The Vegetative Key to the British Flora

A new approach to naming British vascular plants based on vegetative characters

John Poland
&
Eric J. Clement

Illustrations by

Rosalind Bucknall
Sue Nicholls
Delf Smith
Geoff Toone
Robin Walls

Digital artwork by

Niki Simpson

Cyperaceae sections by

Colin Smith

2009

ISBN 13: 978-0-9560144-0-5

Published by John Poland, Southampton
in association with the Botanical Society of the British Isles

Front cover design by Niki Simpson, www.nikisimpson.co.uk

Typeset by Gwynn Ellis, Cardiff

Printed by Hobbs the Printers Ltd, Brunel Road, Totton,
Hampshire SO40 3WX

The authors wish to thank the following organisations
for financial support towards the project

Botanical Society of the British Isles

Natural England

Countryside Council for Wales

Scottish Natural Heritage

Environment Agency

ENIMS (Environmental Consultants)

The Botanical Research Fund

The Wild Flower Society

The following individuals have generously
contributed towards the publication costs

Peter Billinghurst

David Nicolle

John Palmer

Elizabeth Young

CONTENTS

PREFACE

This book flagrantly attempts the impossible – or the apparent impossibility of providing the reader with the means to correctly identify the great majority of our native plants (and many aliens) to species level solely from vegetation bearing neither flowers nor fruits. Our many Floras all imply, indeed indoctrinate us with the belief that it cannot be done. Herein we challenge this concept.

One of the authors (EJC) realized the need for a vegetative key almost immediately upon starting his interest in wild flowers back in 1960. Waiting for flowers or fruits to develop was frustrating! – and he has been waiting for a guide such as this for the intervening forty-eight years.

To produce a consummate tome with infinite detail has not been our intention: we have endeavoured to create a workable volume that contains not too many flaws. Any aims at perfection would retard such a work for many years, whereas we know that many field botanists would like a copy for their living years.

Leaves and twigs are normally present throughout many months of the year, the flowers might blossom for only a few weeks, and the fruits may not set, or might be grazed off. We are left with only vegetative characters to name our plant. Without the book that we proffer, the solution would often prove intractable. A few selective accounts to plant groups or habitats have been published previously, but nothing anywhere near to a comprehensive coverage has ever been achieved.

This book should enable the enthusiast to correctly name both common plants and rarities that tantalisingly sport leaves only. But, being a first attempt at such a vast problem, the keys may sometimes fail, or perhaps not allow for the variability existing within some species.

It is intended that use of our Key will not be restricted to botanists. Entomologists who need to name an insect, or its grub, have an additional powerful aid to identification where such a visitor is host-specific. The same is true for mycologists specialising in rusts, or cedicologists with plant galls.

We would be grateful to receive any corrections or suggestions for improvements that readers might care to offer. We are already working on a bigger and better second edition, as well as an interactive version for a hand-held computer.

We hope that you will enjoy using this book and revel in the new powers that it unlocks – to name a plant that was previously unnameable. Good hunting!

J.P. Poland
Southampton, Hampshire

E.J. Clement
Gosport, Hampshire

March 2009

ACKNOWLEDGEMENTS

From the conception of the idea of a book devoted to naming plants vegetatively, we have received a great deal of support from many members of the BSBI, The Wild Flower Society and county flora groups. Debbie Allan requires a special mention since she typed up the original manuscript into a format which allowed for the easy construction of indented keys. One of us (JP) would particularly like to thank Arthur Chater for his mentoring and unfailing enthusiasm in supporting this project over the past few years.

We have made much use of Botanic Gardens; we thank the Keepers and Trustees of Kew, Cambridge, Edinburgh and the National Botanic Garden of Wales. Both Chelsea and Petersfield Physic Gardens were also particularly welcoming and provided a useful source of herbs not otherwise easily available at short notice. National Collections were also used where necessary, notably those holding *Cotoneaster* and *Sorbus*. A few individuals have had their gardens defoliated by our visits; Penny Condry, David Pearman, Jeremy Roberts and Alan Showler were four such supporters. Numerous people have sent us home-grown specimens on request, or shown us around various botanical localities.

We would like to thank our five artists for their dedication throughout the project. Rosalind Bucknall and Sue Nicholls, both members of the Institute of Analytical Plant Illustration, have provided the bulk of the illustrations. Geoff Toone has drawn the ferns whilst Robin Walls has illustrated the grasses. Delf Smith supplied most of the petiole sections of Apiaceae and a few other select illustrations.

Niki Simpson provided over 150 colour images depicting leaf shapes, bases, margins and even winter twigs, to name a selection. Niki only wished we could have allowed her more space to supply even more, although further examples of her splendid work can be found at www.nikisimpson.co.uk. Colin Smith kindly donated his diagnostic images of the stained Cyperaceae leaf sections.

Gwynn Ellis admirably typeset the entire volume in an extremely short time; he also compiled the index for us. Vicki Russell prepared the glossary and aided with the preparation of the colour plates for printing during the panic-stricken weeks leading up to publication. Mike Wilcox proof-read the text and spotted many errors in the keys which we would otherwise have missed.

Many others have helped in other ways; we must mention: Brian and Barbara Ballinger, Linda Barker, Peter Barnes, Ian Bennallick, Peter Billinghurst, Alison Bolton, Ian Bonner, Margaret Bradshaw, Michael Braithwaite, David Brear, Dick Brummitt, Rosalind Bucknall, Phil Budd, Mark Carine, Clive Chatters, Julie Clarke, Stephen Clarkson, Allen Coombes, Arthur Copping, Jane Croft, Cameron Crook, Andy Cross, Sarah Dalrymple, Mary Dean, Natasha deVere, Mark Duffel, Dave Earl, Bob Ellis, Ian and Pat Evans, Stephen Evans, Andrew Fergusson, Jeanette Fryer, Brian Gale, Jim Gardiner, Mary Ghullham, Barry Goater, Richard Gornall, Mike Grant, Ian Green, Jean Green, Paul Green, Jo Haigh, Alan Hale, Angus Hannah, Gordon Hanson, Kathryn Hart, John Hawksford, Rachel Hemming, Michael and Robin Hickey, Peter Hilton, Nick Hudson, Colin Hutchinson, Richard Jefferson, Clive Jermy, Vic Johnstone, Andy Jones, Hywel Jones, Stephen Jury, Mark and Clare Kitchen, Alan Knapp, Roy Lancaster, Brian Laney, Richard Lansdown, Nick Law, Simon Leach, Bob Leaney, Alan Leslie, Viv Lewis, Alex Lockton, Heather McHaffie, Glyndwr Marsh, Roger Maskew, Jim McIntosh, Pete Michna, Peter Mitchell, Tony Mundell, Sue Nicholls, David Nicolle, John Norton, John and Clare O'Reilly, Philip Oswald, Ken Page, Richard Pankhurst, Stephen Parker, Rosemary Parslow, David Patterson, Robin Payne, Sandy Payne, Jackie Pedlow, Barry Phillips, Colin Pope, Robert Portal, Mike Porter, Mark Poswillo, Hannah Powell, Edward Pratt, John Presland, Chris Preston, Richard and Kath Pryce, Martin Rand, Tim Rich, Francis Rose, Fred Rumsey, Alison Rutherford, Kevin Rylands, John Savidge, Vera Scott, Andy Shaw, Colin Smith, Mary Smith, Mervyn Southam, Mark Spencer, Keith Spurgin, Clive Stace, Paul Stanley, Robin

Stevenson, Sarah Stille, Malcolm Storey, David Streeter, Mike Sutcliffe, John Swindells, Tim Sykes, Ian Thirlwell, Maldwyn Thomas, Helen Thompson,, Ken Trewan, John Twibell, Tim Upson, Kevin Walker, Tim Waters, Sarah Whild, Richard Wilford, Julian Woodman, Len Worthington, Elizabeth Young and Jan Zawadzki.

Our sincere apologies go to anyone who has been inadvertently left off this list. Finally, we must thank the many botanists who have sent us material for naming, and also attendees at various identification workshops who have unwittingly aided the key-writing process.

BIBLIOGRAPHY

The following publications (amongst many others) were consulted during the compilation of the keys. In a few cases, illustrations were re-drawn or used as a basis for the artwork contained in this volume.

CLAPHAM, A.R., TUTIN, T.G. & MOORE, D.M. 1987. *Flora of the British Isles*, 3rd edn. Cambridge University Press, Cambridge.

CLEMENT, E.J. & FOSTER, M.C. 1994. *Alien plants of the British Isles*. Botanical Society of the British Isles, London.

CLEMENT, E.J., SMITH, D.P.J. & THIRLWELL, I.R. 2005. *Illustrations of Alien Plants of the British Isles*. Botanical Society of the British Isles, London.

EGGENBERG, S. & MÖHL, A. 2007. *Flora Vegetativa*. Haupt, Berne.

GRAHAM, G.G. & PRIMAVESI, A.L. 1993. *Roses of Great Britain and Ireland*. Botanical Society of the British Isles, London.

HICKEY, M. & KING, C. 2000. *The Cambridge illustrated glossary of botanical terms*. Cambridge University Press, Cambridge.

HILL, M.O., PRESTON, C.D. & ROY, D. B. 2004. *PLANTATT - Attributes of British and Irish Plants: Status, Size, Life History, Geography and Habitats*. Centre for Ecology and Hydrology, Abbots Rippon.

JERMY, C & CAMUS, J. 1991. *The Illustrated Field Guide to ferns and Allied Plants of the British Isles*. Natural History Museum Publications, London.

JERMY , A.C., SIMPSON, D.A., FOLEY, M.J.Y, & PORTER, M.S. 2007. *Sedges of the British Isles*, 3rd edn. Botanical Society of the British Isles, London.

LANSDOWN, R.V. 2008. *Water Starworts of Europe*. Botanical Society of the British Isles, London.

MEIKLE, R.D. 1984. *Willows and Poplars of Great Britain and Ireland*. Botanical Society of the British Isles, London.

PAGE, C.N. 1997. *The Ferns of Britain and Ireland*, 2nd edn. Cambridge University Press, Cambridge.

POLAND, J. 2005-2008. [A series of articles on naming plants vegetatively]. *BSBI News* **98** - **109**.

PRESTON, C.D. 1995. *Pondweeds of Great Britain and Ireland*. Botanical Society of the British Isles, London.

PRESTON, C.D., PEARMAN, D.A., & DINES, T.D. 2002. *New atlas of the British and Irish flora*. Oxford University Press, Oxford.

RICH, T.C.G. 1991. *Crucifers of Great Britain and Ireland*. Botanical Society of the British Isles, London.

RICH, T.C.G. & JERMY, A.C. 1998. *Plant Crib 1998*. Botanical Society of the British Isles, London.

ROSE, F. 1981. *The Wild Flower Key*. Frederick Warne, London.

ROSE, F. & O' Reilly, C. 2006. *The Wild Flower Key*, 2nd edn. Frederick Warne, London.

RYVES, T.B., CLEMENT, E.J. & FOSTER, M.C. 1996. *Alien grasses of the British Isles*. Botanical Society of the British Isles, London.

SELL, P.D. & MURRELL, G. 1996. *Flora of Great Britain and Ireland, Vol. 5 Butomaceae-Orchidaceae*. Cambridge University Press, Cambridge.

SELL, P.D. & MURRELL, G. 2006. *Flora of Great Britain and Ireland, Vol. 4 Campanulaceae-Asteraceae*. Cambridge University Press, Cambridge.

STACE, C.A. 1997. *New Flora of the British Isles*, 2nd edn. Cambridge University Press, Cambridge.

TUTIN, T.G. 1980. *Umbellifers of the British Isles*. Botanical Society of the British Isles, London.

INTRODUCTION

Coverage

Nearly 3,000 taxa are covered, included all native vascular plants (except some of the highly critical apomictic genera such as *Hieracium*, *Rubus* and *Taraxacum*) as well as the vast majority of aliens and many widely planted, but not yet naturalised, species. Hybrids and infraspecific taxa are mostly omitted – to include them would make this first edition cumbersome and impractical for use in the field.

The whole of Great Britain and Ireland (including the Channel Islands and the Isle of Man) is covered. The inclusion of many aliens originating from all over the world will provide interest to botanists in Europe and beyond.

Nomenclature

The nomenclature and taxonomy generally follows Stace's *New Flora* edn. 2 (1997). Occasionally, we have updated the nomenclature where a change of name appears to be inevitable. For example, *Carex ovalis* has been replaced by *C. leporina* following the publication of *Sedges of the British Isles* ed. 3 (Jermy *et al*, 2007). Unfortunately, recent molecular DNA studies are strongly indicating many more name changes in the near future.

Previous publications

We believe that this key is the first of its kind to give a comprehensive treatment of our flora using vegetative characters. Dr Francis Rose was, perhaps, one of the first British botanists to realise the potential of vegetative keys. Indeed, those first published in *The Wild Flower Key* (1981) almost 30 years ago have provided invaluable help for many botanists, but they were all strictly habitat-based (and some habitats were missed!). More recently, the publication of *Flora Vegetativa* (2007) indicates that our European counterparts also recognise non-flowering keys as an essential part of a botanist's library.

All the keys have been written from personal observation of fresh material. Literature has not been copied blindly as this inevitably results in the perpetuation of errors, e.g. virtually all Floras say that the vegetative parts of *Glaucium flavum* (Yellow-horned Poppy) contain yellow latex but this is simply not true. Of course, we have been guided by the published observations of others and we have relied heavily on the contents of Clapham *et al*. (1987), Sell & Murrell (1996 & 2006) and Stace (1997). The Plant Crib (1998) deserves a special mention, not least for its frequent emphasis on vegetative characters.

Our keys

Our keys have been carefully constructed in an attempt to avoid features that we have disliked in the works of others. We have divided up the species into *Divisions* (e.g. A, B), which are then further divided into *Groups* (e.g. AA, AB). Initial choices are normally easy to make, usually involving very obvious characters. This modular approach allows an experienced user to quickly enter the keys at any selected point.

The keys that we provide are unlike those in most modern Floras in that they are not strictly dichotomous (i.e. giving a choice of just two options). Three or more options are often more appropriate (polychotomous); occasionally there is only one option (a monochotomy) that is indented to allow rapid reading of the text. The keys are indented rather than bracketed, allowing the overall structure to be viewed at a glance. We strongly believe this allows our keys to be far quicker than conventional keys.

This approach has another great bonus - most keys rarely extend beyond a double-page spread. No frantic page-swapping is ever required! In most cases, three turns of the page are all that are needed to provide an identification.

As a visual aid, the initial choices in each key are preceded by the '■' symbol to help users to quickly see all of the possible choices, and prevent any from being overlooked. Care should be taken not to miss widely-spaced options, particularly later in the key.

Always remember the following three points when using the keys:

- Choose a typical specimen (see notes below)
- Read all the choices in the key carefully before a decision is made. In most cases more than one character is given, in order to make identification more certain.
- Always use a hand lens for studying fine detail (x10, or preferably x20)

Naturally, the keys will become much easier to use with experience! To gain confidence, try keying out a plant that has already been identified.

Choosing your plant

For herbs, choose a basal or lower stem leaf (unless absent or otherwise instructed). By default, leaf descriptions refer to those which are basal or on the lower stem; those on the mid-stem can be totally different, the upper leaves may show more variation still, often reducing in size up the stem to become bracts. When selecting a plant for naming, where possible, look at several to confirm that your plant is representative. For trees and shrubs, choose a typical first-year shoot. Avoid sucker or very vigorous growth (which may be atypical).

The keys are not designed to identify seedlings and at least one typical well-developed leaf must be present for them to work. Seedlings normally produce one or two cotyledons (none in *Vicia* and some tuberous plants) before the true leaves appear. Cotyledons usually wither quickly, but they are occasionally mentioned in the descriptions as they can be highly diagnostic. Conversely, it should be noted that some characters (notably the presence of latex) can be hard to interpret correctly in old or dying plants.

Characters

Although the majority of users will be familiar with existing Floras, this volume contains much new information. In the leads, we have always aimed to give simple indicative characters as well as more technical confirmatory ones. The latter may include gross anatomy such as the number of vascular bundles and the presence of stomata. These terms, and many more, are covered in the *Glossary* and *Explanatory notes* . Do not forget that a positive character is far more significant than a negative one. For example, odours may often be difficult to detect by some people and latex can be very sparse or absent from older tissues. Remember that plants are variable, so no Flora or key can reasonably hope to cover all eventualities. We have occasionally found that a 'vegetative ID' is not feasible, and, in the interest of producing a comprehensive volume, reluctantly resorted to including floral or fruiting characters. Flower bud characters are also sometimes used, as buds can precede true flowering characters by as much as several weeks.

Strongly diagnostic characters and the species that exhibit them are also given in the *Explanatory notes*. Many of these characters are characteristic of just a few families, genera or species, and can act as a rapid route to identification, especially if incomplete or inadequate material is being studied.

Spot Characters

Many of the keys incorporate 'spot characters' (marked by ❶, ❷, etc). These are unusual characters, often species-specific, which act as a shortcut to identification. They should be read *before* commencing the relevant key: if one (or more) of these characters fits a specimen, the key can be circumvented, and the reader can move immediately to those very few species marked with the relevant number(s). Of course, the key may also be worked through in the traditional manner.

Selected groups

A few genera or aggregates of species form natural groups that key out in more than one place. In order to prevent unnecessary repetition, keys to these groups have been removed to a 'Selected Groups' section near the back of the book. For example, all the species of *Callitriche* are keyed out in **CAL** (the first three letters being the typical choice for the group name), and the preceding keys, at the relevant points, instruct the user to 'Go To **CAL**', rather than mention a specific species. If the user knows with certainty the genus, an entry can be made immediately at **CAL**.

Measurements

For every species we give an indication of size, usually at least the length of a typical leaf blade (the petiole is not included in the length of the leaf, contrary to some books). In most cases a stem height is also included. To make such measurements more indicative of each species, we give the typical upper limit. Although every effort has been made to ensure accuracy, these sizes are not absolute, and your specimen may not conform. Hence a plant growing out of its typical habitat, e.g. in nutrient-rich soil or deep shade, may well exceed the upper limit that we indicate.

Measurements without qualification are lengths (e.g. lvs 5-10cm); those separated by a multiplication sign (e.g. lvs 5-10 x 6cm) are the lengths and widths respectively of the leaf blade. In this second example, the width is the normal maximum and not an average. Further explanation is required for measurements applying to stems. The phrase 'Stem to 20cm, erect', refers to the height of the erect stem. In cases where ambiguity may exist, such as in creeping or prostrate herbs, the description may be more carefully worded to say 'Stem to 5cm tall' as the stem may be considerably longer, but not taller. Numbers (or characters) found in brackets indicate that this number/feature is much less common. Hence 'lfts 5-7(9)' means that the leaflets are usually 5 to 7 in number, but rarely up to 9.

Definition of terms

We have used technical botanical terms sparingly and only where they enable us to avoid repeating long descriptive phrases. Unfortunately, botanists cannot agree on the precise meaning of even some common terms. The *Glossary* should always be referred to in order to check our interpretation of terms used. Hickey & King (2000) is a splendid all-round glossary with clear textual and illustrative explanations of almost all the terms that we use, and many more in addition.

Plant anomalies and observer errors

The keys do not accommodate for plant abnormalities – adding them would increase the bulk and complexity of the keys for rare rewards. The classic example of a four-leaved clover (*Trifolium* sp) is not allowed for here. Nonetheless, the keys allow for likely errors of observation or interpretation that we can predict.

Line illustrations

A selection of illustrations has been included to show various characters, from overall habit to simple anatomical details. Even with close collaboration between botanist and artist, it should be remembered that artists draw what they observe. For this reason, characters prefixed 'usually' or 'occasionally' in the text may thus be absent from the corresponding illustration. It is intended that more illustrations will be added to a future edition. A comprehensively illustrated volume would be our ultimate goal and we welcome more botanical artists to get involved in this ambitious project. Scale bars are intentionally omitted as typical measurements are already included within the keys.

Colour plates

In addition to the eight pages of diagnostic alphabetically-arranged Cyperaceae leaf sections, approximately 150 leaves of herbs, trees and shrubs were selected for illustrating a broad range of shapes and other characters. The arrangement within the colour plates has been by overall outline alone, from simple and linear to orbicular and compound, and not by the order in which they appear in the keys.

Equipment

A hand lens (or loupe) is an essential piece of equipment for any naturalist. A x10 lens is usually adequate, but we recommend a x20 or x30 lens available from www.eyemagnify.com. A lens with a graticule (a measuring loupe) can give precise measurements in the field without the need for a microscope. Such loupes are available from www.summerfieldbooks.com. Very rarely, microscopic characters are mentioned which require a magnification of x60-100.

Conservation (of plants and botanists!)

Most readers will know that it is against the law to uproot any wild plant in Britain without the permission of the landowner. A few plants are rare enough to warrant inclusion in Schedule 8 of the Wildlife & Countryside Act 1981 (as amended) making it illegal to gather any part, even one leaf thereof. These species are marked 'Sch8' to help to prevent the unsuspecting learner unintentionally breaking the law. Most of these extreme rarities now exist solely within the protection of nature reserves, where reminders not to pick wild plants are conspicuously placed, so accidental picking is not likely to occur. Elsewhere, picking the odd leaf is unlikely to damage any plant population.

This book is a field guide and, as such, we strongly advise that the book is taken to the plant. This does not mean a 'hands off' approach. The BSBI has a sensible and pragmatic *Code of Conduct*. Picking and examining small amounts of widespread plant material for identification and study is essential if botanical expertise is to be developed. It should be noted that photography, if carried out carelessly, may do significantly more harm. Sadly, two of the biggest threats to plants and botany today are the continuing loss of habitat and the lack of botanical expertise through diminishing higher education courses.

Key to Major Divisions

(Based primarily upon the increasing degree of leaf dissection)

For herbaceous plants choose a lower or basal lf unless otherwise instructed. To separate lvs from lfts, lfts never have a bud in the axil. 'Lvs' or 'lf' refers to leaf blade (excluding the petiole). See also notes overleaf (and Introduction)

■Horsetail........ **A**
■Fern........ **B**
■Clubmoss........ **C**
■Conifer........ **D**
■Flowering plant (monocots, dicots)
 Lvs absent or stem-like (occ scale-like or reduced to a short blade)........ **E**
 Lvs present, not stem-like
  Plant with submerged or floating lvs (obligate water plant)........ **F**
  Plant with aerial or emergent lvs (usu land plant)
   Lvs simple (not composed of lfts)
    Lf margin entire (or lvs cylindrical)
     Lvs parallel-veined, usu >3 veins visible (stomata often in parallel rows). Usu monocots (may have bulb, corm or viscid sap)
      Ligule or auricles present (if lvs onion-scented Go to **JF**)
       Ligule adnate (typically sedges)........ **G**
       Ligule free or present as a ring of hairs (typically grasses)........ **H**
       Ligule absent but auricles present (some rushes)........ **I**
      Ligule and auricles both absent........ **J**
     Lvs pinnate-veined (or 0-3 parallel veins), occ obscurely so or palmately veined. Usu dicots (v rarely with bulb, corm or viscid sap)
      Lvs alt........ **K**
      Lvs opp........ **L**
      Lvs whorled (or pseudowhorled) at stem nodes, but not in fascicles........ **M**
    Lf margin toothed (but not lobed)
     Lvs alt........ **N**
     Lvs opp (or whorled)........ **O**
    Lf margin lobed (lobes often toothed, or lvs peltate); lobes palmate or pinnate
     Lvs alt........ **P**
     Lvs opp (or whorled)........ **Q**
   Lvs compound (composed of lfts)
    Lvs with 3-17 lfts (usu palmately arranged)........ **R**
    Lvs 1-pinnate (occ 2 lfts only)........ **S**
    Lvs 2-4-pinnate (or 2-ternate)
     Lfts all opp (mostly *Apiaceae*)........ **T**
     Lfts mostly alt (at least above)........ **U**

Notes on using the keys

Vascular plants are divided into five major classes; horsetails, ferns, clubmosses, conifers and flowering plants. Illustrations of typical members are given in each of the *Divisions*. If uncertainty remains as to which category your plant belongs, consult any botanical book to make this initial choice. Where observer error is likely to occur, species are keyed out in more than one class. For example, the leaves of *Ophioglossum vulgatum* (a fern) may be mistaken for those of a flowering plant. Although the main description of this species is in the fern key, another entry (with a reduced description) is given in the relevant flowering plant key.

Leaf arrangement (phyllotaxy)
A distinction between opposite and alternate leaves is required early in the keys. This is very obvious when the leaves are on a stem but the arrangement is also easily recognised in rosette leaves. Opposite rosette leaves can be paired off (i.e. the leaves will be of equal length where each pair grow from the same node) whilst the leaves of alternate leaves cannot be paired as none of the leaves are of equal length. Opposite-leaved species are usu decussate, giving rise to a neat cross-like appearance. Alternate leaves are typically spirally arranged, giving rise to a more untidy rosette.

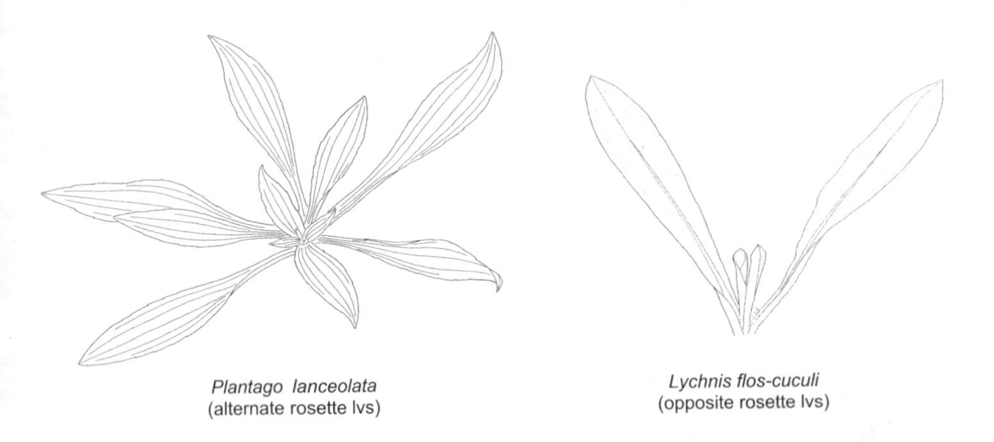

Plantago lanceolata
(alternate rosette lvs)

Lychnis flos-cuculi
(opposite rosette lvs)

Phenology
Amongst the characters that we occasionally use are phenological ones, i.e. months of leafing, flowering or fruiting. Unqualified months (e.g. Mar-Oct) refer to when the leaves are visible in a living state (i.e. excluding dead stems and shrivelled seed heads!).

Deciduous and evergreen leaves (exc some species of *Cotoneaster*) can be separated all year round. Evergreen species usu have older leaves on 2nd year twigs. Evergreen leaves, in addition to a thickened cartilaginous margin, do not have raised 2° veins on the lower surface (and few species are net-veined). In contrast, deciduous trees and large shrubs have the 2° veins raised below (and are usu net-veined, and lack an obvious cartilaginous margin). Evergreen leaves are usu tougher and more leathery than deciduous leaves.

Key to Groups in Division A

(Horsetails Equisetum*)*

Rhizomatous per herbs with ridged jointed stems sheathed at each node. The number of sheath teeth equals the number of stem ridges. A 'tough' stele refers to a stele (vascular cylinder) which remains unbroken when the stem is pulled apart. Apr-Oct (unless otherwise stated)

■Stems with whorls of branches
  Branches with (3)4 angles in TS (check 2nd internode of branch)........ **AA**
  Branches with 5 or more angles in TS (check 2nd internode of branch)........ **AB**
■Stems without whorls of branches........ **AC**

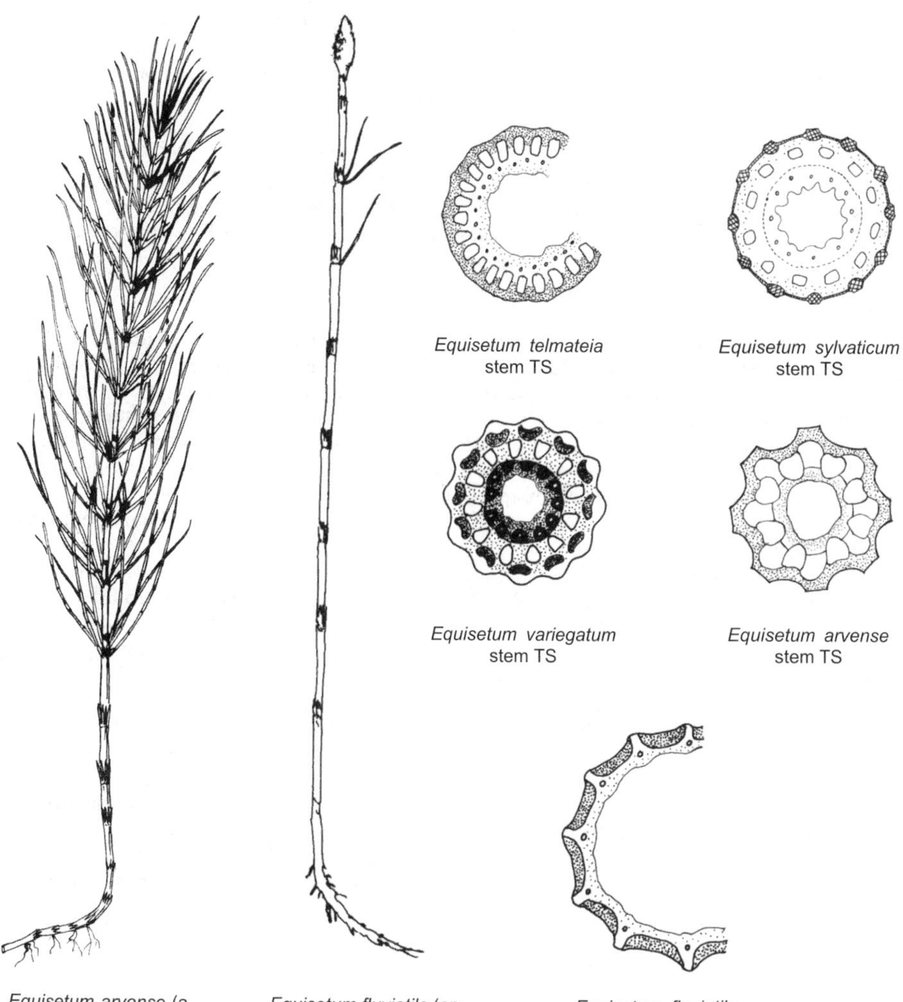

Equisetum telmateia stem TS

Equisetum sylvaticum stem TS

Equisetum variegatum stem TS

Equisetum arvense stem TS

Equisetum arvense (a branched horsetail)

Equisetum fluviatile (an unbranched horsetail)

Equisetum fluviatile stem TS

Group AA – *Stems with whorls of branches, erect (occ decumbent in* E. arvense*). Branches with (3)4 angles.* ❶ *Branches 3-angled.* ❷ *Sheath teeth 2-ribbed.* ❸ *Stems without a tough stele (easily broken).* ❹ *Sheaths loose*

■Stems rough, with ridges 2-angled and long-papillate

Branches 3-angled, usu unbranched

Branches with lowest internode < adjacent stem sheath at lower nodes but longer at upper nodes, 0.5-1mm diam, wavy, with acute teeth not spreading (often brown tipped). Stems 1-2mm diam, with 8-20 ± deep grooves, with hollow <½ diam. Sheath teeth (3)6-20, occ slightly fused, dark with whitish scarious margins. ❶. Uplands, often base-rich, N Br ....*Shady Horsetail* **Equisetum pratense**

Branches 4-angled, repeatedly branched

Branches with lowest internode > adjacent stem sheath, with slightly spreading teeth, 0.5mm diam, drooping, pale green. Stems 1-4mm diam, with 10-18 weak grooves, with hollow ¼ -½ diam. Sheath teeth 6-20, fused into 3-6 obtuse lobes, brown with whitish scarious margins ....*Wood Horsetail* **Equisetum sylvaticum***

Branches with lowest internode < adjacent stem sheath, with spreading teeth, 0.5mm diam, drooping or horizontal, mid-green. Stems 1-4mm diam, with 16 low grooves, with hollow ½ diam. Sheath teeth 6 (joined in groups of 2-3), with brown scarious margins. *E. pratense* x *sylvaticum*...*Milde's Horsetail* **Equisetum** x **mildeanum**

■Stems rough or smooth, with obscurely 2-angled ridges and v short papillae (occ absent)

Branches with spreading teeth. Stems with c14 v weak grooves, with hollow ⅓-½ diam and surrounded by smaller hollows. Sheath teeth 12 (some fused in groups of 2-3), 5-7mm, 2-ribbed at base, brown. ❷. VR. *E. sylvaticum* x *telmateia*......*Bowman's Horsetail* **Equisetum** x **bowmanii**

■Stems smooth, ridges without angles or papillae

Stems with hollow >½ diam, occ orange nr base. ❸

Branches with adpressed teeth

Branches with lowest internode ≤ adjacent stem sheath, developing late or absent, occ 5-ridged, with small hollows around larger central hollow. Stems (2)4-7(10)mm diam, with 10-20(30) v fine shallow grooves, with hollow ¾-4/5 diam. Sheath teeth black, without obvious white margins, unribbed. Usu aquatic ....*Water Horsetail* **Equisetum fluviatile***

Branches with lowest internode usu ± equal to adjacent stem sheath, occ absent. Stems 2-3mm diam, with 14 rounded to ± acute shallow grooves, with hollow ½-⅔ diam. Sheath teeth black, occ with v narrow white margins, 1-ribbed. *E. fluviatile* x *palustre* ....*Dyce's Horsetail* **Equisetum** x **dycei**

Branches with spreading teeth

Branches with lowest internode often ± equal or slightly longer than adjacent stem sheath. Stems without branches nr curved whip-like apex, dull grey-green, with (6)9-15(20) rounded grooves, with hollow >⅔ diam. Sheath teeth black, with indistinct narrow scarious margins. *E. arvense* x *fluviatile*...*Shore Horsetail* **Equisetum** x **litorale**

Stems with hollow <½ diam, never orange nr base

Branches with lowest internode ≥ adjacent stem sheath, 4-angled

Branches with spreading acute green teeth, 1-2mm diam, wavy, solid. Stems 2.5-4mm diam, occ slightly rough, with (6)9-13(18) shallow rounded ridges, with hollow ⅓ diam and surrounded by smaller hollows. Sheath teeth black. ❹. Usu dry habs ....*Field Horsetail* **Equisetum arvense***

Branches with lowest internode < adjacent stem sheath, 4(5)-angled

Stems with c9 ridges and small hollows in TS. VR. *E. arvense* x *palustre* ....*Ditch Horsetail* **Equisetum** x **rothmaleri**

Group AB – *Stems with whorls of branches. Branches with 5 or more angles.* ❶ *Sheaths loose.* ❷ *Sheath teeth 2-ribbed at base*

■Stems usu prostrate, occ branched, rough, with ridges 2-angled and papillate (papillae often low or obscure)

Branches 5-angled, with lowest internode << adjacent stem sheath, solid. Stems to 10(30)cm tall, 1-2mm diam, rarely branched at base, green, with (4)6-10 shallow grooves, with hollow <⅓ diam (± equal to surrounding hollows). Sheath teeth persistent, 1-3(6), with dark central line, with hair-point when young. Sheaths with black band at base, occ orange below. All yr. Dune slacks, uplands...*Variegated Horsetail* **Equisetum variegatum***

Branches 6-7(10)-angled, with lowest internode < adjacent stem sheath, hollow. Stems to 80cm tall, 1-3mm diam, branched, greyish-green, with 10-11(20) rounded grooves, usu with hollow ≥⅔ diam (>> surrounding hollows). Sheath teeth persistent, 3-6mm, black with narrow whitish margins and hair-point. Sheaths occ with black band. All yr. VR, Lincs, Somerset. Sch8 ....*Branched Horsetail* **Equisetum ramosissimum**

■Stems erect (occ decumbent), unbranched, smooth, without ridges 2-angled or papillate

Stems with hollow <¼(⅓) diam

Branches 5-angled (angles v rounded), with lowest internode << adjacent stem sheath, solid. Stems with irregular whorls of branches, 1-3mm diam, greyish-green, with 4-10 rounded ridges, with central hollow ± equal to surrounding hollows. Sheath teeth not ribbed, black, with obvious white margins. ❶..*Marsh Horsetail* **Equisetum palustre***

Stems with hollow ≥½ diam

Stems ivory-white, brittle, fleshy. Branches with lowest internode < adjacent stem sheath

Sheath teeth 2-ribbed at base, 3-8mm, v narrow, brown. Stems 5-13mm diam, with 18-40(60) faint grooves, with hollow (⅓)⅔ diam. Branches 8-angled. ❷. Base-rich clay ....*Great Horsetail* **Equisetum telmateia***

Stems green, not brittle or fleshy. Branches with lowest internode usu equal to stem sheath

Sheath teeth 1-ribbed

Branches well-developed, (4)5-6-angled. Stems 4-10mm diam, never orange nr base, with 18 grooves, with hollow >¾ diam. Sheath teeth black-tipped. VR. *E. fluviatile* x *telmateia* ....*Wilmot's Horsetail* **Equisetum** x **wilmotii**

Branches often absent or ill-developed, (4)5-angled. Stems 2-3mm diam, occ orange nr base, with 14 rounded to ± acute shallow grooves, with hollow ¾-3/5 diam. Sheath teeth black, occ with v narrow white margins. *E. fluviatile* x *palustre* ....Dyce's Horsetail **Equisetum** x **dycei**

Sheath teeth not ribbed

Branches often absent or ill-developed, with adpressed black-tipped teeth. Stems (2)4-7(10)mm diam, often orange below, with 10-20(30) v shallow grooves, with hollow >¾ diam. Sheath teeth black, pale-edged (but without obvious white margins). Usu aquatic ....*Water Horsetail* **Equisetum fluviatile***

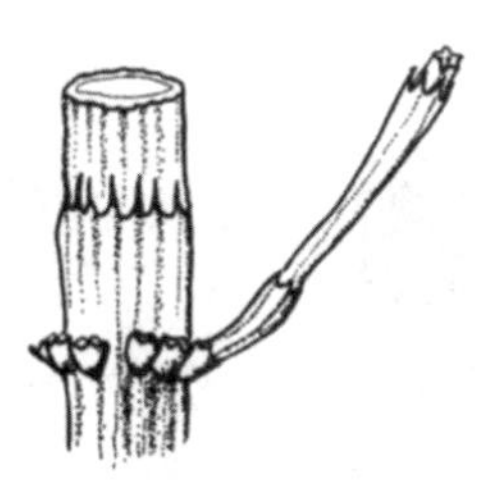

Equisetum fluviatile sheath and branch

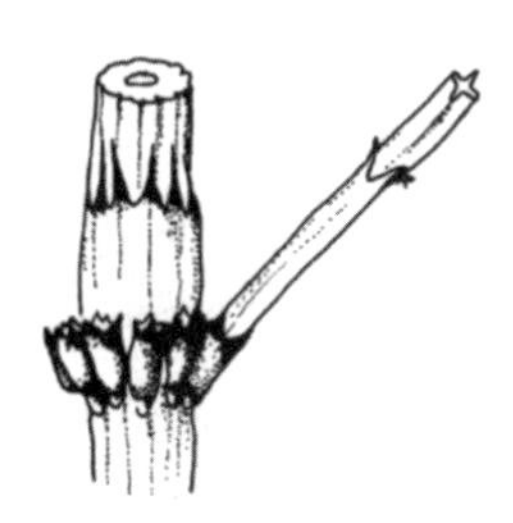

Equisetum arvense sheath and branch

Equisetum palustre sheath and branch

Group AC – *Stems without whorls of branches.* ❶ *Sheath teeth not persistent*

■Stems usu rough, with ridges 2-angled and papillate (often low or obscure)

Sheath teeth persistent

Stems erect, unbranched. Sheath teeth without hair-points

Sheath teeth (3)6-20, occ slightly fused, acute, dark, with scarious margins. Stems 1-2mm diam, soon producing branches, pale green, with 8-20 angular grooves, with hollow c½ diam. Uplands, often base-rich, N Br..*Shady Horsetail* **Equisetum pratense**

Sheath teeth 6-20, fused into 3-6 obtuse lobes, brown, with whitish scarious margins. Stems 1-4mm diam, soon producing branches, green, with 10-18 weak grooves, with hollow ¼ -½ diam...*Wood Horsetail* **Equisetum sylvaticum***

Stems usu prostrate, branched. Sheath teeth with hair-points at least when young

Stems with hollow <⅓ diam (± equal to surrounding hollows), rarely branched at base, 1-2mm diam, green, with (4)6-10 shallow grooves. Sheath teeth 1-6mm, broadly triangular, with dark central line. Sheaths with black band at base, occ orange below. Dune slacks, uplands ....*Variegated Horsetail* **Equisetum variegatum***

Stems with hollow usu ≥⅔ diam (>> surrounding hollows), branched, 1-3mm diam, greyish-green, with 10-11(20) rounded grooves. Sheath teeth 3-6mm, acute, black with narrow whitish margin. Sheaths occ with black band. All yr. Lincs, Somerset. Sch8 ....*Branched Horsetail* **Equisetum ramosissimum**

Sheath teeth not persistent, reduced to crenate margins ❶

Stems 4-6mm diam, grey-green, with slightly swollen internodes, with (10)18-21(30) ridges, with hollow ⅔-¾ diam. Sheaths 4-6mm, often reduced to a 2-3mm black band. All yr ....*Rough Horsetail* **Equisetum hyemale***

■Stems smooth, without ridges 2-angled or papillate

Stems ivory-white or pinkish (fertile stems visible only Apr-May)

Stems with hollow (⅓)⅔ diam, with 18-40 or more v faint grooves. Sheath teeth 2-ribbed at base, 3-8mm, v narrow, brown. Sheaths tight, inflated. Cones 4-8cm. Base-rich clay ....*Great Horsetail* **Equisetum telmateia***

Stems with hollow ⅔-¾ diam, with (6)9-13(18) shallow rounded ridges. Sheath teeth not ribbed, 2.5mm, narrow, black. Sheaths loose. Cones 1-4cm. Usu dry habs ....*Field Horsetail* **Equisetum arvense***

Stems green, often orange nr base

Stems (2)4-7(10)mm diam, with hollow ≥¾ diam, with 10-20(30) v shallow grooves. Sheath teeth 1.5mm, not ribbed, black, without obvious white margins. Sheaths tight. Usu aquatic ....*Water Horsetail* **Equisetum fluviatile***

Stems 2-3mm diam, with hollow ≤¾ diam, with c14 rounded to ± acute shallow grooves. Sheath teeth 2.5mm, 1-ribbed, black, occ with v narrow white margins. Sheaths tight to loose. VR. *E. fluviatile* x *palustre*..*Dyce's Horsetail* **Equisetum** x **dycei**

Equisetum hyemale sheath

Equisetum variegatum sheath

Key to Groups in Division B

(Ferns Pteropsida*)*

Lvs circinate (coiled into a crozier) when young (exc Ophioglossum *and* Botrychium*), with stomata below only (but totally absent in filmy ferns and present on both sides in* Botrychium*). Most species have a narrow cartilaginous margins to the pinnae. Pinnae are usu alt and often hay-scented when fresh*

The examination of vascular bundles (vb's) in a transverse-section (TS) of the petiole often provides the quickest and most reliable shortcut to the identification of a genus. If the plant is considered to be locally rare or protected (Sch8), the alternative characters should be selected. The vb's should not be confused with black strands of phlobaphene (a substance formed from tannins) which lie adjacent to the vb's (e.g. in Phyllitis*) or surround the vb's (e.g. in* Dryopteris*)*

■Lvs entire...**BA**
■Lvs deeply lobed (pinnatisect, pinnae adnate to rachis)...**BB**
■Lvs 1-pinnate (pinnae stalked) OR lvs forked (often repeatedly)...................................**BC**
■Lvs 1-pinnate with the pinnae pinnatisect (or the lowest lobe occ stalked)
  Lvs hairy at least below (hairs occ sparse or confined to rachis)..................................**BD**
  Lvs hairless (may have scales)..**BE**
■Lvs 2-4-pinnate
  Lvs v translucent, 1-cell thick (filmy ferns)..**BF**
  Lvs opaque or ± so, >1-cell thick
    Petiole with 1 or 2 equal vb's..**BG**
    Petiole with 3 or more unequal vb's..**BH**

Group BA – *Lvs entire*

■Lvs 15-75cm, leathery, with distinct midrib and opaque forked veins, with scattered narrow clathrate scales below. Tufted

Lvs strap-shaped, cordate at base, shiny green, each vein ending in a submarginal hydathode. Petiole long, with scales, with 2 C-shaped vb's (fusing into X-shape nr lf), often with 3 black strands. All yr..........*Hart's-tongue* **Phyllitis scolopendrium***

■Lvs <10cm, fleshy, without midrib but with obscure opaque netted veins, scales absent. Rhizomatous

Lvs erect, arising singly, with flat margins

Lvs 4-5(10) x 1.7-3(4)cm, broadly ovate, obtuse, cuneate to cordate at base (rolled into petiole), often channelled, with free vein endings. Petiole 2-4cm, soon hollow. Apr-Sept ....*Adder's-tongue* **Ophioglossum vulgatum***

Lvs ± erect, arising in prs or triplets, with recurved margins

Lvs 2-4cm, lanc to ovate, usu acute, cuneate at base, usu dull green, with free vein endings. Apr-Aug. Short turf, dunes, R..........*Small Adder's-tongue* **Ophioglossum azoricum**

Lvs adpressed to ground, arising singly or in prs, with recurved margins

Lvs <2cm, lanc to linear-lanc, acute or obtuse, cuneate at base, often shiny green, without free vein endings. Oct-May. Short turf, VR, Scilly Is, Channel Is. Sch8 ....*Least Adder's-tongue* **Ophioglossum lusitanicum**

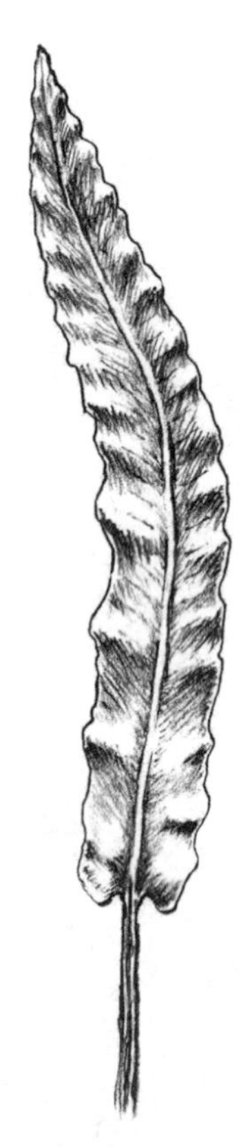

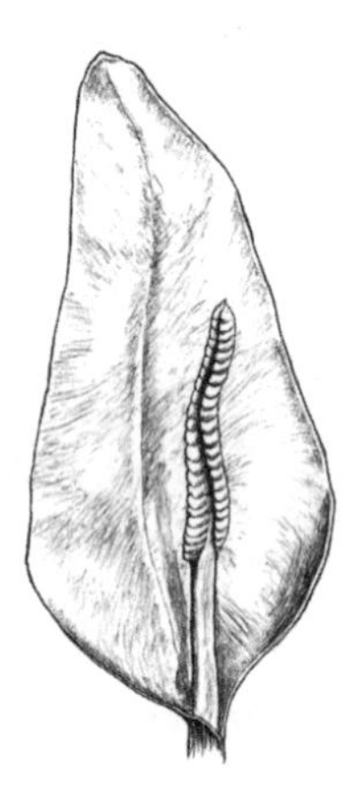

Ophioglossum vulgatum

Phyllitis scolopendrium

Group BB – *Lvs deeply lobed (pinnatisect), lobes not stalked (attached by a flange of tissue).* ❶ *Petiole unchannelled and lvs fleshy with veins obscure*

▪Petiole with 1 or 3 vb's

Petiole ≤⅓ total lf length, with brown scales at base, with 3 vb's. Lobes with obscure veins ending in a submarginal hydathode. All yr

Tufted. Lvs 10-60cm, ± leathery, shiny dark green above; lobes 34-60 prs, 10-20 x 3-7mm, longest nr middle of lf, linear-oblong, acute or apiculate, entire or ± so; rachis green. Acidic habs..........*Hard-fern* **Blechnum spicant***

Rhizomatous. Lvs to 20cm, ± leathery, shiny dark green above; lobes ≤25 prs, 7-10 x 4-7mm, often all ± equal, oblong, acute, entire or ± so; rachis brown. Hortal, VR alien ....*Little Hard-fern* **Blechnum penna-marina**

Petiole >⅓ total lf length, with scales only at extreme base, with 1or 3 vb's. Lobes with translucent free veins ending either in a submarginal hydathode or in a hydathode c½ way to margin. Rhizomatous, with lvs arising singly from a scaly rhizome (often above substrate). All yr

Lvs to 45cm; lobes to 4.5cm, often obtuse, entire or hardly serrate, lowest prs not reflexed and often shorter; sori maturing in summer. (May) Jun-Apr (Jul).......*Polypody* **Polypodium vulgare**

Lvs to 60cm; lobes (2)4-7(9)cm, acute or obtuse, usu slightly serrate, lowest prs ± reflexed, longest nr middle of lf; sori maturing summer to autumn. (Jun) Aug-May (Jul) ....*Intermediate Polypody* **Polypodium interjectum**

Lvs to 50cm; lobes to 7cm, acute, usu serrate, lowest pr reflexed; sori maturing early spring. (Jul) Sept-Jul (Aug)..........*Southern Polypody* **Polypodium cambricum**

▪Petiole with 2 vb's

Lf veins translucent, netted, anastomosing. Rhizomatous, with lvs often emerging singly

Lvs to 100cm, ± pinnately lobed (occ pinnate nr base), light green, with v sparse pale scales on veins below (croziers hairy), with veins to margin. Petiole > lf, blackish at base. Rhizomes with spreading non-clathrate scales. Apr-Oct..........*Sensitive Fern* **Onoclea sensibilis**

Lvs to 60cm, pinnately lobed to entire, mid-green, with v sparse scales below, with obscure netted veins. Petiole > or < lf. Rhizomes with adpressed clathrate scales. All yr. VR alien ....*Kangaroo Fern* **Phymatosorus diversifolius**

Lf veins opaque, netted, obscurely anastomosing. Tufted, with several lvs in a rosette

Lvs 3-20cm; lobes to 2cm, ovate or oblong, obtuse, leathery, dull green above, entire or crenate, with obscure hydathodes, with dense rusty scales below (colourless when young). Petiole <¼ lf, with spreading clathrate scales. All yr. Calc habs..........*Rustyback* **Ceterach officinarum**

▪Petiole with >3 obscure vb's

Lobes 4-7 prs, orb to fan-shaped, fleshy, undulate, dull green, entire (to shallowly crenate), without midrib but with obscure free veins. Petiole weak, without scales. Tufted but lvs usu solitary (often with remains of last yr lvs at base). ❶. Apr-Sept.....*Moonwort* **Botrychium lunaria***

Blechnum spicant petiole TS

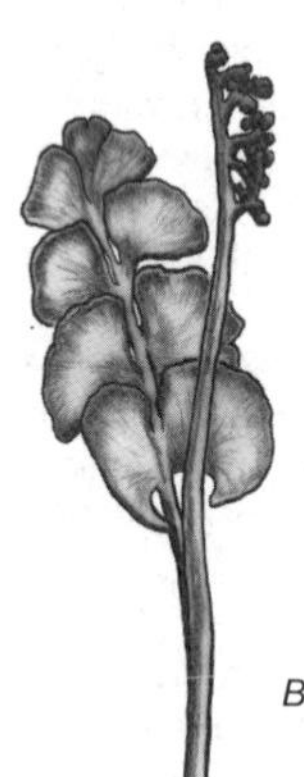

Botrychium lunaria

Group BC – *Lvs 1-pinnate with stalked pinna OR lvs forked (often repeatedly). Pinnae entire or toothed (*Onoclea sensibilis *may key out here).* ❶ *Lf veins netted*

■Petiole usu with 1 vb (vb usu green), scales clathrate. Lvs usu <15cm. Tufted

Pinnae narrowly linear, with veins obscure

Lvs 1-3x forked (often unequally so), 4-15cm, dark dull green, toothed nr apex to ± entire. Petiole >> lf, channelled, blackish at base with dark narrow scales when young. All yr. Acidic rocks, R..*Forked Spleenwort* **Asplenium septentrionale**

Pinnae fan-shaped to broadly obovate, with veins distinct but midrib absent

Lvs forked; pinnae 1cm wide, long-stalked, thin, dull green. Petiole shiny dark brown, wiry, ± ridged, with few scales at base, occ with 2 vb's. Often all yr. Calc rocks ....*Maidenhair Fern* **Adiantum capillus-veneris***

Pinnae ovate, with veins obscure

Rachis black-brown, unwinged. Pinnae to 5mm, ± thin, dull or shiny above. Petiole black, wiry, scales with a central dark stripe. All yr...........*Maidenhair Spleenwort* **Asplenium trichomanes***

Rachis green, narrowly winged. Pinnae 10-40mm, leathery, shiny above. Petiole green, blackish nr base, with black narrow wispy triangular scales at base. All yr. Coastal ....*Sea Spleenwort* **Asplenium marinum***

Rachis green, unwinged. Pinnae to 5mm, thin, dull green above. Petiole green, purplish-black at extreme base, with sparse scales. All yr. Basic rocks....*Green Spleenwort* **Asplenium viride**

■Petiole with 2 vb's, scales not clathrate. Lvs usu <30cm. Rhizomatous

Lvs with terminal pinna >> lateral pinnae, v serrate, with free veins; rachis not winged

Lvs 1-pinnate. Apr-Nov. VR alien..*Ladder Brake* **Pteris vittata**

Lvs 1-2-pinnate or palmately divided. All yr (usu), R alien..................*Ribbon Fern* **Pteris cretica**

■Petiole with ≥3 vb's (black bordered), scales not clathrate. Lvs often >30cm. Tufted or rhizomatous

Pinnae with an enlarged acroscopic basal lobe ('thumb') and stiff spines. Tufted. All yr

Lvs to 1.2m, always 1-pinnate; pinnae 35-55 prs, 4-7cm, dark green above, leathery, serrate, lowest pr hardly shorter than longest. Petiole to ¼ as long as lf, with dense scales nr base (often with dark streaks), with 5 vb's. Hortal............*Western Sword-fern* **Polystichum munitum**

Lvs to 1m, often 2-pinnate (at least nr base); pinnae 25-40 prs, 2-5(7)cm, dark green above, leathery, serrate. Petiole to ¼ as long as lf, with dense ± ovate brown scales, with 4-7 vb's ....*Hard Shield-fern* **Polystichum aculeatum**

Lvs to 30(60)cm, linear; pinnae 20-40 prs, 1-3cm, longest nr middle and often overlapping, shiny mid-green above, leathery, serrate with straight spine-pointed teeth (no longer bristle at pinnae apex), with scales on the rachis and lower surface. Petiole to 1/6 lf, with dense dark reddish-brown scales with wispy apex, with 3-4 vb's. Basic mtn rocks ....*Holly-fern* **Polystichum lonchitis**

Pinnae without an acroscopic lobe or stiff spines

Pinnae with translucent forked free veins

Pinnae usu with undulate margins, 4-15 prs, 6-10 x 2.5cm, leathery, ± cordate at base, ± entire. Petiole channelled, with brown scales nr base, with 7-11 vb's. Shortly rhizomatous, may form small trunk. All yr. Hortal.............................*Chilean Hard-fern* **Blechnum cordatum**

Pinnae with recurved margins, 10-20 prs, 5-10cm, thin to ± leathery, not cordate, serrate. Petiole channelled, with sparse broad scales, with 4-8 vb's. Tufted or shortly rhizomatous. Apr-Nov. Fens, R...*Crested Buckler-fern* **Dryopteris cristata**

Pinnae with opaque or obscure anastomosing netted veins ❶

Pinnae shiny green above, 5-12 prs, ovate, weakly serrate to ± entire, minutely brown hairy below, with terminal pinna similar to laterals. Petiole channelled, with dense scales at base often darker in centre, with 3-6 vb's. Rhizomatous, lvs usu solitary. All yr. R alien ....*House Holly-fern* **Cyrtomium falcatum***

Pinnae dull green above, 5-10 prs, ovate, weakly serrate to ± entire, ± hairless below, with terminal pinna similar to laterals. Petiole channelled, with scales at base rarely darker in centre, with 3 vb's. Tufted, with several lvs in rosette. All yr. Hortal ....*Japanese Holly-fern* **Cyrtomium fortunei***

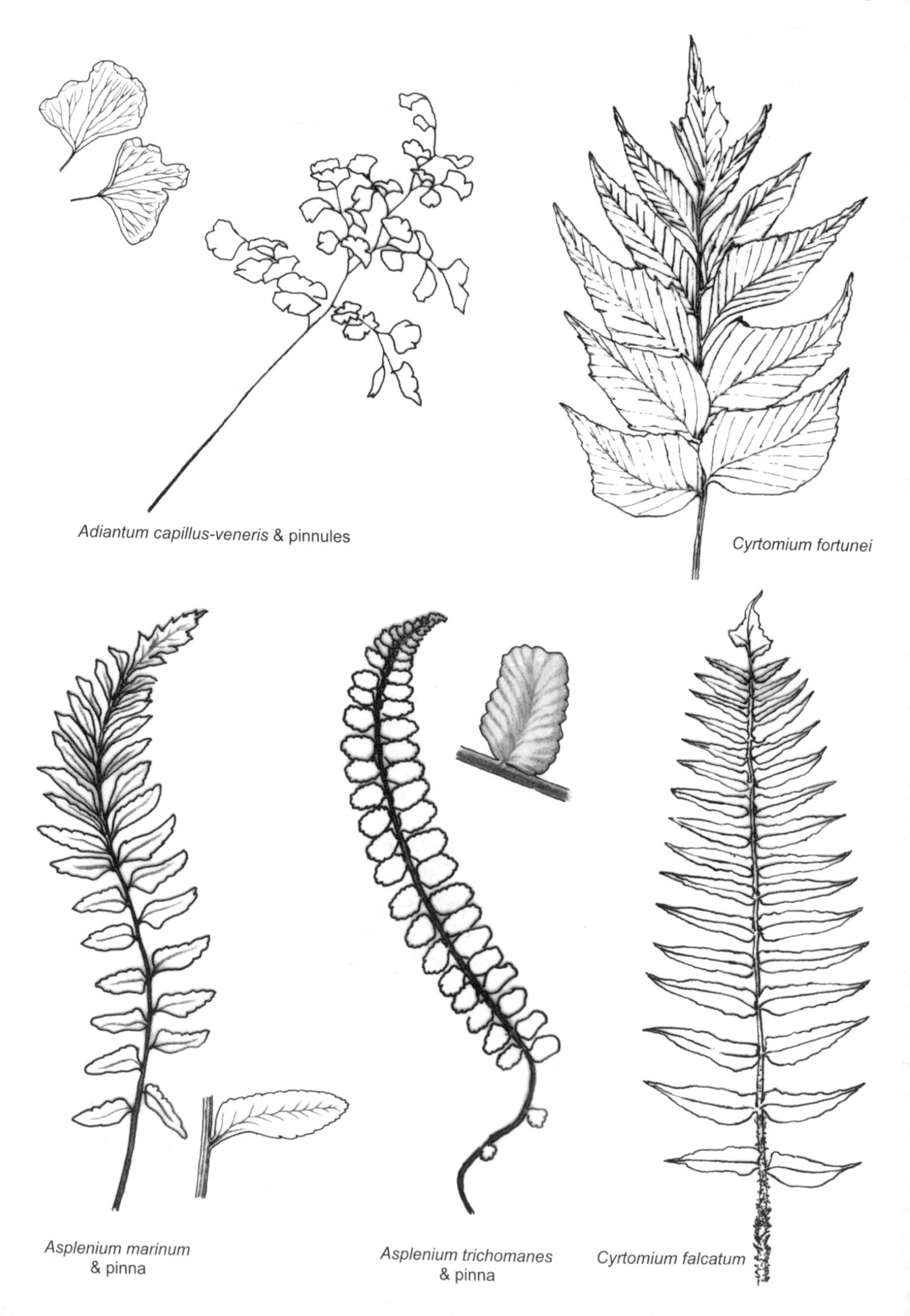

Adiantum capillus-veneris & pinnules

Cyrtomium fortunei

Asplenium marinum & pinna

Asplenium trichomanes & pinna

Cyrtomium falcatum

Group BD – *Lvs 1-pinnate-pinnatisect, hairy at least below (hairs occ sparse or confined to rachis). Petiole channelled (often weakly so).* ❶ *Lvs lemon-scented.* ❷ *Hairs septate*

■Rhizomatous, lvs often emerging singly or in prs.

Lowest pr of pinnae longest, reflexed, with free veins. Rachis winged. Lobes ± acute, straight. Petiole pale at base, with few scales, with 2 vb's fused into U-shape. Apr-Oct ....*Beech Fern* **Phegopteris connectilis**

Lowest pr of pinnae usu shorter than prs above, not reflexed, with free veins. Rachis not winged. Lobes ± acute, curved. Petiole blackish at base, scales soon lost, with 2 vb's fused into U-shape. May-Oct..*Marsh Fern* **Thelypteris palustris**

■Tufted, lvs often forming a 'shuttlecock'

Lvs >30cm. Petiole without a knee-joint

Lvs with yellow sessile glands esp below, with white scales when young, light green, lowest pr of pinnae often shorter than prs above. Petiole with 2 vb's fused into U-shape, green, with few greyish scales. Stolons absent. ❶. Apr-Oct.......*Lemon-scented Fern* **Oreopteris limbosperma**

Lvs without glands, light green, with shorter pinnae nr base. Petiole with 2 vb's not fused, occ purplish, blackish at base, with few scales. Stolons occ present. Apr-Oct ....*Ostrich Fern* **Matteuccia struthiopteris***

Lvs <10cm. Petiole with a knee-joint at base (abscission layer). ❷. Mtn rocks, VR. Sch8

Lvs with pinnae to 15mm; mid-lf pinnae much divided, with (3)6(9) prs of lobes, with dense linear-lanc scales below, with free veins; rachis densely covered with scales and hairs. Apr-Dec (usu)..*Oblong Woodsia* **Woodsia ilvensis**

Lvs with pinnae to 8mm; mid-lf pinnae less divided, with only 1-2(3) prs of lobes, with few or no scales below, with free veins; rachis sparsely covered with scales and hairs, occ ± hairless. Apr-Dec (usu)..*Alpine Woodsia* **Woodsia alpina**

Matteuccia struthiopteris

Group BE – *Lvs 1-pinnate-pinnatisect, hairless (may have scales). Petiole channelled (or weakly so)*

■Petiole with 2 vb's. Lvs to 2m, rooting at tips from a scaly bud, with translucent netted veins
Pinnae serrate. Apr-Nov. VR alien..*Chain-fern* **Woodwardia radicans**

■Petiole with 3-8 vb's. Lvs usu <1m, never rooting at tips, with free veins
Pinnae with an acroscopic basal lobe ('thumb') and stiff spinulose teeth
Lvs usu shiny green, stiff, with basal pinnae much shorter than middle pinnae; pinnules with acute angle at base, those nr rachis usu sessile and decurrent from an acute angled base, tapered to a v acute spinulose apex. All yr.................*Hard Shield-fern* **Polystichum aculeatum**
Pinnae without an acroscopic basal lobe ('thumb') and spinulose teeth
Pinnae with a dark spot below at junction with rachis. Petiole with dense dark-based reddish scales. Lvs to 1.5m, with pinnae margins flat, mid- to yellow-green above, sessile glands usu absent. Apr-Nov (all yr)...*Scaly Male-fern* **Dryopteris affinis** agg
Pinnae without dark spot below at junction with rachis. Petiole with scales usu without dark base
Petiole <¼ of lf. Lvs forming 'shuttlecock', with pinnae shortest nr base
Lvs to 120cm, with pinna margins ± flat or recurved and with acute teeth at apex, green above, usu without sessile glands. Petiole with pale brown wispy scales (occ with darker base). Apr-Nov (usu)...*Male-fern* **Dryopteris filix-mas***
Lvs to 50(80)cm, with pinnae margins crisped upwards and with obtuse teeth at apex, grey- to mid-green above, with sessile glands. Petiole with v pale grey-brown acute scales. Apr-Oct. Uplands...*Mountain Male-fern* **Dryopteris oreades**
Petiole ⅓-¼ of lf. Lvs not or hardly forming 'shuttlecock' (shortly rhizomatous), with pinnae hardly shorter nr base
Lvs to 60cm, with pinnae margins recurved. Petiole with sparse broad acuminate scales. Apr-Nov. Fens, R..*Crested Buckler-fern* **Dryopteris cristata**

Dryopteris filix-mas
petiole TS & scale

Group BF – *Lvs (1)2-4-pinnate, translucent, 1-cell thick, hairless, without stomata. All yr. Shady high-humidity habs, mostly W & N Br*

■Lvs 15-45cm, 2-3-pinnate, broadly triangular, dark green. VR. Sch8
....*Killarney Fern* **Trichomanes speciosum**

■Lvs to 15cm, some often 1-pinnate
Lvs ovate-oblong, blue-green, with veins not reaching truncate toothed apex, with ± square cells. Petiole >⅓ of frond length, with 1 vb. Plant often hanging over substrate
....*Tunbridge Filmy-fern* **Hymenophyllum tunbrigense***
Lvs usu linear, olive-green, with veins reaching rounded (usu entire) apex, with ± rectangular cells. Petiole <⅓ of frond length, with 1 vb. Plant arching away from substrate
....*Wilson's Filmy-fern* **Hymenophyllum wilsonii***

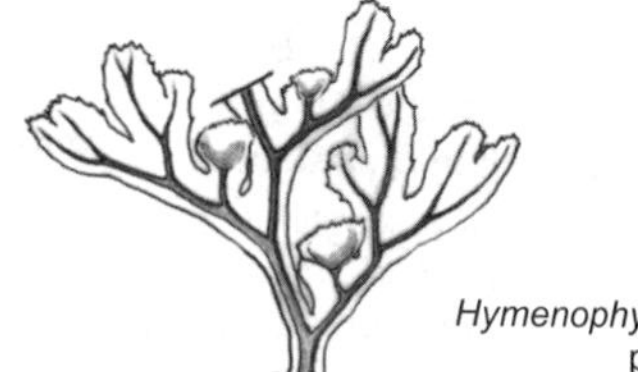

Hymenophyllum tunbrigense
pinnae

Hymenophyllum wilsonii
pinnae

Group BG – *Lvs 2-4-pinnate. Petiole with 1 or 2 equal vb's. (Woodsia may key out here).* ❶ *Ann*

■Large fern with lvs ≥1m

Petiole channelled, with scales and hairs nr base, with involute vb. Lvs 2-pinnate, with lowest pr of pinnae usu shortest, dark green, with tough plastic-feel, usu with brown hairs when young, with veins forked once to tooth apex. Trunk-forming. Apr-Oct. VR alien ....*Tree-fern* **Dicksonia antarctica**

Petiole not channelled, without scales (may have brown hairs), with involute vb. Lvs 2-pinnate, with lowest pr of pinnae often longest, light green, thin, with brown hairs when young, with veins forked to slightly serrate sinus. Tussock-forming. Apr-Nov................*Royal Fern* **Osmunda regalis**

■Medium or small fern with lvs <75cm. Petiole channelled (occ weakly so)

Pinnules with main vein or midrib absent or obscure. Lvs 3-foliate when young

Pinnules rhombic, dark green, thick, all veins obscure. Petiole green, with few scales, with 1-2 vb's. Tufted. All yr...*Wall-rue* **Asplenium ruta-muraria***

Pinnules fan-shaped, dark green, thin, veins distinct but midrib absent. Petiole shiny dark brown, with few scales, with 1-2 vb's. Tufted. Usu all yr. . *Maidenhair Fern* **Adiantum capillus-veneris***

Pinnules wedge-shaped and deeply lobed, yellow-green, thin, with all veins obscure. Petiole green, with few hair-like scales, with 1-2 vb's. Tufted. ❶. Mar-Jun. Channel Is ....*Jersey Fern* **Anogramma leptophylla***

Pinnules with main vein or midrib present (may be opaque). Lvs not 3-foliate when young

Lowest pr of pinnae shorter than prs above. Petiole with 2 vb's fusing into U-shape. Lowest pr pinnae occ reflexed

Petiole often reddish, with few scales. Lvs yellow-green, limp. Sori curved. Tufted. May-Oct ....*Lady-fern* **Athyrium filix-femina***

Petiole rarely reddish, with few scales. Lvs yellow-green, limp. Sori round. Shortly rhizomatous. May-Oct. Mtns >500m, Scot..............*Alpine Lady-fern* **Athyrium distentifolium**

Lowest pr of pinnae longest

Pinnae all opp, curving towards apex. Petiole with 2 vb's never fusing

Lvs with apple-scented ± sessile glands (esp on margins), 2-pinnate-pinnatisect, held horizontal, with acute lower pinnae on 2cm stalk, bright green, rigid; rachis densely glandular-hairy. Petiole to 30cm, dark green, with scales nr base. Rhizomatous. Apr-Oct. Basic rocks...*Limestone Fern* **Gymnocarpium robertianum**

Lvs usu without glands, not apple-scented, 2-pinnate-pinnatisect, held horizontal, with obtuse lower pinnae on 2cm stalk, bright green, limp; rachis sparsely glandular-hairy to hairless. Petiole to 20cm, blackish-green, with scales ± absent. Rhizomatous. Apr-Oct. Usu humus-rich acidic habs...*Oak Fern* **Gymnocarpium dryopteris**

Pinnae mostly alt, occ curving towards apex

Petiole with 2 vb's never fusing. Rhizomatous

Lvs arising singly, 3-4-pinnate, held horizontal on vertical petiole, with lower pinnae on <1.5cm stalk and curving towards apex. Petiole with gland-tipped clathrate scales. May-Oct. VR, Scot mtns.......................................*Mountain Bladder-fern* **Cystopteris montana**

Petiole with 2 vb's fusing into U-shape. Tufted

Lvs parsley-like, limp; pinnules with veins ending before rounded teeth apices and slightly expanded to form sunken hydathodes. Petiole with v few or no scales. Acidic rocks and scree...*Parsley Fern* **Cryptogramma crispa***

Petiole with 2 vb's fusing into X-shape nr lf. Tufted

Lvs shiny dark green above; pinnules ovate-elliptic with acute teeth, with obscure hydathodes. Petiole blackish nr base, with sparse clathrate scales. All yr ....*Black Spleenwort* **Asplenium adiantum-nigrum***

Lvs shiny yellow-green above; pinnules narrowly lanceolate with fine acuminate teeth, with obscure hydathodes. Petiole with sparse clathrate scales. All yr. VR, Ire ....*Irish Spleenwort* **Asplenium onopteris**

Petiole with 2 vb's fusing into X-shape. Lowest pr pinnae usu reflexed

Pinnules ± crenate with obtuse weakly mucronate teeth. Tufted. All yr. Usu coastal, W Br ....*Lanceolate Spleenwort* **Asplenium obovatum***

Petiole with 2 vb's never fusing. Lowest pr pinnae rarely reflexed

Petiole with gland-tipped scales (often v sparse), brittle. Lvs usu light green and opaque

Pinnules with veins ending in teeth, the ultimate segments usu with acute apices, the margins not overlapping. Tufted. Apr-Nov. Basic rocks ....*Brittle Bladder-fern* **Cystopteris fragilis***

Pinnules usu with most veins ending in retuse teeth, the ultimate segments usu with acute (but not tapered) apices, the margins often overlapping. Tufted. Apr-Oct. Rock ledges, sea caves, VR, Scot. Sch8....................................*Dickie's Bladder-fern* **Cystopteris dickieana**

Petiole usu without scales (not gland-tipped if present), brittle. Lvs usu dark green and translucent

Pinnules with most or all veins ending in retuse teeth. Tufted. All yr. Base-poor habs, VR, Cornwall, Ire..*Diaphanous Bladder-fern* **Cystopteris diaphana***

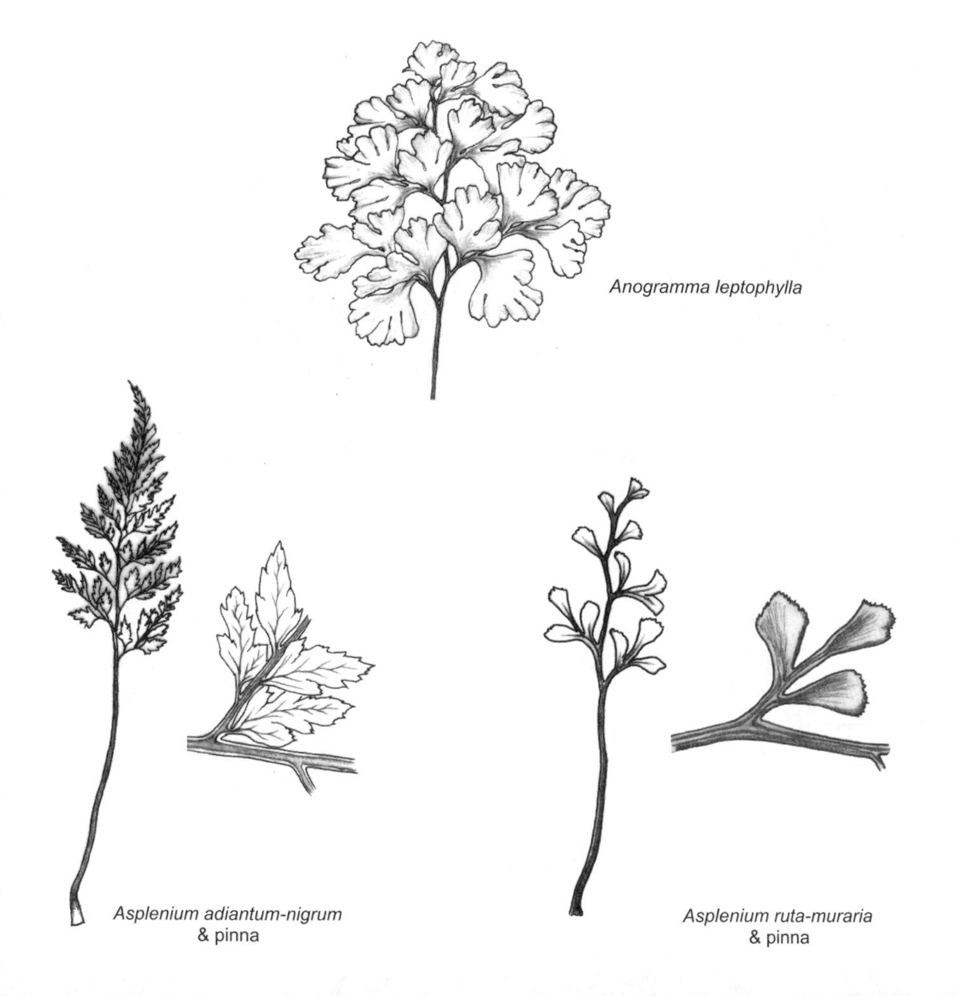

Anogramma leptophylla

Asplenium adiantum-nigrum & pinna

Asplenium ruta-muraria & pinna

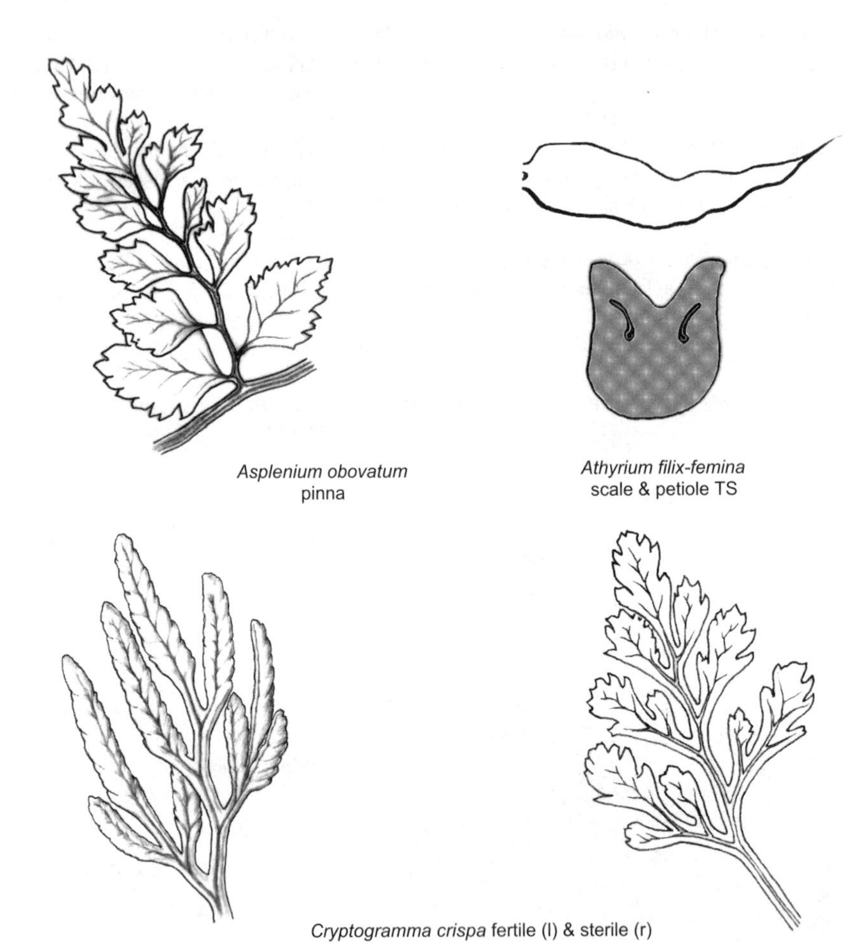

Asplenium obovatum
pinna

Athyrium filix-femina
scale & petiole TS

Cryptogramma crispa fertile (l) & sterile (r)

Cystopteris diaphana pinnule

Cystopteris fragilis pinnule

Group BH – *Lvs 2-3(4)-pinnate, often with obscure 2º veins. Petiole channelled, with 3-7 (occ more) unequal vb's.* ❶ *Petiole with extra-floral nectaries nr pinnae.* ❷ *Scales on petiole with a dark central stripe*

▪Pinnae with an acroscopic basal lobe ('thumb'), usu with ± stiff terminal spines, without visible hydathodes. Lvs without glands, with basal pinnae shorter than middle

Pinnules nr rachis usu stalked into an obtuse angled base, tapered to an obtuse hair-tipped apex (0.7-1.5mm). Lvs forming 'shuttlecock', shiny bright or dark green, limp. Petiole with dense pale grey- to dark reddish-brown scales at base. All yr (usu)...*Soft Shield-fern* **Polystichum setiferum**

Pinnules nr rachis usu sessile, decurrent from an acute angled base, tapered to an acute spinulose apex. Lvs shiny green, stiff. Petiole with dense dark scales (also on rachis) esp nr base. All yr...*Hard Shield-fern* **Polystichum aculeatum**

▪Pinnae without an acroscopic basal lobe nor stiff terminal spines, with elongate submarginal hydathode

Lvs usu with hay-scented (±) sessile glands, with lowest pr pinnae often longest

Lvs grey-green, with flat or recurved margins, with stalked yellow (or colourless) glands above and on margin, weakly scented when fresh. Petiole not purplish or pruinose, with long pale brown scales esp at base. Tufted. May-Dec. Limestone habs ....*Rigid Buckler-fern* **Dryopteris submontana**

Lvs yellow-green, with upwardly crisped margins, with colourless sessile glands (often sparse), strongly scented when dry (often odourless when fresh). Petiole occ purplish (esp on underside) or pruinose, with pale brown scales (usu with dark band at base). Tufted. All yr (usu). Usu acidic habs...*Hay-scented Buckler-fern* **Dryopteris aemula**

Lvs usu without glands (or sparse and scentless if present)

Petiole without scales, minutely reddish hairy at base. Lvs arising singly, with the lowest pr pinnae often longest

Lvs 3-pinnate-pinnatisect, with reddish-brown hairs when young (often retained in axils at maturity), odorous; pinnules with recurved broadly hyaline ciliate margins. Petiole blackish at base, with extra-floral nectaries nr pinna, with several vb's (occ fusing). Strongly rhizomatous. ❶. May-Oct...*Bracken* **Pteridium aquilinum**

Petiole with scales with a dark central stripe (not young plants), hairless. Lvs in a 'shuttlecock', with the lowest pr pinnae usu longest

Lvs usu without glands, dark bluish-green; pinnules with recurved margins, with short (0.2mm) spinulose-tipped teeth. Petiole green (occ brown), with (3)4-7 vb's. ❷. Mar-Dec (usu) ....*Broad Buckler-fern* **Dryopteris dilatata**

Petiole with scales usu without a dark stripe, hairless. Lvs in a loose 'shuttlecock', with the lowest pr of pinnae usu shorter than prs above.

Lvs pale to yellowish green; pinnules with flat or slightly recurved margins, with long incurving spinulose-tipped teeth; basal pinnae without an enlarged basioscopic lobe. Petiole with 3(7) vb's, with pale grey-brown (occ hair-pointed) scales. Occ rhizomatous. Mar-Dec. Damp woods, fens..*Narrow Buckler-fern* **Dryopteris carthusiana**

Lvs light green; pinnules with recurved margins, with short spinulose-tipped teeth. Petiole with 3-8 vb's, with sparse broad acuminate scales; basal pinnae without an enlarged basioscopic lobe. Tufted or shortly rhizomatous. Apr-Nov. Fens, R ....*Crested Buckler-fern* **Dryopteris cristata**

Lvs mid-green; pinnules with flat margins, with short ≤0.5mm spinulose-tipped teeth; basal pinnae with an enlarged basioscopic lobe. Petiole with 4-7 vb's, with abruptly acuminate ginger-brown scales (occ with dark central stripe). Tufted. May-Nov. Mtn scree ....*Northern Buckler-fern* **Dryopteris expansa**

Key to Group in Division C

(Clubmosses Lycopodiopsida*)*

Group CA – *Plants hairless. Stems <20cm tall, branched dichotomously, creeping and rooting adventitiously (exc* Huperzia*). Lvs scale-like, green, overlapping (imbricate), never dropping (no abscission layer), with 1 vein (usu not visible). Evergreen per*

■Lvs opp, decussate (4-ranked), all ± same size

Lvs 2-6mm, acute, ± entire, occ twisted. Stems long, with tufts of ± erect forked branches to 8cm

Plant glaucous. Shoots ± square or cylindrical in TS. Lvs 0.5-1mm wide, lanc ....*Alpine Clubmoss* **Diphasiastrum alpinum***

Plant yellow-green. Shoots ± flattened in TS. Lvs 1-1.3mm wide, elliptic-lanc ....*Issler's Clubmoss* **Diphasiastrum complanatum**

■Lvs alt, spirally arranged in many rows, all ± same size

Lvs with long hair-point (to 4mm) even when mature

Lvs to 5 x 0.5mm, lanc, acute, entire to toothed. Hths, mtn moors ....*Stag's-horn Clubmoss* **Lycopodium clavatum***

Lvs without hair-point when mature

Plant tufted, all stems erect and forking into branches of equal length

Lvs 5-7 x 1-1.5mm, linear-lanc, acute, entire to minutely toothed nr apex. Usu uplands ....*Fir Clubmoss* **Huperzia selago***

Plant patch forming, at least some stems prostrate and all forking into branches of unequal length

Stems v slender, <10cm

Lvs 2-5 x 1mm, ovate to lanc, acute, spreading to ± adpressed, spiny-ciliate, with minute ligule when young. Stems with small central stele surrounded by tiny hollows. Cone shoots erect, deciduous. Damp base-rich habs.............*Lesser Clubmoss* **Selaginella selaginoides**

Stems stout, often >10cm

Lvs entire, to 5 x 1mm, lanc, acute, without a hair-point, curved upwards. Stem without obvious resting stage (all lvs ± same size). Bare peat or damp sand ....*Marsh Clubmoss* **Lycopodiella inundata**

Lvs weakly toothed to ± entire, to 6 x 1mm, lanc, acuminate, with a fragile hair-point to 0.5mm when young, adpressed to spreading. Stems with obvious elongated resting stage (with shorter lvs). Plant like *L. clavatum* in habit. Mtn moors, N Br ....*Interrupted Clubmoss* **Lycopodium annotinum**

■Lvs alt, arranged in 4 rows, of 2 sizes

Lvs minutely toothed, acute, strongly flattened; dorsal lvs (overlying stem) 1-1.5mm, ovate-lanc, unequal at base with rounded auricle, adpressed to stem; lateral lvs 2.5mm, oblong-ovate, acute, rounded at base, spreading from stem. Stems with 2 small steles nr branches. R alien ....*Krauss's Clubmoss* **Selaginella kraussiana***

Diphasiastrum alpinum & stem TS

Lycopodium clavatum & stem TS

Huperzia selago & stem TS

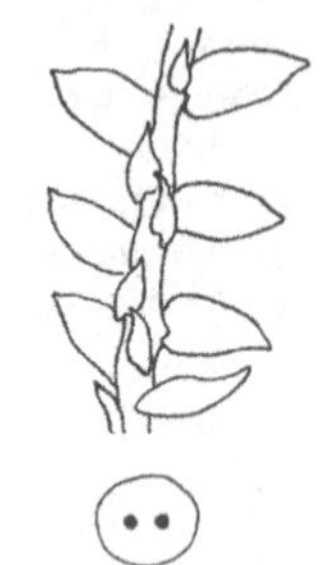

Selaginella kraussiana & stem TS

Key to Groups in Division D

(Conifers Pinopsida*)*

Coniferous tree or shrub. Lvs scale- or needle-like (occ subulate), usu rigid. Branches with monopodial growth, usu in whorls (exc Taxus, Taxodium*) and resinous (exc* Taxus*)*

■Lvs in clusters of 2, 3 or 5, needle-like...**DA**
■Lvs in dense clusters of >10 on short shoots (lf-cushions), needle-like..**DB**
■Lvs 3-whorled around twig, subulate, sharp and usu glaucous above...**DC**
■Lvs single along twig, falling with age
 Young twigs ridged, grooved or slightly angled (with pegs or oblique projection)
  Lvs flat, with stomata below only...**DD**
  Lvs 4-angled or laterally flattened, with stomata all sides...**DE**
 Young twigs round, not ridged or grooved (without pegs). Lvs flat..**DF**
■Lvs imbricate along twig (sessile and at least partly adnate to twig), never falling
 Lvs alt, spirally arranged, 3-50mm, without resin glands..**DG**
 Lvs opp, decussate (4-ranked), usu 1.5-4mm, often with resin gland (usu sunken)....................**DH**

DA

Pinus sylvestris bud
(showing reflexed scales)

Pinus sylvestris lf TS

Pinus pinaster lf TS

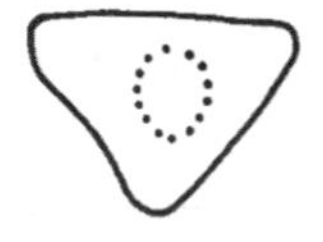

Pinus wallichiana lf TS

Group DA – *Lvs in clusters of 2, 3 or 5, needle-like, falling together within a sheath*

■Lvs in prs, semi-cylindrical (flat above, rounded below), weakly serrulate, with 2 vb's

Lvs short, mostly to 8cm

Lvs glaucous, with fine white fibres when broken. Buds rarely resinous, scales free or reflexed

Lvs 2-8cm x 1-2mm, stiff, twisted, with 8-11 rows of stomata both sides, with 5-10 marginal resin canals. Lf sheaths 8-12mm. Buds 5-12mm, ovoid, acute, reddish-brown. Bark reddish-brown (at least above), fissured into plates................................*Scots Pine* **Pinus sylvestris***

Lvs green, not showing fibres when broken. Buds v resinous, scales adpressed

Tree to 25m, usu with single trunk. Lvs dark green, 1.2-1.4mm wide (1.7mm in var *latifolia*), stiff, twisted, with rows of stomata both sides, with median resin canals. Buds 10-20mm, acute to obtuse. Lf sheaths 2-4(6)mm. Bark dark reddish-brown, deeply fissured into plates ....*Lodgepole Pine* **Pinus contorta**

Shrub or tree to 4m, multi-stemmed or many-branched. Lvs dark grey-green, 1.2-1.5mm wide, stiff, twisted, with rows of stomata both sides, with marginal resin canals. Buds 6-12mm, acute to obtuse. Lf sheaths (5)10-15mm. Bark grey-pink, cracked into squares ....*Dwarf Mountain-pine* **Pinus mugo**

Lvs long, mostly >10cm

Buds resinous or not, usu with adpressed scales. Lf sheaths 12mm when young

Lvs (8)10-17cm x 1-2mm, dark grey-green, sharply acute, with 6-8 rows of stomata above and <11 below, with 3-17 median resin canals. Buds 1.2-2.5cm, oblong-ovoid or cylindrical, acuminate, light brown. Bark dark brown or blackish, with scaly plates flaking ....*Corsican Pine* **Pinus nigra**

Buds not resinous, with reflexed ciliate scales. Lf sheaths 20mm when young

Lvs usu 12-20cm x 1.5-3mm, dark or bright green, sharply acute, with 10-11 rows of stomata above and 11-15 below, with 6 median resin canals. Buds 2-3cm, cylindrical, acute, light brown. Bark dark reddish-brown, v fissured..............................*Maritime Pine* **Pinus pinaster***

■Lvs in clusters of 3, flattened triangular (keeled above, rounded below), serrulate, with 2 vb's

Lvs (8)10-15(25)cm x 0.8-1.2mm, bright green (basal 5mm white), flexible, twisted, sharply acute (not apiculate), with 8 rows of stomata below, weak fibres occ present on breaking, with 2 marginal resin canals, aniseed-scented. Buds to 1.5cm, acute, reddish-brown, resinous, with adpressed scales. Lf sheaths (5)10-15mm, reddish-brown, wrinkled. Twigs pale greyish, hairless. Bark deeply fissured..*Monterey Pine* **Pinus radiata**

Lvs 13-25cm x 1.4-1.6mm, dark grey-green to deep yellow-green, stiff, often twisted, sharply long-apiculate, with 4-5 rows of stomata below, fibres absent on breaking, with 2 median resin canals, pine-scented. Buds to 4cm, acute, reddish-brown, resinous, with adpressed scales. Lf sheaths (10)20-30mm. Twigs reddish-brown, hairless. Bark fissured into plates ....*Western Yellow-pine* **Pinus ponderosa**

■Lvs in clusters of 5, trigonous (rounded above, keeled below), serrulate, with 1 vb

Young twigs minutely hairy nr apex (and below the lf clusters). Buds ± not resinous

Lvs 6-14cm x 0.5-0.7mm, with ± obtuse but serrulate apex, pale green, flexible, occ pendent, with 2 rows of stomata along each of the 2 lower sides (none above), with marginal resin canals. Buds ovoid to conical, usu acuminate, brownish. Twigs brownish, not pruinose, with short reddish-brown hairs...*Weymouth Pine* **Pinus strobus**

Young twigs hairless. Buds ± not resinous

Twigs usu pruinose. Lvs (10)13-20cm x 0.5-0.7mm, with ± obtuse but serrulate apex, pale bluish-green, flexible, usu pendent, serrulate along length, with 2-4 rows of stomata along each of the 2 lower sides, with marginal resin canals. Buds greyish with black-tipped scales, sharply acuminate..*Bhutan Pine* **Pinus wallichiana***

Twigs not pruinose. Lvs 7-12cm x 0.7-0.9mm, with ± obtuse but serrulate apex, dark bluish-green to bright green, flexible, often pendent, serrulate in distal ½, with 4 rows of stomata along each of the 2 lower sides (none above but occ 2 whitish lines), with marginal resin canals. Buds grey-brown without black-tipped scales, sharply acuminate (to obtuse) ....*Macedonian Pine* **Pinus peuce**

Group DB – *Lvs in dense clusters of >10 on short shoots (lf-cushions) (occ spirally arranged single lvs on long shoots), needle-like, falling singly (without sheaths). Buds ovoid*

▪Lvs deciduous, 1-2mm wide, flat, with stomata below only. Twigs hairy or hairless

Lvs 30-40 per short shoot, sessile, flexible, obtuse but apiculate on long shoots (not sharply so). Short shoots 3-7mm

Young twigs straw-yellow, hairless. Lvs 12-30 x 1mm, bright green, with indistinct greenish bands below, with midrib raised below. PL 20............................*European Larch* **Larix decidua**

Young twigs usu pinkish-brown, usu hairless. Lvs 40-50 x 2mm, glaucous, often with pale bands below, with midrib hardly raised below. *L. decidua x kaempferi* ....*Hybrid Larch* **Larix** x **marschlinsii**

Young twigs pruinose, soon reddish, sparsely hairy to hairless. Lvs 15-40 x 1mm, v glaucous at least when young, with 2 obvious broad whitish stomatal bands below, with midrib flat to slightly raised both sides..*Japanese Larch* **Larix kaempferi**

▪Lvs evergreen, to 1mm wide, 3-5-angled (weakly trigonous), with stomata all sides. Twigs minutely hairy

Lvs 3-6cm, flexible. Leading shoot (and young branches) drooping or pendent

Lvs grey-green to bright green, ± gradually tapered to a sharp colourless translucent tip to 0.4mm...*Deodar* **Cedrus deodara**

Lvs usu <2.5cm, stiff. Leading shoot (esp of young trees) curved or erect

Lvs mostly 1.5-3cm, abruptly tapered to a sharp reddish 0.2mm translucent tip, dark green or (rarely) glaucous. Young branches mostly horizontal..............*Cedar-of-Lebanon* **Cedrus libani**

Lvs mostly <1.8cm, gradually tapered to a sharp yellowish 0.3-0.5mm translucent tip, dark green or glaucous (more so than *C. libani*). Young branches mostly ascending (occ pendent) ....*Atlas Cedar* **Cedrus atlantica***

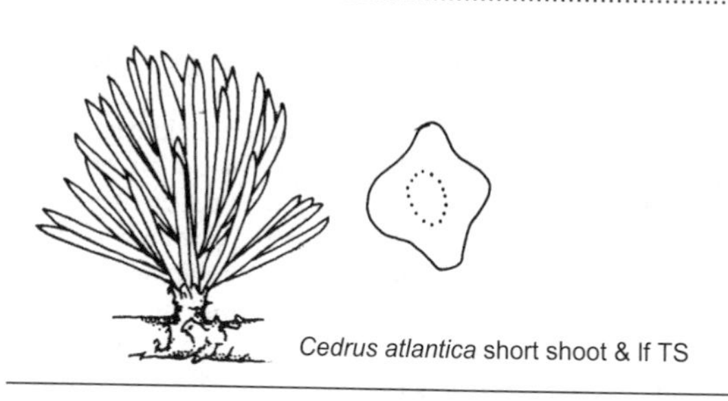

Cedrus atlantica short shoot & lf TS

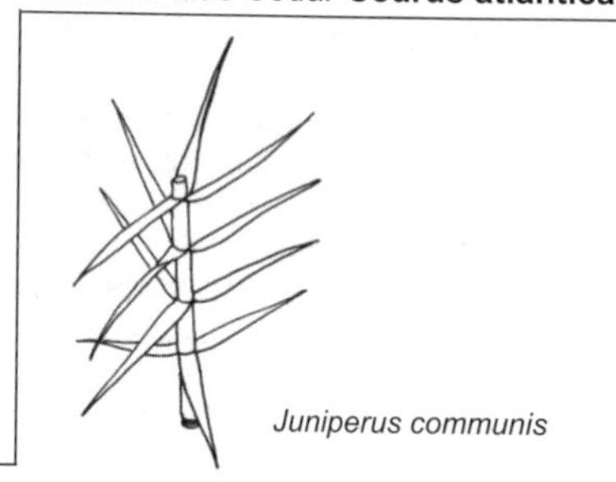

Juniperus communis

Group DC – *Juvenile lvs 3-whorled around twig (rarely opp), subulate, spreading or ascending, sharp, sessile, ± flat, usu glaucous above, entire. Adult lvs (if present) scale-like, adpressed, opposite, decussate*

▪All lvs subulate (juvenile foliage)

Lvs entirely glaucous above, green below

Lvs (5)10-20 x 1mm, linear-subulate, acute, with a sharp translucent 0.2mm apex, flat above, weakly keeled below, with stomata above only. Twigs ridged, hairless, sandlewood-scented. Bark brown, fibrous, flaking in strips. Shrub or small tree, to 4(10)m, multi-stemmed, erect to prostrate...*Common Juniper* **Juniperus communis***

Lvs with 2 glaucous bands above, green or glaucous below

Lvs (2.5)5-8(10) x 1-1.5mm, with 0.5mm spine at apex, decurrent at base, spreading to ± adpressed, straight or slightly curved, slightly concave, weakly keeled below. Small tree or shrub, usu 1-stemmed, to 12m, erect to ± prostrate........*Meyer's Juniper* **Juniperus squamata 'Meyeri'**

▪Few lvs subulate (juvenile foliage), often in prs, mostly small adpressed scale lvs (adult foliage)

Adult lvs usu acute with a 0.1mm translucent point, with stomata totally absent. Tree ....*Pencil Cedar* **Juniperus virginiana***

Adult lvs usu obtuse without a translucent point, with sparse stomata along pale proximal margins. Tree or shrub...*Chinese Juniper* **Juniperus chinensis***

Group DD – *Twigs ridged, grooved or slightly angled, often with raised pegs or oblique projections at lf bases. Lvs evergreen, stiff, oblong, linear or needle-like, flat (dorso-ventrally), with stomatal bands below only. ❶ Lvs occ twisted (turned over, exposing lowerside above) and overlying shoot*

■Young twigs green, hairless, without pegs (but with oblique projections)

Lvs with midrib raised both sides, 10-25(30) x 2-3mm, acute, shortly stalked, dark green above, paler yellowish-green below with 2 indistinct stomatal bands, entire, with recurved margins, odourless, without resin canals. Buds to 3.5mm, green, not resinous. Trunk often with epicormic buds. Bark plate-like

Main branches spreading. Lvs spreading in 2 lateral rows along twig, straight, dull dark green above..*Yew* **Taxus baccata**

Main branches fastigiate. Lvs spiralling all around twig, strongly curved, dull or shiny blackish green above...*Irish Yew* **Taxus baccata 'Fastigiata'**

Lvs with midrib flat above, 6-18 x 2-3mm, acute, ± sessile, shiny dark green above, ± glaucous below with 2 obvious stomatal bands, entire, with flat margins, with sour parsley scent, with 3 resin canals. Buds to 3.5mm, green, not resinous. Trunk without epicormic buds. Bark thick, spongy, red-brown..*Coastal Redwood* **Sequoia sempervirens**

■Young twigs brown, with obvious 0.3-1.5mm pegs

Lvs minutely serrulate, those overlying shoot occ twisted ❶

Lvs 2mm wide, obtuse, dark green above, grapefruit-scented, with midrib slightly sunken above (not raised below). Buds globose-ovoid, obtuse. Leading shoot and branches often pendent at tips

Lvs 5-20mm, v unequal (lateral lvs longer than upper lvs), ± parallel-sided, with 2 broad whitish bands below. Young twigs densely hairy with long and short hairs ....*Western Hemlock-spruce* **Tsuga heterophylla**

Lvs to 10mm, unequal, slightly tapered to apex, with 2 indistinct whitish bands below. Young twigs shaggy-hairy with long and short hairs.......*Eastern Hemlock-spruce* **Tsuga canadensis**

Lvs entire, those overlying shoot (if present) never twisted over

Petiole to 1mm. Lvs usu spreading either side of twig (leaving a parting), with midrib slightly sunken above. Twigs weakly ridged; lf scars on small pegs, <20° to twig. Trunk deeply gnarled

Lvs 2-3(4)cm x 1.5-1.8mm, ± equal, acute to retuse, shiny light to dark green above, with 2 white to pale green bands below, tangerine-scented. Twigs yellowish-brown, minutely hairy. Buds 5-10mm, slender, v acute, shiny reddish-brown, not or hardly resinous, with sparsely ciliate adpressed scales...*Douglas Fir* **Pseudotsuga menziesii***

Petiole 0-0.5mm. Lvs usu spreading around twig (some lvs above, obscuring twig), with midrib slightly raised above. Twigs distinctly ridged; lf scars on small pegs, 30-90° to twig. Bark scaly

Young twigs hairless, pale brown. Lvs 10-25 x 1mm, usu sharply acute, ± forward-pointing, dark green above, with 2 whitish bands below (rarely 1-2 rows of stomata above), with sweet fruity banana scent. Buds 6mm, ovoid, acute, brown, not resinous, without free scale tips. Branches usu drooping...*Sitka Spruce* **Picea sitchensis***

Young twigs hairy, pale brown. Lvs 10-30 x 1.5-2mm, ± obtuse, often upswept, yellowish- or bluish-green above, with 2 broad white bands below, orange-scented. Buds 6mm, ovoid, ± obtuse, brown, hardly resinous, with free scale tips. Branches drooping but upswept at apices (like ski-jumps!)...*Serbian Spruce* **Picea omorika**

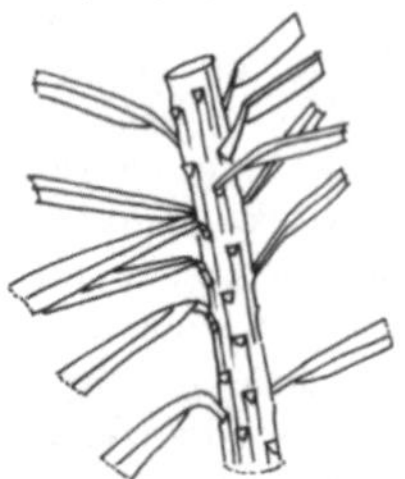

Pseudotsuga menziesii

Picea sitchensis lf TS

Group DE – *Twigs ridged, grooved or slightly angled, with raised pegs or oblique projections. Lvs evergreen, not in 2 lateral rows, stiff, oblong, linear or needle-like, 4-angled (rhombic, at least as thick as wide) in TS, or laterally flattened, entire, with stomata all sides. Buds not or hardly resinous, ovoid.* ❶ *Branches pendent.* ❷ *Buds with reflexed scales*

■Lvs 3-5cm

Young twigs usu hairless, whitish-brown. Lvs 1mm wide, acute, shiny green, curved torwards twig apex, tangerine-scented. Buds 3-4mm, ± obtuse, brown, shiny but not viscid. ❶ ....*Morinda Spruce* **Picea smithiana**

■Lvs 1.2-2.5cm

Young twigs hairless, pale brown. Lvs 10-20 x 1.2mm, acute or obtuse, curved, bluish-green, with 2 ± obvious whitish stomatal bands each side, fetid or tangerine-scented. Buds 6mm, acute, brown. Bark grey-brown, ± smooth...*White Spruce* **Picea glauca**

Young twigs usu hairless, reddish orange-brown (occ yellowish). Lvs 10-20 x 0.8mm, acute-apiculate, straight or curved (all erecto-patent with upper ranks pointing forwards), green, with 1-3 faint whitish stomatal bands below, resin-scented. Buds 6mm, acute, brown. Bark reddish, ± smooth...*Norway Spruce* **Picea abies***

Young twigs hairy (occ sparsely so), dull greyish. Lvs 13-20(25) x 0.8-1.2mm, often sharply acute, straight, bluish-green, with 2 ± obvious white bands below, fetid or tangerine-scented. Buds 5mm, obtuse, brown. Bark grey to reddish-brown, ± smooth. ❷..*Engelmann Spruce* **Picea engelmannii**

Young twigs hairy, yellow-brown. Lvs 10-26(30) x 1.2mm (often unequal), ± obtuse, straight but jumbled around twig, glaucous both sides or shiny green above with indistinct dull grey-green bands below, tangerine-scented. Buds 3-4mm, ovoid, brown. Bark dark grey to reddish, scaly, fissured...*Mountain Hemlock-spruce* **Tsuga mertensiana**

■Lvs 0.6-1(1.2)cm

Young twigs hairy, whitish, soon reddish-brown. Lvs to 1mm wide, ± obtuse, shiny green brown, weakly fetid or orange-scented. Buds 2mm, ovoid, acute, pale or red-brown. Bark brown, scaly ....*Oriental Spruce* **Picea orientalis**

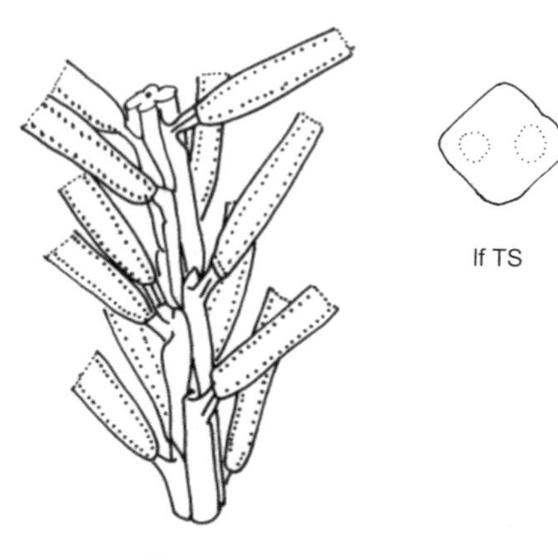

Picea abies

Group DF – *Twigs round, not ridged or grooved. Lvs linear, ± flat. (*Tsuga *may key out here)*

■Lvs deciduous, flexible, on deciduous branchlets (each with 20-35 prs of lvs). Buds 1-3mm, ovoid, obtuse

Lvs light green, obtuse to acute, entire, spreading in 1 plane on opp sides of twigs (but actually spirally arranged), with stomata below only (obscure). Buds pale brown, scales 4-ranked with raised midrib. Branchlets reddish, hairless, leaving circular scar with ring of vb's. Bark fibrous, stringy

Lvs alt, 10-17 x 2mm. Buds alt, 1mm, not superposed and developing adjacent to branchlet scar. Roots often above ground as lumps (pneumatophores) ....*Swamp Cypress* **Taxodium distichum**

Lvs mostly opp, 10-25 x 2mm. Buds opp, 3mm, superposed and developing below branchlet scar. Roots below ground...............................*Dawn Redwood* **Metasequoia glyptostroboides**

■Lvs evergreen, stiff. Buds 2-5mm, ovoid, (±) obtuse

Lvs on twigs parted so twig is completely visible from above

Lvs with stomata above (occ confined to midrib)

Buds brown, often resinous. Lvs 5-8cm x 2.2mm, notched, with 2 v pale greenish bands below, strongly orange-scented, with resin canals marginal. Twigs orange-brown, with hairs <0.25mm. Buds brown, often resinous. Trunk with resin blisters ....*White Fir* **Abies concolor** var **lowiana**

Lvs without stomata above, arranged in 2 sets (lateral and upper), the upper much shorter

Buds purplish (pearl-like), resinous. Lvs 2-6cm x 2.5mm, notched, shiny green above, glaucous below with 2 indistinct stomatal bands, tangerine-scented. Twigs olive-green to rufous-brown, with hairs <0.25mm. Trunk with few resin blisters........*Giant Fir* **Abies grandis***

Buds brown, not or hardly resinous. Lvs 1.5-3.5cm x 2-2.5mm, entire or notched, shiny dark green above, with 2 greyish (to white) bands below, grapefruit-scented. Twigs grey, usu with hairs >0.25mm. Trunk without resin blisters..............................*European Silver-fir* **Abies alba**

Lvs on twigs not parted (upper lvs overlying twig) so twig is mostly hidden from above

Lvs with stomata both sides (4-6 stomatal rows above)

Lvs 1-2(3.5)cm x 1.5-1.7mm, v dense, curved outwards, entire or slightly notched, glaucous above, lateral lvs spreading, upper lvs curving upwards along shoot. Twigs reddish, hairy. Buds purple, resinous. Trunk without resin blisters......*Noble Fir* **Abies procera**

Lvs without stomata above

Buds not resinous, scales with 1(3) ribs

Lvs 1.5-4cm x 2-2.2mm, petiolate, straight, notched, shiny dark (to bright) green above, 2 obvious white bands below, orange odour, persisting longer than *A. alba* (>3 yrs), 2 median resin canals. Twigs olive-brown, with 0.3mm hairs. Buds brown. Trunk without resin blisters, white...*Caucasian Fir* **Abies nordmanniana**

Buds resinous, scales not ribbed

Trunk often hollowed below branches, occ with resin blisters

Buds purple-brown, resinous. Lvs 2.5cm x 1.7mm, often curved up, notched, shiny dark green above, with 2 broad white stomatal bands below, paint-scented, with 2 median resin canals. Twigs orange-brown, hairy, with few lvs beneath. VR planted ....*Veitch's Silver-fir* **Abies veitchii**

Trunk never hollowed below branches, with resin blisters

Lvs 2-3cm x 2.2mm, linear, straight and flat on twig above (few or no lvs below) and spreading at wider angle than most spp (sprays wide but shallow), notched, shiny dark green above, with 2 broad white stomatal bands below, tangerine-scented. Twigs grey to orange, with hairs ≤0.4mm. Buds grey (pearl-like). Trunk silvery, with few fissures. Large tree..*Red Fir* **Abies amabilis**

Lvs 1-1.3(1.8)cm x 2mm, narrowed to base, curved up, notched, shiny dark green above (often white nr apex), with 2 broad white stomatal bands below, resin-scented. Twigs pale grey, often hairless. Buds reddish-brown. Trunk grey. Small tree (rarely to 15m). R hortal ....*Korean Fir* **Abies koreana**

Group DG – *Lvs scale-like, imbricate, at least partly adnate to twig, sessile, alt, spirally arranged, 3-50mm, without resin glands. Unless stated, 'lvs' refers to the dorsal facial scale lvs. Twigs usu reddish-brown*

■Lvs totally adpressed to twig, not decurrent, with sunken salt glands. Twigs (branchlets) deciduous. Not a conifer...(*Tamarix*) Go to **KC**

■Lvs at least partly spreading from twig, decurrent, without glands

Lvs with free part 25-50 x 10-14mm, flat, ovate-lanc, dark yellow-green, sharply spine-tipped, spiralling in 4-5 ranks, v stiff, with stomata in many parallel lines both sides, ± odourless. Buds green, without scales. Branches often falling. Bark dark grey ....*Monkey-puzzle* **Araucaria araucana**

Lvs with free part 5-20 x 1.5mm, laterally flattened (± 4-sided and usu as thick as broad), subulate, bright green, ± acute (exc narrow juv foliage), spiralling in 3 ranks, flexible, with 2 stomatal bands each side, resin-scented. Buds green, with scales. Branches often branched at 60º. Bark stringy, red-brown...*Japanese Red-cedar* **Cryptomeria japonica**

Lvs with free part 3-7 x 1mm, scale-like, linear, grey-green, acute (v sharp translucent point), spiralling in 3 ranks, stiff, with 2 bands of scattered stomata each side (exc along midlines), leather-scented. Buds green, without scales. Branches often downcurved. Bark red-brown, v spongy...*Wellingtonia* **Sequoiadendron giganteum**

DF

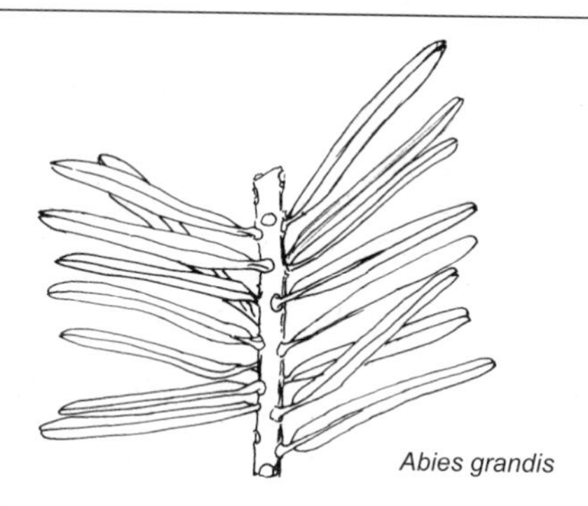

Abies grandis

DH

Thuja plicata

Chamaecyparis lawsoniana

Chamaecyparis nootkatensis

Group DH – *Lvs scale-like, imbricate, at least partly adnate to twig, sessile, opp, decussate (4-ranked), usu 1.5-4mm, often with translucent resin gland (usu sunken). Unless stated, 'lvs' refers to the dorsal facial scale lvs*

■Branchlets in 1-plane, flat in TS

Lvs strongly pineapple-scented, with pale stomata-covered patches below

Lvs with obvious oblong translucent raised resin gland, not keeled, ± cuspidate or acuminate, parallel-sided in outline (no sinus between lateral lvs), shiny dark green above; apical lateral lvs incurved and longer than inner facial lvs (often obscured). Branchlets pendent .... *Western Red-cedar* **Thuja plicata***

Lvs parsley-scented, without stomata visible either side (whitish marks usu indistinct or absent below)

Lvs with v obvious translucent oval resin gland (occ linear or absent), not keeled, ± acute, with strong sour parsley scent, green to glaucous above; apical lateral lvs usu incurved. Branchlets often pendent. Bark red-brown, v spongy. Terminal male fls purple. Many cultivars exist! .... *Lawson's Cypress* **Chamaecyparis lawsoniana***

Lvs with weakly translucent long-linear resin gland, keeled, acute to acuminate, sweetly parsley-scented, dark green above, paler below without white markings; apical lateral lvs divergent and much longer than facial lvs. Branchlets usu pendent. Bark orange-brown. Terminal male fls yellow.. *Nootka Cypress* **Chamaecyparis nootkatensis***

Lvs with rancid odour, or odourless

Lvs with large bright white patches below

Lvs with weakly translucent linear resin gland (often obscure), not keeled, obtuse, stiff, paint-scented; apical lateral lvs slightly divergent and equal to inner facial lvs .... *Hiba* **Thujopsis dolabrata**

Lvs without bright white patches below

Lvs with weakly translucent linear resin gland, not or weakly keeled, acute to acuminate, rancid to scentless, often with stomata, dull mid-green both sides, the facial lvs equal to the lateral lvs; apical lateral lvs parallel and equal to inner facial lvs, acute to acuminate but not sharp. Branchlets often drooping. Bark cinnamon-brown, fibrous .... *Chinese Thuja* **Platycladus orientalis***

Lvs with obscure (or invisible) resin gland, not or weakly keeled, rarely ± acute, sweetly or resin-scented, without stomata, shiny dark green above, with whitish scale edges below, the 1-1.5mm facial lvs much shorter than the 1-3mm lateral lvs; apical lateral lvs incurved. Branchlets drooping. Bark reddish-brown, peeling in strips .... *Hinoki Cypress* **Chamaecyparis obtusa**

■Branchlets mostly not in 1-plane, usu ± square, round or triangular in TS

Lvs strongly parsley-scented

Lvs with oblong-oval translucent resin gland (occ obscure), indistinctly keeled (flat or ± so), acute, green (may be glaucous above), usu with grey bases and pale margins, with stomata absent at least below. *Chamaecyparis nootkatensis* x *Cupressus macrocarpa* .... *Leyland Cypress* x **Cupressocyparis leylandii**

Lvs strongly orange-scented

Lvs strongly glaucous, adpressed

Lvs with ovate opaque resin gland (often resin encrusted), without visible stomata .... *Smooth Arizona-cypress* **Cupressus glabra**

Lvs green and/or spreading

Lvs with linear translucent resin gland (often obscure in cultivars), not keeled, acute to ± acuminate, orange-or parsley-scented, pale or bright dark green, with pale markings occ visible above (inside lateral scales). Branchlets in ascending sprays, divergent, often with crisped appearance (esp in cultivars that retain juvenile lvs such as 'Plumosa', 'Squarrosa'). Terminal male fls purple.. *Sawara Cypress* **Chamaecyparis pisifera***

Lvs strongly lemon-scented

Lvs bright yellowish or dark green, obtuse to ± acute (occ with v minute blunt or acute translucent tip), usu with resin gland visible (occ obscure), with stomata visible or not; juvenile lvs absent or soon lost.. *Monterey Cypress* **Cupressus macrocarpa***

Lvs with rancid odour (like paint) or weakly scented (to odourless)
Lvs usu acute with a 0.1mm translucent point (sharp to touch when fingers run backwards along shoot)
Lvs 1.5-3mm, deep green, with resin gland nr base (often obscure), with rancid scent, with stomata totally absent; juvenile subulate lvs often present in prs (or 3-whorled) nr shoot tips. Tree..*Pencil Cedar* **Juniperus virginiana***
Lvs usu obtuse without a translucent point (not sharp to touch)
Subulate juvenile lvs often present nr shoot tips (3-whorled or in prs)
Lvs 1.5-3mm, deep green, not keeled, obtuse, with obscure resin gland at mid-point, with rancid scent, with sparse stomata along pale proximal margins. Tree or shrub ....*Chinese Juniper* **Juniperus chinensis***
Subulate juvenile lvs absent
Lvs 1-1.5mm, grey-green both sides, not keeled, ± acute, with weakly translucent or obscure narrow-oblong resin gland, weakly or not scented, often with scattered stomata; lateral lvs 1-1.5mm, strongly adpressed. Fastigiate tree ....*Italian Cypress* **Cupressus sempervirens***

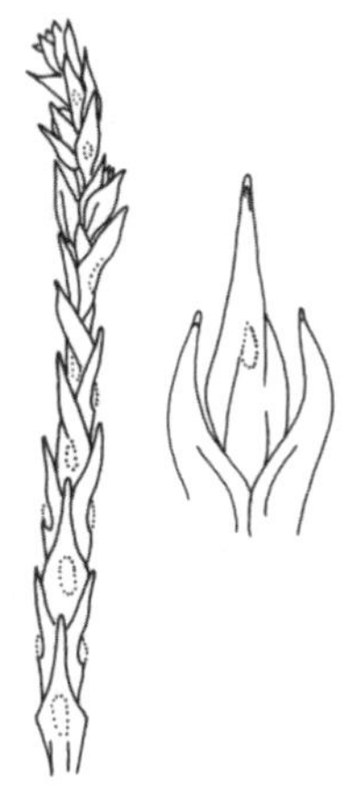

Chamaecyparis pisifera

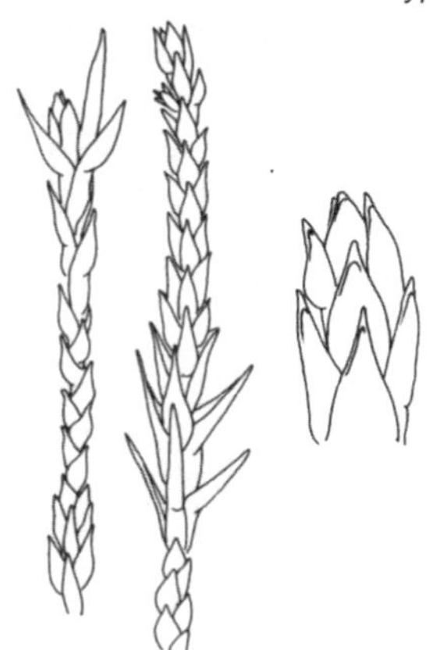

Juniperus virginiana

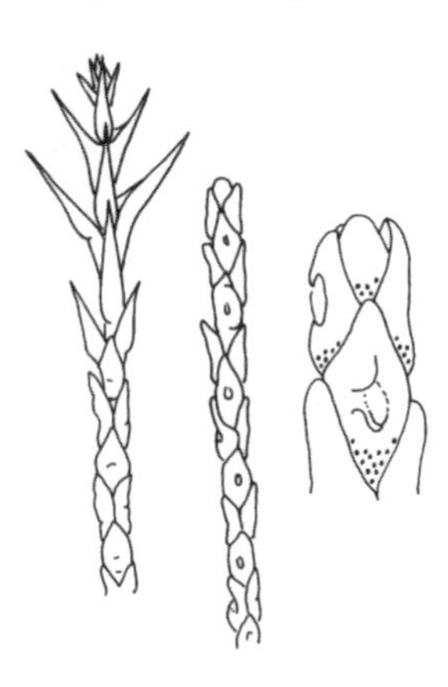

Juniperus chinensis

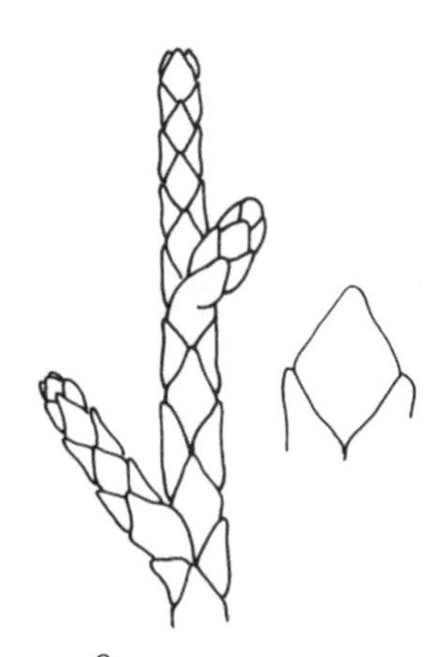

Cupressus macrocarpa

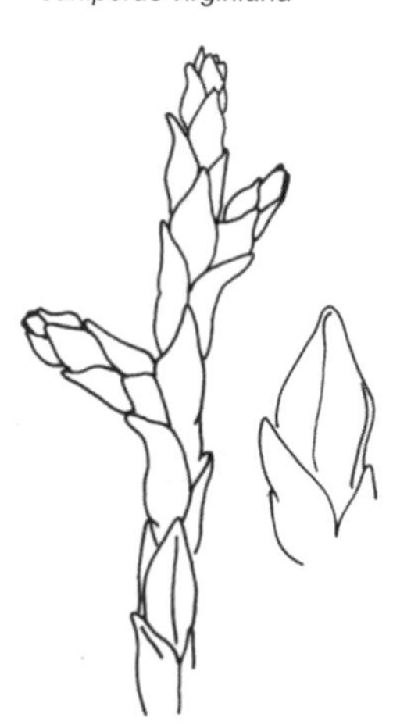

Cupressus sempervirens

Platycladus orientalis

Key to Groups in Division E

(Lvs absent or stem-like, occ scale-like or reduced to a short blade)

Reduced lvs may be present at the base of the plant in the form of sheaths that are usu bladeless (cataphylls)

■Shrub, spiny or spineless....................**EA**
■Herb, spineless
  Plant green (with chlorophyll)
    Plant with fleshy stems (lvs absent), may be woody near base. Saltmarshes....................**EB**
    Plant with rush-like cylindrical, flattened or triangular stems/lvs, never fleshy or woody. Wet habs (occ saline)
      Stems usu ≥5mm diam....................**EC**
      Stems usu 1.5-5mm diam....................**ED**
      Stems <1.2mm diam....................**EE**
    Plant with tubular insectivorous pitchers. Bogs....................**EF**
  Plant not green (without chlorophyll), saprophytic (myco-heterotrophic) or parasitic
    Stems absent, buds clustered at ground level. Plant parasitic on *Salix*, *Populus* or *Alnus* roots
    ....**EG**
    Stems twining, parasitic on stems of other plants....................**EH**
    Stems erect. Plant saprophytic or parasitic on roots of other plants....................**EI**

Group EA – *Shrub. Twigs green at least when young*

▪Plant spiny. Buds small, usu superposed

Spines simple or with terminal portion much longer than laterals

Shrub to 2.5m, usu erect, grey-green. Spines 1-3cm, stout, v rigid, furrowed, with 2mm point. Fls mostly in spring, dead fls visible autumn..*Gorse* **Ulex europaeus***

Shrub 0.3-1m, usu erect, dark green. Spines 1-2.5cm, stout, rigid, weakly furrowed or striate, with 1-1.5mm point. Fls late summer (mid-Jul), dead fls visible spring ....*Western Gorse* **Ulex gallii**

Shrub 0.1-0.5(0.7)m, prostrate, dark green. Spines 0.8-1.5cm, ± weak, flexible, weakly furrowed or striate, with 1-1.3mm point. Fls late summer (late Jul), dead fls visible spring with frs green throughout winter... *Dwarf Gorse* **Ulex minor***

Spines much-branched, with lateral portions ± equal to terminal portion

Shrub to 0.8(1.2)m. Spines slightly furrowed, with 1mm reddish point. Lvs present on young growth (Jul-Sept), simple, ± sessile, adpressed-hairy. Stems v densely branched, slightly furrowed, ± densely adpressed hairy.........*Spanish Gorse* **Genista hispanica** ssp **occidentalis**

▪Plant spineless. Buds (if present) small, not superposed

Twigs round (but with 20-50 striations, rough to touch when dry)

Lvs 1-3 x 0.4-0.7cm, few, soon falling, oblong-lanc, apiculate, thickish, (±) hairless above, silky-hairy below, with translucent midrib and indistinct 2° veins, with stomata both sides. Petiole 3mm. Twigs 2.5-4.5mm diam, branched (often opp or subopp), straight, flexible, with stomata all around, pith-filled. Shrub to 3m..*Spanish Broom* **Spartium junceum**

Twigs with 15 grooves

Lvs to 10mm, hairy, falling Aug. Twigs pendent, occ minutely hairy, without visible stomata. Shrub (occ small tree) to 6m...*Mount Etna Broom* **Genista aetnensis**

Ulex europaeus with stem TS

Ulex minor with stem TS

Group EB – *Stems fleshy, may be woody near base, jointed, often branched, hairless, with salty taste. 'Lvs' enveloping stem to form opp connate segments, with stomata (exc on the scarious margins of segments). Saltmarshes*

▪Rhizomatous per. Stems ± woody below, ± prostrate, often several, rooting at nodes. All yr
Lvs 5-6 x 3mm, with basal segments often keeled, often yellowish or orange
....*Perennial Glasswort* **Sarcocornia perennis**

▪Ann. Stems not woody, ± erect, usu 1 (often many close together), not rooting at nodes. Jul-Nov
Fertile segment with 1 fl. Upper saltmarshes................*One-flowered Glasswort* **Salicornia pusilla**
Fertile segment with 3 fls. Mid to lower saltmarshes
Central fl much larger than laterals. Stem segments with distinctly convex sides
....**Salicornia europaea** agg
Central fl ± equal to laterals. Stem segments with straight or slightly convex sides
....**Salicornia procumbens** agg

Group EC – *Herb with rush-like stems and/or lvs. Wet habs. Stems usu >5mm diam, with continous pith. Basal sheaths may have a short blade often adpressed to stem (absent from* J. pallidus*).* ❶ *Tufted (all other spp are rhizomatous)*

▪Stem sharply triangular (triquetrous) along length
Lvs absent or reduced to short blade. Stems 3-8mm diam. Apr-Oct. Tidal rivers. Sch8
....*Triangular Club-rush* **Schoenoplectus triqueter**
Lvs 2-3, to 30cm x 3-5.5mm, linear. Stems to 3.5mm diam. Apr-Oct. Dune slacks, Lancs
....*Sharp Club-rush* **Schoenoplectus pungens**

▪Stem triangular (trigonous) above midpoint but round nr base
Stems 3-8mm diam, acutely 3-angled at apex, spongy. Apr-Oct. *S. tabernaemontani* x *triqueter*
....**Schoenoplectus** x **kuekenthalianus**

▪Stems round along length
Stems dull glaucous-grey
Stems 3-9mm diam, to 1.5m, spongy. Basal sheaths closed, ladder-fibrillose, reddish-brown to purplish black, with cross-veins. Lvs short, tapered, acute, channelled to flat. Underwater linear lvs v rarely present. Apr-Oct. Brackish or freshwater, usu in shallow water
....*Grey Club-rush* **Schoenoplectus tabernaemontani**
Stems dull or shiny green
Stems 10-15mm diam, to 3m, dull or shiny green, spongy. Basal sheaths closed, ladder-fibrillose, reddish-brown to purplish black, with cross-veins. Lvs short, tapered, acute, channelled to flat. Underwater linear lvs occ present. Apr-Oct. Freshwater, often in deep water
....*Common Club-rush* **Schoenoplectus lacustris**
Stems 5-10mm diam, to 2m, shiny bright green, hardly spongy. Basal sheaths open, not ladder-fibrillose, warm-brown, without cross-veins. Underwater lvs never present. ❶. All yr. Freshwater, in shallow water, R alien.......................................*Great Soft-rush* **Juncus pallidus**

ED

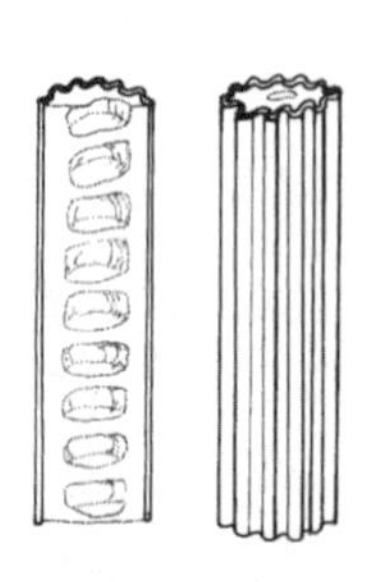

Juncus inflexus

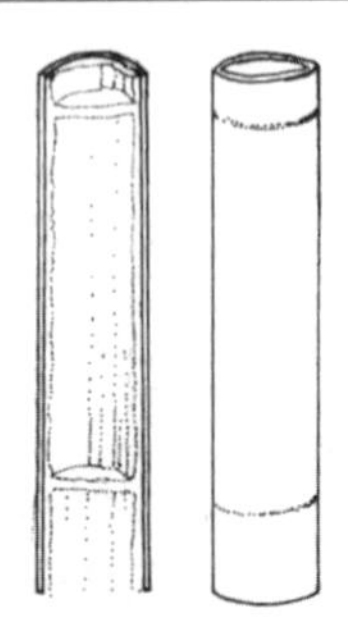

Juncus acutiflorus

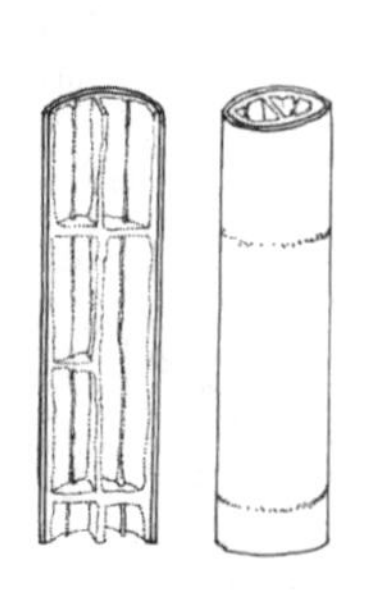

Juncus subnodulosus

Group ED – *Stems (or stem-like lvs) usu 1.5-5mm diam. Wet habs (occ saline)*

▪Stem-like lvs with horizontal and vertical partitions (septa)

Basal sheaths greenish, pale orange-brown at extreme base, open. Lvs 2-3mm diam, bright green, smooth. Stems to 1.2m, occ purplish above, not compressible, solid to hollow, with scattered vb's. Rhizomatous. All yr.........................*Blunt-flowered Rush* **Juncus subnodulosus***

▪Stem-like lvs with horizontal partitions only

Lvs with 1(2) septa per 5cm. Basal sheaths often reddish, open

Lvs 3mm diam, smooth, slightly curved, oval, slightly flattened, shiny green, hollow. Auricles 0.5-2mm. Stems to 100cm tall, ± erect. Rhizomatous. All yr
....*Sharp-flowered Rush* **Juncus acutiflorus***

Lvs with 5-10 septa per 5cm. Basal sheaths often reddish, open

Lvs 1.5-2.5mm diam, slightly ribbed, curved, ± oval to strongly flattened, dull green, pith-filled or hollow. Auricles 1.5mm. Stems to 50cm tall, ± prostrate at base. Tufted to strongly rhizomatous, occ rooting at lower nodes. Often all yr...................................*Jointed Rush* **Juncus articulatus**

Lvs to 1mm diam, smooth, curved, oval, slightly flattened, dull green, pith-filled, with 5-10 septa per 5cm. Auricles 2mm. Stems usu <7cm tall, ± prostrate at base. Shortly rhizomatous. All yr. Wet mtn habs...*Alpine Rush* **Juncus alpinoarticulatus**

▪Stem-like lvs or true stems without partitions

Stems ridged or grooved (at least to touch)

Stems with continuous pith (occ interrupted in *J.* x *diffusus*). Tufted

Stems with (15)20-32 ridges, 3-5mm diam, dull green to grey-green. Basal sheaths red-brown to olive-green. All yr...*Compact Rush* **Juncus conglomeratus**

Stems with 30-40(45) ridges, 3-5mm diam, dull greyish-green. Basal sheaths reddish-brown. *J. conglomeratus* x *effusus*..**Juncus** x **kern-reichgeltii**

Stems with (18)28-45 shallow ridges (not or hardly visible in TS), 1.5-2.5mm diam, dull pale green. Basal sheaths reddish-black. All yr. *J. effusus* x *inflexus*.................**Juncus** x **diffusus**

Stems with interrupted pith

Tufted, may form low tussocks. Stems with 15-20(25) distinct ridges, 1-2.5mm diam, to 1m tall, dull glaucous. Basal sheaths shiny blackish-purple. All yr...........*Hard Rush* **Juncus inflexus***

Shortly rhizomatous, with stems arising singly. Stems with 10-15(25) ridges (often hardly visible in TS), (1)1.5-2.5mm diam, to 1m tall, green. Basal sheaths shiny blackish. Apr-Oct. VR, S Lancs...**Juncus balticus** x **inflexus**

Stems smooth (not ribbed), with continuous pith. (*Schoenus nigricans* may key out here)

Basal sheaths closed, membranous, ladder-fibrillose. Tufted

Stems all yr, 2-2.5(4)mm diam, to 1m tall, pale green, oval. Basal sheaths whitish, often reddish, mostly without blades. Shoots slightly flattened. Lvs visible Apr-Aug, 1-2, all nr base, 1-1.2mm wide, shorter than stem, obtuse, rigid, U-shaped, with scabrid margins, whitish above (exposed pith), with stomata at least below. Coastal, R
....*Round-headed Club-rush* **Scirpoides holoschoenus***

Basal sheaths closed, membranous, not ladder-fibrillose, without aristate tip, convex (slightly concave on other side). Rhizomatous, with tufts of c6 stems. May-Oct

Stems with (16)23-30 vb's around margin, 1.5-4mm diam, to 60cm tall, oval, mid-green, reddish at base, pliable, often with 4-5 large hollows (and small hollows in outer margin). PL 8
....*Common Spike-rush* **Eleocharis palustris***

Stems with 15-23(25) vb's around margin, 1.5-4mm diam, to 60cm tall, round or ± flattened, yellow-green, hardly reddish at base weak, brittle, pith-filled or ± hollow. R, N Br
....*Northern Spike-rush* **Eleocharis mamillata** ssp **austriaca**

Basal sheaths open, tough, not ladder-fibrillose, with fragile aristate tip

Rhizomatous, with shoots arising singly. Apr-Oct

Stems (1)1.5-2.5mm diam, to 60cm tall, shiny green, not striate. Basal sheaths pale yellow-brown. Mostly dune slacks, R, N Br..*Baltic Rush* **Juncus balticus**

Stems 1.5-2(3)mm diam, to 1.2m tall, dull grey-green, weakly striate, occ with interrupted pith. Basal sheaths dark orange-brown. Dune slacks, VR, W Lancs.....**Juncus balticus** x **inflexus**

Tufted, clump forming. All yr

Stems single, shiny bright green (stem-like lvs absent)

Stems (1.5)3-5mm diam, to 1.2m tall, pliable, acute. Basal sheaths dull reddish-brown to black (darker nr base), without distinct blade..............................*Soft-rush* **Juncus effusus***

Stem-like lvs 2(3), shiny dark green. Coastal

Lvs 2-3.5mm diam, to 1.5m tall, v stiff, extremely sharply acute. Basal sheaths greenish-brown, without distinct blade. Plant forming dense tussocks. R ....*Sharp Rush* **Juncus acutus**

Lvs 1.5-2mm diam, to 1m tall, stiff, sharply acute. Basal sheaths shiny reddish-brown, often with short green blade..*Sea Rush* **Juncus maritimus**

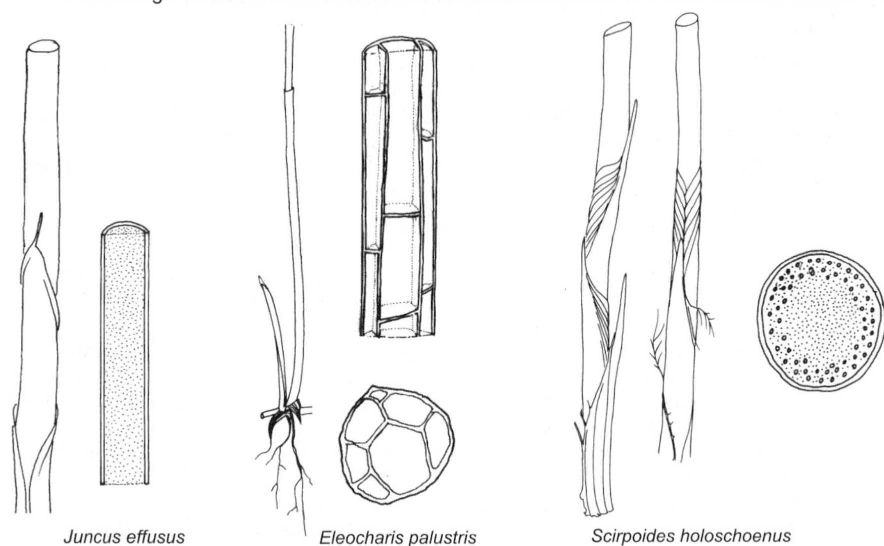

Juncus effusus *Eleocharis palustris* *Scirpoides holoschoenus*

Group EE – *Stems or stem-like lvs <1.2mm diam, ± obtuse (acute in* Pilularia *and* Juncus filiformis*). Basal sheaths (cataphylls) usu lacking a blade*

▪Basal sheaths absent (a fern)

Lvs arising in 1-3's from rhizomes, to 8cm tall, 1-1.5mm diam, wavy, circinate when young (in croziers), green, soon purplish, hay-scented, with opaque clathrate pattern (HTL), without stomata, with 10-12 hollows around central stele. Rhizomes with pill-like swellings when fertile. Jun-Oct..*Pillwort* **Pilularia globulifera***

▪Basal sheaths tough, open, with a fragile aristate tip (occ with a short blade)

Stems arising singly every 3-4mm, to 30cm tall, 0.5-1mm diam, light green, obscurely ridged, hollow or with loose pith. Basal sheaths reddish-brown, several, upper often with a short green blade. Rhizomatous. Mar-Oct. Northern, lake and reservoir margins ....*Thread Rush* **Juncus filiformis**

▪Basal sheaths ± tough, closed, with short blade(s) to 5mm

Stems furrowed, 0.7-1mm diam, round, ± solid, stomata all around (exc in sunken cartilaginous grooves). Ligule minute, adnate. Basal sheaths shiny pale-brown (occ reddish). Densely tufted per. Mar-Nov. Peaty moors and bogs.........................*Deergrass* **Trichophorum cespitosum** agg

Stems not furrowed, 0.5-1mm diam, oval, ± solid, with stomata all around. Ligule absent. Basal sheaths with purplish or dark red veins. Tufted ann to per. Apr-Oct (all yr)

Spikelets 1(3). Terminal bract usu < infl. Nutlet smooth.........*Slender Club-rush* **Isolepis cernua**

Spikelets (1)2-4. Terminal bract >> infl. Nutlet ridged.............*Bristle Club-rush* **Isolepis setacea**

■Basal sheaths v thin, closed, without blade or aristate tip, convex on one side (concave on the other)
Stems mostly >10cm
Rhizomatous, with stems 10-60cm
Basal sheaths reddish. Stems 0.9-1.2mm diam, cylindrical to ± oval, with hollows in TS. Usu saline gsld..*Slender Spike-rush* **Eleocharis uniglumis**
Densely tufted (v short rhizomes), with stems 10-35cm
Basal sheaths orange-brown (occ purple). Stems 0.5-1.1mm diam, obscurely ridged, with 6(8) hollows (often obscure) in TS. All yr. Bogs...*Many-stalked Spike-rush* **Eleocharis multicaulis**
Stems mostly <10cm. Rhizomatous (occ weakly so, or with stolons)
Basal sheaths orange-brown to reddish. Stems 0.5-1mm diam, to 10cm tall, round, pith-filled, occ with 6 large hollows. Damp habs...........*Few-flowered Spike-rush* **Eleocharis quinqueflora**
Basal sheaths colourless (occ brownish at apex) with purple veins. Stems 0.2-0.9mm diam, to 8cm tall (underwater stems rarely to 50cm), round to 3-4-angled, with 3 hollows (occ pith-filled). Often submerged..*Needle Spike-rush* **Eleocharis acicularis**
Basal sheaths whitish. Stems 0.3-0.5mm diam, usu to 3(5)cm tall, ± oval, translucent, with sparse transverse septa, with 2-3(4) hollows, with few stomata in longitudinal lines. Rhizomes with small white tubers (forming short turf). Estuarine, VR. Sch8
....*Dwarf Spike-rush* **Eleocharis parvula**

Pilularia globulifera

Group EF – *Lvs basal, consisting of a tubular insectivorous pitcher (inflated petiole) and small hood (lf) at entrance (with stiff adpressed unicellular hairs on inner surface), narrowed to a short stalk-like base, with large flange on dorsal side and scattered stomata. Per. Bogs*
■Pitchers to 30cm, decumbent, curved, green; hood erect, red-veined. R alien
....*Pitcherplant* **Sarracenia purpurea**
■Pitchers 30-60cm, straight, erect, green; hood arching, rarely red-veined. VR alien
....*Trumpets* **Sarracenia flava**

Group EG – *Plant not green (without chlorophyll), parasitic on* Salix, Populus *or* Alnus *roots. Stems and lvs absent. Buds at ground level*
■Rhizome scales at or just below ground, opp, whitish, decussate, ovate, fleshy, hairless. Fl buds clustered, purplish (corolla bright purple). Damp shady habs..*Purple Toothwort* **Lathraea clandestina**

Group EH – *Stems not green (without chlorophyll), long, matted, lfless, twining anti-clockwise, parasitic on aerial stems of other plants (rootless but attached to host by suckers), with coloured sap. Ann (occ per)*
■Stems reddish (rarely yellowish). Jul-Oct
Stems (0.1)0.4-0.5(0.8)mm diam, round, with yellow or orange sap. Often on *Calluna*, *Ulex* and *Thymus*. Dry habs...*Dodder* **Cuscuta epithymum**
Stems 0.8-1.5mm diam, round, with orange sap. Usu on *Urtica* and *Humulus*. By rivers, R
....*Greater Dodder* **Cuscuta europaea**
■Stems yellowish (occ pale green at nodes)
Stems 0.7-0.8mm diam, weakly ribbed (with 4 ribs slightly more prominent than others), with yellow or orange sap. On crop or garden plants esp carrot, tomato, beetroot and lucerne. R alien
....*Yellow Dodder* **Cuscuta campestris**

Group EI – *Plant not green (without chlorophyll), usu with scales (scale-like lvs), saprophytic (myco-parasitic) or parasitic on roots of other plants.* ❶ *Stems branched.* ❷ *Fls clove-scented*

▪Stems hairless or hairy above (occ glandular). Tufted or rhizomatous per

Stems jointed, usu ridged, without scales..*Equisetum* Go to **Division A**

Stems unjointed, never ridged, often with scales

Stems 8-30cm, white or pale pinkish, sparsely hairy above, few whitish ovate scales nr base. Tufted per. Usu on roots of *Corylus* or *Ulmus*. Apr-Jun. Shady, usu calc habs ....*Toothwort* **Lathraea squamaria**

Stems 8-30cm, waxy, pale creamy-yellow or ivory-white pale, ± fleshy, brown when dry, usu hairy above, solid, with disinfectant odour, numerous scale lvs to 12mm. Shortly rhizomatous. Jun-Aug. Shady habs or dune slacks..........................*Yellow Bird's-nest* **Monotropa hypopitys**

Stems 7-25cm, slender, yellowish-green, hairless or sparsely hairy, with 2-4 long brown-veined scales to ½ way up stem. Rhizomes coral-like. May-Jul. Mossy woods, dune slacks or bogs. N Eng, Scot...*Coralroot Orchid* **Corallorhiza trifida**

Stems rarely produced, 5-15cm, turgid, weak, ± translucent, pinkish-white, streaked with red, hairless, without scales, usu slug-eaten. Rhizomatous. Jun-Sept. V dark beech woods, S Eng (please inform authors if found!). VR. Sch8.........................*Ghost Orchid* **Epipogium aphyllum**

Stems 20-45cm, 4-7mm diam, robust, yellow-brown to tan, hairless below, minutely sparsely glandular-hairy above, slightly angled, with numerous scales below. Scales 3 x 1cm, strongly veined. May-Jun. Humus-rich calc (esp beech) woods.....*Bird's-nest Orchid* **Neottia nidus-avis**

▪Stem usu glandular-hairy along length, with 5-25mm brown scales esp nr base, pith-filled. Ann or tuberous per. Jun-Aug (but mid-May in *Orobanche alba*)

Stems blue or purple

Stems 15-45cm x 4-6mm, bluish. Usu on *Achillea millefolium* ....*Yarrow Broomrape* **Orobanche purpurea**

Stems 10-50cm x 4-8mm, dark purplish. On *Hedera helix*. Mostly coastal ....*Ivy Broomrape* **Orobanche hederae**

Stems red or reddish-brown

Stems 15-70cm x 10-15mm, yellowish-brown (rarely red). Usu on *Centaurea scabiosa*. Calc turf ....*Knapweed Broomrape* **Orobanche elatior**

Stems 10-40cm x 4-8mm, yellowish-red, occ purplish at base. On many hosts, esp legumes ....*Common Broomrape* **Orobanche minor**

Stems 4-25cm x 5-10mm, purplish-red. Usu on *Thymus*. Often coastal ....*Thyme Broomrape* **Orobanche alba**

Stems pale or yellowish

Stems usu branched nr base, yellowish-white or bluish. ❶. On crop plants, VR alien ....*Branched Broomrape* **Orobanche ramosa**

Stems usu simple

Stems 20-80cm x 10-15mm, yellowish. Usu on *Ulex* or *Cytisus* ....*Greater Broomrape* **Orobanche rapum-genistae**

Stems 15-70cm x 10-15mm, stout, yellowish-brown (-red), yellow glandular-hairy, slightly ridged. Scales with a dark central line. Usu on *Centaurea scabiosa*. Calc turf ....*Knapweed Broomrape* **Orobanche elatior**

Stems 10-60cm x 8-13mm, pale yellowish, often purple tinged, occ ± hairless. On *Picris* and *Crepis*. Coastal. Sch8....................*Oxtongue Broomrape* **Orobanche artemisiae-campestris**

Stems 15-50cm, pale, often purplish below. On *Carduus* and *Cirsium*. NW Yorks. Sch8 ....*Thistle Broomrape* **Orobanche reticulata**

Stems 15-40cm x 3.5-9mm, yellowish or white (occ pinkish). Scales mostly below, yellow or purple-brown. ❷. On *Galium*. Sch8............*Bedstraw Broomrape* **Orobanche caryophyllacea**

Key to Groups in Division F

(Obligate water plant)

Plant submerged or floating, with no leafy stem out of water, hairless, without stomata on submerged parts. Ann to per, never woody

■Lvs submerged (rarely emergent or exposed at low water level)
  Lvs whorled or pseudowhorled (spiraling) on stem
    Lvs simple..**FA**
    Lvs lobed or compound (1-pinnate, with pinnae usu alt and thread-like)..................................**FB**
  Lvs opp on stem...**FC**
  Lvs alt or all basal, occ free-floating. Fragments often washed up on shoreline
    Lvs simple and unlobed (margins entire or toothed)
      Sea water (*Zostera*)..**FD**
      Freshwater or brackish
        Lvs usu >10mm wide...**FE**
        Lvs <10mm wide
          Basal lvs absent (with lvs all on stem)..**FF**
          Basal lvs present (without lvs on stem)...**FG**
    Lvs compound (1-pinnate etc) or lobed
      Lvs with translucent bladders (for trapping micro-fauna)..**FH**
      Lvs without bladders..**FI**
■Lvs floating (rarely emergent)
  Lvs >10mm wide
    Lvs orb or ± so; lf veins usu pinnate, palmate, or obscure...**FJ**
    Lvs elliptic-lanc to broadly ovate; lf veins parallel..**FK**
  Lvs <7mm wide
    Plant free-floating. Lvs usu orb..**FL**
    Plant rooted to substrate. Lvs linear..**FM**

Group FA – *Lvs submerged, whorled or pseudowhorled, without distinct petiole, simple. Stems round.* ❶ *Ann.* ❷ *Turions in lf axils or at end of stolons.* ❸ *Midrib obscure*

■Lvs pseudowhorled in spiral of 3-4 around stem (rarely whorled), minutely toothed

Lvs 2-2.5cm x 3mm, strongly recurved, not twisted, dense at shoot apices, acute, with translucent midrib and dark adjacent longitudinal lacunae strips. Stems 2-2.5mm diam, with aerenchyma around central stele. All yr. Still water................................*Curly Waterweed* **Lagarosiphon major**

■Lvs 3(4)-whorled, toothed (often minutely so) at least nr apex

Stems with several hollows around central stele (cartwheel-like). Per

Lvs 5-15 x 1-1.5mm, narrowly linear, acute, strongly recurved, usu twisted, with midrib visible. All yr..*Nuttall's Waterweed* **Elodea nuttallii***

Lvs 5-10 x 2.5-4mm, ± broadly-linear or oblong, ± obtuse, not strongly recurved, not twisted, with midrib visible. Often all yr..................................*Canadian Waterweed* **Elodea canadensis***

Stems solid. Ann ❶. Apr-Oct. VR. Sch8

Stems smooth. Lvs 0.3-1mm wide, apiculate or acute, translucent, minutely toothed to entire, midrib not scabrid. Lakes, N Br, Ire..*Slender Naiad* **Najas flexilis**

Stems spiny. Lvs 1-4mm wide, acute, opaque, deeply spiny toothed, midrib scabrid below. Norfolk Broads..*Holly-leaved Naiad* **Najas marina**

■Lvs 4-5-whorled, minutely toothed (occ entire)

Lvs 4-whorled, to 30 x 5mm, narrowly oblong to linear, abruptly acute, recurved, not twisted, dark green but soon colourless and v translucent, reddish-streaked, midrib not scabrid below. Stem with stele not purple. Often all yr....................................*Large-flowered Waterweed* **Egeria densa**

Lvs (3)4-5(6)-whorled, to 20(40) x 2-2.5(5)mm, linear-lanc, acute, straight, not twisted, green, midrib often scabrid below. Stem with purple stele. ❷. Apr-Oct. VR, Scot, Ire ....*Esthwaite Waterweed* **Hydrilla verticillata**

■Lvs 6-11-whorled, entire

Lvs to 20 x 2mm, linear, with hard (±) acute apex, with stomata both sides when above water (absent from submerged lvs). Stems often emergent, solid, with aerenchyma around central stele. ❸. Apr-Oct..*Mare's-tail* **Hippuris vulgaris***

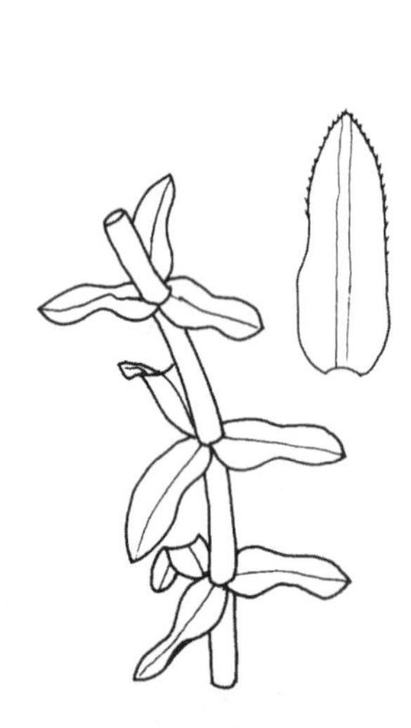

Elodea canadensis

Elodea nuttallii

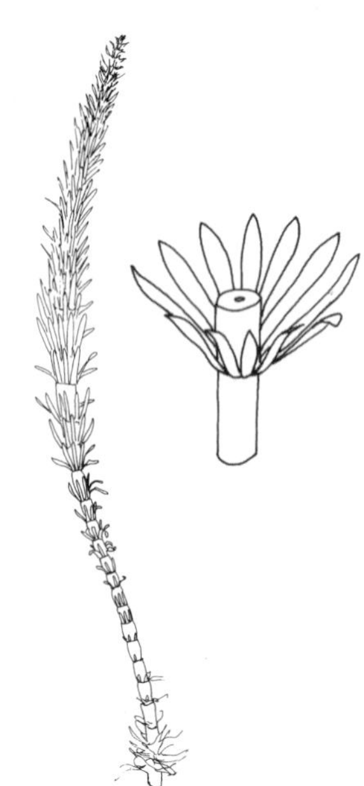

Hippuris vulgaris

Group FB – *Lvs submerged, whorled or pseudowhorled, without distinct petiole, lobed or compound.* ❶ *Turions in lf axils (detachable club-shaped bud of tightly clustered lvs, produced in autumn).* ❷ *Roots absent*

■Lvs pseudowhorled (appearing 2-6 whorled); lobes linear, untoothed, without bristles

Lvs crowded at stem apex into rosette just below water surface, pinnately lobed, petiolate, flat, pale green, with obscure veins. Stems round, floating and often rooting in shallow water, with odourless (±) sessile glands when emergent, aerenchyma-filled. All yr .... *Water-violet* **Hottonia palustris**

■Lvs (3)4-whorled; lobes thread-like (to 0.3mm wide), untoothed, without bristles

Lvs with 6-12 lobes per side, 1-2.5cm, the uppermost may be opp or alt. Stems 0.8-1.2mm diam, with 8-10 hollows around central stele (often reddish). Fl stems emergent, with tip drooping in bud. Plant often pinkish. Often all yr. Usu acidic water .... *Alternate Water-milfoil* **Myriophyllum alterniflorum***

Lvs with 11-18 lobes per side, 1.5-3cm. Stems 1-3mm diam, pinkish, with 9-12 hollows around reddish central stele. Fl stems emergent, with tip erect in bud. Plant often encrusted in marl. Apr-Oct (all yr). Usu eutrophic or base-rich water .... *Spiked Water-milfoil* **Myriophyllum spicatum**

■Lvs 5-6-whorled; lobes thread-like, untoothed, without apical bristles

Stems submerged, branched, 1-4mm diam, with 12-20 hollows around central stele. Lvs usu 5-whorled, with (12)16-20 lobes per side, dark green, without glands. Turions in leaf axils ❶. Often all yr. Usu base-rich water................................ *Whorled Water-milfoil* **Myriophyllum verticillatum**

Stems emergent, unbranched, usu 3-4mm diam, with c25 hollows around central stele. Lvs 4-6-whorled, with 4-15 lobes per side, feathery, pale blue-green, with abundant sessile glands. All yr. Usu eutrophic water. Invasive alien............................. *Parrot's-feather* **Myriophyllum aquaticum***

■Lvs (7)9-11-whorled, forked, thread-like (usu 0.3mm wide), minutely toothed at least nr apex, with 2-4 bristles at apex (and on teeth). ❷. All yr

Lvs 1(2)-forked, the terminal fork usu with 3 teeth, each tooth with 2 bristles; upper lvs often with thicker and wider (1-1.5mm) lobes. Stems without hollows around stele .... *Rigid Hornwort* **Ceratophyllum demersum**

Lvs 3-forked, the terminal fork with 1-3 teeth, each tooth with 2-4 bristles; upper lvs never with thicker lobes. Stems without hollows around stele. Usu brackish water, or hortal .... *Soft Hornwort* **Ceratophyllum submersum***

lf

apex of terminal fork

Ceratophyllum submersum

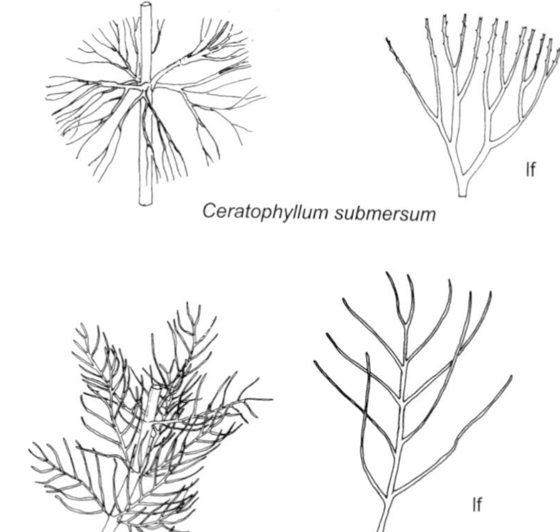

Myriophyllum alterniflorum

Myriophyllum aquaticum

Group FC – *Lvs submerged, opp on stem. Stems round*

■Lvs simple. Stems smooth

Lf margin entire

Lvs usu with notched apex. Stipules absent. Lvs without hollows, ± linear (when submerged), usu connate at base, thin. Ann to per..(*Callitriche*) Go To **CAL**

Lvs usu with entire apex

Stipules tubular, translucent, soon falling. Lvs with 2 hollows in TS

Lvs 15-50(100) x 0.3-2mm, linear to thread-like, tapered, with v fine minutely mucronate apex, flat, translucent, with midrib visible as wide dark green stripe, often with cross-veins. Stems 0.5mm diam, whitish. Apr-Oct (all yr)......................*Horned Pondweed* **Zannichellia palustris**

Stipules free, translucent, toothed. Lvs without hollows

Lvs to 8mm, occ 4-whorled, spathulate, obtuse, not connate, usu opaque, midrib usu obscure. Stems rooting at nodes (many roots per node), round, with 6-10 hollows (cartwheel-like)

Fl stalk > fl bud length. Summer ann on mud, short-lived per underwater. Lakes, ponds, wet mud..*Six-stamened Waterwort* **Elatine hexandra***

Fl stalk < fl bud length (to absent). Ann. Ponds, small lakes, R ....*Eight-stamened Waterwort* **Elatine hydropiper**

Stipules absent. Lvs without hollows

Lvs 4-15(20)mm, linear to ovate-lanc, acute, connate at base, fleshy, with all veins usu obscure. Stems simple or branched, with dark ring below node (often slightly constricted). Per. All yr. Invasive alien.................................*New Zealand Pigmyweed* **Crassula helmsii***

Lf margin toothed

Lvs minutely toothed, 10-40 x 5-15mm, 2-ranked, bright green, sessile, clasping, flaccid, recurved, often undulate, midrib obvious, with 1(3) veins per side. Stems smooth, branched, rooting at nodes, with dark line below nodes, with minute aerenchyma around central stele. Stipules usu absent. Per.................................*Opposite-leaved Pondweed* **Groenlandia densa**

Lvs with dark-tipped spines, 15-25 x 1-4mm, 3-ranked, dark green, sessile, not clasping, stiff and brittle, not recurved or undulate, veins obscure (midrib rarely visible). Stems spiny, often forking, not rooting at nodes, ± solid, with rectangular cells visible. Stipules absent. Ann. Norfolk Broads. Sch8..*Holly-leaved Naiad* **Najas marina**

■Lvs repeatedly forked. Stems usu spiny-ciliate

Lvs clustered at stem apices, often recurved, palmately dissected (first division 4-5-palmately divided, each division forking 2(3)-chotomously several times); lobes 10 x 1mm, flat, obtuse, stiffly ciliate, veins obscure. Petiole 1-2(3)cm. Stems 2mm diam, with aerenchyma. Per. All yr. VR alien, Basingstoke Canal...*Carolina Water-shield* **Cabomba caroliniana**

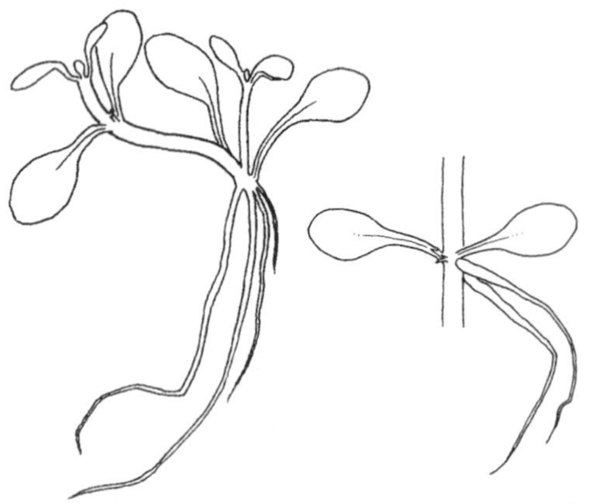

Elatine hexandra & close-up showing minute stipules

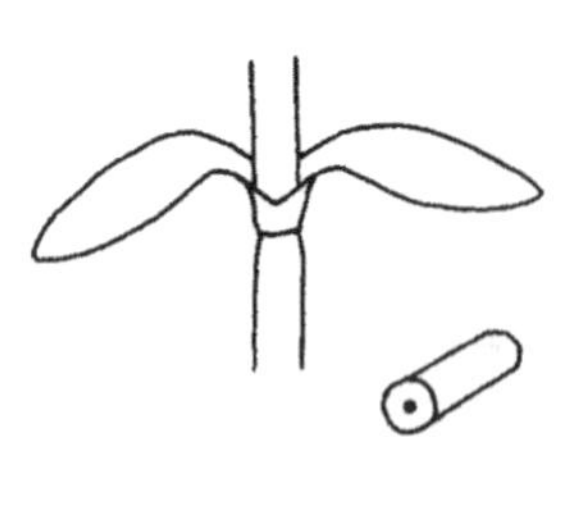

Crassula helmsii & stem TS

Group FD – Zostera. *Lvs submerged, alt, simple, entire, dark green, translucent, with 3-5 equal parallel veins, with cross-veins (often subopp), often with hollows and spiral fibres on tearing, without stomata. Sheaths flattened. Rhizomatous per. All yr (but with short narrow lvs in winter). Sea water*

■Lvs 4-10mm wide, with (3)5 veins. Sheaths closed..................................*Eelgrass* **Zostera marina**

■Lvs (1)1.5-2.5mm wide, with 3 veins. Sheaths closed ....*Narrow-leaved Eelgrass* **Zostera angustifolia**

■Lvs 0.5-1mm wide, with 3 veins. Sheaths open................................*Dwarf Eelgrass* **Zostera noltei**

Group FE – *Lvs mostly submerged, alt, simple and unlobed, usu >10mm wide, without stomata. Fresh or brackish water. ❶ Lf apex hooded. ❷ Stems flattened and ± dumbbell-shaped (round in other spp)*

■Lvs with long petiole (>20mm) (*P. polygonifolius* and *P. epihydrus* may key out here)

Stipules to 10cm, strongly 2-keeled

Lvs 10-20 x 3.5cm, ovate-elliptic, ± obtuse to acute, usu with excurrent midrib at apex, cuneate at base, green, involute when young, minutely undulate, with entire margins, with 3-6 opaque veins per side. Petiole (1.5)2-7cm, flattened. Floating lvs with stomata above. Apr-Oct. *P. lucens* x *natans*............**Potamogeton** x **fluitans**

Stipules absent

Lvs 8-35cm, broadly ovate, fan-shaped, crumpled or involute when young, thin (like semi-translucent lettuce lvs), shiny green, sinuate-crenate, with strong midrib and opaque-bordered pinnate veins forking towards margin. Petiole with equal sized aerenchyma (like those of floating lvs). Apr-Oct............(submerged lvs) *Yellow Water-lily* **Nuphar lutea**

■Lvs with short petiole (usu 5-15mm)

Lf margin minutely denticulate. Floating lvs absent

Lvs 8-20 x 3-6cm, oblong-lanc, obtuse, cuspidate or apiculate (occ acuminate), with excurrent midrib at apex (often long), cuneate at base, short-petiolate but decurrent and appearing ± sessile, shiny green, minutely undulate, with 2-3 main veins per side, with cross-veins raised above. Petiole to 12mm. Stipules to 6cm, obtuse, strongly 2-keeled, persistent. Apr-Oct ....*Shining Pondweed* **Potamogeton lucens***

Lf margin entire. Floating lvs with stomata above

Lvs 2-7 x 1.5-5cm, ovate-elliptic (occ narrowly so), ± obtuse, without excurrent midrib at apex, truncate to cuneate at base, shiny green or pinkish, slightly undulate, with 4-10 opaque veins per side, with cross-veins raised above. Petiole to 20mm. Stipules 2-5cm, obtuse, weakly 2-keeled, ± persistent. All yr. Fens............*Fen Pondweed* **Potamogeton coloratus***

■Lvs sessile (occ short-petiolate on uppermost lvs in *P. gramineus*)

Stipules absent. Lvs all basal

Lf margins strongly spiny-serrate

Lvs 15-50 x 1-1.6cm, aloe-like, acute, brittle, becoming tough and opaque, with many indistinct opaque veins and cross-veins, with midrib raised below. Rosettes usu floating, joined by whitish stolons. All yr............*Water-soldier* **Stratiotes aloides**

Lf margins entire

Lvs 7-10(18)mm wide............*Branched Bur-reed* **Sparganium erectum***

Lvs 4-5(10)mm wide............*Unbranched Bur-reed* **Sparganium emersum**

Stipules present at least when young. Lvs on stems

Lf margin entire

Lvs rounded at base, ± clasping, 6-18 x 2-5cm, strap-shaped to oblong, obtuse, involute when young, slightly undulate, dull green, with (1)3 main veins per side (often reddish). Stipules 0.5-6cm, obtuse, not 2-keeled, usu persisting as fibres. Floating lvs absent. ❶. Often all yr ....*Long-stalked Pondweed* **Potamogeton praelongus***

Lvs cuneate at base, ± sessile or shortly stalked, usu 6-15 x 1-2.5cm, narrowly oblong-elliptic, ± obtuse, involute when young, flat, shiny green or reddish, with 6-10 veins per side (plus c3 crowded veins adjacent to midrib). Stipules 2-6cm, usu obtuse, not 2-keeled, occ persisting. Floating lvs occ present, thicker, often with short petiole. Apr-Oct ....*Red Pondweed* **Potamogeton alpinus***

Lf margin minutely toothed (at least when young)

Lvs strongly clasping with auricles at cordate base

Lvs 2-10 x 1.3-6cm, usu ovate, obtuse (v rarely mucronate), involute when young, shiny green, slightly undulate, with 2-4 main veins per side (c10 total per side), veins never reddish. Stipules to 1cm, obtuse, not 2-keeled, rarely persisting. Apr-Oct ....*Perfoliate Pondweed* **Potamogeton perfoliatus***

Lvs ± auriculate or rounded at base

Lvs 5-10 x 0.8-1.5cm, lanc to linear-lanc, obtuse (rarely mucronate), applanate when young, shiny green, usu strongly undulate, 1 main vein per side (and faint submarginal vein), veins often reddish. Stipules 1-2cm, obtuse, not 2-keeled, soon torn. ❷. Often all yr .... *Curled Pondweed* **Potamogeton crispus***

Lvs cuneate at base

Lvs 3-20 x 1.5-4cm, oblong, ± clasping, ± obtuse, involute when young, denticulate, with 4-8 main veins per side. Stipules 2-3cm, obtuse, weakly 2-keeled, persistent. Floating lvs absent. Apr-Oct. *P. lucens* x *perfoliatus*...................................... **Potamogeton** x **salicifolius**

Lvs 3-10 x 0.3-1.5cm, linear-lanc to elliptic-oblong, not clasping (uppermost shortly stalked), acuminate or cuspidate, involute when young, denticulate or minutely serrate, with 3-5 main veins per side. Stipules 2-5cm, acute, not 2-keeled, persistent. Floating lvs few or absent, long-petiolate, ± rounded at base, leathery, with cross-veins (HTL). Apr-Oct .... *Various-leaved Pondweed* **Potamogeton gramineus**

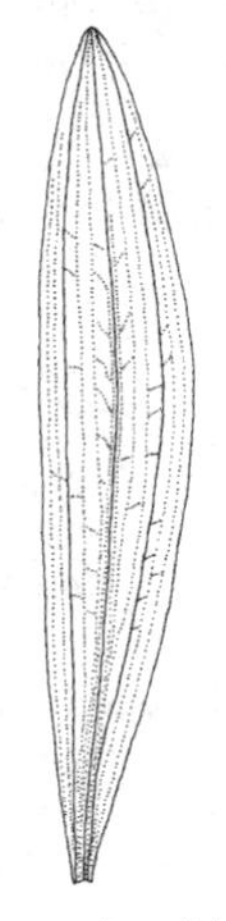

Potamogeton alpinus

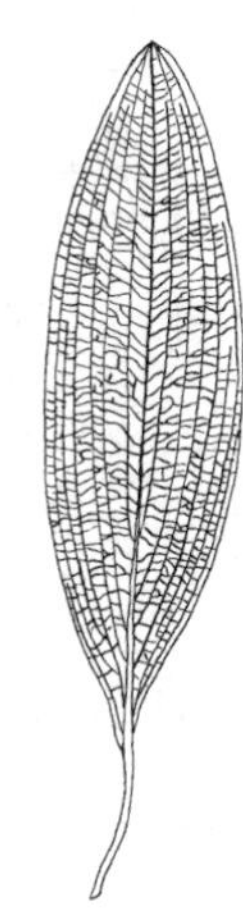

Potamogeton coloratus

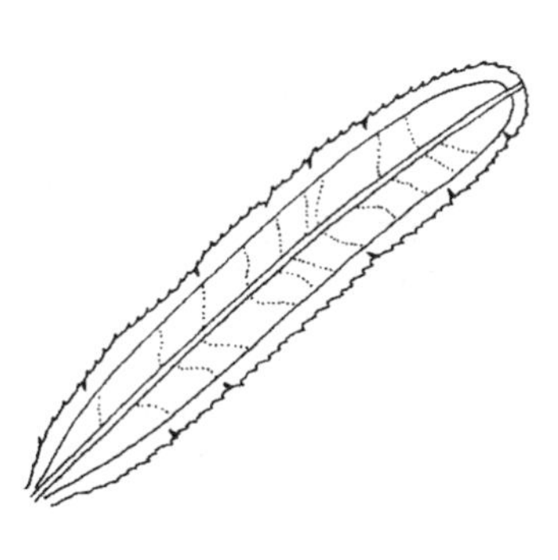

Potamogeton crispus

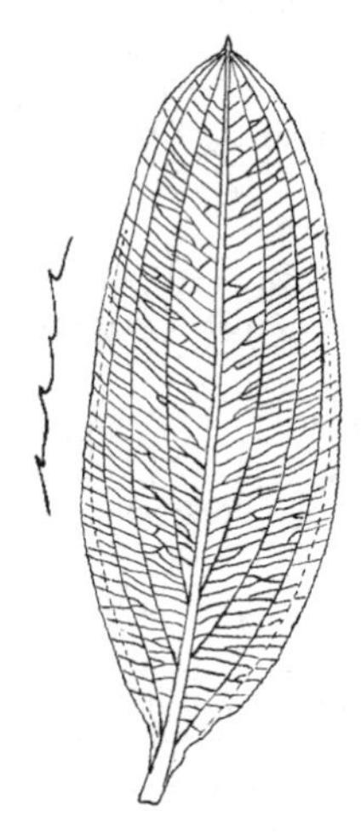

Potamogeton lucens & lf margin

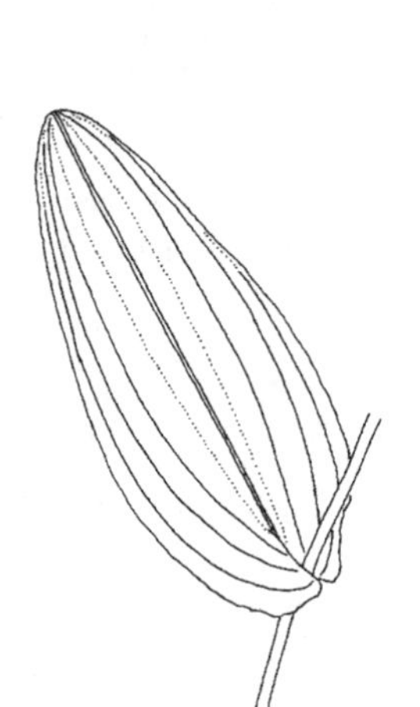

Potamogeton perfoliatus

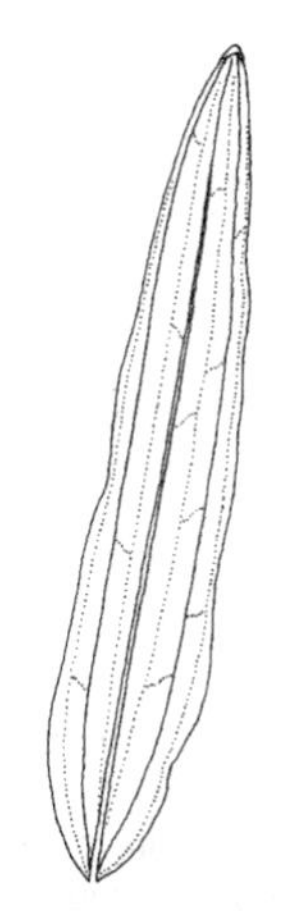

Potamogeton praelongus

Group FF – *Lvs all on submerged stems, <10mm wide, simple and unlobed. Stems may have 'glands' at nodes (actually root primordials, from where 1-2 roots may develop). Fresh or brackish water*

■Stem round or v slightly flattened. Lvs usu <2mm wide

Lf margin minutely toothed nr apex

Lvs 0.3-1mm wide, channelled, often opaque, with dark opaque wide midrib not reaching apex, with 2 hollows in TS. Sheaths without a ligule, loose, open to base, v thin. Stems 0.3-0.7mm diam, brittle, solid, with central stele. Ann to per. Apr-Oct (all yr). Brackish or saline water

Infl stalks usu <2.5cm. Lvs acute, light green................*Beaked Tasselweed* **Ruppia maritima**

Infl stalks >4cm. Lvs obtuse, dark green...........................*Spiral Tasselweed* **Ruppia cirrhosa**

Lf margin entire

Sheaths with a ligule 1.5-15mm

Lvs 0.3-1mm wide, ± translucent, usu with 2 hollows in TS. Stems 0.7mm diam, with c10 hollows around central stele (cartwheel-like). Plant stiff or flaccid. Brackish or freshwater

Sheaths open, overlapping, acute. Lvs usu acute, dark green, with 0-5 veins. Ligule 1.5-3(15)mm. Stems v branched. Apr-Oct (all yr)..*Fennel Pondweed* **Potamogeton pectinatus***

Sheath closed, not overlapping, obtuse. Lvs obtuse, yellowish-green, with 1(3) veins (laterals faint and marginal). Ligule 5-15mm. Stems mostly unbranched. Apr-Oct. N Br ....*Slender-leaved Pondweed* **Potamogeton filiformis***

Sheaths without a ligule

Sheaths closed at least when young (soon splitting), not overlapping

Stipules absent

Lvs 30-70 x 0.5-1mm, tapered to an acute apex, often opaque, solid (no aerenchyma) or with 2 unequal hollows, ± flat or weakly channelled above, light green, with 3(4) opaque veins, occ with sparse cross-veins. Stems to 1mm diam, occ channelled, with large aerenchyma and central vb. All yr...............................*Floating Club-rush* **Isolepis fluitans**

Stipules soon falling

Lvs 1.5-5(10)cm x 0.3-2mm, linear to thread-like, tapered, with v fine minutely mucronate apex, flat, translucent, with 2 hollows in TS, with midrib visible as wide dark green band, often with cross-veins. Stipules tubular, translucent. Stems 0.5mm diam, whitish. Apr-Oct (all yr)...*Horned Pondweed* **Zannichellia palustris**

Stipules ± persistent.

Lvs 10-50 x 0.3-1(3)mm, abruptly acute, translucent, rarely with a central silvery stripe, solid or with aerenchyma, flat, mid- to dark green, with 3(5) veins joining midrib at narrow angle 2-3 lf widths below apex. Stipules 2-17mm. Stems rarely with 'glands' at nodes. Apr-Oct..*Lesser Pondweed* **Potamogeton pusillus***

Sheaths open, overlapping

Sheaths loose. Lvs 0.5-1mm wide, thread-like, occ with 2 ± equal hollows

Lvs often opaque, weakly channelled. Stems with 'glands' at nodes. Apr-Oct. *P. pectinatus* x *vaginatus*..**Potamogeton** x **bottnicus**

Sheaths not loose. Lvs 0.5-1mm wide, thread-like, without distinct hollows

Lvs <0.5mm wide, stiff, not adhering together when removed from water

Lvs 20 x 0.5(1)mm, ± acute, saucer-shaped in TS, with thick midrib visible as groove above (dark borders to midrib equal to ⅓ lf width), with indistinct lateral veins (occ 3). Stipules 7-12mm, ± acute, not keeled. Apr-Oct ....*Hairlike Pondweed* **Potamogeton trichoides***

Lvs 1-2mm wide, limp, adhering together when removed from water

Lvs 20-60 x 0.5-2mm, obtuse to acute (abruptly contracted), usu shortly mucronate, usu with central silvery stripe (lacunae strip along midrib), with 3 lateral veins nr margin (joining midrib ± at right angles and ending well before apex (0.5-1x lf width)), often with cross-veins. Stem usu with 'glands' at nodes. Stipules 3-15mm, blunt, 2-keeled, not persisting. Apr-Oct....................................*Small Pondweed* **Potamogeton berchtoldii***

■Stems flattened. Lvs usu >2mm wide (0.5-1mm in *P. rutilus*). Lf margin entire

Lvs with 1 main lateral vein per side, without faint lateral veins, applanate when young

Lvs 2-4mm wide

Lvs v acute to gradually acuminate or long cuspidate, 5-13cm, green, with lateral veins (often weak) joining green midrib at wide angle. Stems 1.5-2(4)mm wide. Stipules 1.5-2.5cm, acute, strongly parallel-veined, not 2-keeled, ± persistent (soon fibrous). Apr-Oct ....*Sharp-leaved Pondweed* **Potamogeton acutifolius***

Lvs usu obtuse and minutely apiculate, 3-9cm, reddish later in yr, usu with distinct lateral veins joining reddish midrib at wide angle (occ >90°). Stems to 1.2(1.5)mm wide. Stipules 1.3-2cm, obtuse, faintly parallel-veined, not 2-keeled, ± persistent. Apr-Oct ....*Blunt-leaved Pondweed* **Potamogeton obtusifolius***

Lvs 0.5-1mm wide

Lvs v acute to acuminate, bright green to reddish, with strong marginal vein. Stems much-branched, slender. Stipules 1-2cm, closed, strongly parallel-veined, ± fibrous and persistent. Scot..*Shetland Pondweed* **Potamogeton rutilus***

Lvs with 2 main lateral veins per side (1 strong and 1 weak) or with many faint lateral veins, applanate when young

Lvs (±) acute, often mucronate, 4-8(10)cm x 2-3mm, green (but veins can be reddish), with lateral veins joining midrib at (¼)½ lf width below apex, without faint lateral veins, with cross-veins. Stems usu 0.5-1mm wide, not winged. Stipules 0.7-1.5cm, obtuse, not 2-keeled, soon jagged. Nodal glands obvious. Apr-Oct.................*Flat-stalked Pondweed* **Potamogeton friesii***

Lvs obtuse, cuspidate (occ acuminate), 9-20cm x (2)3-4mm, dark green, with main lateral veins joining midrib immediately below apex, with many other faint lateral veins, with scattered cross-veins. Stems 3-6mm wide, ± winged. Stipules 2.5-3.5cm, obtuse, not 2-keeled, soon jagged. Nodal glands absent. Apr-Oct.................*Grass-wrack Pondweed* **Potamogeton compressus***

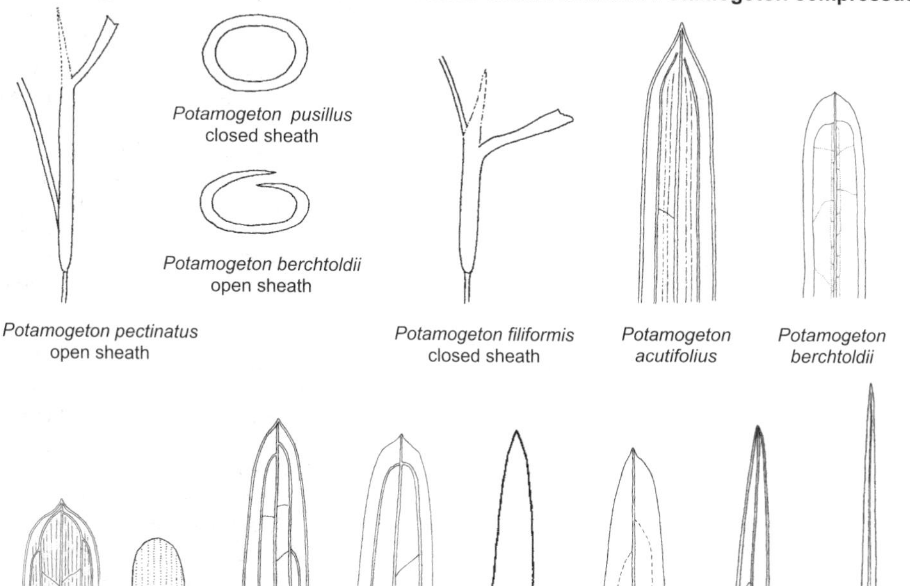

Group FG – *Lvs mostly submerged, all basal, alt, simple and unlobed, <10mm wide. (Young plants of* Damasonium alisma *may key out here).* ❶ *Auricles present.* ❷ *Ligule present*

■Lvs with 4 large hollows, without latex. Roots whitish, hollow. Plant often washed up on shoreline

Lvs to 12cm x 2-3mm, ± cylindrical, with flattened expanded base partly sheathing swollen corm, brittle (snapping audibly), with cross-veins. All yr

Lvs stiff, curved, ± parallel-sided, ± obtuse, dark green, the epidermal cells usu 8-10x as long as wide. Megaspores netted..*Quillwort* **Isoetes lacustris***

Lvs limp, straight, tapered to a slender acute apex, bright green, the epidermal cells usu 3-4x as long as wide. Megaspores spiny to netted.................*Spring Quillwort* **Isoetes echinospora***

■Lvs with 2 large hollows, with white latex (occ sparse). Roots whitish, solid

Lvs 3-8cm x 2-4mm, strap-shaped, ± flattened, obtuse, ± recurved, without cross-veins. Stem lfless, hollow. All yr. Upland lakes..*Water Lobelia* **Lobelia dortmanna***

■Lvs solid or with indistinct hollows/aerenchyma. Roots whitish, solid

Lvs translucent, brittle, snapping audibly, with distinct aerenchyma blocks and cross-veins

Lvs with white latex confined to margins in TS. Roots not septate. Apr-Oct

Lvs 8-30(100) x 0.2-2(3)cm, occ floating or emergent, linear, narrowly obtuse, flat, often developing a short (15-40mm) linear-lanc or narrowly oblong blade. Stolons absent. VR, Worcs. Sch8...*Ribbon-leaved Water-plantain* **Alisma gramineum***

Lvs to 30cm, submerged, linear, narrowly obtuse, semi-cylindrical in TS. Stolons absent ....*Water-plantain* **Alisma plantago-aquatica**

Lvs with white latex scattered throughout in TS. Roots weakly septate. Apr-Oct

Stolons absent. Lvs submerged, 5-20 x 1-2cm, linear, with 3-many veins joining midrib before apex. Petiole indistinct. Pl 14...*Arrowhead* **Sagittaria sagittifolia**

Stolons present. Lvs floating (rarely emergent), to 5 x 1cm, elliptic-oblanc, with 3 veins. Petiole long, semi-cylindrical. VR alien............*Narrow-leaved Arrowhead* **Sagittaria subulata**

Lvs without latex. All yr

Lvs with strong coriander odour. Stolons rarely present

Lvs 4-5mm wide, ± semi-cylindrical tapered to flat ± acute apex (<1mm wide), with 10-15 green veins, with large aerenchyma in TS...*Lesser Water-plantain* **Baldellia ranunculoides**

Lvs odourless

Stolons often abundant. Roots not segmented. Lvs 5-20(60)cm x 1-8mm, linear to sword-shaped, acute, flat, green to whitish, with 6-10 green veins, with aerenchyma or small hollows. Sch8...*Floating Water-plantain* **Luronium natans***

Stolons absent. Roots segmented. Lvs 3.5(10)cm x 4-5mm, tapered from base, finely acute, ± flat, green, with 6-10 green veins, solid. Stem 2mm diam, sheathing at base, with 8-10 hollows around central stele. W Scot, Ire............................*Pipewort* **Eriocaulon aquaticum**

Lvs ± opaque, not brittle or snappable

Lvs in rosette or with stolons, <3mm wide

Sheaths with 1(2)mm overlapping auricles and cross-veins (HTL, occ indistinct)

Lvs to 30cm x 0.5-1.5mm, obtuse, flattened, with 2-4 obscure hollows, Plant often reddish, with stolons, irreg clustered from a bulbous base. ❶. All yr *Bulbous Rush* **Juncus bulbosus**

Sheaths without auricles or cross-veins

Stoloniferous per, often turf-forming. Lvs 30-70 x 2-3mm, ± cylindrical to flat, obtuse to acute (hydathode often purplish), usu with a few obscure hairs at sheathing base, ± rigid, with spongy aerenchyma and tiny central vb. All yr.......................*Shoreweed* **Littorella uniflora***

Tufted bi or ann. Lvs 20-40 x 1-1.5mm, trigonous to semi-cylindrical, finely acute, hairless even at dilated base, flaccid, solid but with hollows at base. Often all yr. Mtn lakes ....*Awlwort* **Subularia aquatica**

Lvs not in a rosette, without stolons, usu >3mm wide

Lvs acute, linear, tapered, flat, with adnate obtuse ligule at base (to 6mm), with opaque veins, with cross-veins. Sheathing base flattened. All yr

Lvs to 1.2m x 3-5mm, finely acute, with margins rarely scabrid, with 7 opaque veins. ❷ ....(submerged lvs) *Common Club-rush* **Schoenoplectus lacustris**

Lvs obtuse, linear, tapered, weakly keeled below (flat nr apex), without ligule, with opaque veins, usu with cross-veins. Sheathing base flattened. Apr-Nov

Lvs to 0.3(0.5)m

Lvs 2-6(7)mm wide, ± all floating, flat but curved, with 15 opaque veins, with stomata both sides, with 10 hollows. Organic-rich pools....................*Least Bur-reed* **Sparganium natans**

Lvs 1-2m

Lvs 7-10(18)mm wide..*Branched Bur-reed* **Sparganium erectum***

Lvs 2-7mm wide

Lvs to 2m x 4-5(10)mm, usu floating or submerged, flattened triangular, usu with >10 veins. Stems often absent..........................*Unbranched Bur-reed* **Sparganium emersum**

Lvs to 1.5m x 2-5mm, usu all floating, flat or flattened triangular, with 7-10 opaque veins. Stems round, solid, with scattered vb's. Peaty mtn lakes ....*Floating Bur-reed* **Sparganium angustifolium**

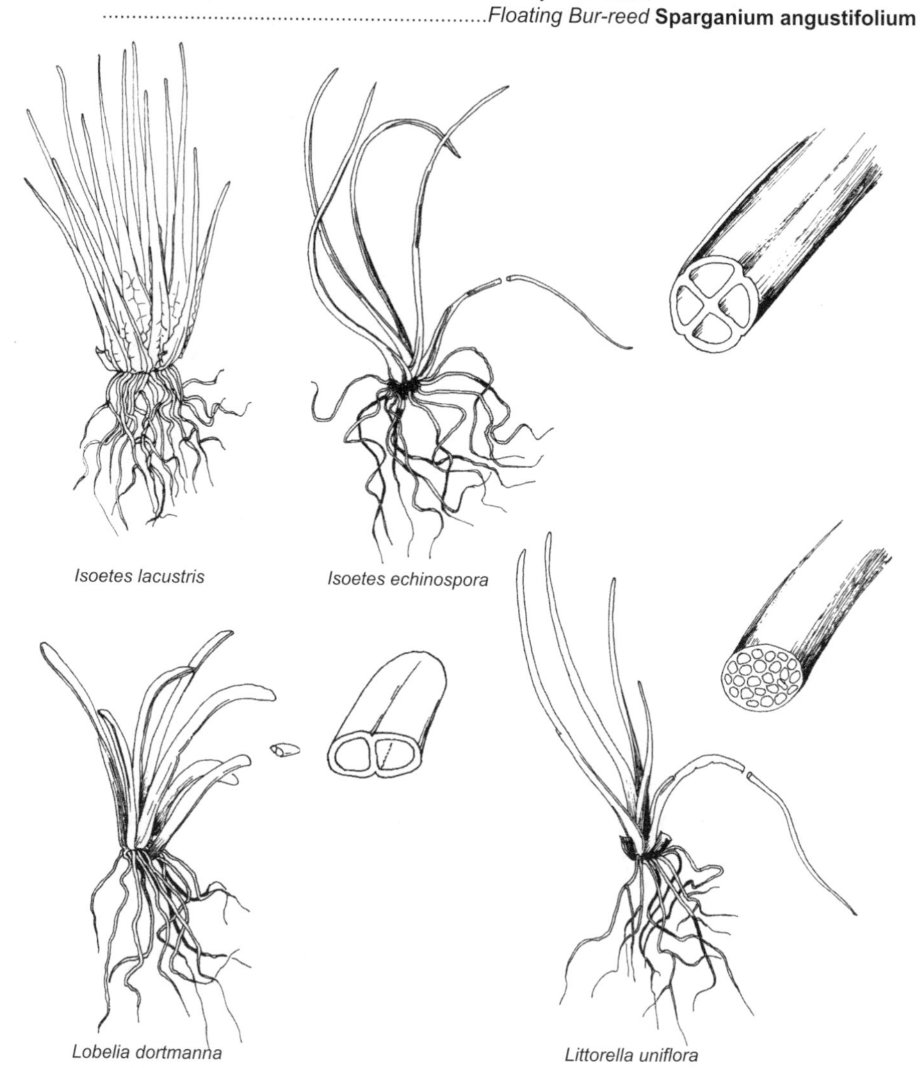

Group FH – Utricularia. *Lvs submerged, alt, usu 2-pinnate, with translucent bladders (for trapping micro-fauna); segments thread-like to linear (with main vein visible), with apical bristles. Stems rarely >40cm, buried or floating in shallow water, green or colourless, round, striate, with hollows around central stele (cartwheel-like). Roots absent. Apr-Oct*

■Lvs pinnately divided, mostly on same stems as bladders. Lf segments minutely toothed

Lf teeth (and lobe apices) with 1-2 bristles. Lvs 1-4cm. Stems 0.5-0.8mm diam, with 12 hollows. Acidic water........*Bladderwort* **Utricularia australis**

Lf teeth (and lobe apices) with >2 bristles. Lvs 1-3cm. Stems 0.5-0.7mm diam, with 12 hollows. Base-rich water........*Greater Bladderwort* **Utricularia vulgaris**

■Lvs palmately divided, mostly on separate stems to bladders

Lf segments entire, with single bristle at apex. Lvs 0.3-2cm. Stems 0.4-0.6mm diam, with 7 hollows........*Lesser Bladderwort* **Utricularia minor**

Lf segments minutely toothed, with 1-2 bristles on each tooth and apex. Lvs 0.3-1.5mm. Stems 0.5-0.7mm diam, with 7 hollows........*Intermediate Bladderwort* **Utricularia intermedia** agg

---

FG

Luronium natans (submerged lvs only)

Group FI – *Lvs submerged, alt or all basal, compound (1-pinnate etc) or lobed, without translucent bladders.* Lemna *consists of a thallus (a plant body not differentiated into stems and lvs). For the purpose of this key, we refer to those parts resembling leaves as leaves (lvs)*

■Lvs (thalli) 3-lobed, joined in 'chains'. Plant suspended nr surface, with 1 root per lf

Lvs <15mm diam, translucent, elliptic-lanc, obscurely 3-veined, tapered at base to 7mm stalk, ± acute, usu minutely serrate at apex. All yr............................*Ivy-leaved Duckweed* **Lemna trisulca**

■Lvs 1-pinnate (or pinnately lobed). Petiole without sheathing base. Plant rooted to substrate ....*Water-violet* **Hottonia palustris**

■Lvs 2-3(6) pinnate. Petiole with sheathing base. Plant rooted to substrate

Lf lobes (segments) with 1-4 apical bristles. Lvs trichotomously branched. Plant odourless

Lf lobes round, thread-like, ± alt, ± obtuse. Sheathing base not inflated. Floating lvs, if present, ± orb...(*Ranunculus aquatilis* agg*) Go to **RAN-BAT**

Lf lobes without apical bristles. Lvs dichotomously branched.

Plant with parsley odour when crushed ....(submerged lvs) *Fine-leaved Water-dropwort* **Oenanthe aquatica**

Plant with sweet celery odour when crushed

Petiole v hollow, often with v sparse white latex. Stolons developing from mid-summer ....(submerged lvs) *Tubular Water-dropwort* **Oenanthe fistulosa**

Petiole hollow, without latex. Stolons present, rooting at nodes

Lfts obtuse, flat, 0.4mm wide, cuneate, toothed nr apex, with main veins visible only. Petiole channelled, not swollen at sheathing base, often with small pithy hollow. Stems >2mm diam, hollow. Usu all yr. Fast-flowing calc rivers...........*River Water-dropwort* **Oenanthe fluviatilis**

Lfts acute, flat or cylindrical, 1mm wide, sessile, entire or lobed but not toothed, with veins obscure. Petiole ± round, ± swollen at sheathing base, hollow. Stems 1.5-2mm diam, hollow. All yr. Muddy pond margins..*Lesser Marshwort* **Apium inundatum**

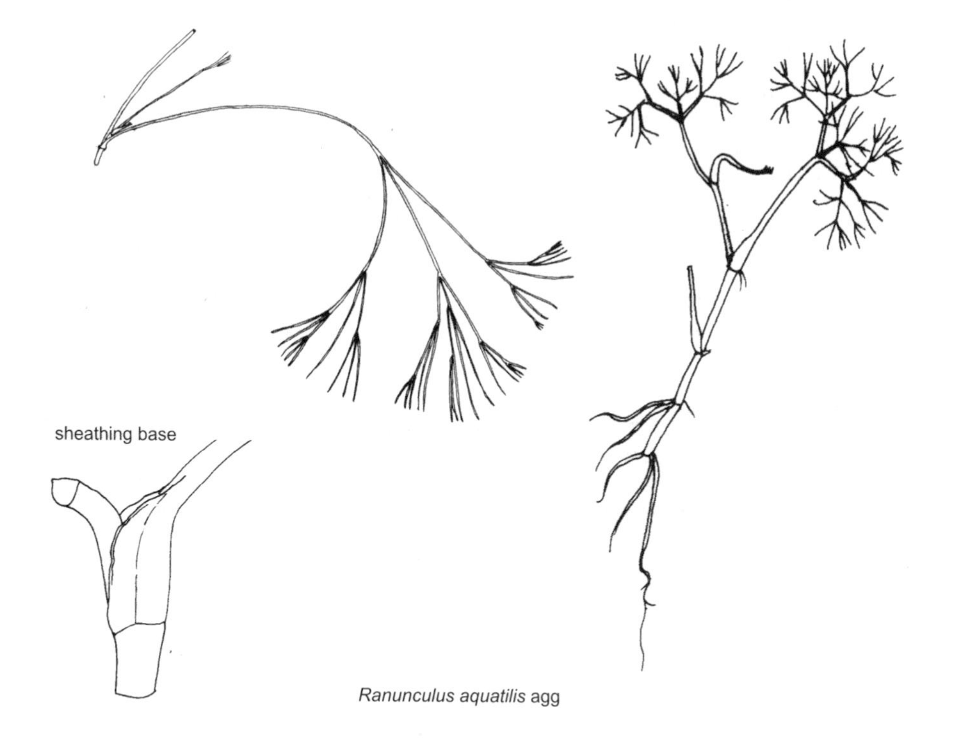

Ranunculus aquatilis agg

Group FJ – *Lvs floating on surface (often emergent in* Nuphar advena *and* Nymphaea marliacea*), >10mm wide, usu broadly ovate to orb. Lf veins pinnate, palmate, or obscure*

■Lvs with mostly pinnate veins along midrib. Rhizomatous

Lvs ± orb, with forking veins, involute when young. Underwater lvs semi-translucent. Stipules absent

Lvs 12-40 x 8-30cm, ovate to ± orb, with broad triangular basal sinus and overlapping to divergent (often ≥90°) lobes c⅓ lf length, leathery, shiny green both sides, rarely with sparse hairs, with 20-38 distinct 2° veins forking 3-5x to margin. Petiole ± round to oval, with >20 vb's in aerenchyma. Apr-Oct.. *Spatter-dock* **Nuphar advena**

Lvs 12-40 x 8-30cm, ovate-oblong, with deep basal sinus and divergent lobes c⅓ lf length, leathery, shiny green above, midrib occ hairy below, with 23-28 distinct 2° veins forking 3-4x towards margin. Petiole trigonous, >10mm diam, with 17-23 vb's in aerenchyma. Apr-Oct .... *Yellow Water-lily* **Nuphar lutea***

Lvs 4-8(14) x 3.5-7(13)cm, broadly oval, with deep basal sinus and ± divergent lobes c⅓ lf length, leathery, shiny green above, often discoloured, with 11-18 indistinct 2° veins forking 3x towards margin. Petiole flattened, with 3(10) vb's in aerenchyma. Apr-Oct. Highland lakes, R .... *Least Water-lily* **Nuphar pumila**

Lvs lanc, with anastomosing veins, revolute when young. Underwater lvs absent. Stipules fused into ochreae

Lvs 6-15 x 2-4(6)cm, adpressed hispid-hairy above (to hairless), pitted above, with stomata at least above. Petiole 2-4cm, reddish. Ochreae adpressed hispid-hairy (when not submerged). Apr-Oct...(floating lvs) *Amphibious Bistort* **Persicaria amphibia***

■Lvs with mostly palmate veins (radiating from petiole junction). Rhizomatous, stoloniferous, or tufted

Floating lvs deeply palmately lobed

Thread-like submerged lvs often present. Floating lvs with stomata above only. Tufted, occ stoloniferous...(*Ranunculus aquatilis* agg*) Go to **RAN-BAT**

Thread-like submerged lvs never present. Floating lvs with stomata both sides. Stoloniferous. Invasive alien..*Floating Pennywort* **Hydrocotyle ranunculoides**

Floating lvs entire

Lvs with purplish tubercles below, involute when young. Stipules absent but auriculate sheathing base present

Lvs 3-11cm diam, ± orb, ± angular, cordate at base with rounded lobes, entire or sinuate, crisped upwards, thin, shiny dark green above, with veins occ obscure. Petiole 5-10mm, slender, limp, with 1(3) vb's in aerenchyma around central stele. Fl stems with opp lvs. Rhizomatous. May-Nov...*Fringed Water-lily* **Nymphoides peltata***

Lvs without purplish tubercles, involute when young. Stipules present

Petiole with c14 unequal hollows, ± round. Lvs floating,10-30cm diam, ± orb, with closed basal sinus ⅓-½ lf length (lobes rarely overlapping), shiny dark green above, often reddish below, with veins forking 5-6x to margin. Rhizomatous. May-Nov. PL 24 .... *White Water-lily* **Nymphaea alba**

Petiole with 4 main equal hollows, ± round. Lvs often emergent, 10-45cm diam, ± orb, with basal sinus open or closed, shiny dark green above, often reddish below, with submarginal vein. Rhizomatous. May-Nov.. **Nymphaea** x **marliacea***

■Lvs with 2-3 ± parallel veins converging at apex (connected by pinnate veins). Stoloniferous

Lvs to 4cm diam, ± orb, cordate at base, often with basal lobes overlapping, reddish esp below. Petiole round, with hollows. Stipules large, ovate, scarious, often reddish-veined. Plant free-floating, with feathery roots. May-Oct (overwinters as buds on stolons) .... *Frogbit* **Hydrocharis morsus-ranae***

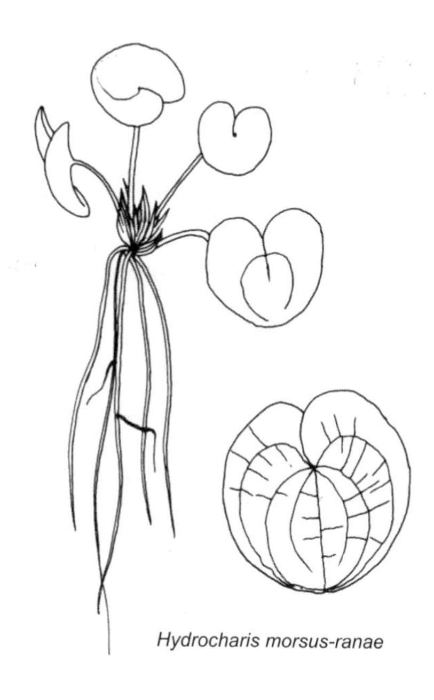

Hydrocharis morsus-ranae

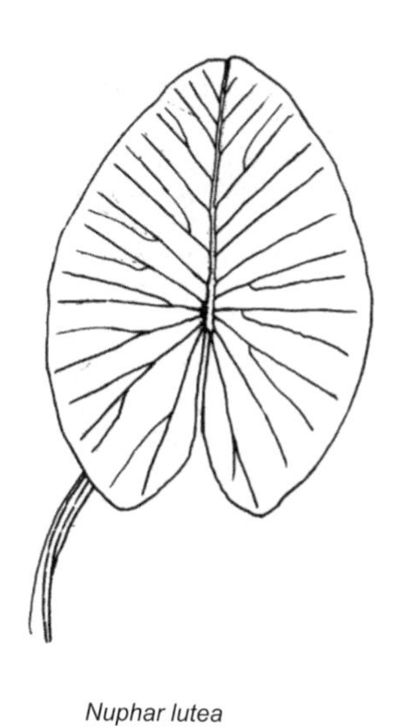

Nuphar lutea

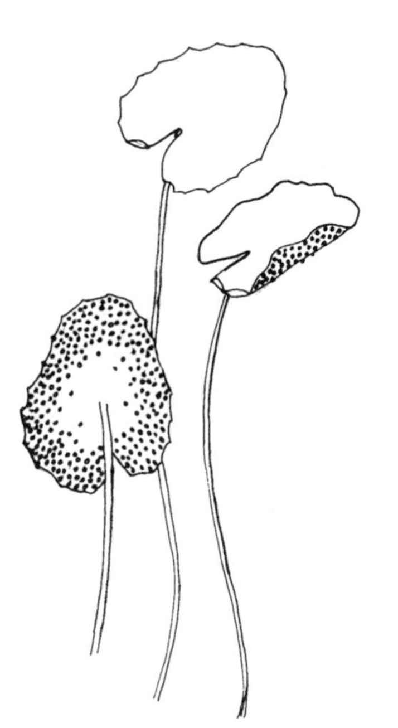

Nymphoides peltata

Nymphaea x marliacea

Group FK – *Lvs floating on surface (rarely emergent or submerged), >10mm wide, usu elliptic-lanc to broadly ovate, those of* Potamogeton *and* Aponogeton *are involute when young (with stomata above only). Lf veins parallel. Stems and stolons round. Roots solid.* ❶ *Lvs densely septate-hairy.* ❷ *Petiole with white latex*

■Stipules present at least when young

Petiole with discoloured flexible joint below lf. Lvs with all veins translucent. Stems with interlacunae vb's

Lvs to 12 x 7cm, elliptic to ovate-lanc, widest at or below middle, ± acute, usu rounded to ± cordate at base, leathery, with 9-14 veins per side, with cross-veins. Petiole > lf, flattened, with 3(5) vb's. Stipules 4-20cm x 8-12mm, 2-keeled. Submerged lvs (phyllodes) 9-20cm x 1-3mm, linear, rarely with small blade..........................*Broad-leaved Pondweed* **Potamogeton natans***

Petiole without discoloured flexible joint. Lvs usu with all veins opaque. Stems without interlacunae vb's

Lvs with oblique cross-veins (HTL)

Petiole 5-25cm

Lvs 6-20 x 2.5-6cm, oblong-elliptic or ovate, shortly acute, cuneate, leathery. Petiole 2.5mm diam, round, with 3(5) vb's. Stipules 4-10cm, lanc, 2-keeled. Stems stout, unbranched. Submerged lvs 10-20 x 1.5-4cm, lanc to ovate, ± cuspidate, thin, translucent, net-veined, entire, often rotting away..................................*Loddon Pondweed* **Potamogeton nodosus***

Petiole to 5cm (usu ½-1x lf length)

Floating lvs ± leathery

Submerged lvs (if present) entire. Floating lvs (4)8-20 x 1-4cm, broadly elliptic to lanc, acute, cuneate to cordate at base, with 6-10 veins per side. Stipules 1-5cm, obtuse, often with 2 stronger veins. Bogs.........................*Bog Pondweed* **Potamogeton polygonifolius***

Submerged lvs usu denticulate. Floating lvs 3-10 x 0.3-1.5cm, cuneate, with 5-10 veins per side. Petiole rarely to 5x lf length. Stipules 0.5-2.5cm, obtuse to ± acute (often rolled and appearing acute), many veined, keeled or not ....*Various-leaved Pondweed* **Potamogeton gramineus**

Floating lvs v thin and translucent (often pinkish/coppery and marl-encrusted)

Submerged lvs (if present) entire. Floating lvs 2-7 x 1.5-5cm, ovate-elliptic (occ narrowly so), ± obtuse, truncate to cuneate at base, with 4-10 opaque veins per side. Stipules 2-5cm, obtuse, weakly 2-keeled, ± persistent. All yr. Fens ....*Fen Pondweed* **Potamogeton coloratus***

Lvs without cross-veins (exc on translucent thin submerged lvs). Occ submerged lvs only

Lvs to 12 x 1.5-2.5cm, oblong to elliptic, obtuse, cuneate and tapered into petiole, leathery. Submerged lvs 8-20cm x 3-8mm, linear, obtuse or (±) acute, sessile and tapered, with 3-5(7) veins, midrib bordered with strip of hollows. Petiole with 3 vb's. Stems slender, mostly unbranched, flattened in submerged growth, round in floating growth. Stipules to 3.5cm, broad, obtuse. VR, Outer Hebs (native), canals in N Eng (alien). Apr-Oct ....*American Pondweed* **Potamogeton epihydrus**

■Stipules absent

Plant free-floating rosette with feathery roots

Lvs 5-15cm, orb to ovate or wider than long, v obtuse, usu cordate at base, shiny dark green, entire, often undulate, with numerous parallel veins, with stomata both sides. Petiole 2-30cm, floating, v inflated, spongy, with minute stomata. Stolons round, with ± solid aerenchyma and spiral fibres. Roots blackish. Stoloniferous per. All yr (but frost-sensitive). VR alien ....*Water-hyacinth* **Eichhornia crassipes**

Lvs to 5(10)cm, fan-shaped to obovate, retuse apex occ recurved, cuneate at base, dull pale grey-green, entire, undulate, ± net-veined, pleated with parallel forked veins strongly raised below, with stomata above. Petiole indistinct. Stolons round, solid, without spiral fibres. Roots green. Stoloniferous per. ❶. All yr (but frost-sensitive). VR alien. *Water-lettuce* **Pistia stratiotes**

Plant rooted in substrate

Petiole with latex canals confined to margin (and around vb's). Roots not septate. ❷

Lvs 8-30(100) x 0.2-2(3)cm, occ floating or submerged, linear, hardly discernible from petiole, often developing a short (15-40mm) linear-lanc or narrowly oblong blade. Not tuberous. VR, Worcs. Sch8..*Ribbon-leaved Water-plantain* **Alisma gramineum***

Petiole with latex canals throughout aerenchyma. Roots weakly septate. ❷

Petiole sharply ± triangular, with sheathing base and large aerenchyma. Lvs usu acute (with hydathode obscure), without submarginal veins. Tuberous per. PL 14 ....*Arrowhead* **Sagittaria sagittifolia**

Petiole without latex. Roots not septate

Lvs >5cm, without pore-like hydathode

Lvs 6-25 x 8cm, elliptic-lanc to oblong-elliptic, obtuse to acute, ± rounded to cuneate at base, involute when young, with abundant small reddish glandular streaks below, with entire cartilaginous margin, with 5-9 parallel opaque veins and opaque ladder-like cross-veins, with midrib only raised below, with stomata above only. Petiole to 1m, with sheathing base, ± round (slightly channelled above), with large spongy aerenchyma. Tuberous per. Apr-Oct. VR alien...*Cape-pondweed* **Aponogeton distachyos**

Lvs <5cm, with pore-like hydathode at apex below

Ann. Lvs 2-5 x 1-3cm, ovate to oblong, usu obtuse, rounded to cordate at base, yellow-green, with 5 opaque veins (not prominent) and submarginal vein, with stomata above only. Petiole to 6cm x 2mm, with sheathing base, oval with 2 weak opp angles, with cross-veins, with hollows each side of central vb. Nov-Aug. VR. Sch8........*Starfruit* **Damasonium alisma**

Stoloniferous per. Lvs 1-3 x 0.3-0.7cm, ovate-elliptic, obtuse, rounded-cuneate at base, green, with midrib and submarginal vein converging at apex (connected by pinnate 2° veins at ± 90° to midrib), with red bordered stomata above only. Petiole to 10cm x 1-2mm, with sheathing base, flattish, with aerenchyma around central vb. R, mostly Montgomery-Manchester region. Sch8....................................*Floating Water-plantain* **Luronium natans***

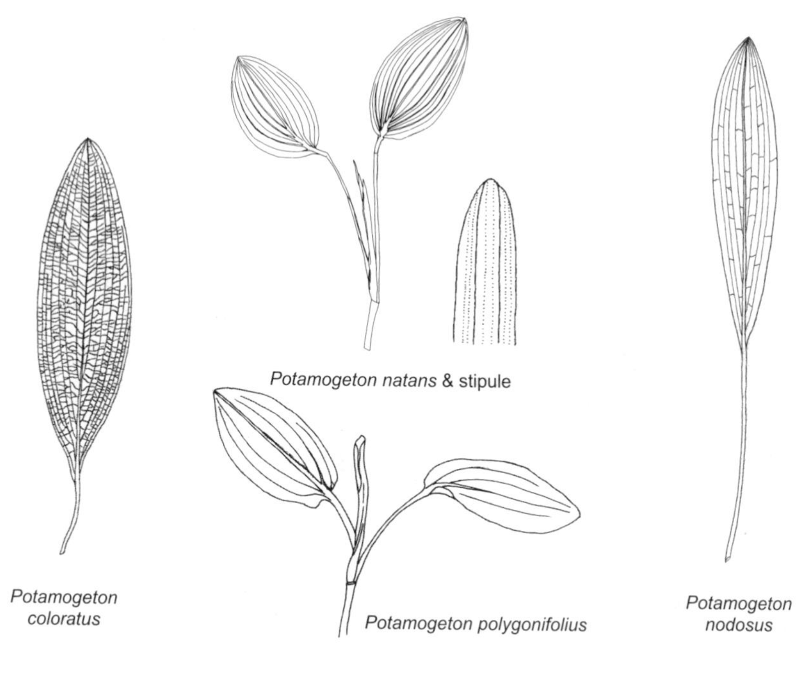

Potamogeton natans & stipule

Potamogeton coloratus

Potamogeton polygonifolius

Potamogeton nodosus

Group FL – *Lvs free-floating (often growing in large masses), <7mm diam, orb (or ± so), with stomata above.* Lemna, Spirodela *and* Wolffia *constitute a thallus (a plant body not differentiated into stems and lvs). For the purpose of this key, we refer to those parts resembling leaves, as leaves (lvs). Turions are small daughter plants (often rootless)*

■Roots absent

Lvs floating on or nr surface, 0.5-1(1.6)mm diam, ovoid to ellipsoid (occ ± globose), ± acute, smooth, green, veins absent. Turions developing. Apr-Oct (sinks to bottom in winter) ....*Rootless Duckweed* **Wolffia arrhiza**

■Roots 1 per lf

Lvs overlapping in 2-ranks, imbricate (like fish-scales), 2-lobed

Lvs with alt lobes 1-1.5mm, ovate-triangular, blue-green, turning red, with broad hyaline margins often reddish, with few short papillae and microscopic unicellular hairs both sides. Apr-Nov (occ all yr but usu sinking in winter)....................................*Water Fern* **Azolla filiculoides***

Lvs not overlapping in 2-ranks, entire

Lvs v swollen below (often flat in turions plants, or after fl)

Lvs 3-5mm, obscurely 3(5)-veined, usu with opaque reticulations. Turions usu absent. Root to 6cm...*Fat Duckweed* **Lemna gibba**

Lvs ± flat (slightly convex below)

Lvs bright shiny green, (2)3-4(5) x 3(4)mm, obscurely 3-veined, not ridged, usu obovate, obtuse never with apiculus. Turions usu absent. Root to 6(15)cm. All yr (but frost-damaged plants often sink in winter)..*Common Duckweed* **Lemna minor***

Lvs dull dark olive-green, usu 1.5-3 x 1.3-2mm, 1-veined, ridged (often obscure) or with low ridge of papillae, usu elliptic, obtuse, occ with v minute apiculus. Turions absent. Root to 1.5cm. Often all yr...*Least Duckweed* **Lemna minuta**

Lvs olive-green, often reddish below, 2-5mm, obscurely 3-veined, ridged above with line of papillae, usu obovate, obtuse, without apiculus. Turions often present, <1.5mm. Roots to 6(15)cm. VR alien..*Red Duckweed* **Lemna turionifera**

■Roots 7-12 per lf

Lvs 4-7mm diam, orb, usu purplish below, usu with 5-12 veins. Turions present. Roots to 3cm ....*Greater Duckweed* **Spirodela polyrhiza***

Azolla filiculoides

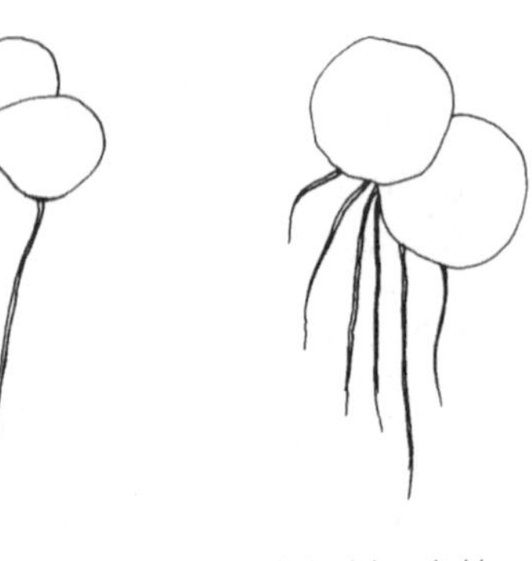

Spirodela polyrhiza (with turion)

Lemna minor (with turion)

Group FM – *Lvs floating on surface (rarely emergent), <7mm wide, linear. Plant rooted to substrate. Stems round, solid, with scattered vb's. (Other* Sparganium *spp and* Limosella aquatica *may key out here)*

■Lvs to 0.3(0.5)m

Lvs 2-6(7)mm wide, ± all floating, flat but curved, with 15 opaque veins, with stomata both sides, with 10 hollows. Organic-rich pools..*Least Bur-reed* **Sparganium natans**

■Lvs 1-2m

Lvs 7-10(18)mm wide...*Branched Bur-reed* **Sparganium erectum***

Lvs 2-7mm wide

Lvs to 2m x 4-5(10)mm, usu floating or submerged, flattened triangular, usu with >10 veins. Stems often absent..*Unbranched Bur-reed* **Sparganium emersum**

Lvs to 1.5m x 2-5mm, usu all floating, flat or flattened triangular, with 7-10 opaque veins. Stems round, solid, with scattered vb's. Peaty mtn lakes ....*Floating Bur-reed* **Sparganium angustifolium**

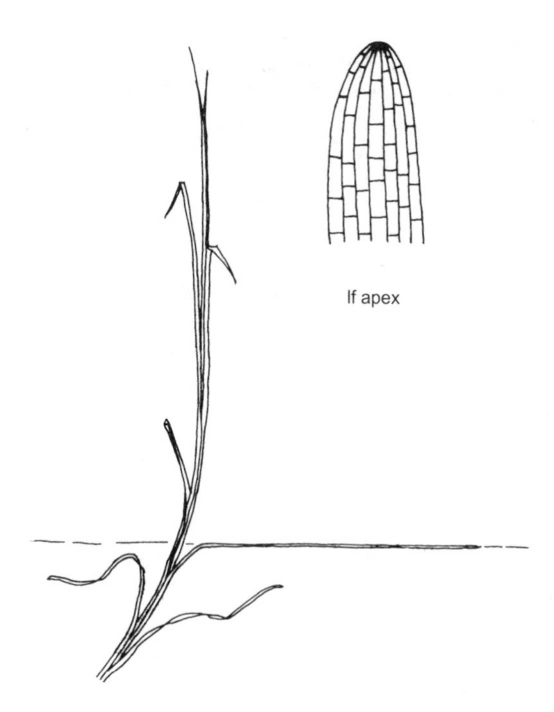

Sparganium erectum

Key to Groups in Division G

(Cyperaceae. *Ligule present, ± adnate to lf)*

Lvs usu all basal and 3-ranked, with margins scabrid at least nr apex, parallel-veined, with the number of veins equal to the number of stomatal bands (and number of hollows in TS). Ligule adnate to lf, exc for a narrow free margin. Sheaths closed, usu obscurely parallel-veined (but pinnate-veined in ladder-fibrillose spp). Stems often trigonous, without nodes (exc where stated), ± without sap (exc Blysmus *spp). Shoots forming false-stems (consisting of tightly rolled lvs and sheaths) in some species. Hairs, when present, unicellular (but papillate in* Carex hirta*). All yr (unless otherwise stated)*

Plates 1-8 illustrate (in alphabetical order) the TS of all the Carex *spp, together with examples of* Eleocharis, Eriophorum *and* Kobresia

■Lvs ≤1mm wide.......... **GA**
■Lvs >1mm wide
 Lvs and/or sheaths distinctly hairy.......... **GB**
 Lvs and sheaths hairless or obscurely hairy
  Lvs with stomata above (often dull above).......... **GC**
  Lvs with stomata below only
   Lvs with distinct hollows in TS at midpoint, often spongy and U-shaped.......... **GD**
   Lvs solid (occ with hollows in TS nr base), usu thin and V- to M-shaped
    Lvs dull glaucous below.......... **GE**
    Lvs shiny green below
     Ligule >2mm
      Basal sheaths reddish, often ladder-fibrillose. Lvs occ hairy.......... **GF**
      Basal sheaths not reddish, rarely ladder-fibrillose. Lvs never hairy.......... **GG**
     Ligule <1.5(2)mm
      Basal sheaths reddish/orange (or with dark red veins) AND/OR ligule minutely fimbriate. Lvs occ sparsely or minutely hairy.......... **GH**
      Basal sheaths not reddish or orange. Ligule never fimbriate. Lvs never hairy
       Lvs with trigonous tip.......... **GI**
       Lvs without trigonous tip.......... **GJ**

Group GA – *Lvs ≤1mm wide.* ❶ *Shoots intravaginal, fasciculate.* ❷ *Shoots 'bulbous' at base*

■Lvs >20cm, with hollows in TS. Basal sheaths not fibrous

Plant tussock-forming. Lvs often with cross-veins

Lvs 0.7-1mm wide, triangular, rounded at apex with 0.5-1mm blackish tip, with sparsely scabrid margins nr apex, with 3-7 stomatal bands below. Ligule to 1mm, obtuse. Sheaths with concave apex. Basal sheaths pinkish. Shoots trigonous. Stems round below, trigonous above. Bogs, wet hths. PL 8..*Hare's-tail Cottongrass* **Eriophorum vaginatum**

Plant rhizomatous. Lvs always with cross-veins

Lvs 1-2.5mm wide, U- or V-shaped, thick, with long whip-like acute trigonous tip, with scabrid margins, grey-green, with 4-8 veins each side of midrib (best viewed below), with 10 hollows in TS. Ligule 1mm, obtuse, brown-scarious. Sheaths with ± acute U- to V-shaped brown-scarious apex. Basal sheaths dark reddish-brown, ladder-fibrillose. Stems round at base. Bogs. PL 4 ....*Slender Sedge* **Carex lasiocarpa**

Lvs 0.8-1.1mm wide, triangular, channelled above, with rounded black scabrid apex, with smooth margins, with 5-7 veins on 2 of the 3 sides. Ligule 1mm, obtuse. Sheaths with concave apex. Basal sheaths pale reddish-brown, not ladder-fibrillose. Stems triquetrous. V wet bogs. Sch8..*Slender Cottongrass* **Eriophorum gracile**

■Lvs <10cm, without hollows in TS

Basal sheaths fibrous, persistent, usu with dead lvs. Lvs tough, without cross-veins. Tufted

Dry calc gsld, S Eng. Lvs 0.5-1(2)mm wide, U-shaped (but outer lvs ± flat), with acute trigonous tip (often dead and wispy), often twisted nr sheath, with ± long (often retrorse) scabrid cilia nr ligule, with 1-2 veins each side of midrib (midrib most translucent vein) but 4-6 unequal stomatal bands below. Ligule 0.5mm, v obtuse, minutely fimbriate. Sheaths with concave apex. Basal sheaths ladder-fibrillose, with blackish-brown veins. ❶. PL 4.*Dwarf Sedge* **Carex humilis**

Calc mtn flushes, N Eng, Scot. Lvs 0.7-1mm wide, folded or deeply V-shaped, with v acute long trigonous tip (often >2cm), with scabrid margins esp nr apex, shiny dark green, with 2-4 obscure veins each side of midrib. Ligule to 0.5mm, obtuse, obscurely fimbriate, brown-scarious. Sheaths with concave apex. Basal sheaths pale brown, soon splitting. ❷. PL 8 ....*False Sedge* **Kobresia simpliciuscula**

Exposed mtn rocks, Scot. Lvs 0.5-1.5mm wide, deeply U- to shallowly V-shaped, with a long trigonous tip (often dead and curled), with scabrid margins esp nr apex, mid-green, with 3-6 obscure veins per lf (midrib translucent only). Ligule 1mm, retuse, entire. Sheaths with concave apex. Basal sheaths dark brown. PL 7..*Rock Sedge* **Carex rupestris**

Basal sheaths not fibrous or persistent. Lvs not tough, without cross-veins. Shortly rhizomatous

Lvs with 2 stomatal bands below

Lvs to 40 x 0.6-0.8mm, channelled (semi-cylindrical), with 3-5 veins per lf. Ligule 0.1mm. Stems 0.5mm diam, round, hollow. Bogs. PL 2.........................*Dioecious Sedge* **Carex dioica**

Lvs with 4 stomatal bands below

Basal lvs present. Stems 0.5mm diam, round, solid

Lvs deeply channelled (± as wide as thick), 0.3-0.6mm wide, with margins scabrid nr apex only (not to touch). Ligule <0.5mm. Sheaths occ scarious. Bogs. PL 6 ....*Flea Sedge* **Carex pulicaris**

Basal lvs absent (stem lvs only). Stems 1mm diam, trigonous, solid

Lvs to 40 x 1-1.2(2)mm, flat to shallowly channelled, wider than thick, with rounded apex, with ± smooth margins (occ scabrid esp distally), shiny green both sides. Ligule <0.5mm, rounded. Upland boggy habs. PL 6...........................*Few-flowered Sedge* **Carex pauciflora**

Lvs 30-70 x 0.5-1mm, channelled distally, almost as wide as thick, with rounded apex, with smooth margins, usu shiny green both sides. Ligule <0.5mm. Mtn bogs, VR, Scot. PL 5 ....*Bristle Sedge* **Carex microglochin**

Group GB – *Lvs (and/or sheaths) v hairy, without trigonous tip, with scabrid margins*

■Lvs on stems (stems often vegetative but with nodes nr base). Lvs often with cross-veins. Rhizomatous

Hairs papillate (x60). Lvs 2-6mm wide, V- or weakly M-shaped, greyish-green. Ligule 2-5(7)mm, rounded to ± acute, retuse. Sheaths hairy, ± straight to concave at apex. Basal sheaths occ reddish, weakly ladder-fibrillose. Shoots trigonous

Lvs with long hairs both sides, without hollows. Ligule usu long-ciliate and hairy. Dry sunny habs. PL 4..*Hairy Sedge* **Carex hirta** var **hirta**

Lvs sparsely hairy to hairless, often with hollows. Ligule often hairless. Wet sunny habs ....*Hairy Sedge* (var) **Carex hirta** var **sublaevis**

■Lvs ± all basal. Lvs with cross-veins obscure or absent. Tufted

Hairs smooth. Lvs 3-5mm wide, flat to M-shaped, pale green, sparsely hairy below. Ligule 3-4mm, acute, shortly ciliate-hairy. Sheaths V-shaped at apex, hairy. Basal sheaths brown, rarely reddish, not ladder-fibrillose. Shoots trigonous. Damp or shady habs. PL 6 ....*Pale Sedge* **Carex pallescens**

Group GC – *Lvs with stomata above (caution - the whitish cuticle cells of* Carex divisa *and* Blysmus *spp may resemble stomata). Sheaths with concave apex (obliquely so in* C. aquatilis *and* C. recta*). Shoots (±) round*

■Lvs often >40cm, with hollows at midpoint. Rhizomatous

Lvs with long (3-10cm) trigonous tip. Shoots usu forming false-stems. Sheaths with aerenchyma blocks to 6mm

Lvs 2.5-4(7)mm wide, flat to U-shaped, folding on drying, glaucous at least above, with scabrid margins nr apex, with 7-8 veins each side of midrib, with obvious cross-veins, with stomata occ in broken lines below esp distally. Ligule 2-3mm (shorter than wide), usu retuse, with 0.5mm free margin. Basal sheaths reddish, ladder-fibrillose. Apr-Oct. PL 7 .... *Bottle Sedge* **Carex rostrata**

Lvs without trigonous tip. Shoots not forming false-stems. Sheaths with aerenchyma blocks 8-12mm

Basal sheaths usu reddish, ladder-fibrillose

Lvs 4-6mm wide, M-shaped (or weakly so), with scabrid margins, usu shiny green both sides, with 8 veins each side of midrib, with obvious cross-veins, occ with sparse stomata below. Ligule (2)4-6mm, acute. *C. rostrata x vesicaria*...**Carex** x **involuta**

Basal sheaths reddish, not ladder-fibrillose

Lvs (3)5-8mm wide, M-shaped, involute on drying, dull grey-green above, bright shiny green below, 7-11 veins each side of midrib, with stomata both sides or above only. Ligule 5-10mm, usu acute, with wide (1mm) free margin. PL 1............................*Water Sedge* **Carex aquatilis**

Lvs 4-6mm wide, M-shaped, folding on drying, rough above, dull mid-green above, shiny green below, with 7-10 veins each side of midrib (with 1 raised opaque vein each side), with tiny hollows (not spongy), usu with stomata above only. Ligule 6-9mm, acute. Estuarine, VR, Scot. PL 7...*Estuarine Sedge* **Carex recta**

Basal sheaths whitish, not ladder-fibrillose

Lvs 4-5mm wide, weakly M- or V-shaped, glaucous both sides, with scabrid margins nr apex, with 7 veins each side of midrib, with weak cross-veins, with stomata both sides. Ligule 8-10mm, acute. Estuarine, VR, Scot. PL 7..................................*Saltmarsh Sedge* **Carex salina**

■Lvs mostly <40cm, solid (occ hollows nr ligule or in sheaths)

Rhizomes long to short. Lvs usu green above (occ glaucous both sides). Ligule minutely fimbriate (x30)

Lvs (1)2-4(5)mm wide, usu weakly M-shaped, with or without trigonous tip, with smooth margins in proximal ½, dull below, with translucent midrib, with 4-7 less translucent veins each side of midrib (occ dark-bordered, or with 1 raised vein each side), usu with cross-veins, often with stomata both sides. Ligule 1-3mm, not longer than wide, ± acute. Basal sheaths brown. Apr-Oct. PL 5..*Common Sedge* **Carex nigra**

Rhizomes short to absent. Lvs whitish-green to glaucous above. Ligule entire

Lvs 2-3.5mm wide, V- to M-shaped (occ flat), with or without trigonous tip, usu with scabrid margins, whitish-glaucous above, dull below, with 5-7 veins each side of midrib, with or without cross-veins, with stomata both sides. Ligule 1.5-4mm, occ longer than wide, usu acute. Basal sheaths whitish-brown. PL 2...*White Sedge* **Carex canescens**

Lvs 2.7-3mm wide, flat (occ weakly M-shaped), without trigonous tip, with smooth margins proximally, dark green or glaucous above, dull below, with 3-4 veins each side of midrib (1-2 more clearly translucent), with v weak cross-veins, with broken stomatal lines below. Ligule 1.5-3mm, ± acute, retuse. Basal sheaths whitish-brown. *C. canescens* x *lachenalii* ....**Carex** x **helvola**

Group GD – *Lvs U- or V-shaped with hollows in TS, spongy when pressed (exc trigonous tips).* ❶ *Stems full of sap*

■Lvs usu <3mm wide

Lvs >20cm, with scabrid margins (often strongly so)

Tussock-forming. Lvs 1.5-2.5mm wide, U-shaped, with whip-like apex, with c6 veins each side of midrib, with cross-veins. Ligule 1-2mm, obtuse, retuse, with free margin <0.5mm. Sheaths with concave to ± straight apex, often scarious. Basal sheaths orange to dark brown, v fibrous. Shoots round at base. Fens. PL 1........................*Fibrous Tussock-sedge* **Carex appropinquata**

Tufted or shortly rhizomatous, rarely tussock-forming. Lvs 2-3(3.5)mm wide, flatter than *C. appropinquata*, shiny green both sides, with 3 translucent and 3 opaque veins each side of midrib, with cross-veins. Ligule to 4mm, acute. Sheath with straight to convex apex (forming a long tubular section). Basal sheaths dark brown or blackish, not fibrous. Shoots round at base. Occ all yr. Fens. PL 2..*Lesser Tussock-sedge* **Carex diandra**

Rhizomatous. Lvs 1-2.5mm wide, U- or V-shaped, thick, with long whip-like acute trigonous tip, with scabrid margins, grey-green, with 4-8 veins each side of midrib (best viewed below), with cross-veins, with 10 hollows in TS. Ligule 1mm, obtuse, brown-scarious. Sheaths with ± acute U- to V-shaped brown-scarious apex. Basal sheaths dark reddish-brown, ladder-fibrillose, not fibrous. Shoots round at base. Bogs. PL 4...............................*Slender Sedge* **Carex lasiocarpa**

Lvs <20cm, with margins not scabrid (exc nr apex)

Rhizomatous. Lvs 1-2mm wide, strongly U-shaped, with rounded apex, with white cells and cross-partitions above, with obscure opaque veins. Ligule ≤1mm, rounded, ± entire, tubular. Sheaths with concave apex. Stems 1-2mm diam, round, with small hollows around large irreg hollow, with cross-partitions. ❶. Saltmarshes, N Br, Wales, Ire ....*Saltmarsh Flat-sedge* **Blysmus rufus**

■Lvs >3mm wide

Lvs usu >30cm long

Tussock-forming. Shoots forming round false-stems

Lvs 3-7mm wide, U-shaped, with v scabrid margins, smooth above, dull dark green (shiny when young), with 10-12 veins each side of midrib, with obvious cross-veins. Ligule 1-3(5)mm, v obtuse, occ asymmetrical. Sheaths with concave apex. Basal sheaths blackish. Fens. PL 6 ....*Greater Tussock-sedge* **Carex paniculata**

Rhizomatous. Shoots not forming false-stems

Lvs 8-20mm wide

Lvs M-shaped, with hard rounded apex (hydathode easily seen and felt), often with trigonous tip to 1.5cm, with scabrid margins (and raised midrib below), rough and often slightly puckered above, with 18-23 veins each side of midrib (occ opaque), with obvious cross-veins, with small hollows in TS. Ligule 2-20mm (absent from some lvs), occ obtuse or strongly asymmetric, often minutely fimbriate, often with minute free margin. Sheaths oblique at apex, with 5-12mm aerenchyma blocks nr ligule (longer below), herbaceous on stem lvs. Shoots often several from each node on rhizome, whitish, weakly trigonous to ± round at base. Stems trigonous, with aerial nodes.........*Wood Club-rush* **Scirpus sylvaticus**

Lvs <10mm wide

Lvs (4)6-8.5mm wide, M-shaped, acute, without trigonous tip, with scabrid margins, shiny yellowish-green both sides, with cross-veins, with midrib raised below, often with dark-bordered veins. Ligule 5-10(15)mm, acute, entire. Sheaths with asymmetric concave apex. Basal sheaths reddish, ladder-fibrillose. Shoots trigonous at base. PL 8 ....*Bladder-sedge* **Carex vesicaria**

Lvs (2)3-6(10)mm wide, flat or shallowly channelled, obtuse (x10), with 5-10cm triquetrous tip (often blackish-red), with slightly scabrid margins at least nr apex, shiny dark green above (dying red or with red spots), paler below with sunken translucent midrib, with cross-veins, with c10 hollows each side of midrib in TS. Ligule 0.5mm, minutely fimbriate. Sheaths with concave apex. Basal sheaths pink, not ladder-fibrillose. Stems round below, pinkish, solid, with scattered vb's and tiny hollows around margin. Usu all yr. Bogs, fens. PL 8 ....*Common Cottongrass* **Eriophorum angustifolium**

Lvs usu <30cm long

Lowlands

Dune slacks, VR, Lancs. Stems to 100cm, triquetrous, ± solid. Lvs to 30cm x 3-5.5mm, deeply V-shaped, with trigonous tip, with smooth margins, whitish above due to cuticle cells, yellow-green below, with midrib raised below, with 8 opaque or weakly translucent veins, with cross-veins, with 8 hollows. Ligule 1.5-3mm, rounded, retuse (auricle-like), entire. Sheaths with concave apex (often scarious). Basal sheaths rarely ladder-fibrillose, occ reddish beneath dead brown sheaths. Shoots ± round or weakly trigonous at base. Shortly rhizomatous. Sch8 .... *Sharp Club-rush* **Schoenoplectus pungens**

Wet saline habs. Stems to 40cm, round, solid. Tufted. Lvs to 40cm x 3-6mm, usu U-shaped, with long trigonous tip, with smooth margins (occ scabrid nr apex), dark glaucous-green above, ± shiny green below, dying yellow, with 3-5 translucent veins each side of midrib (occ dark-bordered), with cross-veins (occ obscure). Ligule 0.5-2mm, obtuse. Sheaths with shallowly concave apex. Basal sheaths never ladder-fibrillose or reddish. Shoots ± forming round false-stems. PL 3........ *Long-bracted Sedge* **Carex extensa**

Fens, marshes, occ coastal. Stems to 30cm, round, solid. Rhizomatous. Lvs to 30cm x 2-4mm, U-shaped, with long trigonous tip (to touch), often with smooth margins, whitish above due to cuticle cells, with midrib barely discernible, mid-green below, often without midrib raised below, with opaque veins, with cross-veins, occ solid. Ligule 1mm, obtuse or retuse, entire. Sheaths with concave apex. Basal sheaths never ladder-fibrillose or reddish. Shoots forming round false-stems. Apr-Oct........ *Flat-sedge* **Blysmus compressus**

Mtns

Lvs to 15cm x 3-3.5(4)mm, U-shaped to ± flat, with trigonous tip to 5cm, with scabrid margins (occ weakly so), dark green, with (5)6-7 ± translucent to opaque veins each side of midrib, with 12-14 hollows in TS, with cross-veins. Ligule entire, with 0.3mm free margin. Sheaths with concave apex. Basal sheaths purple. Shoots round at base, with obvious cataphylls. Stems trigonous, rough nr infl. Rhizomatous

Ligule 1-2mm, obtuse. Stigmas 2. PL 7........ *Russet Sedge* **Carex saxatilis**

Ligule 2.5-3(6)mm, obtuse to acute, often v slightly retuse. Stigmas 3. PL 4 .... *Graham's Sedge* **Carex** x **grahamii**

Group GE – *Lvs dull glaucous (or ± so) at least below*

■Lvs >40cm, M-shaped, without trigonous tip

Plant tufted. Lvs with cross-veins weak or visible only nr ligule

Lvs 14-21mm wide, with rounded M-angles in TS, revolute on drying, shiny green above, with weakly scabrid margins. Ligule 15-50mm, acute. Sheaths obliquely concave at apex, with 1-3mm aerenchyma blocks below ligule. Basal sheaths reddish, not ladder-fibrillose. Shoots trigonous at base. All yr. Damp shady habs. PL 6....................*Pendulous Sedge* **Carex pendula**

Lvs (3)6-8mm wide, with ± acute M-angles in TS, revolute on drying, dark bluish-green above, with v scabrid margins. Ligule 5-12(20)mm, acute, with 0.6-1mm free margin. Sheaths obliquely concave at apex, with (1)4-5(6)mm aerenchyma blocks. Basal sheaths whitish, weakly ladder-fibrillose. Shoots trigonous at base. Apr-Oct. Wet habs. PL 3..............*Tufted-sedge* **Carex elata**

Plant rhizomatous. Lvs with obvious cross-veins

Ligule obtuse, often much wider than long

Lvs 5-24mm wide, usu folding on drying, bluish-green above, with v scabrid margins. Ligule (2)5-10(15)mm, with 0.6-0.8mm free margin. Sheaths concave at apex, with 1.5-3.5mm aerenchyma blocks (clearly visible for some distance above ligule). Basal sheaths occ reddish, ladder-fibrillose or obscurely so. Shoots trigonous at base. PL 7 ....*Greater Pond-sedge* **Carex riparia**

Ligule acute, always longer than wide

Lvs 7-10(15)mm wide, usu folding on drying, bluish-green above, with v scabrid margins. Ligule (5)7-10(15)mm, with 0.3-0.4mm free margin. Sheaths concave at apex, never wrinkled, with 3(8)mm aerenchyma blocks. Basal sheaths reddish, ladder-fibrillose. Shoots trigonous at base. PL 1...*Lesser Pond-sedge* **Carex acutiformis**

Lvs 4-6(7)mm wide, revolute on drying, dark green above, with v scabrid margins. Ligule 4-6mm, with 0.8-1.2mm free margin. Sheaths concave at apex, often wrinkled, with 5-10mm, aerenchyma blocks. Basal sheaths not reddish or ladder-fibrillose. Shoots trigonous at base. PL 1...*Slender Tufted-sedge* **Carex acuta**

■Lvs usu <40cm, weakly M- or V-shaped

Lvs 1.5-2.5mm wide, with or without trigonous tip

Plant with aerial stolons. Shoots with bladeless sheaths (cataphylls). Bogs

Lvs to 25cm x 1-3.5mm, V- to weakly M-shaped, usu without trigonous tip, with v scabrid margins (and raised midrib below), glaucous both sides, with translucent midrib and 6 translucent veins each side of midrib, with distinct cross-veins. Ligule 2-5mm, (±) acute. Sheaths with acute V-shaped apex, soon splitting. Basal sheaths brown to reddish, ladder-fibrillose. Shoots round at base, occ forming false-stem, usu with persistent dead lvs or sheaths. Stolons branched, with 25 elongated hollows (cartwheel-like) in TS. Roots densely orange-hairy. Mar-Oct. PL 4..*Bog-sedge* **Carex limosa**

Lvs to 15cm x 0.5-2.5mm, deeply V-shaped, with trigonous tip, with weakly scabrid margins, ± glaucous (to green) both sides, with translucent midrib and 3 ± opaque veins each side of midrib, with cross-veins (often obscure). Ligule 1-2mm, ± acute, with free margin <0.5mm. Sheaths with concave apex. Basal sheaths pale brown, not ladder-fibrillose. Shoots round at base, often forming false-stem, with persistent dead lvs. Roots not orange-hairy. Apr-Oct. R, Scot. PL 2...*String Sedge* **Carex chordorrhiza**

Plant without aerial stolons. Shoots without cataphylls

Ligule 5-8mm, acute, not retuse, entire, with 3mm free margins. Lvs to 6cm x (1)1.5-2mm, without trigonous tip, ± flat, curved, with margins scabrid nr apex, glaucous both sides, with 3-4 veins each side of midrib, with sparse cross-veins. Sheaths concave or V-shaped at apex, soon splitting. Basal sheaths shiny reddish-black to purple. Rhizomes long. Roots densely orange-hairy. Mtn bogs. PL 6..*Mountain Bog-sedge* **Carex rariflora**

Ligule 1.5mm, rounded to ± acute, retuse, occ minutely toothed. Lvs to 10cm x 1.5-2.5mm, with trigonous tip to 25mm, flat, straight, with margins scabrid nr apex, ± pale glaucous-green both sides, with 4 veins (1 translucent) each side of midrib (the midrib most translucent), with weak cross-veins. Sheaths concave or V-shaped at apex, soon splitting. Basal sheaths brown. Rhizomes short. Roots not hairy. Wet mtn ledges. PL 4....*Hare's-foot Sedge* **Carex lachenalii**

Lvs 2-6(9)mm wide

Lvs with trigonous tip

Sheaths convex at apex, often split

Lvs 2-4mm wide, with 5-15mm trigonous tip (short or absent on upper lvs), mid-green above, with v translucent midrib, with 4-8 dark-bordered veins each side of midrib, with cross-veins. Ligule 0-1mm, retuse, entire. Basal sheaths usu dead, brown, fibrous. Mar-Oct (all yr). PL 4 .... *Tawny Sedge* **Carex hostiana**

Sheaths concave (or obliquely so) to ± straight at apex

Lvs (2)3-5mm wide, with trigonous tip to 10mm, v glaucous above (usu less so below), with midrib raised below (more translucent and wider than other veins) and 6-12 unequal irregularly spaced veins each side of midrib (3-4 per mm, unlike *C. flacca*), with cross-veins ± absent (exc nr ligule). Ligule 1.5-2mm, obtuse, retuse, entire. Basal sheaths not reddish. Stems trigonous, with wide hollow. Rhizomes short or absent. PL 6 .... *Carnation Sedge* **Carex panicea**

Sheaths concave to convex at apex

Lvs 2-6.5(9)mm wide, with scabrid margins (and raised midrib below) at least nr apex, green to ± glaucous both sides, with 7-11 unequal veins each side of midrib, with weak to distinct cross-veins. Ligule 2-4(10)mm, ± obtuse to acute, retuse, often asymmetric. Basal sheaths reddish. Stems trigonous, ± solid. Rhizomes short or absent. Mar-Oct (all yr). Mtns, VR. PL 1 .... *Black Alpine-sedge* **Carex atrata**

Lvs without trigonous tip

Ligule 1-2.5mm, retuse, minutely fimbriate (often obscure)

Lvs 2.5-5mm wide, folding on drying, v glaucous below (occ also above), often rough above, with midrib v translucent, with (6)8-12 ± equal regularly spaced veins each side of midrib (5 per mm, unlike *C. panicea*), without cross-veins. Basal sheaths occ reddish, occ weakly ladder-fibrillose, often with persistent dead lvs and sheaths. Stems trigonous, usu with small hollow. Rhizomes often long. PL 3.......... *Glaucous Sedge* **Carex flacca**

Ligule >3mm, acute, entire

Roots densely orange-hairy. Rhizomes often long. Lvs 2-4mm wide. Bogs, N Br, Wales, N Ire. PL 5.......... *Tall Bog-sedge* **Carex magellanica** ssp **irrigua**

Roots not hairy. Rhizomes short, ± tufted. Lvs 2.5-3.5mm wide, weakly M- or V-shaped, with scabrid margins esp nr apex (often retrorsely so nr ligule), often rough above nr apex, mid- to dull grey-green, with cross-veins weak (to absent). Ligule 3-6(10)mm. Sheaths with asymmetrically concave apex. Basal sheaths ladder-fibrillose, usu reddish, persistent. Shoots forming short false-stems, trigonous at base. Fens, Scot. PL 1 .... *Club Sedge* **Carex buxbaumii**

Group GF – *Ligule usu >2mm. Basal sheaths reddish or purplish, often ladder-fibrillose*

■Plant with long rhizomes, or shoots arising in linear line

Lvs v sparsely hairy when young (look carefully!). Lowland calc habs

Lvs 1.5-4mm wide, without trigonous tip, pale to dark green both sides, occ dull below, with v translucent narrow midrib and 8 unequal veins each side of midrib (1 main vein each side), with cross-veins nr base. Ligule 1.5-3mm, obtuse to acute (occ cuspidate), minutely fimbriate, with free margin <1mm. Sheaths obliquely concave at apex. Basal sheaths reddish, ladder-fibrillose. Usu all yr. PL 3...*Downy-fruited Sedge* **Carex filiformis**

Lvs always hairless

Lowland wet habs (usu). PL 8..*Bladder-sedge* **Carex vesicaria**

Upland or mtn habs

Lvs 5-17(25)cm x (2.5)3-6mm, ± flat, rarely abruptly acute, often with short trigonous tip, with smooth margins proximally, rough above distally, dark bluish-green to ± glaucous above, with 6-9 veins each side of midrib, with cross-veins at least nr ligule. Ligule 2-4mm, acute to obtuse, rarely asymmetric, occ obscurely fimbriate. Sheaths concave at apex, soon splitting. Basal sheaths often reddish. Stems triquetrous, ± winged, rough at least above, solid. Rhizomes occ short, with scale lvs occ purplish. Apr-Oct. PL 1.....*Stiff Sedge* **Carex bigelowii**

Lvs 15-35cm x (3)4-5(6)mm, weakly M-shaped, often abruptly acute (often dead), without trigonous tip, with smooth margins proximally, occ puckered above, often dark bluish-green to ± glaucous above, with 6-8 veins each side of midrib (occ dark-bordered), with cross-veins. Ligule 4-7mm, usu acute, asymmetric, entire. Sheaths asymmetrically concave at apex, soon splitting, often wrinkled. Basal sheaths brownish-white, with dead lvs persisting as fibres. Stems ± trigonous, not winged. Rhizomes always long, with scale lvs occ purplish. Apr-Oct. PL 8..*Sheathed Sedge* **Carex vaginata**

■Plant tufted or with short rhizomes

Dry or shady habs. Lvs to 45(60)cm

Lvs 2-4mm wide, usu V-shaped, without trigonous tip, with smooth margins, smooth and not puckered above, dark green above, with 3 main veins each side of midrib, with weak cross-veins (stronger in sheaths). Ligule 3-8mm, ± acute to rounded, entire. Sheaths concave at apex. Basal sheaths reddish, fibrous at extreme base. Shoots forming weakly trigonous false-stem. PL 7..*Spiked Sedge* **Carex spicata**

Lvs 4-6mm wide, shallowly M-shaped, without trigonous tip, with v scabrid margins (and raised midrib below), rough and puckered above (at least with age), dull green above, with 4-6 main veins each side of midrib, with v weak cross-veins. Ligule 2-4(6)mm, obtuse to ± acute, minutely fimbriate. Sheaths with concave apex. Basal sheaths reddish, not fibrous. Shoots ± round at base, not forming false-stems. Shady habs, VR, S Eng, Ire. Sch8. PL 2 ....*Starved Wood-sedge* **Carex depauperata**

Fens or boggy habs. Lvs to 35cm

Roots densely orange-hairy. Lvs 2-4mm wide, weakly M-shaped, without trigonous tip, with scabrid margins, usu smooth above, bright green to ± glaucous above, with 5 veins each side of midrib, without cross-veins. Ligule 6-7mm, acute, entire. Sheaths with acute V-shaped apex, often asymmetric. Basal sheaths occ reddish, not ladder-fibrillose. Shoots trigonous at base, not forming false-stems. Rhizomes often short, with >20 hollows (cartwheel-like). PL 5 ....*Tall Bog-sedge* **Carex magellanica** ssp **irrigua**

Roots not hairy. Lvs 2.5-3.5mm wide, weakly M- or V-shaped, without trigonous tip, with scabrid margins esp nr apex (often retrorsely so nr ligule), often rough above nr apex, mid-green to dull grey-green, with 4-6 veins each side of midrib, with weak cross-veins (occ absent). Ligule 3-6(10)mm, acute, entire. Sheaths with asymmetrically concave apex. Basal sheaths usu reddish, ladder-fibrillose, persistent. Shoots forming short false-stems, trigonous at base. Rhizomes short, solid. Fens, Scot. PL 1...*Club Sedge* **Carex buxbaumii**

Group GG – *Ligule >2mm. Basal sheaths not reddish.* ❶ *Ligule often asymmetrically W-shaped.* ❷ *Sheaths often ladder-fibrillose.* ❸ *Sheath (and ligular area of lf) with white spots.* ❹ *Stem lvs with 2 minute glands where lf joins sheath*

■Shoots arising from long rhizomes

Lvs 8-20mm wide, M-shaped. Shoots often several from each node on rhizome

Lvs with hard rounded apex (hydathode easily seen and felt), often with trigonous tip to 1.5cm, with scabrid margins (and raised midrib below), rough and often slightly puckered above, with 18-23 veins each side of midrib (occ opaque), with obvious cross-veins, with small hollows in TS. Ligule 2-20mm (absent from some lvs), occ obtuse or strongly asymmetric, often minutely fimbriate, often with minute free margin. Sheaths oblique at apex, with 5-12mm aerenchyma blocks nr ligule (longer below), herbaceous on stem lvs. Shoots whitish, weakly trigonous to ± round at base. Stems trigonous, with aerial nodes. ❶........*Wood Club-rush* **Scirpus sylvaticus**

Lvs <6mm wide, flat or U- to V-shaped. Shoots solitary, not at every node on rhizome

Shoots forming false-stems, or all lvs on stems

False-stems ± round, without nodes, with short lvs often adpressed at base. Sheaths hyaline, concave at apex. Lvs to 45cm x 2-3.5mm, U- or V-shaped, without trigonous tip, often with smooth margins, with whitish (cuticle cells) above, with midrib not or hardly raised below, with 7 unequal veins each side of midrib, with cross-veins (often indistinct). Ligule 2-3mm, obtuse, not retuse, entire, not brown-scarious. Usu saline habs. PL 3.......*Divided Sedge* **Carex divisa**

True stems trigonous to triquetrous, with nodes (often visible as a dark line). Sheaths herbaceous exc for hyaline apex. Lvs to 45(60)cm x 2-6mm, V-shaped to flat, without trigonous tip, with scabrid margins at least distally, shiny green both sides, with midrib raised below, with dark-bordered veins, with obvious cross-veins. Ligule 2-8mm, acute, retuse, entire, often brown-scarious. ❷. Usu freshwater habs. PL 2.......*Brown Sedge* **Carex disticha**

Shoots occ forming false-stems (never with aerial nodes), some lvs basal

Lvs (1.5)2-4mm wide, weakly U-shaped, without trigonous tip, tough, ± shiny dark green above, with v translucent midrib and weakly translucent or opaque veins, often with weak cross-veins. Ligule 1.5-2.5mm, acute, minutely fimbriate, thick, soon brown-scarious. Sheaths shallowly concave at apex, tough, soon brown-scarious. Shoots arising singly from every 4th node along rhizome. Rhizomes occ exposed by wind erosion, with brown scale lvs persisting as fibres. Sandy habs. PL 1..*Sand Sedge* **Carex arenaria**

■Shoots forming raised tussock, none forming false-stems. Basal sheaths not fibrous

Lvs to 80cm x 4-5mm, weakly M-shaped, with scabrid margins (and midrib below) at least nr apex, usu puckered above, with 3-5 main translucent veins each side of midrib, with cross-veins. Ligule (2)4-8mm, ± obtuse to acute. Sheaths concave at apex. Basal sheaths chestnut-brown, soon splitting. Shoots trigonous at base. ❸. PL 3...............................*Elongated Sedge* **Carex elongata**

■Shoots densely tufted (may form small tussock in *C. remota*), all forming false-stems

Basal sheaths fibrous, often blackish. Dry sunny habs.....................(*Carex muricata* agg) Go to **GJ**

Basal sheaths not fibrous or blackish. Damp shady habs. PL 7.........*Remote Sedge* **Carex remota**

■Shoots tufted, often loosely so

Woodland or shady habs

At least some lvs >9mm wide

Lvs to 30(40)cm x 9-13mm, with 1(2) strongly raised veins each side of midrib forming prominent ribs, without trigonous tip (apex often dead and whip-like), with scabrid margins, shiny green above, with c18 veins each side of midrib, with obscure cross-veins. Ligule (5)8-10(13)mm, acute, often asymmetric. Sheaths asymmetrically concave or oblique at apex. Basal sheaths rarely reddish. PL 7...........................*Thin-spiked Wood-sedge* **Carex strigosa**

Lvs to 60cm x (5)8-11mm, without strongly raised veins, without trigonous tip, yellow-green, with margins scabrid distally, with c10 veins each side of midrib, with ± obscure cross-veins (obvious nr sheath). Ligule (4)8-12mm, as long as wide (much longer than wide in *C. sylvatica*), ± acute, occ asymmetric, with 0.7-1.3mm free margin. Sheaths convex to concave at apex, with green veins and cross-veins. PL 4.........*Smooth-stalked Sedge* **Carex laevigata**

All lvs <8mm wide

Lvs to 60cm x (4)6-7mm, M-shaped, without trigonous tip, occ rough above nr apex, without puckering, dark green above, paler below, with 8-13 translucent veins each side of midrib, with cross-veins weak or absent. Ligule 4-7(16)mm, longer than wide, acute, entire, usu asymmetric. Sheaths concave at apex (often asymmetric). Shoots trigonous at base. PL 7 .... *Wood-sedge* **Carex sylvatica**

Lvs to 30cm x 4-5mm, weakly M- (to -V) shaped, often with short trigonous tip, rough above nr apex, often with puckering, ± shiny bright green both sides, with veins occ dark-bordered, with obvious cross-veins. Ligule to 6mm, usu longer than wide, obtuse to ± acute, occ minutely fimbriate, often asymmetric. Sheaths asymmetrically V- to U-shaped at apex. Shoots weakly trigonous at base. VR, N Eng. PL 4.. *Large Yellow-sedge* **Carex flava**

Sunny wet or marshy habs (occ saline)

Lvs mostly >6mm wide

Shoots usu forming sharply triangular false-stems

Sheaths wrinkled, with green veins. Lvs (5)9-12mm wide, (±) M-shaped, with 3-4cm trigonous tip, usu with smooth margins, ± dull green both sides, smooth above but puckered with sunken cross-veins, with c10 veins each side of midrib, with obvious cross-veins. Ligule 4-12mm, ± obtuse. Stems triquetrous, strongly winged. VR. PL 8 *True Fox-sedge* **Carex vulpina**

Sheaths smooth, with green veins, with concave apex (often asymmetric but occ straight or convex on upper lvs). Lvs 5-8(10)mm wide, (±) M-shaped, occ with short trigonous tip, often with smooth margins proximally, shiny mid- to yellow-green both sides, smooth above but puckered with sunken cross-veins, with c10 veins each side of midrib, with obvious cross-veins. Ligule 5-10mm, ± acute or obtuse. Stems triquetrous, hardly winged. PL 5 .... *False Fox-sedge* **Carex otrubae**

Shoots not forming false-stems

Lvs (6)8-15mm wide, M-shaped, without trigonous tip, with scabrid margins (and raised midrib below), rough or minutely puckered above, shiny yellow-green both sides, with dark-bordered veins, with distinct cross-veins. Ligule 10-13mm, acute, with free margin usu ≤1mm. Sheaths smooth. Basal sheaths whitish, ladder-fibrillose (often obscure). Shoots trigonous at base, spongy. PL 6.. *Cyperus Sedge* **Carex pseudocyperus**

Lvs <6mm wide

Shoots not forming false-stems. Lvs 4-5.5mm wide, weakly M-shaped, usu without trigonous tip, often with smooth margins proximally, smooth above, yellow-green both sides, with c8 veins each side of midrib, with distinct cross-veins. Ligule 4-8(10)mm, ± acute, usu minutely fimbriate, with free margin 0.6-1mm. Sheaths concave at apex. Basal sheaths pale brown, not fibrous. Usu coastal. PL 6... *Dotted Sedge* **Carex punctata**

Shoots usu forming weakly triangular false-stems. Lvs (2.5)3-3.5mm wide, flat or weakly M-shaped (folding on drying), often with trigonous tip, often with ± smooth margins, smooth above, (±) shiny dark or pale green both sides, with weak cross-veins often obscure or absent. Ligule (1)2-3(4)mm, ± acute, entire (fimbriate on upper lvs). Sheaths concave to ± straight at apex, with cross-veins. Basal sheaths brown, occ fibrous and persistent. ❹. PL 4...................... *Oval Sedge* **Carex leporina**

Upland or mtn habs

Lvs 5-17(25)cm x (2.5)3-6mm, ± flat, rarely abruptly acute, often with short trigonous tip, with smooth margins proximally, rough above distally, dark bluish-green to ± glaucous above, with 6-9 veins each side of midrib, with cross-veins at least nr ligule. Ligule 2-4mm, acute to obtuse, rarely asymmetric, occ obscurely fimbriate. Sheaths concave at apex. Basal sheaths often reddish. Stems triquetrous, ± winged, rough at least above, solid. Rhizomes occ short, with scale lvs occ purplish. Apr-Oct. PL 1.. *Stiff Sedge* **Carex bigelowii**

Lvs 15-35cm x (3)4-5(6)mm, weakly M-shaped, often abruptly acute (often dead), without trigonous tip, with smooth margins proximally, occ puckered above, often dark bluish-green to ± glaucous above, 6-8 veins each side of midrib (occ dark-bordered), with cross-veins. Ligule 4-7mm, usu acute, asymmetric, entire. Sheaths asymmetrically concave at apex, often wrinkled. Basal sheaths brownish-white, with dead lvs persisting as fibres. Stems ± trigonous, not winged. Rhizomes always long, with scale lvs occ purplish. Apr-Oct. PL 8 .... *Sheathed Sedge* **Carex vaginata**

Group GH – *Lvs green both sides, shiny at least below. Ligule <1.5(2)mm. Basal sheaths reddish/orange or with reddish-black veins.* ❶ *Roots turpentine-scented (odourless in other spp)*

■Lf margin antrorsely scabrid (to smooth) throughout. Ligule minutely fimbriate

Lvs 4-8mm wide. Sheaths with convex (to straight) fimbriate apex

Lvs often >15(30)cm, usu without trigonous tip, often with smooth margins, dark green to ± glaucous above, often turning orange, with indistinct cross-veins. Ligule 0.5-1(2)mm, truncate to obtuse, with 0.5mm free margin. Basal sheaths usu orange with red-black veins. Shoots trigonous at base. PL 1..*Green-ribbed Sedge* **Carex binervis**

Lvs <4mm wide. Sheaths with concave (to straight) entire apex

Rhizomatous, shoots arising in linear lines

Lvs (1.5)2-4mm wide. Ligule 1.5-2.5mm, acute, opaque, turning brown. Sandy habs. PL 1 ....*Sand Sedge* **Carex arenaria**

Tufted, patch-forming

Lvs 5-12(20)cm x (1.5)2-2.5mm, often with trigonous tip, with weakly scabrid to smooth margins, often puckered above, shiny yellow-green both sides, often with 1-3 v translucent veins each side of midrib (more translucent than midrib), with cross-veins obscure or absent. Ligule 0.2-1mm, truncate. Sheaths with ± straight apex. Basal sheaths reddish, or pale with reddish veins, ladder-fibrillose. Shoots occ prostrate. ❶. Acidic habs. PL 6 ....*Pill Sedge* **Carex pilulifera**

Lvs to 7cm x 1.5-3.5mm, without trigonous tip, with weakly scabrid margins, never puckered above, shiny (yellow) green both sides, with dark-bordered veins, with obvious cross-veins. Ligule 0.5-1mm, obtuse. Basal sheaths shiny reddish-purple, not ladder-fibrillose. Shoots never prostrate. Mtns, VR. PL 5.........................*Close-headed Alpine-sedge* **Carex norvegica**

Lvs 0.5-1(2)mm wide. Sheaths with concave (to straight) entire apex

Lvs U-shaped, with midrib the most translucent vein (unlike *C. pilulifera*), without cross-veins. Ligule 0.5mm. Basal sheaths ladder-fibrillose. Calc gsld, S Eng. PL 4 ....*Dwarf Sedge* **Carex humilis**

■Lf margin retrorsely scabrid below midpoint

Ligule minutely fimbriate. Shoots usu forming false-stems. Basal sheaths ladder-fibrillose. Apr-Oct

Lvs sparsely hairy above when young, 1.5-2.5mm wide, often with short trigonous tip, ± smooth above, pale green, with 5-7 veins each side of midrib, without cross-veins. Ligule <1mm, obtuse to ± acute. Sheaths U- to V-shaped at apex (occ asymmetric), hairless. Basal sheaths reddish. Rhizomes short. ❶. Usu hths or limestone gsld.......................*Soft-leaved Sedge* **Carex montana**

Ligule entire. Shoots never forming false-stems. Basal sheaths not ladder-fibrillose

Lvs sparsely hairy above at least nr ligule when young, 3-5mm wide, usu without trigonous tip, bright yellow-green to dull bluish grey-green, with 5 main veins each side of midrib, without cross-veins. Ligule 0.5-1(2)mm, obtuse to truncate. Sheaths concave at apex. Basal sheaths purplish or with purplish veins. Rhizomes absent. Limestone habs. PL 2 ....*Fingered Sedge* **Carex digitata**

Lvs hairless, 2-5mm wide, with trigonous tip, light green to dull dark green, with 1-3(4) main veins each side of midrib, without cross-veins. Ligule 0.5-1mm, obtuse to truncate. Sheaths concave at apex. Basal sheaths purplish or with purplish veins. Rhizomes absent. Limestone habs. PL 5..*Bird's-foot Sedge* **Carex ornithopoda**

Group GI – *Lvs shiny green at least below, with trigonous tip. Ligule <1.5(2)mm, entire. Basal sheaths not reddish (rarely pinkish or with purplish veins when dying)*

■Lvs with obvious cross-veins. Wet or damp habs (*Carex chordorrhiza* may key out here)

Rhizomes long (but still forming tufts). Uppermost lvs without a ligule

Lvs (3)4-6(8)mm wide, shallowly M- or V-shaped, obtuse, with trigonous tip to 5cm, with scabrid margins at apex, yellow-green, turning orange nr apex, with 8-11 veins each side of midrib, with cross-veins occ weak, with 10-16 minute hollows in TS. Ligule 0.5mm, rounded, those on stem lvs to 2mm, turning scarious, sinuate or concave, all with 0.2mm free margin. Sheaths with concave apex, white or yellowish. Shoots round at base. Stems triquetrous. Usu all yr ....*Broad-leaved Cottongrass* **Eriophorum latifolium**

Rhizomes short or absent (tufted). All lvs with a ligule

Sheaths with convex apex, often split. Mar-Oct

Lvs 2-4mm wide, with 5-15mm trigonous tip (short or absent on stem lvs), with v translucent midrib, with 4-8 dark-bordered veins each side of midrib. Ligule 0-1mm, retuse, entire. Basal sheaths usu brown, fibrous. PL 4...*Tawny Sedge* **Carex hostiana**

Sheaths with concave (to ± straight) apex, often split. Lvs shiny bright green both sides, with the translucent midrib raised below (the most translucent vein), usu with dark-bordered veins

Lvs (5)10-30cm x 3-4(5)mm, often with rough margins. Ligule 0.5-2mm, ± as long as wide, obtuse. Fens, calc mires. PL 8 ....*Long-stalked Yellow-sedge* **Carex viridula** ssp **brachyrrhyncha**

Lvs 5-20cm x 2-4(5)mm, often with smooth margins. Ligule 0.5-2mm, wider than long, v obtuse to ± truncate. PL 8...................*Common Yellow-sedge* **Carex viridula** ssp **oedocarpa**

Lvs 3-8cm x (1.5)2-2.5(3)mm, with smooth margins (exc in distal ¼). Ligule 0.5-1mm, wider than long, ± truncate. PL 8..................*Small-fruited Yellow-sedge* **Carex viridula** ssp **viridula**

■Lvs without cross-veins (or obscure if present)

Bogs or bare peat

Basal sheaths lfless but with short fragile aristate tip, tough, rush-like, with 1(3) buds in their axils. Lvs 7-20cm x 0.7-2(2.5)mm, shorter or ± equal to stems, channelled, with rounded apex, with rough or smooth margins (always rough nr apex), yellow-green, with 2-3 veins each side of translucent midrib visible below. Sheaths often convex at apex. Stems trigonous to round, hollow. Tufted or with short rhizomes. Apr-Oct. Bogs....*White Beak-sedge* **Rhynchospora alba**

Basal sheaths mostly with short blades, without overwintering buds in their axils. Lvs 4-15cm x 0.7-2(2.5)mm, usu much shorter than stems, channelled, yellow-green, with 3-4 veins each side of midrib visible below. Sheaths often convex at apex. Stems trigonous to round, with irreg hollow (partially divided into c3) in TS. Rhizomes long. Apr-Oct. Bare peat and bogs, R ....*Brown Beak-sedge* **Rhynchospora fusca**

Dry or damp habs

Shortly rhizomatous. Lvs (1.5)2-4mm wide. Ligule 1(2)mm, v obtuse, often retuse. Lvs 3-15cm, dark green, with scabrid margins at least nr apex and ligule, rough above, with wide translucent midrib, with 1-5 veins each side of midrib (occ indistinct). Sheaths with shallowly concave apex (occ asymmetric). Basal sheaths with purple veins when dying, persisting as fibres

Lvs with 3-4mm trigonous tip. Infl pale or reddish brown. Usu calc gsld. PL 2 ....*Spring-sedge* **Carex caryophyllea**

Lvs often with 5-15mm trigonous tip. Infl dark brown or blackish. Calc gsld, R. PL 3 ....*Rare Spring-sedge* **Carex ericetorum**

Tufted. Lvs 1.5-3mm wide. Ligule <0.5mm, truncate, usu retuse. Lvs to 5(10)cm, flat to weakly M-shaped, with 5mm trigonous tip, with scabrid margins (at least nr ligule and apex), green or grey-green, with 1(3) main veins each side of midrib. Sheaths ± straight at apex. Basal sheaths occ with purple sheen, veins not purple, persisting as fibres. Calc upland gsld and flushes. PL 2 ....*Hair Sedge* **Carex capillaris**

PLATE 1

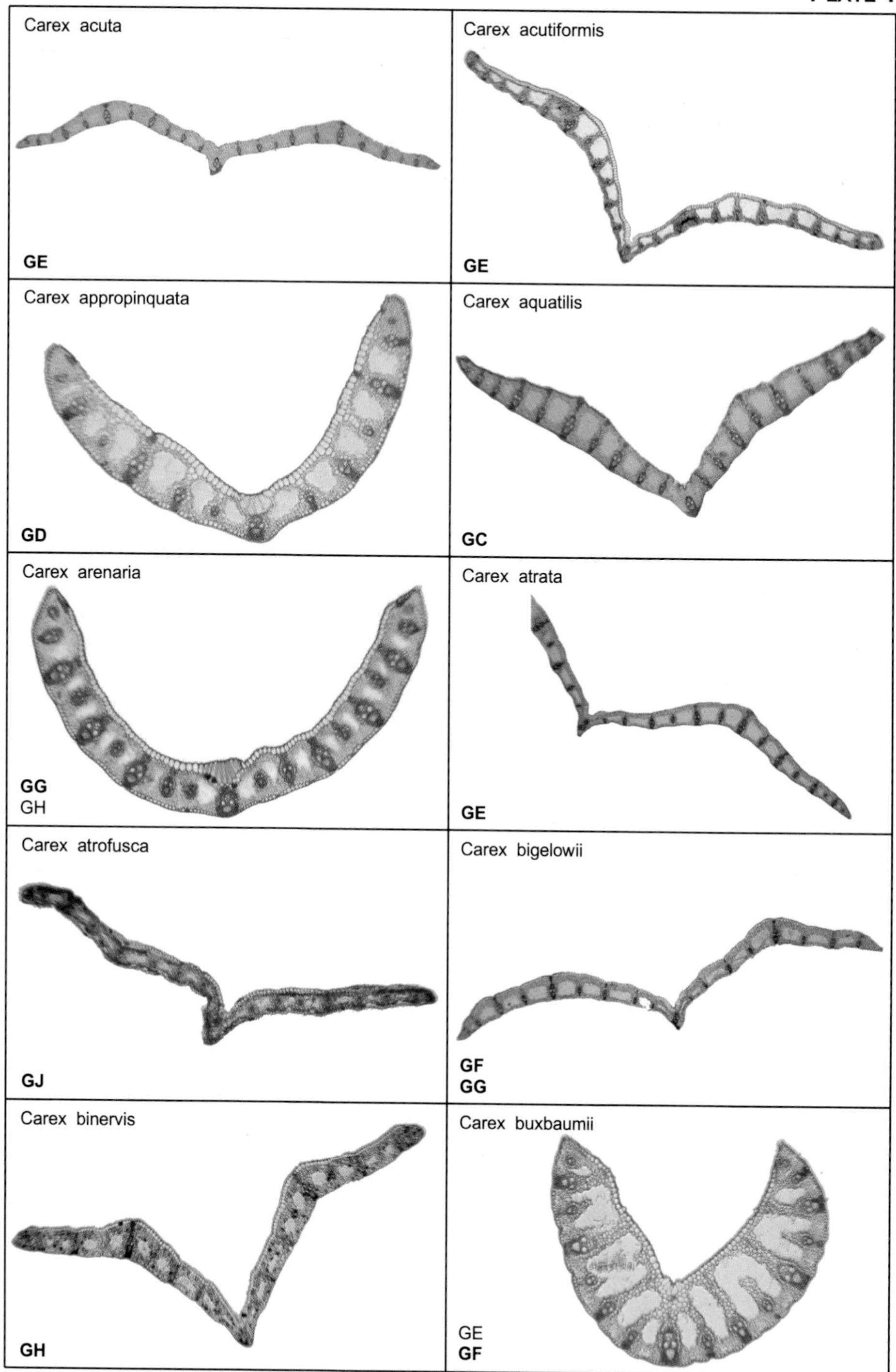
Carex acuta
GE
Carex acutiformis
GE
Carex appropinquata
GD
Carex aquatilis
GC
Carex arenaria
GG
GH
Carex atrata
GE
Carex atrofusca
GJ
Carex bigelowii
GF
GG
Carex binervis
GH
Carex buxbaumii
GE
GF

PLATE 2

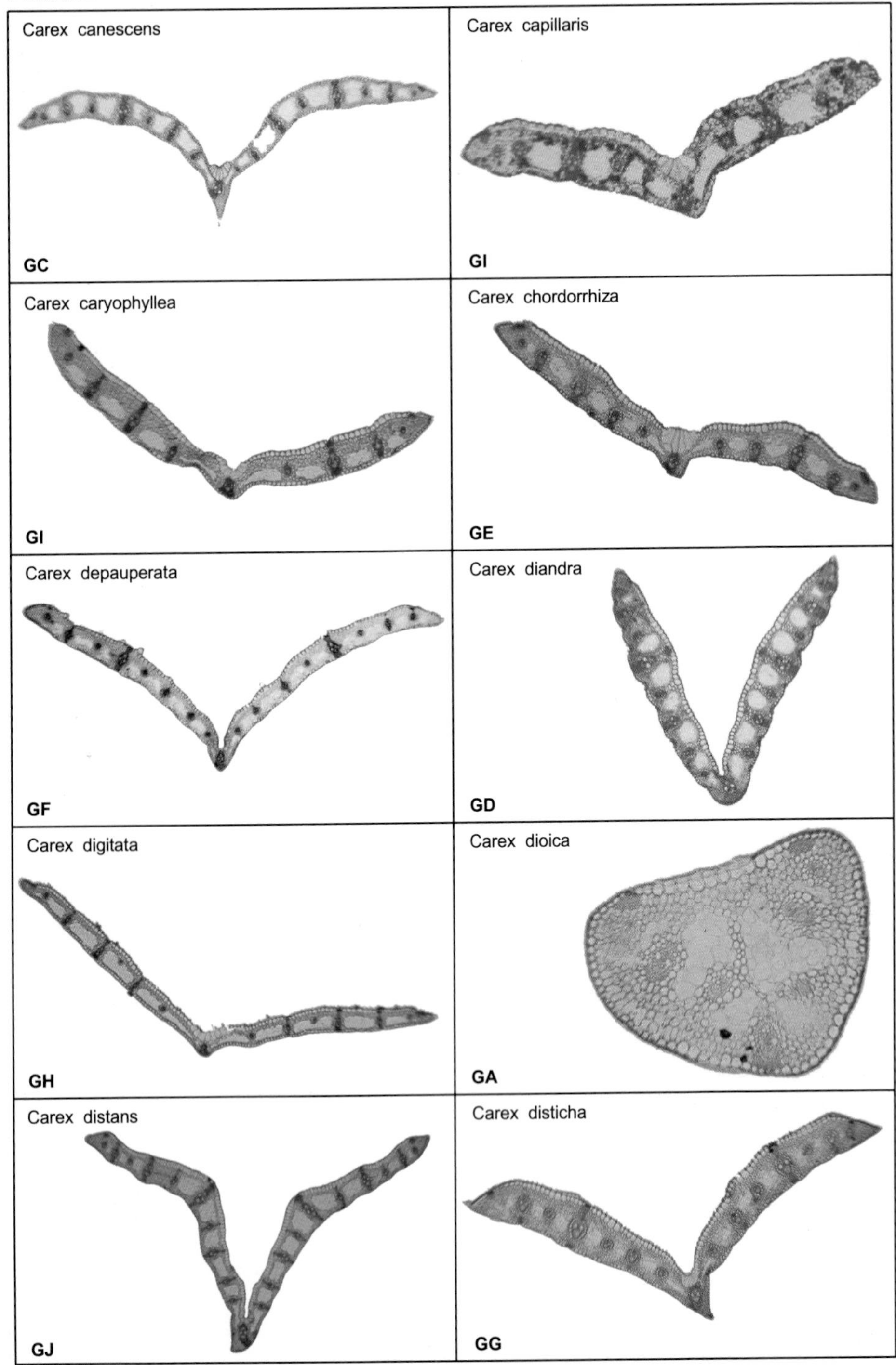

Carex canescens
GC
Carex capillaris
GI
Carex caryophyllea
GI
Carex chordorrhiza
GE
Carex depauperata
GF
Carex diandra
GD
Carex digitata
GH
Carex dioica
GA
Carex distans
GJ
Carex disticha
GG

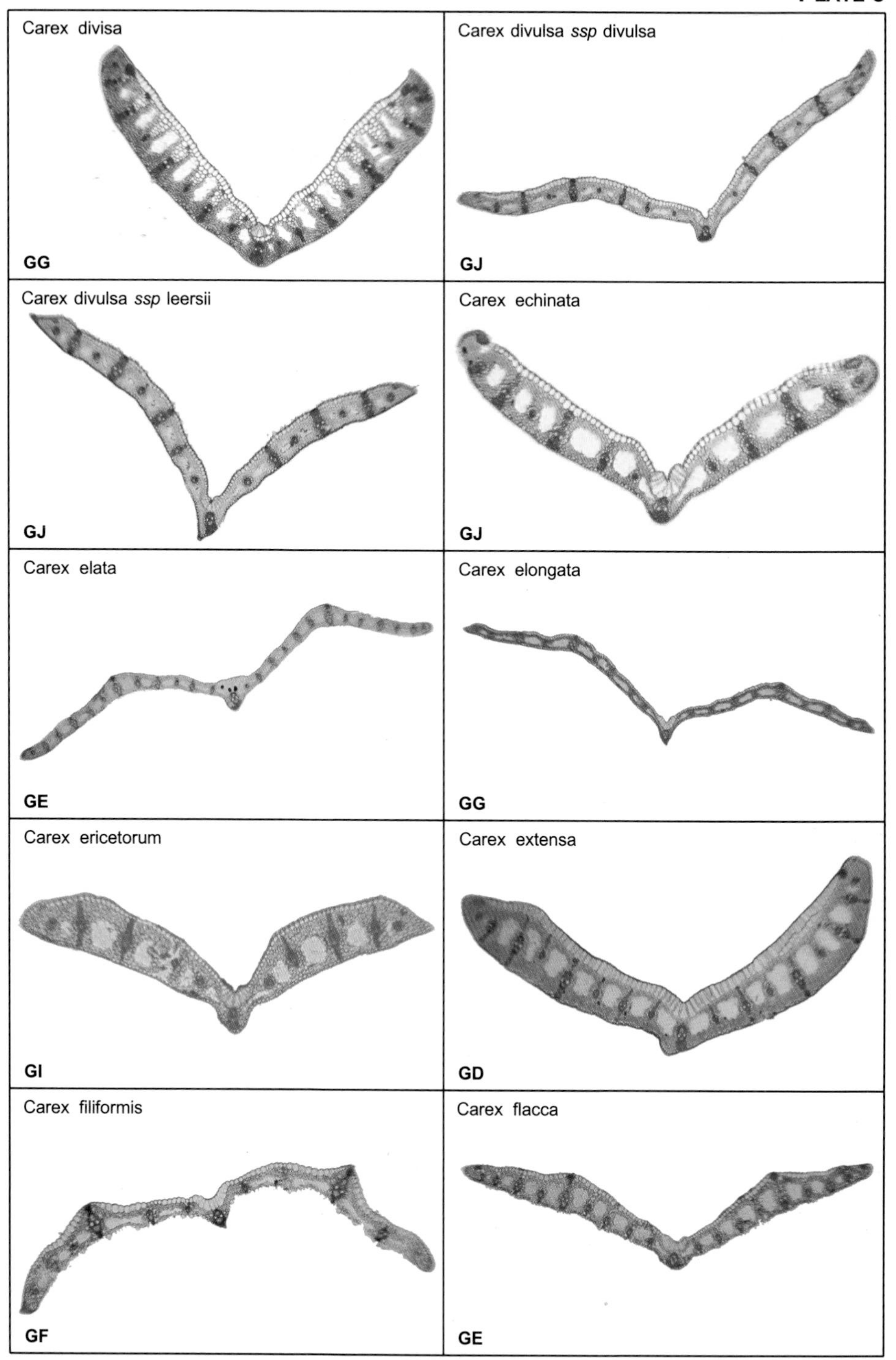
Carex divisa
GG
Carex divulsa *ssp* divulsa
GJ
Carex divulsa *ssp* leersii
GJ
Carex echinata
GJ
Carex elata
GE
Carex elongata
GG
Carex ericetorum
GI
Carex extensa
GD
Carex filiformis
GF
Carex flacca
GE

PLATE 4

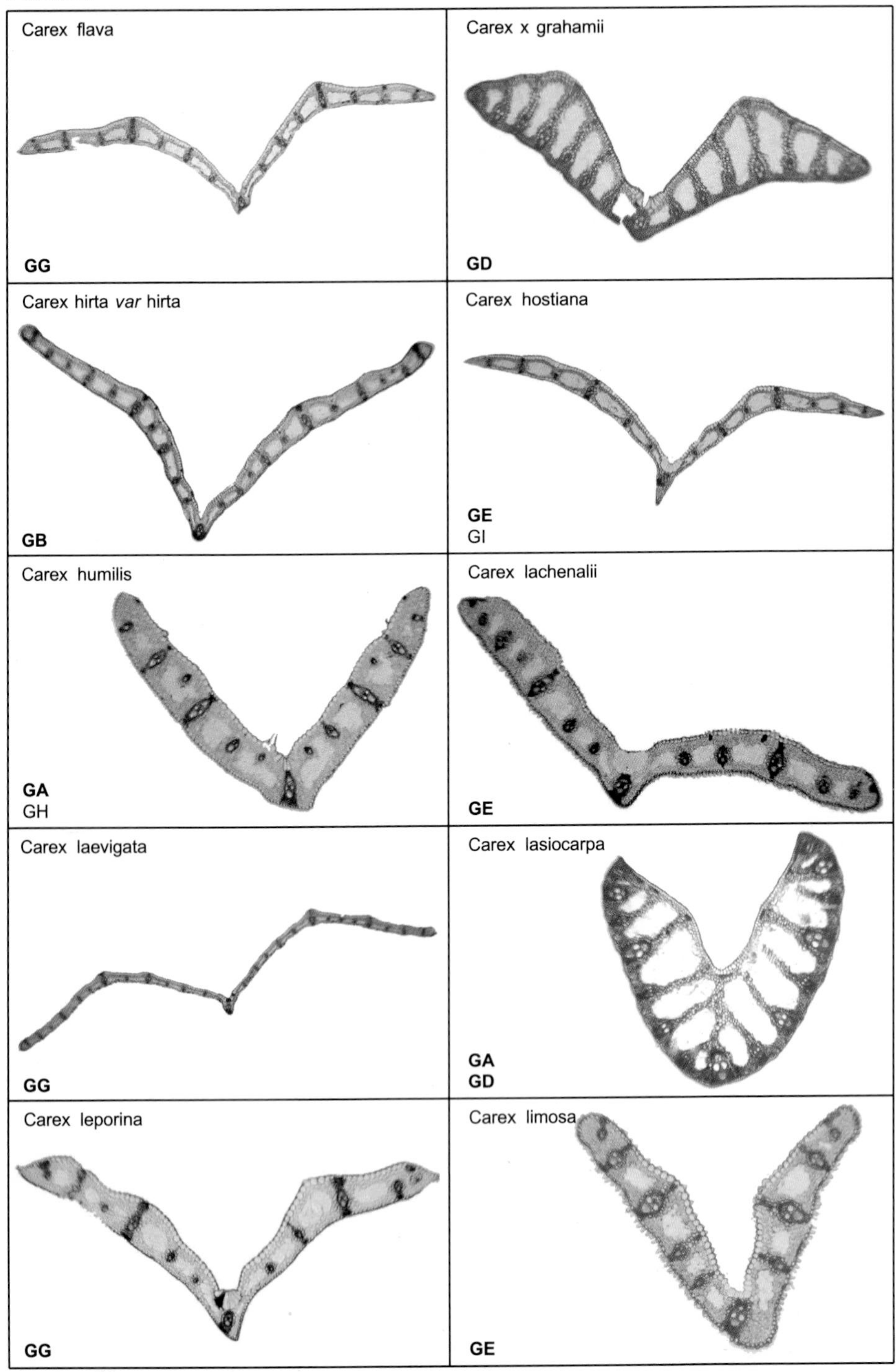
Carex flava
GG
Carex x grahamii
GD
Carex hirta *var* hirta
GB
Carex hostiana
GE
GI
Carex humilis
GA
GH
Carex lachenalii
GE
Carex laevigata
GG
Carex lasiocarpa
GA
GD
Carex leporina
GG
Carex limosa
GE

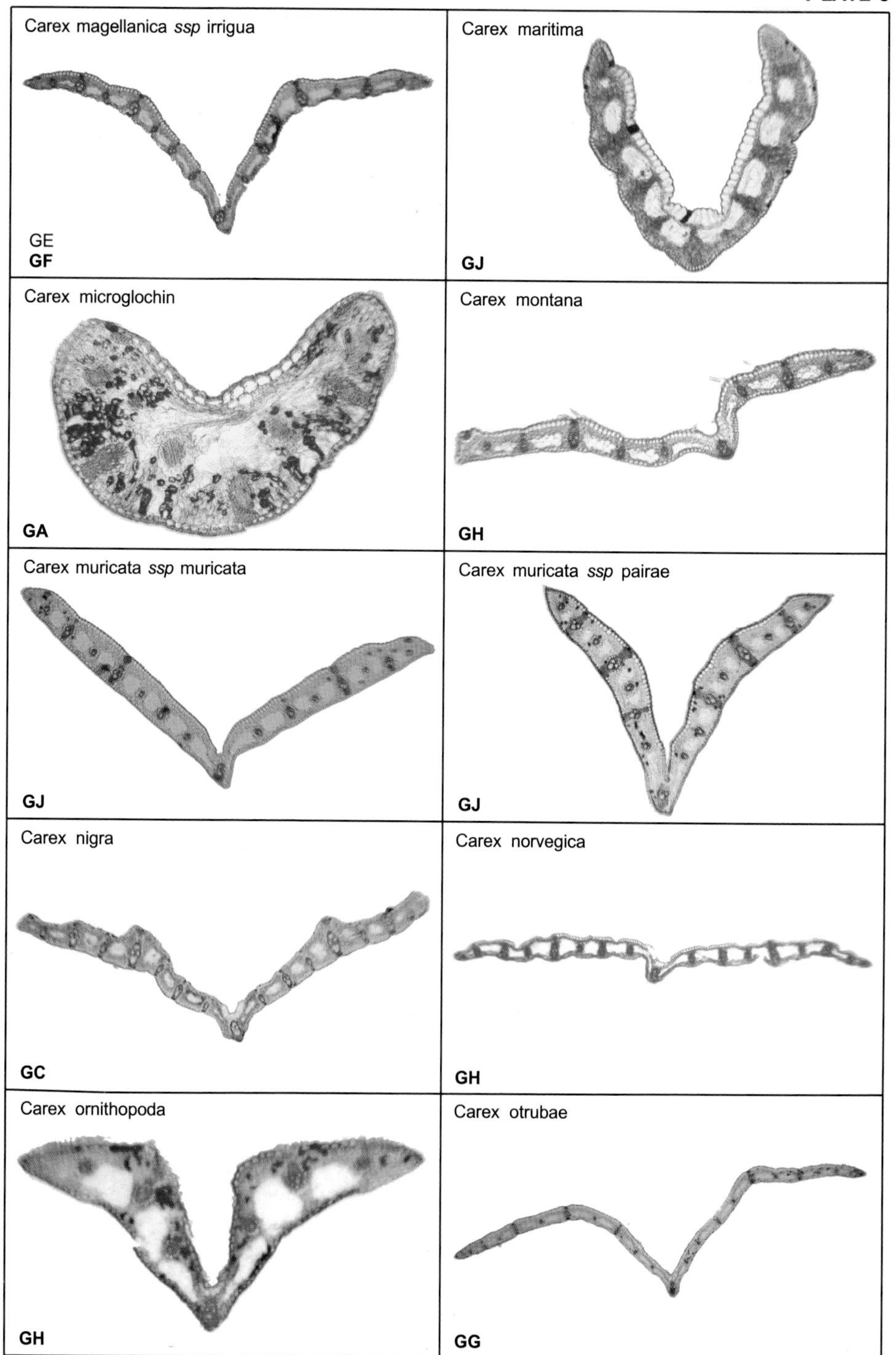
Carex magellanica *ssp* irrigua
GE
GF
Carex maritima
GJ
Carex microglochin
GA
Carex montana
GH
Carex muricata *ssp* muricata
GJ
Carex muricata *ssp* pairae
GJ
Carex nigra
GC
Carex norvegica
GH
Carex ornithopoda
GH
Carex otrubae
GG

PLATE 6

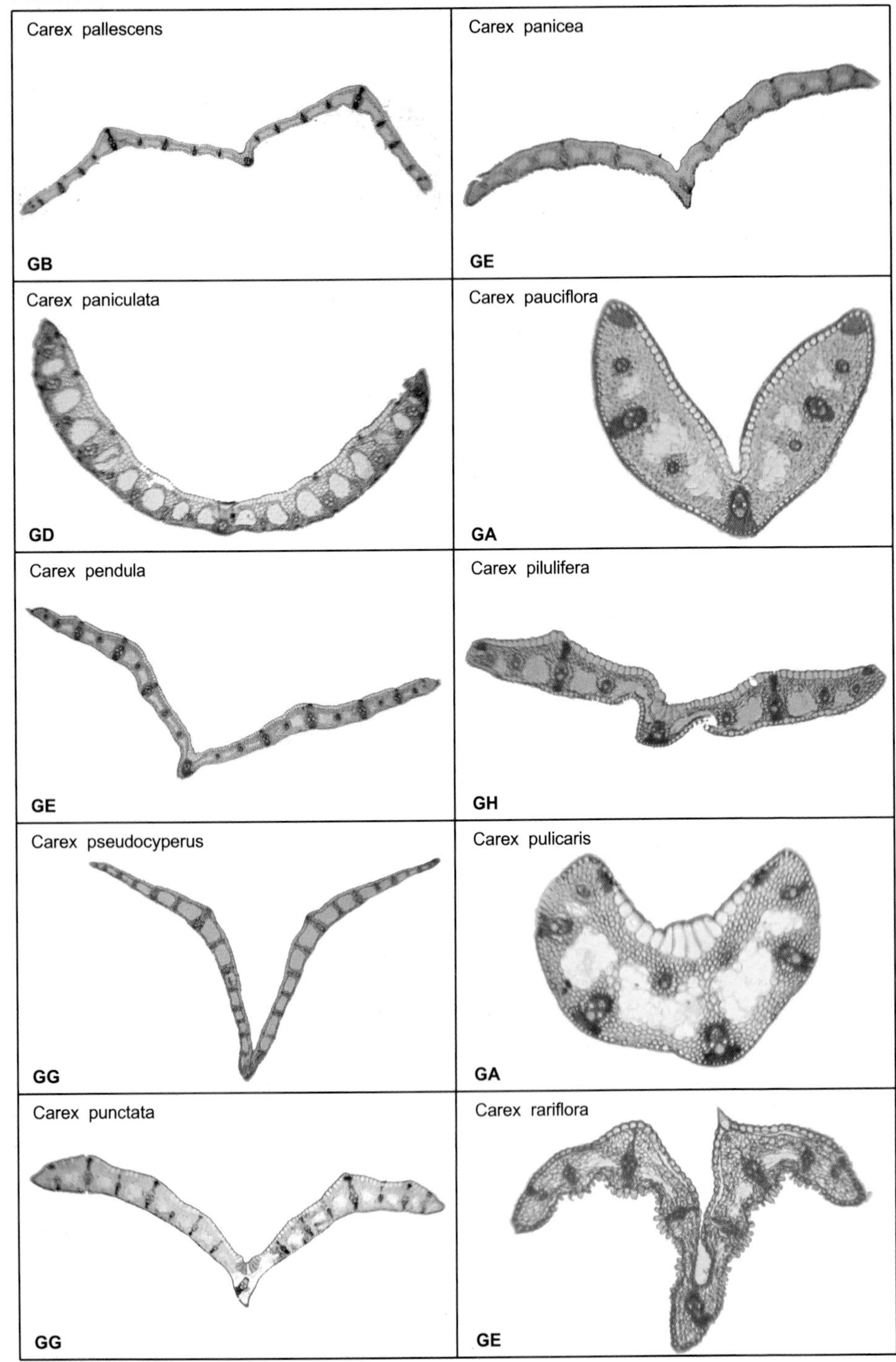

Carex pallescens
GB
Carex panicea
GE
Carex paniculata
GD
Carex pauciflora
GA
Carex pendula
GE
Carex pilulifera
GH
Carex pseudocyperus
GG
Carex pulicaris
GA
Carex punctata
GG
Carex rariflora
GE

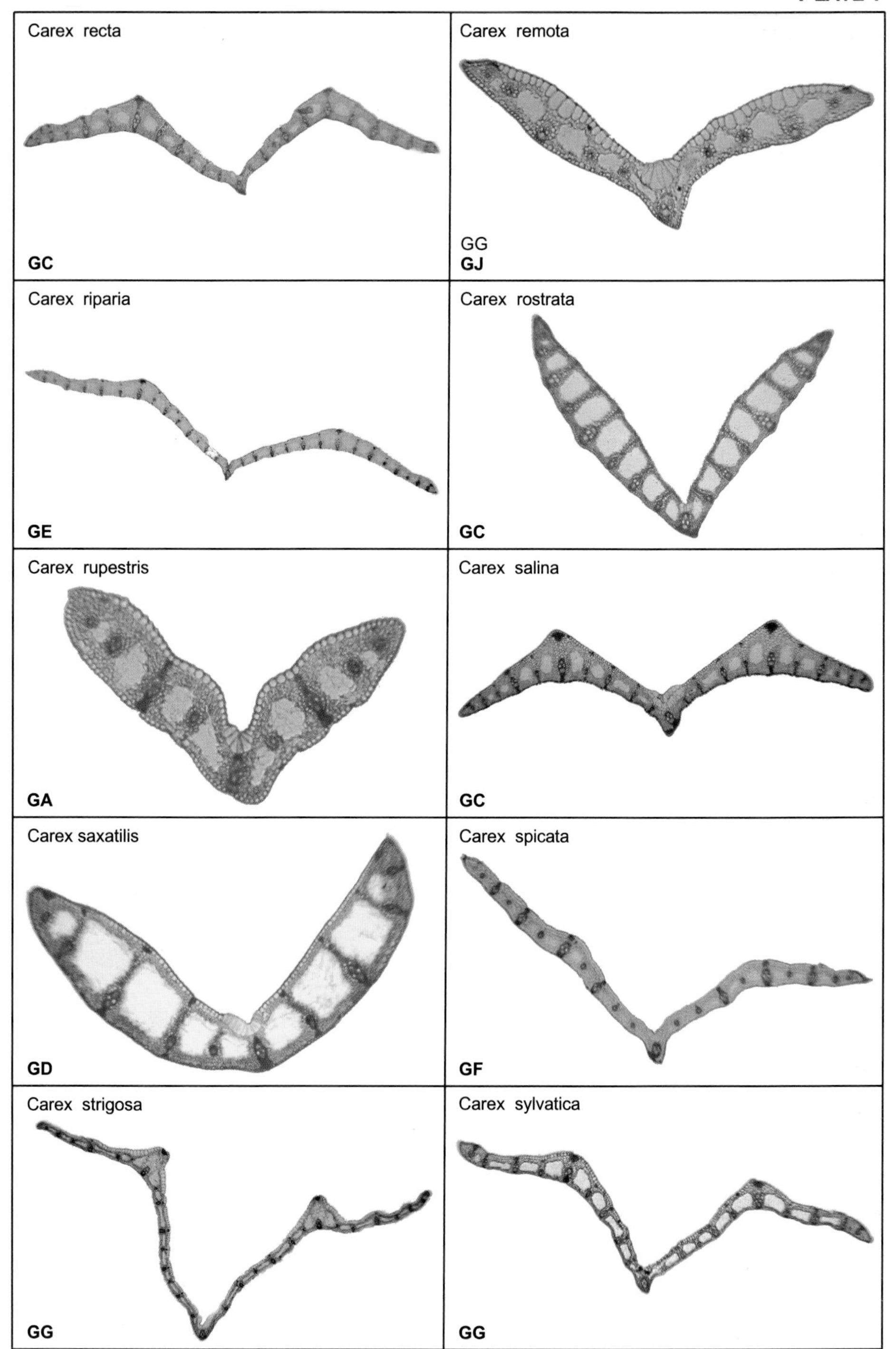
Carex recta
GC
Carex remota
GG
GJ
Carex riparia
GE
Carex rostrata
GC
Carex rupestris
GA
Carex salina
GC
Carex saxatilis
GD
Carex spicata
GF
Carex strigosa
GG
Carex sylvatica
GG

PLATE 8

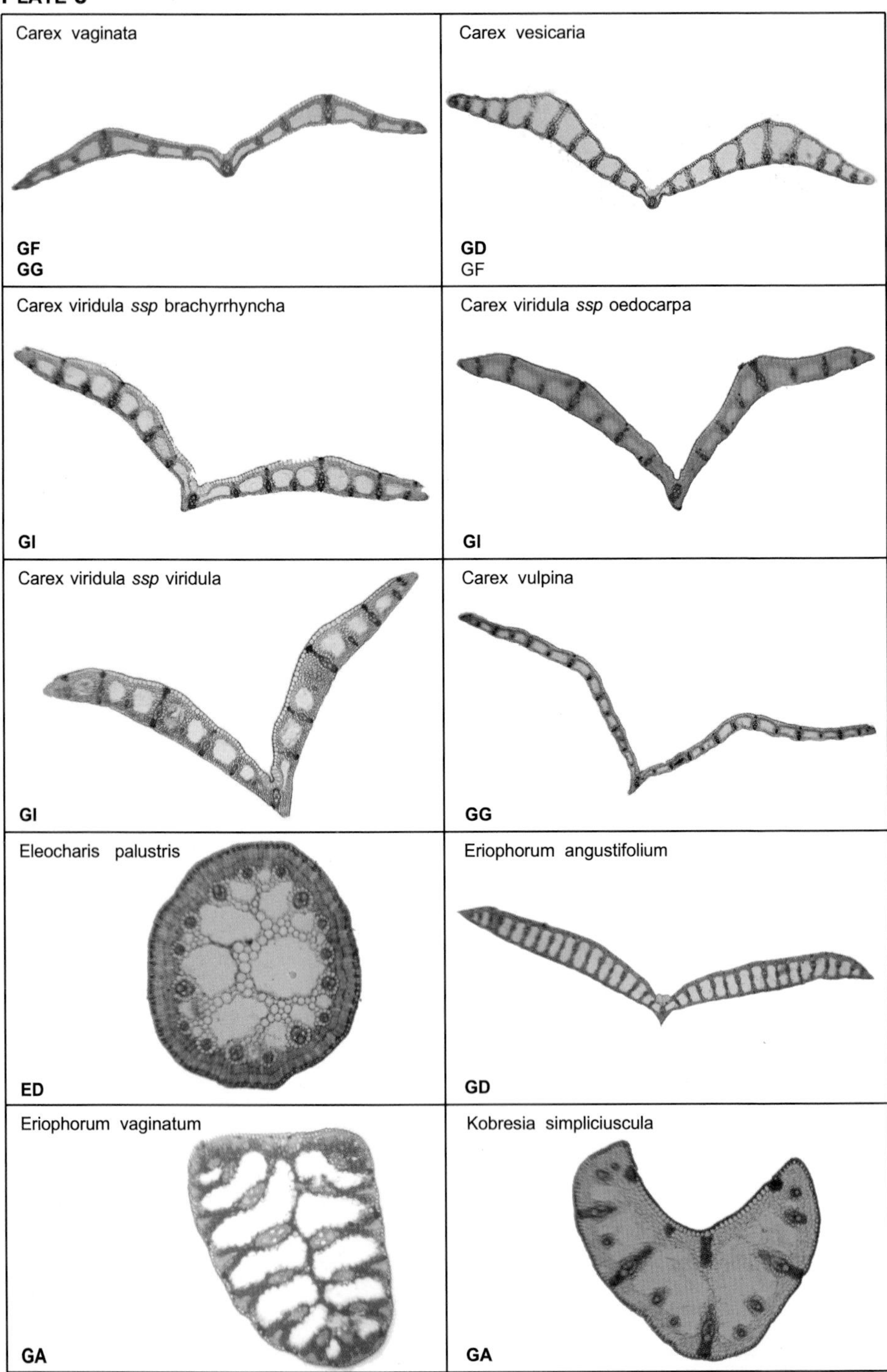

Carex vaginata
GF
GG
Carex vesicaria
GD
GF
Carex viridula *ssp* brachyrrhyncha
GI
Carex viridula *ssp* oedocarpa
GI
Carex viridula *ssp* viridula
GI
Carex vulpina
GG
Eleocharis palustris
ED
Eriophorum angustifolium
GD
Eriophorum vaginatum
GA
Kobresia simpliciuscula
GA

Group GJ – *Lvs shiny green at least below, without trigonous tip. Ligule <1.5(2)mm, entire. Basal sheaths not reddish or orange.* ❶ *Rhizomes branched, with persistent scale lvs*

■Lvs usu M-shaped (occ weakly so). Shoots not forming false-stems

Sheaths convex at apex, with cross-veins. Lvs 3-5(6)mm wide, with weakly scabrid margins, with 7-9 veins each side of midrib (1 often raised), with cross-veins often obscure. Ligule 0-2(3)mm, obtuse to ± acute, with 0.2-0.5mm free margin. Basal sheaths dark brown, without persistent fibres. Stems round. All yr. Usu coastal. PL 2.................................*Distant Sedge* **Carex distans**

Sheaths concave at apex, with cross-veins. Lvs 2-5mm wide, with weakly scabrid margins, with 3 main veins each side of midrib. Ligule 0.5-1.5(2.5)mm, ± obtuse. Basal sheaths pale brown, with persistent fibres. Stem ± trigonous. Calc mtn flushes. PL 1 ....*Scorched Alpine-sedge* **Carex atrofusca**

■Lvs usu U- or V-shaped. Shoots forming false-stems (occ so in *C. echinata*)

Rhizomatous, shoots arising in linear lines

Lvs 6-8cm x 0.8-2mm, U-shaped (no keel below), solid, with (±) smooth margins, shiny green both sides, with 4-7 opaque veins per lf, with obvious cross-veins. Ligule 0.3-1mm, truncate. Sheaths concave at apex. False-stems round. Stem weakly trigonous, hollow. ❶. Sandy coasts, R, Scot. PL 5...*Curved Sedge* **Carex maritima**

Loosely tufted, never forming tussocks

Lvs to 12cm x (1)2-3mm, shiny green both sides, V- (or U)-shaped, with scabrid margins at least distally, occ rough above, with 3-5 less weakly translucent veins each side of v translucent midrib (occ dark-bordered), usu with weak cross-veins. Ligule (0.5)1-3mm, ± acute to obtuse. Sheaths with concave apex. False-stems with 1(2) reduced lvs below. Stems 1mm diam, trigonous, with pithy hollow. Bogs. PL 3................................*Star Sedge* **Carex echinata**

Densely tufted, often forming small tussocks with age

Sheaths acutely V-shaped at apex, never wrinkled, without red-black dots

Lvs to 40(60)cm x 2-3mm, ± V-shaped, margins usu rough, yellowish-green, with translucent midrib weakly raised below, the other veins indistinct, with cross-veins. Ligule to 1(2)mm, retuse. False-stems trigonous. Stems hollow. Damp shady habs. PL 7 ....*Remote Sedge* **Carex remota**

Sheaths convex at apex, strongly wrinkled, with tiny red-black dots

Lvs to 25cm x 3-3.5mm, V-shaped, with weakly scabrid margins at least in distal ½, without midrib raised below, with 3-7 unequal veins each side of midrib, with cross-veins at least nr ligule. Ligule 0.5mm, rounded. Stems ± round, hollow. VR alien ....*American Fox-sedge* **Carex vulpinoidea**

Sheaths concave to convex at apex, rarely wrinkled, without red-black dots

Lvs to 60(75)cm x 1.5-3.5(5)mm, ± flat or weakly M- or V-shaped, with scabrid margins, with midrib raised below, with 3-8 veins each side of midrib, often with cross-veins. Ligule (1)1.5-3.5(5)mm, shorter than wide (occ ± equal), rounded to ± acute. False-stems ± round to weakly trigonous. Stems round to ± triangular, pith-filled..........................**Carex muricata** agg

Infl 5-18cm. Utricles 3-4(4.5)mm. PL 3......................*Grey Sedge* **Carex divulsa** ssp **divulsa**

Infl 4-8cm. Utricles 4-4.5(4.8)mm. Calc habs. PL 3 ....*Many-leaved Sedge* **Carex divulsa** ssp **leersii**

Infl (1)2-3(4)cm. Utricles 3.5(4.5)mm. Calc habs, R. PL 5 ....*Prickly Sedge* **Carex muricata** ssp **muricata**

Infl (1)2-3(4)cm. Utricles 2.6-3.5(4)mm. PL 5 ....*Prickly Sedge* **Carex muricata** ssp **lamprocarpa**

Key to Groups in Division H

(Grasses, Triglochin, Scheuchzeria*)*

Grasses or grass-like monocots. Lvs 2-ranked. The ligule descriptions refer to lower ones of vegetative shoots; those on culms (esp close to the infl) are often much longer. All yr (unless otherwise stated)

- ■Pseudopetiole present AND/OR lvs with obvious white midrib above. Culms often woody.......... **HA**
- ■Pseudopetiole absent. Culms never woody
  - Ligule thick, not membranous (*Triglochin*, *Scheuchzeria*).......... **HB**
  - Ligule absent, or present as a ring of hairs
    - Lf margins smooth. Saltmarshes (*Spartina*).......... **HC**
    - Lf margins serrate (often minutely so) at least nr apex.......... **HD**
  - Ligule thin, membranous, occ fringed with hairs
    - Lvs <1.2mm wide, bristle-like
      - Ann. Nov-Jul.......... **HE**
      - Per. All yr
        - Ligule >0.3mm.......... **HF**
        - Ligule <0.1mm.......... **HG**
    - Lvs >1.2mm wide, rarely bristle-like
      - Young lvs folded. Shoot usu flattened
        - Sheaths and/or lvs hairy.......... **HH**
        - Sheaths and lvs hairless (lvs may be shortly scabrid above)
          - Lower sheaths open (with margins overlapping).......... **HI**
          - Lower sheaths closed (may split but margins never overlapping)
            - Lvs and sheaths with cross-veins or hollows. Wet habs.......... **HJ**
            - Lvs and sheaths without cross-veins or hollows. Mostly dry habs
              - Lvs ribbed above, tramlines absent, apex not hooded.......... **HK**
              - Lvs not ribbed above, tramlines usu present, apex hooded.......... **HL**
      - Young lvs rolled. Shoot usu cylindrical or oval (rarely flattened or 4-angled)
        - Auricles present at junction of lf and sheath (often poorly developed)
          - Lvs and/or sheaths hairy.......... **HM**
          - Lvs and sheaths hairless.......... **HN**
        - Auricles absent
          - Lvs hairy at least above (occ sparsely or minutely so)
            - Sheaths hairy to densely so.......... **HO**
            - Sheaths hairless or with sparse scattered hairs
              - Sheaths closed at least below when young.......... **HP**
              - Sheaths open (with margins overlapping).......... **HQ**
          - Lvs hairless (occ ciliate)
            - Ligule <1.5mm, always shorter than broad, (±) truncate.......... **HR**
            - Ligule >1.5mm, usu longer than broad, truncate to acute
              - Lvs ribbed above in TS.......... **HS**
              - Lvs not ribbed above in TS.......... **HT**

Group HA – *Pseudopetiole present AND/OR lvs with obvious white midrib above. Culms often woody. Sheaths open. Ligule present.* ❶ *Culms solid*

■Culms woody. Pseudopetiole <2mm, distinct from lf. Lvs with stomata below only

Lvs evergreen, finely tessellate-veined.. *Bamboos* Go to **BAM**

■Culms herbaceous

Pseudopetiole >10mm, tapered into lf. Lvs with stomata both sides. Rhizomatous (patch-forming)

Lvs 2-3.5cm wide, yellow-green. Wet habs

Lvs widest nr middle, hairless, smooth both sides, with scabrid margins, not or hardly ribbed above in TS, with 4-5 main veins (9-10 minor veins between the main veins), with midrib raised below, with obscure cross-veins. Ligule 1-2cm, acute, minutely fimbriate, hairless, with purple triangular blotch either side, distinctly parallel-veined. Pseudopetiole with large aerenchyma. Shoots spongy, oval. VR alien.................. *Manchurian Wild-rice* **Zizania latifolia**

Lvs 1-2cm wide, green. Dry habs

Lvs widest nr middle, hairless, smooth both sides, with thick hyaline margins v scabrid distally, not or hardly ribbed above in TS, with (2)4 translucent main veins each side of midrib merging with midrib nr middle (with c20 semi-opaque wavy minor veins between main veins), with white midrib raised below in proximal ½. Ligule 3mm, obtuse (to ± truncate), hairless, long-ciliate, apex often reddish. Sheaths with cross-veins, smooth, hairless. Pseudopetiole with tiny aerenchyma (solid TNE). R alien....................... *Johnson-grass* **Sorghum halepense**

Pseudopetiole absent. Lvs with stomata below only. Shortly rhizomatous (clump-forming). ❶

Lvs hairless or long-hairy, with scabrid or smooth margins, with white midrib raised below. Ligule truncate, long-ciliate, hairy, surrounded by whiskers. Apr-Nov. Biocrop

Culms to 4m. Lvs 40-120 x 1-2(4)cm. Sheaths reddish at base. Ligule to 1.5mm. Fls late Oct onwards. *M. sacchariflorus* x *sinensis*....................... *Elephant-grass* **Miscanthus** x **giganteus**

Culms 0.8-2(4)m. Lvs 18-75 x 0.3-2(4)cm. Sheaths not reddish at base. Ligule 0.5-4mm. Fls Sept onwards... *Chinese Silver-grass* **Miscanthus sinensis**

Group HB – Juncaginaceae. *Ligule thick, not membranous, running down sheath. Lvs often aromatic, obtuse and flattened at apex, hairless, with smooth margins. Sheaths open, not overlapping, the base with persistent remains of dead lvs. Rhizomatous (*Triglochin palustre *has auricles instead of a ligule and may key out here in error)*

■Lvs aromatic (soapy odour), semi-cylindrical but channelled above, without pore nr apex. Saline habs

Lvs 15-60cm x (1)2-3mm, with obscure veins, with elongate stomata in rows both sides, snapping audibly, with 3 vb's within spongy aerenchyma. Ligule 2.5-6mm, obtuse to acute. Basal sheaths flattened, whitish (occ purple). Stems flattened nr base. Apr-Oct ....*Sea Arrowgrass* **Triglochin maritimum***

■Lvs odourless, semi-cylindrical, with large pore nr apex. Bogs, VR, Scot

Lvs 5-30cm x (1)2mm, with 8 veins forming strong ribs and 8 bands of elongate stomata. Ligule 3-5mm, obtuse (often inrolled). Basal sheaths flattened, whitish. Stems ± round ....*Rannoch-rush* **Scheuchzeria palustris***

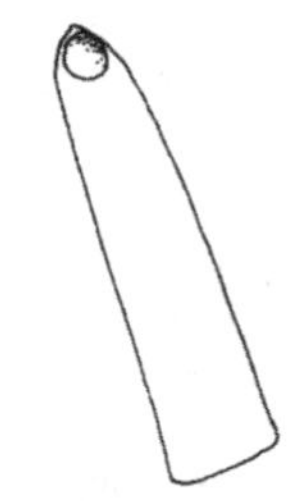

Scheuchzeria palustris lf apex

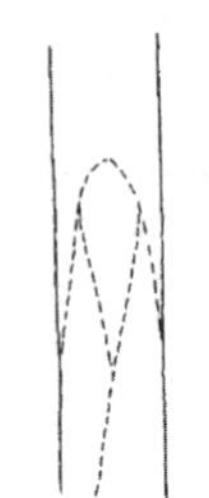

Triglochin maritimum ligule

Group HC – *Ligule present as a fringe of matted hairs (rarely a membrane). Lvs widest at base, smooth shiny yellow-green below, with smooth cartilaginous margins (v rare in grasses), hairless (minutely hairy above in* S. patens*), with deep flat ribs above in TS, without discernible midrib either side (not raised below), with stomata obscure above but in rows bordering veins below. Sheaths smooth, hairless (ciliate in* S. alterniflora*). Shoots round. Rhizomatous. Saltmarshes*

■Lvs >3mm wide, with (20)30-50 ribs (often wavy) above. Sheaths with distinct cross-veins

Ligule hairs 0.5-3mm

Lvs 6-12(15)mm wide, acute, flat or inrolled, tough, ± smooth above. Lower and upper saltmarshes

Anthers protruding, healthy. Pollen fertile..........*Common Cord-grass* **Spartina anglica**

Anthers not protruding, withered. Pollen sterile or absent. *S. alterniflora* x *maritima* ....*Townsend's Cord-grass* **Spartina** x **townsendii***

Ligule hairs 0.5(1)mm

Lvs 5-10mm wide, with 20-25 ribs above. Sheaths minutely ciliate on both margins. Upper saltmarses, VR alien..........*Smooth Cord-grass* **Spartina alterniflora**

Ligule hairs <0.5mm

Lvs c4mm wide, often purplish. Rhizomes short. Upper saltmarshes, R ....*Small Cord-grass* **Spartina maritima**

■Lvs 1-2.5mm wide, with 7 ribs above. Sheaths with cross-veins obscure or absent

Ligule hairs 0.5mm. Lvs to 30cm, channelled or inrolled, grey-green and minutely hairy above (x20). Upper saltmarses, VR alien, Sussex..........*Thorney Island Cord-grass* **Spartina patens**

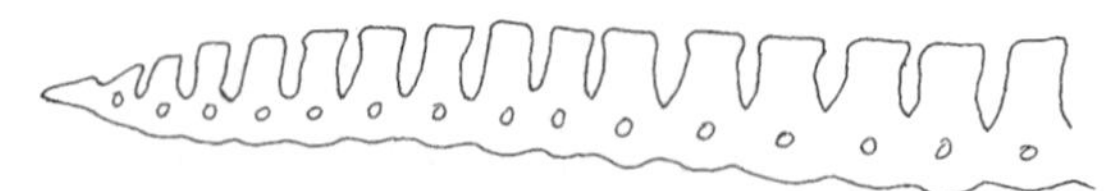

Spartina x *townsendii* lf TS

Group HD – *Ligule absent or present as a fringe of hairs. Sheaths open (exc lower sheaths of* Eleusine indica*).* ❶ *Often cleistogamous.* ❷ *Ligule absent (but pale collar where ligule should be).* ❸ *Lf margins papillate*

■Lvs usu >60cm, with strongly serrate sharp cutting margins, twisted, without stomata below. Ligule a dense fringe of silky hairs 2.5-4mm. Tussock-forming (often >1m diam). Wet or dry habs, not saline (but often nr coast)

Lvs 0.8-2.8m x 10-30mm, folding on drying (scabrid to touch), tearing easily, with ± equal regular veins (not forming strong ribs either side), with midrib raised below. Ligule with long whiskers on collar to 6mm. Sheaths hairless or with sparse long antrorse hairs, with obvious cross-veins ....*Pampas-grass* **Cortaderia selloana***

Lvs 0.6-1.2m x 6-15mm, inrolling on drying (smooth and cylindrical to touch), v tough, with unequal veins (every 4-8th vein thicker and forming a strong rib), with deep flat ribs above, without midrib raised below. Ligule without whiskers on collar. Sheaths hairless, with cross-veins obscure to absent. Mostly W Br...*Early Pampas-grass* **Cortaderia richardii***

■Lvs usu <60cm, with weakly serrate (to ± smooth) margins

Lvs folded when young, ± parallel-sided, hooded or not at apex

Lvs 2-4mm wide, flat, stiff, hooded at apex, dull grey-green to pruinose above, shiny green below, hairless or with sparse long hairs, occ ciliate, with tramlines, with stomata absent or sparse below. Ligule 0.5mm, truncate, with long whiskers on collar. Sheaths flattened, hairless or with spreading hairs. Per, with rhizomes short or absent. ❶. Usu acidic gsld, rarely chalk gsld..*Heath-grass* **Danthonia decumbens***

Lvs 5-7mm wide, strongly folded, stiff, not hooded at apex, mid-green above, shiny green below, with v sparse long hairs above, not ciliate, with every 4-6th vein thicker, with minute stomata both sides. Ligule 0.5mm, truncate, with long whiskers on collar. Sheaths strongly flattened, hairless, long-ciliate, soon splitting. Ann. VR alien...........................*Yard-grass* **Eleusine indica**

Lvs rolled when young, tapered at one or both ends, not hooded at apex

Rhizomatous per

Lvs 60-100cm x 5-15mm, widest at base, dull ± glaucous above, shiny green below, hairless, v tough, strongly involute on drying, with flat-topped low ridges above, with tough unequal veins, with midrib slightly raised below, without cross-veins, with stomata both sides. Ligule to 4mm. Sheaths round, not swollen at nodes, hairless, with cross-veins. Culms to 2m. VR alien ....*Prairie Cord-grass* **Spartina pectinata**

Lvs 20-60cm x (3)10-45mm, narrowed at base (and again ⅓-½ along lf), dull grey-green above, ± glaucous below, hairless or with sparse adpressed hairs, not tough, rolled on drying, hardly ribbed in TS, with 8 main veins (and 3-7 minor veins between main veins), with midrib hardly raised below, without cross-veins, with v minute stomata both sides. Ligule 1mm. Sheaths round, not swollen at nodes, hairless to hairy, ciliate along one margin when young, with cross-veins. Culms to 4(6)m. May-Oct. Wet habs ....*Common Reed* **Phragmites australis**

Lvs to 20cm x 4mm, widest at base, acute to acuminate, dull green both sides, flat (usu rolled at apex), stiff, rough above, with scabrid margins, hairless to hairy above, not ribbed in TS, with 3-4 main veins (and 5 minor veins between main veins), without cross-veins, with minute stomata both sides. Ligule <0.5mm, with long (4mm) whiskers on collar. Sheaths round or ± slightly flattened, often ± swollen at nodes, hairy or hairless, without cross-veins. Cataphylls to 4cm, tough, white. Culms to 30cm tall, but often with prostrate stolons ....*Bermuda-grass* **Cynodon dactylon**

Tufted per, often forming raised tussocks

Lvs (2)4-10mm wide, flat or inrolled, thin, dull grey-green above, hardly shiny below, with sparse long hairs above, with v low ribs above in TS, with every 4th vein stronger, with stomata absent or obscure below. Ligule <0.5mm, with long whiskers on collar. Sheaths often with cross-veins. Shoots oval, usu with swollen clavate base. Only deciduous grass with abscission layer...*Purple Moor-grass* **Molinia caerulea***

Ann, stems often branched from base. Subtropical ruderals usu germinating in Jun
Sheaths hairy (hairs long, often piercing to touch)
Lvs 13-18mm wide, occ wrinkled, ± hairless, ciliate nr base, with c55 equal veins each side of midrib, not ribbed. Ligule c1.5mm, truncate, shortly membranous but with long fringe of hairs. Shoots slightly flattened. Bird-seed casual...........*Common Millet* **Panicum miliaceum**
Lvs 7-12mm wide, occ wrinkled, with patent hairs both sides, long-ciliate nr base, with c28 equal veins each side of midrib, not ribbed. Ligule 1mm, truncate, shortly membranous but with long fringe of hairs. Shoots ± round. R alien...................*Witch-grass* **Panicum capillare**
Sheaths hairless but ciliate along at least one margin. Lf collar usu with purple blotches
Lvs 4-13mm wide, widest nr base, often rolled at acute apex, occ folded at maturity, hairless, rough (esp above), with minutely scabrid and often wrinkled margins, not ribbed. Ligule to 2mm. Sheaths round (lowest flattened)
Inf bristles antrorsely scabrid...*Green Bristle-grass* **Setaria viridis**
Infl bristles retrorsely scabrid (barbed), v rough.......*Rough Bristle-grass* **Setaria verticillata**
Sheaths hairless, not ciliate. Lf collar without purple blotches
Ligule absent ❷
Lvs (8)13-16mm wide, often rolled at acute apex, rough above nr apex, occ wrinkled, with scabrid cartilaginous margins (occ smooth nr base), with 4-5 main veins each side of midrib, with many crowded minor veins, with stomata both sides. Sheaths flattened, keeled. ....*Cockspur* **Echinochloa crus-galli**
Ligule present
Lvs 3-30cm x 4-10mm, with scabrid margins, occ folded at maturity, with scattered long hairs nr ligule, with 4 whitish main veins each side of midrib, with stomata both sides. Ligule 0.5mm, usu with long whiskers on collar. Sheaths keeled, the lower flattened and purplish at base...*Yellow Bristle-grass* **Setaria pumila**
Lvs 2-12cm x 2(4)mm, with distinctly papillate margins. Ligule 0.5mm, with long whiskers to 2mm. Sheaths round, the lower often purplish at base, occ with sparse long hairs. ❸. VR alien..*Small Love-grass* **Eragrostis minor**

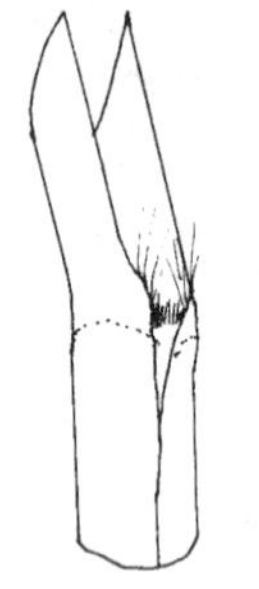
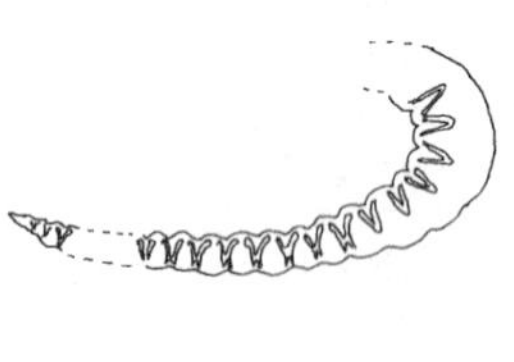

Cortaderia selloana ligule & lf TS

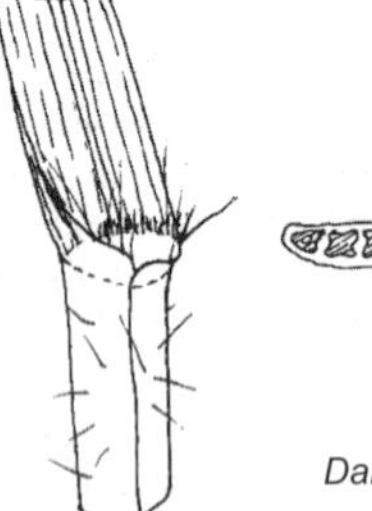

Danthonia decumbens ligule & lf TS

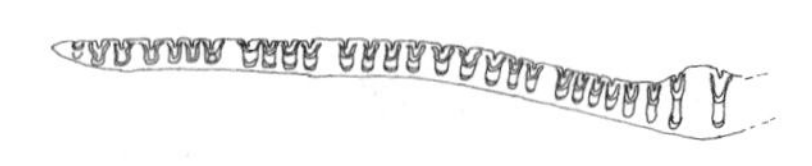

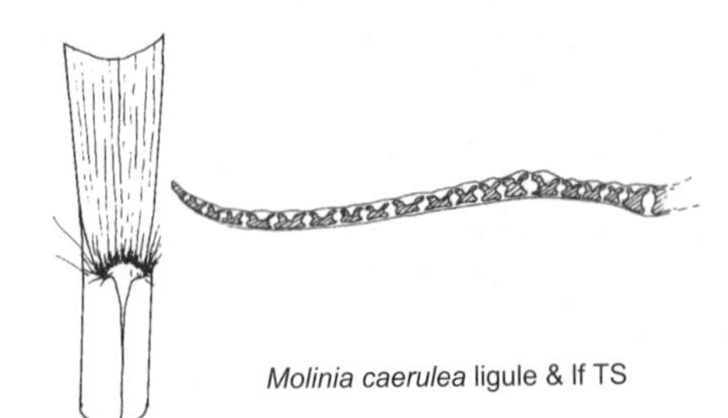

Cortaderia richardii lf TS

Molinia caerulea ligule & lf TS

Group HE – *Ann, without persistent fibrous sheath remains. Lvs <1.2mm, rolled when young, becoming flat, folded or inrolled. Sheaths smooth, round. Nov-Jul*

▪Lvs minutely hairy above

Lvs 0.5-3mm wide, flat to inrolled or folded, dull green below, strongly ribbed above, rough above at least nr apex, smooth and hairless below, with scabrid margins, usu with stomata sparse or absent below. Sheaths open (exc v lowest), hairless

Lvs with c17 ribs above, green, obtuse. Ligule c1mm. Sheaths inflated. Sand dunes ....*Dune Fescue* **Vulpia fasciculata**

Lvs with 9-11 ribs, green, acute. Ligule c1mm, minutely toothed, ciliate. Sheaths not inflated, purple at base...*Rat's-tail Fescue* **Vulpia myuros**

Lvs with c9 ribs, green, acute. Ligule 0.5-1mm, entire but deeply retuse, with auricles. Sheaths not inflated...*Squirreltail Fescue* **Vulpia bromoides***

Lvs with c7 ribs, green or reddish, acute. Ligule 0.5-1mm, entire. Sheaths ± inflated ....*Bearded Fescue* **Vulpia ciliata** ssp **ambigua**

Lvs with c7 ribs, green or purplish, acute to obtuse. Ligule 0.5-1mm, toothed. Sheaths not inflated...*Mat-grass Fescue* **Vulpia unilateralis**

▪Lvs hairless above

Ligule 1-2(5)mm, acute, sparsely minutely hairy (occ hairless), decurrent down sheath. Lvs without ribs above (veins obscure) but weak tramlines often present, U-shaped, with cartilaginous oblique or obtuse apex, with weakly scabrid margins, with stomata above only. Sheaths open, not reddish at base, with stomata

Ligule (±) entire. Lvs 0.3-0.5mm wide. Culms usu to 12cm..........*Early Hair-grass* **Aira praecox***

Ligule distinctly toothed. Lvs 0.3mm wide. Culms to 50cm...*Silver Hair-grass* **Aira caryophyllea**

Ligule <1mm, truncate to obtuse

Lvs ribbed above in TS, strongly U-shaped, usu 1-2 per culm, short, curved when young, with cartilaginous truncate apex, rigid, dark green (turning reddish), hairless, with smooth margins, with obscure opaque veins (translucent between veins), usu with 2 wider tramlines, with stomata above only. Sheaths usu reddish at base, closed when young

Lvs 1.5-2mm wide, usu with 5 ribs above (often with raised cartilaginous ridges). Ligule 0.3-1mm, fimbriate to jagged, decurrent down sheath. Sheaths hairless. Coastal ....*Sea Fern-grass* **Catapodium marinum***

Lvs 0.5mm wide, with 5 ribs above, soon minutely rough. Ligule 0.5mm, jagged, decurrent down sheath. Sheaths occ minutely hairy on veins. Dry habs ....*Fern-grass* **Catapodium rigidum**

Lvs not ribbed above, usu inrolled (occ flat), mostly basal, to 2cm x 0.5mm, obtuse, rolled when young, pale green to purple, smooth above, with margins bluntly scabrid or papillate (like minute beads). Ligule 0.5-1mm, entire, obtuse. Sheaths whitish, open and overlapping, with low blunt papillae. Culms to 6cm. Sand dunes, R.................*Early Sand-grass* **Mibora minima***

Vulpia bromoides lf TS

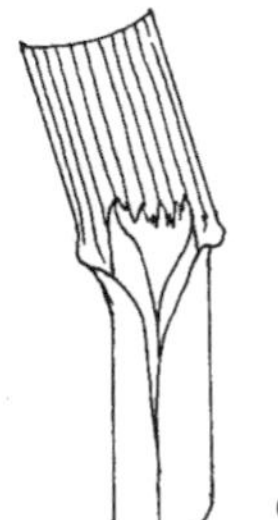

Catapodium marinum ligule & lf TS

Aira praecox lf TS

Mibora minima lf TS

Group HF – *Per, often with persistent fibrous remains. Lvs <1.2mm wide, bristle-like, inrolled or folded (occ flattened). Ligule >0.3mm*

■Ligule acute, usu >1.5mm. Sheaths open. Tufted

Lvs 20-40cm, green, hairless, acute, with v scabrid margins, without stomata below

Lvs 0.3-0.4mm wide, channelled, with 4 vb's. Ligule 1.5-2mm, hairless. Sheaths rough, hairless ....*Feather-grass / Pony-tail Grass* **Stipa tenuissima**

Lvs <20cm, glaucous or grey-green, hairless, acute, with scabrid margins, without stomata below

Lvs 0.3-0.5(1)mm wide, papillate. Ligule 3-4mm, minutely hairy. Sheaths usu purplish, retrorsely scabrid, with veins forming ribs running down sheath. Prophylls to 15mm, retrorsely scabrid on ribs. Culms with blackish nodes. Mainly East Anglia and coastal ....*Grey Hair-grass* **Corynephorus canescens***

Lvs 0.3mm wide, not papillate. Ligule 1.5-2.5(4)mm, minutely antrorsely (or retrorsely) hairy, often torn. Sheaths whitish, retrorsely scabrid, without veins clearly running down sheath. Prophylls to 9mm, retrorsely scabrid in proximal ½ and antrorsely scabrid distally. Dry hths, SW Eng..........*Bristle Bent* **Agrostis curtisii***

Lvs 0.3mm wide, not papillate. Ligule 2mm, minutely antrorsely hairy. Sheaths whitish, smooth to retrorsely scabrid, with veins clearly running down sheath. Prophylls to 9mm, retrorsely scabrid in proximal ½ and antrorsely scabrid distally. Damp hths and wet peaty habs, R ....*Bog Hair-grass* **Deschampsia setacea***

■Ligule truncate to obtuse, usu <1.5mm

Sheaths closed (at least below). Tufted

Lvs often >30cm x 1-2.5mm, green, with tramlines above in TS (but no ribs), hooded at apex, folded when young, minutely scabrid or hairy above, not ciliate, usu with smooth margins, with stomata both sides. Ligule 0.5-2mm, entire. Sheaths hairless, flattened, keeled, closed. Basal sheaths not persisting..........*Narrow-leaved Meadow-grass* **Poa angustifolia**

Lvs 5-25cm x 0.3-1(4)mm, glaucous, with 3(6) deep ribs above in TS, not hooded at apex, folded when young, usu smooth and sparsely hairy both sides (esp above nr ligule), sparsely ciliate, often with rough margins, with stomata above only. Ligule 0.3-1mm, truncate, ciliate. Sheaths densely hairy (lowest occ hairless), with 0.1-0.5mm spreading or slightly retrorse hairs, round, not keeled, closed (but the upper open and overlapping). Basal sheaths persisting as dense fibrous remains..........*Crested Hair-grass* **Koeleria macrantha***

Sheaths open

Stoloniferous. Culms solid. Saltmarshes

Lvs ± glaucous, with smooth margins, snapping audibly when mature. Sheaths hairless, with stomata..........*Common Saltmarsh-grass* **Puccinellia maritima***

Tufted. Culms hollow

Lvs usu held at c90° from pale flexible sheath junction, 4-30cm x 0.2-1mm, acute, inrolled with groove above, greyish-green to green, minutely hairy above or hairless, smooth to strongly ridged below, occ with sparse stomata below. Ligule 0.3-2mm, obtuse, obscurely antrorsely hairy. Basal sheaths hairless, smooth, shiny, white, tough. Cataphylls similar to sheaths, strongly veined, persisting as fibres. Hths..........*Mat-grass* **Nardus stricta***

Lvs usu held at an acute angle from sheaths, 5-30cm x 0.5mm, obtuse, ± cylindrical (weakly 5-6 angled) with indistinct groove, green, hairless, smooth below, without stomata visible below. Ligule 0.5-1mm, truncate or retuse, minutely hairy. Sheaths minutely hairy or hairless, usu rough. Cataphylls obscure or absent. Hths, acidic gsld ....*Wavy Hair-grass* **Deschampsia flexuosa***

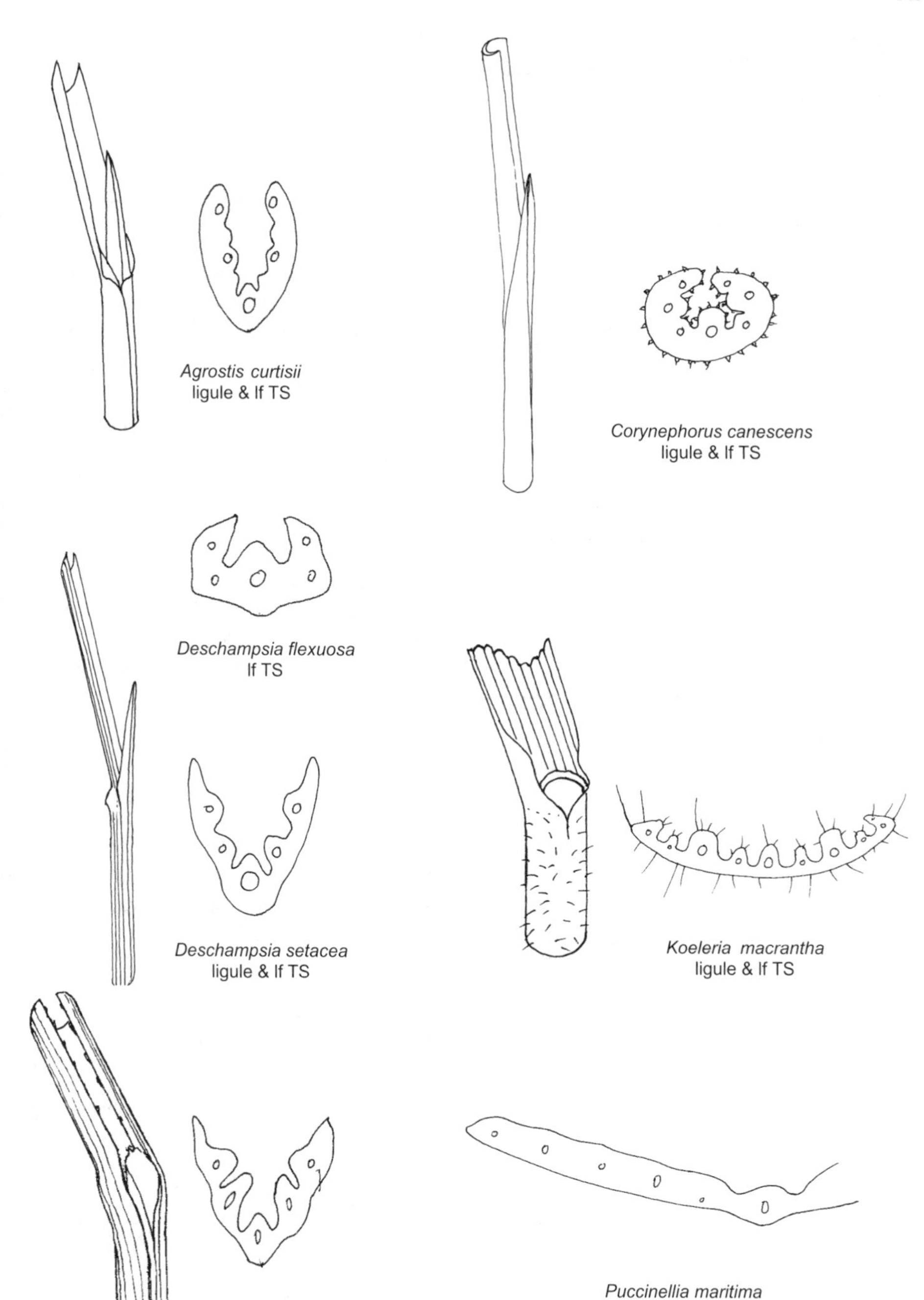
Agrostis curtisii
ligule & lf TS
Corynephorus canescens
ligule & lf TS
Deschampsia flexuosa
lf TS
Deschampsia setacea
ligule & lf TS
Koeleria macrantha
ligule & lf TS
Nardus stricta
ligule & lf TS
Puccinellia maritima
lf TS

Group HG – *Lvs <1.2mm wide, bristle-like, rolled or folded, without stomata below. Ligule ± absent (<0.1mm), usu with tiny rounded auricles. Basal sheaths fibrous, persistent. (Narrow lvs of* Nardus stricta *may key out here)*

■Sheaths closed

Usu rhizomatous, often with stolons. Shoots extravaginal, usu reddish at base. Lvs usu smooth to touch

Lvs to 40cm x (0.3)0.5-1mm, oval, usu acute, 6-angled, ± shiny green, minutely hairy above on 3-9 ribs, with 5 distinct and 2 weaker vb's. Culm lvs >1mm wide, flat. Basal sheaths often minutely retrorsely hairy

Lvs green, not fleshy. Widespread................................*Red Fescue* **Festuca rubra** ssp **rubra***

Lvs green, fleshy. Saltmarshes...........................*Red Fescue* (ssp) **Festuca rubra** ssp **litoralis**

Lvs green, not fleshy. Sand dunes..............................*Rush-leaved Fescue* **Festuca arenaria***

Lvs dull glaucous or pruinose, not fleshy. Coastal ....*Red Fescue* (ssp) **Festuca rubra** ssp **juncea**

Tufted. Shoots intravaginal, dark brown at base. Lvs rough to touch

Lvs 0.3-0.5(0.6)mm wide, 3(4)-angled, hairless. Sheaths minutely retrorsely hairy or hairless. Clumps with persistent white dead lvs around outer edge. Culm lvs 2-4mm wide, flat, hairy above on ribs. Woods..*Various-leaved Fescue* **Festuca heterophylla***

■Sheaths open. Tufted. Shoots intravaginal, retaining old lvs

Lvs usu rough at least nr apex (often smooth in *F. vivipara*)

Lvs glaucous or bluish-green. Basal sheaths occ pinkish

Lvs 0.5-1(2)mm wide, ± obtuse, stiff, strongly folded (rarely flat), usu hairless, with 7(9) unequal veins and 4(6) grooves above each side of midrib, with midrib weakly raised below. Sheaths smooth or rough due to minute retrorse stiff hairs, or minutely hairy ....*Hard Fescue* **Festuca brevipila**

Lvs usu dark green (occ glaucous or greyish). Basal sheaths whitish

Lvs mostly >20cm

Lvs 0.3-0.4mm wide, usu 5-veined, hairless, with 3 vb's. Auricles small, rounded, with prominent auricles. Sheaths minutely hairy to hairless. Clumps with intermixed persistent dead lvs...*Fine-leaved Sheep's-fescue* **Festuca filiformis***

Lvs mostly <10cm

Lvs 0.3-0.6(0.8)mm wide, 5-7-veined, often minutely hairy, with 3 prominent and 4 smaller vb's. Auricles small, rounded. Culm lvs like basal lvs (unlike *F. rubra*)

Auricles hardly prominent. Lvs green to glaucous..................*Sheep's-fescue* **Festuca ovina**

Auricles prominent. Lvs green, often smooth. Upland ....*Viviparous Sheep's-fescue* **Festuca vivipara**

Lvs usu smooth, glaucous or bluish-green. Basal sheaths whitish

Lvs (0.5)0.8-1.2mm wide (when folded), acute, ± stiff, pruinose, 7-veined, with 3-5 ribs above, with midrib weakly raised below. Ligule ciliolate. Sheaths smooth. Auricles distinct

Lvs hairless. E Anglian Brecks.............................*Blue Fescue* (ssp) **Festuca longifolia** ssp **A**

Lvs minutely ciliate nr sheath. Coastal cliff-tops, Isle of Wight, Devon, Channel Is ....*Blue Fescue* (ssp) **Festuca longifolia** ssp **B**

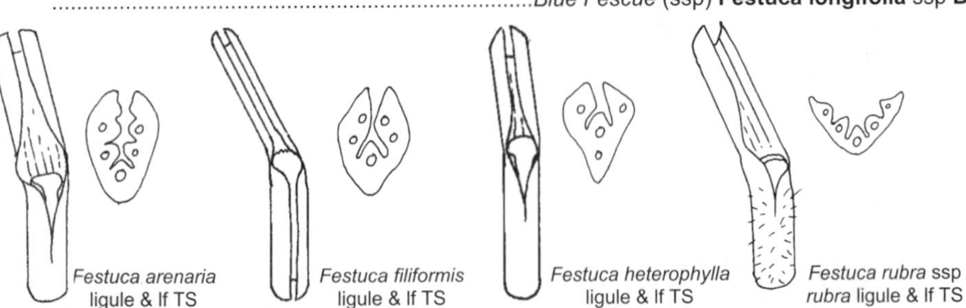

Festuca arenaria ligule & lf TS

Festuca filiformis ligule & lf TS

Festuca heterophylla ligule & lf TS

Festuca rubra ssp *rubra* ligule & lf TS

Group HH – *Tufted or rhizomatous per. Lvs folded when young (occ rolled in* Koeleria *and* Bromopsis*). Lvs or sheaths hairy (or ciliate). Sheaths closed at least below.* ❶ *Lvs not hooded at apex (hooded in all other taxa)*

■Lvs with 3 deep ribs each side of midrib in TS (stripy HTL), often glaucous. Shoots round

Lvs 5-25cm x (0.3)1-4mm, usu rolled when young (occ folded), later flat or inrolled, narrowed to base, shortly ciliate, with longer cilia nr ligule, with ribs densely hairy above (occ hairless), often hairless below, with stomata above only. Ligule 0.3-1mm, truncate, usu with long whiskers to 2mm on collar. Sheaths with dense short patent or retrorse hairs. Basal sheaths hairless, whitish, fibrous. Tufted...*Crested Hair-grass* **Koeleria macrantha***

■Lvs with 5-6 shallow ribs each side of midrib in TS (hardly stripy HTL), green. Shoots usu round

Lvs 10-25(50)cm x 2-3.5mm, occ rolled when young, acute, widest at or nr base, often hairy above, with long antrorse cilia to 2mm (occ absent), with midrib slightly raised below, with 1(3) main veins each side of midrib, with stomata absent or v sparse below. Ligule 0.2(3)mm, truncate, minutely fimbriate to entire, hairless. Sheaths hairless or with long hairs. Basal sheaths whitish or purplish, persisting as fibres. Tufted. ❶....................................*Upright Brome* **Bromopsis erecta***

■Lvs not ribbed above but tramlines may be present (not stripy HTL), green. Shoots flattened

Lvs with tramlines. Sheaths hairless. Rhizomatous, often with dead lvs at base

Lvs 15-45cm x 1-2.5mm, ± stiff, flat or folded, hairless or hairy above, not ciliate, usu with smooth margins, with 5 veins each side of midrib (midrib raised below), with stomata both sides or above only. Collar often retrorsely hairy. Ligule 0.5-2mm, truncate, minutely hairy to hairless. Dry habs ....*Narrow-leaved Meadow-grass* **Poa angustifolia**

Lvs 5-30cm x 2-5mm, ± limp, flat or channelled, occ hairless above, long-ciliate nr ligule, usu with smooth margins, with 2 veins each side of midrib (midrib raised below), with stomata both sides. Collar usu hairless. Ligule <0.5(2)mm, truncate, minutely hairy to hairless. Usu gsld ....*Smooth Meadow-grass* **Poa pratensis***

Lvs without tramlines (but midrib in deep furrow). Sheaths with patent to reflexed hairs. Tufted, often with dead lvs at base

Lvs 4-30cm x (2)3-6(8)mm, obtuse, hairless or with sparse long hairs both sides, with retrorse to antrorse cilia, with midrib strongly raised below, with 4 veins each side of midrib, with stomata sparse below. Ligule 0.2mm (3-8mm on culm). Sheaths often purplish below. Calc gsld..*Downy Oat-grass* **Helictotrichon pubescens***

Poa pratensis lf TS

Helictotrichon pubescens lf TS

Bromopsis erecta ligule & lf TS

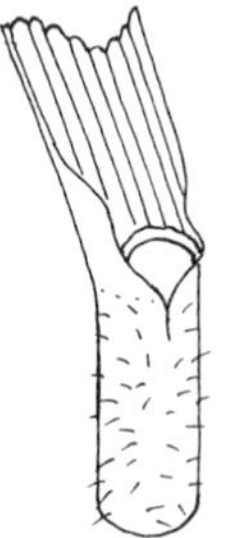

Koeleria macrantha ligule & lf TS

Group HI – *Sheaths and lvs hairless, not ciliate (exc nr ligule), glaucous or dark green. Sheaths open, without cross-veins*

■Lvs ribbed above in TS

Lvs with tramlines above, hooded at apex, flat or folded (occ inrolled). Ligule 1-2mm, obtuse

Lvs 1.5-2.5(5)mm wide, with 2-6 low ribs above each side of midrib, glaucous and rough above, with stomata both sides. Ligule c2mm, often split, hairless. Shoots round. Culms solid or with minute hollow. Wet saline habs ....*Borrer's Saltmarsh-grass* **Puccinellia fasciculata***

Lvs 1-2.5mm wide, with 2-4 deep ribs above each side of midrib, grey-green both sides, smooth or rough above, with stomata both sides. Ligule 1mm, entire, hairless. Shoots ± round or oval. Culms solid. Damp saline habs (inc road verges) ....*Reflexed Saltmarsh-grass* **Puccinellia distans***

Lvs without tramlines, not hooded at apex, U-shaped. Ligule 1-5mm, acute

Stoloniferous. Damp to wet acidic habs..*Velvet Bent* **Agrostis canina***

Shortly rhizomatous. Dry acidic habs...*Brown Bent* **Agrostis vinealis***

■Lvs not (or hardly) ribbed above, tramlines present, hooded at apex

Coastal saline habs. Culms (and stolons) solid

Stolons present (at least in summer), often reddish. Lvs to 20cm x 1-3mm, ± parallel-sided, grey-green above, ± fleshy (occ snapping audibly), often minutely white-hairy above, smooth below, with smooth margins, with 1-3 weak veins each side of midrib (less translucent than midrib), with stomata both sides. Ligule 1mm, ± obtuse, hairless, often brownish and split. Upper sheaths not inflated. Saltmarshes........*Common Saltmarsh-grass* **Puccinellia maritima***

Stolons absent. Lvs to 5(10)cm x 1.8-3mm, tapered from base, grey-green, not fleshy but stiff, always hairless, often rough above, smooth below, with scabrid margins, with 3 weak veins each side of midrib, with stomata both sides. Ligule 1-1.5mm, ± acute or cuspidate to ± obtuse, hairless. Upper sheaths loosely inflated. Winter-wet bare saline habs ....*Stiff Saltmarsh-grass* **Puccinellia rupestris**

Chalk gsld. Culms hollow

Lvs to 50cm x 3-4mm, stiff, pruinose or glaucous above, usu rough above, occ hairy and ciliate nr ligule, with stomata scattered below. Ligule 0.3(3)mm, often ciliolate, minutely hairy. Sheaths usu hairless...*Meadow Oat-grass* **Helictotrichon pratense***

Shady or mtn habs. Culms hollow..*Wood Meadow-grass* **Poa nemoralis**

Puccinellia fasciculata lf TS

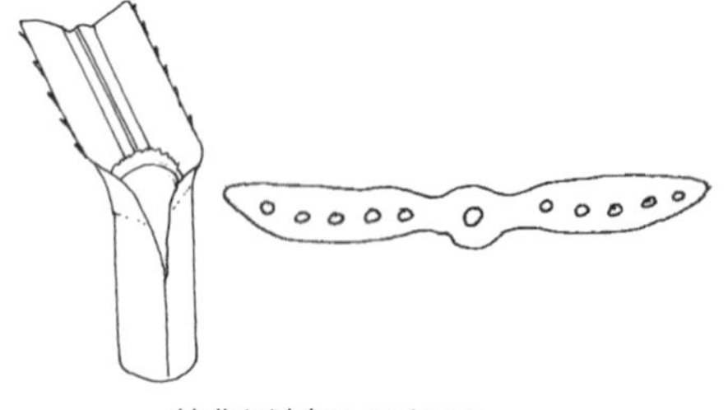

Helictotrichon pratense ligule & lf TS

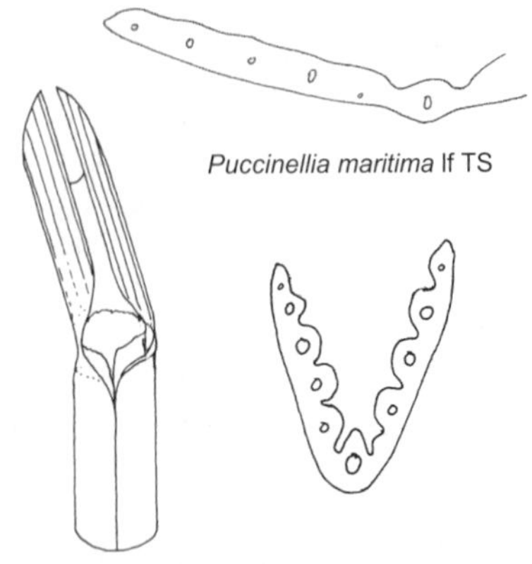

Puccinellia maritima lf TS

Puccinellia distans ligule & lf TS

Group HJ – *Sheaths and lvs hairless. Sheaths closed. Lvs with cross-veins, or hollows in sheaths. Some lvs usu emergent as well as floating, often ± parallel-sided, with translucent veins.* ❶ *Lvs glaucous, with cross-veins absent or obscure*

▪Lvs 10-16mm wide, usu emergent, with stomata both sides. Rhizomatous

Lvs apiculate, hooded at apex, occ wrinkled, smooth or rough above, grey-green above, shiny green below, not ribbed in TS, with 14-17 veins each side of midrib. Ligule 3-4mm, shorter than wide, cuspidate. Sheaths often reddish at base, keeled, rough. Floating shoots all yr, aerial shoots usu Apr-Nov..*Reed Sweet-grass* **Glyceria maxima***

▪Lvs usu <10mm wide, floating or emergent, with stomata above only on floating lvs. Stoloniferous

Ligule 1-3mm, usu shorter than wide, hairless, toothed. Lvs slightly hooded at apex (occ asymmetrically notched), with tramlines often indistinct

Lvs 7-10mm wide, semi-emergent or floating, bright green, gradually tapered, limp, often wrinkled, with cross-veins absent or obscure, not ribbed, with 6-7 indistinct veins each side of midrib. Sheaths often reddish at base, oval, smooth, with hollows ....*Whorl-grass* **Catabrosa aquatica***

Ligule 6-10mm, longer than wide, usu minutely hairy, entire (occ torn). Lvs hooded at apex, with distinct tramlines

Lvs usu glaucous or grey-green, not or hardly ribbed above in TS, ± without cross-veins

Lvs 3-5(9)mm wide. Ligule 6-8mm, v sparsely minutely hairy to ± hairless, often with 3 green veins. Sheaths not reddish, smooth. ❶. Usu acidic muddy or shallow water ....*Small Sweet-grass* **Glyceria declinata**

Lvs yellow- to dark green, with 10-14 ribs above each side of midrib in TS, with distinct cross-veins

Lvs 5-10(15)mm wide, yellow-green, deeply ribbed, often rough at least below. Sheaths not reddish, rough above. Usu in shallow water...................*Plicate Sweet-grass* **Glyceria notata***

Lvs 5-10mm wide, dull dark green, variably ribbed, rough or smooth both sides. Sheaths not reddish, often rough. *G. fluitans* x *notata*.............*Hybrid Sweet-grass* **Glyceria** x **pedicellata***

Lvs 4-6(10)mm wide, fresh green, shallowly ribbed, occ rough above. Sheaths often reddish at base, rough. Usu in deep water..............................*Floating Sweet-grass* **Glyceria fluitans***

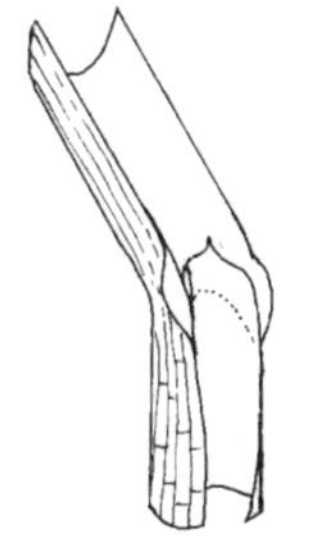

Glyceria maxima
(sheath opened to show ligule)

Glyceria notata ligule & lf TS

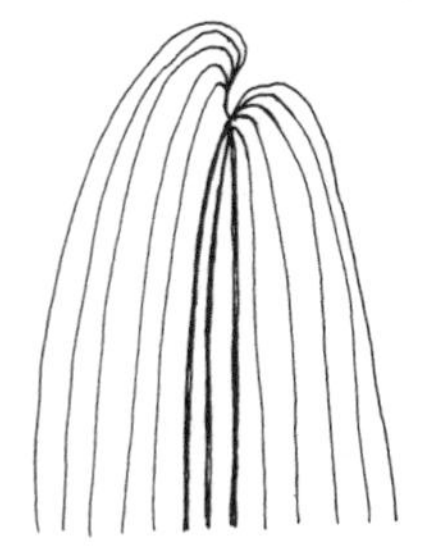

Catabrosa aquatica
lf apex

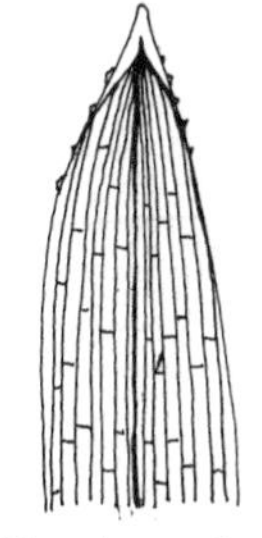

Glyceria x *pedicellata*
lf apex

Glyceria fluitans lf TS

Group HK – *Lvs without cross-veins, ribbed above in TS, without stomata below (occ below on culm lvs of* Cynosurus cristatus*). Sheaths closed at least below, hairless, without hollows. Tufted per*

■Lvs v rough above

Lvs (10)20-60cm x (2)3-5mm, dark grey-green, inrolling on drying, with 6 deep translucent furrows above, hairless, antrorsely scabrid above, smooth below. Ligule 3-10mm, acute, with 3 main veins. Sheaths strongly flattened, keeled...................*Tufted Hair-grass* **Deschampsia cespitosa***

■Lvs ± smooth both sides

Sheaths purplish at base. Auricles often small, hooked

Lvs 3-20cm x 2-6mm, rarely rolled when young, dull green to grey-green above, shiny green below, with 7-8 deep ribs each side of midrib (stripy HTL), with midrib slightly raised below. Ligule 0.7-1.2mm, entire. Sheaths flattened, the upper open ....*Perennial Rye-grass* **Lolium perenne***

Sheaths yellowish at base. Auricles absent

Lvs 5-15cm x (1)2-4mm, often held 90° to sheath, rarely rolled when young, tapered to apex, shiny green below, with smooth margins (exc nr apex), with 2-4 v low ribs each side of midrib (stripy HTL), with midrib slightly raised below. Culm lvs usu held 45º to sheath, weakly ribbed above. Ligule 0.5-0.8mm (to 1.5mm on culm), truncate, denticulate, often minutely fimbriate, often brownish, hairless to minutely hairy. Sheaths round, the upper open ....*Crested Dog's-tail* **Cynosurus cristatus***

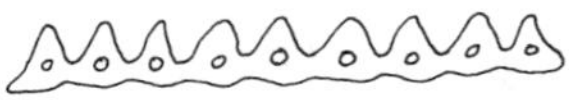

Deschampsia cespitosa lf TS

Lolium perenne lf TS

Cynosurus cristatus lf TS

Group HL – *Lvs not ribbed above, tramlines present (exc* Dactylis glomerata*), hooded at apex (obscure in* Poa nemoralis*). Ligule often decurrent down sheath. Sheaths hairless, closed (but often open and overlapping above).* ❶ *Lvs occ hairy above.* ❷ *Culms with blackish nodes*

■Lvs strongly glaucous or pruinose above, stiff. (*Helictotrichon pratense* may key out here in error but the sheaths are open)

Lvs to 20cm x (2.5)3.5-4(8)mm, flat, hairless, usu smooth both sides, with distinct hyaline (0.1mm) scabrid margins, with midrib raised below, with stomata both sides or above only. Ligule 0.1-0.3mm, ciliolate. Shoots flattened, with persistent dead sheaths at base. Usu limestone gsld, N Br ....*Blue Moor-grass* **Sesleria caerulea**

Lvs 4-8cm x 3-6mm, flat, hairless, ± smooth both sides, with scabrid margins nr apex, often wrinkled (on culm), with 7-8 main veins each side of midrib, with stomata both sides. Ligule 0.3-0.5mm, obtuse, obscurely ciliolate, hairless. Shoots flattened, with fibrous remains at base. Bulbs occ present. Mtns............*Alpine Meadow-grass* **Poa alpina**

Lvs to 4cm x 1.5-2mm, ± inrolled, hairless or with obscure minute hairs, ± smooth both sides, with scabrid margins at least nr ligule and apex, with 1-2 main veins each side of midrib (others obscure), with stomata at least above. Ligule 0.2-0.5mm, truncate, obscurely ciliolate. Shoots slightly flattened, the lower often minutely rough. Mtns.........*Glaucous Meadow-grass* **Poa glauca**

■Lvs green to grey-green both sides, stiff or limp

Ann (per). Lvs usu wrinkled

Lvs (1)3-4mm wide, limp, with ± indistinct tramlines, often with 5 indistinct veins each side of midrib (occ v translucent vein in outer ⅓ nr margin), with midrib slightly raised below, with stomata both sides. Ligule 1-4mm, obtuse, ± entire, hairless. Sheaths ± flattened, smooth. Shoots intravaginal

All yr, often perennating. Lvs 1-10cm, yellow-green, often folded. Culms branched at lowest nodes............*Annual Meadow-grass* **Poa annua**

Dec-May, never perennating. Lvs mostly <1.2cm, pale yellow-green, often flat. Culms unbranched............*Early Meadow-grass* **Poa infirma**

Per (often weak in *Poa trivialis*). Lvs not wrinkled

Lvs usu >5mm wide. Sheaths sharply keeled, strongly flattened

Lvs dull grey-green both sides, 5-14mm wide, v acute to aristate, without tramlines (but midrib sunken above), rough or smooth above, with scabrid margins (retrorsely so nr ligule), with stomata both sides. Ligule 0.5-12mm, truncate to acute, hairless. Sheaths rough or smooth, fleshy. Tufted, often with persistent fibrous sheaths................*Cock's-foot* **Dactylis glomerata**

Lvs shiny bright green both sides, 5-10mm wide, with abruptly apiculate apex, with tramlines above, smooth above, with scabrid margins, with stomata both sides. Ligule to 1.5mm, truncate, hairless. Sheaths rough or smooth, not fleshy. Tufted. Shady habs ....*Broad-leaved Meadow-grass* **Poa chaixii**

Lvs usu <5mm wide. Sheaths not sharply keeled, occ flattened

Plant with bulbs at base. Shoots intravaginal

Lvs to 5cm x 1-2mm, dark green, stiff, with obscurely scabrid-ciliate margins, with midrib translucent only (at least when young), with stomata above only. Ligule 1-3mm, usu ± acute. Sheaths purplish at base. Cataphylls 6mm, acute, with ciliate keel. Oct-Jul. Open gsld or sandy habs, often nr coast..*Bulbous Meadow-grass* **Poa bulbosa***

Plant without bulbs at base. Shoots intravaginal or extravaginal

Lf margins ciliolate at extreme base (on collar). Rhizomatous

Lvs 15-45cm x 1-2.5mm, flat or folded, green, usu with smooth margins, minutely scabrid or hairy above (esp distally), with stomata both sides. Ligule 0.5-2mm, truncate to obtuse, not ciliolate. Sheaths smooth....................*Narrow-leaved Meadow-grass* **Poa angustifolia**

Lvs to 6cm x 1.5-4mm, flat or channelled, dark blue-green (occ ± glaucous), with scabrid margins at least nr apex, hairless, with stomata both sides or above only. Ligule 0.3-1(1.5)mm, truncate, ciliolate, usu hairless (minutely hairy on culm lvs). Sheaths smooth ....*Spreading Meadow-grass* **Poa humilis**

Lf margins hairless at extreme base (occ scabrid). Rhizomatous or tufted

Ligule <1mm

Lvs shiny light green below

Lvs 4-20cm x 1-4(6)mm, limp, occ rough above, hairless, usu with smooth margins, with tramlines occ indistinct, with opaque midrib raised below, with stomata above only. Ligule 0-3(7)mm, acute to ± absent, minutely hairy, occ greenish. Sheaths v flattened, smooth below, rough above esp below nodes. Loosely tufted or weakly stoloniferous ....*Rough Meadow-grass* **Poa trivialis***

Lvs 5-30cm x 2-5mm, ± limp, occ rough above, occ hairy above, long-ciliate nr ligule, usu with smooth margins, with translucent midrib raised below, with stomata both sides. Ligule <0.5(2)mm, truncate, minutely hairy to hairless. Sheaths slightly flattened, rough above. Rhizomatous. ❶. Usu gsld...........*Smooth Meadow-grass* **Poa pratensis***

Lvs shiny dark green below

Lvs 4-12cm x 1-3mm, limp, occ rough above, with smooth margins (retrorsely scabrid when young), with tramlines visible only nr apex, with midrib v weakly raised below, with stomata above only (occ v sparse below). Ligule <0.6mm, truncate, minutely ciliolate, obscurely hairy. Sheaths ± round, smooth, rarely pruinose or reddish at base. Loosely tufted. ❷. Shady habs, mtns...................*Wood Meadow-grass* **Poa nemoralis**

Lvs dull grey-green below

Lvs 4-12cm x (1)3-4mm, stiff, usu smooth above, with scabrid margins nr apex, with distinct tramlines and 6-7 (1-2 main) veins each side of midrib, with midrib raised below, with stomata both sides. Ligule 0.3-0.7mm (to 2mm on culm lvs), truncate, ciliolate (often obscure), minutely hairy. Sheaths often reddish at base, occ with cilia on collar (like P. humilis). Shoots flattened. Culm flattened, geniculate, with flattened nodes. Rhizomatous......................................*Flattened Meadow-grass* **Poa compressa**

Ligule >1mm

Rhizomatous

Ligule 0.5-2mm, truncate, minutely hairy. Lvs 5-30cm x 2-5mm, obtuse, flat or channelled, dull green, with midrib raised below, with stomata both sides. Collars dark brown to blackish below. ❶...............................*Smooth Meadow-grass* **Poa pratensis***

Tufted

Upper sheaths rough...*Rough Meadow-grass* **Poa trivialis***

All sheaths smooth. Sheaths without fibrous remains at base (unlike *P. alpina*)

Damp lowland habs, R. Ligule 2-5mm, obtuse to truncate, ± entire to jagged. Lvs 1.5-4mm wide, flat, with ± scabrid margins, with tramlines visible only nr apex. Sheaths ± round, covering internodes...............................*Swamp Meadow-grass* **Poa palustris**

Mtns, VR. Ligule 1-4mm, acute, ± jagged. Lvs 1-2mm wide, ± inrolled, with an obliquely mucronate apex, often with smooth margins, with tramlines distinct to obscure. Sheaths flattened, the uppermost shorter than its lf ....*Wavy Meadow-grass* **Poa flexuosa**

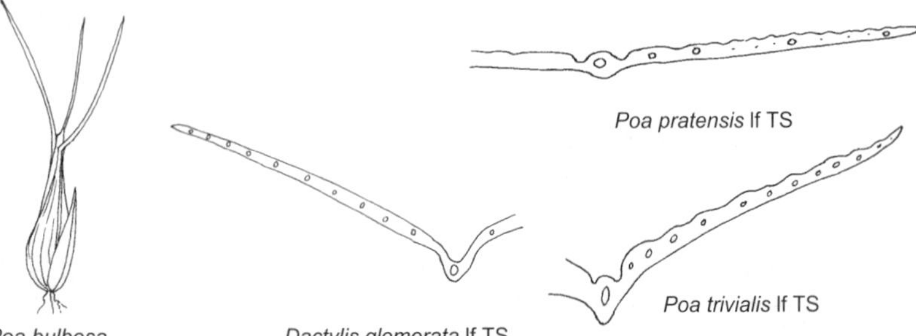

Poa pratensis lf TS

Poa trivialis lf TS

Poa bulbosa

Dactylis glomerata lf TS

Group HM – *Auricles present. Lvs (and/or sheaths) hairy, not (or hardly) ribbed in TS. Shoots ± round (slightly flattened in* Ceratochloa*).* ❶ *Lvs with long hairs nr ligule (whiskers)*

■Ligule >1mm, toothed, ciliolate or not, hairless or minutely hairy

Sheaths always open, margins overlapping

Tufted per. Lvs 1-6mm wide, dull yellow-green above, (±) shiny below, sparsely hairy and ciliate (occ hairless), usu smooth above, hay-scented, with v bitter taste, with stomata sparse or absent below. Auricles rounded to acute, often absent. Ligule 1-4mm, ± truncate, fimbriate, hairless. Sheaths hairless or shortly hairy. ❶....*Sweet Vernal-grass* **Anthoxanthum odoratum***

Ann. Lvs 5-10mm wide, dull dark bluish-green or glaucous/pruinose both sides, rarely with long hairs below, rough above, not hay-scented, tasteless, with stomata at least above. Auricles tiny, not overlapping. Ligule 2mm, truncate, ciliolate, hairless. Sheaths hairless or with long hairs. Crop relic, R..*Rye* **Secale cereale**

Sheaths closed (at least below when young)

Sheaths soon splitting (margins never overlapping). Woods

Lvs 10-14mm wide, narrowed at both ends, often twisted, rough both sides, green, with long hairs above and on margins (usu only on midrib below), with midrib raised below, with 15 unequal veins each side of midrib (3-4 main veins), with stomata both sides but sparse below. Ligule 2.5mm, hairless, scarious, often brown, with veins

All sheaths with long ± patent or retrorse hairs (occ sparse on upper). Culms densely hairy below 2nd node...*Hairy-brome* **Bromopsis ramosa**

Upper sheaths hairless or v shortly hairy with few or no long hairs. Lower sheaths with adpressed or retrorse hairs. Culms ± hairless below 2nd node. Dry calc woods, R ....*Lesser Hairy-brome* **Bromopsis benekenii**

Sheaths not splitting (often strongly fused ± to apex). Shoots. Ruderal

Lower sheaths retrorsely hairy or hairless, ciliate, ± smooth, with cross-veins weak to absent, the upper hairless and open. Lvs 5-14mm wide, widest nr base, rough at least above, occ fleshy, usu dark green, with sparse long hairs on one or both sides, with translucent midrib raised below, with c10 veins each side of midrib, with stomata more abundant above. Ligule (1)2-4mm, long-hairy to hairless, often green or scarious. Auricles small or absent ....*Rescue Brome* **Ceratochloa cathartica**

All sheaths hairless but may be ciliate and hairy at apex, ± smooth, with cross-veins weak to absent. Lvs 6-14mm wide, widest nr base, occ rough below, occ fleshy, usu dark green, ± hairless to hairless, often ciliate (cilia to 1mm), with opaque midrib raised below, with c15 veins each side of midrib, with stomata more abundant above. Ligule 1-3.5mm, long-hairy to minutely hairy or hairless, often green or scarious. Auricles small or absent ....*California Brome* **Ceratochloa carinata**

■Ligule 0.5-0.7(1)mm, truncate, ciliolate (to minutely fimbriate), hairless

Sheaths closed (strongly fused). Rhizomatous per

Lvs 5-10mm wide, v slightly narrowed to base, with minute hair-tuft at sheath junction, hairy to hairless, smooth or hardly rough above, scabrid below on raised veins (but midrib smooth), with scabrid margins, with 5-6 main veins (occ unequal) and 1-3 minor veins between main veins. Ligule 0.5-1mm. Auricles small to absent. Sheaths hairless or with short retrorse hairs, with weak cross-veins. Culms occ with hairs nr nodes....*Hungarian Brome* **Bromopsis inermis**

Sheaths open above (but closed below)

Rhizomatous per

Sheaths not ciliate. Rarely sand dunes

Lvs 3-5mm wide, widest in middle, dull green or glaucous, twisted, inrolling on drying, rough below at least distally, with sparse long hairs esp above (occ v shortly hairy above or hairless), with scabrid margins, with 8-16 v low ribs, with stomata both sides or above only. Ligule ≤0.3mm. Auricles occ v small. Basal sheaths often shortly retrorsely hairy, rarely ciliate..*Common Couch* **Elytrigia repens***

Sheaths ciliate on 1 margin (soon rubs off). Coastal (often sand dunes)

Lvs with 11-25 distinct broad ribs above, 2-7(10)mm wide, sharply acute, stiff, twisted, inrolling tightly on drying, grey-green or glaucous both sides, densely minutely antrorsely hairy above, with smooth margins, with stomata above only. Ligule 0.5mm. Sheaths purplish at base, without cross-veins, the lowest closed. Auricles ± absent. Young dunes ....*Sand Couch* **Elytrigia juncea***

Lvs with 6-10 distinct narrow ribs above, minutely short-hairy above. *E. atherica* x *juncea* ....**Elytrigia** x **obtusiuscula**

Lvs with 6-10 indistinct narrow ribs above, with sparse long hairs above. *E. atherica* x *repens*..........**Elytrigia** x **oliveri**

Tufted per

Auricles distinct, without purplish collar. Lvs 4-14mm wide, narrowed to base, often twisted and shredded nr apex, ± shiny green below, often slightly rough above, ± hairless to hairy above, with 14-18 veins each side of midrib (3 main veins, every 4th vein thicker), with midrib asymmetric esp proximally and raised below, with stomata above only. Ligule 0.5-0.7(1)mm. Sheaths with weakly retrorse hairs (c1mm, often unequal), rough on culm, soon splitting. Calc woods, R..........*Wood Barley* **Hordelymus europaeus***

Auricles indistinct (occ absent), with purplish collar. Lvs 4-9mm wide, slightly narrowed to base, shiny bright to dark bluish-green below, densely to sparsely hairy above (some hairs to 1mm), hairless and smooth below, with 8-11 v low ribs, with sparse stomata below. Ligule <0.5mm. Sheaths with minute (<0.1mm) retrorse hairs. Shady habs ....*Bearded Couch* **Elymus caninus***

Auricles indistinct to absent, without purplish collar. Lvs 2-5mm wide, involute on drying, green, with long or short hairs above or hairless, ± rough or smooth below, with stomata sparse or absent below. Ligule 0.5mm. Lower sheaths with dense short (<0.5mm) hairs, the lowest closed with 1mm unequal patent hairs. Mid and upper sheaths hairless, open and not inflated. Gsld, esp on clay..........*Meadow Barley* **Hordeum secalinum**

Ann

Auricles often long, narrow, spreading or overlapping. Lvs green, 2-15cm x 2.5-8mm, widest at base, twisted, sparsely long-hairy both sides (occ hairless), limp, often rough both sides, hardly ribbed above, with 1-3 main veins each side of midrib (every 3rd-4th vein v translucent), with stomata both sides, midrib raised below. Ligule 0.3-1mm, minutely fimbriate. Sheaths hairless and inflated above (only lowest v hairy and closed), smooth. Ruderal ....*Wall Barley* **Hordeum murinum***

Auricles v small. Lvs bluish-green, to 3cm x 1-3.5mm, widest at base, twisted, minutely hairy both sides (occ hairless), stiff, smooth (rarely rough) both sides, with 4 low ribs above each side of midrib, with translucent or opaque veins, with stomata sparse or absent below. Ligule 0.5mm, obscurely ciliolate, usu minutely jagged, hairless. Sheaths densely retrorsely minutely hairy below, hairless and slightly inflated above, without ciliate margins. Coastal ....*Sea Barley* **Hordeum marinum**

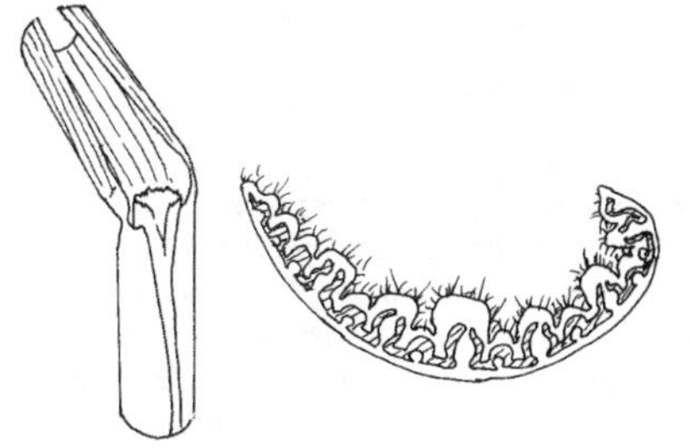

Elytrigia juncea ligule & lf TS

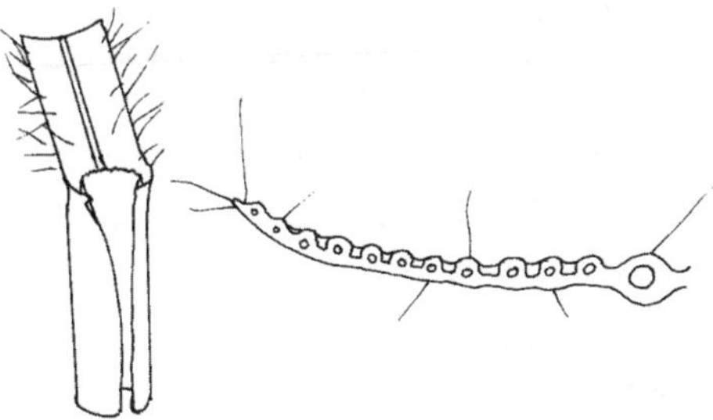

Elymus caninus ligule & lf TS

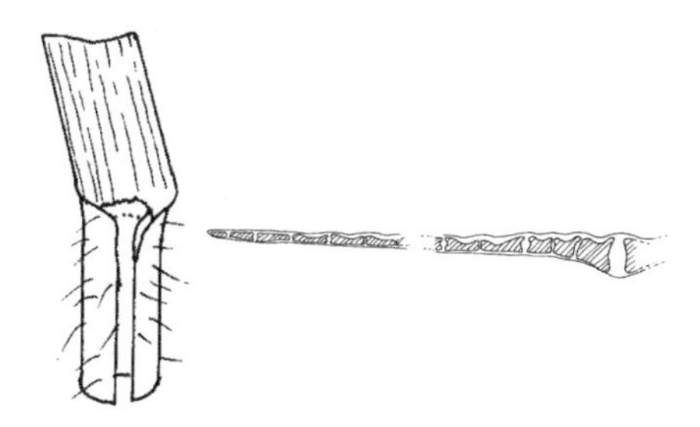

Hordelymus europaeus ligule & lf TS

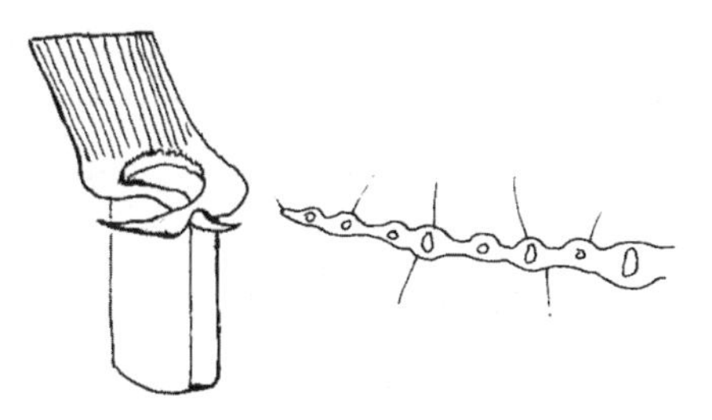

Hordeum murinum ligule & lf TS

HN

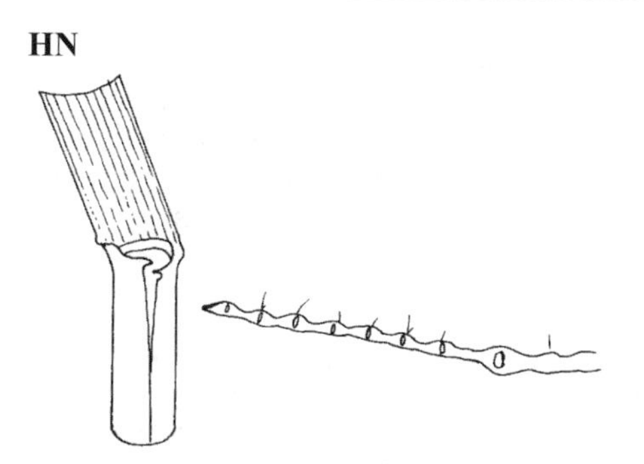

Elytrigia repens ligule & lf TS

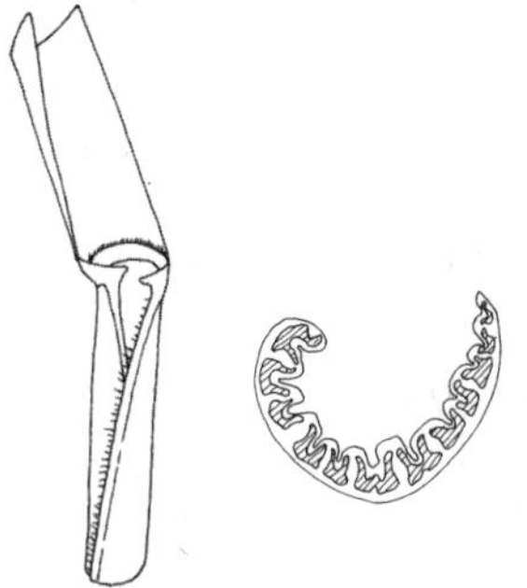

Elytrigia atherica ligule & lf TS

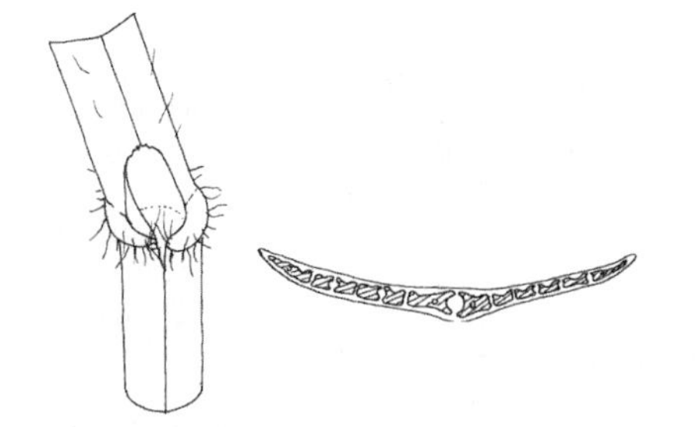

Anthoxanthum odoratum ligule & lf TS

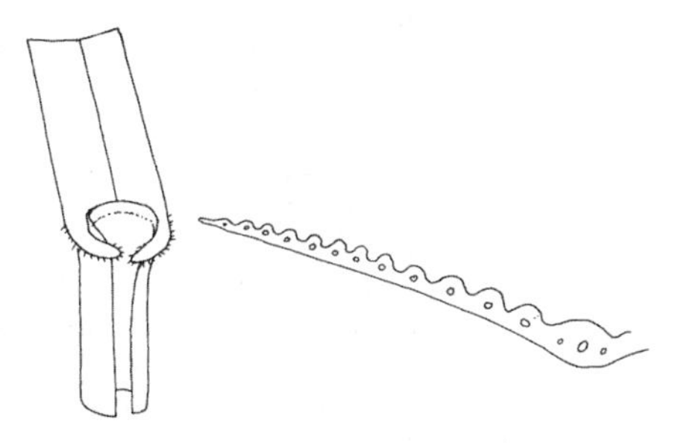

Festuca arundinacea ligule & lf TS

Group HN – *Auricles present. Lvs (and/or sheaths) hairless, widest nr base. Sheaths open*

■Auricles (or collar) ciliate at least when young

Per

Auricles shortly ciliate (at least when young), long, acute, creamy-white. Lvs 3-12(16)mm wide, shiny green below, smooth or slightly rough above, often with v scabrid margins (occ retrorse nr ligule), never ciliate, with 8-10 flat deep ribs each side of midrib in TS, with stomata sparse or absent below. Ligule 0.2-1(2)mm, minutely toothed or fimbriate-ciliate. Sheaths round, hairless, with cross-veins, the lowest purplish and rough............*Tall Fescue* **Festuca arundinacea***

Auricles hairless (but long whiskers nr ligule), short, rounded to acute, often absent. Lvs 1-6mm wide, (±) shiny yellow-green below, usu smooth above, without ribs, occ ciliate, hay-scented, with v bitter taste, with stomata sparse or absent below. Ligule to 4mm, ± truncate, jagged, fimbriate. Sheaths round, hairless (below) or hairy, without cross-veins, never purplish or rough ....*Sweet Vernal-grass* **Anthoxanthum odoratum***

Ann. Crop relic

Auricles long-ciliate, long, often overlapping. Lvs 8-14mm wide, occ hairy above, dull ± glaucous above, ± shiny below, slightly rough above, smooth below, hardly ribbed in TS, with 7 main veins each side of midrib (every 3rd-4th vein v translucent), with stomata both sides. Ligule 1.5-2mm, truncate, ciliolate, jagged, hairless. Sheaths smooth......*Bread Wheat* **Triticum aestivum**

Auricles sparsely short-ciliate or hairless, shorter, spreading, not overlapping. Lvs 12-15mm wide, hairless, dull glaucous (pruinose) both sides, slightly rough above, smooth below, not ribbed in TS, with stomata both sides. Ligule to 2mm, truncate, ciliolate, jagged, hairless. *Secale* x *Triticum*...x **Triticosecale**

■Auricles (and collar) not ciliate

Lvs v shiny green below, usu with stomata above only. Basal sheaths usu purplish

Lvs rolled when young

Per

Lvs 6-18mm wide, weakly ribbed with 15-20 ribs above in TS (occ absent), widest nr base, slightly rough at least above, with scabrid margins (often retrorse nr ligule), twisted, with opaque midrib raised below. Ligule 0.7-2.5mm, entire, brownish, hairless. Auricles usu purplish, overlapping. Sheaths rough below. Culms with purple nodes. Shady habs ....*Giant Fescue* **Festuca gigantea***

Lvs 3-8mm wide, strongly ribbed with 15-20 flat-topped ribs above in TS, widest nr base, with scabrid margins. Ligule 0.5-1mm, entire, hairless. Auricles small, weak. Gsld (usu damp)...*Meadow Fescue* **Festuca pratensis**

Ann (often short-lived per in *L. multiflorum*)

Lvs (3)6-8.5(10)mm wide, strongly ribbed with 15-25 ribs above in TS, widest nr base, with scabrid margins. Ligule 0.6-2.5(4)mm. Auricles long, narrow, overlapping ....*Italian Rye-grass* **Lolium multiflorum**

Lvs (3)9-13mm wide, strongly ribbed with c23 rough ribs above in TS, widest nr base, with scabrid margins. Ligule to 2mm, truncate or obtuse, hairless. Auricles narrow, overlapping (occ spreading). VR alien..*Darnel* **Lolium temulentum**

Lvs partly rolled and partly folded when young. Per (often short-lived)

Damp gsld. Lvs 4-10mm wide. Ligule to 1mm. *Festuca pratensis* x *Lolium perenne* ....*Hybrid Fescue* x **Festulolium loliaceum**

Dry gsld, often sown. Lvs 3-6.5mm wide, with 13 low rounded ribs each side of midrib, with midrib raised below. Auricles often overlapping. *L. multiflorum* x *perenne* ....**Lolium** x **boucheanum**

Lvs dull green or glaucous below, with stomata both sides. Basal sheaths not or rarely purplish

Auricles overlapping, clasping, acute. Lvs 6-20mm wide, widest nr base, not twisted, often ± pruinose, usu smooth both sides, with shallow ribs, with stomata both sides. Ligule 0.7-1mm, truncate, ciliolate, hairless. Sheaths smooth, often ± pruinose. Ann. Crop relic

Fertile spikelets arranged in 2 vertical rows.....................................*Barley* **Hordeum distichon**

Fertile spikelets arranged in 6 vertical rows. R...................*Six-rowed Barley* **Hordeum vulgare**

Auricles small, not overlapping

Lvs >8mm wide

Auricles distinct. Lvs 10-20mm wide, glaucous, stiff, sharply acute, shortly scabrid above, smooth below, with scabrid margins (occ smooth), with 10-20 flat ribs each side of midrib. Ligule to 1mm, truncate, fimbriate. Sheaths hairless, smooth, round, with cross-veins. Sand dunes..*Lyme-grass* **Leymus arenarius***

Auricles indistinct. Mar-Nov

Rhizomatous per. Usu wet habs........................*Reed Canary-grass* **Phalaris arundinacea***

Tufted per (rhizomes short or absent). Dry habs...*Bulbous Canary-grass* **Phalaris aquatica**

Ann. Dry arable or disturbed habs

Anthers 3mm...*Canary-grass* **Phalaris canariensis**

Anthers 1.5-1.8mm. Spikelets all alike, not in clusters, disarticulating above the persistent glumes.. *Lesser Canary-grass* **Phalaris minor**

Anthers 1.5-1.8mm. Spikelets falling in clusters of 7, a single fertile spikelet encircled by 6 reduced sterile spikelets.................................*Awned Canary-grass* **Phalaris paradoxa**

Lvs <8mm wide

Sheath margins hairless. Lvs with 8-16 indistinct v low flat ribs above in TS ....*Common Couch* **Elytrigia repens***

Sheath margins ciliate on 1-side (soon rubs off)

Lvs with 13-16 distinct broad flat ribs above in TS, 2-6mm wide, glaucous at least above, acute, rigid, snapping audibly, often rough above, smooth below, usu with scabrid margins. Ligule 0.3mm, truncate, ciliolate. Auricles acute, occ v small. Sheaths occ purplish at base..*Sea Couch* **Elytrigia atherica***

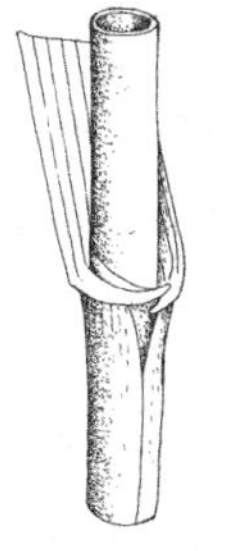

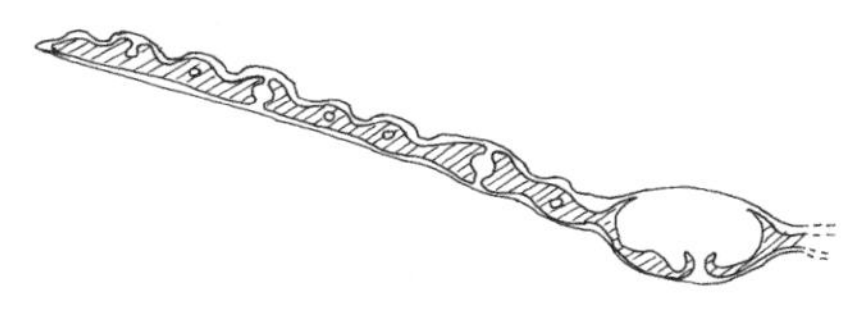

Festuca gigantea ligule & lf TS

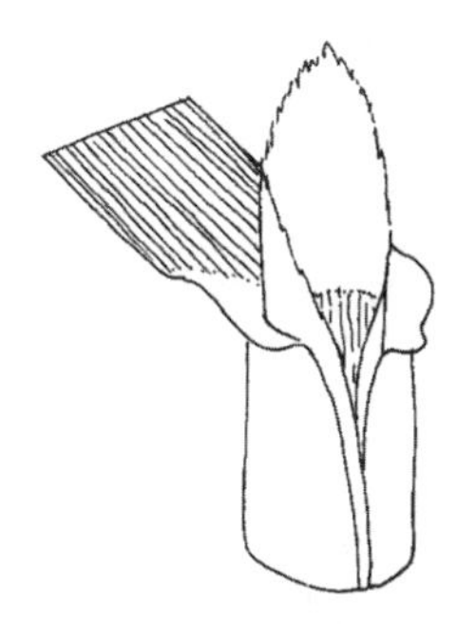

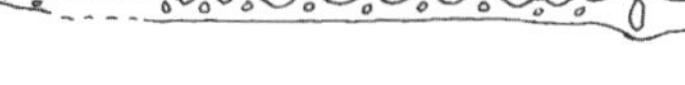

Phalaris arundinacea ligule & lf TS

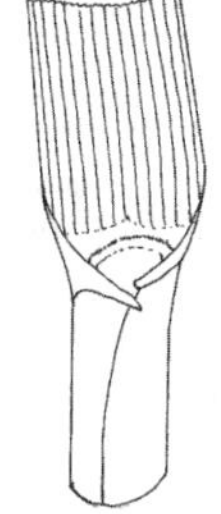

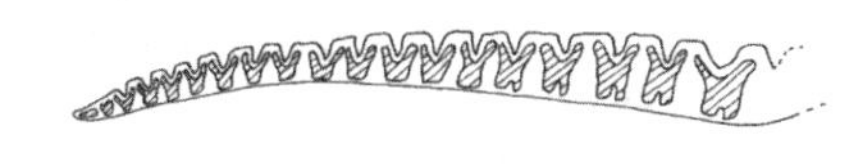

Leymus arenarius ligule & lf TS

Group HO – *Lvs hairy (rarely hairless). Sheaths densely hairy*

■Lvs strongly narrowed to base (widest nr middle)

Rhizomatous. Lvs with low flat ribs above in TS, 3-8mm wide, stiff, strongly and neatly inrolled on drying, yellow-green, rough above, sparsely hairy above, sparsely scabrid-ciliate, with 10(15) veins each side of midrib (4 main veins each side of midrib, with 3-4 minor veins between), with stomata usu above only. Ligule 1-2(6)mm, truncate to obtuse, ciliolate (occ long-ciliate), shortly hairy. Sheaths open, hairless or retrorsely hairy. Cataphylls tough, white, v ciliate. Calc gsld ....*Tor-grass* **Brachypodium pinnatum**

Tufted. Lvs not ribbed in TS, 3-9(13)mm wide, limp, twisted, drooping, inrolling on drying, dark yellow-green, often rough esp below, sparsely hairy below, ciliate, with c20 veins each side of midrib (3-4 main veins each side of midrib, with 5-6 minor veins between), with stomata both sides or above only. Ligule 1-2(5)mm, jagged, brown, long-ciliate (long and short hairs), hairy. Sheaths closed below, weakly retrorsely or patent-hairy. Cataphylls absent or obscure. Often shady habs..*False Brome* **Brachypodium sylvaticum**

■Lvs not strongly narrowed to base

Ligule <1mm, truncate

Lvs ribbed above in TS, often stripy HTL, without translucent veins

Lvs glaucous, 1-3mm wide, often inrolled, usu densely short-hairy above, scabrid-cilia longer nr ligule, with 3 deep ribs each side of midrib. Ligule to 0.5mm, fimbriate toothed, v minutely ciliolate. Sheaths minutely hairy (upper occ hairless), the lower closed (with concave apex) and ladder-fibrillose. Shoots ± bulbous and fibrous at base. Per. Limestone gsld, Somerset ....*Somerset Hair-grass* **Koeleria vallesiana**

Lvs green or glaucous, 1-4mm wide, often inrolled, densely short-hairy above, ciliate, with 3 deep ribs each side of midrib. Ligule to 1mm, erose, ciliolate. Sheaths densely short-hairy, the lower closed. Shoots not bulbous or fibrous at base. Per. Calc gsld ....*Crested Hair-grass* **Koeleria macrantha***

Lvs green, 2-4mm wide, usu flat, long-hairy esp below, long-ciliate to 1.5mm, with >3 ribs each side of midrib, with stomata above only or sparse below. Ligule 0.3(0.7)mm, minutely toothed, not ciliolate. Sheaths closed, with dense hairs often interwoven. Shoots not bulbous or fibrous at base. Ann or weak per. Calc or clay gsld, R, S Eng ....*French Oat-grass* **Gaudinia fragilis**

Lvs not ribbed above, not stripy HTL, usu with translucent veins

Per

Lvs 1.5-4(7)mm wide, yellow-green both sides, hairy above, sparsely hairy or hairless below, slightly rough above, with stomata both sides. Ligule to 1mm, fimbriate-ciliate. Sheaths with retrorse hairs or hairless, often ciliate along one margin, open. Gsld ....*Yellow Oat-grass* **Trisetum flavescens**

Lvs 2-5mm wide, green, with long or short hairs above or hairless, ± rough or smooth below, with stomata sparse or absent below. Ligule 0.5mm, ciliolate, hairless. Sheaths with dense short (<0.5mm) hairs below (the lowest closed, with 1mm unequal patent hairs), the mid and upper sheaths hairless, open and not inflated. Gsld, esp on clay ....*Meadow Barley* **Hordeum secalinum**

Ann

Lvs densely velvety short-hairy both sides, light green, 2-12mm wide, not wrinkled, with stomata both sides. Ligule usu <1mm, minutely toothed or erose, densely hairy. Sheaths densely long-hairy, inflated above. Shoots round. Culms with shortly hairy nodes. Oct-Jul. Usu sand dunes..*Hare's-tail* **Lagurus ovatus**

Lvs sparsely long-hairy at least above, green to purplish (esp nr ligule), 3-13mm wide, occ wrinkled, with stomata both sides. Ligule 1mm, minutely jagged, hairless. Sheaths rough with swollen-based patent hairs to 3mm, not inflated above. Shoots flattened. Culms with long-hairy nodes. Jun-Nov. Ruderal......................*Hairy Finger-grass* **Digitaria sanguinalis**

Ligule >1mm, often toothed. Lvs not or hardly ribbed above in TS

Ligule hairy

Per

Ligule with short hairs, 1.5-2(3)mm, truncate (to obtuse), finely toothed to jagged. Lvs pale green, 2-8mm wide, softly hairy at least above, often twisted, often shortly ciliate, not or hardly ribbed above, with stomata both sides. Sheaths rarely with purplish veins (occ purplish at base), long-hairy or v shortly retrorsely hairy, the lower closed, the upper open above and ± inflated. Culm nodes long-hairy. Rhizomatous, vegetative shoots also rooting at lowest nodes..*Creeping Soft-grass* **Holcus mollis**

Ligule with long and short hairs, (1)2-3(4)mm, truncate, finely toothed to jagged. Lvs greyish-green, 6-10mm wide, softly hairy at least above, twisted, ciliate, not or hardly ribbed above, with stomata both sides. Sheaths with purplish veins, with slightly retrorse (or spreading) hairs to 1mm, the lower closed, the upper overlapping and inflated. Culm nodes as hairy as rest of sheath or culm. Tufted...............................*Yorkshire-fog* **Holcus lanatus**

Ann

Lower sheaths with long patent hairs only...(*Avena*) Go to **AVENA**

Lower sheaths with short retrorse hairs only

Ligule to 1mm, hairy. Lvs 2-5mm wide, green, softly hairy. Sheaths soon splitting, the upper sparsely hairy to hairless..................................*Slender Soft-brome* **Bromus lepidus**

Lower sheaths with sparse long (2mm) patent hairs and dense short (0.3mm) retrorse hairs

Ligule truncate, minutely toothed. Lvs densely long-hairy (to 2mm) both sides, ciliate, 4-8mm wide, greyish-green, with midrib, with stomata both sides. Lower sheaths often purplish. Upper sheaths (and culm) shortly hairy, with long hairs often absent

Ligule 0.5-2.5mm, long-hairy, ciliate. Lvs with midrib often translucent and not raised below...*Soft-brome* **Bromus hordeaceus**

Ligule 0.5-1mm, sparsely long-hairy, not ciliate. Lvs with midrib hardly visible HTL and not raised below. VR...*Interrupted Brome* **Bromus interruptus**

Ligule hairless. Ann

Sheaths with long (1-1.5mm) ± patent hairs and often with short retrorse hairs

Lowest sheaths with reddish veins. Ligule 2-2.5(5)mm, toothed to jagged. Lvs 6-11mm wide, widest nr base, softly sparsely hairy with long hairs both sides, with midrib not v translucent but raised below..*Great Brome* **Anisantha diandra**

Lowest sheaths without reddish veins. Ligule 1.5mm (length ≥ width), shortly toothed. Lvs 6-11mm wide, widest nr base, with sparse short hairs below. Culms with v minute retrorse hairs and longer unequal (TNE) 1.5mm hairs. Culm nodes with longer hairs ....*Smooth Brome* **Bromus racemosus**

Sheaths with short (<0.3mm) retrorse ± equal hairs (mid and upper sheaths usu hairless)

Lowest sheaths with purplish veins

Ligule 1.5-4mm, jagged. Lvs 2-7mm wide, softly hairy both sides (hairs long and short above but minute below), ± twisted, with 1-3 main translucent veins each side of midrib, with indistinct cross-veins, with stomata both sides. Sheaths with 0.3mm hairs (equal TNE) or with sparse long hairs, soon splitting...........................*Barren Brome* **Anisantha sterilis**

Ligule 1.5-4mm, truncate to jagged. Lvs with minute (≤0.2mm) hairs both sides. Sheaths with minute (0.2mm) retrorse hairs. R alien..........*Compact Brome* **Anisantha madritensis**

Lowest sheaths without purplish veins. R

Ligule 3-5mm, jagged. Lvs 2-4mm wide, softly hairy. Upper sheaths hairless. Infl drooping ....*Drooping Brome* **Anisantha tectorum**

Ligule 2-4mm, toothed. Lvs 2-6mm wide, with long hairs. Infl erect ....*Field Brome* **Bromus arvensis**

Ligule 1.5-3mm, shortly toothed. Lvs 2-6mm wide, with short hairs. Culms smooth to slightly rough, middle nodes with some longer hairs (like rest of culm). Infl erect ....*Meadow Brome* **Bromus commutatus**

Group HP – *Lvs hairy (occ sparsely so), not ribbed above. Sheaths closed (at least the lower), hairless or with sparse scattered hairs. (*Elytrigia juncea *may key out here as the lowest sheath is closed).* ❶ *Lower sheaths purplish.* ❷ *Ligule margins wrap around to make a 1-4mm bristle-like anti-ligule (visible as strong vein when young)*

■Shoots 4-angled. Lvs with raised veins below. ❶

Ligule 0.5-1(2)mm, entire, hairless. Lvs 3-6mm wide, narrowed to base, bright green, usu with long scattered hairs above, with stomata above only. Sheaths rough, hairless or with scattered retrorse hairs. ❷. Shady habs..*Wood Melick* **Melica uniflora***

Ligule 0.3-0.5mm, (±) entire, hairless. Lvs 2-6mm wide, slightly narrowed nr base, mid-green, sparsely hairy or hairless above, with stomata above only. Sheaths rough, usu hairless but often hairy nr ligule. Limestone habs...*Mountain Melick* **Melica nutans**

■Shoots round or slightly flattened. Lvs without raised veins below (exc midrib)

All sheaths soon splitting. Tufted per

Shady habs. Lvs 5-15mm wide, widest nr middle, occ wrinkled, dull whitish- or grey-green above, shiny green below, hairless, never ciliate, smooth or only slightly rough above, with stomata above only (occ sparse below nr veins). Ligule 0.5-4(10)mm obtuse, entire to shredded, hairless, occ pruinose. Sheaths hairless. ❶.................*Wood Millet* **Milium effusum***

Arable. Lvs 4-10mm wide, widest nr base, never wrinkled, green both sides, with long hairs above (occ sparse), shortly hairy below, occ ciliate, rough above, with stomata both sides. Ligule 1mm, truncate, hairless. Sheaths ± hairless. Culms hairless, with minutely hairy nodes ....*Rye Brome* **Bromus secalinus**

Lower sheaths remaining closed

Rhizomatous per

Lvs 5-10mm wide, occ hairless, with minute hair-tuft at junction with sheath, (±) smooth above, often rough below (exc midrib), with 5-6 main veins (occ unequal) with 1-3 minor veins between. Ligule 0.5-1mm, minutely fimbriate, hairless. Auricles small to absent. Sheaths hairless or with short retrorse hairs, with weak cross-veins. Culms occ with some hairs nr nodes ....*Hungarian Brome* **Bromopsis inermis**

Tufted short-lived per

Lower sheaths ± retrorsely hairy or hairless, ciliate, ± smooth, with cross-veins weak to absent, the upper hairless and open. Lvs 5-14mm wide, widest nr base, rough at least above, occ fleshy, usu dark green, with sparse long hairs esp above, with translucent midrib raised below, with c10 veins each side of midrib, with stomata more abundant above. Ligule (1)2-4mm, long-hairy to hairless, often green or scarious. Auricles small or absent ....*Rescue Brome* **Ceratochloa cathartica**

All sheaths hairless but may be ciliate and hairy at apex, ± smooth, with cross-veins weak to absent. Lvs 6-14mm wide, widest nr base, occ rough below, occ fleshy, usu dark green, ± hairless to hairless, often ciliate (cilia to 1mm), with opaque midrib raised below, with c15 veins each side of midrib, with stomata more abundant above. Ligule 1-3.5mm, long-hairy to minutely hairy or hairless, often green or scarious. Auricles small or absent ....*California Brome* **Ceratochloa carinata**

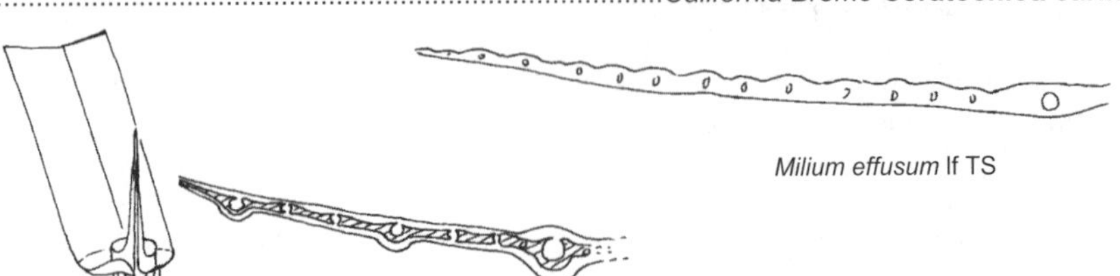

Milium effusum lf TS

Melica uniflora ligule & lf TS

Group HQ – *Lvs hairy (occ sparsely so). Sheaths open, hairless or with sparse scattered hairs. (*Holcus mollis *may key out here)*

■Lvs strongly narrowed to base. Wet habs

Strongly rhizomatous, forming extensive patches. Culms often >1m, branched

Culms 3mm diam, stout. Lvs 30-60cm x (3)5-7mm, greyish- or bluish-green with long hairs above, ± shiny hairless green below, rough esp above, usu inrolling on drying, weakly ribbed above (and often below), without midrib raised below, with stomata both sides. Ligule 2mm (upper to 10mm), ± truncate, torn, hairy, turning brown. Sheaths hairless, smooth, occ pinkish below. Mar-Oct (all yr). R............................*Scandinavian Small-reed* **Calamagrostis purpurea**

Culms 1.5mm diam, slender. Lvs 20-40cm x 2.5-6mm, greyish with scattered long hairs above, ± shiny hairless green below, rough both sides, inrolling on drying, not or weakly ribbed above, slightly ribbed below, with midrib slightly raised below, with stomata above only. Ligule 1(5)mm, obtuse, split or entire, ciliolate, hairy, turning brown. Sheaths hairless, smooth, occ purplish below. Mar-Oct. Fens..*Purple Small-reed* **Calamagrostis canescens***

Shortly rhizomatous, forming compact tufts. Culms to 1m, not branched

Lvs 2-5mm wide, shiny green below, slightly rough above, smooth below, ribbed above in TS, not ribbed below, with 5 veins each side of midrib (1 main vein each side, slightly thicker than midrib), with stomata sparse or absent below. Ligule 1(3)mm, ± truncate, jagged, ciliolate, densely hairy. Sheaths hairless to densely minutely hairy, smooth to minutely rough, occ pinkish below. Mar-Oct (all yr)

Lvs with long hairs above (occ hairless). VR, Scot *Scottish Small-reed* **Calamagrostis scotica**

Lvs usu with short hairs above. R............................*Narrow Small-reed* **Calamagrostis stricta**

■Lvs slightly or not narrowed to base. Dry or saline habs

Ligule <1mm, truncate

Rhizomatous per (shoots arising singly from long rhizomes). Lvs with 6-13 broad ribs above

Lvs 2-7(10)mm wide, sharply acute, stiff, twisted, inrolling tightly on drying, grey-green or glaucous both sides, densely minutely antrorsely hairy above, with smooth margins, with stomata above only. Ligule 0.5mm, fimbriate, hairless. Sheaths purplish at base, the lowest closed. Auricles ± absent. Young dunes..................................*Sand Couch* **Elytrigia juncea***

Tufted per. Lvs not or hardly ribbed above

Lvs (1)1.5-4(7)mm wide, yellow-green both sides, hairy above, sparsely hairy or hairless below, slightly rough above, with 1-3 main veins each side of midrib, with stomata both sides. Ligule to 1mm, fimbriate-ciliate. Sheaths with retrorse hairs or hairless, often ciliate along one margin. Gsld..*Yellow Oat-grass* **Trisetum flavescens**

Ann

Ruderal. Lvs with 10 medium obtuse ribs above

Lvs 1-2.5mm wide, long- or short-hairy above, hairless below, without midrib raised below, stripy HTL, with stomata v sparse below. Ligule 0.5-1mm, deeply retuse, with auricles ....*Squirreltail Fescue* **Vulpia bromoides***

Ruderal. Lvs not or hardly ribbed above

Lvs to 10(15)cm x 1-8mm, widest nr base, pale green, sparsely hairy above only, with c12 unequal irreg veins (3-5 more translucent), without discernible midrib, with stomata both sides. Ligule c1mm, deeply fimbriate, hairless. Lowest sheaths sparsely hairy, densely ciliate along one margin. VR alien....................*Mediterranean Hair-grass* **Rostraria cristata**

Saline habs. Lvs strongly ribbed above

Anthers 1.5-4mm. Damp habs..*Hard-grass* **Parapholis strigosa**

Anthers 0.5-1mm. Dry habs.....................................*Curved Hard-grass* **Parapholis incurva**

Ligule >1mm, obtuse to acute

Rhizomatous per

Sand dunes. Lvs to 90cm x 6mm, widest nr base, odourless, strongly inrolled, sharply acute, stiff, glaucous, densely minutely white-hairy above, smooth below, with smooth margins, with 6-8 unequal ribs above, with stomata above only. Auricles absent. Ligule 10-16(30)mm, acute, hairless. Sheaths hairless, smooth, purplish at base, the upper ± inflated. Occ rooting at lower nodes...*Marram* **Ammophila arenaria***

Wetlands. Lvs to 40cm x (4)5-6(10)mm, widest nr middle, hay-scented, flat (but inrolling on drying), twisted, with long fine apex, stiff, green above, shiny bright green below, sparsely minutely hairy or hairless above, rough above, with v scabrid margins, with c18 v low indistinct ribs above, with sparse stomata below. Auricles rarely present, obscure. Ligule 1-3mm (±) acute, often torn, minutely hairy. Sheaths hairless, ± smooth, pinkish nr base. R, N Br, Ire...*Holy-grass* **Hierochloe odorata**

Tufted per

Lvs with long whiskers nr ligule, 1-6mm wide, not twisted, usu sparsely hairy, usu ciliate, (±) shiny yellow-green below, dull above, hay-scented esp when dried, with v bitter taste, with stomata sparse or absent below, not ribbed above. Ligule to 4mm, ± truncate, jagged, fimbriate, hairless. Sheaths smooth, hairy, the lowest hairless ....*Sweet Vernal-grass* **Anthoxanthum odoratum***

Lvs without whiskers nr ligule, 4-8mm wide, twisted, usu with sparse long hairs above, rarely ciliate, often with retrorsely scabrid margins nr ligule, dull greyish-green both sides, odourless, tasteless (but culms with bitter taste), not or hardly ribbed above, with stomata both sides or above only. Ligule 1-2(3)mm, obtuse, minutely toothed to jagged, ciliolate, minutely retrorsely hairy, occ greenish. Sheaths hairless or with sparse long hairs, slightly keeled, the lowest closed and purplish at base

Basal internodes swollen into 2(3) globose corms (often at or below ground-level) ....*Onion Couch* **Arrhenatherum elatius** var **bulbosum**

Basal internodes not swollen................*False Oat-grass* **Arrhenatherum elatius** var **elatius***

Ann

Lvs 3-7cm wide, not or rarely twisted...*Maize* **Zea mays**

Lvs to 3cm wide, twisted...(*Avena*) Go to **AVENA**

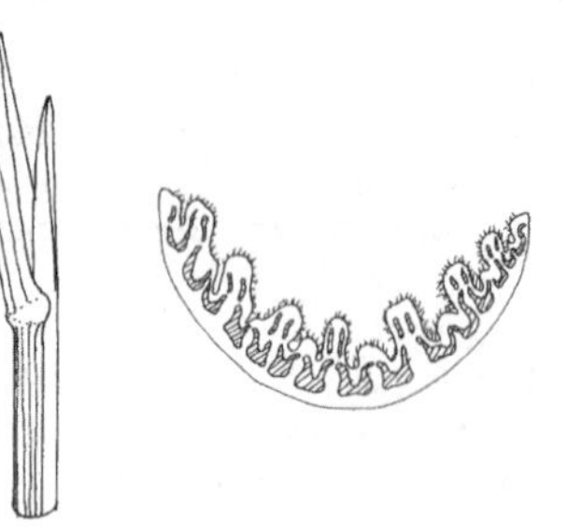

Ammophila arenaria ligule & lf TS

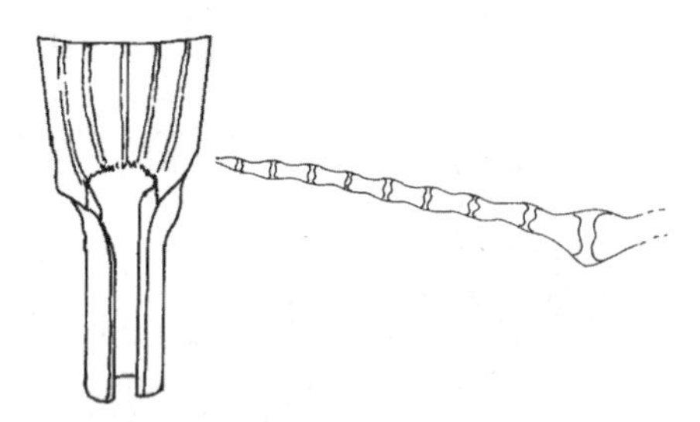

Arrhenatherum elatius var *elatius* ligule & lf TS

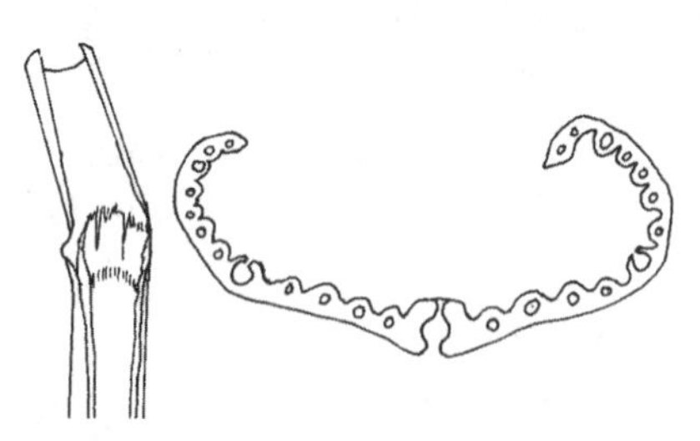

Calamagrostis canescens ligule & lf TS

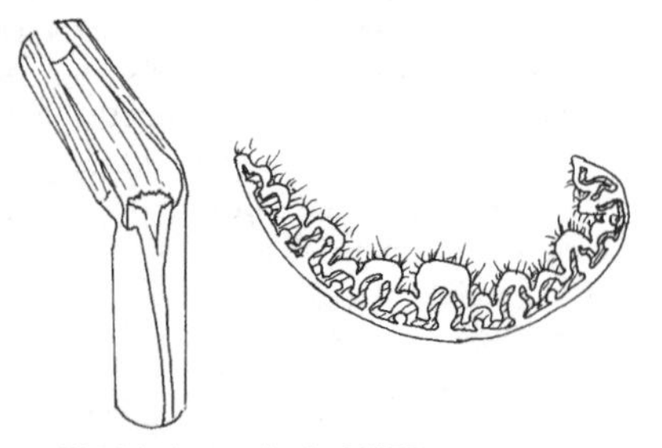

Elytrigia juncea ligule & lf TS

Group HR – *Lvs hairless. Ligule (±) truncate (collar-like), usu <1.5mm, entire or toothed, always shorter than wide. (*Cynosurus cristatus *may key out here)*

■Ann

Arable. All sheaths open. Lvs not (or weakly) ribbed above in TS

Lvs 25-100cm x 30-70mm, widest nr middle, with scabrid-ciliate margins esp distally, inrolling on drying, smooth both sides, yellow-green, hairless to hairy above (the hairs with minutely swollen bases), with midrib raised below, with stomata both sides. Ligule 2-3mm, minutely fimbriate or long-ciliate, hairless. Sheaths often with long hairs nr lf, the basal purplish. Shoots single, erect, slightly oval. Crop relic...*Maize* **Zea mays**

Lvs to 12cm x 2-7mm, often purplish (esp nr ligule), with scabrid margins. Ligule 0.5-2mm. Shoots tufted at base, ± prostrate, flattened.............*Smooth Finger-grass* **Digitaria ischaemum**

Coastal. Lowest sheaths closed, mid and upper open. Lvs ribbed above in TS

Lvs >2mm wide...*Sand Cat's-tail* **Phleum arenarium**

Lvs <2mm wide

Lvs with stomata above only

Lvs 1-5(10) x 1.5-2mm, occ rolled or folded, rigid, shiny dark green below, turning reddish, with opaque veins (stripy HTL), with wide tramlines, often with raised silica ridges on ribs above. Ligule 0.3-1mm, obtuse to truncate, fimbriate or jagged. Sheaths often reddish at base...*Sea Fern-grass* **Catapodium marinum***

Lvs with stomata both sides

Anthers 1.5-4mm. Lvs 1.5mm wide, dark green, minutely rough above, with 3-4 ribs each side of midrib (stripy HTL). Ligule 0.3-1mm, fimbriate to weakly jagged. Sheaths not inflated above, the lower hairy on keel. Stems often >10cm, erect or ascending, usu branched below, ± straight. Damp saline habs...................................*Hard-grass* **Parapholis strigosa**

Anthers 0.5-1mm. Lvs 1mm wide, dark green, minutely rough above, with 3-4 ribs each side of midrib (stripy HTL). Sheaths inflated above, the lower hairless. Ligule 0.3-1mm, fimbriate to weakly jagged. Stems usu <10cm, prostrate to ± ascending, usu branched below, curved. Stony saline habs...*Curved Hard-grass* **Parapholis incurva**

■Per

Mtns

Ligule 1-1.5(3)mm, occ obtuse, not ciliolate, minutely hairy. Lvs 6-17cm x (3)5-7mm, flat or inrolled, green or glaucous, v rough above, smooth below, with weakly translucent midrib not or hardly raised below, with 6 deep flat ribs each side of midrib, with stomata both sides. Sheaths smooth, the basal fibrous and occ reddish, the upper inflated ....*Alpine Foxtail* **Alopecurus borealis**

Ligule 0.5(2)mm, occ obtuse, minutely ciliolate (x20), hairless. Lvs 4-6(15)cm x 5(6)mm, flat or inrolled, slightly grey-green, smooth both sides, with scabrid margins (occ retrorse nr ligule), long-ciliate at extreme base (x20), with weakly translucent midrib not or hardly raised below, with c8 v low obtuse ribs each side of midrib. Sheaths smooth, the basal fibrous and dark brown, the upper inflated..*Alpine Cat's-tail* **Phleum alpinum**

Lowlands, may climb mtns

Sheaths open (even lowest)

Rhizomatous per. Calc gsld..*Tor-grass* **Brachypodium pinnatum**

Tufted per

Lvs not ribbed above in TS

Ligule 0.3-1(2)mm, minutely fimbriate or ciliolate, hairless. Lvs usu <10cm x (1)2.5-4.5mm, with scabrid margins (occ slightly retrorse), often rough above, greyish green, unribbed, with midrib not raised below, with stomata both sides. Sheaths open, smooth, the lower purplish, the upper slightly inflated. Culms often purplish. Densely tufted, occ fibrous at base...*Purple-stem Cat's-tail* **Phleum phleoides**

Lvs deeply ribbed above in TS

Shoots usu bulbous at base

Ligule 0.5-1mm (2-5mm on culm), truncate to ± acute, toothed, usu ciliolate, hairless, soon brownish. Lvs 10-45cm x 2-4(10)mm, with retrorsely scabrid margins at least nr ligule, smooth or slightly rough both sides, pale or grey-green, occ sparse hairs nr base, with translucent midrib not or weakly raised below, with stomata both sides. Sheaths brown and ± fibrous at base, the upper ± inflated. Infl spike 6-15cm ....*Timothy* **Phleum pratense***

Ligule (1)1.5(4)mm, ± acute to obtuse, toothed, usu ciliolate, hairless, soon brownish. Lvs 3-5(12)cm x (2)3(5)mm, with retrorsely scabrid margins at least nr ligule, smooth or slightly rough both sides, pale or grey-green, occ sparse hairs nr base, with translucent midrib not or weakly raised below, with stomata both sides. Sheaths brown (occ purplish) and ± fibrous at base, the upper ± inflated. Infl spike 1-6(8)cm ....*Smaller Cat's-tail* **Phleum bertolonii**

Shoots not bulbous at base

Ruderal. Tufted, to 1m, with long false-stems and whitish cataphylls like *Nardus stricta*

Lvs to 35cm x 4.5mm, widest nr base, inrolling on drying, tough, with v scabrid margins, ± smooth above, occ rough below, dull greyish-green above, shiny dark green below, often minutely hairy nr ligule above, without any veins raised below, with c20 deep narrow flat-topped ribs above, with translucent veins (midrib not discernible), with stomata above only. Ligule 0.5(1)mm, with auricles, occ ciliate. Sheaths hairless but one margin densely ciliate, smooth. VR alien ....*Pheasant's-tail Grass* **Anemanthele lessoniana***

Wet habs. Rhizomatous, to 1m

Lvs 5-10mm wide, with sharply scabrid margins (retrorse nr ligule), yellowish-green, rough to smooth both sides, with many dark-bordered veins. Ligule <1mm, ± jagged, with teeth ± awned, hairless, with slight auricles. Sheaths open, with indistinct cross-veins, the lower smooth, the upper rough (or minutely hairy). Culm nodes swollen, the lower ± hairless, the upper long-hairy. Sch8......................*Cut-grass* **Leersia oryzoides**

Gsld

Rhizomatous. Lvs 2-2.5mm wide, smooth to obscurely rough above. Ligule 0.5-2mm ....*Common Bent* **Agrostis capillaris**

Tufted. Lvs (1)1.5-4(7)mm wide, slightly rough above, yellow-green both sides, usu hairy at least above, with 1-3 main veins each side of midrib, with stomata both sides. Ligule to 1mm, fimbriate-ciliate. Sheaths with retrorse hairs or hairless, often ciliate along one margin..*Yellow Oat-grass* **Trisetum flavescens**

Sheaths closed at least below

Lvs ribbed above in TS

Lvs with 3 deep acute ribs each side of midrib in TS. Dry calc gsld ....*Crested Hair-grass* **Koeleria macrantha***

Lvs with 6 shallow flat ribs each side of midrib in TS. Usu damp gsld

Lvs 2.5-8mm wide, widest nr base, rough above, ± smooth below, shiny dark green below, with midrib hardly discernible (stripy HTL) and not raised below, with stomata both sides or above only. Ligule 0.5-1mm (1-3mm on culm), brownish, occ ± acute, stiff, minute retrorsely hairy, ciliolate, often ± with auricles. Basal sheaths closed, soon splitting, dark brown (rarely purplish). Upper sheaths inflated. Culms ± geniculate at lowest node(s) ....*Meadow Foxtail* **Alopecurus pratensis**

Lvs not (or weakly) ribbed above in TS

Shady habs (occ mtns in *P. nemoralis*). Tufted per

Lvs 1-3mm wide, widest nr base, never wrinkled, shiny dark green below, occ rough above, with smooth margins (retrorsely scabrid when young), with tramlines visible only nr apex, with midrib v weakly raised below, with stomata above only (occ v sparse below). Ligule <0.6mm, minutely ciliolate, obscurely hairy. Sheaths rarely pruinose or reddish at base ....*Wood Meadow-grass* **Poa nemoralis**

Lvs 5-15mm wide, widest nr middle, occ wrinkled, dull whitish- or grey-green above, shiny green below, smooth or slightly rough above, with scabrid margins, without tramlines, with midrib raised below, with stomata above only (occ sparse below nr veins). Ligule 0.5-4(10)mm, obtuse, entire to shredded, hairless, occ pruinose. Sheaths purplish at base ....*Wood Millet* **Milium effusum***

Dry gsld. Tufted

Lvs 5-15cm x (1)2-4mm, acute, often held 90º to sheath, usu folded when young, shiny green below, smooth both sides, with smooth margins (exc nr apex), stripy HTL, with midrib slightly raised below, with 2-4 v low ribs each side of midrib, with stomata above only (both sides on culm lvs). Ligule 0.5-0.8mm (to 1.5mm on culm), denticulate, often minutely fimbriate, often brownish, hairless to minutely hairy. Sheaths smooth, the basal closed, yellowish, the upper open.........................*Crested Dog's-tail* **Cynosurus cristatus***

Lvs 2.5-8(15)cm x 2-5(10)mm, slightly hooded or rolled at acute apex, held 90° to sheath (± flat on ground, v strongly 2-ranked), always rolled when young, pale green, smooth both sides, with scabrid margins often retrorse nr ligule, stripy (HTL), occ with v low ribs, with midrib hardly or not raised below, usu with stomata both sides. Ligule 0.5(1.5)mm, entire (but wavy and v slightly retuse), hairless. Sheaths usu smooth, the basal closed, white to dark reddish-brown and fibrous, the upper open......................*Quaking-grass* **Briza media***

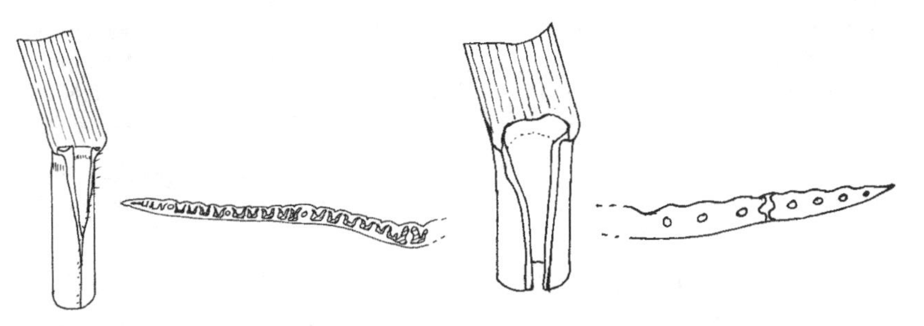

Anemanthele lessoniana ligule & lf TS

Briza media ligule & lf TS

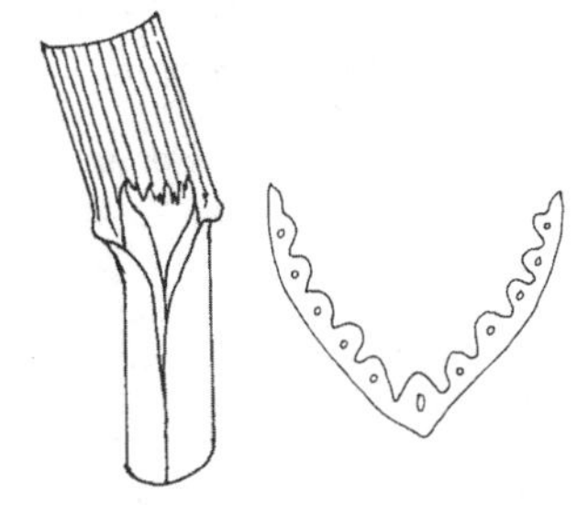

Catapodium marinum ligule & lf TS

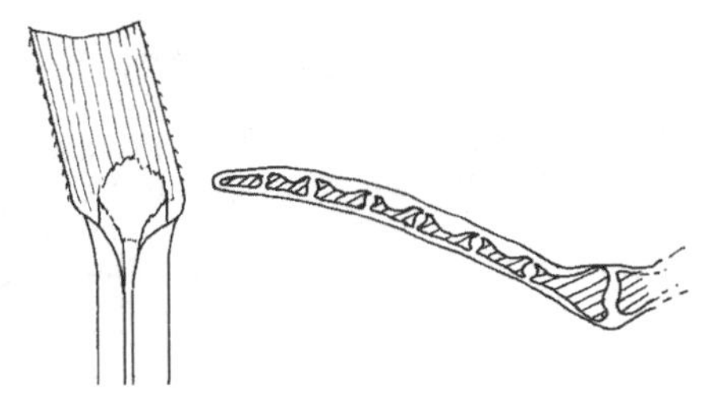

Phleum pratense ligule & lf TS

Group HS – *Lvs hairless, ribbed in TS. Ligule >1.5mm, obtuse to acute, usu longer than wide, minutely retrorsely hairy.* ❶ *Hay-scented.* ❷ *Lvs channelled when young.* ❸ *Sheaths and lvs often pruinose*

▪Rhizomatous per (occ shortly stoloniferous in *Agrostis*)

Lvs strongly narrowed to base, with midrib raised below

Calc gsld. Lvs with low flat ribs above, 3-8mm wide, odourless, stiff, not twisted, strongly and neatly inrolled on drying, yellow-green, rough above only, with sparsely scabrid-ciliate margins, with stomata usu above only. Ligule 1-2(6)mm, ± truncate or obtuse, shortly hairy, ciliolate, occ fringed with longer hairs. Sheaths hairless or with retrorse hairs, not pinkish. Cataphylls tough, white, strongly ciliate..*Tor-grass* **Brachypodium pinnatum**

Wetlands. Lvs with v low indistinct ribs above, (4)5-6(10)mm wide, hay-scented, stiff, twisted, inrolling on drying, mid-green above, shiny bright green below, rough above, often rough on raised midrib below, with v scabrid margins, with sparse stomata below. Ligule 1-3mm, (±) acute, often torn, minutely hairy. Sheaths hairless, pinkish nr base. Cataphylls absent. ❶. R, N Br, Ire...*Holy-grass* **Hierochloe odorata**

Lvs not or hardly narrowed to base, with midrib not or hardly raised below, with 10 ± low ribs above in TS, usu with opaque veins, with stomata both sides

Ligule shorter than wide. Rhizomes or stolons to 5cm............*Common Bent* **Agrostis capillaris**

Ligule longer than wide

Lvs U-channelled or weakly folded ❷

Lvs 0.6-3mm wide, occ bristle-like, greyish-green, rough at least above, stripy HTL, with translucent midrib. Ligule 0.7-2.5(5)mm, acute (rarely obtuse), toothed. Sheaths smooth. Infl contracted in fr. Densely tufted per with rhizomes to 10cm. Hths, mtns ....*Brown Bent* **Agrostis vinealis***

Lvs flat or rolled

Terminal spikelet with hairless lemma (occ with hairs at base). Rhizomes to 25cm. Lvs v rough both sides, 2-8mm wide, dull green. Ligule 1.5-2(6)mm, truncate, jagged, turning brown. Sheaths antrorsely rough, often smooth above............*Black Bent* **Agrostis gigantea**

Terminal spikelet with hairy lemma (usu with longer hairs at base). Rhizomes to 10(40)cm. Lvs rough above, 1.5(4.5)mm wide, bluish-green. Ligule 1-3mm, truncate to obtuse, ciliolate. Sheaths usu smooth. Lowest panicle branches bare of spikelets ....*Highland Bent* **Agrostis castellana**

▪Stoloniferous per or ann, rooting at least the lowest nodes. Lvs widest nr base, without midrib raised below, with stomata both sides

Ligule fimbriate and ciliolate

Lvs (0.5)2.5-5mm wide, often grey-green, rough above, with c8 low ribs each side of midrib. Ligule 1-2(4)mm, truncate. Sheaths smooth. Stolons <5cm. Plant resembling *Agrostis capillaris*. Ruderal...*Water Bent* **Polypogon viridis***

Ligule entire or ± so (occ torn)

Lvs U-channelled ❷

Lvs 1(3)mm wide, bright grey-green, with 3-9 ribs above. Ligule (1)1.5-5mm, acute to obtuse, ± entire, occ purplish. Sheaths usu smooth. Infl contracted in fr. Stolons often with several shoots at a node giving a knotted (spider-plant) appearance. Damp gsld ....*Velvet Bent* **Agrostis canina***

Lvs flat or rolled

All sheaths open, not pruinose or inflated

Lvs smooth or rough above, smooth below, (0.5)3-5.5mm wide, occ stripy, with 5-6 low ribs each side of midrib, with midrib (HTL) v narrowly translucent or indistinct. Ligule (1)1.5-2.5(7)mm, acute (to truncate), entire or erose, occ purplish. Sheaths not pruinose, smooth or rough. Culms often purplish at base or at nodes. Stolons 5-100cm ....*Creeping Bent* **Agrostis stolonifera***

Lowest sheaths closed, the upper pruinose and inflated. ❸

Per. Lvs usu with 5-8 deep acute ribs above each side of midrib, rough above, below, 3-4mm wide, stripy (HTL), without translucent veins. Ligule 2-5mm, acute to often torn, with overlapping margins. Sheaths smooth. Culms strongly geniculate, rooti nodes, the nodes often purplish..............................*Marsh Foxtail* **Alopecurus geniculatu**

Ann (per). Lvs with 6-7 low (to deep) ribs above each side of midrib, rough above, smooth below, 2.5-5mm wide, stripy (HTL), without translucent veins. Ligule 1-4mm, acute (to truncate), often torn. Sheaths smooth. Culms geniculate, rooting and branching at nodes, often purplish at base..*Orange Foxtail* **Alopecurus aequalis**

■Tufted per. Lvs widest at base, without midrib raised below, with stomata both sides

Mtns >600m. Lvs to 17cm x (3)5-7mm, green or glaucous, with 6 deep flat ribs each side of midrib, v rough above, smooth below. Ligule 1-1.5(3)mm, truncate (occ obtuse). Sheaths smooth, the upper inflated. Shoots fibrous and occ reddish at base. Culms erect. ❸ ....*Alpine Foxtail* **Alopecurus borealis**

Grassy saltmarshes. Lvs to 12cm x 2-2.5mm, mid-green, with 3 deep ribs each side of midrib (esp when young), usu smooth above, smooth below. Ligule 1-3(5)mm, acute (occ torn), the lowest obtuse. Sheaths ± inflated above. Shoots fibrous at base, usu with two swollen basal internodes. Culms often geniculate..*Bulbous Foxtail* **Alopecurus bulbosus**

■Ann. Lvs widest at base, usu without midrib raised below, with stomata both sides

Saltmarshes (or ruderal). Ligule 3.5-8(10)mm, acute, jagged

Lvs 3-4(8)mm wide, with 10 medium ribs each side of midrib, minutely rough above. Sheaths smooth or minutely rough, ± inflated above......*Annual Beard-grass* **Polypogon monspeliensis**

Arable

Ligule 2-4(5)mm, obtuse to truncate, jagged, not wrapped around culm, with veins obscure. Lvs (2)3-5(8)mm wide, with ± deep ribs (occ acute) above, rough or slightly so above, smooth or with low ribs below. Auricles absent. Sheaths ± inflated above, hairless, the lower often slightly rough. Oct-Jul (all yr)..*Black-grass* **Alopecurus myosuroides**

Ligule 1-1.5(3)mm, acute to truncate, entire, often wrapped around culm, with veins clearly visible. Lvs 2-3(4)mm wide, with medium ribs above, usu rough above, smooth below. Auricles v reduced. Sheaths ± inflated above, hairless, smooth. Nov-Jun. R ....*Nit-grass* **Gastridium ventricosum**

Agrostis stolonifera ligule & lf TS

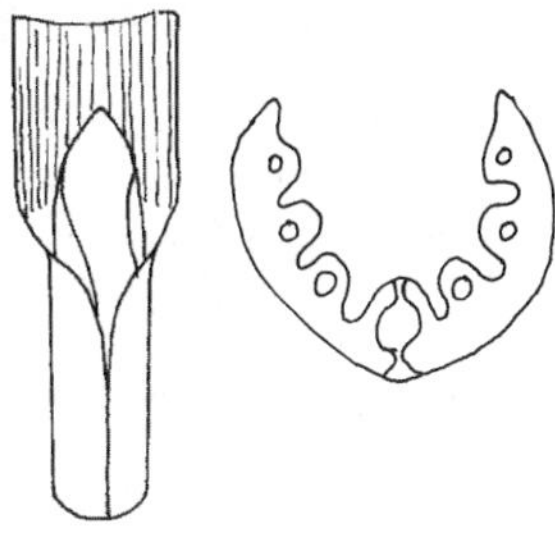

Agrostis vinealis ligule & lf TS

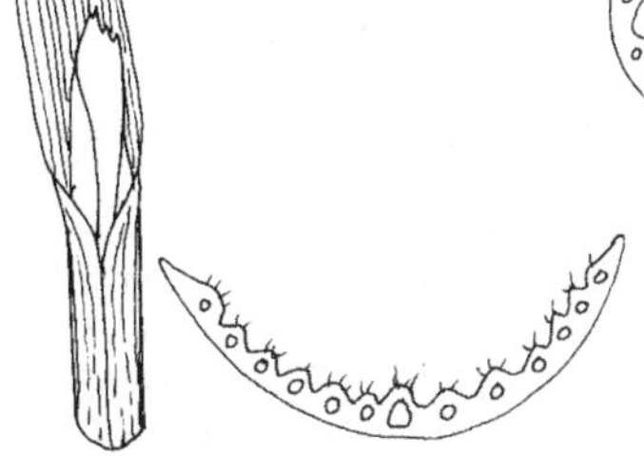

Alopecurus geniculatus ligule & lf TS

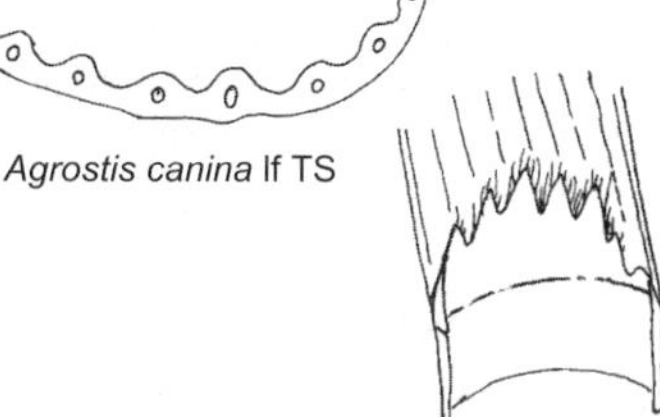

Agrostis canina lf TS

Polypogon viridis ligule & ligule margin

...s, not ribbed in TS. Ligule >1.5mm, obtuse to acute, always longer than
...cc present in Phalaris *and* Ceratochloa. *(Hairless forms of* Arrhenatherum

...t to dry habs

...ightly narrowed to base, slightly hooded and twisted at apex, ± limp, with ...rorse nr ligule), smooth both sides, mid-green, with midrib slightly raised below, without ribs, usu with cross-veins (esp nr base), with stomata both sides. Ligule 6-10mm, obtuse, slightly jagged, fimbriate, sparsely ciliolate, hairless to minutely hairy. Sheaths open, smooth, usu with cross-veins and hollows. Shoots slightly bulbous at base, often fibrous. Often rooting at lower nodes. Mar-Nov. Usu wet habs..........*Reed Canary-grass* **Phalaris arundinacea***

Lvs (4)7-8(10)mm wide, strongly narrowed to base, acute, stiff, with scabrid margins, ± smooth above, rough below, pale green above, ± shiny dark green below, rarely hairy, with v low unequal ribs above, with midrib hardly raised below, without cross-veins, with stomata both sides (often sparse below). Ligule 2-4(12)mm, obtuse to acute, torn, stiff, ciliolate, often brown, hairless to obscurely hairy. Sheaths open, occ slightly rough, without cross-veins or hollows. Shoots not bulbous at base. All yr. Often damp habs..................*Wood Small-reed* **Calamagrostis epigejos***

■Tufted per (rhizomes short or absent). Dry habs

Shady habs

Cataphylls usu large, tough, pale brown. Lvs 4-14mm wide, twisted, inrolling on drying, ± glaucous above, shiny bright green below, usu smooth both sides, with scabrid margins, with midrib raised below, with stomata above only (occ sparse below). Ligule 1-2.5(5)mm, obtuse, minutely toothed, fringed, hairless, never pruinose. Sheaths not purplish at base, smooth or occ rough. Usu humid wooded valleys, mostly W Br................*Wood Fescue* **Festuca altissima**

Cataphylls absent. Lvs 5-15mm wide, not twisted, occ wrinkled, dull whitish- or grey-green above, shiny green below, smooth both sides or slightly rough above, with scabrid margins, with midrib raised below, with stomata above only (occ sparse below nr veins). Ligule 0.5-4(10)mm, obtuse, entire to shredded, hairless, occ pruinose. Sheaths purplish at base ....*Wood Millet* **Milium effusum***

Gsld or ruderal

Lvs >5mm wide

Shoots bulbous at base. Ligule obscurely minutely hairy

Lvs 25-40cm x 7-16mm, the lower narrowed to base, smooth or rough above only, with midrib raised below, with stomata both sides. Ligule 6mm, not ciliolate. Anthers 3-3.5mm. Shoots bulbous at base (bulbs often in prs). R alien.............*Bulbous Canary-grass* **Phalaris aquatica**

Shoots never bulbous at base. Ligule hairy to hairless

Lower sheaths retrorsely hairy or hairless, ciliate, ± smooth, with cross-veins weak to absent, the upper hairless and open. Lvs 5-14mm wide, rough at least above, occ fleshy, usu dark green, usu with sparse long hairs esp above, with translucent midrib raised below, with stomata more abundant above. Ligule (1)2-4mm, long-hairy to hairless, often green or scarious..*Rescue Brome* **Ceratochloa cathartica**

All sheaths hairless but may be ciliate and hairy at apex, ± smooth, with cross-veins weak to absent. Lvs 6-14mm wide, occ rough below, occ fleshy, usu dark green, ± hairless to hairless, often ciliate (cilia to 1mm), with opaque midrib raised below, with stomata more abundant above. Ligule 1-3.5mm, long-hairy to minutely hairy or hairless, often green or scarious..*California Brome* **Ceratochloa carinata**

Lvs mostly <5mm wide. Shoots usu bulbous at base. Ligule hairless

Ligule 0.5-1.5mm (2-5mm on culm), truncate to ± acute, minutely toothed, usu ciliolate, soon brownish. Lvs retrorsely scabrid at least nr ligule, slightly rough both sides or smooth, pale or grey-green, occ sparse hairs nr base, with translucent midrib not or weakly raised below, with 1-3 main veins each side of midrib (6-8 total), with stomata both sides. Basal sheaths brown (occ purplish), ± fibrous, with few obscure cross-veins nr apex. Upper sheaths ± inflated

Lvs 10-45cm x 2-4(10)mm. Infl spike 6-15cm..............................*Timothy* **Phleum pratense***

Lvs 3-5(12)cm x (2)3(5)mm. Infl spike 1-6(8)cm............*Smaller Cat's-tail* **Phleum bertolonii**

■Ann

Sand dunes

Lvs to 3cm x 2-5mm, inrolling nr apex, with scabrid-papillate margins, rough above, yellow-green, with midrib not discernible or raised below, with translucent veins in distal ⅓ (rarely visible proximally). Ligule (1.5)2-7mm, acute, occ with green cells, jagged at apex (uppermost ± entire), overlapping at back. Sheaths open, smooth, the basal occ purplish, the upper inflated. Nov-Jun.......*Sand Cat's-tail* **Phleum arenarium**

Dry arable or disturbed habs

Lvs usu >10mm wide. Ligule hairless.......(*Avena*) Go to **AVENA**

Lvs <10mm wide. Ligule usu minutely hairy

Ligule minutely hairy (occ obscure)

Anthers 3mm. Lvs 10-25cm x 5-10mm, green or grey-green, rough above. Sheaths ± smooth, the upper ± inflated. Ligule 3-8mm, toothed......*Canary-grass* **Phalaris canariensis**

Anthers 1.5-1.8mm. Lvs 3-15cm x 3-9mm, shiny green, smooth. Ligule 3-8(12)mm, truncate to rounded, toothed to jagged. Sheaths ± smooth, the upper inflated ....*Lesser Canary-grass* **Phalaris minor**

Anthers 1.5-1.8mm. Lvs 5-15cm x 5(7)mm, grey-green. Ligule (3)5(8)mm, toothed, entire to torn. Sheaths ± smooth, the upper inflated. Spikelets falling in clusters of 7, a single fertile spikelet encircled by 6 reduced sterile spikelets (in other *Phalaris* the spikelets are alike, not in clusters and disarticulate above the persistent glumes) ....*Awned Canary-grass* **Phalaris paradoxa**

Ligule hairless

Ruderal

Lvs (4)6-12mm wide, widest nr base, twisted, with scabrid margins (occ weak), smooth both sides, pale grey-green, shiny green below, with discernible midrib raised below, with c8 v low ribs each side of midrib, with stomata both sides. Ligule 2-5mm, ± acute to rounded, ± toothed or torn. Sheaths open (but the v lowest closed), purplish at base, smooth, the upper not inflated. Sunny disturbed habs.......*Greater Quaking-grass* **Briza maxima**

Lvs 3-10mm wide, widest at base, occ twisted, often slightly rough above, smooth below, dull green above, shiny below, with midrib hardly discernible (HTL) and ± not raised below, with stomata both sides. Ligule 4-10mm, acute, entire. Sheaths open, not purplish, smooth, the upper inflated.......*Rough Dog's-tail* **Cynosurus echinatus**

Arable

Ligule 6-15mm, obtuse, toothed. Lvs 6-18(25)cm x 2-5(10)mm, rough both sides or smooth below. Sheaths often purplish. Culms usu several, 20-80(120)cm, weakly branched, hollow, with 3-5 nodes, with internodes shorter or longer than sheaths ....*Loose Silky-bent* **Apera spica-venti**

Ligule 3-6(8)mm, acute, entire, not decurrent. Lvs 3-7(14)cm x (3)5-10mm, twisted, rough above, smooth below. Sheaths smooth. Culms usu several, 10-50cm, unbranched, usu solid, with 2 nodes.......*Lesser Quaking-grass* **Briza minor**

Ligule (1.5)2-5mm, acute to truncate, toothed to erose, decurrent down sheath. Lvs 4-12cm x 0.3-4mm, usu inrolled, rough above (rarely minutely hairy), smooth below exc nr apex. Sheaths smooth, often purplish. Culms 1-several, 10-50(75)cm, occ weakly branched nr the base, hollow, with several nodes, the internodes usu longer than sheaths. E Anglia ....*Dense Silky-bent* **Apera interrupta**

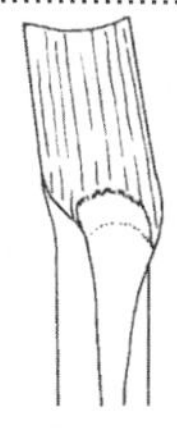

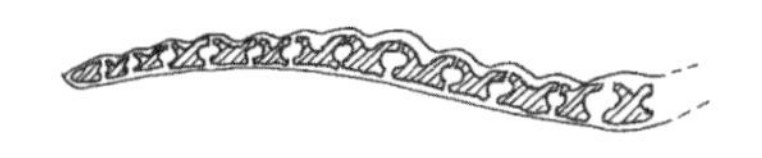

Calamagrostis epigejos ligule & lf TS

Key to Groups in Division I

(Auricles present, occ resembling a ligule)

Juncus, Triglochin. *Monocot. Lvs 2-ranked at least below, with smooth margins, with stomata all sides*

■Lvs (semi-) cylindrical, with entire apex, AND/OR aromatic..**IA**
■Lvs flat or channelled, with minutely bifid or trifid apex (occ obscure or broken), not aromatic.......**IB**

Group IA – *Lvs cylindrical. Sheaths open (and overlapping in* Juncus*).* ❶ *Lvs aromatic, with soapy odour.* ❷ *Basal sheaths with purplish veins.* ❸ *Shoots bulbous at base*

■Lvs with distinct cross-partitions detectable by touch (run fingers firmly along lf)

Lvs with longitudinal partitions. Basal sheaths greenish, pale orange-brown at extreme base

Lvs 2-3mm diam, bright green, smooth. Stems to 120cm, occ purplish above, not compressible, solid to hollow, with scattered vb's. Rhizomatous per. All yr ....*Blunt-flowered Rush* **Juncus subnodulosus***

Lvs without longitudinal partitions. Basal sheaths often reddish

Lvs with 1(2) septa per 5cm

Lvs 3mm diam, smooth, slightly curved, oval, slightly flattened, shiny green, hollow. Auricles 0.5-2mm. Stems to 100cm, ± erect. Rhizomatous per. All yr ....*Sharp-flowered Rush* **Juncus acutiflorus***

Lvs with 5-10 septa per 5cm

Lvs 1.5-2.5mm diam, slightly ribbed, curved, ± oval to strongly flattened, dull green, pith-filled or hollow. Auricles 1.5mm. Stems to 50cm, ± prostrate at base. Tufted to strongly rhizomatous per, occ rooting at lower nodes. Often all yr.....*Jointed Rush* **Juncus articulatus**

Lvs to 1mm diam, smooth, curved, oval, slightly flattened, dull green, pith-filled. Auricles 2mm. Stems usu <7cm, ± prostrate at base. Shortly rhizomatous per. All yr. Wet mtn habs ....*Alpine Rush* **Juncus alpinoarticulatus**

■Lvs with obscure cross-partitions (often visible HTL), not detectable by touch

Lvs solid

Lvs to 15cm x 0.8-1.5(2.5)mm, 2-ranked, semi-cylindrical, channelled above nr base, obtuse, with veins not visible. Auricles <1mm, obtuse, overlapping to look like a ligule. Shoots flattened at white sheathing base. Stems 1.2mm diam, round, solid. Rhizomatous per. ❶. Apr-Oct. Freshwater marshes..*Marsh Arrowgrass* **Triglochin palustre***

Lvs with 1 hollow

Lvs 20-100cm x 2-3mm, mostly on stem, cylindrical to oval, obtuse, shiny green, without cross-partitions, hollow or pith-filled. Auricles to 2mm, obtuse. Stems to 100cm, channelled. Strongly rhizomatous per. Saltmarshes, VR alien, Somerset, Scot....*Somerset Rush* **Juncus subulatus**

Lvs 3-8cm x 0.5-1.5mm, all basal, cylindrical, obtuse, shiny green, with cross-partitions (HTL), with 1-3 hollows at base. Auricles to 0.5mm, obtuse. Stems 5-12cm, grooved along one side. Tufted per. Mtn flushes, often mica-rich, Scot...................*Two-flowered Rush* **Juncus biglumis**

Lvs 0.5-3cm x 0.5mm, mostly basal, cylindrical, obtuse, green or purplish, with indistinct cross-partitions, hollow or pith-filled. Auricles to 0.5mm, obtuse. Stems to 3cm, round. Ann. ❷. Mar-Jun. Winter-wet habs, VR, Lizard..*Pigmy Rush* **Juncus pygmaeus**

Lvs with 2 distinct hollows (soon breaking up)

Lvs 1.5-10cm x 0.5-1(1.5)mm, usu all (±) basal, flattened (occ cylindrical), obtuse (red-tipped), usu channelled above nr base, channelled below, with obscure cross-partitions, occ with 5 hollows at base. Auricles 0.7-1.5mm, ± acute to obtuse. Basal sheaths ± shiny (reddish) brown. Stems 5-20cm, weakly 3-ridged, hollow. Tufted per. Mtn flushes, often base-poor, N Br ....*Three-flowered Rush* **Juncus triglumis**

Lvs with 2-several indistinct hollows

Lvs 2-10cm x 0.7-1.2mm, usu all basal, ± cylindrical but channelled above, with indistinct cross-partitions. Auricles 1-2mm. Basal sheaths with c2mm fragile aristate point. Per with bulb-like swellings at base, often rooting at nodes. ❸. All yr.................*Bulbous Rush* **Juncus bulbosus**

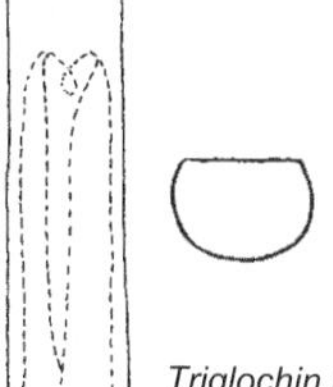

Triglochin palustre auricles & lf TS

Group IB – *Lvs flat or channelled, with minutely bifid or trifid apex (occ obscure or broken)*

■Auricles entire. Lvs with smooth margins

Tufted per, or in compact rosette. All yr

Lvs 6-12, in a basal rosette, usu held ± 90° to sheath, 5-30cm x 1-2(3)mm, U-shaped, stiff, curved, shiny green both sides, occ whitish above, often orange or brown in winter, with >5 bands of stomata below, with vb's obscure. Auricles to 0.5mm. Basal sheaths tough, white to brown, with reddish veins, usu without aristate tip. Shoots flattened at base. Hths, moors ....*Heath Rush* **Juncus squarrosus**

Lvs (3)6, on false-stems, often held erect, 4-20cm x 1-1.5mm, weakly channelled, flexible, curved, usu shiny green, with 3 veins and 5 bands of stomata below. Auricles (1)3mm, obtuse, rarely overlapping. Basal sheaths warm-brown, occ reddish, with fragile aristate tip. Shoots flattened at base. Bare gravely or sandy habs

Auricles usu longer than wide, whitish.........................*Slender Rush* **Juncus tenuis** var **tenuis**

Auricles wider than long, yellow-brown. VR, Scot.......*Dudley's Rush* **Juncus tenuis** var **dudleyi**

Rhizomatous per (occ shortly so). Apr-Oct

Lvs 2-4, on stem, the lowest 2-ranked, the upper 3-ranked, 5-30cm x 1-1.5(3)mm, dull dark green, weakly ridged below, with midrib translucent below only, with obscure cross-veins, with 5 bands of stomata below. Auricles to 1mm, overlapping. Basal sheaths greenish to warm-brown, with fragile aristate tip. Stems solid, round, sap-filled

Saltmarshes. Rhizomes long, usu >5cm...............................*Saltmarsh Rush* **Juncus gerardii**

By freshwater (usu). Rhizomes short, usu <5cm......*Round-fruited Rush* **Juncus compressus**

■Auricles deeply laciniate. Lvs with minutely scabrid margins (often obscure). May-Oct

Lvs basal (some reduced to sheaths) or on stems (as long bracts), to 8cm x 0.5-0.7mm, slightly channelled, tapered to v fine apex. Auricles to 2(4)mm. Basal sheaths brown with 10mm fragile aristate point. Shoots with dead fibrous remains at base. Stems to 15cm, 0.8mm diam, round, with small hollow. Tufted per, occ with v short rhizomes. Mtns ....*Three-leaved Rush* **Juncus trifidus**

Key to Groups in Division J

(Ligule and auricles both absent)

Lvs parallel-veined, usu >3 veins visible (stomata often in parallel rows). Usu monocots (may have bulb, corm or viscid sap)

■Evergreen tree or shrub (beware young plants) with sharply acute lvs.......... **JA**

■Herb. Lvs equitant (*Iridaceae*, *Araceae*, *Liliaceae*).......... **JB**

■Herb. Lvs not equitant

Lvs with distinct petiole

Plant emergent from water. Petiole may have white latex.......... **JC**

Plant not emergent from water (may be found in damp habitats). Petiole without latex.......... **JD**

Lvs sessile or narrowing to indistinct petiole

Lvs with white latex drying red.......... **JE**

Lvs without latex

Lvs with odour of onion, garlic or leek (*Allium* and allies).......... **JF**

Lvs vanilla- (rarely goat-) scented (*Orchidaceae*)

Lvs with raised veins below (other than midrib), all on stem (± basal in *Listera*).......... **JG**

Lvs without raised veins below (midrib may be raised), often basal

Lvs spotted or blotched.......... **JH**

Lvs not spotted or blotched

Lvs with scattered stomata above (often in single rows each side of veins).......... **JI**

Lvs with stomata absent above (or confined to margins)

Lvs with hyaline margins (±) entire.......... **JJ**

Lvs with hyaline margins minutely crenate or papillate

Lvs >6cm.......... **JK**

Lvs <5cm.......... **JL**

Lvs not vanilla- or goat-scented

Lvs opp or whorled on stems, may be ribbed below by veins.......... **JM**

Lvs alt (occ single or in prs)

Lf margins ciliate with long hairs (at least nr base).......... **JN**

Lf margins rough or scabrid/papillate (at least nr apex), never ciliate.......... **JO**

Lf margins smooth and (±) entire, never ciliate

All lvs on stems.......... **JP**

At least some lvs basal

Lvs with a white stripe along midrib above.......... **JQ**

Lvs without a white stripe along midrib above (may have a glaucous stripe above)

Lvs 1-4 mm wide, rarely 4-angled.......... **JR**

Lvs >4mm wide, never 4-angled

Lvs triangular at least nr base, or semi-cylindrical (usu aquatic).......... **JS**

Lvs flat or U-shaped

Lf veins (other than midrib) strongly raised below. Dicot, never bulbous.......... **JT**

Lf veins (except midrib) not or weakly raised below. Monocot, usu bulbous

Lvs 1-2 per shoot (lvs emerging singly or in prs).......... **JU**

Lvs 3 or more per shoot

Lvs glaucous above.......... **JV**

Lvs green above.......... **JW**

Group JA – *Evergreen tree or shrub (beware young plants). Lvs usu tough, with sharp apex, entire*

■Lvs 20-100cm, spirally arranged in rosettes at stem apices

Lvs with obvious translucent veins, with stomata in parallel bands both sides

Lvs (1)3-6cm wide, linear-lanc, tapered to sharp slender point, cartilaginous margins whitish (without reddish speckles), flat, erect when young, later drooping, mid-green. Tree or shrub. Hortal......*Cabbage-palm* **Cordyline australis**

Lvs with indistinct opaque veins (inc midrib), with scattered stomata both sides

Lvs 3-5cm wide, broadly lanc, rolled when young, slightly channelled, thick, dull green to grey-green, with yellowish cartilaginous margins soon reddish-speckled. Woody shrubby per

Lvs straight, erect. Fls spring. Hortal......*Spanish-dagger* **Yucca gloriosa**

Lvs usu drooping nr apices. Fls late summer/autumn. Sand dunes, hortal ......*Curved-leaved Spanish-dagger* **Yucca recurvifolia**

■Lvs to 10cm, spirally arranged along stems

Lvs ± sessile, twisted at base, with indistinct veins, with stomata both sides (the green 'lvs' are actually cladodes, the scarious 'stipules' are the true lvs). Stems striate. Shortly rhizomatous

Cladodes 1-4 x 0.4-1cm, alt, spine-tipped to 2mm, rigid, tough, shiny dark green. PL 9 ......*Butcher's-broom* **Ruscus aculeatus**

Cladodes 3-10 x 1.3-3cm, occ 3-4-whorled or opp, acute but not spine-tipped, soft, shiny pale green. Hortal, R alien......*Spineless Butcher's-broom* **Ruscus hypoglossum**

JB

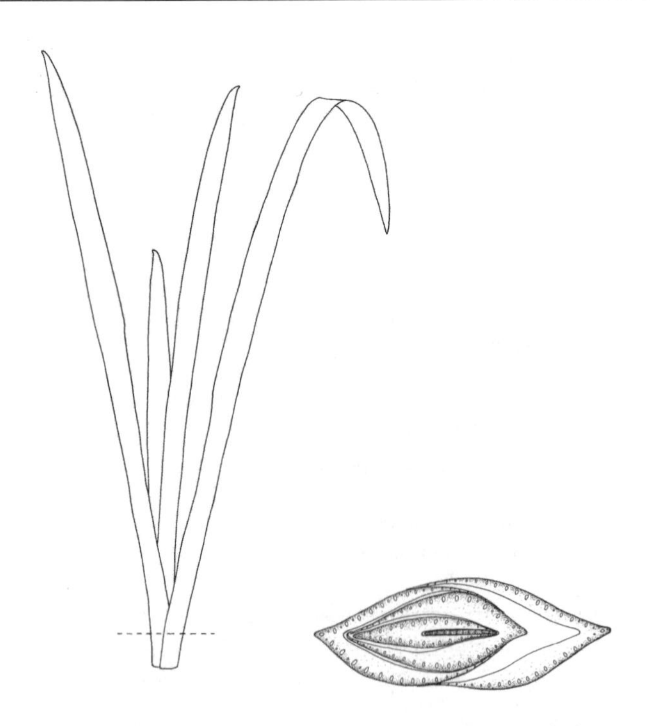

Iris pseudacorus with lvs in TS (showing equitant vernation)

Group JB – *Lvs equitant, acute (often with curved apex), with cartilaginous margins. Unlike most plants, the true upper surface of the lvs is not visible as the lvs are folded laterally along the midrib with the faces joined together (just the true lower side visible).* ❶ *Lvs non-equitant distally*

■Lvs <5mm wide

Lvs odourless, with scattered stomata

Rhizomatous per. Lvs green. Stems round, not winged

Lvs 5-30cm x 2-5mm, often curved, entire, usu with 5 veins. Mar-Oct. Bogs ....*Bog Asphodel* **Narthecium ossifragum**

Lvs 1-4cm x 1-2mm, hardly curved, serrate-ciliate along one margin, with 3(5) veins (often obscure). May-Oct. Wet upland habs, N Br..........*Scottish Asphodel* **Tofieldia pusilla**

Tufted per. Lvs green to glaucous (occ reddish at base). Stems flattened, narrowly winged

Lvs 20cm x 2-5mm, entire to minutely serrate, with 5-7 indistinct veins. Mar-Oct (all yr). Damp to dry habs..........*Blue-eyed-grass* **Sisyrinchium bermudiana**

Lvs with weak sweet citrus odour when crushed, with stomata in bands. Rhizomatous per

Lvs 5-45cm x 2-4(6)mm, shiny green or variegated, reddish at base, entire. Damp habs, VR alien..........*Slender Sweet-flag* **Acorus gramineus**

■Lvs >5mm wide, with scattered stomata

Lvs with sweet citrus odour when crushed. Rhizomatous. Apr-Oct

Lvs to 120cm x 10-25mm, acuminate, usu with transverse wrinkling, shiny bright green, with (30)80 veins, with thick mid-vein (often eccentric) raised both sides. Aquatic ....*Sweet-flag* **Acorus calamus**

Lvs with fetid (fresh meat) odour when crushed

Tufted. Lvs with strong odour. All yr

Lvs to 80cm x 20-25mm, rarely with transverse wrinkling, shiny dark green, serrate nr apex, flat but ribbed with 10-20 slightly raised veins, cross-veins present (HTL)*Stinking Iris* **Iris foetidissima**

Rhizomatous. Lvs with weak odour. Apr-Oct

Lvs to 100cm x 10-20(25)mm, usu light green to slightly glaucous, ± thick, serrate nr apex, flat but ribbed with slightly raised veins, cross-veins absent..........*Turkish Iris* **Iris orientalis**

Lvs to 100cm x (6)10-20mm, dark green to slightly glaucous, ± thin, entire, flat, hardly ribbed, cross-veins absent..........*Blue Iris* **Iris spuria**

Lvs odourless

Lvs glaucous, often strongly so

Rhizomatous per. Lvs usu >20mm wide, weakly pruinose

Dry habs. Lvs 20-60mm wide, occ purplish at base, all veins weakly raised (mid-vein not more strongly raised than others), without cross-veins, solid, without spiral fibres. Shoots fan-like. Rhizomes brown. Mar-Oct. R alien..........*Bearded Iris* **Iris germanica**

Wet habs. Lvs 15-50mm wide, purplish at base, with veins slightly raised, with mid-vein (2 ribbed) strongly raised, with cross-veins, with aerenchyma and several square hollows, with spiral fibres weak or absent. Rhizomes pink. Jan-Jul (all yr). Aquatic..........*Yellow Iris* **Iris pseudacorus***

Tufted or cormous per. Lvs <20mm wide

Lvs 8-15(20)mm wide, strongly glaucous-pruinose, reddish at base, rarely twisted, entire, with broad cartilaginous margins often reddish, with slightly sunken veins, without cross-veins, with spiral fibres when torn. Cataphylls absent. Stems v flattened, narrowly winged. Tufted. All yr..........*Pale Yellow-eyed-grass* **Sisyrinchium striatum**

Lvs (5)6-8(10)mm wide, glaucous, twisted, entire, with narrow cartilaginous margins never reddish, with mid-vein raised both sides and at least 1 raised translucent vein, without cross-veins, without spiral fibres when torn. Cataphyll 1. Cormous. Apr-Jul. New Forest. Sch8 ....*Wild Gladiolus* **Gladiolus illyricus**

Lvs (2)5-5.5mm wide, usu glaucous, rarely twisted, obscurely serrate, with narrow cartilaginous margins never reddish, with obscure veins, without raised mid-vein, without cross-veins, with spiral fibres when torn. Cataphylls absent. Stems v flattened, distinctly winged. Tufted or with short rhizomes..*American Blue-eyed-grass* **Sisyrinchium montanum**

Lvs green or weakly glaucous

Lvs strongly pleated with 7-10 strongly raised veins, concertina-like. Strongly rhizomatous

Lvs 3-7cm wide, pale green, entire, with stomata in rows. Basal sheaths reddish. Mar-Oct

Tough. Widespread..*Aunt-Eliza* **Crocosmia paniculata**

Tender, R, mostly Cornwall, Ire..........................*Giant Montbretia* **Crocosmia masoniorum**

Lvs flat, may have slightly raised veins

Strongly rhizomatous per (with corms), rapidly forming large patches

Lvs (5)15-17(20)mm wide, yellow-green, entire, with raised midrib and at least 2 raised translucent veins, often with some minor veins slightly raised. Cataphylls 1-3. *Crocosmia aurea* x *pottsii*. Mar-Oct.......................................*Montbretia* **Crocosmia** x **crocosmiiflora**

Rhizomatous per (without corms), forming small patches

Lvs 40-100cm x 6-18mm, green or slightly glaucous, not purplish at base, minutely serrate, without any raised veins (but with 30-50 translucent veins), with cross-veins and spiral fibres weak or absent. Apr-Oct. Damp habs..................................*Siberian Iris* **Iris sibirica**

Lvs 50-80cm x 25-30mm, green, purplish at base, entire, occ with 1-several slightly raised veins, usu without cross-veins or spiral fibres. Apr-Oct. Aquatic. *I. versicolor* x *virginica* ....*Windermere Iris* **Iris** x **robusta**

Lvs 40-70cm x 15-21mm, light green (occ slightly glaucous), purplish at base, slightly serrate nr apex, without raised veins (double mid-vein not or hardly raised), with weak cross-veins, with v sparse spiral fibres. Mar-Oct. Aquatic ....*Japanese Water Iris* **Iris laevigata**

Lvs 30-80cm x (8)15(25)mm, occ ± glaucous, purplish at base, entire, without raised veins (not even mid-vein), without cross-veins or spiral fibres. Mar-Oct. Aquatic ....*Purple Iris* **Iris versicolor**

Tufted per

Lvs 100-300cm, sharply acute. Dry habs

Lvs 5-12cm wide, v tough, often dull ± glaucous below, with reddish cartilaginous margins, with many narrow translucent veins, with stomata in rows (often obscure) at least above. Tussock-forming. ❶. All yr....................*New Zealand Flax* **Phormium tenax**

Lvs 30-90cm, not sharply acute. Dry habs

Lvs (8)12-24mm wide, bluish-green, twisted, with midrib raised both sides (more so than other veins). Cormous. May-Aug ....*Eastern Gladiolus* **Gladiolus communis** ssp **byzantinus**

Lvs 6-12mm wide, dark green (occ glaucous), twisted, with all veins equally raised. Densely tufted (weakly tussock-forming) with v short rhizomes. All yr ....*Chilean-iris* **Libertia formosa**

Lvs to 15(30)cm, not sharply acute. Damp habs

Lvs 5-8.5mm wide, bluish-green, occ purple at base, usu entire, with all veins equally raised, with cross-veins weak or absent, without spiral fibres when torn. Mar-Oct (all yr) ....*Yellow-eyed-grass* **Sisyrinchium californicum**

Group JC – *Lvs with distinct petiole, usu emergent, entire, with stomata both sides. Petiole with sheathing base. Aquatic. (*Nuphar advena *may key out here).* ❶ *Plant with strong coriander odour*

■Petiole with latex (often sparse). Apr-Oct

Lvs sagittate at base (cuneate in *S. ridiga*), with obscure hydathode at apex. Petiole with latex canals throughout aerenchyma. Roots weakly septate

Petiole with purple-black spotting nr base. Lvs with distinct 4° veins

Petiole ± round (many-sided), with small air-spaces in TS. Lvs 5-20 x 3-10cm, hooded at apex, the acute basal lobes occ absent, often with purplish margin. Tuberous and stoloniferous..........*Duck-potato* **Sagittaria latifolia**

Petiole without purple-black spotting. Lvs with obscure 4° veins

Petiole ± sharply triangular, with large air-spaces in TS. Lvs 5-20 x 3-10cm, ovate or elliptic, sagittate at base, with 2-3 parallel veins each side of midrib, with stomata both sides. Submerged lvs, if present, linear. Tuberous. PL 14..........*Arrowhead* **Sagittaria sagittifolia**

Petiole triangular, usu with large air-spaces in TS. Lvs 5-15(20) x 6cm, ovate or elliptic, cuneate at base, with 2-3 parallel veins each side of midrib, with stomata both sides. Submerged lvs ± linear, ± stiff. Stoloniferous. VR alien ....*Canadian Arrowhead* **Sagittaria rigida**

Lvs not sagittate, with dark hydathode below at apex. Petiole with latex canals confined to margin (and around vb's). Roots not septate

Lvs 8-30 x 4-10cm, elliptic-ovate, rounded to ± cordate at base, rolled when young, with midrib raised below, with (2)3-4 opaque 2° veins each side of midrib (slightly raised below), with oblique translucent 3° veins (4° veins obscure). Petiole semi-cylindrical, spongy, with ± equal small (3-5mm) aerenchyma hollows..........*Water-plantain* **Alisma plantago-aquatica**

Lvs 3-25 x 3-7cm, lanc, long-cuneate at base, 3-4 opaque veins each side of midrib, rolled when young, with midrib raised below, with obscure cross-veins. Petiole with irreg large (>5mm) aerenchyma hollows..........*Narrow-leaved Water-plantain* **Alisma lanceolatum**

Lvs 8-30(100) x 0.2-2(3)cm, often developing a short (15-40mm) linear-lanc or narrowly oblong blade, occ floating or submerged, rolled when young, ribbon-like, hardly discernible from petiole. VR, Worcs. Sch8..........*Ribbon-leaved Water-plantain* **Alisma gramineum***

■Petiole without latex

Lvs 5-25 x 5-15cm, triangular-ovate, ± acute with hydathode below, cordate at base, with 20-30 main veins and c100 fine veins fading out distally (not raised below), with stomata both sides. Petiole long, with sheathing base, round, spongy, with 25-60 scattered vb's and spiral fibres. Stems ± prostrate or floating, to 1m. Rhizomatous per. Apr-Oct ....*Pickerelweed* **Pontederia cordata***

Lvs 4-7(10) x 0.5-2cm, usu lanc, acute, cuneate at base, with 3 opaque veins raised below, with cross-veins distinct to obscure, with stomata both sides. Tufted or stoloniferous per. ❶. All yr ....*Lesser Water-plantain* **Baldellia ranunculoides**

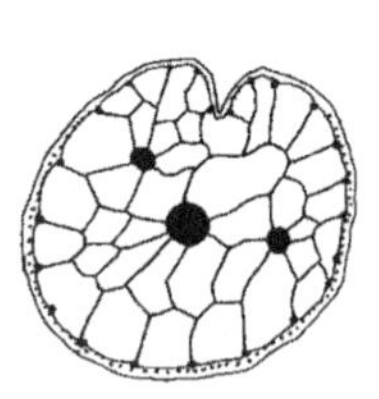

Alisma gramineum petiole TS

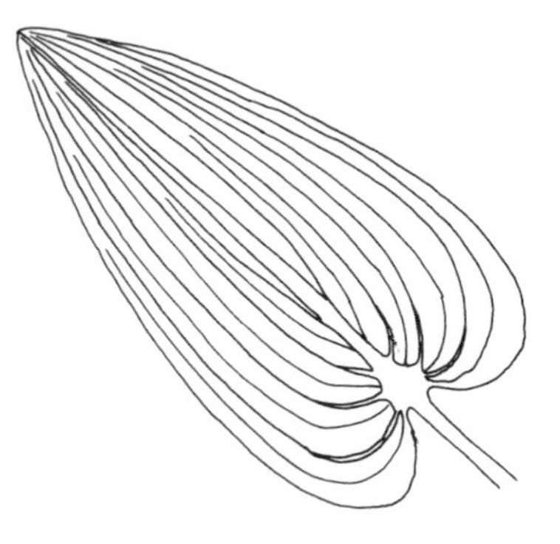

Pontederia cordata

Group JD – *Lvs with distinct petiole. Plant hairless (sparse hairs on* Maianthemum bifolium*). Damp or dry hab. (*Tulipa sylvestris *may key out here)*

■Lvs with onion odour, cuneate (occ rounded) at base

Lvs 2(3) per shoot, 10-20 x 1.5-5.5cm, ± elliptic, acute, twisted, shiny green above, dull below, with crenulate-erose hyaline margins, with opaque veins, with midrib raised below, with stomata below only. Petiole to 7cm, semi-cylindrical, with 7-11 vb's in shallow arc. Bulbs joined by short rhizomes. Mar-Jul..*Ramsons* **Allium ursinum**

■Lvs odourless, cordate

Basal lvs solitary. Dry habs

Lvs ovate, acute to acuminate, shiny green esp below, with minutely serrate-scabrid margins, with 30-50 parallel veins converging at apex, with stomata below only. Stems round, often channelled above, with 2 scale lvs at base and 2 foliage lvs nr apex. Apr-Aug. Rhizomatous

Lvs 5-10 x 10cm, hairless. Petiole >10cm. Stems to 40cm, hairless. Hortal, R ....*False Lily-of-the-valley* **Maianthemum kamtschaticum**

Lvs 4-6 x 2.5-6cm, sparsely hairy (esp below when young). Petiole 5-12cm. Stems to 20cm, usu hairless. Mostly N Eng..*May Lily* **Maianthemum bifolium**

Basal lvs in prs (occ in 3's). Dry habs

Lvs to 12 x 3-4(5)cm, elliptic to ovate-lanc, acute, rolled when young, ± dull green above, shiny green below, pruinose when young, with hyaline margins, with many translucent parallel veins, with midrib raised below, without cross-veins and viscid sap, with stomata both sides. Lower lf with closed sheath forming a false-stem. Cataphylls and false-stem purple-spotted. Rhizomatous. Apr-Aug...*Lily-of-the-valley* **Convallaria majalis**

Basal lvs several, usu in a rosette. Damp habs

Lvs 10-30cm, elliptic-ovate, cuspidate, with many parallel veins converging at apex, with entire margins, with stomata below only. Petiole 10-30cm, with long sheathing base, weakly channelled, with many vb's. Stem lvs absent. Stems round, solid. Rhizomes stout, jointed, scaly. Apr-Oct...*Bog Arum* **Calla palustris**

Lvs 1.5-4cm, ovate, obtuse to ± acute, with few parallel veins converging at apex (occ obscure), with minutely crenulate hyaline margins, with stomata below only. Petiole to 7cm, without sheathing base, channelled, with 1 vb. Stem lvs ± clasping. Stems 5-ridged, solid. Rhizomes absent. Apr-Sept...*Grass-of-Parnassus* **Parnassia palustris**

Group JE – *Lvs with white latex (drying red), usu with a white stripe along midrib above, sparsely woolly above when young (esp nr base), soon hairless, hollow at least nr base*

■Lvs linear-lanc, sessile, with a white stripe along midrib above, often resembling leeks (*Allium*)

Lvs basal and/or on stems, 20-40 x 0.6-1cm, with cylindrical acute apex (rounded x20), sheathing at dilated base, weakly clasping, often with scabrid hyaline margins (at least nr apex), with 3-4 translucent veins each side of midrib, with raised midrib below (2° veins rarely raised), with stomata both sides. Stems to 80cm, branched, striate, hollow. Fl colour usu visible when in bud

Fls yellow. Stems not or hardly dilated below fls. Lvs grey-green, with a straight apex. Short-lived per (occ ann) with taproot and dead fibrous lf remains at base. Often all yr. PL 24 .... *Goat's-beard* **Tragopogon pratensis**

Fls purple. Stems dilated below fls. Lvs ± glaucous, with a recurved to coiled apex. Bi (ann). Alien, often coastal. All yr.. *Salsify* **Tragopogon porrifolius**

Fls pink. Stems strongly dilated below fl. Lvs grey-green, with a straight apex. Ann. VR casual .... *Slender Salsify* **Tragopogon hybridus**

■Lvs narrowly elliptic to linear-lanc, tapered to long petiole-like base, often with a whitish stripe along midrib above nr base, resembling *Plantago lanceolata*

Lvs basal, to 25 x 0.9-1.5cm, with minutely hooded or obtuse apex, with ± entire hyaline margins, soon hairless, with 5 veins raised below (occ pinnate veins), with scattered stomata both sides. Petiole narrowly winged, sheathing at base, hollow. Stems woolly, soon hairless. Per. Apr-Oct. Damp gsld, VR, Dorset, S Wales. Sch8.. *Viper's-grass* **Scorzonera humilis**

Group JF – *Lvs with odour of onion, garlic or leek, often 2-ranked, with pale terminal hydathode at apex, with sheathing base, with obscure opaque veins. Many spp have a short adnate ligule at the base of the lf. Bulbous.* ❶ *Bulb with honeycomb-like (reticulate) tunic.* ❷ *Lf margins ciliate*

■Lvs cylindrical or semi-cylindrical, often hollow, with stomata all sides

Lvs solid (at least distally) or with small hollow nr base, not compressible. Ligule absent

Lvs 2-3(4)mm wide, tapered to whitish obtuse apex, semi-cylindrical, distinctly ribbed with 3-5 papillate raised veins both sides (at least nr base). Shoots round. Bulb with sessile bulbils. Oct-Jul............*Field Garlic* **Allium oleraceum** var **oleraceum**

Lvs 1-3(4)mm diam, tapered to pale reddish obtuse apex, ± cylindrical but channelled above, obscurely ridged with long papillae nr base. Shoots round. Bulb with stalked bulbils. Feb-Jul. Sch8............*Round-headed Leek* **Allium sphaerocephalon**

Lvs with large hollow, readily compressible. Ligule present

Lvs >5mm diam, green to grey-green, semi-cylindrical, not or weakly ridged, without papillae

Bulb usu solitary. All yr............*Onion* **Allium cepa**

Bulbs several. All yr............*Welsh Onion* **Allium fistulosum**

Lvs <5mm diam, grey-green

Lvs ± cylindrical, channelled, 2-5mm diam, narrowed to v acute apex, with (5)10-18 distinct ridges (occ papillate). Bulb ovoid, with bulbils. Jan-Aug............*Wild Onion* **Allium vineale**

Lvs cylindrical, rarely channelled, 1-4mm diam, narrowed to v acute (or obtuse) apex, with c15 obscure ridges (sparsely papillate when young). Bulbs elongate, clustered on short rhizome. Feb-Jul............*Chives* **Allium schoenoprasum**

■Lvs flat to triquetrous (never cylindrical), solid, with stomata both sides (exc *A. paradoxum*, *A. triquetrum*)

Basal lvs 1(2) per bulb, rolled when young. Lf margin entire

Lvs dull glaucous both sides, 6-35mm wide, linear-lanc, abruptly tapered to wide acute apex, flat, without wrinkling, with midrib raised below. Shoots round. Mar-Jul ....*Yellow Garlic* **Allium moly**

Lvs dull grey-green above, shiny green below, 6-15(18)mm wide, linear-lanc, abruptly tapered to wide acute apex, triquetrous, occ with transverse wrinkling nr base, with midrib raised below, with stomata above only (a few along midrib below). Shoots slightly flattened, keeled. Feb-Jul ....*Few-flowered Garlic* **Allium paradoxum**

Basal lvs ≥2 per bulb

Lf margin (and keel if present) papillate-scabrid at least nr base

Lvs triquetrous or strongly keeled below (usu V-shaped at maturity). Ligule present

Lvs 40-60mm wide, glaucous both sides, tapered to obtuse apex, folded when young. Shoots with fleshy basal sheaths forming false-stem. Bulb without bulbils. Feb-Aug. Hortal ....*Leek* **Allium porrum**

Lvs 12-35mm wide, glaucous both sides, tapered to broadly acute hooded apex, folded when young. Shoots slightly flattened, with basal rarely thin and not forming false-stem. Bulb with bulbils inside tunic. Oct-Jun. SW Eng, and hortal ....*Babington's Leek* **Allium ampeloprasum** var **babingtonii**

Lvs not or weakly keeled below (may be shallowly V- or U-shaped at maturity)

Ligule present

Lvs 12-35mm wide, glaucous both sides, tapered to broad acute hooded apex, with long-papillate margins esp nr base. Shoots slightly flattened. Bulb with yellowish bulbils. Oct-Jun. SW Eng, also hortal............*Wild Leek* **Allium ampeloprasum** var **ampeloprasum**

Lvs 10-24mm wide, ± dull green both sides, tapered to obtuse apex, with long-papillate margins esp nr base. Shoots slightly flattened. Bulbs with purplish bulbils. Mar-Jul ....*Sand Leek* **Allium scorodoprasum**

Ligule absent

Lvs leek-scented

Lvs 5-25mm wide, tapered to v acute apex, shallowly V-shaped, shiny green both sides, often with long-papillate margins nr base, with midrib raised below. Shoots ± round. Oct-Jun..*Neapolitan Garlic* **Allium neapolitanum**

Lvs onion-scented

Lvs ribbed below. Bulb with non-reticulate tunic

Lvs 3-5mm wide, shallowly U-shaped or ± flat above, abruptly tapered to obtuse apex, dull grey-green both sides, with long-papillate margins (at least nr base). Shoots flattened to round, occ reddish. Oct-Jun........................*Keeled Garlic* **Allium carinatum**

Lvs 2-4mm wide, channelled above, ribbed below, abruptly tapered to acute apex, dull grey- or mid-green both sides, with long-papillate margins (and often ribs below). Shoots round, never reddish.....*Field Garlic* (var) **Allium oleraceum** var **complanatum**

Lvs not ribbed below. Bulb with reticulate tunic

Lvs 5-12mm wide, shallowly U-shaped, tapered to v acute apex, dull grey-green above, shiny green below, with minute papillae (mostly nr base), without veins raised below (the midrib hardly raised). Shoots round, occ reddish at base. Bulb with many bulbils. ❶. Oct-Jun..*Rosy Garlic* **Allium roseum**

Lf margin always entire

Lvs strongly keeled by midrib below

Lvs 40-60cm x 15-30(50)mm, blunt or acute (never finely so), ± dull green both sides, triquetrous or strongly V-shaped, with unpleasant sickly odour. Ligule absent. Shoots round. Stems round, solid. Jan-Jun...................................*Honey Garlic* **Nectaroscordum siculum**

Lvs to 30cm x 10-14mm, blunt or acute (never finely so), dull green above, ± shiny below, ± triquetrous esp nr base, with onion odour, with stomata scattered above but confined to keel area below. Ligule absent. Shoots round below, trigonous above. Stems triquetrous, solid. Oct-Jun...*Three-cornered Garlic* **Allium triquetrum**

Lvs flat, not or weakly keeled below (*A. ursinum* would key out here but lvs are petiolate)

Lf margins ciliate with unicellular hairs to 1.5mm ❷

Lvs 4-5mm wide, tapered to acute apex, channelled proximally, flattened distally, pale or mid-green. Shoots round. Nov-May.................................*Hairy Garlic* **Allium subhirsutum**

Lf margins hairless

Lvs (20)30-60mm wide. Ligule absent

Lvs to 60cm, weakly leek-scented, with erose margins. Mar-Jul ....*Broad-leaved Leek* **Allium nigrum**

Lvs <25mm wide

Ligule present. Lvs garlic-scented

Lvs 5-25mm wide, tapered to ± acute narrow hooded apex, flat or weakly channelled, with spiral fibres when torn. Shoots flattened. Feb-Jul.................*Garlic* **Allium sativum**

Ligule absent. Lvs onion-scented

Lvs to 25cm x 4-11mm, obtuse, flat to weakly channelled, pale grey-green, with erose to entire margins, without spiral fibres. Shoots round. Bulb with non-reticulate tunic. Nov-Jun...*Spring Starflower* **Tristagma uniflorum**

Lvs to 20(40)cm x 6mm, obtuse, flat, slightly greyish-green, entire, without spiral fibres. Shoots flattened. Bulb with reticulate tunic. ❶. Apr-Sept. VR alien ....*Chinese Chives* **Allium tuberosum**

on stems, with >3 raised veins below. Stems with scale-like lvs nr base, solid
). ❶ Lvs hairy
/ate, sessile, with stomata below only

.., broadly ovate (to elliptic), with upcurved apiculus, broadly cuneate to ± cordate at base, usu shiny green both sides, with minutely erose margins, opaquely net-veined, with 1-3 raised veins each side of midrib. Stems 20-60cm, ± square and hairless below lvs, round and glandular-hairy above. May-Jul. Lowland................................*Common Twayblade* **Listera ovata**

Lvs 1-2cm, triangular-ovate, with rounded hard mucro, broadly cuneate to ± cordate at base, slightly translucent, shiny green both sides (paler below), with ± entire to erose-crenulate margins, with 1 main vein each side of midrib (weaker veins also present). Stems 6-20cm, furrowed, hairless below, glandular-hairy above. Jun-Aug. Mostly N Br...*Lesser Twayblade* **Listera cordata**

■Lvs >2, alt (largest nr mid-stem)

Lf margin with papillae (often long)

Lvs with stomata scattered above, with veins slightly papillate both sides

Lvs 3-8(10) x 1.5-5cm, acute, elliptic to ovate, dark green above, purplish or green below. Stem usu 1, 15-30(60)cm, violet below, ± hairy (densely so above). Limestone habs ....*Dark-red Helleborine* **Epipactis atrorubens**

Lvs with stomata below only

Lf veins distinctly papillate at least above

Lvs 5-10 x 2-5cm, broadly ovate to broadly lanc, acute, dull yellow-green or dark green, ± undulate, with papillate margins, with 7-9 main veins raised below. Stems 1-several, 20-70cm, ± hairy (esp above), usu green. Often woods (esp beech) ....*Narrow-lipped Helleborine* **Epipactis leptochila** var **leptochila**

Lvs 4-8 x 3cm, ovate-lanc to ovate-oblanc, usu acute, usu yellow-green, undulate, with papillate margins (papillae often long, unequal, occ absent), with 7-9 main veins raised below. Stems 1(3), to 60cm, violet and hairless below, shortly hairy above. Coastal dunes ....*Dune Helleborine* **Epipactis leptochila** var **dunensis***

Lf veins minutely papillate both sides

Lvs 4-10 x 2-5cm, ovate-lanc to lanc, acute to shortly acuminate, grey-green to purplish (esp below), with irreg longish papillate margins, with 1-3 main veins raised below each side of midrib. Stems (1)3-5(10), 20-70cm, violet below, shortly hairy above (often sparsely so) ....*Violet Helleborine* **Epipactis purpurata**

Lf veins not or obscurely papillate either side (occ papillate in *E. helleborine*)

Lvs 5-17 x 2.5-6(10)cm, broadly ovate to ovate-lanc, acute or shortly acuminate, dull deep green, with long unequal papillate margins, with c15-36 veins each side of midrib, with c5 main veins raised below (occ papillate). Stems 1-3(5), 25-80cm, often violet below, whitish with short hairs above..............................*Broad-leaved Helleborine* **Epipactis helleborine***

Lvs 3-7 x 1.5-3cm, elliptic, acuminate, green, often undulate, with unequal long-papillate margins, with 14 veins each side of midrib. Stems 1(3), 10-40(65)cm, green, hairless to sparsely hairy above............................*Green-flowered Helleborine* **Epipactis phyllanthes***

Lf margin crenulate (to entire). Lf veins not or obscurely papillate above

Lvs 0.8cm wide, usu folded at maturity

Lvs 5-10cm, spirally arranged, ovate-lanc to linear-lanc, v acute. Scale lvs 2-4, brownish, occ green-tipped. Stem 20-50cm, striate, glandular-hairy above. Sch8 ....*Red Helleborine* **Cephalanthera rubra**

Lvs 1.5-1.8cm wide, usu folded at maturity

Lvs 7-10(20)cm, 2-ranked, elliptic-lanc to lanc, acute, with crenulate margins, with 2-3 veins unequal regular veins each side of midrib. Scale lvs 2-4, whitish, occ green-tipped. Stems 20-60, ± hairless, slightly ridged above. Dry calc woods ....*Narrow-leaved Helleborine* **Cephalanthera longifolia**

Lvs >1.8cm wide, not or hardly folded at maturity

Lvs shortly hairy both sides ❶

Lvs (5)12-15 x 5-7cm, spirally arranged, ovate-oblong, acute or acuminate (occ obtuse), light green, flat to weakly channelled, ciliate, with 5-9 main veins raised below (30-40 minor veins each side of midrib), usu with stomata below only. Stems 1-3, 15-45cm, shortly hairy. VR, N Eng. Sch8..*Lady's-slipper* **Cypripedium calceolus**

Lvs hairless

Shady habs. Lvs shiny dark green, 5-8(10) x 1.5-2.5cm, ± 2-ranked, elliptic-lanc to ovate-oblong, rarely with minutely papillate margins, with 2 main veins raised below each side of midrib (16 veins in total). Scale lvs 2-3, brown, the upper often green-tipped. Stems 15-50cm, usu hairless but papillate nr infl........*White Helleborine* **Cephalanthera damasonium**

Sunny damp calc habs. Lvs yellow-green (the lowest purplish), 5-15 x 1.8-3.8cm, spirally arranged, oblong-ovate to elliptic-lanc, acute to obtuse, with erose to crenulate (rarely papillate) margins, with 2-3 veins raised below each side of midrib (30-60 veins in total). Scale lvs 2-4, often purplish. Stems 15-45(60)cm, hairy above ....*Marsh Helleborine* **Epipactis palustris***

JG

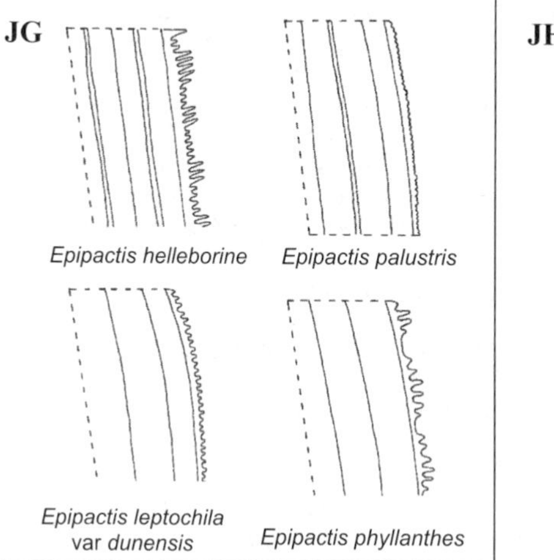

Epipactis helleborine *Epipactis palustris*

Epipactis leptochila var *dunensis* *Epipactis phyllanthes*

JH

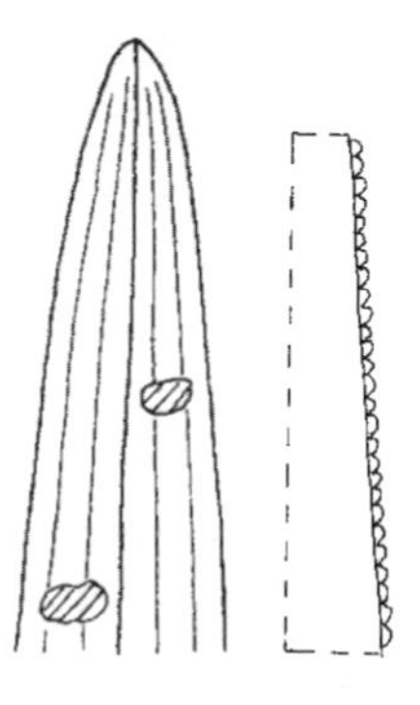

Dactylorhiza maculata
lf apex & lf margin

Group JH – *Lvs spotted, spirally arranged in rosette or false rosette. Scale lvs whitish*

■Lvs with stomata below only (or confined to margins above), with crenulate or papillate margins. Stems solid

Lvs shiny green both sides

Lvs 3-8 in rosette, 5-20 x (0.5)2-4cm, oblanc-oblong, ± obtuse with apiculus occ recurved, with transverse blotches, with 3-5 main veins each side of midrib. Feb-May. Calc habs (esp woods) ....*Early-purple Orchid* **Orchis mascula**

Lvs usu dull green above

Dry calc turf. Lvs 2-3(6), 7-12 x 1.5-2.5cm, elliptic to obovate-oblong, ± obtuse, with stomata below only, often with purple crenulate (often acutely so) margins, with transverse blotches, with 4-5 veins each side of midrib. Feb-Jul..........*Common Spotted-orchid* **Dactylorhiza fuchsii**

Damp acidic turf. Lvs 2-3(6), 7-10 x 0.8-1.5cm, often oblanc, acute to obtuse, usu with stomata nr margins above, often with purplish crenulate margins, with transverse blotches, with 3-8 veins each side of midrib. Mar-Jul......................*Heath Spotted-orchid* **Dactylorhiza maculata***

■Lvs with stomata scattered above (not confined to margins), with entire margins. Stems hollow (often <½ diam). R

Lvs 3-12 x 1-1.5cm, heavily spotted or blotched, weakly hooded at ± acute apex (not reflexed). Apr-Aug. NW Scot. Sch8.......................................*Lapland Marsh-orchid* **Dactylorhiza lapponica**

Lvs 5-10(12) x 2-3cm, rarely spotted, weakly hooded at ± acute reflexed apex. Apr-Aug. N Scot, W Wales, Ire..*Western Marsh-orchid* **Dactylorhiza majalis**

Group JI – *Lvs unspotted, with scattered stomata above (often in single rows each side of veins), with hyaline margins (±) entire, rarely crenulate or papillate*

■Dry short gsld (*Dactylorhiza praetermissa* may rarely key out here)

Lvs 4-9, in flat rosette, 2.5-4.5 x 1.8cm, stiff, trullate-ovate, acute, dull bluish-green, with up to 5 veins each side, with stout midrib narrowing to apex (hardly raised below). Stems 7-20cm, round, glandular-hairy (esp above), with several pale green adpressed bract-like scale lvs. Sept-Jul ....*Autumn Lady's-tresses* **Spiranthes spiralis**

■Damp habs

Lvs mostly ≥10mm wide

Lf margin (±) entire

Stem hollow >½ diam. Lvs (4)7-12(20) x 1.3-2cm, elliptic-lanc, narrowly hooded at apex, not reflexed, usu shiny yellow-green above (rarely with faint spots). Apr-Aug ....*Early Marsh-orchid* **Dactylorhiza incarnata***

Stem hollow usu <½ diam. Lvs 3-10(12) x 1-2cm, green (occ faintly spotted), narrowly oblong-lanc or linear-lanc, weakly hooded at obtuse to ± acute apex, not reflexed. Apr-Aug. Calc fens, dune-hollows......................................*Narrow-leaved Marsh-orchid* **Dactylorhiza traunsteineri**

Stem hollow usu <½ diam. Lvs 5-10(12) x 2-3cm, elliptic-lanc, green, rarely spotted, weakly hooded at ± acute reflexed apex. Apr-Aug. R, N Scot, W Wales, Ire ....*Western Marsh-orchid* **Dactylorhiza majalis**

Lf margin minutely crenulate or papillate. Stems often with small hollow <½ diam

Basal lvs usu 5-8, 10-13 x 0.9-2.5cm, not or hardly hooded at ± acute apex, green, with minutely crenulate to entire margins, with stomata in single rows each side of veins. Apr-Aug. Eng, Wales..*Southern Marsh-orchid* **Dactylorhiza praetermissa***

Basal lvs usu 5-8, 7-15 x 1.2-3cm, elliptic-lanc, not or hardly hooded at ± acute apex, green, with minutely crenulate to entire margins, with 4-8 veins each side of midrib, with stomata in single rows each side of veins. Apr-Aug. Ire, N Eng, Scot, Wales ....*Northern Marsh-orchid* **Dactylorhiza purpurella**

Lvs mostly <8mm wide. Lf margin always entire

Lvs basal, 5-6, held erect, 5-10(15)cm x 4-7(10)mm, lanc to linear-oblanc, hooded at acute apex, U-shaped, with silvery sheen above (due to cuticle cells), with midrib raised below, with 1(2) veins each side of midrib. Uppermost lvs bract-like, acuminate, often spreading. Stems 12-25cm, bluntly 3-angled, sparsely hairy above. All yr. VR, Scot, Ire. Sch8 ....*Irish Lady's-tresses* **Spiranthes romanzoffiana**

Lvs ± basal, 5-6, held ± erect, 5-12cm x 5-9mm, lanc, hooded at obtuse apex, flat to channelled, bright shiny green both sides, without midrib raised below. Uppermost lvs bract-like, adpressed. Stems 10-40cm, angled, ± glandular-hairy above. Jul-Sept. VR, New Forest (now re-introduced alien!)..*Summer Lady's-tresses* **Spiranthes aestivalis**

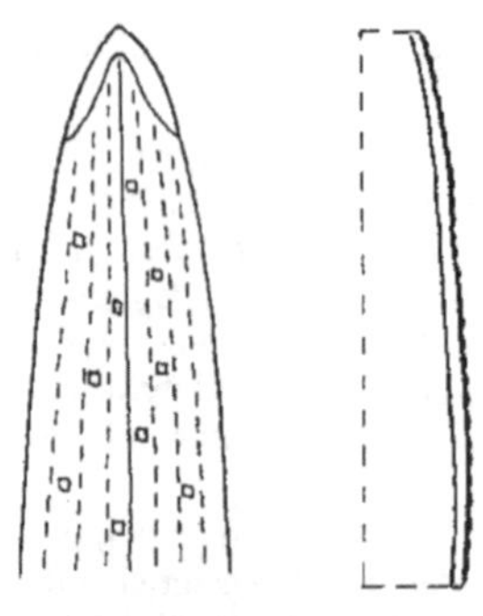

Dactylorhiza incarnata
lf apex & lf margin

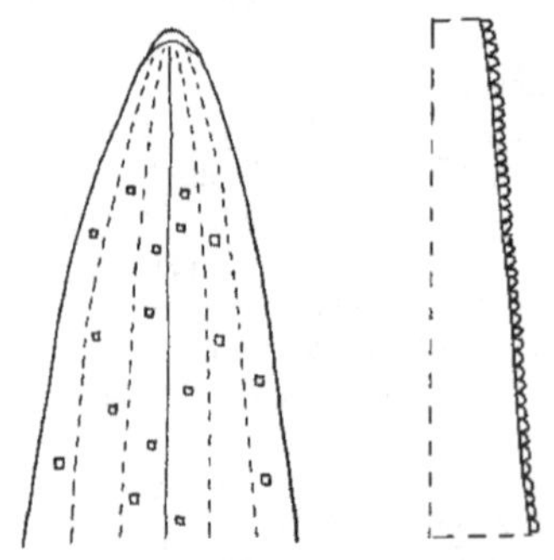

Dactylorhiza praetermissa
lf apex & lf margin

Group JJ – *Lvs unspotted, with stomata absent above, with hyaline margins (±) entire*

■Lvs >10cm

Lvs with broad entire hyaline margins, unpleasantly goat-scented, 6-22 x 3-5cm, elliptic-oblong, ± obtuse, ± shiny to dull green above, with 3-6 main veins each side of midrib. Stems 20-40(90)cm, stout, occ purple-mottled, hairless. Feb-Jun (all yr). Calc habs, S Eng. Sch8. ....*Lizard Orchid* **Himantoglossum hircinum**

Lvs with narrow ± entire hyaline margins, pleasantly vanilla-scented, 10-15(25) x 3-5cm, elliptic to ovate-oblong, obtuse to apiculate, shiny green above, paler with silvery sheen below, with 3-4(8) main veins each side of midrib. Feb-Jul. S Eng.................................*Lady Orchid* **Orchis purpurea**

■Lvs <10cm, with narrow hyaline margin

Dry habs. Lvs elliptic-oblong

Lvs 2(3), ± basal, with entire to erose margins. Shady calc habs

Lvs 4-8(12) x 1-2cm, ± acute, shiny pale green. Stems 15-60cm, ± hairless ....*Fly Orchid* **Ophrys insectifera**

Lvs 2-4(6) in rosette, entire. Calc gsld. (*Orchis morio* may key out here)

Lvs usu dark bluish-green

Lvs 3-8(10) x 0.5-2.5cm, ± acute, with upturned apiculus. Sept-Jul ....*Bee Orchid* **Ophrys apifera**

Lvs 4-10 x 0.5-2.5cm, ± acute, with upturned apiculus. Sept-Jul. Calc gsld. Sch8 ....*Late Spider Orchid* **Ophrys fuciflora**

Lvs pale grey-green

Lvs 4-10 x 0.5-1.8cm, hooded at apex. Sept-Jul. Sch8 ....*Early Spider Orchid* **Ophrys sphegodes**

Fens, dune slacks. Lvs ovate-elliptic to ± orb

Lvs 2, ± opp, 2.5-4(7) x 2-4cm, usu obtuse, greasy-looking, with entire margins. Stem 6-20cm, hairless, strongly 3(5)-angled above, with 2-3 sheathing basal scales enclosing pseudobulb. Jun-Oct. Sch8..*Fen Orchid* **Liparis loeselii**

Group JK – *Lvs >6cm, unspotted, with stomata absent above (or nr margins in* Gymnadenia conopsea*), with minutely crenate or papillate margins*

■Lvs 2, elliptic to elliptic-oblanc, obtuse, with translucent longitudinal streaks, with minutely crenulate margins, dull above with greasy-feel, shiny below, often odourless, with 5-7 ± opaque veins each side of midrib, with midrib raised below. Stems usu 20-40cm, hairless, ± angled above. Apr-Jul

Lvs (5)8-21 x 3-7cm. Fls with strongly curved spur............*Butterfly-orchid* **Platanthera chlorantha**

Lvs 4-10(15) x 0.7-4cm. Fls with weakly curved spur....*Lesser Butterfly-orchid* **Platanthera bifolia**

■Lvs (2)3-8

Lvs transversely wrinkled nr base, not waisted

Lvs 3-6, spirally arranged in rosette, 4-12 x 1.5-3cm, oblong to oblong-lanc, obtuse to ± acute, with apiculus recurved or upturned (often withered), shiny light green both sides, with low crenulate margins (occ papillate), with 3-5 main veins. Mar-Jun ....*Man Orchid* **Aceras anthropophorum**

Lvs occ wrinkled, waisted (strongly channelled) in middle, occ squarrose nr apex

Lvs 3-8, spirally arranged in rosette, 3-11 x 0.5-2cm, elliptic-oblong to lanc, acute with mucro, shiny dark green both sides, with low crenulate margins, with 3-5 main veins and irreg opaque cross-veins. Oct-Jun...*Green-winged Orchid* **Orchis morio**

Lvs never wrinkled or waisted

Lvs strongly folded (even at maturity), 2-ranked in rosette

Lvs with apiculus, often with obtusely crenulate margins, with stomata absent above

Lvs 7-15 x 0.8-1.5(2)cm, narrowly oblong-lanc, acute at slightly hooded apex, soon dying as fls appear, with (3)5-8 main veins each side of midrib. Stems slightly angled above. Jan-Jul ....*Pyramidal Orchid* **Anacamptis pyramidalis***

Lvs without apiculus, often with acutely crenulate margins, often with stomata nr margins above

Lvs (3)6-15 x 0.5-2(3)cm, narrowly oblong-lanc, U- or V-shaped, blunt or ± acute at slightly hooded apex, similar to *Anacamptis* but thicker, stiffer and more persistent (even after fl), shiny yellow- to grey-green above, veins often obscure. Stems 15-50cm, occ purplish above. Apr-Aug

Fls early Jun to mid-Jul. Dry calc turf ....*Chalk Fragrant-orchid* **Gymnadenia conopsea** ssp **conopsea**

Fls late Jun to late Jul. Damp acidic to neutral turf, usu N & W Br ....*Heath Fragrant-orchid* **Gymnadenia conopsea** ssp **borealis**

Fls early Jul to Aug. Damp calc turf ....*Marsh Fragrant-orchid* **Gymnadenia conopsea** ssp **densiflora**

Lvs flat, usu spirally arranged in rosette

Lvs with apiculus

Lvs 3-8, 5-20 x (0.5)2-4cm, oblanc-oblong, ± obtuse, with apiculus occ recurved, bright or dark green, usu with transverse blotches, with margins crenulate or papillate, with 3-5 main veins each side of midrib. Feb-May. Calc habs (esp woods) ....*Early-purple Orchid* **Orchis mascula**

Lvs without apiculus

Lvs in a basal rosette. Chalk gsld, VR, S Eng. Sch8

Lvs 2(3), (5)8-20 x 3-5cm, ovate-elliptic, usu obtuse, dull grey-green above with greasy-feel, silvery below, with crenate-papillate margins, with 3-4 main veins each side of midrib ....*Monkey Orchid* **Orchis simia**

Lvs usu 3-5, (5)8-12(18) x 3-4cm, elliptic, acute, ± shiny bright green above, silvery below, with erose to slightly crenulate margins, with 4 main veins each side of midrib ....*Military Orchid* **Orchis militaris**

Lvs not in a basal rosette. Wet gsld, Channel Is

Lvs 3-8, 7-18 x 0.8-1.5(2)cm, lanc or linear-lanc, acute, ± folded, with low crenulate margins...*Loose-flowered Orchid* **Orchis laxiflora**

Group JL – *Lvs <5cm, unspotted, with stomata absent above, with minutely crenate margins.* ❶ *Lvs usu with small brown spots in interrupted lines.* ❷ *Plant rhizomatous with petiolate lvs and pinnate veins.* ❸ *Lf apex distinctly fringed with tiny bulbils.* ❹ *Lvs occ with stomata above nr margins*

■Lowland calc gsld habs

Lvs 2-5, hardly forming rosette, 1.5-6(10) x 1-3cm, ± orb or broadly oblong to obovate-elliptic, acute to obtuse, without apiculus, flat, dull grey- or yellow-green above, shiny below, with 3-5 veins each side of midrib (often 1(2) veins more translucent). (Mar) May-Jul ....*Frog Orchid* **Coeloglossum viride**

Lvs 2(4) in rosette, 4-5 x 1-1.5cm, elliptic-lanc, acute or obtuse, without apiculus, slightly channelled, mid- or yellowish-green, with 5 veins each side of midrib (1 v obvious). May-Jul ....*Musk Orchid* **Herminium monorchis**

Lvs 2-3(6) in rosette, 2-5 x 0.5-1.5cm, elliptic-oblong, obtuse or acute, without apiculus, weakly channelled to flat, bluish-green, with 4-7 veins each side of midrib. Oct-Aug (at least some plants) ....*Burnt-tip Orchid* **Orchis ustulata**

Lvs 3-6 in rosette, 3-7 x 0.5-1.5cm, elliptic-oblong, obtuse, mucronate, weakly channelled to flat, mid- to yellow-green. Mar-Jun. ❶. VR, Ire....................*Dense-flowered Orchid* **Neotinea maculata**

■Lowland coniferous woods

Lvs 5-6 in flat rosette, 1.5-3.5cm, ovate to ovate-lanc, without apiculus, abruptly narrowed to 1cm petiole, rolled when young, dark green, often mottled with pale green, with 2 main veins each side of midrib, with apparently pinnate 2° veins (occ with cross-veins). Stems 10-25cm, round below but angled above, glandular-hairy esp above. Scale lf 1, whitish, occ green-tipped. ❷ All yr ....*Creeping Lady's-tresses* **Goodyera repens**

■Bogs

Lvs 2-3(5), ± basal, 0.5-1cm x 5-8mm, ovate, broadly obtuse, bright or pale green, deeply channelled, with 3(7) veins. Stems 3-12cm, 3-5 angled above, hairless. Plant with pseudobulb enclosed by lvs. ❸....................*Bog Orchid* **Hammarbya paludosa**

■Upland habs

Lvs with translucent longitudinal streaks (HTL), usu 4(6), 2.5-8 x 0.6-1.7cm, oblanc, apiculate, shiny green above, ± with silvery sheen below, with c5 obscure veins each side of midrib. ❹ ....*Small-white Orchid* **Pseudorchis albida**

Lvs without longitudinal streaks....................*Frog Orchid* **Coeloglossum viride**

JK

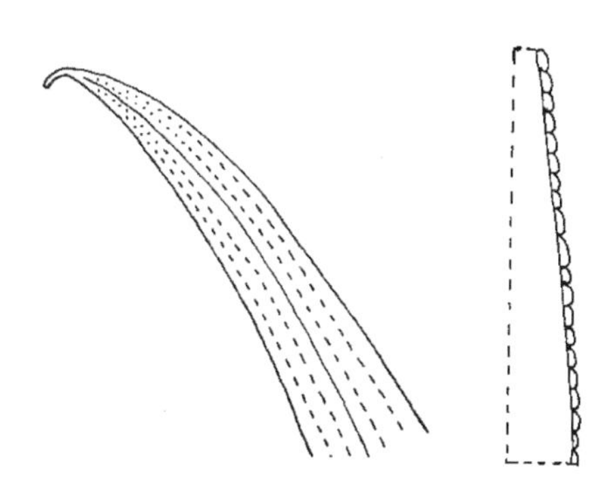

Anacamptis pyramidalis lf apex & lf margin

Group JM – *Lvs opp or whorled.* ❶ *Emergent stems and lvs hairy (hairless in all other spp)*

■Lvs opp. Dicot

Plant with bitter taste, with rhizomes short or absent. Lvs (1)3-5-veined, applanate when young ....(Gentianaceae) Go to **GENT**

Plant without bitter taste

Stoloniferous. Lvs 5(7)-veined, applanate when young. ❶. Bogs ....*Marsh St John's-wort* **Hypericum elodes**

Rhizomatous. Lvs 3(5)-veined, obvolute when young. Dry habs

Lvs 5-10cm, broadly ovate to elliptic, acute, with minutely crenate or scabrid-ciliate cartilaginous margins (often recurved), occ with pinnate veins but not net-veined, with stomata both sides. Petiole indistinct, with 3 flat vb's. Stems (and reduced lvs) purplish nr base, round. Apr-Oct..*Soapwort* **Saponaria officinalis**

■Lvs (3)5-whorled. Monocot

Lvs 4-8(12) x 0.5-0.8cm, linear-lanc, narrowly obtuse (hydathode easily felt), sessile, light green both sides, with scabrid-papillate margins, with 5-15 translucent veins, without cross-veins, with stomata below only. Stems to 80cm, slender, red-spotted, slightly rough with papillae, ± round, striate, fetid. Rhizomatous. May-Aug. Sch8..*Whorled Solomon's-seal* **Polygonatum verticillatum**

■Lvs (4)5-11-whorled below, alt above (occ single lf below). Monocot

Lvs 8-14 x 3-6cm, obovate-lanc, narrowed to short petiole, shiny green at least below, with minutely crenate-papillate margins, with parallel 2° veins sunken above and raised below, with pinnate cross-veins, with stomata below only. Stems round, often yellowish at extreme base, with scattered vb's. Bulbous. Apr-Aug...*Martagon Lily* **Lilium martagon**

Group JN – *Lvs long-ciliate, sessile, translucent veins not raised below (exc midrib in* Tradescantia*). Sheaths closed, reddish-purple or with purplish veins. Stems solid. Hairs unicellular. Per*

■Lvs mostly >6mm wide

Lvs with midrib raised below. Plant with abundant viscid sap

Lvs on stem, 2-ranked, 3-6 x 1-2cm, ovate, with rounded apex, with cartilaginous margins often purplish, shiny green above, not ribbed above, hairless, densely shortly ciliate, with long hairs at sheathing base, with spiral fibres absent or v sparse. Stems often >100cm (but usu <20cm tall), rooting at nodes, purplish, round, brittle, often with single line of short hairs, with smaller cataphylls. Stoloniferous. Usu all yr............................*Wandering-jew* **Tradescantia fluminensis**

Lvs often basal, 2-ranked below, spirally arranged above, 15-35 x 1.2-2.5cm, linear, with rounded apex, with cartilaginous margins, dark green, slightly ribbed above (not in TS), usu hairless both sides, ciliate at least nr base, with stomata both sides, with spiral fibres absent or v sparse. Stems to 0.5m, ± erect, rarely rooting at lowest nodes, often purplish, round, brittle, usu hairless. Shortly rhizomatous. Often all yr....................*Spiderwort* **Tradescantia virginiana**

Lvs without midrib raised below. Plant with sparse non-viscid sap

Lvs parallel-sided, or widest at base (may taper to apex)

Basal lvs 10-50cm x 6-15(20)mm, broadly linear, tapered to v acute apex, shiny bright green, sparsely hairy, with cilia usu >8mm, with midrib not discernible from 10-12 main veins, with cross-veins, with stomata below only. Rhizomatous and stoloniferous, often forming tussocks, with persistent dead brown lvs. All yr. Woods, mtns...........*Great Wood-rush* **Luzula sylvatica**

Lvs tapered at both ends, widest nr middle

Basal lvs 10-14cm x 5-11mm, linear-lanc, with swollen bulbous apex, bright green, with cilia to 10mm, often with opaque cross-veins. Stem lvs to 6cm. Tufted. All yr. Shady habs ....*Hairy Wood-rush* **Luzula pilosa**

■Lvs all ≤6mm wide, usu tapered at both ends and widest nr middle, midrib not raised below

Lf apex usu swollen, with short mucro. Shady habs

Basal lvs 4-6mm wide. Stem lvs to 6cm. *L. forsteri* x *pilosa*...............................**Luzula** x **borreri**

Basal lvs 2-4mm wide, shiny dark green above, with 1-3 main (and 2-3 obscure) veins each side of midrib, occ with stomata along margins above. Stem lvs often >6cm. All yr ....*Southern Wood-rush* **Luzula forsteri***

Lf apex rounded, not swollen, without mucro. Sunny habs

Shortly rhizomatous, in crowded clumps. Lvs occ with stomata along margins above

Basal lvs 2-5mm wide, often with reddish hydathode at apex, bright green, with sparse 5-10mm cilia, with 1-3 main veins each side of midrib, with obscure cross-veins. All yr ....*Field Wood-rush* **Luzula campestris***

Tufted. Lvs without stomata above

Acidic gsld, hths. Basal lvs 3-30cm x (4)6mm, often with reddish hydathode at apex, bright green, with cilia to 12mm, with 1-3 main veins each side of midrib, with cross-veins obscure or absent. Stems to 60cm. All yr.......................................*Heath Wood-rush* **Luzula multiflora**

Fens, VR. Basal lvs to 10cm x 3(6)mm wide, with hydathode not reddish at apex, pale green, with cilia to 5mm (mostly nr base), with 1-2 main veins each side of midrib, with cross-veins obscure or absent. Stems to 20cm. All yr. Fens, VR.............*Fen Wood-rush* **Luzula pallidula**

Lf apex v acute. Sunny habs

Lvs 8-20(30)cm. Usu lowlands

Basal lvs 3-6mm wide, widest nr base or ± parallel, channelled when young, inrolling on drying, shiny green, with cilia 1-5mm, with 5-7 veins each side of midrib, without cross-veins. Shoots often forming false-stem when young. Tufted, often with short stolons. All yr ....*White Wood-rush* **Luzula luzuloides**

Lvs <10cm. Usu mtns

Basal lvs 4-9cm x 1.5-3mm, widest at base, recurved, (±) channelled, sparsely hairy, with cilia 4-10mm (longest nr base), with midrib only translucent (others obscure). Stems <20cm. Infl compact, nodding. Tufted, with short stolons and persistent lf sheaths. All yr ....*Spiked Wood-rush* **Luzula spicata**

Basal lvs 2-5(7)cm x 1.5(2)mm, widest at base, recurved, stiff, deeply U-shaped, shiny green and hairless both sides, with cilia 2-8mm (longest nr base), with midrib only translucent (occ 1 vein each side of midrib). Stems <8cm. Infl diffuse, not nodding. Tufted, with short stolons or rhizomes and persistent lf sheaths. All yr...........................*Curved Wood-rush* **Luzula arcuata**

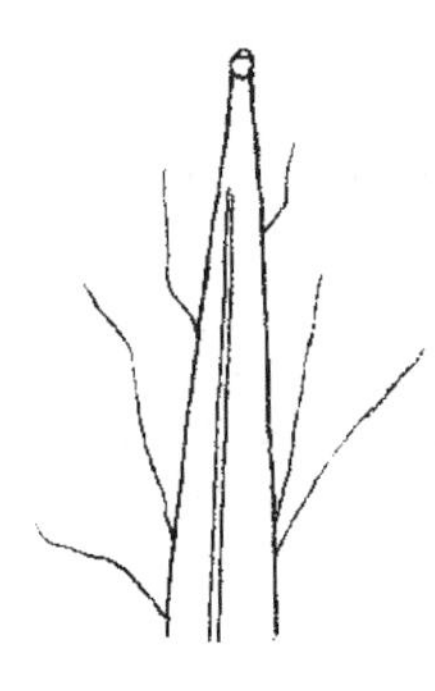

Luzula campestris lf apex

Luzula forsteri lf apex

Group JO – *Lvs with scabrid or papillate margins (at least at apex), never ciliate.* ❶ *Ligule occ present, 1mm, retuse (absent in all other spp)*

■Lvs >2mm wide

Lvs mostly or all basal

Lvs 2-ranked. Strongly rhizomatous per, with tuberous roots

Lvs obtuse, with slightly hooded apex and pale hydathode, folded when young, V-shaped to flat at maturity, with margins minutely scabrid-papillate or crenulate but smooth to touch, dark or light green, with midrib raised below, with other translucent veins v weakly raised below, with indistinct cross-veins, occ with c16 hollows in TS, with stomata below only or sparsely scattered above. Shoots flattened, whitish. Mar-Oct

Lvs to 90cm x 12-37mm. Fls dull orange, net-veined......*Orange Day-lily* **Hemerocallis fulva**

Lvs to 40(65)cm x 7-15mm. Fls yellow, parallel-veined ....*Yellow Day-lily* **Hemerocallis lilioasphodelus**

Lvs spiralling or 3-ranked. Weakly rhizomatous or tufted per, without tuberous roots

Lvs mostly >2cm wide, solid, with spiral fibres when torn

Lvs 10-30, 40-90(200) x 2-6cm, ± acute, weakly V-shaped, spongy, dull green to bluish-green both sides, with scabrid-papillate margins (and raised midrib below), with indistinct veins, without cross-veins, with scattered elongate stomata both sides. Stems 1-3m. Bulb absent. Tufted. All yr...*Red-hot-poker* **Kniphofia uvaria** agg

Lvs 4-8, 20-40 x 1.8-2.5(3)cm, *Hyacinthoides*-like, long-acute, channelled at least proximally, not spongy, green or weakly glaucous, with scabrid-papillate margins, with obscure veins, without cross-veins, with scattered elongate stomata both sides. Stems 15-60cm. Bulb 3-3.5 x 2.5-3cm, ovoid, with pink tunic. Mar-Jul.................*Tassel Hyacinth* **Muscari comosum**

Lvs <2cm wide, with hollows, without spiral fibres

Lvs 100-200 x 1-2cm, V-shaped, acute, with whip-like trigonous apex, with v sharply serrate margins (and raised midrib below), dull glaucous, with >15 translucent veins each side of translucent midrib, without cross-veins (exc at base), with 20 hollows in TS, with stomata below only. Sheaths concave at apex, the basal yellow-brown. False-stems absent. Stems to 3m, 1-4cm diam, round, hollow. Rhizomatous. All yr. Fens ....*Great Fen-sedge* **Cladium mariscus**

Lvs to 35 x 0.5-1.5cm, V-shaped (folding on drying), acute, with scabrid margins nr apex, shiny green both sides, with 8-12 translucent veins each side of opaque midrib (raised below), with cross-veins, with 8-10 flattened small hollows (non-compressible), with stomata in bands (between veins) below and along margins above. Sheaths oblique or concave at apex, soon splitting. False-stems often present. Stems to 60cm, 2-4mm wide, trigonous, occ reddish at base, solid. Tufted or shortly rhizomatous. All yr. R alien (increasing) ....*Pale Galingale* **Cyperus eragrostis**

Lvs all on stem (or false-stem)

Lvs 2-ranked on stem. Ann. *Poaceae*

Lvs usu (8)13-16mm wide, acute (apex often rolled), often with transverse wrinkling, with rough thickened cartilaginous margins (often smooth nr base), with midrib raised below, with 4-5 main veins each side of midrib (not visible when HTL), with v crowded minor veins, with stomata both sides. Sheaths open, flattened and keeled below. Culms to 150cm. The only grass without a ligule or ring of hairs...................................*Cockspur* **Echinochloa crus-galli**

Lvs 3-ranked or spirally arranged. Per. *Liliaceae*

Dry (often shady) habs, often hortal. Lvs 7-10 x 0.8cm, linear-lanc, acute, ± flat, shiny green both sides, with long-papillate margins (not scabrid to touch), with 1-3 veins each side of midrib (raised below with thicker midrib), ± without pinnate cross-veins. Stems to 100cm ....*Pyrenean Lily* **Lilium pyrenaicum**

[cont'd]

Wet saline habs. Lvs (3)5-8(10)mm wide, ± folded or weakly M-shaped, with acute trigonous apex, thick, with plastic-feel, shiny green both sides, with margins (and keel) of lower lvs smooth but those of upper lvs rough, with c12 equal translucent veins each side of midrib, usu with cross-veins obscure or absent, with stomata both sides, occ with hollows in TS. Sheaths closed, herbaceous, with convex hyaline apex, with cross-veins. Ligule absent but with ± acute pale mark 4-10mm. Stems triquetrous, rough nr apex. Shoots trigonous, arising from bulbous tubers along rhizome. Apr-Oct.........*Sea Club-rush* **Bolboschoenus maritimus**

Wet, usu freshwater habs. Lvs (4)6-10mm wide, M-shaped, acute (apex often dead), slightly rough above esp nr apex, ± shiny green both sides, with scabrid margins, with veins translucent to opaque, with indistinct cross-veins, solid, with stomatal bands below only (less obvious nr margins and midrib). Sheaths straight to convex at apex. Ligule absent but with pale mark. Shoots weakly trigonous to round, dead sheaths at base often reddish. Stems trigonous, solid. Rhizomes not or hardly swollen. Mar-Nov..*Galingale* **Cyperus longus**

■Lvs <2mm wide, never truly basal

Sheaths closed, ladder-fibrillose with age, whitish or reddish

Lvs visible Apr-Aug, 1-2, all nr base, 1-1.5mm wide, to 80cm, U-shaped, obtuse, rigid, whitish above (due to exposed pith), with scabrid margins, with stomata at least below. Stems all yr, to 100cm, 2-2.5(4)mm diam, pale green, oval, smooth, loosely pith-filled. Basal sheaths mostly without blades. Tufted. Coastal, R..........*Round-headed Club-rush* **Scirpoides holoschoenus***

Sheaths closed, never ladder-fibrillose, whitish to pale brown

Ann

Lvs to 4cm x 1-5mm, usu shorter than stems, dull grey-green, soon withering, with cross-veins. Stems to 5(10)cm, triquetrous, smooth. Jul-Oct. Muddy habs, VR. Sch8 ....*Brown Galingale* **Cyperus fuscus**

Per. ❶

Lvs 0.7-2(2.5)mm wide, channelled, with rounded apex, with scabrid or smooth margins (always scabrid nr apex), yellow-green, with translucent midrib, with 2-3 veins each side of midrib visible below, without cross-veins, with stomata below only. Sheaths often convex at apex, without cross-veins. Stems trigonous or round, hollow. Apr-Oct

Basal sheaths lfless but with short fragile aristate tip, tough, with 1(3) buds in their axils. Lvs 7-20cm, shorter or ± equal to stems. Tufted or with short rhizomes. Bogs ....*White Beak-sedge* **Rhynchospora alba**

Basal sheaths mostly with short blades, without overwintering buds in their axils. Lvs 4-15cm, usu much shorter than stems. Rhizomes long. Bare peat and bogs, R ....*Brown Beak-sedge* **Rhynchospora fusca**

Sheaths open, never ladder-fibrillose, shiny blackish

Lvs 2-3(6), usu 15-40cm x 0.5-1.5mm, channelled (semi-cylindrical with involute margins), acute, solid, with minutely scabrid margins, with stomata all sides. Basal sheaths with 9-11 veins, with tough adnate auricles. Stems 15-75cm, oval to round, striate, solid, with stomata. Tussock-forming. All yr. Usu base-rich bogs......................*Black Bog-rush* **Schoenus nigricans**

Lvs 2-3(6), 1-15cm x 0.5mm, not channelled, acute, solid, with minutely scabrid margins, with stomata all sides. Basal sheaths usu with 7 veins, ± without auricles. Stems to 12(40)cm, oval to round, striate, solid, with stomata. Tussock-forming. All yr. Base-rich flushes, VR, Perth ....*Brown Bog-rush* **Schoenus ferrugineus**

Group JP – *All lvs on stem (basal rosette absent).* ❶ *Dicot with stipules to 1mm, subulate (all other spp are monocots without stipules)*

■Lvs with main veins (and midrib) raised below. Rhizomatous

Lvs minutely hairy esp below, ± shiny green below, narrowed to short indistinct petiole, ± 2-ranked, with stomata below only. Stems arching, with distinct odour (not fetid)

Stem to 90cm, v weakly ridged, solid, with many scattered vb's (and around margin). Lvs 8-18 x 4-8cm, elliptic, minutely ciliate. Basal sheaths long-ciliate ....*False Solomon's-seal* **Smilacina racemosa**

Lvs hairless, usu glaucous/pruinose below, sessile, ± 2-ranked, with stomata below only. Stems arching, with fetid (meat) odour

Stems round throughout, to 70cm. Lvs 5-12 x 3-4cm, ± elliptic, acute, rolled when young, shiny green above, with erose to v minutely crenulate margins, with many translucent veins. Basal sheaths white or purplish. Fls 1-6 per lf axil..............*Solomon's-seal* **Polygonatum multiflorum**

Stems (weakly) angled above lowest lf (often round below), to 90cm. Lvs 5-12 x 3-4cm, ± elliptic, acute, rolled when young, with erose or minutely crenulate margins, shiny green above, with many translucent veins. Basal sheaths white or purplish. Fls 1-6 per lf axil. *P. multiflorum* x *odoratum*...*Garden Solomon's-seal* **Polygonatum** x **hybridum**

Stems distinctly angled or ridged throughout, to 20cm. Lvs to 5(10) x 2-3cm, ± ovate, bluntly cuspidate, rolled when young, with erose margins, dull greyish-green above, with many translucent veins. Fls 1 per lf axil. Limestone habs. W Br ....*Angular Solomon's-seal* **Polygonatum odoratum**

■Lvs without main veins (exc midrib) raised below

Rhizomatous per. Lvs M- or V-shaped

Lvs to 30cm x 3-5.5mm, tapered to acute trigonous apex, deeply V-channelled (± folded), <0.5mm thick, with whitish cells above, yellow-green below, smooth, with midrib raised below, with opaque or weakly translucent veins, with cross-veins, with 8 hollows in TS. Ligule 1.5-3mm, obtuse, retuse (auricle-like), with 0.5mm free margin. Sheaths closed, with concave apex often scarious. Basal sheaths occ reddish beneath dead brown sheaths, rarely ladder-fibrillose. Shoots ± round or weakly trigonous at base. Stems to 100cm, triquetrous, ± solid. Shortly rhizomatous. Dune slacks. VR, Lancs. Sch8..........*Sharp Club-rush* **Schoenoplectus pungens**

Lvs to 40cm x (3)5-8(10)mm, ± folded or weakly M-shaped, with acute trigonous apex, thick, with plastic-feel, shiny green both sides, with margins (and keel) of lower lvs smooth but those of upper lvs rough, with c12 equal translucent veins each side of midrib, usu with cross-veins obscure or absent, with stomata both sides, occ with hollows in TS. Sheaths closed, herbaceous, with convex hyaline apex, distinctly veined, with cross-veins. Ligule absent but with ± acute pale mark 4-10mm. Stems to 100cm, triquetrous. Shoots trigonous, arising from bulbous tubers along rhizome. Wet saline habs.......*Sea Club-rush* **Bolboschoenus maritimus**

Tufted ann or per. Lvs flat to weakly U-shaped

Stems ± square. Ann

Lvs (actually phyllodes) to 15cm x (2)4-8mm, linear-lanc, grass-like, with smooth margins (erose-crenulate), with 3-4 parallel veins each side of midrib slightly raised below, with weak pinnate veins, with stomata both sides. Stems to 70cm, sparsely minutely hairy when young, hollow. Apr-Oct. ❶...*Grass Vetchling* **Lathyrus nissolia**

Stem round (occ furrowed). Per (ann)

Lvs usu 1, nr base of stems. Plant not bulbous

Lvs to 5 x 0.5mm, obtuse. Stems 15-50cm, 0.7-1mm diam, furrowed, round, ± solid, with stomata all around (exc in sunken cartilaginous grooves). Ligule minute, adnate. Basal sheaths closed, shiny pale-brown (occ reddish). Densely tufted per. Mar-Nov. Peaty moors, bogs..*Deergrass* **Trichophorum cespitosum** agg

[cont'd]

Lvs to 5 x 0.7mm, obtuse. Stems to 15cm, 0.5-1mm diam, not furrowed, oval, ± solid, with stomata all around. Ligule absent. Basal sheaths closed, with purplish or dark red veins. Tufted ann or per. Apr-Oct (all yr)

Spikelets 1(3). Terminal bract usu ≤ infl. Nutlet smooth.. *Slender Club-rush* **Isolepis cernua**

Spikelets (1)2-4. Terminal bract >> infl. Nutlet ridged...... *Bristle Club-rush* **Isolepis setacea**

Lvs 2, on stem. Plant bulbous

Lvs 5-15 x 5-6mm, linear, ± flat to U-shaped, obtuse at slightly hooded apex (with pale hydathode), green, with stomata both sides. Stem 1, to 20cm. Jan-Apr. Hortal .... *Alpine Squill* **Scilla bifolia**

Lvs 6-15cm x 4-10mm, linear, ± flat to U-shaped, acute at hooded apex (with pale hydathode), green or ± glaucous, with stomata both sides. Stem 1, to 40cm. Winter-flooded calc gsld or hortal... *Fritillary* **Fritillaria meleagris**

Lvs >5, spirally arranged on stems. Plant bulbous

Lvs 7-10 x 0.8cm, linear-lanc, acute, ± flat, shiny green both sides, usu with papillate margins (not rough to touch), with 1-3 veins each side of midrib raised below. Stems to 100cm. May-Sept. Hortal, often shady habs...................... *Pyrenean Lily* **Lilium pyrenaicum**

Group JQ – *Lvs linear, with a central whitish stripe above*

■Lvs flat or U-shaped, hooded at apex with pale hydathode, with 8-20 weakly raised veins below (midrib hardly raised), with viscid sap, with scattered elongate stomata both sides. Basal sheaths absent or obscure. Bulbous. Feb-Jun

Lvs 30-60cm x (5)8-15mm. Scape to 60cm

Bulb with numerous bulbils, progressively renewed over 3-4 yrs ....*Drooping Star-of-Bethlehem* **Ornithogalum nutans**

Lvs 20-30cm x 3-6(8)mm. Scape to 40cm

Bulb with few elongate bulbils, renewed each yr. Lvs 4-6(9), with c8 minute hollows in TS ....*Star-of-Bethlehem* **Ornithogalum angustifolium***

Bulb with numerous globose bulbils, renewed each yr. Lvs 4-6(9), with c8 minute hollows in TS. VR alien..*Nap-at-noon* **Ornithogalum umbellatum**

■Lvs with revolute margins, truncate at apex and lacking visible hydathode, with wide midrib strongly raised below only, without viscid sap, with stomata below only. Basal sheaths 3-5, large, scarious, white or brown. Cormous. Feb-May

Lvs appearing in spring. Fls autumn

Lvs 2-4mm wide, 3-5 per shoot. Fls purplish.........................*Autumn Crocus* **Crocus nudiflorus**

Lvs appearing in spring with fls

Lvs 4-7mm wide, (2)3-4 per shoot. Fls purple to white..................*Spring Crocus* **Crocus vernus**

Lvs 2-4mm wide, 3-4 per shoot. Fls purple to white.........*Early Crocus* **Crocus tommasinianus***

Lvs 1-4mm wide, 4-6(8) per shoot. Fls yellow. *C. angustifolius* x *flavus* ....*Yellow Crocus* **Crocus** x **stellaris**

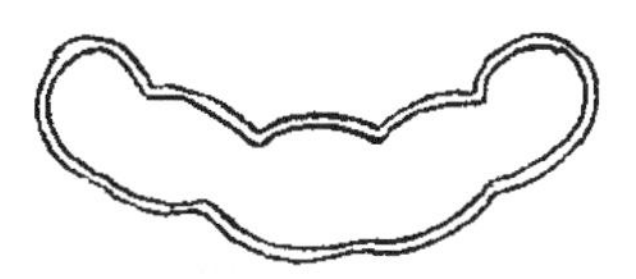

Ornithogalum angustifolium lf TS

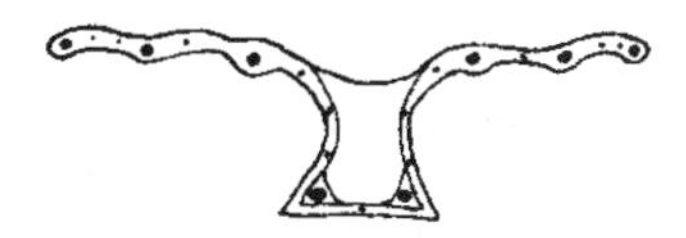

Crocus tommasinianus lf TS

Group JR – *Lvs 1-4mm wide, occ bristle- or thread-like, often twisted and fragile.* ❶ *Lvs ± square in TS*

▪Plant with bulb or corm

Lvs without stomata visible either side, all basal, hollow

Lvs (1)2-6 per corm, 4-6(10)cm x 0.5-1(2)mm, recurved, often twisted, green, ± square in TS with 4 grooves (2 on each of the upper and lower sides). ❶. Oct-May. Coastal turf, VR, SW Eng. Sch8..*Sand Crocus* **Romulea columnae**

Lvs with stomata both sides, all basal, solid

Lvs ≥2mm wide

Lvs 2-6 per bulb, to 13cm x 2-5mm, flat to channelled or ± cylindrical, wavy, green, without spiral fibres but with c5 obscure vb's and few viscid sap strands, with scattered stomata all around. Bulb ovoid, with bulbils. Oct-May. Coastal, mostly W Br......*Spring Squill* **Scilla verna**

Lvs 3-6 per bulb, (10)15-30(40)cm x 2-5(8)mm, strongly channelled (to ± cylindrical), wavy, slightly glaucous above, dark blue-green below, with spiral fibres, with stomata scattered above but mostly in rows below. Bulb ovoid, usu with bulbils. Nov-Jun. E Anglian Brecks or R hortal..*Grape-hyacinth* **Muscari neglectum**

Lvs 1-2mm wide

Coastal habs. S or SW Eng, Channel Is

Lvs 4-8 per bulb, to 12cm x 1-2mm, semi-cylindrical, obscurely ridged below, shiny green, dying yellowish, with stomata scattered above but in 5-8 bands below, with spiral fibres and viscid sap. Oct-Jul. S Eng, Channel Is................................*Autumn Squill* **Scilla autumnalis**

Lvs c12 per corm-like stock, 2-3cm x 0.7-1mm, semi-cylindrical, blunty triangular to flattened, with weak cartilaginous ridges, shiny green, turning orange, with 4 rows of stomata, without spiral fibres or viscid sap. Oct-May. VR, SW Eng (Lizard), Channel Is ....*Land Quillwort* **Isoetes histrix**

Mtns or inland rocky habs. Wales

Lvs usu 2 per bulb, 10-20cm x 1mm, obtuse to ± acute, semi-cylindrical, twisted, hairless, with stomata both sides. Stem lvs similar to basal lvs. Stems 5-15cm, hairless. Bulb 1, in tunic, producing slender stolons. May-Jul. Mtns, Snowdon. Sch8 ....*Snowdon Lily* **Lloydia serotina**

Lvs (1)2 lvs per non-fl bulbil (2 to 4 per fl bulb), 4-10(20)cm x 0.5-1mm, acute, semi-cylindrical, slightly sinuous, occ shortly hairy, with elongate stomata both sides. Stem lvs appearing basal, 15-40 x 2-4mm, narrowly lanc, ciliate. Stems 1-3cm, ± hairless. Bulbs 2, in same tunic, producing bulbils. Oct-Apr. VR, Radnor. Sch8 ....*Early Star-of-Bethlehem* **Gagea bohemica**

▪Plant without bulb or corm

Lvs hollow or strongly 4-angled, not pitted, with stomata scattered all sides

Lvs emerging singly (with cataphyll) or on false-stem, to 120cm x 4.5mm, acute, 4-angled, greyish-green, hollow. Stems 1-3, 20-40cm. Rhizomes absent but rootstock branched into 2-4 tubers. Nov-Jul. Hortal. R alien. ❶.......................*Snake's-head Iris* **Hermodactylus tuberosus**

Lvs 3-5, some on stem (with long sheathing base), 5-10(20)cm x 0.5-4(6)mm, obtuse, ± channelled above with involute margins, with 1-many hollows in TS, with stomata below only. Stem 1, 8-40cm, 1-2(3.5)mm diam, round. Rhizomes present. Mtn flushes ....*Chestnut Rush* **Juncus castaneus**

Lvs solid, never 4-angled, pitted, with stomata both sides

Lvs in dense rosette, linear-lanc, ± fleshy, flat to ± semi-cylindrical, green, reddish at base, 3-5(7)-veined, with pale green sunken midrib. Rootstock woody. Coastal, Channel Is and VR alien..*Jersey Thrift* **Armeria arenaria**

Lvs solid, never 4-angled, not pitted, with stomata below only (in bands)

Lvs 1.7-2.5(3)mm wide, acute with 0.2mm aristate tip, bright green, slightly channelled, occ inrolled, with obscure veins and cross-veins. Stems round, solid, with branches diverging at >90°. Ann (per). Apr-Aug (all yr). Damp habs..............................*Leafy Rush* **Juncus foliosus**

Lvs 0.6-1.5mm wide, acute with minute (<0.1mm) mucro, mid-green, slightly channelled, occ inrolled, with obscure veins and cross-veins. Stems round, solid, with branches diverging at <90°. Ann. Apr-Aug. Damp habs

Basal sheaths occ reddish. Plant reddish ....*Toad Rush* **Juncus bufonius**

Basal sheaths reddish. Plant light brown. Brackish muddy habs ....*Frog Rush* **Juncus ambiguus**

Basal sheaths with greenish-brown veins (plant may be reddish). (*Isolepis fluitans* may key out here but is usu aquatic)

Sheaths closed. Lvs 1-2, nr base of stems, to 2cm x 0.5-0.7mm, ± obtuse, ± flat but channelled above, usu shorter than stems, with indistinct cross-veins, with obscure hollow(s) in TS. Stems to 15cm, oval, ± solid, with stomata all around. Tufted ann or per, usu remaining green. Apr-Oct (all yr)

Spikelets 1(3). Terminal bract usu ≤ infl. Nutlet smooth....*Slender Club-rush* **Isolepis cernua**

Spikelets (1)2-4. Terminal bract >> infl. Nutlet ridged........*Bristle Club-rush* **Isolepis setacea**

Group JS – *Lvs >4mm wide, triangular at least nr base, or semi-cylindrical, 2-ranked, sheathing at base, with translucent veins. Rhizomatous. Apr-Nov. Often growing in shallow water*

■Lvs sharply triangular at midpoint, flat nr acute apex, usu reddish at base, with opaque veins, with abundant spiral fibres when torn

Lvs all basal, rarely submerged or floating, 4-15mm wide, twisted, green, with stomata all sides, with aerenchyma in TS. Stems to 1.5m, round.....................*Flowering-rush* **Butomus umbellatus**

■Lvs triangular nr base only, keeled below by midrib at midpoint, flat nr obtuse apex, usu whitish at base, with translucent veins (opaque in floating lvs), with sparse spiral fibres when torn

Lvs all basal, rarely floating, usu twisted, with aerenchyma in TS, with stomata both sides (above only on floating lvs). Sheathing base flattened. Rhizomatous. Apr-Nov

Lvs erect, sharply keeled below

Lvs 50-150cm x 10-25mm, bright yellow-green. Floating lvs usu absent. Rivers and ponds ....*Branched Bur-reed* **Sparganium erectum***

Lvs 20-50(80)cm x 4-10(12)mm, bright yellow-green. Floating lvs usu present. Rivers and ponds (not acidic)..*Unbranched Bur-reed* **Sparganium emersum**

Lvs not erect, often floating, weakly keeled below

Lvs to 150cm x 2-5mm. Peaty mtn lakes...........*Floating Bur-reed* **Sparganium angustifolium**

Lvs to 30(50)cm x 2-6(7)mm. Organic-rich pools...............*Least Bur-reed* **Sparganium natans**

■Lvs semi-cylindrical, without midrib raised below, obtuse, usu whitish at base, usu twisted, with sparse or no spiral fibres when torn

Lvs all nr base, with margins forming a cylindrical open overlapping sheath nr base (occ with shoulder), with large aerenchyma (easily compressed), with stomata both sides

Lvs 12-20mm wide, glaucous when young..*Bulrush* **Typha latifolia**

Lvs usu 4-8mm wide, dark green. Often coastal....................*Lesser Bulrush* **Typha angustifolia**

JT

Plantago lanceolata *Plantago media*

Group JT – *Lvs with main veins (and midrib) raised below, with stomata both sides. (*Plantago maritima *and* Gagea lutea *may key out here). Dicots.* ❶ *Latex present*

■Lvs with some hairs at least when young, all basal, without obvious cartilaginous margin

Lvs with white latex turning red, with cottony hairs when young. ❶ ....*Viper's-grass* **Scorzonera humilis**

Lvs without latex, with septate hairs and silky hairs at extreme base

Lvs acute or apiculate. Fl stalk ribbed. Shortly rhizomatous per, with 1-3 rosettes

Lvs 5-30cm, linear-lanc (to elliptic), sessile and narrowing gradually to base, often concave, not or hardly undulate, hairy to ± hairless, with long thin (>1 x 0.05mm) hairs, with 3-5 translucent main veins slightly sunken above. Petiole not purplish at base, with long white silky hairs in rosette centre. PL 9..................................*Ribwort Plantain* **Plantago lanceolata***

Lvs obtuse or ± so. Fl stalk round

Shortly rhizomatous per, with 1-3 rosettes. Lvs ± sessile or with short petiole, usu hairy. All yr

Lvs 5-8cm, ovate, hairy, with medium thick (<1 x 0.2mm) hairs, thick, convex, undulate, with 3-5(7) opaque main veins (HTL), without cross-veins. Petiole often purplish at base, with brown silky hairs in rosette centre.......................................*Hoary Plantain* **Plantago media***

Tufted per or ann, with single rosette. Lvs with distinct petiole, often ± hairless. Mar-Nov

Lvs 7-15cm, elliptic to ovate, often undulate, thin, green, often with shortish thick (to 0.6 x 0.1mm) hairs, with (3)5-9 translucent main veins. Petiole not purplish at base, with white (occ brown) hairs in rosette centre

Lvs ± hairless, cuneate to ± cordate at base, usu dark green, 5-9-veined. Per. Usu dry habs..*Greater Plantain* **Plantago major** ssp **major**

Lvs minutely hairy, cuneate at base, usu pale green, 3-5-veined. Usu ann. Damp or saline habs...*Greater Plantain* (ssp) **Plantago major** ssp **intermedia**

■Lvs hairless, basal or on stems (often clasping), with distinct cartilaginous margin

Per (bi), with 1-several stems from stout rootstock. Upper lvs ± clasping without auricles

Basal lvs 4-7cm, oblanc to broadly ovate, acute to obtuse, petiolate, often ± falcate, dull dark green above, paler below, with minutely scabrid to smooth margins, 5-6-veined, with opaque network of minor veins. Stem lvs oblanc to linear-lanc, acute, often ± falcate. Stems to 1m, erect, hollow. VR, Essex...................................*Sickle-leaved Hare's-ear* **Bupleurum falcatum**

Ann, usu with 1 stem from taproot

Upper stem lvs clasping with rounded auricles

Basal (and lower stem) lvs to 10cm, obovate, obtuse, cuneate, ± sessile, ± fleshy, ± glaucous, occ slightly papillate. Upper stem lvs obovate-oblong to ± orb, obtuse. Stem 1, 10-30(60)cm, ± glaucous. Occ bi. Oct-Jul (all yr). VR alien.............*Hare's-ear Mustard* **Conringia orientalis**

Upper stem lvs perfoliate

Upper stem lvs ± orb, mostly <2x as long as wide, ± glaucous. Basal (and lower stem) lvs 2-6 x 1.5-4cm, broadly oblanc to ± orb, apiculate, narrowed to petiole or sessile, with minutely scabrid to smooth margins (often purplish), occ without raised veins below. Stems 15-30cm, often purple-tinged, hollow. Oct-Sept. PL 24...............*Thorow-wax* **Bupleurum rotundifolium**

Upper stem lvs ovate to narrowly so, mostly >2x as long as wide, ± glaucous. Basal (and lower stem) lvs 2-7 x 1.5-3cm, oblanc to broadly so, apiculate, narrowed to petiole, or sessile, with minutely scabrid to smooth margins (often purplish), occ without raised veins below. Stems 15-30cm, often purple-tinged, hollow. Oct-Sept ....*False Thorow-wax* **Bupleurum subovatum**

Group JU – *Bulb. Lvs 1-2 per shoot (emerging singly or in prs), linear to narrowly oblanc, with veins not raised below (exc midrib).* ❶ *Fls autumn (others fl in spring).* ❷ *Lf veins usu translucent (opaque or obscure in other spp)*

■Lvs 4-8cm wide, emerging singly or in prs

Lvs elliptic, with short hooded apex, rolled into short petiole, rolled when young, glaucous or pruinose, with entire cartilaginous margins, often hairy above, with stomata both sides, with sparse spiral fibres. Bulb tunic often with sparse hairs................*Garden Tulip* **Tulipa gesneriana**

■Lvs <3cm wide

Lvs emerging singly (rarely in prs) from bulb, without viscid sap forming strands (occ with spiral fibres)

Lf 12-25mm wide. Bulb tunic often with sparse hairs

Basal lf glaucous or pruinose above, narrowed to base and rolled into petiole (often below ground). Stem lvs 1-3, all nr base, to 30cm, apiculate at hooded apex, shallowly channelled, with entire hyaline margins, shiny green above, with red spots at apex, with stomata both sides, with sparse spiral fibres when torn. Mar-Jun.......................*Wild Tulip* **Tulipa sylvestris**

Lf <12mm wide. Bulb tunic hairless

Basal lf green or ± glaucous above, 6-15cm x 4-10mm, acute at hooded apex (with pale hydathode), rolled when young, ± flat to slightly channelled, without ridges below, with stomata both sides, without spiral fibres or viscid sap. Stem lvs 3(5), 2-ranked. Winter-flooded calc gsld or hortal. Apr-Jun...*Fritillary* **Fritillaria meleagris**

Basal lf green above, 10-40cm x 5-12mm, hooded with long inrolled apex (to 15mm), ± curled like a crozier when young, channelled, with 1-5 ridges below (visible as dark lines HTL), with stomata both sides, with non-viscid sap and v sparse spiral fibres. Stem lvs reduced to bracts, 3-ranked. Damp shady calc habs. Feb-May.................*Yellow Star-of-Bethlehem* **Gagea lutea**

Lvs emerging in prs from bulb (within membranous cataphyll), often several prs per bulb, with viscid sap forming strands, (±) obtuse at apex with pale hydathode. Dec-May

Lvs flat when young (occ with slightly revolute margins at maturity)

Lvs glaucous both sides, (5)7-8(10)mm wide, flat or slightly hooded at apex, with stomata sparse above but abundant below. Jan-May.............................*Snowdrop* **Galanthus nivalis***

Lvs green with a central glaucous band above, green or glaucous below, 3-9mm wide, flat or slightly hooded at apex. ❶........................*Queen Olga's Snowdrop* **Galanthus reginae-olgae**

Lvs explicative (lf margins sharply revolute) when young

Lvs ± glaucous above, pruinose-glaucous below, to 25cm x 6-20mm, with stomata abundant below but confined to nr midrib above.........................*Pleated Snowdrop* **Galanthus plicatus***

Lvs rolled when young (with flat margins at maturity). ❷

Lvs glaucous both sides, to 15(30)cm x 6-15(35)mm, often hooded at apex, with c15 veins each side of midrib (occ shallowly furrowed), rarely with raised cross-veins, with stomata abundant below but absent or confined to nr midrib above, with weak spiral fibres ....*Greater Snowdrop* **Galanthus elwesii**

Lvs shiny green at least above, to 8(16)cm x (5)15-25mm, not hooded at apex, with c15 veins each side of midrib (often shallowly furrowed), with raised cross-veins, with stomata abundant below but absent or confined to nr midrib above, with weak spiral fibres ....*Green Snowdrop* **Galanthus woronowii**

Galanthus nivalis
young lf TS

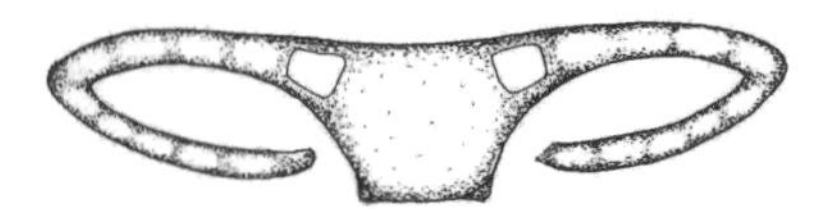

Galanthus plicatus
young lf TS

Group JV – *Lvs ≥3 per shoot, dull glaucous (or slightly so) above, linear to linear-lanc, all basal (exc* Tulipa sylvestris*, which may have some stem lvs). Bulb (exc* Simethis planifolia *which is shortly rhizomatous)*

▪Lvs with viscid sap when torn, spiral fibres v sparse or absent

Lvs 3-4, 20-60cm x 5-17mm, obtuse with pale hydathode, flat or weakly channelled (esp nr base), twisted, tough, ± spongy, with entire hyaline margins, with several obscure veins each side of midrib, with 'double' midrib raised below (visible as 2 dark lines HTL), with hollows in TS, with elongate stomata both sides. Cataphylls membranous. Scape with 2 raised lines, hollow. Bulb with brownish scales. Jan-May. Woods and hortal...............*Daffodil* **Narcissus pseudonarcissus**

▪Lvs with viscid sap and/or spiral fibres when torn

At least some lvs >10mm wide, with stomata scattered both sides

Lvs 5-8, 30-75cm x (3)10-15mm, slightly hooded at apex, slightly channelled, weakly ribbed below, with weakly scabrid-crenulate hyaline margins, with viscid sap and spiral fibres when torn. Scape c5mm diam, tough, round, glaucous-pruinose, pith-filled. Bulb with a whitish or brown tunic. Feb-Jun..............................*Spiked Star-of-Bethlehem* **Ornithogalum pyrenaicum**

All lvs <10mm wide, with stomata scattered above but mostly in rows below

Lvs 6-18 per bulb, 10-50cm x 5-9mm, with slightly hooded apex, channelled, wavy, with 10-15 faint veins visible as ribs below. Bulb ovoid, usu with bulbils, with light brown tunic. Nov-Jun. Hortal..*Garden Grape-hyacinth* **Muscari armeniacum**

Lvs 3-6 per bulb, (10)15-30(40)cm x 2-5(8)mm, with slightly hooded apex, channelled (to ± cylindrical), wavy, with up to 10 faint ribs. Bulb ovoid, usu with bulbils, with dark brown tunic. Nov-Jun. E Anglian Brecks or R hortal................................*Grape-hyacinth* **Muscari neglectum**

▪Lvs without viscid sap or spiral fibres when torn

Lvs 3, to 30cm x (6)12-25mm, shallowly channelled. Bulb enclosed in a tunic (the tunic often with sparse hairs on the inner side). R alien..*Wild Tulip* **Tulipa sylvestris**

Lvs >3, 15-45 x 2-5cm, ± channelled, grass-like. Rhizomes short, clothed with fibrous lf bases. Coastal hths, Kerry, Cork...*Kerry Lily* **Simethis planifolia**

Group JW – *Lvs 3 or more per shoot, all basal (exc* Juncus planifolius, *which may have some stem lvs), green above, never glaucous. Usu bulbous or tuberous.* ❶ *Lf veins translucent (± obscure and opaque in all other spp).* ❷ *Bulbs ann without collar lines*

▪Lvs usu with erose to crenulate hyaline margins (x20)

Rhizomatous, with tuberous roots. Lvs 2-ranked, with slightly hooded apex, folded when young, with sap not viscid. ❶

Fls dull orange, net-veined...*Orange Day-lily* **Hemerocallis fulva**

Fls yellow, parallel-veined....................................*Yellow Day-lily* **Hemerocallis lilioasphodelus**

Bulbous. Lvs 3-ranked, rolled or channelled when young, with clearly hooded apex, with viscid sap forming strands

Lvs slightly ribbed both sides (visible as darker lines HTL)

Lvs often 3, to 20cm x 11-18mm, channelled (at least proximally), shiny green both sides, with entire to erose hyaline margins (often purplish), with stomata both sides. Bulb per, with dark brown tunic...*Turkish Squill* **Scilla bithynica**

Lvs slightly ribbed below only (often visible as darker lines HTL)

Lvs 2-4, 10-15cm x 5-20mm, ± channelled or weakly so, with crenulate margins, with c8 ribs each side of midrib below. Scapes 1-4, to 20cm. Bulb per, with purplish tunic ....*Siberian Squill* **Scilla siberica**

Lvs not ribbed either side

Lvs 9-15 per bulb, often dull green both sides, with tiny purple-black spots below

Lvs 25-60(80) x 2-3cm, slightly channelled at base, with erose margins, with translucent midrib (not raised below), with scattered elongate stomata both sides. Bulb per, with purplish tunic...*Portuguese Squill* **Scilla peruviana**

Lvs 3-8 per bulb, shiny green both sides, without spots below

Lvs 20-50cm, flat to channelled, with crenulate hyaline margins (occ weak), with midrib raised below (occ visible as 2(3) ribs or opaque lines below), with scattered elongate stomata both sides, with viscid spiral fibres when torn. Cataphyll 1. Bulb ann, without collar lines, whitish, without tunic. ❷

Lvs 4-8, 2-4cm wide. Raceme not 1-sided. Fls remaining erect at least distally, not fragrant; stamens equal; anthers blue. Hortal.................*Spanish Bluebell* **Hyacinthoides hispanica**

Lvs 3-8, 1-3cm wide. Raceme not 1-sided. Fls nodding or erect distally, fragrant; stamens unequal; anthers white to blue. Hortal. *H. hispanica* x *non-scripta* ....*Hybrid Bluebell* **Hyacinthoides** x **massartiana**

Lvs 3-6, 0.7-1.6cm wide. Raceme 1-sided. Fls all soon nodding, fragrant; stamens unequal; anthers cream. (Jan) Mar-Jun. Woods, cliff-tops in W Br ....*Bluebell* **Hyacinthoides non-scripta**

▪Lvs with entire hyaline margin

Lvs >2cm wide

Rhizomes tuber-like, forming dense clumps of lvs

Lvs 6-7 per shoot, to 45(60) x 2.5-5cm, oblanc, obtuse (not hooded at apex), channelled nr base, ± dull green both sides, occ with slightly recurved margins, with midrib strongly raised below, with other veins weakly raised below, occ with weak cross-veins, with scattered elongate stomata both sides, with viscid sap and spiral fibres. All yr ....*African Lily* **Agapanthus praecox** ssp **orientalis**

Bulb or corm, not forming dense clumps of lvs

Lvs mostly >4cm wide

Lvs 5-12, present at fl, 60-120 x 4-10cm, linear to elliptic-oblong, slightly hooded at apex, rolled when young, channelled, not ribbed, ± shiny green above, with entire (occ erose) hyaline margins turning reddish, with midrib usu raised below, with indistinct cross-veins, with 30-40 small hollows in TS, with spiral fibres and abundant sap, with scattered square stomata both sides. Cataphylls absent. Bulbous. All yr. Hortal. *C. bulbispermum* x *moorei* ....*Powell's Cape-lily* **Crinum** x **powellii**

Lvs 2-5, absent at fl, 15-30 x 2-6(7)cm, oblong-lanc, slightly hooded at apex, rolled when young, channelled, slightly ribbed both sides, shiny yellow-green both sides, with entire hyaline margins remaining colourless, with midrib raised below, without cross-veins, without hollows, ± without fibres or sap, with scattered elongate stomata both sides. Shoots round, yellowish at base. Cataphylls absent. Cormous. Feb-Jul. Damp gsld or woods on basic soils, or hortal..*Meadow Saffron* **Colchicum autumnale**

Lvs all <4cm wide

Lf surface slightly ribbed both sides

Lvs 2-4cm wide, hooded at apex, often ± flat, shiny or dull dark green, without midrib raised below, occ with hollows in TS, with viscid sap forming strands, with scattered elongate stomata both sides. Bulbous. Jan-Jun. Hortal......*Hyacinth* **Hyacinthus orientalis**

Lvs <2cm wide

Bulbous

Lvs (4)5, linear-lanc, obtuse (with pale hydathode), weakly channelled esp nr base, thick, shiny bright green, without viscid sap, with elongate scattered stomata both sides. Cataphylls herbaceous. Dec-Jun

Lvs 30-50cm x 10-18mm, with 15-20 hollows in TS

Scape angled, with 2 sharp edges entire throughout. Hortal ....*Summer Snowflake* (ssp) **Leucojum aestivum** ssp **pulchellum**

Scape angled, with 2 sharp edges remotely and often obscurely denticulate. By rivers, S Eng, Ire, R native, occ hortal..........*Summer Snowflake* **Leucojum aestivum** ssp **aestivum**

Lvs 10-30cm x 8-12mm, with up to 15 hollows in TS. Scape round. VR alien and hortal ....*Spring Snowflake* **Leucojum vernum**

Non-bulbous

Lvs not linear, pitted at least above. Dicot. Often ❶.................................(*Limonium*) Go to **LIM**

Lvs linear, not pitted. Monocot

Lvs numerous, 3-ranked, to 10cm x (2)5-13mm, widest nr base (stem lvs widest nr middle), minutely mucronate at acute apex, flat, soon slightly involute, thin, translucent, green, often resembling those of *Luzula*. Basal sheaths open, often purplish or with purple veins, with dead fibrous lvs at base. Rhizomes ± absent. ❶. R alien, Galway ....*Broad-leaved Rush* **Juncus planifolius**

Key to Groups in Division K

(Lvs entire, alt)

■Tree or tall shrub >1.2m
  Lvs with peltate scales or stellate hairs....................**KA**
  Lvs with simple or woolly hairs, or hairless
    Lvs net-veined (often strongly so)....................**KB**
    Lvs not net-veined....................**KC**
■Climbing, sprawling or twining vine-like shrub or herb (may have latex)....................**KD**
■Low shrub or subshrub to 1.2m (often much less), erect, trailing or mat-forming
  Lvs revolute when young, usu with distinctly revolute margins at maturity....................**KE**
  Lvs not revolute when young, usu with flat margins at maturity....................**KF**
■Herb, occ ± woody at base, never climbing. (Many bulbous plants may key out here in error)
  Latex present
    Lvs in a basal rosette, OR clasping stem with auricles....................**KG**
    Lvs all on stems but never clasping with auricles (*Euphorbia*)....................**KH**
  Latex absent
    Lvs spiny or with spine-tipped apex....................**KI**
    Lvs not spiny (may be bristly)
      Lvs (or stems) hairy or ciliate (look carefully!)
        Hairs branched, stellate, or medifixed....................**KJ**
        Hairs woolly or cottony (at least on lower side of lvs)....................**KK**
        Hairs bladder-like, often silvery-white (*Chenopodium*, *Atriplex*)....................Go to **CHAT**
        Hairs glandular....................**KL**
        Hairs simple
          Lvs with some minute hooked hairs at least below (adhering to woolly clothing!)....................**KM**
          Lvs without hooked hairs
            Hairs v bristly or sharply hispid (may pierce skin) ....**KN**
            Hairs not v bristly (may have swollen bases which feel rough)
              Lvs with 2° veins translucent....................**KO**
              Lvs 0-1-veined or 2° veins opaque....................**KP**
      Lvs (and stem if present) totally hairless (aquatic *Myosotis* spp may key out here)
        Lvs with (0)1-3(5) ± parallel veins (midrib occ absent)
          Lvs in basal rosette (at least initially)....................**KQ**
          Lvs all on stem (no basal rosette)
            Lvs fleshy (often strongly so)....................**KR**
            Lvs not fleshy....................**KS**
        Lvs with pinnate veins (occ obscure or palmate)
          Lvs in basal rosette, or plant stoloniferous in freshwater habs....................**KT**
          Lvs all on stem (no basal rosette). Plant not stoloniferous....................**KU**

Group KA – *Tree or tall shrub >1.2m. Lvs with peltate scales or stellate hairs. Twigs without short shoots*

■Lvs with peltate scales. Shrub

Lvs 1.5-5cm wide, ± undulate

Lvs strongly net-veined, 4-7cm, ovate-oblong, with slightly recurved or erose non-cartilaginous margins, shiny green above, whitish or with buff peltate scales below, with 2° veins not sunken above and hardly raised below. Petiole to 1cm. Twigs angled, minutely hairy, brown, or whitish with scales. Unarmed. Evergreen, not suckering..............................*Akiraho* **Olearia paniculata**

Lvs not (or obscurely) net-veined, 3-10cm, elliptic-ovate, dark green above, often with silvery scales turning brown below, with ± entire margins, with 2° veins not sunken above and hardly raised below. Petiole to 1cm. Twigs angled, silvery with scales. Spines often present. Deciduous or semi-evergreen, suckering..................*Spreading Oleaster* **Elaeagnus umbellata**

Lvs 0.6-1cm wide, not undulate

Lvs 4-6cm, linear-lanc, ± 1-veined, often ± hairless and dull green above, silvery below, with fimbriate scales turning brownish, with midrib raised below. Petiole v short. Twigs angled, with brownish scales. Thorns usu present. Buds 2.5mm, red-brown, with 2 scales. Deciduous, suckering. Coastal, esp sand dunes. PL 9.....................*Sea-buckthorn* **Hippophae rhamnoides**

■Lvs with stellate hairs

Evergreen tree to 20m. Lvs 2-6(9) x 3cm, ovate to elliptic, occ spine-tipped when young (or with spiny teeth on vigorous shoots), usu with flat margins, dark green and hairless above, stellate-hairy below, odourless, net-veined, with stomata below only. Petiole to 15mm. Twigs densely stellate-hairy when young, soon ± hairless, round. Buds with several scales ....*Evergreen Oak* **Quercus ilex**

Deciduous shrub to 4m. Lvs 8 x 1.5cm, narrowly elliptic to linear, obtuse to acute, often with recurved margins, dark green above, hairless above, white stellate-hairy below, weakly aromatic, net-veined, with stomata below only. Petiole 1-3mm. Twigs stellate-woolly or green when young, soon orange and hairless, weakly ridged, arching, brittle. Buds naked ....*Alternate-leaved Butterfly-bush* **Buddleja alternifolia**

Group KB – *Tree or tall shrub >1.2m. Lvs with simple or woolly hairs, or hairless, net-veined, with stomata below only (both sides on some* Salix, Bupleurum fruticosum *and* Pittosporum tenuifolium*). (*Pyrus communis *may key out here). ❶ Lvs bay-scented. ❷ Twigs monopodial*

■Stipules present, soon falling

Buds naked (occ with 2 parted scales in *Cotoneaster*)

Deciduous or evergreen. Twigs odourless. Short shoots present..........(*Cotoneaster*) Go to **COT**

Deciduous. Twigs with odorous inner bark. Short shoots absent.

Lvs 2-7 x 3cm, obovate, bluntly apiculate, ± undulate, odourless, folded when young, without cartilaginous margins, shiny green above, turning yellow-red in autumn, brown-hairy esp below, soon hairless, ciliate, net-veined, with 7-8(10) prs of ± parallel 2° veins anastomosing at margin. Petiole to 2cm, hairy, with 1 vb. Stipules to 4 x 0.5-1mm, red nr apex, hairless. Twigs sparsely minutely hairy, round, with weakly raised long lenticels, with strongly odorous lime-green inner bark; lf scars with 3 traces. Buds adpressed, 3-5mm, brown-hairy when young. Shrub to 5m..*Alder Buckthorn* **Frangula alnus***

Buds with 1 scale. Deciduous...(*Salix*) Go to **SAL**

Buds with (1)2 or more overlapping scales. Deciduous

Evergreen. Twigs with almond-scented inner bark

Lvs 5-10 x 3cm, lanc, acute, occ undulate, with almond odour when crushed, folded and purplish when young, with narrow cartilaginous margins, shiny green above, ± glaucous below, with reddish veins in autumn, hairless, often indistinctly net-veined, with 7-9 prs of anastomosing 2° veins flat or sunken above. Petiole 1-7cm, hairy, with 1 vb. Stipules to 5 x 0.5mm, lanc. Twigs with adpressed *Cotoneaster*-like hairs when young, round, with lenticels. Buds adpressed, 5-8mm, reddish, hairless. Shrub or tree to 3(8)m. Hortal. R alien ....*Stranvaesia* **Photinia davidiana**

Deciduous

Lvs 10-15 x 7-10cm, broadly ovate, acute to emarginate, rounded to ± cordate at base, ± folded when young, flat, ± hairless above, woolly below when young (soon sparsely so), with 10 prs of anastomosing 2° veins. Petiole to 3cm, reddish, channelled. Stipules to 2cm, ovate, soon falling. Twigs shiny dark olive-green, hairless, with lenticels. Buds adpressed to twig, 2-5mm, purplish-brown, woolly when young, with (1)2 scales. Small tree to 3(6)m ....*Quince* **Cydonia oblonga**

Lvs 4-9 x 3-6cm, ovate-elliptic, acute, cuneate or rounded (often asymmetrically) at base, folded-plicate when young, ± undulate, soon hairless exc for silky buff hairs along veins below, long-ciliate, occ sinuate-toothed, with 6-9 prs of straight 2° veins usu anastomosing at margins (or in teeth). Petiole to 1cm. Stipules to 2cm, linear, soon falling. Twigs reddish-brown, with sparse long hairs when young, with lenticels, often with short shoots. Buds spreading, 1-2cm, spindle-shaped, acute, reddish-brown, sparsely hairy, ciliate, with the lowest 4 prs of scales opp in 4-ranks. Tree to 46m. ❷. PL 12, 21........*Beech* **Fagus sylvatica***

Lvs to 8 x 1.5cm, elliptic to linear-lanc, acute, cuneate at base, often undulate, silvery silky-hairy (esp below), rarely glandular toothed, with midrib raised below. Petiole to 1.5cm, not channelled. Stipules 5mm, subulate. Twigs pendent (weeping), silky-hairy, round. Buds tiny. Tree to 5m. R hortal...*Willow-leaved Pear* **Pyrus salicifolia**

■Stipules absent

Lvs white-felted below

Lvs to 8cm, oblong-elliptic, entire to remotely denticulate. Buds naked. Evergreen shrub to 1.5m. Hortal...*Shrub Ragwort* **Brachyglottis** x **jubar***

Lvs not white-felted below

Lvs (and twigs) in lax pseudowhorls nr twig apices (not on short shoots), revolute when young

Evergreen

Twigs hairless, round. Lvs (6)8-20 x 4cm, eliptic to oblong, acute, cuneate at base, occ ± undulate, dark green above, paler below, hairless or ± so (viscid-hairy when young), with midrib raised below. Petiole to 1.8cm, channelled, with arc vb. Buds terminal and lateral; terminal (fl) buds to 35mm; lateral buds 3mm, with several ciliate scales. PL 9 ....*Rhododendron* **Rhododendron ponticum**

Twigs minutely hairy, round. Lvs to 8 x 3(4)cm, oblong, obtuse to retuse, cuneate at base, undulate, leathery, shiny green above, hairless both sides (midrib occ hairy above), with midrib raised below. Petiole to 2cm. Buds short, ovoid, with several scales .... *Japanese Pittosporum* **Pittosporum tobira**

Deciduous

Twigs bristly glandular-hairy when young, ± round or angled. Lvs 8-17 x 3-5cm, oblong-lanc, mucronate or with short aristate point, cuneate at base, occ slightly undulate, aromatic, yellow-green both sides, hairy above and at least on main veins below, often serrate, glandular-ciliate, with midrib and 2° veins raised below. Petiole 0.8-1.5cm, channelled. Buds terminal and lateral; terminal (fl) buds to 20mm; lateral buds 1mm, with 2-3 ciliate scales .... *Yellow Azalea* **Rhododendron luteum**

Lvs on long or short shoots, not revolute when young

Twigs with spines (often 3-partite), with yellow inner bark......................... (*Berberis*) Go to **BER**

Twigs without spines

Evergreen. Lvs bay-scented ❶

Lvs 6-12 x 2-5cm, oblong-elliptic, usu acute, cuneate at base, undulate, shiny dark green above, paler below, with thick cartilaginous margins, hairless, with anastomosing 2° veins not sunken above or raised below. Petiole 1cm, often red, with 1 vb. Twigs green or red. Buds c3mm, narrowly ovoid, red, occ superposed, the 2 outer scales ± valvate. PL 9 .... *Bay* **Laurus nobilis**

Evergreen. Lvs odourless

Lvs ± sessile and weakly clasping, to 7 x 2cm, oblong to narrowly obovate, cuneate at base, with flat or revolute margins, with narrow cartilaginous margins, thick, shiny dark green above, dull pale green below, hairless, with stomata below and along midrib above. Twigs hairless, ± round, green to grey-brown; lf scars with 3(5) traces. Buds naked .... *Shrubby Hare's-ear* **Bupleurum fruticosum**

Lvs petiolate to 1cm, not clasping, to 7 x 2.5cm, elliptic-oblong, rounded at base, undulate, with thick cartilaginous margins, thin, shiny dark green above, paler below, ± hairless at maturity, with midrib raised below, with stomata both sides (esp abundant nr midrib above). Petiole minutely hairy above, with 3 vb's. Twigs hairless, ± round, blackish-brown. Buds 1-3mm, globose, green to purple, with several scales..... *Kohuhu* **Pittosporum tenuifolium**

Deciduous

Twigs densely hairy to hairless, round, occ green with raised patterning when young. Lvs 3-9 x 4cm, ovate or elliptic, mucronate, rolled when young, with weakly recurved margins, leathery, reddish in autumn, soon hairless above, hairy below, odourless, ± entire to minutely toothed, ± net-veined, with anastomosing 2° veins. Petiole 2-3mm, reddish. Buds 5-6mm, reddish, with v acuminate scales..................... *Blueberry* **Vaccinium corymbosum**

Twigs hairless, round, greenish to reddish, with lenticels, odorous. Lvs 4-9 x 7cm, broadly ovate, often slightly retuse, folded when young, with flat cartilaginous margins, thin, occ purple, hairless above, sparsely hairy to hairless below, odorous, ± net-veined, with 2° veins at 60-90° to midrib and forking 3-4x to margin. Petiole to 50mm. Buds tiny, green-black, apiculate. PL 12, 23.. *Smoke-tree* **Cotinus coggygria**

Brachyglottis x *jubar* lf & petiole TS

Fagus sylvatica

Frangula alnus

Group KC – *Tree or tall shrub >1.2m. Lvs with simple or woolly hairs, or hairless, not net-veined. Stipules absent (if present* Go to **COT**). *Magnolias (no description) key out here if ochreae (fused stipules) present.* ❶ *Lvs silvery due to raised crystalline cells both sides.* ❷ *Lvs scale-like*

■Evergreen (occ semi-evergreen in *Atriplex halimus*)

Lvs white or buff short-woolly below

Lvs occ opp or whorled, to 8cm, obovate to oblanc, obtuse, cuneate at base, with revolute margins, leathery, shiny dark green above, ± net-veined, without 2° veins raised below, with stomata below only. Petiole to 2cm. Twigs shortly woolly, round. Buds short, ovoid, v hairy, with several scales..*Karo* **Pittosporum crassifolium**

Lvs hairless but dotted with translucent glands

Lvs occ pseudowhorled, to 9 x 3cm, ovate-lanc, ± acute, thick (snapping audibly), often with cartilaginous margins slightly recurved, dark green and, pitted above with weakly aromatic glands, paler with flat glands below, with obscure 2° veins not raised or sunken either side, with stomata below only. Petiole to 1cm, often reddish, hardly channelled. Twigs green, round, with raised glands. Hortal. PL 9...*Skimmia* **Skimmia japonica**

Lvs hairless but silvery due to raised crystalline cells both sides

Lvs with Kranz venation (veins visible as a dark network - scratch surface to check), to 4 x 2.5cm, oblong to elliptic-ovate, acute to obtuse, cuneate at base, ± fleshy, with ± sinuate or erose margins, with few opaque 2° veins. Petiole to 0.3(1)cm. Twigs much-branched, purplish, angled to ± round, brittle, with 1 axillary shoot per axil. ❶....*Shrubby Orache* **Atriplex halimus***

Lvs hairless, not gland-dotted or silvery

Lvs with spine-tipped apex (occ spiny teeth on vigorous shoots), 5-9(12) x 5cm, ovate-oblong, ± symmetrically rounded at base, leathery, with narrow cartilaginous margins, shiny dark green above, dull pale green below, with 2° veins few and hardly raised or sunken either side, with stomata below only. Petiole 1-2cm, with 1 vb. Twigs green, round. *I. aquifolium* x *perado* ....*Highclere Holly* **Ilex** x **altaclerensis**

Lvs with obtuse apex, 5-11 x 3-8cm, broadly ovate to ± orb, asymmetrically rounded at base, leathery, with broad cartilaginous margins, shiny yellow-green above, dull pale green below, with 2° veins few or obscure (rarely net-veined) and hardly raised or sunken either side, with stomata below only. Petiole 1-2.5cm, with ciliolate sheathing base, with 1 vb. Twigs yellow-green, round. PL 12..*New Zealand Broadleaf* **Griselinia littoralis***

■Deciduous

Stems (branchlets) deciduous. ❷

Lvs narrowly triangular-ovate, acute to acuminate, sessile, scale-like, adpressed, imbricate, spirally arranged, hairless, pitted with salt glands. Twigs not in 1-plane reddish-brown, hairless, round, with scale lvs often arising every 4-5th node (sparse with age). Buds with scales

Lvs 1-1.5(4)mm. Infl axis and bracts hairless. Usu coastal...............*Tamarisk* **Tamarix gallica***

Lvs 1.5-3mm. Infl axis and bracts papillate. VR alien..........*African Tamarisk* **Tamarix africana**

Stems not deciduous, arching, branching at 90°

Lvs to 6(10)cm, alt or in fascicles of 3-7, cuneate at base, ± fleshy, occ undulate, hairless (exc sparse ± sessile glandular hairs when young), with few obscure 2° veins (weakly or not raised below), with stomata both sides. Petiole to 1cm, indistinct. Twigs ± brittle, greyish-white, occ puplish, often ± ridged. Buds without scales, often obscure. Suckering. Hortal, ruderal

Thorns usu present. Lvs lanc (widest nr middle), grey-green ....*Duke of Argyll's Teaplant* **Lycium barbarum***

Thorns usu absent. Lvs narrowly ovate or rhombic (widest below middle), bright green. R ....*Chinese Teaplant* **Lycium chinense**

Stems not arching or branching at 90°

Twigs with spines (often 3-partite)...(*Berberis*) Go to **BER**

Twigs without spines

Twigs with 20-50 fine striations (rough to touch when dry)

Lvs 1-3 x 0.4-0.7cm, few, soon falling, oblong-lanc, apiculate, ± thick, (±) hairless above, silky-hairy below, with translucent midrib, with 2° veins indistinct, with stomata both sides. Petiole 3mm. Twigs 2-5mm diam, branched (often opp or subopp), straight, green, round, flexible, filled with solid pith, with stomata all around. Shrub to 3m ....*Spanish Broom* **Spartium junceum**

Twigs with 15 ridges

Lvs to 1cm, hairy both sides, soon falling. Twigs drooping. Shrub (occ small tree) to 6m ....*Mount Etna Broom* **Genista aetnensis**

Twigs with 6-7 ridges

Lvs to 2.5 x 0.5cm, silvery silky-hairy below, with indistinct opaque veins (but midrib raised below), with stomata both sides. Twigs silvery silky-hairy when young. Tall shrub or small tree to 7m. VR alien..*Madeira Broom* **Genista tenera**

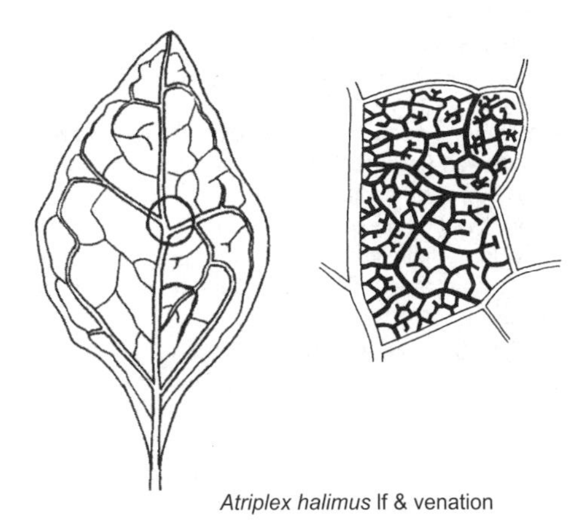

Atriplex halimus lf & venation

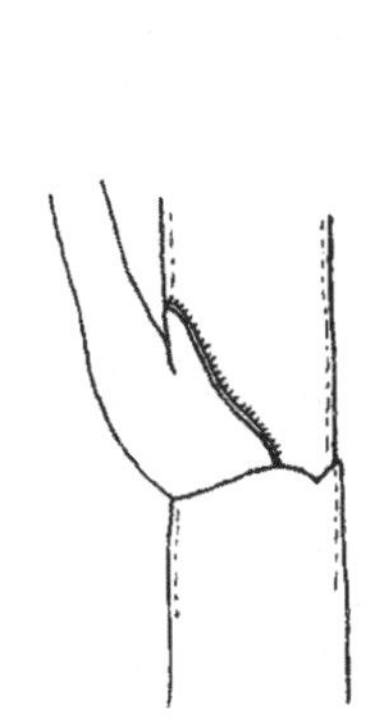

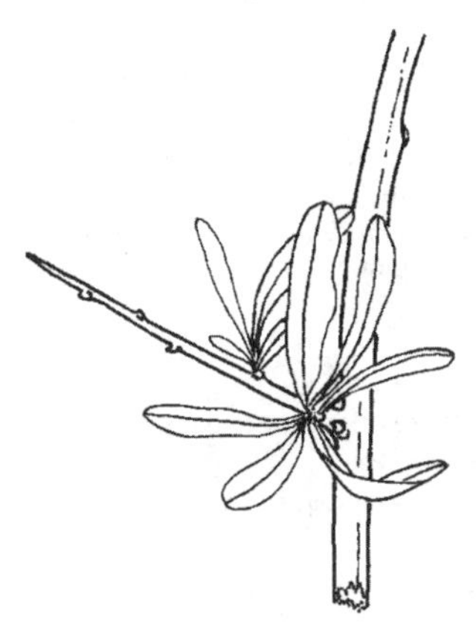

Griselinia littoralis petiole base

Lycium barbarum

Tamarix gallica branchlet

Group KD – *Climbing, sprawling or twining vine-like shrub or herb*

■Petiole (and stems) with white latex. Rhizomatous per. Lvs rolled when young, ± palmate-veined. Stems climbing, trailing, or sprawling, twining anti-clockwise

Lvs with stomata all over the upper surface, 1.5-5cm

Lvs fleshy, reniform, obtuse (rarely mucronate), hairless, not ciliate, pitted both sides, not net-veined, with anastomosing veins hardly raised below (and fading before margin). Petiole 3-8cm, channelled. Stems trailing, angled, with erose cartilaginous angles. Apr-Oct. Sandy or shingle shores..*Sea Bindweed* **Calystegia soldanella***

Lvs thin, broadly ovate to lanc, obtuse with mucro, usu hastate-sagittate at base, with erose margins, hairy when young, often ciliate, not pitted, weakly net-veined, with anastomosing veins raised below. Petiole 0.3-2cm, channelled. Stems climbing, 3-5-ribbed. PL 14 ....*Field Bindweed* **Convolvulus arvensis***

Lvs with stomata confined to vein margins on the upper surface, 8-15cm. Stems climbing

Stem and petioles hairless. Lvs hairless to sparsely hairy

Lvs with ± V-shaped acute basal sinus (lowermost vein does not lie on sinus margin), net-veined. Petiole usu < lf, channelled. Stems 5-ribbed ....*Hedge Bindweed* **Calystegia sepium** ssp **sepium***

Lvs with basal sinus rounded (lowermost vein partially lies on sinus margin), net-veined. Stems 5-ribbed. PL 14..*Large Bindweed* **Calystegia silvatica***

Stems and petioles sparsely hairy (at least distally). Lvs shortly hairy both sides

Lvs with ± V-shaped acute basal sinus (lowermost vein does not lie on sinus margin), net-veined. Often coastal.........................*Hedge Bindweed* (ssp) **Calystegia sepium** ssp **roseata**

Lvs with stomata absent from the upper surface, 5-12cm. Stems climbing

Stems and petioles sparsely hairy (at least distally). Lvs dull green above (much less shiny than other spp), with ± square wide sinus (lowermost vein partially lies on sinus margin), net-veined. Apr-Nov..*Hairy Bindweed* **Calystegia pulchra***

■Petiole (and stems) without latex. Herb or woody vine

Adventitious aerial roots present between stem nodes. Ochreae absent. Evergreen ....(*Hedera*) Go to **IVY**

Adventitious roots absent

Ochreae present

Evergreen. Lvs widely-spaced along stems, 0.4-1.2(2)cm, usu orb (slightly angular), leathery, dark green above, ± entire (occ lobed), with indistinct anastomosing 2° veins, with stomata sparse or absent above. Petiole 4-6mm, without extra-floral nectary. Stems 1-1.5(3)mm diam, red-black, tough, round, interwoven, minutely scabrid (x20). Ochreae soon disintegrating, brownish, scabrid..*Wireplant* **Muehlenbeckia complexa**

Deciduous. Lvs often in fascicles, 3.5-6 x 4cm, ovate, acuminate, rounded to cordate at base (occ sagittate), revolute when young, slightly undulate, with reddish erose cartilaginous margins, green, weakly net-veined, with anastomosing 2° veins raised below, with stomata below only. Petiole 1-3cm, with 1 extra-floral nectary at base (below abscission layer), channelled, with 5-7 vb's. Stems to 5m, clockwise-climbing, woody, green or reddish, soon ashy grey, ± round, with faint lenticels and striations, solid to hollow. Ochreae persistent, greenish, soon brownish...*Russian-vine* **Fallopia baldschuanica**

Ochreae absent. Deciduous

Lvs not cordate at base, pinnate-veined..............(*Solanum dulcamara*, *Salpichroa*) Go to **SOL**

Lvs cordate at base, palmate-veined. Plant hairless

Lvs shiny green below, dark green above, 5-10(15) x 4-11cm, broadly ovate, acuminate (with reddish mucro or 'runner tip' when young), odourless, hardly net-veined, with (3)7-9 palmate veins curving to apex, with stomata below only. Petiole 5-15cm, with 2 curved stipoid glands at base, round, hollow or solid, with 7 vb's and slightly viscid sap. Stems to 4m, twining clockwise, slender, angled to round, unbranched, solid. Buds absent. Rootstock tuberous, often forked. May-Oct. PL 13..............................*Black Bryony* **Tamus communis***

Lvs dull grey-green below, dark or greyish-green above, to 15 x 15cm, broadly ovate, obtuse or retuse, folded (but rolled together) when young, fetid, with erose to sparsely papillate cartilaginous margins, strongly net-veined, with ≥5 palmate anastomosing veins (with thick vein bordering sinus), with stomata below only. Petiole 5-20cm, narrowly channelled, solid, with 5 vb's. Stems to 0.8m, scrambling or trailing (rarely erect), weakly furrowed, unbranched, ± solid. Buds superposed (visible after lf fall). Rhizomatous. Mar-Oct. PL 13 ....*Birthwort* **Aristolochia clematitis**

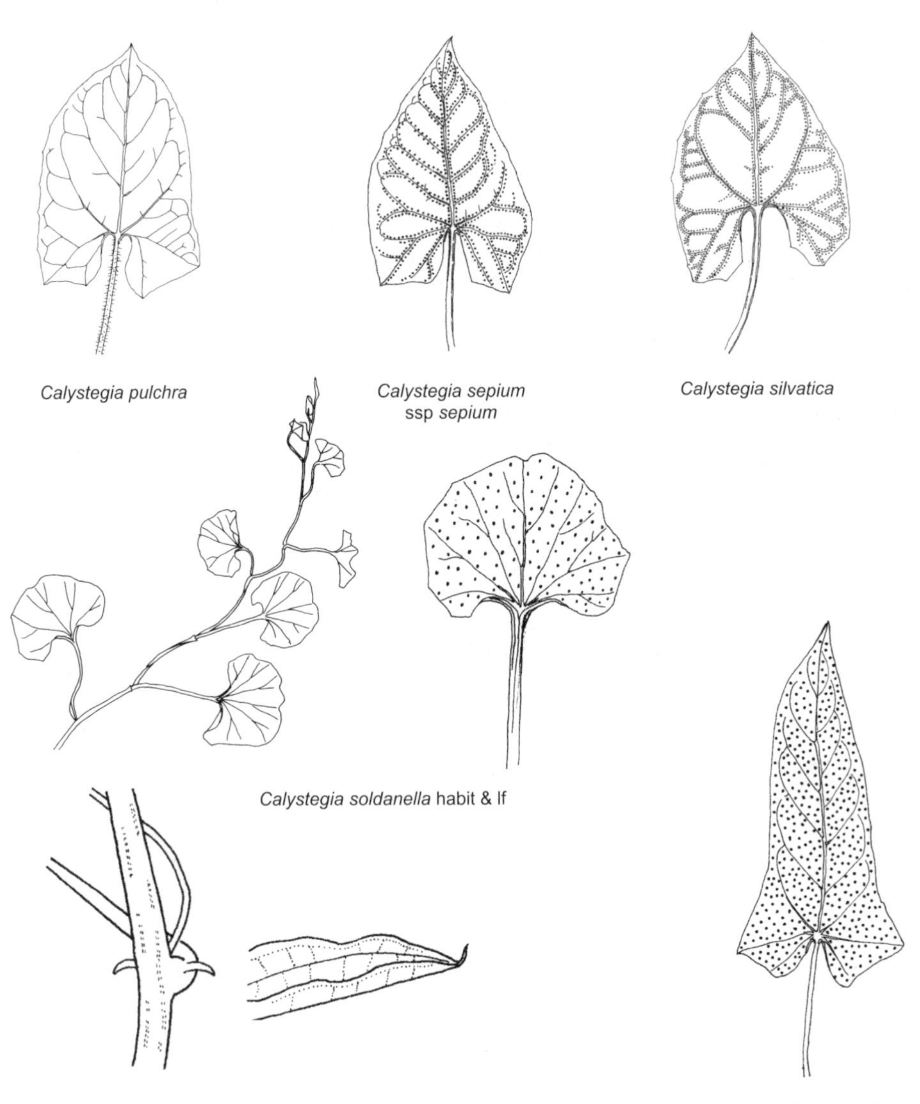

Calystegia pulchra

Calystegia sepium ssp *sepium*

Calystegia silvatica

Calystegia soldanella habit & lf

Tamus communis stipules & lf apex

Convolvulus arvensis

Group KE – *Evergreen low shrub <1.2m (often much less). Lvs revolute when young, usu with distinctly revolute margins, leathery, with stomata below only. Stipules absent. Stems (±) round. Buds minute (larger in* Ledum palustre*)*

■Lvs strongly revolute and ± cylindrical (underside visible only as white line)

Lvs 4-6 x 0.8-1.2mm, crowded, obtuse, hairless, with glandular cilia when young, not net-veined. Petiole v short. Twigs often reddish, minutely glandular-hairy when young. Shrub to 45cm. Mtn moorland, hths...*Crowberry* **Empetrum nigrum***

■Lvs orange-woolly below

Lvs dark green with aromatic yellow sessile glands above when young, 2-5 x 0.5-2cm, linear to elliptic-oblong, leathery, with midrib raised below. Petiole c3mm. Twigs rusty-woolly. Buds with c3 scales. Shrub to 50(100)cm. Raised bogs.......................................*Labrador Tea* **Ledum palustre**

■Lvs white-woolly below (midrib hairless)

Lvs dark green above with scattered glandular hairs, 5-12 x 4-5mm, elliptic or elliptic-linear, acute with protruding rounded terminal hydathode (turning reddish), with 2° veins anastomosing (usu indistinct), with translucent midrib. Petiole v short. Twigs with long red-tipped glandular hairs. Buds without scales. Rhizomatous shrub to 50cm. R, W Ire
....*St Dabeoc's Heath* **Daboecia cantabrica**

■Lvs hairless pale green and gland-dotted below

Lvs shiny dark green and hairless above, 1.8 x 0.8cm, obovate to elliptic, slightly retuse with terminal hydathode, leathery, with thick cartilaginous margins occ remotely crenulate by hydathode-teeth, with translucent midrib slightly raised below, with indistinct anastomosing veins. Petiole to 3mm. Twigs hairy when young, green. Buds with 2 scales. Rhizomatous shrub to 30cm. Moors...*Cowberry* **Vaccinium vitis-idaea**

■Lvs hairless with white bloom below, not gland-dotted

Stems erect, not rooting at nodes, >1mm diam. Rhizomatous shrub to 30cm

Lvs hairless with white bloom below, dark green and hairless above, to 3.5 x 0.3-0.6cm, elliptic-linear, acute or apiculate with terminal hydathode, ± net-veined, with anastomosing 2° veins. Petiole to 1.5mm. Twigs hairless, pale brown. Buds with 2 scales. Bogs, N of Humber-Severn line...*Bog-rosemary* **Andromeda polifolia**

Stems trailing, rooting at all nodes, ≤1mm diam. Subshrub to 15cm

Lvs well-spaced, with reddish terminal hydathode, dark green above, with wavy midrib sunken both sides or raised below, with obscurely anastomosing 2° veins. Stems often reddish. Buds with 2 scales

Lvs (6)10-11(18)mm, oblong or elliptic-oblong, obtuse, with flat to slightly revolute margins, often turning reddish above. Petiole 1-2mm. Stems 1mm diam. Frs with apple-like taste. Boggy ground. R alien.....................................*American Cranberry* **Vaccinium macrocarpon**

Lvs (4)5-8mm, oblong-ovate, ± acute, with slightly revolute margins. Petiole 0.5mm. Stems 0.5-0.7mm diam, with sparse crisped hairs when young. Frs with sharp acidic taste. Wet bogs
....*Cranberry* **Vaccinium oxycoccos**

Lvs 3-7mm, ovate, ± acute, with revolute margins. Petiole 0.5mm. Stems 0.3mm diam, hairless. Frs with sharp acidic taste. Bogs (drier parts), Scot. May be a var of *V. oxycoccos*.
....*Small Cranberry* **Vaccinium microcarpum**

Empetrum nigrum shoot & lf TS

Group KF – *Evergreen or deciduous low shrub <1.2m (often much less). Lvs not revolute when young, usu with flat margins. (*Polygala, Ruscus *and* Cytisus scoparius *may key out here).* ❶ *Lvs v fleshy*

▪Stipules present at least when young. (If buds adpressed with 1 scale (*Salix*) Go to **SAL C)**

Stipules herbaceous, usu reddish, entire, soon falling...............................(*Cotoneaster*) Go to **COT**

Stipules herbaceous, green, entire, persistent

Deciduous. Lvs with stomata both sides, 18-37 x 5-8mm, oblong-lanc, acute to apiculate, ± sessile, often held erecto-patent, often with recurved margins, shiny green both sides, soon hairless, adpressed-ciliate, with c4 prs of weak 2° veins anastomosing at margins (occ fading). Stipules 1-2mm, subulate. Stems to 60cm, 1.5mm diam, green, sparsely hairy or hairless, striate or weakly angled when young, with prominent lf pegs, with stomata on non-cartilaginous ridges...*Dyer's Greenweed* **Genista tinctoria**

Stipules membranous, silvery-white, fringed, usu persistent

Lvs usu 2-6mm, broadly ovate to obovate, appearing alt due to one lf of opp pr smaller or absent (esp nr shoot apices), strongly ciliate, with obscure veins, with stomata both sides. Stipules ovate-acuminate, best seen on underside of stems. Stems to 5cm tall, prostrate, woody nr base, usu only hairy on upperside, round. All yr. Coastal

Lvs strongly ciliate, usu 2-6mm, often with 1 lf of opposite pr reduced or abortive (esp above) thus appearing alt, cuneate at base, with obscure veins, with stomata both sides. Stems prostrate, woody at base, with stele in TS. Per

Stems only usu hairy on upperside. Lvs broadly ovate to obovate. Cornwall, Alderney and Guernsey..*Fringed Rupturewort* **Herniaria ciliolata** ssp **ciliolata**

Stems hairy all round. Lvs narrowly elliptic. Jersey ....*Fringed Rupturewort* (ssp) **Herniaria ciliolata** ssp **subciliata**

▪Stipules always absent

Plant with spines

Dense shrub. Lvs few, soon falling, 6-9 x 4mm, obtuse, elliptic, green, hairless above, silky-hairy below, ciliate, with obscure 2° veins, with stomata both sides. Stems to 80(120)cm, green, ridged, adpressed-hairy, without lf pegs....*Spanish Gorse* **Genista hispanica** ssp **occidentalis**

Weak shrub. Lvs numerous, 4-9 x 2-3.5mm, ovate, acute or apiculate, channelled, ± glaucous grey-green below, hairless or with adpressed whitish hairs below (sparse above), occ ciliate, with obscure 2° veins, with stomata both sides. Stems to 50cm, green when young, round, adpressed-hairy or hairless, with prominent lf pegs..........................*Petty Whin* **Genista anglica**

Plant without spines

Lvs v fleshy ❶

Lvs ± cylindrical, glaucous

Lvs 5-18 x 1-1.5mm, rounded or shortly apiculate, (±) sessile, convex above with weak lateral keels, hairless, often with translucent midrib. Evergreen (sub)shrub 40-120cm, much-branched. Coastal, S & E Eng....................................*Shrubby Sea-blite* **Suaeda vera**

Lvs ± semi-cylindrical, green

Lvs all along stems, to 3 x 0.2cm, linear, sessile, sparsely adpressed-hairy, with obscure veins, with stomata all around. Stems to 30cm, woody nr base, reddish-brown, hairless, slightly ridged, brittle, with lf scars. Mar-Oct (all yr). Coastal dunes, W Br ....*Dune Wormwood* **Artemisia campestris** ssp **maritima**

Lvs flat, green

Lvs mostly nr stem apices, flattened, 4-7 x 1.3cm, elliptic-ovate, sessile, shiny green above, hairless, with obscure veins, with stomata both sides. Stems to 75cm, woody, green, hairless, round, brittle. Evergreen.................*Greater Mexican-stonecrop* **Sedum praealtum**

Lvs not v fleshy (occ ± fleshy)

Lowlands. Lvs often all nr top of stems (lf scars clearly visible below)

Evergreen

Lvs 5-12 x 2.5cm, obovate-lanc, ± acute, cuneate at base, shiny dark green above, paler below, leathery, with entire cartilaginous margins slightly rolled below, hairless, with indistinctly anastomosing 2° veins, with stomata below only. Stems to 100cm, round, flexible, weakly odorous. Calc woods. PL 9..........................*Spurge-laurel* **Daphne laureola**

Lvs to 5 x 0.2-0.4cm, ± linear, obtuse, cuneate at base, ± shiny green above, paler below, ± fleshy (usu snapping audibly), with erose cartilaginous margins, hairless, with obscure 2° veins, with stomata both sides. Stems to 60cm, round but ridged by decurrent lf pegs, snapping audibly, weakly cress-scented.............*Perennial Candytuft* **Iberis sempervirens**

Deciduous

Lvs 3-10 x 2.2cm, oblanc, acute (with brown scale-tips when young), thin, light green above, paler below, hairless, ciliate when young, opaquely net-veined, with 2° veins not or hardly raised below, with stomata below only. Petiole short, with 1 vb. Stems to 100cm, round, flexible, hairless or minutely hairy when young, odorous, with tiny lenticels. Lf buds 1.5mm, black, not ciliate, with 4-6 scales. Fl buds to 8mm, with dark-tipped green scales. Calc woods, R. PL 9...*Mezereon* **Daphne mezereum**

Mtns (often coastal cliff-tops in *Genista pilosa*). Lvs usu spaced along stems

Lvs strongly net-veined

Evergreen. Lvs 1-2cm, obovate-elliptic, obtuse or ± acute, cuneate at base, leathery, dark green above, paler below, hairless, wavy-ciliate, with 2° veins sunken both sides, with stomata below only. Petiole to 3.5mm. Twigs reddish at least on upperside, minutely wavy-hairy, round. Buds minute, with c3 scales. Stoloniferous shrub to 20cm. Upland hths and moors...*Bearberry* **Arctostaphylos uva-ursi**

Deciduous. Lvs 1.5-4cm, ovate-obovate, obtuse, cuneate at base, often with 1 basal gland per side, thin, dull blue-green above, paler below, hairless (to hairy), not ciliate, with 2° veins ± raised below, with stomata below only. Petiole to 2mm. Twigs brown, hairless or minutely hairy, round. Buds 3-4mm, with 3 scales. Rhizomatous shrub to 40cm. Upland hths and blanket bogs, N Br...*Bog Bilberry* **Vaccinium uliginosum**

Lvs not net-veined

Evergreen. Lvs densely crowded, 5-12 x 6mm, obovate-spathulate, recurved nr obtuse apex, cuneate at base, leathery, shiny dark green above, hairless, with indistinct 2° veins (midrib translucent but fading nr apex), with stomata below only. Stems ± woody at base, 1mm diam, obscured by lvs, with dead lvs persisting at base. Buds minute, naked. Cushion-forming shrub (easily mistaken for a herb) to 5cm. Scot. Sch8 ....*Diapensia* **Diapensia lapponica**

Deciduous. Lvs few to many, to 8 x 3mm, oblong-ovate, obtuse, cuneate at base, thin, mid-green, hairless above, densely adpressed-hairy below, with veins obscure or ± so, with stomata both sides. Lateral lfts vestigial, reduced to extra-floral nectaries. Stems woody at base, 1mm diam, green, silky-hairy when young, ridged, with stomata all around. Buds small, naked. Low or prostrate shrub to 40cm. R, W Br....*Hairy Greenweed* **Genista pilosa**

Group KG – *Herb. Latex present. Lvs in a rosette, or clasping stem with auricles, with stomata both sides (confined to midrib area above in* Gazania*) (*Centaurea montana *occ has sparse orange latex). (Some* Hieracium *spp (no description) key out here due to the presence of scabrid hairs)*

■Lvs white-hairy below, with some long scabrid hairs above, in basal rosette

Lvs with stellate hairs below, the lowest rosette lvs may be ± hairless in winter, rolled when young, 1-6 x 0.8-1.5cm, dark green above, obscurely toothed; latex sparse. Per. All yr

Lvs with short (<6mm) scabrid hairs above (occ sparse or confined to margin). Stolons long, slender...*Mouse-ear-hawkweed* **Pilosella officinarum**

Lvs with long (6-10mm) scabrid hairs above. Stolons short, thick. Usu calc sea-cliffs, S Eng ....*Shaggy Mouse-ear-hawkweed* **Pilosella peleteriana**

Lvs with simple hairs below, revolute when young; latex often abundant

Lvs to 20 x 1.4cm, linear-lanc, cuneate at base, with recurved margins at maturity, shiny dark green above, antrorsely scabrid-ciliate, with translucent midrib raised below, with obscure 2° veins. Petiole indistinct, to 8cm, with 5 vb's. Per (ann). All yr. Walls, rocky habs nr sea ....*Treasureflower* **Gazania rigens**

■Lvs not white-hairy below

Stem lvs always absent. Lvs hairless. Aquatic *Narrow-leaved Water-plantain* **Alisma lanceolatum**

Stem lvs always absent. Lvs with some long scabrid hairs. Dry habs

Rhizomatous (occ stoloniferous). Basal lvs persisting at fl, 5-15(25)cm, oblanc, apiculate, narrowed to a winged petiole (occ indistinct), rolled when young, with bristly scabrid hairs to 6mm both sides (occ with purple swollen bases), with sparse tiny stellate hairs below, sinuate-toothed, with 2° veins weakly visible, with opaque reticulations, with stomata both sides. Stems occ purplish at extreme base, with stellate hairs and yellow-tipped black glandular hairs that have swollen black bases, hollow...*Fox-and-cubs* **Pilosella aurantiaca**

Stoloniferous, stolons often purplish, with swollen-based hairs. Basal lvs persisting at fl, 5-15cm, oblanc, apiculate, narrowed to a winged petiole (occ indistinct), rolled when young, with bristly scabrid hairs 4-6(10)mm both sides, sinuate-toothed, with 2° veins weakly visible, with opaque reticulations, with stomata both sides. Stems reddish at base, with swollen-based hairs, hollow. R native (Shetland) or alien.......................*Shetland Mouse-ear-hawkweed* **Pilosella flagellaris**

Stem lvs clasping with auricles. Lvs hairless to hairy (hairs never scabrid)

Stem lvs with smooth midrib below, green, not waxy, hairless or hairy, not net-veined

Basal lvs 5-10cm, oblanc, obtuse, narrowed to a long winged petiole, occ minutely sinuate-toothed. Stem lvs oblong or oblong-spathulate, sessile, ± clasping with rounded auricles. Stems often >1, 30-60cm, hairless or hairy. Rootstock black, premorse. Per. Apr-Oct (all yr). R, N Eng, Scot...*Northern Hawk's-beard* **Crepis mollis**

Stem lvs with ± spiny midrib below, usu glaucous or waxy, hairless, net-veined

Basal lvs 7-15(30) x 1-3(8)cm, unlobed to sinuate. Stem lvs usu held erect, often in 1-plane, occ runcinate-lobed with acute terminal lobe and a few distant prs of narrowish acute lateral lobes backcurved distally, the upper often entire with clasping acute auricles. Stem usu 1, to 150cm, whitish-green (rarely red nr base), often spiny, round, solid. Ann (bi). Often all yr. Ruderal...*Prickly Lettuce* **Lactuca serriola***

Basal lvs 4-7 x 0.5-2(5)cm, ± entire to narrowly acutely lobed (occ runcinate) with long slender terminal lobe. Stem lvs usu held erect and in 1-plane, occ lobed and clasping (occ sagittate), the upper linear-lanc and usu entire. Stem usu 1, to 100cm, whitish, hairless or bristly below, round, solid. Ann. May-Oct. Shingle and sea-walls, SE Eng. Sch8 ....*Least Lettuce* **Lactuca saligna**

Group KH – *Lvs all on stems, sessile but never clasping with auricles, spirally arranged, rolled when young (may appear folded), without Kranz venation, usu with translucent midrib raised below and 2º veins obscure. Stems ± erect, round, often lfless below with lf scars. Stipules absent.* ❶ *Lvs with stomata both sides (below only in all other taxa)*

■Lvs hairy, often sparsely so. Stems with bluish latex

Rhizomatous per, often strongly so

Stems hairless, scaly or not at base

Lvs 3-11 x 1-3.5cm, oblong to obovate or linear, occ long-apiculate, ± leathery, ± shiny dark green above, dull pale green below, ± hairless or shortly patent-hairy below, ciliate (hairs turning brown). Stems bi, 30-60cm, reddish. All yr. Hortal ....*Leathery Wood-spurge* **Euphorbia amygdaloides** ssp **robbiae**

Stems hairy

Lvs reddish-purple

Lvs 3-5 x 1.5cm, oblong, obtuse, (±) hairless above, patent-hairy below. Stems 30-60cm, persistently scaly at base, sparsely hairy with hairs to 0.5mm. Apr-Oct ....*Sweet Spurge* (cv) **Euphorbia dulcis 'Chameleon'**

Lvs greenish

Lvs with dense patent hairs below to 1mm, (±) hairless above, 3-5 x 1.5cm, oblong, obtuse, green below, often serrate nr apex. Stems 30-60cm, persistently scaly at base, sparsely hairy with hairs to 0.5mm. Apr-Oct......................................*Sweet Spurge* **Euphorbia dulcis**

Lvs sparsely hairy below, hairless above, 5-10 x 0.6-1.2cm, oblong to oblanc-oblong, obtuse, slightly glaucous below, often serrate in distal ½. Stems 30-60cm, not scaly at base, reddish, patent-hairy with hairs to 1mm. Apr-Oct.........*Balkan Spurge* **Euphorbia oblongata**

Tufted per, or with v short rhizomes, not scaly at base

Lvs with minute adpressed hairs above

Lvs to 10 x 1cm, oblanc, acute, drooping or reflexed, dark bluish-green above, ± patent-hairy below, crisped-ciliate. Stems to 150cm, reddish, minutely hairy. Per. All yr ....*Mediterranean Spurge* **Euphorbia characias**

Lvs patent-hairy or hairless

Shady habs

Lvs 5-12 x 1.5-2.5cm, densely clustered towards stem apices, acute (often with short mucro), not leathery, dull dark green above, paler or ± glaucous below, often purplish, densely patent-hairy at least below (hairs usu 0.8-1mm). Stems 30-80cm, often reddish, with long patent hairs at least when young (often turning brown). All yr ....*Wood-spurge* **Euphorbia amygdaloides** ssp **amygdaloides**

Lvs 5-10 x 0.6-2.2cm, along stems, oblong to oblanc-oblong, retuse (to obtuse), not leathery, yellow-green, weakly glaucous below, turning red, hairless above, sparsely hairy below (esp nr midrib). Stems 30-60cm, reddish, usu hairless. Apr-Oct. SW Br, Ire ....*Irish Spurge* **Euphorbia hyberna**

Sunny habs

Lvs 5-8 x 2.5cm, oblong, obtuse, bluish-green above, paler below, often with purplish margins, softly hairy with 1mm patent hairs both sides, usu finely serrulate nr apex. Stems to 80cm, reddish, hairless. Apr-Oct. R alien................*Coral Spurge* **Euphorbia corallioides**

Lvs 4-6 x 1.5cm, oblong, obtuse, mid-green above, paler below, turning reddish in autumn, softly hairy with 0.5mm patent hairs both sides, usu finely serrulate nr apex. Stems to 45cm, green, patent-hairy. Apr-Oct. VR alien...................*Cushion Spurge* **Euphorbia polychroma**

■Lvs hairless (stems may be hairy above). Stems with bluish or white latex

Lvs 1-2(3)mm wide. Plant green to glaucous

Lvs 1.5-2.5(3)cm, narrowly linear, usu acute or apiculate, with stomata below only (occ along midrib above). Stems 10-30cm, hairless, round. Strongly rhizomatous with thin rhizomes. Apr-Oct..*Cypress Spurge* **Euphorbia cyparissias**

Lvs >2mm wide

Plant green. Ruderal

Per with long thin horizontal rhizomes. ❶

Lvs (2)5-8.5cm x (4)7-11mm, linear-oblanc (broadest above middle), mucronate, dark bluish-green both sides, with narrow (0.1mm) entire cartilaginous margins, hairless. Stems 20-60cm, green. Apr-Oct..*Leafy Spurge* **Euphorbia esula**

Lvs 2-8cm x 4-5mm, linear-lanc (± parallel-sided), mucronate, dark bluish-green both sides, with narrow (0.1mm) entire cartilaginous margins, occ sparsely hairy below when young. Stems 30-80cm, green. Apr-Oct. *E. esula* x *waldsteinii* ....*Twiggy Spurge* **Euphorbia** x **pseudovirgata**

Ann, with vertical taproot

Lvs 0.5-3 x 0.5-2cm, ovate to obovate, obtuse to retuse, short-petiolate, entire to obscurely minutely crenulate, with stomata below only. Stems 10-30cm, simple or 2-5-forked into branches, hairless, with bluish latex. All yr...........................*Petty Spurge* **Euphorbia peplus**

Plant ± glaucous. Ann. Bare calc habs, often arable

Lvs 0.5-3cm x 1.5-6mm, linear, usu acute or mucronate (rarely obtuse), sessile, with stomata below only. Stem 5-20cm. Apr-Oct..*Dwarf Spurge* **Euphorbia exigua**

Plant glaucous. Tufted per. Coastal

Lvs with stomata above only, dull glaucous above, usu shiny green below, 1-2cm x 2.5-5mm, ovate to oblong, ± acute (to retuse), often fleshy, ± without cartilaginous margins, usu with obscure midrib not raised below. Stems 20-40cm, often reddish, hairless. All yr. Sand dunes, drift-lines..*Sea Spurge* **Euphorbia paralias**

Lvs with stomata below and along midrib above, dull glaucous both sides, 1-2cm x 3-5(10)mm, obovate to oblanc, mucronate or tri-mucronate, not fleshy, with erose cartilaginous margins, with translucent midrib raised below. Stems 5-30cm, often reddish, minutely hairy above. All yr (but less hardy than *E. paralias*). Mostly exposed coasts ....*Portland Spurge* **Euphorbia portlandica**

Group KI – *Lvs spiny or spine-tipped*

■Lvs fleshy, first lvs opp

Stems to 60cm, woody below, brittle above, branched, purplish or with pale green or reddish cartilaginous stripes, often papillate. Lvs 1-4cm x 3-5mm, ± semi-cylindrical, with spine-tipped apex to 1.5mm, sessile (with slightly scarious sheathing base), opaque but with translucent channel above nr base, papillate on side keels and keel below, with stomata all around. Ann. Jun-Oct. Sandy beaches...*Prickly Saltwort* **Salsola kali**

■Lvs not fleshy, all lvs alt...(Thistles and thistle-likes) <u>Go to</u> **THIS**

Group KJ – *Plant hairy. Hairs branched, stellate or medifixed*

■Hairs medifixed (2-partite and strongly adpressed)

Lvs 5-10cm x 3-12mm, oblong-lanc, ± acute, narrowed to short petiole, shiny dark green above, paler below, sparsely hairy, occ with obscure teeth, with ± obscure 2° veins, with stomata below only. Stems 20-60cm, stout, usu unbranched below, ridged. Per. All yr. Walls and rocky habs (esp calc)............ *Wallflower* **Erysimum cheiri**

Lvs 2-4cm x 1.5-5(8)mm, occ opp or subopp, linear-lanc, ± acute, narrowed to short petiole, dull grey-green to purplish with both sides, densely greyish-hairy, entire, with obscure 2° veins, with stomata both sides. Stems 10-30cm, slender, branched, angled. Ann to per. All yr. Coastal, ruderal............ *Sweet Alison* **Lobularia maritima**

■Hairs stellate, usu sessile (v sparse in *Artemisia dracunculus* and *Erophila verna* agg)

Plant aromatic

Lvs 3-6.5 x 0.2-0.7cm, linear-lanc, acute, indistinctly petiolate, ± fleshy, dark blue-green, with sparse stellate wool when young, occ 3-lobed at apex, with translucent midrib occ slightly raised both sides, with 2° veins obscure, with stomata both sides. Stems 20-100(150)cm, woody at base, ± round to weakly angled, striate with weak cartilaginous ridges. Shortly rhizomatous per. Apr-Oct. VR alien............ *Tarragon* **Artemisia dracunculus**

Plant odourless

Stems stout, usu >1.5mm diam

Lvs 7-12 x 1-2cm, linear-oblong to lanc, obtuse, cuneate at base, thick, ± involute when young, with slightly upcurved or undulate margins at maturity, grey-green, densely hairy with short stellate hairs, with midrib raised at least below, with ± obscure opaque 2° veins. Petiole short, with 3(5) vb's. Stems 30-70cm, stellate-hairy, forming aerial rosettes in 2nd yr, with lf scars. Per. All yr. Mostly sea cliffs............ *Hoary Stock* **Matthiola incana**

Stems slender, usu ≤1.5mm diam

Stem lvs usu absent. Ann, usu <10cm............ *Common Whitlowgrass* **Erophila verna** agg

Stem lvs present (*Alyssum saxatile* and *Tuberaria guttata* may rarely key out here)

Bi (occ ann or per). Stem lvs 5-20 x 6mm, broadly linear-oblong, sessile, grey-green, with adpressed stellate hairs above, with stellate and simple adpressed hairs below, with midrib raised below. Basal rosette lvs 3-8(10)cm, oblanc, obtuse to acute, broadly petiolate, entire to sinuate-toothed. Stems to 60(80)cm, 1.5mm diam, purplish, densely adpressed stellate-hairy (hairs often with <6 rays), with some simple hairs............ *Hoary Alison* **Berteroa incana**

Ann. Stem lvs 5-18(30) x 1(3)mm, oblanc, sessile, grey-green, densely stellate-hairy (occ simple-hairy) both sides, with midrib raised below. Basal rosette lvs dead by fl. Stems to 30cm, 1mm diam, greyish, densely adpressed stellate-hairy (hairs with 6-10 rays). May-Aug. Arable or sandy habs, VR, SE Eng. Sch8............ *Small Alison* **Alyssum alyssoides**

Group KK – *Lvs cottony or woolly. Young shoots of* Aster linosyris *may key out here*

■Lvs with (±) prominent hydathodes along margin

Lvs with long petiole, rugose..*Burdock* **Arctium minus** agg

Lvs with short or indistinct petiole, clasping or decurrent

Lvs rugose, often with revolute margins, odorous

Rhizomatous per. Lvs to 5(8) x 1-2.5cm, oblong-lanc to oblanc, cordate at base, clasping, with revolute margins, long-hairy above, with cottony hairs at least below, without glands below, obscurely toothed with hydathodes. Stems to 50cm, woolly or nonglandular-hairy, ridged. Damp habs..*Common Fleabane* **Pulicaria dysenterica**

Ann. Lvs to 4 x 0.5-1.5cm, elliptic-lanc, oblanc, (±) not cordate at base, weakly clasping, with undulate margins, sparsely hairy both sides, with cottony hairs at least below, with yellow sessile glands below, occ weakly toothed with hydathodes. Stems to 40cm, glandular-hairy, ridged. May-Oct. VR, New Forest. Sch8.............................*Small Fleabane* **Pulicaria vulgaris**

Lvs not rugose, without revolute margins, odourless

Basal lvs flat on ground, persistent at fl

Per (bi). Lvs 2-5(10) x 1.5-3(5)cm, broadly ovate to ± orb, abruptly narrowed to short winged petiole, leathery, septate-hairy both sides (hairs with withered tips above), white-cottony esp below when young, occ remotely dentate, with indistinct 2° veins. Stem(s) 7-30cm, cottony when young. Apr-Oct (all yr). Calc turf.........*Field Fleawort* **Tephroseris integrifolia**

Basal lvs held ± erect or absent, not persistent at fl

Per. Lvs 10-15cm, broadly oblanc. All yr...............*Perennial Cornflower* **Centaurea montana**

Ann (bi). Lvs 5-20cm, linear or oblanc, narrowed to ± sessile base, greyish with cottony hairs (esp above), occ distantly hydathode-toothed or pinnately lobed, with midrib raised below, often with opaque ≥3° veins raised both sides. Stems 20-90cm, usu branched, sparsely cottony, grooved or angled, with cartilaginous ridges. Apr-Sept (all yr). Arable, wildflower mixes..*Cornflower* **Centaurea cyanus**

■Lvs without hydathodes along margins. (If ochrea present and lvs revolute when young <u>Go to</u> **DOCK**)

Per with stolons rooting at nodes. Lvs obscurely pinnately veined

Lvs mostly in rosettes at end of short stolons, to 3cm, obovate-spathulate, usu apiculate, narrowed to petiole, green and hairless or sparsely hairy above, white-woolly below with adpressed long hairs, with stomata above. Stems (and stolons) white-woolly, to 15cm. All yr ....*Mountain Everlasting* **Antennaria dioica**

Per without stolons, with abundant non-fl shoots. Lvs (1)3-veined

Lowlands

Rhizomatous. Lvs basal or on stem, to 12(15) x 1-2cm, narrowly lanc to elliptic, acute, ± clasping at base, with slightly revolute margins, soon hairless above, white-woolly below, with at least midrib raised below, with stomata below only. Stems usu 1, to 1m, adpressed white-woolly or thinly so. Mar-Oct (all yr).......................*Pearly Everlasting* **Anaphalis margaritacea**

Tufted. Lvs basal and on stem, 2-8cm x 2-8mm, oblanc to linear, acute, narrowed to long petiole-like base, with flat margins, soon green and hairless above, whitish-woolly below, with midrib raised below, with stomata below only. Stems often several, (8)20-60cm, woolly. Apr-Oct..*Heath Cudweed* **Gnaphalium sylvaticum**

Mtns, usu >450m. Tufted

Stems 8-30cm, woolly. Lvs basal and on stem, 50-80(120) x 3-10(18)mm, broadly lanc, whitish-woolly, 3(5)-veined. VR Scot................*Highland Cudweed* **Gnaphalium norvegicum**

Stems 2-10cm, thinly woolly. Lvs often all basal, 5-15 x 2mm, linear-oblanc, woolly or silky-hairy with adpressed long wavy hairs, 1-veined..........*Dwarf Cudweed* **Gnaphalium supinum**

Ann, usu without non-fl shoots. Lvs 1(3)-veined

Winter-wet habs, v common (but often avoiding calc soils)

Lvs 1-5cm x 1-5mm, linear-oblong to oblanc, ± acute, cuneate at base, sessile, often undulate, adpressed-cottony both sides, 1(3)-veined. Stems 4-20cm, usu with many decumbent to ascending branches from nr the base, adpressed-woolly or -cottony. Apr-Oct (all yr) ....*Marsh Cudweed* **Gnaphalium uliginosum**

Usu dry or well-drained habs, uncommon to VR

Plant greyish- or white-woolly

Stem usu 1

Stems 5-30(45)cm, erect or ascending, simple or branched at base, densely white-woolly. Lvs 10-20(30)mm, erect, lanc, obtuse to acute, usu undulate, often with distinct midrib. Oct-Aug (usu). Mainly acidic to neutral soils..................*Common Cudweed* **Filago vulgaris**

Stem 2-15(30)cm, erect or ascending, branched, greyish-woolly. Lvs 4-10mm, oblong to linear-lanc, erect or adpressed to stem, acute, not undulate, with midrib indistinct. Apr-Oct. Acidic to neutral soils..*Small Cudweed* **Filago minima**

Stem 4-30(45)cm, erect, usu with ascending branches from nr the base, simple below but corymbosely branched above, densely white-woolly (occ silky-hairy). Lvs 15-50(70)mm, oblanc to oblong, usu obtuse (the upper acute), ± clasping, undulate, with recurved margins, white-woolly both sides (hairs more adpressed below), with translucent midrib raised below. Oct-Jul. Often ruderal. Sch8........*Jersey Cudweed* **Gnaphalium luteoalbum**

Stems 1-several

Stems 5-30(40)cm, decumbent or ascending, divaricately branched above. Lvs (5)10-15 x 3-4mm, narrowly oblong to ± broadly spathulate, apiculate, hardly undulate, with midrib raised below. Oct-Aug. Chalky soils, VR, SE Eng. Sch8 ....*Broad-leaved Cudweed* **Filago pyramidata**

Plant ± grey-green with short silky hairs. Stems 1-several

Stems 5-20(25)cm, erect or ascending, usu branched below, dichasially branched above. Lvs 8-25mm, held weakly erect, usu linear, ± long-acute. Essex, Suffolk, VR ....*Narrow-leaved Cudweed* **Filago gallica**

Plant yellowish-woolly. Stems 1-several

Stems 10-25cm. Lvs 3-6mm, broadly oblong-lanc to spathulate, apiculate, hardly undulate. Oct-Aug (usu) or spring ann. Acidic to neutral soils, VR, mainly SE Eng. Sch8 ....*Red-tipped Cudweed* **Filago lutescens**

Group KL – *Lvs with at least some sessile or stalked glandular hairs*

■Lvs in a basal rosette. Bogs or peaty habs. Plant insectivorous

Lvs with long ± flattened petiole, viscid with reddish patent tentacle-like glandular hairs above, circinate when young, with stomata both sides. Stipules scarious, partly adnate to petiole, laciniate

Petiole hairy. Lvs often adpressed to ground, abruptly narrowed to petiole

Lvs (3)7-10mm diam, ± orb, with red hairs to 6mm (shortest nr middle of lf, often papillate or with ciliate gland-tips). Petiole to 4cm. Stipules 5-7-fid. Winter bud to 2.5mm, purplish. Apr-Oct. Damp to wet bogs......................................*Round-leaved Sundew* **Drosera rotundifolia**

Petiole hairless or with sparse sessile glands. Lvs usu held erect, narrowing gradually to petiole

Lvs 0.4-2cm x 3mm, obovate to spathulate. Petiole 1.5-3.5cm, hairless or with sparse sessile glands. Stipules 7-fid, often absent at maturity. Winter bud to 3mm, purplish. Apr-Oct. Damp peaty hths and moors...*Oblong-leaved Sundew* **Drosera intermedia**

Lvs (1.5)2.5-4cm x 2-5mm, narrowly obovate to linear-oblong, green with long red tentacles. Petiole 1-7cm, hairless. Stipules c12-fid. Winter bud to 10mm, green. Apr-Oct. Wet sphagnum bogs..*Great Sundew* **Drosera anglica**

Lvs ± sessile, viscid with 0.3mm (and sessile) glandular hairs above, with involute margins, with stomata both sides (often v sparse above nr midrib). Stipules absent

Lvs 2.5-5(9) x 0.9-2.5cm, yellow-green (occ pinkish below), opaque, slightly fleshy. May-Sept

Lvs with more stellate arrangement than *P. grandiflora* when young, ovate-oblong, often with obscure midrib. Winter bud to 10mm, green, with 4-5 scales. Basic bogs, mostly N & W Br ....*Common Butterwort* **Pinguicula vulgaris**

Lvs ovate-oblong, with distinct midrib. Winter bud to 10mm, green, with 4-5 scales. Basic bogs, SW Ire (native), Derbs (alien)...........*Large-flowered Butterwort* **Pinguicula grandiflora**

Lvs to 2 x 1cm, pale olive-green with reddish veins, translucent, v thin. All yr

Lvs ovate, narrowed to short indistinct petiole, strongly involute, with scattered short red-tipped glandular hairs (and longer septate hairs along midrib) above, hairless below, with anastomosing 2° veins. Acidic bogs, mostly W Br.........*Pale Butterwort* **Pinguicula lusitanica**

■Lvs in a basal rosette. Dry sandy or calc habs. Plant not insectivorous

Basal lvs soon dying, to 1cm, obovate-spathulate, with a distinct pore nr apex above, cuneate at base, reddish or yellow-green, with red-tipped glandular hairs esp on margins, with obscure veins. Ann...*Rue-leaved Saxifrage* **Saxifraga tridactylites**

■Lvs on stem, or in an aerial rosette. Mostly dry habs. Plant not insectivorous

Stems trailing or prostrate, usu 20-50cm

Plant densely glandular-hairy. Lvs 2.5-6cm, ovate to ± orb, obtuse, rounded or ± cordate at base, shortly petiolate, grey-green, hairy both sides, occ slightly toothed, 3(5)-pli-veined from base, with stomata both sides. Stems c1mm diam, branched from base, with patent long nonglandular and short glandular hairs, round. Ann. May-Oct ....*Round-leaved Fluellen* **Kickxia spuria**

Plant (sparsely) hairy, hardly glandular. Lvs 1.5-3 x 1-2cm, ovate, ± obtuse to acute, hastate at base (at least upper and middle lvs), shortly petiolate, grey-green, hairy both sides, occ with 4 teeth per side, with few anastomosing veins (occ fading nr margins), with stomata both sides. Stems c1mm diam, branched from base, with unequal spreading to patent nonglandular septate hairs to 2mm (the longest hairs often widely spaced) and short glandular hairs, round. Ann. May-Oct...*Sharp-leaved Fluellen* **Kickxia elatine***

Stems erect or ± so

Lvs >3cm (*Misopates orontium* may key out here)

Lf margin with prominent hydathodes. Odorous or weakly so

Lvs with some cottony hairs. Sch8...................................*Small Fleabane* **Pulicaria vulgaris**

Lvs without cottony hairs. Hortal......................................*Pot Marigold* **Calendula officinalis**

Lf margin without hydathodes

Odorous ann or per..(*Solanaceae*) Go to **SOL**

Weakly aromatic ann. Lvs to 6 x 1.5-3cm, ovate to ovate-elliptic, obtuse, narrowed to a short petiole, dull grey-green, densely septate-hairy both sides (minutely glandular), with 4 prs of 2° veins (anastomosing or fading nr margins) often raised below. Petiole with flat arc vb. Stems to 60cm, densely glandular-hairy. Apr-Oct. *P. axillaris* x *integrifolia* ....*Petunia* **Petunia** x **hybrida**

Weakly aromatic per. Lvs (on stem) 2.5-3.5cm, obovate-oblong, ± obtuse, sessile, dull grey-green, with v tiny glandular hairs, long-ciliate esp at base, with opaque or weakly translucent veins (fading nr margins) often raised both sides. Stems to 50cm, reddish at base, with septate hairs and tiny glandular hairs. All yr......*Seaside Daisy* **Erigeron glaucus**

Lvs ≤2.5cm. Ann (*Sedum album* and *S. villosum* may key out here but are per with fleshy lvs)

Lvs opp below, alt above, 10-25 x 2.5-3.5mm, linear to oblong-lanc, ± obtuse, narrowed at base to a short petiole, with flat margins, dark grey-green to ± glaucous, glandular-hairy, with translucent midrib raised below, with obscure 2° veins (occ 3-pli-veined). Stem 8-25cm, erect, branched, usu glandular-hairy, round. Ruderal, arable ....*Small Toadflax* **Chaenorhinum minus**

Lvs 3-whorled or pseudowhorled below, alt above, 3-10 x 1.5mm, lanc or oblanc, ± acute, narrowed to short petiole, with flat (or revolute) margins, light green, viscid-hairy, veinless. Stems to 15cm, often ascending-erect, much-branched, viscid-hairy, round. Sand dunes, VR, SW Br...*Sand Toadflax* **Linaria arenaria**

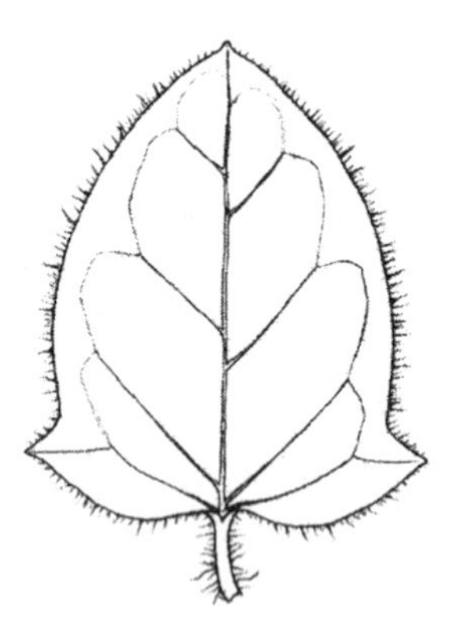

Kickxia elatine

Group KM – Symphytum, Parietaria. *Lvs with some minute hooked hairs at least below (often minute but adhering to woolly clothing), rolled when young (revolute in* Parietaria*), with 1-4° veins raised below. All hairs simple and unicellular.* ❶ *Lvs occ with black-based hairs*

■Lvs with abundant hooked hairs both sides, cuneate at base

Lvs with stomata both sides, to 10 x 4cm, elliptic to narrowly ovate, dark green, with short hispid hairs above. Petiole to 7cm, with 3 vb's. Stems 30-50cm, not purplish, hispid. Tubers ± globose, occ connected by thin rhizomes. Jan-Apr......................*Bulbous Comfrey* **Symphytum bulbosum**

Lvs with stomata below only, to 17 x 7cm (mid-stem lvs larger than basal lvs), elliptic to ovate, dark green, with additional slightly curved hispid hairs above, with veins occ purplish below. Petiole to 7cm, with 3(5) vb's. Stems 30-50cm, purplish or with purple hair-bases, sharply hispid. Rhizomes tuberous, short, stout, with thin portions. Feb-Jul ....*Tuberous Comfrey* **Symphytum tuberosum***

■Lvs with hooked hairs below only (occ also v sparse above), cuneate to cordate at base

Stems rooting at lower nodes, unwinged. Patch-forming, with stolons. ❶

Lvs 4-8cm, ovate, acuminate, usu cordate at base, usu undulate and rugose, dark green, v rough, with v short curved hairs to 0.2mm above. Petiole 2-8cm, decurrent down stem, ± unwinged, with 3(5) vb's. Stems usu 10-25cm tall, ± prostrate, rough with retrorsely curved hairs. All yr. Often damp habs............................*Creeping Comfrey* **Symphytum grandiflorum**

Lvs 8-18cm, ovate, acuminate, ± cordate to rounded at base, hardly undulate or rugose, dark green, rough, with weakly curved hairs to 0.5mm above. Petiole 6-15cm, ± decurrent down stem, ± unwinged, with 3(5) vb's. Stems usu 25-70cm tall, ascending, rough with antrorse hooked hairs and v stout short straight or retrorse hispid hairs. All yr. *S. grandiflorum* x *?uplandicum*..*Hidcote Comfrey* **Symphytum 'Hidcote Blue'**

Stems not rooting at nodes, erect, strongly winged to unwinged. Tufted, with taproot

Hairs usu sharply hispid (esp in *S.* x *uplandicum*)

Plant usu nr water or in winter-wet habs. Stems 2(4)-winged (due to broad long-decurrent petioles)

Basal lvs 15-30cm, narrowly triangular, acute, rounded-cuneate at base, with weakly adpressed hairs above to 1.5mm. Petiole strongly and broadly decurrent, with 5(9) vb's. Stems to 150cm, with long weakly hispid retrorse hairs. Apr-Oct ....*Common Comfrey* **Symphytum officinale**

Plant in damp to dry habs. Stems not winged (petioles narrowly and shortly decurrent on stem)

Basal lvs cordate at base, 5-15cm, elliptic-lanc, roughly hairy. Petiole unwinged. Stems to 180cm tall, with short stout hooked bristles. Calyx-teeth obtuse. Apr-Oct (all yr). R alien ....*Rough Comfrey* **Symphytum asperum**

Basal lvs not cordate at base, 8-18cm, elliptic-lanc, with spreading hairs 0.5-1.2mm above. Petiole narrowly winged. Stems to 1m tall, with short stout hooked bristles and longer retrorse hispid hairs. Calyx-teeth acute. Apr-Oct. *S. asperum* x *officinale* ....*Russian Comfrey* **Symphytum** x **uplandicum**

Hairs not sharply hispid. Dry habs

Lvs >10cm, rugose

Stem lvs shortly decurrent, to 20 x 8cm, ovate to lanc, ± acute, cuneate to rounded at base, the lower petiolate, the upper sessile, slightly grey-green, roughly hairy (but not piercing), with 0.1mm hooked hairs below. Petiole to 15cm, with 5 vb's. Stems to 90cm, rough. Fls blue. All yr...*Caucasian Comfrey* **Symphytum caucasicum**

Stem lvs not or hardly decurrent, to 20 x 10cm, ovate or oblong, ± acute, cordate to rounded at base, pale green, softly hairy, with 0.2mm hooked hairs below. Petiole to 20cm, with 5 vb's. Stems to 150cm, rough, hardly ridged, with ± retrorse hairs. Fls white. Oct-Jun ....*White Comfrey* **Symphytum orientale**

Lvs <8cm, not rugose

Lvs 4-7cm, often >4x petiole length, ovate, widest below middle, acuminate, revolute when young, shiny yellow-green above, ciliate, ± net-veined, with 1-3° veins raised below. Petiole to 3cm, channelled, with 3 vb's. Stems to 60cm, ± erect, stout, reddish, round, semi-translucent when young, softly hairy. Bracts fused. Apr-Oct. VR alien ....*Eastern Pellitory-of-the-wall* **Parietaria officinalis**

Lvs to 4(6)cm, to 3x petiole length, ovate to lanc, widest in middle, often acuminate, revolute when young, shiny dark green above, ciliate, weakly net-veined, with 1-3° veins raised below. Petiole to 1cm, channelled, with 3 vb's. Stems to 40cm, ascending, reddish, round, semi-translucent when young, softly hairy. Bracts free. All yr. Often on walls ....*Pellitory-of-the-wall* **Parietaria judaica**

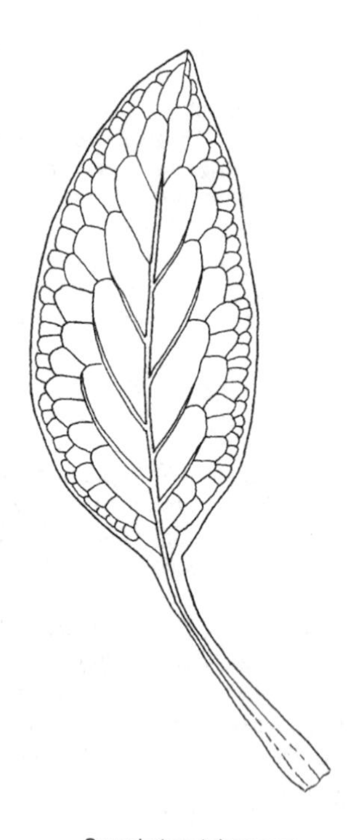

Symphytum tuberosum

Group KN – Boraginaceae. *Plant with v bristly or sharply hispid unicellular hairs (young lvs often softly hairy)*

■Lvs net-veined or at least 3° veins visible (lvs usu petiolate, with at least 2° veins raised below)

Lvs appearing distinctly toothed due to crimped margins, rugose

Per with short stout rhizome. Basal lvs persistent, 4-15 x 5cm, lanc to ovate or obovate, obtuse, rounded to cuneate at base, rugose, with strong cucumber odour, shiny dark green above, with stomata both sides. Petiole short, often indistinct, winged, with 1(3) vb's, with TS turning brown. Stems to 25cm. All yr..*Slender Borage* **Borago pygmaea**

Ann. Basal lvs soon dying or absent, 5-15(20) x 8cm, ovate, obtuse to acute, ± cordate at base, rugose, with weak cucumber odour, dark or grey-green above, with stomata both sides. Petiole 3-7cm, narrowly winged nr lf and decurrent down stem, with 1-5 vb's. Stems to 100cm, usu with swollen-based hairs, angled to ± round, soon hollow, with abundant sap, with TS turning brown..*Borage* **Borago officinalis**

Lvs entire (occ obscurely toothed in *Pentaglottis sempervirens*), not rugose (but ± rugose in *Omphalodes verna*)

Lvs cuneate at base

Lvs all on stem, 15-45 x 8cm, linear-lanc to ovate-lanc, acute, sharply hispid, often with red-based hairs, entire, with stomata both sides (esp nr veins above). Petiole often indistinct, short, winged, not channelled, with flat arc vb and several smaller vb's along lower margin. Stem 1, to 4m, usu unbranched, v hispid, round. Monocarpic per. All yr ....*Giant Viper's-bugloss* **Echium pininana**

Lvs mostly basal, to 16 x 7cm, ovate-lanc, minutely aristate when young, hispid both sides, with white swollen-based hairs above, with stomata below only. Petiole distinct, to 10cm, winged, channelled, with 5(7) vb's. Stems 1-several, to 1m, usu branched, v hispid with whitish retrorse hairs, round. Per. All yr................*Green Alkanet* **Pentaglottis sempervirens***

Lvs cordate to rounded at base

Plant tufted or shortly rhizomatous

Basal lvs rarely >10cm diam, broadly ovate to orb, cordate at base, usu hispid both sides, with stomata below only. Petiole long, hispid, with 3(5) vb's turning brown, sap with cucumber odour...*Great Forget-me-not* **Brunnera macrophylla**

Plant strongly rhizomatous or stoloniferous

Basal lvs 15-50 x 25cm, ovate, ± obtuse, cordate at base, rolled when young, shortly hispid both sides (or shortly patent-hairy below), obscurely toothed nr base, with stomata below only. Petiole to 15cm, roughly hairy, V-channelled, with 5(7) vb's turning brown, with cucumber odour. Rhizomes stout, with persistent dead petioles. Stolons never developing. Feb-Oct..*Abraham-Isaac-Jacob* **Trachystemon orientalis**

Basal lvs to 10(15) x 7cm, ovate, often acuminate, rounded to cordate at base, rolled when young, with short sparse white hispid hairs (± adpressed to patent) both sides, often with stomata above. Petiole to 10cm, sparsely ciliate to hairless, V-channelled, with 1(3) vb's turning brownish, odourless. Rhizomes slender, with persistent dead petioles. Stolons often developing in spring. Oct-Jul. Shady habs.....................*Blue-eyed-Mary* **Omphalodes verna**

■Lvs not net-veined (3° veins not visible)

Basal or lower stem lvs with distinct petiole (may be winged), without 2° veins raised below. Lvs usu broadly ovate, rolled when young, usu with swollen-based hairs, with indistinct opaque 2° veins, with stomata both sides. Petiole to 15cm, winged, with 1(3) vb's

Lvs distinctly white-spotted, to 15(20)cm

Lvs rounded at base, acute-acuminate, not undulate, usu with sparse minute glandular hairs. All yr..*Lungwort* **Pulmonaria officinalis**

Lvs cuneate at base, acute-acuminate, undulate, without glandular hairs. All yr. VR alien ....*Red Lungwort* **Pulmonaria rubra**

Lvs unspotted to weakly white-spotted, to 15cm, lanc to oblong, acute, long-cuneate at base, with long hairs to 1mm, without minute or glandular hairs. Feb-Dec. Usu shady habs, R, S Eng ....*Narrow-leaved Lungwort* **Pulmonaria longifolia**

Lvs unspotted or with faint green spots, to 8(10)cm, ovate, acute-acuminate, rounded at base, with long hairs to 1mm and minute hairs <0.1mm, occ with glandular hairs. Apr-Sept. Woods, VR, E Suffolk...*Suffolk Lungwort* **Pulmonaria obscura**

Basal or lower stem lvs sessile or with indistinct petiole, OR 2° veins raised below

Lvs apparently toothed, with margins undulate or recurved. Ann (easily uprooted)

Basal lvs often <10, in a rosette, usu dead at fl, 7-15 x 0.8-1.7cm, oblanc to linear-oblong, obtuse or apiculate, with unequal ± translucent or white swollen-based hairs (occ flattening), with obscure 2° veins, with stomata both sides. Petiole with flat arc vb. Stems to 50cm, erect, v hispid (hairs not purple-based), ± round. Taproot slender, brown, without fibrous remains. Apr-Oct (all yr)..*Bugloss* **Anchusa arvensis**

Lvs entire, with margins undulate

Bi, with stout black taproot, and with dense fibrous lf bases (firmly rooted)

Basal lvs often >20, in adpressed rosette, dying as stem develops, to 15 x 1.5(2)cm, linear-lanc to oblong, often with recurved margins, with long white hispid hairs (often swollen-based or bent at 90°), with translucent midrib, with opaque indistinct or obscure 2° veins (visible on larger lvs). Petiole with flat arc vb (and 2 rib bundles). Stems to 90cm, with long hispid hairs (with purple swollen bases) and short hairs......*Viper's-bugloss* **Echium vulgare**

Ann (easily uprooted)

Basal lvs to 8cm, linear to oblanc, v hispid with patent white hairs both sides, with obscure 2° veins. Stem(s) to 70cm. Mar-Oct....................*Common Fiddleneck* **Amsinckia micrantha**

Lvs entire, with margins flat or ± so

Rhizomatous per, with prostrate stems rooting at tips, to 60cm tall

Lvs 4-8cm x 6-15mm, narrowly lanc, acute, rough with ± antrorse adpressed hairs above, ± sessile, dark green above, paler below, 1-veined, with translucent midrib raised below, with 2° veins indistinct or opaque. Stems always present, rough with swollen-based spreading hairs at least below (often antrorsely adpressed). Mar-Nov. Shady calc habs or coastal ....*Purple Gromwell* **Lithospermum purpureocaeruleum**

Tufted per, with erect stems to 60cm

Lvs 12-20cm, elliptic-lanc, narrowed to petiole, roughly hispid with minutely swollen-based 1mm hairs, with c4 prs of translucent 2° veins. Petiole indistinct, with 1(3) vb's ....*False Alkanet* **Cynoglottis barrelieri**

Ann or bi, with erect stems to 60cm

Lvs to 14cm, ovate, petiolate, often pustulate above, with antrorse hispid hairs, with translucent 2° veins raised below. Stems without purple based hairs ....*Purple Viper's-bugloss* **Echium plantagineum**

Ann, with erect stems to 50cm

Lvs to 14cm, ovate-lanc to lanc, greyish-green, softly hairy when young (hairs 1mm), with obscure 2° veins, with stomata both sides.................................*Alkanet* **Anchusa officinalis**

Pentaglottis sempervirens

Group KO – *Plant not sharply hispid (stems and lvs may have rough swollen-based hairs). Lvs with 2º veins visible. Hairs unicellular but septate in* Aster, Amaranthus *and some* Solanaceae. *(*Parietaria *will key out here if the hooked hairs are overlooked).* ❶ *Mat-forming, to 1cm tall.* ❷ *Lvs with mouse-like odour*

■Stipules fused into ochreae. Lvs revolute when young........................(*Polygonaceae*) Go to **DOCK**

■Stipules reduced to minute glands. Aquatic...........................*Water-primrose* **Ludwigia grandiflora**

■Stipules absent

Stem lvs clasping, always present

Stem lvs green, net-veined, to 6 x 1cm....................................(*Aster novae-angliae*) Go to **AST**

Stem lvs glaucous, not net-veined, 7-15 x 1.5-4cm, sagittate, softly hairy or hairless, occ ciliate. Basal lvs in 1st yr, lanc, narrowed to long petiole, often obscurely sinuate-toothed. Stems 60-120cm. Bi or per. All yr..*Woad* **Isatis tinctoria***

Stem lvs not clasping, or absent

Basal rosette often present. Plant hairy

Lvs softly hairy with ± silky weakly adpressed hairs both sides, to 14 x 6cm, lanc to ovate, the basal dying as stem develops, usu acute, rolled when young, rugose or bullate, greyish, not or hardly net-veined, with anastomosing 2º veins raised below, with stomata both sides. Upper lvs sessile, undulate. Petiole to 18cm, not channelled, with 3(5) vb's. Short-lived per, with 1-4 rosettes per rootstock. ❷. Mar-Oct. Calc habs ....*Hound's-tongue* **Cynoglossum officinale**

Lvs (±) hairless above, roughly hairy with short spreading swollen-based hairs below, to 14 x 5cm, lanc to ovate, narrowed to petiole, rolled when young, undulate, shiny dark green above, weakly net-veined, with stomata below and occ above. Petiole to 8cm, sparsely hairy, channelled, with 3(5) vb's. Stems with retrorse swollen-based hairs. Bi to per, with 1 rosette per rootstock. All yr. R, S Eng. Sch8........*Green Hound's-tongue* **Cynoglossum germanicum**

Basal rosette absent

Mat-forming per, with prostrate stems rooting at the nodes ❶

Lvs 2-7mm, ± orb, ± sessile to shortly petiolate, dark green, paler below, translucent-dotted (or at least pitted above), sparsely hairy to hairless, with ± indistinct veins. Stems <1mm diam, pale reddish, hairy or hairless, with elastic stele. All yr ....*Mind-your-own-business* **Soleirolia soleirolii**

Tufted per

Lvs petiolate...(*Solanaceae*) Go to **SOL**

Lvs sessile, 5-8 x 1.5cm, lanc, acute, often with revolute margins, dark green above, antrorsely ± adpressed-hairy above (hairs on tiny swollen bases), with short patent hairs below, with midrib raised below (rough due to swollen-based hairs), with c3 prs of 2º anastomosing veins (often fading nr margins or raised below), with stomata both sides. Stems to 80cm, 1-several, erect, simple or branched above, with antrorsely adpressed (rarely ± patent) swollen-based white hairs. Mar-Oct ....*Common Gromwell* **Lithospermum officinale**

Ann. Lvs petiolate

Lvs with Kranz venation (occ obscure). Stems with scattered vb's. Apr-Oct ....(*Amaranthus*) Go to **AMAR**

Lvs without Kranz venation. Stems with a ring of vb's. Jun-Oct. (*Solanum nigrum*) Go to **SOL**

Group KP – *Plant not sharply hispid (stems and lvs may have rough swollen-based hairs). Lvs 0-1-veined or 2° veins opaque. Hairs unicellular or septate. If ochreae present (*Polygonaceae*)* Go to **DOCK**. (Pulmonaria *may key out here)*

■Lvs with swollen white hydathode below at apex (dark when mature), folded when young, ciliate or not, with translucent midrib. Hairs unicellular, white, swollen-based.................(*Myosotis*) Go to **MYO**

■Lvs with sunken apical pore above nr apex (HTL), not folded when young, usu ciliate at least nr base, with all veins obscure. Hairs septate (differentiating multiseriate from uniseriate hairs requires a microscope (x100) but multiseriate hairs often have a ± swollen base)

Lvs with an obvious apical pore nr apex (occ lime-encrusted)

Lvs 6-20 x 1-2mm, oblong-linear, acute or minutely apiculate, sessile, thick, flat, hairless, usu remotely ciliate with stiff hairs at least nr base (often resembling teeth), with stomata both sides. Stems rooting at nodes, often reddish and sparsely hairy, round. Hairs uniseriate. Stoloniferous. Mar-Oct (all yr). Mtn wet flushes..*Yellow Saxifrage* **Saxifraga aizoides**

Lvs with an indistinct pale apical pore (never lime-encrusted)

Lvs 10-40 x 2-4mm, elliptic-lanc to oblanc, obtuse, sparsely ciliate nr base, with stomata below only. Petiole ≥ lf on lower lvs. Stems not rooting at nodes, often reddish, often sparsely long-hairy, round. Hairs multiseriate. Rhizomatous. Mar-Oct. Sch8 ....*Marsh Saxifrage* **Saxifraga hirculus**

Lvs 6-10 x 0.5-1.5mm, ± linear (some usu 3-5-lobed), acute with arista to 0.5mm, sparsely ciliate esp nr base with long wavy nonglandular hairs, with stomata below only. Petiole ± absent. Stems rooting at nodes, usu with axillary bulbils, hairless or sparsely wavy-hairy (occ glandular), round. Hairs uniseriate. All yr........................*Mossy Saxifrage* **Saxifraga hypnoides**

■Lvs without swollen white hydathode or apical pore, not folded when young

Stem always present, erect, rough with swollen-based unicellular hairs at least below (usu antrorsely adpressed)

Per. Lvs 5-8 x 1.5cm, lanc, acute, sessile, often with revolute margins, dark green above, ± adpressed-hairy above (hairs with swollen-bases), with short patent hairs below, with midrib raised below, with c3 prs of 2° anastomosing veins (often fading nr margins or raised below), with stomata both sides. Stems 1-several, to 80cm, white-hairy (hairs rarely ± patent). Taproot ± swollen, without purple dye. Mar-Oct................*Common Gromwell* **Lithospermum officinale**

Ann. Lvs 2-3(4)cm x 3-6mm, linear-oblong, acute, indistinctly petiolate or ± sessile, antrorsely adpressed-hairy above, with translucent midrib raised below, with indistinct 2° veins, with stomata both sides. Stem usu 1, to 50cm. Taproot not swollen, with purple dye. Arable ....*Field Gromwell* **Lithospermum arvense**

Stem, if present, not rough with swollen-based hairs

Lvs with 2° veins raised above (only midrib strongly raised below). Hairs septate

Per

Basal lvs 1.5-3 x 1cm, narrowly oblanc, narrowed to long winged petiole, densely hairy, ciliate nr base, occ with remote hydathode-teeth. Stem lvs few, linear-oblong, sessile, ± clasping. Stems 7-20cm, erect, usu unbranched, hairy. All yr. Mtns, R. Sch8 ....*Alpine Fleabane* **Erigeron borealis**

Basal (and lower) lvs 1-3(5) x 1.5cm, obovate, cuneate at base, sparsely adpressed-hairy to ± hairless, often 3-lobed or with 3-5 mucronate teeth at apex. Upper lvs to 1.5 x 0.3cm, linear-lanc, acute, sparsely hairy to ± hairless, adpressed-ciliate, ± entire, with translucent midrib raised below, with other veins ± obscure. Stems to 50cm, erect or sprawling, branched, slightly angled, sparsely hairy, solid. Often on walls. All yr ....*Mexican Fleabane* **Erigeron karvinskianus**

Ann (bi)

Basal lvs persisting at fl, to 5 x 1cm, obovate-oblanc to spathulate, obtuse with purple hydathode, narrowed to indistinct petiole, rolled when young, occ undulate, weakly aromatic, hairy both sides, ciliate, occ 3-pli-veined, with reddish midrib, with opaque anastomosing 2° veins, with stomata both sides. Stem lvs linear-lanc, sessile or narrowing to petiole, ± clasping. Petiole purplish, flat or channelled. Stem 1, purple, hairy, ± round to weakly ridged. Ann or bi (per). Often all yr..*Blue Fleabane* **Erigeron acer**

Basal lvs often absent at fl. Stem lvs 3-8 x 0.4-1(2)cm, oblanc to linear-lanc, acute with hydathode, ± sessile or with indistinct petiole, weakly aromatic to odourless, patent- to adpressed-hairy both sides (occ roughly so), shortly ciliate, entire to weakly toothed, with translucent midrib raised below, other veins obscure and opaque. Stems (3)10-100cm, shortly hairy, weakly ridged. R...............................*Argentine Fleabane* **Conyza bonariensis**

Basal lvs often absent at fl. Stem lvs 1-5 x 0.2-1cm, obovate-lanc to linear, acute with hydathode, ± sessile or with indistinct petiole, weakly aromatic, dull pale yellow-green both sides, roughly adpressed-hairy both sides (± strigose), with long widely spaced cilia to 10(15)mm, entire to weakly toothed. Stems (3)50-100cm, roughly hairy, ridged. Ann (bi). Oct-Aug (all yr)...*Canadian Fleabane* **Conyza canadensis**

Lvs without veins raised above. Hairs unicellular

Stems (or lf margins) with short antrorsely crisped hairs, often slightly woody at base. Lvs basal and/or on stems, the upper longer than the lower, linear to obovate, ± sessile or narrowed to indistinct petiole, applanate when young, midrib often obscure or opaque. Stems 0.7-1mm diam, weakly angled to ± round. Per. All yr

Lvs in rosette slightly above ground level (false aerial rosette) at end of stock (visible until late summer), all alt, with stomata both sides

Lvs 5-20mm, elliptic-lanc to obovate, ± obtuse, often thick, often with crisped hairs above. Stem lvs oblanc to elliptic-oblanc, ± acute. Stems to 10cm. Chalk gsld, S Eng ....*Chalk Milkwort* **Polygala calcarea**

Lvs in basal rosette at ground level, or on stems only

Lvs all alt, with stomata both sides

Lvs in basal rosette or on stems, 5-23 x 5mm, elliptic-lanc, ± acute, tasteless, occ with crisped hairs on midrib above, occ ciliate. Stems usu 3-8, to 25cm, much-branched, often visible from previous yr................................*Common Milkwort* **Polygala vulgaris**

Lvs usu all in basal rosette, 5-20 x 5mm, obovate, obtuse, bitter-tasting. Stem lvs narrowly obovate to lanc, acute. Stems often 1, to 10cm, hardly branched, not visible from previous yr. Calc gsld, VR, Yorks, Kent............*Dwarf Milkwort* **Polygala amarella**

Lvs often opp below (but not connate), without stomata above

Lvs in basal rosette or on stems, 3-15 x 4mm, elliptic-lanc, ± obtuse, occ with crisped hairs on midrib above. Stems to 15cm, hardly branched, rarely visible from previous yr ....*Heath Milkwort* **Polygala serpyllifolia**

Stems, if present, without antrorsely crisped hairs, herbaceous at base

At least some lvs on stem

Stem lvs clasping, glaucous, 7-15 x 1.5-4cm, sagittate, softly hairy or hairless, occ ciliate, not net-veined. Basal lvs only in 1st yr, lanc, narrowed to long stalk, often obscurely sinuate-toothed. Stems 60-120cm. Bi or per. All yr...........................*Woad* **Isatis tinctoria***

Stem lvs not clasping, green to grey-green..(*Ranunculus lingua*, *R. flammula*) Go to **RAN A**

Lvs all basal

Winter-wet habs

Stoloniferous per. Lvs 3-7cm x 1-3mm, ± cylindrical to flattened or linear and channelled, blunt to acute (hydathode often purplish), bright green, occ hairy or sparsely ciliate, ± solid with aerenchyma, with 1(3) vb's, with stomata both sides. Roots white. All yr ....*Shoreweed* **Littorella uniflora***

Ann. Lvs 3-5cm x 2-4mm, linear, flat, yellow-green, with adpressed unicellular hispid hairs both sides, with antrorsely adpressed cilia, with translucent midrib, solid, with stomata both sides. All yr. VR alien............*White Forget-me-not* **Plagiobothrys scouleri**

Coastal habs. Tufted per

Lvs 2-22cm x 3-8mm, semi-cylindrical or channelled, ± acute with pale hydathode, fleshy, weakly pitted both sides, green to glaucous, purplish at base, hairless but with long white hairs in rosette centre, with all veins obscure (but with 3(5) vb's), with stomata both sides. Rootstock woody, stout. Apr-Oct..................*Sea Plantain* **Plantago maritima**

Lvs 2-15cm x 1(2)mm, flat and linear to ± semi-cylindrical, acute to obtuse, ± fleshy, pitted both sides, green, reddish at base, occ hairy or ciliate, 1-veined with midrib sunken above, with stomata scattered both sides. Rootstock woody, forming cushion-like tufts..*Thrift* **Armeria maritima** ssp **maritima**

Dry habs or hortal. Tufted per

Lvs petiolate, not imbricate, not persistent when dead

Lvs 15-30 x 8cm, elliptic to obovate, with mucro at apex, with cartilaginous margins. Apr-Sept..(*Limonium latifolium*) Go to **LIM**

Lvs (±) sessile, ± imbricate, persistent when dead

Lvs 20-60 x 10mm, obovate-lanc, mucronate, fleshy, dark green to glaucous, often reddish, shortly ciliate, with stomata both sides. Stems to 40cm, glandular-hairy above. Compact cushion with 1-many rosettes. All yr.......*House-leek* **Sempervivum tectorum**

Lvs 10-20 x 2.5mm, linear, bristle-tipped, not fleshy, mid-green, with strongly wavy white cilia, with midrib raised below (strongly so when dead), with stomata both sides. Stems to 10cm, not glandular-hairy. Compact cushion with many rosettes. All yr. Gower, S. Wales..*Yellow Whitlowgrass* **Draba aizoides**

Group KQ – *Lvs in basal rosette, hairless (exc at extreme base in* Plantago maritima*), 1-3-veined or veins obscure. (Some monocots e.g.* Crocus *may key out here.* Glaux maritima *may key out here if lvs alt).* ❶ *Lvs coriander-scented.* ❷ *Lvs aromatic (soapy).* ❸ *Stipules present*

■Lvs >8mm wide, rarely linear, petiolate

Lvs cuneate at base

Aquatic. ❶..*Lesser Water-plantain* **Baldellia ranunculoides**

Saltmarshes

Lvs with mucro, not fleshy, with cartilaginous margins, usu pitted with salt-glands both sides ....(*Limonium*) Go to **LIM**

Lvs without mucro, fleshy, without cartilaginous margins, not pitted

Lvs elliptic, narrowed to petiole, rolled when young, occ with septate cilia nr base when young, occ with obscure hydathode-teeth, with obscure veins (occ with 1-3 ± parallel veins, rarely appearing pinnate-veined), with stomata both sides. Stem lvs sessile, weakly clasping. Petiole channelled, with 5-7 vb's. Stems weakly ridged. Apr-Oct (all yr) ....*Sea Aster* **Aster tripolium**

Lvs ± rounded at base

Lvs with opaque netted veins, without midrib. (A fern) ....*Adder's-tongue* **Ophioglossum vulgatum***

Lvs with distinct pinnate veins, with midrib. (An orchid) ....*Creeping Lady's-tresses* **Goodyera repens**

Lvs with distinct parallel veins, with midrib................................*Pink Purslane* **Claytonia sibirica**

Lvs with all veins obscure. PL 11.. *Springbeauty* **Claytonia perfoliata**

Lvs cordate at base

Lvs 1.5-4cm, ovate, obtuse to ± acute, with few parallel veins (occ obscure) converging at apex, with minutely crenulate margins, with stomata below only. Stem lvs sessile, ± clasping. Petiole to 7cm, without sheathing base, channelled, with 1 vb. Stems 5-ridged, solid. Tufted per ....*Grass-of-Parnassus* **Parnassia palustris**

■Lvs <8mm wide, usu linear, ± sessile or with indistinct petiole

Per. Lvs with translucent or obscure midrib

Lvs with sparse long white hairs at extreme base (easily overlooked), with obscure midrib

Lvs 2-22cm x 3-8mm, semi-cylindrical or channelled, ± acute with pale hydathode, green to glaucous, purplish at base, weakly pitted both sides, solid, with 3(5) vb's, with stomata both sides. Tufted per, with stout woody rootstock. Apr-Oct. Coastal ....*Sea Plantain* **Plantago maritima**

Lvs 3-7cm x 1-3mm, ± cylindrical to flattened or linear and channelled, obtuse to acute with hydathode often purplish, mid-green, occ sparsely ciliate, not pitted, ± solid with aerenchyma, with 1(3) vb's, with stomata both sides. Stoloniferous per. All yr. Winter-wet habs ....*Shoreweed* **Littorella uniflora***

Lvs hairless even at extreme base

Lvs with translucent midrib

Plant forming 1-several cushion-like tufts, with woody rootstock

Lvs basal, 2-15cm, linear, acute to obtuse, flat to ± semi-cylindrical, ± fleshy, pitted both sides, reddish at base, occ hairy or ciliate, with midrib sunken above, with scattered stomata both sides

Lvs 1(2)mm wide, usu 1-veined. Saltmarshes, sea cliffs ....*Thrift* **Armeria maritima** var **maritima**

Lvs 2mm wide, 3-veined. Mtns......................*Thrift* (var) **Armeria maritima** var **planifolia**

Lvs ≥2mm wide, 3-5(7)-veined. Coastal, Jersey................*Jersey Thrift* **Armeria arenaria**

Plant solitary, with corm-like rootstock

Lvs basal, c12 in rosette, 2-3cm x 0.7-1mm, semi-cylindrical or blunty triangular to flattened, shiny green, turning orange, solid, with stomata in 4 rows. Oct-May. VR, SW Eng (Lizard), Channel Is..*Land Quillwort* **Isoetes histrix**

Lvs without translucent midrib

Lvs aromatic (soapy odour), semi-cylindrical and channelled above. ❷

Lvs 15-60cm x (1)2-3mm. Ligule 2.5-6mm, obtuse to acute. Apr-Oct. Saline habs ....*Sea Arrowgrass* **Triglochin maritimum***

Lvs to 15cm x 0.8-1.5(2.5)mm. Auricles overlapping to look like a ligule, <1mm, obtuse. Apr-Oct. Freshwater marshes.................................*Marsh Arrowgrass* **Triglochin palustre***

Lvs odourless or hay-scented, flat. (A fern)

Lvs 2(3), ± erect, with free vein endings. Apr-Aug. Short turf, dunes, R ....*Small Adder's-tongue* **Ophioglossum azoricum**

Lvs 1(2), adpressed to ground, without free vein endings. Oct-May. Short turf, VR, Scilly Is, Channel Is. Sch8..................................*Least Adder's-tongue* **Ophioglossum lusitanicum**

Ann (usu). Lvs usu with obscure midrib

Stipules to 1mm, acuminate, semi-sagittate, scarious to herbaceous, erose ❸

Lvs 5-30 x 3-5mm, occ ± opp, linear-oblanc, obtuse, cuneate at base, weakly fleshy, with erose cartilaginous margins, ± glaucous, with obscure veins, with stomata both sides. Stems 5-25cm, decumbent, often reddish, round. Occ bi. Jun-Oct (all yr). Lagoon shingle, VR, Devon, also VR alien. Sch8..*Strapwort* **Corrigiola litoralis**

Stipules absent

Lvs 3-14cm x 2-8mm, lanc-spathulate to subulate, obtuse, with dark hydathode below, flat but fleshy (snapping audibly), pitted above, often with narrowly involute margins, green, with stomata both sides. Petiole long, wavy, cylindrical, with a hyaline sheathing base, with 1 vb surrounded by 12 small hollows. Stolons often developing. May-Oct. Wet mud ....*Mudwort* **Limosella aquatica**

Lvs 2-10cm x 1-2mm, oblanc or linear, obtuse, flat to ± fleshy, green, with stomata both sides. Petiole indistinct, dilated at purplish base, without a sheathing base. Stolons absent. Apr-Jun. Muddy winter-wet habs...*Mousetail* **Myosurus minimus**

Lvs 1-2.5cm x 0.5-1mm, linear-subulate, v obtuse, ± cylindrical and fleshy, pitted, light green, with 1 vb surrounded by 12 small hollows (at least nr base), with stomata all around. Petiole indistinct, with a hyaline sheathing base. Stolons occ developing. May-Sept. Estuarine, VR, Wales. Sch8..*Welsh Mudwort* **Limosella australis**

Group KR – *Lvs not in a basal rosette, hairless, 1-3-veined or veins obscure, fleshy (often strongly so).* ❶ *Lvs with turpentine or camphor odour.* ❷ *Stipules present*

■Plant branching only at ground level, often patch-forming..................................(*Sedum*) Go to **SED**

■Plant usu branching aerially, not patch-forming

Per, with stout taproot

Lvs to 50 x 6mm, mainly on stem, linear, sessile, weakly channelled above, green, often ciliate esp proximally, usu 3-toothed at apex, with midrib grooved below, with obscure 2° veins. Stems 15-90cm tall. Herb. ❶. Mar-Oct. Coastal............................*Golden-samphire* **Inula crithmoides**

Lvs 5-18 x 1-1.5mm, all on stem, crowded (esp nr branch apices), ± cylindrical (with v slight lateral keels), rounded or shortly apiculate, ± sessile, glaucous, often with translucent midrib. Stems 40-120cm tall. Evergreen subshrub. Coastal, S & E Eng ....*Shrubby Sea-blite* **Suaeda vera**

Ann, with slender taproot

Stipules absent

Lvs 1-3-pli-veined, with opaque network of minor veins (Kranz venation), 20-50 x 3-7mm, linear-lanc, (±) acute, ± sessile, light green or reddish, occ sparsely ciliate. Stems 30-100cm, branched, reddish, nonglandular-hairy, ± round. Jun-Nov. Ruderal, esp on motorways ....*Summer-cypress* **Bassia scoparia**

Lvs with all veins obscure, 3-30 x 1-4mm, semi-cylindrical, (±) acute, ± sessile, glaucous or reddish, not ciliate, with stomata (and rounded translucent raised cells) all around. Stems to 30cm, branched, green to reddish, rough with low papillae on ridges. Jun-Nov. Saltmarshes ....*Annual Sea-blite* **Suaeda maritima**

Stipules without veins, hyaline, c1mm, ± obtuse

Lvs often ± opp, to 6mm, broadly elliptic to ± orb, shortly narrowed to petiole, often ± fleshy (to 0.4mm thick), weakly pitted above, with translucent midrib fading before apex, with stomata both sides. Stems usu <6cm, weak, wavy, reddish, ± semi-cylindrical. ❷. Ann. Upland, VR, Skye, Mull..*Iceland-purslane* **Koenigia islandica**

Group KS – *Lvs not in a basal rosette, hairless, 1-3-veined or veins obscure, not fleshy.* ❶ *Stipules present, fused into ochreae when young*

■Lvs pseudowhorled to 5-whorled below, alt above

Ann. Lvs ± 3-5-whorled, 1-3cm x 2.5mm, linear. Stems 5-20cm, decumbent to ascending, glandular nr infl. VR alien, SW Br..*Prostrate Toadflax* **Linaria supina**

Per. Lvs ± 5-whorled, to 4cm x 6mm, linear-lanc, ± acute. Stems to 1m, ± erect

Stems to 6mm diam, usu purplish, pruinose at least near base, never glandular towards infl. Lvs 4-6mm wide, glaucous or dark bluish-green, the lower often purplish, with stomata both sides or below only. Fls purple. Tufted...*Purple Toadflax* **Linaria purpurea**

Stems to 2mm diam, green to glaucous, occ pruinose, occ glandular towards infl. Lvs 1.5-5mm wide, pale glaucous-green, with stomata both sides. Fls yellow. Patch-forming, occ extensively, by stems arising from root buds. All yr.................................*Common Toadflax* **Linaria vulgaris**

Stems to 1.5mm diam, green, occ purplish, rarely pruinose. Lvs 1.5-2.5mm wide, glaucous-pruinose, rarely purplish, with stomata below only. Fls pale lilac. Patch-forming (not v extensively) by stems arising from root buds.................................*Pale Toadflax* **Linaria repens**

■Lvs alt, not pseudowhorled

Lvs (actually cladodes) in 1-sided bunches (fascicles) of 4-7(10), to 17 x 0.3-0.5mm, unequal, needle-like, apiculate, linear (to ± cylindrical), veinless, with stomata all sides. Petiole round, solid. True lvs reduced to short stipule-like scarious brown scales. Stems round, solid. Shortly rhizomatous per. Old fl stalks occ mistaken for sucker-tipped stipular glands!

Stems to 150cm, erect, herbaceous. Cladodes green, often wavy. May-Oct. Crop relic or alien ....*Garden Asparagus* **Asparagus officinalis** ssp **officinalis**

Stems to 30cm, ± prostrate, woody below. Cladodes glaucous, stiff. All yr. Western grassy sea cliffs...*Wild Asparagus* **Asparagus officinalis** ssp **prostratus**

Lvs not in bunches, not needle-like (but may be linear)

Per. Calc gsld or coastal

Lvs 30-50mm, gland-pitted

Lvs to 2mm wide, linear, acute, sessile, with minutely scabrid margins, slightly glaucous, cottony when young, soon hairless, 1-veined, with stomata both sides. Stems 10-50cm, slender, erect, woody at base, round, minutely scabrid. Limestone cliffs, VR, W Br. Apr-Nov ....*Goldilocks Aster* **Aster linosyris**

Lvs <30mm, not gland-pitted

Lvs glaucous or ± so

Lvs erecto-patent to adpressed nr stem apices, 5-20 x 0.5-4mm, linear, acute, with entire or minutely scabrid-serrate margins, (1)3-veined at least nr base (occ raised below), with stomata both sides. Stems 0.6-2mm diam, ± woody at base, weakly angled. Short-lived per (bi). All yr..*Pale Flax* **Linum bienne**

Lvs spreading to recurved nr stem apices, 5-20 x 1-4mm, linear, acute, with minutely scabrid margins, mostly 1-veined, with stomata both sides. Stems 1-2mm diam, ± woody at base, weakly angled. Per. All yr. Calc gsld.......................*Perennial Flax* **Linum perenne**

Lvs olive- to yellow-green

Lvs 8-25 x 1-2mm, linear to obovate, acute or obtuse, ± sessile or indistinctly petiolate, rarely fleshy, with erose narrow cartilaginous margins, 1-veined (often obscure or opaque), with midrib occ slightly raised below (rarely above), with stomata both sides. Stems 1mm diam, often woody at base, strongly angled to ± flattened (angles slightly rough), with stomata all around. Hemiparasitic. Mar-Oct. Calc turf ....*Bastard-toadflax* **Thesium humifusum**

Per. Ruderal or hortal

Lvs 4-7(12)cm x 5mm, linear, acute, ± clasping with weak auricles at base, revolute when young, mid-green, not gland-dotted, with sparse septate hairs in lf axils, ± entire to (rarely) pinnately lobed, 1-veined, with midrib raised below, with stomata both sides. Stems 30-60cm, erect in bud, ± hairless, slightly angled. Apr-Oct ....*Narrow-leaved Ragwort* **Senecio inaequidens**

Lvs 5-12cm x 3-9mm, linear to linear-lanc, acute, not clasping at base, rolled when young, dark green, obscurely gland-dotted, rarely septate-hairy, entire but with antrorsely scabrid margins, 3(5)-veined, with midrib raised below, with stomata both sides. Stems to 150cm, nodding in bud, occ hairy, round with cartilaginous ridges. Apr-Oct. VR alien ....*Grass-leaved Goldenrod* **Solidago graminifolia**

Ann

Lvs 1-veined, all alt

Stipules absent. Lvs 25 x 3mm, linear-lanc, ± entire. Stems hairless. Apr-Oct. Hortal ....*Garden Candytuft* **Iberis umbellata**

Stipules parallel-veined, membranous, silvery. ➊. Often ruderal or arable ....(*Polygonum*) Go to **DOCK E**

Lvs 1-veined, opp below (examine lf scars if lvs have dropped)

Stem lvs 10-25 x 2.5-3.5mm, linear to oblong-lanc, obtuse, narrowed to short petiole, dark grey-green to ± glaucous, with translucent midrib raised below, with obscure 2° veins (occ 3-pli-veined), with stomata both sides. Stems 8-25cm, erect, branched, usu glandular-hairy, round. Apr-Oct. Ruderal...*Small Toadflax* **Chaenorhinum minus**

Stem lvs 10-25 x 5mm, oblong-ovate, the upper linear-lanc, obtuse to acute, ± sessile, pale green, with translucent midrib, with obscure 2° veins, with stomata both sides. Stem 10-25cm, erect, unbranched, hairless, with 2 obscure raised lines that are obscurely crenulate. Apr-Oct (all yr). Winter-wet habs. Sch8...........................*Grass-poly* **Lythrum hyssopifolium**

Lvs 3-veined, all alt

Stems 2-3mm diam, with lf scars below. Lvs without cartilaginous margins widening at base

Lvs to 30 x 1.5-4mm, linear, acute, grey-green, with minutely scabrid to entire margins, translucent-dotted esp nr margins, with slightly raised veins both sides, with stomata both sides. Stems to 75cm, erect, branched, round (occ becoming ridged), with stomata all around. Apr-Oct. Crop relic...*Flax* **Linum usitatissimum**

Stems to 1mm diam, without lf scars. Lvs with cartilaginous margins widening to form partially sheathing base

Lvs 15-30(60) x 3mm, narrowly oblanc, acute, grey-green, with minutely scabrid cartilaginous margins (occ obscure), with veins raised below. Stems 15-50cm, erect and wavy or ± prostrate, branched above, ± round with translucent lines. Jul-Oct. Coastal ....*Slender Hare's-ear* **Bupleurum tenuissimum**

Lvs 10-20 x 3mm, spathulate or oblanc, acute, dark green, with minutely scabrid cartilaginous margins, with veins raised below (laterals weakly so). Stems 1-10(25)cm, erect, simple or divaricately branched, ± round. Calc sea cliffs, VR, S Eng, Channel Is. Sch8...*Small Hare's-ear* **Bupleurum baldense**

Group KT – *Lvs in basal rosette (or plant stoloniferous in freshwater habs), hairless, pinnate-veined. Stipules absent but if ochreae (or remains) present (*Polygonaceae*)* Go to **DOCK**. *(*Goodyera repens *may rarely key out here)*

■Basal lvs sagittate or hastate (stem lvs never present). Monocot

Lf apex with long drip-tip (1-10cm x 0.5mm). Tuberous per

Lvs 20-50cm, ovate, cordate-hastate at base (often asymmetrically so), rolled when young, leathery, undulate, with cartilaginous margins, ± shiny darkish green above, all veins sunken above, with midrib raised below, with main 2° veins raised below (not anastomosing but running along margins to apex), with ≥50 sinuate minor veins, with cross-veins, with stomata both sides. Petiole >30cm, semi-cylindrical, weakly channelled, with long sheathing base, with many green vb's scattered throughout aerenchyma. Cataphylls obscure or absent. All yr (but frost-sensitive). Marshy habs or hortal, mostly SW Eng.......*Altar-lily* **Zantedeschia aethiopica***

Lf apex without long drip-tip (may have apiculus <5mm). Rhizomatous per

Aquatic. Cataphylls absent..*Spatter-dock* **Nuphar advena**

Dry, often shady, habs. Lvs rolled when young, shiny dark green, ± net-veined, with veins anastomosing and forming submarginal vein, with stomata both sides. Petiole 10-20cm, with sheathing base, channelled, with vb's scattered throughout aerenchyma. Cataphylls large, white (often smaller cataphylls present in connate prs)

Lvs (2)4-8, with pale whitish colouration along midrib and 2° veins

Lvs Oct-Jul, 10-35cm, rarely with black blotches, with rounded lobes at base. Strongly rhizomatous (patch-forming)...............*Italian Lords-and-Ladies* **Arum italicum** ssp **italicum***

Lvs 2-3(6), completely green

Lvs Oct-Jul, 10-35cm, v rarely black-blotched, with rounded lobes at base. Strongly rhizomatous (patch-forming)...*Italian Lords-and-Ladies* (ssp) **Arum italicum** ssp **neglectum**

Lvs Jan-Jul, to 15cm, often black-blotched, with ± acute lobes at base. Weakly rhizomatous. PL 14..*Lords-and-Ladies* **Arum maculatum**

■Basal lvs not sagittate (may be cordate) or stem lvs sagittate. Dicot (exc *Lysichiton*)

Lvs with fragile bristle at apex..(*Limonium*) Go to **LIM**

Lvs without bristle at apex

Petiole strongly sheathing at base ....(*Ranunculaceae*) Go to **RAN**

Petiole not strongly sheathing at base (may be dilated)

Lvs fleshy. Petiole long

Basal lvs 4-40cm. Petiole to 30cm, channelled above, soon furrowed below, with 3-7 vb's. Lvs broadly triangular, obtuse (often with mucron), cuneate to truncate but decurrent along petiole, revolute when young (occ with sparse septate hairs), shiny green, often with reddish cartilaginous margins, stomata both sides. Per, with stout rootstock. All yr ....*Sea Beet* **Beta vulgaris** ssp **maritima***

Basal lvs <5cm. Petiole to 5cm, channelled above, with 1 vb. Lvs with white (turning dark) hydathode below at apex (or along margins below), cress-scented, shiny green above, with few anastomosing 2° veins (often indistinct), with stomata both sides or below only. Stems round or ridged, often purplish at base

Basal lvs cuneate at base. Stem lvs petiolate, or sessile and clasping

Lvs to 4cm, obovate-ovate, entire or with 1(4) tooth per side. Bi or per. Saltmarshes ....*English Scurvygrass* **Cochlearia anglica**

Basal lvs cordate or rounded at base. Stem lvs sessile, often clasping

Per (bi). All yr

Saltmarshes. Lvs to 5cm, ovate-orb, entire or 1(4) teeth per side ....*Common Scurvygrass* **Cochlearia officinalis**

Mtns. Lvs to 2cm, orb (occ ivy-shaped), entire or 1(4) teeth per side ....*Mountain Scurvygrass* **Cochlearia micacea**

Ann. Oct-Jun. Dry saline habs (inc roadsides). Lvs to 1.5cm, orb to ivy-shaped ....*Danish Scurvygrass* **Cochlearia danica**

Lvs not fleshy. Petiole short or absent

Lvs 30-150cm, without stipoid glands. Wet shady habs

Basal lvs elliptic, obtuse to acute with tiny apiculate hydathode, abruptly narrowed to short petiole, rolled when young, with entire narrow hyaline margins, weakly coffee-scented, dull dark or shiny green above (occ with faint dark green marbling), usu shiny below, with midrib raised below, with stomata below only. Petiole winged, channelled, spongy, with vb's hidden in aerenchyma and weak spiral fibres. Shortly rhizomatous. Mar-Aug ....*American Skunk-cabbage* **Lysichiton americanus**

Lvs 5-10cm, with minute stipoid glands at base. Dry sunny habs (often ruderal)

Basal lvs narrowly oblanc, sessile or narrowing to short petiole, ± bullate, strongly undulate, with minutely crenulate cartilaginous margins often turning orange, shiny dark green above, with 2° veins anastomosing nr margins (hardly raised below), with stomata both sides. Petiole with flat arc vb. Stipoid glands to 1.5mm, shorter on stem lvs, acute, reddish. Stems to 150cm, ribbed, hollow. Bi. All yr..*Weld* **Reseda luteola**

Lvs 1-8cm, without stipoid glands. Wet sunny habs (often saline)

Basal lvs obovate to spathulate, obtuse (without apical hydathode), abruptly narrowed to short petiole, weakly rolled when young, without cartilaginous margins, shiny light green, pitted both sides, few whitish 2° veins raised or flat above (not raised below), with stomata both sides. Petiole with 1 vb. Stems to 45cm, round, with solid aerenchyma. Per (bi). All yr ....*Brookweed* **Samolus valerandi**

Lvs 1-2(3)cm, without stipoid glands. Dry metal-rich soils

Basal lvs spathulate or ovate, contracted abruptly into petiole (1.5cm), ± entire. Stem lvs green, elliptic-lanc, clasping with ± acute auricles, always hairless, entire or sinuate-toothed, not net-veined. Stems 5-40cm. Per (bi) ....*Alpine Penny-cress* **Thlaspi caerulescens**

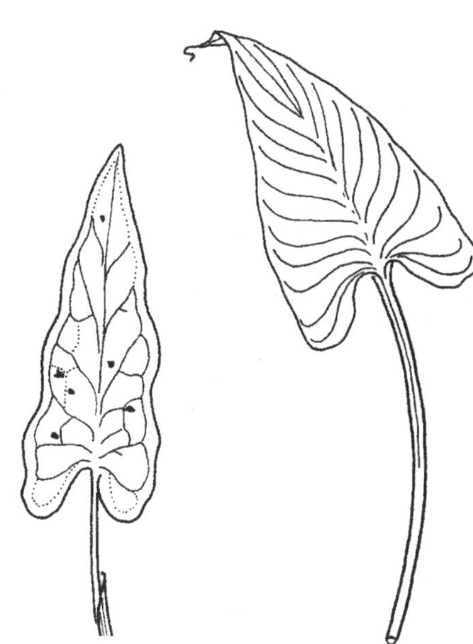

Arum italicum ssp *italicum*

Zantedeschia aethiopica

Beta vulgaris ssp *maritima*

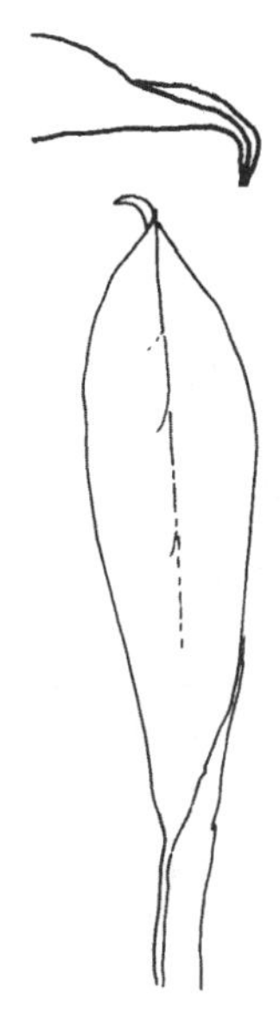

Limonium vulgare

Group KU – *Lvs all on stems, hairless, pinnate-veined. Stem hairs unicellular but septate in* Amaranthus *and some* Solanaceae

■Stem lvs clasping and/or sagittate. Plant may have cress/mustard odour (occ fetid). (*Fagopyrum esculentum* may key out here if the ochreae are overlooked)

Ann

Lower lvs green, to 6cm, oblanc or obovate, ± obtuse, cuneate and narrowed to petiole-like base, without cartilaginous margins, fetid when crushed, rarely persisting to fl. Stem lvs clasping, sagittate. Stems 1-several, 10-60cm. Often all yr..*Field Penny-cress* **Thlaspi arvense**

Lower lvs pruinose, to 10cm, narrowly obovate, obtuse, ± sessile, with v narrow translucent margins, mustard-scented when crushed, rarely persisting to fl. Stem lvs clasping with rounded lobes. Stem 1, 10-30cm. VR alien................................*Hare's-ear Mustard* **Conringia orientalis**

Per

Stem lvs glaucous, 7-15 x 1.5-4cm, clasping and sagittate at base, softly hairy or hairless, occ ciliate, (±) entire, not net-veined. Basal lvs to 30cm, lanc, narrowed to long petiole, often obscurely sinuate-toothed, softly hairy, glaucous, similar to *Lepidium draba*. Stems 60-120cm. Bi or per. All yr..*Woad* **Isatis tinctoria***

Stem lvs green, 1-3cm, elliptic-lanc, clasping with ± acute auricles, always hairless, entire or sinuate-toothed, not net-veined. Basal lvs 1-2(3)cm, spathulate or ovate, contracted abruptly into petiole (1.5cm), ± entire. Stems 5-40cm. Per (bi). Metal-rich soils ....*Alpine Penny-cress* **Thlaspi caerulescens**

■Stem lvs not clasping or sagittate (but may be hastate)

Stipules fused into a closed ochreae..(*Polygonaceae*) Go to **DOCK**

Stipules fused into an open sheath, partly adnate to petiole

Lvs (2)4-10 x (2)2.5-6cm, broadly ovate, shortly acuminate, cordate at base, gland-dotted, with erose cartilaginous margins, orange-scented, often purplish below or along margins, occ hairy in vein axils, ± net-veined, 5(7)-pli-veined (veins slightly raised both sides), with stomata below only. Petiole 1-4cm, channelled, with 7 vb's and spiral fibres. Stipular sheath 0.5-2cm, usually ciliate. Stems (10)30-60cm, ascending, channelled on one side. Rhizomes thin. Hortal. VR alien, increasing...*Chameleon-plant* **Houttuynia cordata**

Stipules absent (lvs not sheathing at base)

Ann

Lvs with mucro...(*Amaranthus*) Go to **AMAR**

Lvs without mucro...(*Solanum nigrum*) Go to **SOL**

Per

Lvs v glaucous or pruinose both sides

Lvs 0.5-6cm, ovate-obovate, with red-brown apical hydathode, the lower long-petiolate, the upper sessile nr infl, fleshy, pitted above, with midrib raised below, with ± indistinct 2° veins, with stomata both sides. Stems ± prostrate, green or purple, round, with tough stele. Apr-Oct. Shingle and rocky beaches, N & W Br..........................*Oysterplant* **Mertensia maritima**

Lvs glaucous below only................................*Rosebay Willowherb* **Chamerion angustifolium***

Lvs not glaucous either side

Plant with unpleasant odour when crushed. Stems hairy to hairless ....(*Atropa belladonna*, *Solanum dulcamara*) Go to **SOL**

Plant odourless. Stems hairy...........................*Common Gromwell* **Lithospermum officinale**

Plant odourless. Stems hairless

Lvs 15-20(35) x 10-18cm, broadly ovate to elliptic, resembling *Atropa belladonna*, ± acute to obtuse, cuneate at base, ± folded when young, with erose narrow cartilaginous margins, not net-veined, with whitish 1-3° veins, with stomata both sides. Petiole 2.5-3(7)cm, occ indistinct, with arc vb (consisting of 3-4 vb's). Stems to 150(200)cm, copperish when emerging. Apr-Nov. R alien......................................*Indian Pokeweed* **Phytolacca acinosa**

Key to Groups in Division L

(Lvs entire, opp)

- ■Tree or non-climbing shrub (occ epiphytic). Lvs often >2cm
  - Lvs with clearly translucent pinnate 2° veins. Evergreen or deciduous
    - Twigs with interpetiolar ridge. Lvs strictly opp.................... **LA**
    - Twigs without interpetiolar ridge. Lvs often subopp.................... **LB**
  - Lvs with 2° veins indistinct or opaque (lvs often thick or aromatic). Evergreen.................... **LC**
- ■Climbing shrub. Lvs often >2cm.................... **LD**
- ■Dwarf or trailing shrub <20(40)cm tall, with thin wiry or ± woody stems. Lvs <2cm.................... **LE**
- ■Herb, occ woody at base (if tendrils present <u>Go to</u> **SH**)
  - Latex present (esp in young stems) (*Vinca*, *Euphorbia*).................... **LF**
  - Latex absent
    - Stipules (occ fused) or stipoid glands present.................... **LG**
    - Stipules absent
      - Lvs with black glands along margins and/or scattered translucent dots, sessile (*Hypericum*) **LH**
      - Lvs with scattered orange glands (or along apex), sessile or petiolate (*Lysimachia*).................... **LI**
      - Lvs without orange glands or obvious translucent dots (rarely scattered black glands below)
        - Lvs <1.2mm wide.................... **LJ**
        - Lvs >1.2mm wide
          - Plant totally hairless, lf margins may be scabrid but never ciliate
            - Lvs with long petiole (>> lf).................... **LK**
            - Lvs with short or indistinct petiole (< lf), or sessile
              - Lvs 3(5)-veined (veins parallel and often strongly raised below).................... **LL**
              - Lvs 0-1(3)-veined or pinnate-veined
                - Lf margin scabrid. Lvs (±) connate at base
                  - Plant glaucous.................... **LM**
                  - Plant not glaucous.................... **LN**
                - Lf margin smooth to erose. Lvs connate or free at base.................... **LO**
          - Plant with at least some hairs, often confined to stems or petioles
            - Lvs with mealy (often sparse), woolly, stellate or medifixed hairs.................... **LP**
            - Lvs with simple hairs (but not woolly)
              - Stems square.................... **LQ**
              - Stems round (occ angled) or absent
                - Stems with 1-2 lines of hairs, always present.................... **LR**
                - Stems without 1-2 lines of hairs (may be hairy), or absent
                  - Lvs usu long-ciliate nr base with long wavy septate hairs (longer than elsewhere on lf), usu spathulate or long-cuneate, obvolute when young.................... **LS**
                  - Lvs without longer cilia nr base (cilia not long and wavy)
                    - Lvs with v long petiole (>2x lf) (*Asarum*).................... **LT**
                    - Lvs with short petiole or sessile
                      - Lvs alt above, weakly pinnate-veined to 1(3)-veined.................... **LU**
                      - Lvs all opp, pinnate-veined (occ weakly so).................... **LV**
                      - Lvs all opp, 0-5-veined
                        - Stem (or lvs) with minute hairs (often retrorse). Hairs usu unicellular, not usu glandular. Lvs (0)1-3(5)-veined.................... **LW**
                        - Stem (or lvs) with spreading hairs (often long, rarely absent). Hairs septate, occ glandular. Lvs 1-veined (*Cerastium*).................... **LX**

Group LA – *Tree or non-climbing shrub. Lvs always exactly opp, with clearly pinnate 2° veins, with stomata below only. Twigs with interpetiolar ridge. Hairs unicellular. (*Lonicera nitida *and* Buxus sempervirens *may key out here).* ❶ *Plant with dendritic hairs.* ❷ *Stipules present, interpetiolar.* ❸ *Lvs with medifixed hairs*

■Petiole <0.4mm (lvs ± sessile)

Lvs <1.5cm wide, revolute when young and with revolute margins at maturity

Lvs 20-40(60) x 3-15mm, rarely 3-whorled, linear to obovate, acute, cuneate at base, odourless, leathery, dark hairless green above, densely white-hairy below. Twigs minutely hairy to hairless, 2-ridged. Evergreen shrub to 0.5m. Hths, moors, VR alien..............*Bog-laurel* **Kalmia polifolia**

Lvs 15-40 x 3.5mm, always opp, linear, obtuse, cuneate at base, rosemary-scented, leathery, shiny dark hairless green above, whitish-woolly below. Twigs minutely hairy, ± round. Evergreen shrub to 2m. Hortal..*Rosemary* **Rosmarinus officinalis***

Lvs 1.5-6cm wide, valvate when young, with flat or weakly revolute margins at maturity

Lvs translucent-dotted (x20), often scented, with hydathode below at apex, ± cordate to clasping at base, with 2° veins weakly raised below. Stems to 1(2)m tall, erect, brittle. Rhizomes short or absent

Lvs net-veined, often yellow-green above, thin. Stems ridged or winged

Stem 2-ridged, with narrow wings distinctly undulate below nodes. Lvs 7-12cm, ovate, obtuse, weakly aromatic (curry- or coffee-scented). Frs fertile. Usu all yr ....*Tutsan* **Hypericum androsaemum***

Stems 2-ridged or v narrowly winged, wings not or hardly undulate below nodes. Lvs 3-9cm, ovate to oblong-lanc, obtuse, weakly goat-scented or aromatic. Frs partially fertile. All yr. *H. androsaemum* x *hircinum*...*Tall Tutsan* **Hypericum** x **inodorum**

Stems 2-4-ridged, wings never undulate below nodes. Lvs 2.5-4(6) x 2cm, ovate-lanc, acute, strongly scented. Frs fertile. Often all yr.......................*Stinking Tutsan* **Hypericum hircinum**

Lvs not net-veined, dark green above, leathery. Stems round

Lvs 3-6 x 2-4cm ovate, obtuse to slightly retuse, with translucent glandular dots and short streaks, citrus-scented. VR alien.................................*Forrest's Tutsan* **Hypericum forrestii**

Lvs not translucent-dotted but with minute glands (x20) nr or attached to free veinlets, odourless, without hydathode below at apex, not cordate or clasping at base, with 2° veins not or hardly raised below. Stems low-growing, to 30cm tall, brittle. Rhizomes often long

Lvs 5-9 x 3cm, oblong to elliptic, obtuse, ± leathery, with slightly recurved margins, dark green above, net-veined. Stems bluntly 2(4)-ridged. All yr...*Rose-of-Sharon* **Hypericum calycinum***

■Petiole >0.6mm (lvs petiolate). Lvs never translucent-dotted

Evergreen. Lvs >2cm, net-veined or not

Lvs with dendritic (occ stellate) hairs both sides. ❶

Lvs 3-9cm, elliptic-ovate, cuneate to truncate at base, obvolute when young, rugose, weakly sage-scented, dull green above, paler below, entire to weakly crenate, net-veined, with stomata below only. Petiole 2.5cm. Twigs yellow-white woolly with dendritic hairs, square. Hortal...*Jerusalem Sage* **Phlomis fruticosa**

Lvs with dense silvery silky hairs below

Lvs 3-7cm, ovate, v obtuse, occ apiculate, ± cuneate at base, with slightly recurved margins, shiny dark green above, entire to weakly toothed, not net-veined, with anastomosing translucent 2° veins hardly raised below, with stomata below only. Petiole 0.5cm. Twigs white-hairy, angled or ridged. Coastal, W Br..................................*Ake-ake* **Olearia traversii**

Lvs hairless both sides exc for hair-tufts in vein axils below

Lvs 4-10 x 3-6cm, ovate-oblong, ± acute, ± folded when young, leathery, with slightly recurved cartilaginous margins, shiny dark green above, paler below, usu long-ciliate, net-veined, with 2° veins hardly raised below, with stomata below only. Petiole 1-2cm, reddish, sparsely roughly hairy to ± hairless, usu ciliate along weak channel. Twigs often reddish, ± hairless, ± round to square, soon developing lenticels. Buds naked, with brown stellate hairs, occ glandular-ciliate. PL 9...*Laurustinus* **Viburnum tinus**

Lvs hairless both sides, with domatia (pits) in vein axils below

Lvs 3-8 x 2-4cm, broadly ovate-oblong, v obtuse, cuneate at base, with recurved cartilaginous margins, shiny dark green above, paler below, net-veined, with 2° veins not raised either side, with stomata below only. Petiole to 1cm. Stipules small, fused at centre of interpetiolar ridge. Twigs green when young, ± square. ❷ .... *Tree Bedstraw* **Coprosma repens***

Evergreen. Lvs 0.5-2cm, not net-veined

Lvs held at 90° to twig (rarely subopp nr apices), without midrib raised above, 0.6-0.8cm wide, ovate, shiny dark green above, paler below with indistinct sunken glands (occ translucent-dotted), hairless, with 3 prs of indistinct anastomosing 2° veins, with stomata below only. Petiole 1mm. Twigs with minute dense antrorse brown hairs, often with longer hairs, round. Buds with persistent scales .... *Wilson's Honeysuckle* **Lonicera nitida***

Lvs held at 45° to twig, with midrib raised above, 0.4-1cm wide, narrowly elliptic, shiny dark green above, paler below with indistinct sunken glands, (±) hairless, with 3 prs of distinct anastomosing 2° veins, with stomata below only. Petiole 1mm. Twigs with ± equal brown hairs (often antrorse), occ with red ± sessile glandular hairs, round. Buds with persistent scales .... *Box-leaved Honeysuckle* **Lonicera pileata***

Deciduous. Lvs 1-16cm, net-veined

Buds spreading, with persistent scales (i.e. with papery scales persisting at twig base). Twigs brittle, soon hollow. Lvs without medifixed hairs

Lvs broadly ovate to ± orb. Plant suckering by rhizomes or tip-rooting stolons

Lvs dull green above, ± glaucous below, with 2° veins hardly sunken above or raised below, usu with stomata below only. Twigs often plagiotropic (in horizontal plane) or arching, pale reddish-brown, ± round. Buds to 2mm, brown, with 2 opp prs of apiculate scales

Twigs hairless. Lvs hairless, sparsely ciliate. Rhizomatous

Lvs 2-6 x 3-6cm, obtuse, occ sinuate-lobed or toothed on vigorous shoots. Petiole 5-10mm. Pl 19 .... *Snowberry* **Symphoricarpos albus**

Twigs hairy when young. Lvs hairy at least below, ciliate. Stoloniferous

Lvs 2-2.5cm, ± acute. Petiole to 5mm. *S. microphyllus* x *orbiculatus* .... *Hybrid Coralberry* **Symphoricarpos** x **chenaultii**

Lvs 1-2cm, ± obtuse. Petiole to 5mm .... *Coralberry* **Symphoricarpos orbiculatus**

Lvs ovate-elliptic. Plant not suckering

Lvs densely short-hairy both sides, with stomata both sides. Twigs shortly hairy when young

Lvs 3-6 x 3cm, plagiotropic (held in horizontal plane), not acuminate, with hairs often turning brown, ciliate, with 6 prs of anastomosing opaque 2° veins. Petiole 5(20)mm, reddish above, shortly hairy. Twigs reddish above, green below, becoming ashy-grey. Buds 6-8mm, hairy distally, long-ciliate .... *Fly Honeysuckle* **Lonicera xylosteum**

Lvs 5-8 x 4cm, not plagiotropic, strongly acuminate, ciliate, with 6 prs of anastomosing translucent 2° veins. Petiole to 8mm. Twigs ashy-grey. Buds to 3(5)mm, hairy. VR alien .... *Maack's Honeysuckle* **Lonicera maackii**

Lvs sparsely hairy to hairless, with stomata below only. Twigs usu hairless

Twigs green, with stomata .... *Himalayan Honeysuckle* **Leycesteria formosa**

Twigs soon brown, without stomata

Lvs 4-10 x 5cm, adpressed-hairy on midrib below (rarely sparsely hairy below), sparsely ciliate, acuminate, slightly glaucous below. Petiole 5mm, with sparse cilia. Twigs hairless. *L. fragrantissima* x *standishii* .... *Purpus's Honeysuckle* **Lonicera** x **purpusii**

Lvs 4-8 x 4cm, hairless (rarely sparsely hairy either side), sparsely ciliate, acuminate, paler below. Petiole 3-8mm, hairless (occ hairy). Twigs hairless (occ minutely hairy when young) green to reddish or brown .... *Tartarian Honeysuckle* **Lonicera tatarica**

Buds adpressed (spreading in *Cornus mas*), naked. Twigs flexible, solid. Lvs with medifixed hairs at least above

Lvs ovate, involute when young, with 2° veins mostly in proximal ½ and curving towards apex, with spiral fibres visible on tearing, with stomata below only. Twigs often with interpetiolar ridge weak to absent. ❷

Lvs with hairs usu spreading below (occ simple) but adpressed above, with sparse 3° veins

Lvs 6-7 x 4.5cm, dark green above, pale green below, usu turning reddish in autumn, cuspidate or acuminate, with 4 prs of 2° veins raised below and sunken above. Petiole 0.8-1.5cm, channelled. Twigs reddish (at least above), hairless or hairy. Buds 6mm, dark brown, hairy. Suckering shrub to 4m. Calc soils. PL 22 ....*Dogwood* **Cornus sanguinea** ssp **sanguinea***

Lvs 5-13 x 2.5-7cm, dark green above, pale green below, often remaining greenish in autumn. VR planted...*Asian Dogwood* **Cornus koenigii**

Lvs (and twigs) with hairs adpressed both sides (all medifixed), with abundant 3° veins

Small tree or large shrub, to 4(8)m. Twigs square

Lvs 5-10 x 3-4cm, undulate, with 4-5 prs of 2° veins. Petiole c7mm, reddish, channelled. Twigs green or purplish, soon hairless. Buds often spreading from twig, to 6mm, flattened, dark brown, hairy...*Cornelian-cherry* **Cornus mas**

Shrub, to 3m. Twigs round

Lvs (8)10-16 x 6-10cm, acuminate, rounded at base, mid- to dark green above, turning yellowish in autumn, pale glaucous below, with 5-7 prs of 2° veins. Petiole (1)1.5-2.5cm. Twigs usu reddish on upperside in 1st yr, greenish-yellow in winter. Suckering or stoloniferous...*Red-osier Dogwood* **Cornus sericea***

Lvs to 10 x 6cm. Suckering. VR alien ....*Southern Dogwood* **Cornus sanguinea** ssp **australis**

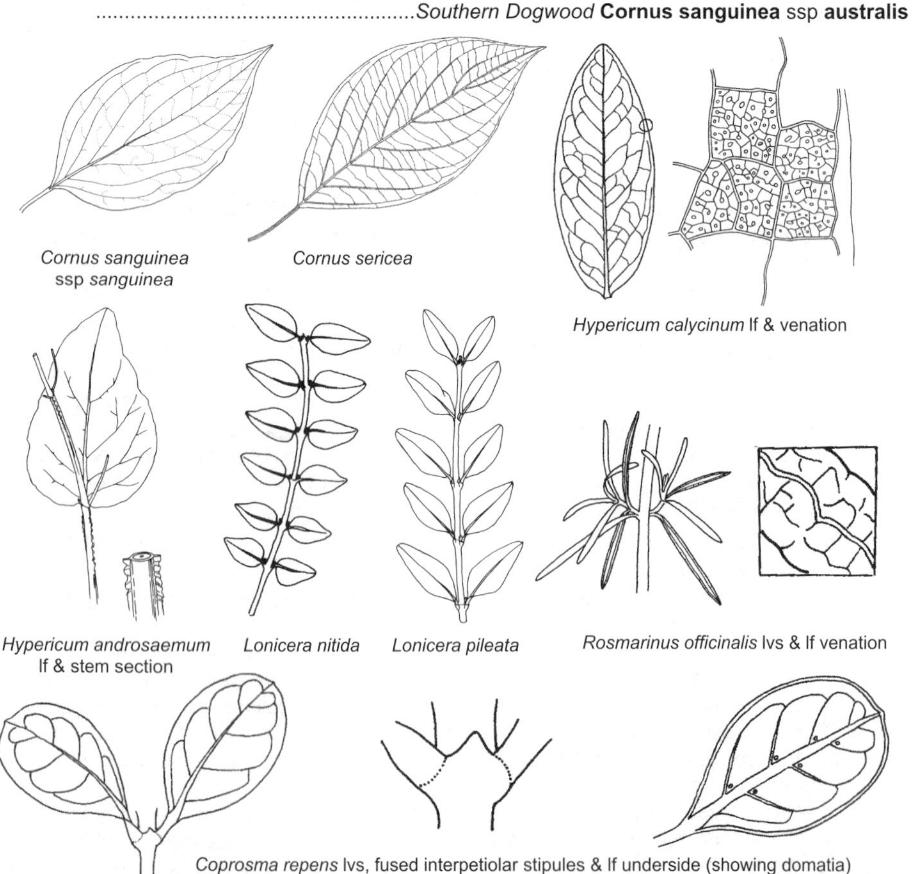

Cornus sanguinea ssp *sanguinea*

Cornus sericea

Hypericum calycinum lf & venation

Hypericum androsaemum lf & stem section

Lonicera nitida

Lonicera pileata

Rosmarinus officinalis lvs & lf venation

Coprosma repens lvs, fused interpetiolar stipules & lf underside (showing domatia)

Group LB – *Tree or non-climbing shrub. Lvs often subopp, with clearly pinnate 2° veins, obvolute when young. Petiole channelled. Twigs without interpetiolar ridge. Buds ovoid. (*Lonicera nitida *and* Buxus sempervirens *may key out here)*

■Lvs 4-12 x 9cm, strongly net-veined, not turning black when dry

Lvs ovate, shortly acuminate, truncate to cordate at base, dull green above, obscurely pitted both sides, hairless, with 6 prs of 2° veins raised below, with stomata both sides. Petiole to 3cm, swollen at base, hairless or with sparse minutely glandular hairs when young, with arc vb (tiny rib bundles also present). Twigs olive-brown, hairless, ± brittle, round (rarely ± 4-angled), striate, with lenticels, pith-filled. Buds in prs at twig apices, to 1cm, green (purplish in winter), glandular-ciliate, with 6-7 scales. Deciduous suckering shrub or small tree to 7m.......*Lilac* **Syringa vulgaris**

■Lvs 3-6 x 1-3cm, not net-veined, turning black when dry

Twigs densely minutely hairy when young, pale olive-brown, brittle, round, with lenticels. Lvs usu broadly lanc, obtuse to acute, cuneate at base, with 4 prs of ± anastomosing 2° veins (often indistinct or opaque), with stomata below only. Petiole to 4mm. Buds 1.5-3mm, obtuse, ciliate, with 4 prs of opp scales, dark purple-brown, or green with dark tips. Deciduous to semi-evergreen shrub. PL 9, 22..*Wild Privet* **Ligustrum vulgare***

Twigs (±) hairless, pale olive-brown, round, brittle, with lenticels. Lvs elliptic-ovate to elliptic-oblong, obtuse to acute, cuneate at base, with 4 prs of ± anastomosing 2° veins (often indistinct or opaque), with stomata below only. Petiole to 4mm. Buds 1.5-3mm, obtuse, ciliate, with 4 prs of opp scales, green, occ with brown spreading tips. Semi-evergreen to evergreen shrub ....*Garden Privet* **Ligustrum ovalifolium**

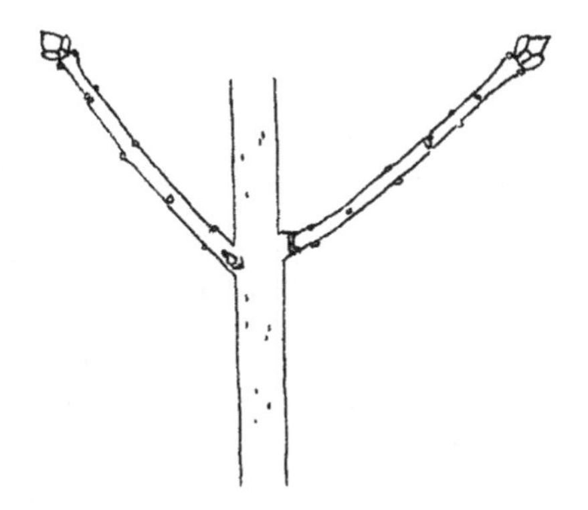

Ligustrum vulgare twig (showing absence of interpetiolar ridge)

Group LC – *Evergreen tree or shrub >30cm. Lvs with 2° veins indistinct, often thick or aromatic. Buds naked (exc* Buxus*). (*Lonicera nitida *may key out here).* ❶ *Lvs 1-2mm.* ❷ *Lvs silvery-grey.* ❸ *Plant parasitic on tree branches*

■Lvs whitish hairy at least below, strongly aromatic

Lvs grey-grey with stellate floccose hairs both sides, 15-70 x 2-4(10)mm, linear, ± obtuse, cuneate at base, with revolute margins, perfume-scented, with stomata both sides. Twigs angled or square, furrowed, with flaky bark..............................*Garden Lavender* **Lavandula angustifolia**

Lvs shiny dark green and hairless above, whitish-woolly below, 15-40 x 3.5mm, linear, obtuse, cuneate at base, with revolute margins, rosemary-scented, with stomata below only. Twigs ± round, with fissured bark...*Rosemary* **Rosmarinus officinalis***

■Lvs not whitish hairy either side, odorous or not

Twigs (±) square. Lvs odorous

Twigs without interpetiolar ridge. Shrub or tree usu >1m

Lvs 1-3 x 1-2.5cm, ovate, obtuse to emarginate, strongly odorous, shiny dark green above, paler below, hairless (sparsely hairy when young, or hairs confined to midrib above), with midrib raised both sides, with obscure herring-bone arrangement of forking 2° veins, with stomata below only (stomata largest nr to midrib). Petiole 1.5-4mm. Twigs green, usu hairy, with 2-4 ridges. Buds 3mm, flat-ovoid, with several scales.............*Box* **Buxus sempervirens***

Twigs with interpetiolar ridge. Low shrub to 60cm

Young twigs green, minutely retrorsely hairy. Lvs to 30 x 1.5-7mm, linear-lanc, obtuse to apiculate, narrowed to short indistinct petiole, dark bluish-green above, gland-pitted esp below, menthol-scented, hairless, minutely scabrid-ciliate, with translucent midrib raised below, with stomata both sides. VR alien...................................*Hyssop* **Hyssopus officinalis**

Young twigs whitish, minutely retrorsely hairy. Lvs 15-30 x 3-4mm, linear-lanc, narrowed to sessile base, acute, mid-green, reddish gland-pitted both sides, sage-scented, hairless, scabrid-ciliate (cilia longest at base), with translucent midrib not raised below, with stomata both sides. Old walls, VR alien..*Winter Savory* **Satureja montana**

Twigs round or angled. Lvs (±) odourless

Lvs strongly revolute (appearing trigonous), peg-like, sessile. ❶.........*Heather* **Calluna vulgaris**

Lvs not revolute

Petiole sinus usu formed by each pr of young lvs

Lvs distinctly pitted below (domatia) nr margins. VR hortal *Townson's Hebe* **Hebe townsonii**

Lvs obscurely pitted above

Lvs 12-16 x 5-6mm, oblong, *Buxus*-like, hairless, with obscure ± parallel 2° veins, with stomata both sides. Petiole sinus minute. Twigs minutely hairy in 2 broad opp lines. VR hortal...*Hooker's Hebe* **Hebe brachysiphon**

Lvs not pitted either side (the following 2 spp below often form backcrossed hybrids)

Lvs 8-15 x 1.5-3cm, 5-8x as long as wide, narrowly lanc, shiny dark green above, dull pale green below, hairy along midrib above, occ minutely ciliate, with translucent midrib raised below, with stomata below only. Petiole forming large circular sinus. Twigs hairless ....*Koromiko* **Hebe salicifolia***

Lvs to 9 x 2.5cm, <5x as long as wide, elliptic, shiny dark green above, dull pale green below, occ minutely hairy along midrib above, occ minutely ciliate, with translucent midrib raised below, with stomata below only. Petiole sinus large, circular. Twigs v minutely hairy in 2 broad opp lines. *H. elliptica* x *speciosa*..............*Hedge Veronica* **Hebe** x **franciscana***

Petiole sinus absent

Plant of coasts and saltmarshes. Twigs much-branched

Lvs to 5cm, elliptic, narrowed to short petiole, obtuse, fleshy, ± warty, silvery or mealy-grey, with midrib raised both sides, with obscure stomata both sides. Twigs decumbent, round below, angled above. Low shrub to 80cm. ❷. All yr...*Sea-purslane* **Atriplex portulacoides**

Plant parasitic on tree branches. Twigs repeatedly forked. ❸

Twigs green, round, brittle, with stomata all around. Lvs 3-8cm, narrowly obovate (often ± curved), narrowed to short petiole, applanate when young, thick, leathery, yellow-green, with stomata both sides. PL 10 .... *Mistletoe* **Viscum album**

Twigs brown, round, brittle, without stomata. Lvs 3-8cm, narrowly obovate (often ± curved), narrowed to short petiole, applanate when young, thick, leathery, yellow-green, with stomata both sides. VR hortal (Kew) .... *Yellow-berried Mistletoe* **Loranthus europaeus**

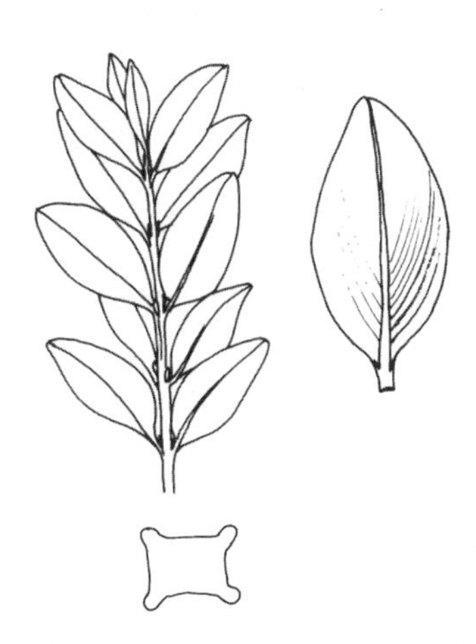

Buxus sempervirens shoot, twig TS & lf

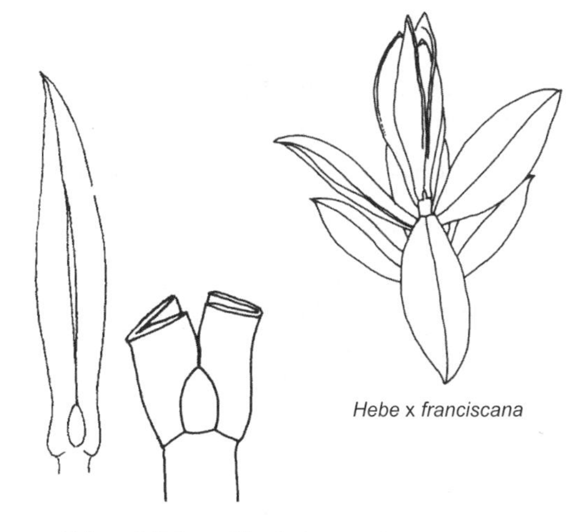

Hebe x *franciscana*

Hebe salicifolia pr of lvs & sinus

Group LD – *Climbing shrub. Lvs often >2cm, rolled when young, net-veined. Stems clockwise-twining, round, becoming hollow. Buds with persistent scales (i.e. with papery scales persisting at shoot base)*

■Lvs (semi-) evergreen, leathery, never glaucous below

Lvs (±) cordate at base, 4-10 x 4cm, broadly lanc-elliptic to ovate, acute to ± acuminate, mid-green above, without purplish margins, hairless exc for midrib below, ciliate, with 2° veins sunken both sides, with stomata below only. Petiole 7-15mm, hairy. Twigs densely brown-hairy when young. Fls terminal. Evergreen .... *Henry's Honeysuckle* **Lonicera henryi**

Lvs not cordate at base, (3)5-9 x 3-5cm, ovate to oblong, often mucronate, shiny yellow-green above, occ with purplish margins, sparsely hairy or hairless (hairy on midrib below), ciliate, rarely deeply lobed (oak-like), with 2° veins weakly raised below, with stomata below only. Petiole 7-15mm, (glandular-) hairy at least when young. Twigs (glandular-) hairy when young. Fls in lateral axils. Semi-evergreen .... *Japanese Honeysuckle* **Lonicera japonica**

■Lvs deciduous (occ overwintering), thin, usu glaucous below

Uppermost pr of lvs free

Lvs 3-7 x 2-3cm, ovate to elliptic, usu acute, dull dark green above, patent-hairy to hairless above, patent-hairy below, ciliate, rarely deeply lobed (oak-like), with 2° veins not or hardly raised below or sunken, with stomata below only (occ along veins above). Petiole to 15mm, long-ciliate. Twigs rarely pruinose, hairless or sparsely patent-hairy. PL 23 .... *Honeysuckle* **Lonicera periclymenum**

Uppermost pr(s) of lvs perfoliate (fused around stem)

Lvs 4-10 x 5cm, ovate to obovate, usu acute, dull dark green above, hairless both sides, not ciliate, with 2° veins not or hardly raised below or sunken, with stomata below only. Petiole to 10mm, hairless. Twigs often pruinose, sparsely hairy to hairless. *L. caprifolium* x *etrusca* .... *Garden Honeysuckle* **Lonicera** x **italica**

Group LE – *Dwarf (sub)shrubs usu <20cm tall, often trailing with thin, wiry or ± woody stems. Lvs <2cm.* ❶ *Lvs <2mm.* ❷ *Deciduous (but lvs at shoot apices wintergreen)*

■Lvs stellate-woolly at least below, margins usu at least slightly revolute, valvate when young, ± 1-veined, with stomata both sides (usu obscured by hairs). Petiole to 2mm

Stipules linear-subulate, often as long as petiole. Lvs 0.5-2cm

Lvs 5mm wide, oblong, with flat or weakly revolute margins, dark green and ± hairless above or with sparse long septate hairs (often 2 hairs originating from same point on minute swollen base), ciliate. Stems reddish when young. Calc turf ....*Common Rock-rose* **Helianthemum nummularium**

Lvs 2-3mm wide, oblong-linear, usu with strongly revolute margins, grey-woolly both sides. Stems greyish-hairy when young. Limestone turf, Devon, N Somerset ....*White Rock-rose* **Helianthemum apenninum**

Stipules absent. Lvs to 1.2cm

Lvs with revolute margins, obovate-elliptic to ovate, greyish-hairy above. Limestone habs, Wales, N Eng..........*Hoary Rock-rose* **Helianthemum oelandicum** ssp **incanum**

Lvs with flat margins, obovate-elliptic to ovate, sparsely greyish-hairy above. Limestone habs, W Ire..........*Hoary Rock-rose* (ssp) **Helianthemum oelandicum** ssp **piloselloides**

■Lvs with simple hairs or hairless, margins flat or revolute

Lvs aromatic or weakly so, translucent-dotted with sunken glands both sides

Stems 20-40cm tall, solid. Petiole with short white septate cilia. Lvs densely white-hairy

Lvs 3-8 x 0.5-2.5mm, occ strongly revolute (appearing cylindrical). Stems erect to decumbent, not or hardly rooting at nodes, purplish, densely white-hairy, ± 4-angled. Old walls, stony banks..........*Garden Thyme* **Thymus vulgaris**

Stems <20cm tall, minutely hollow. Petiole with long white septate cilia. Lvs shiny dark green above

Lvs 6-11 x 3-6mm, with stomata below only, strongly scented, with 2-4 prs of 2° veins. Stems usu ± erect, tufted, rooting sparingly at nodes, reddish

Stems sharply 4-angled, with crisped cilia on angles, 2 opp faces narrow and shortly hairy or hairless, the other 2 opp faces wider and hairless. Lvs hairless, scabrid-ciliate. S Br ....*Large Thyme* **Thymus pulegioides**

Lvs to 4 x 2mm, with stomata both sides, weakly scented, with 2 prs of 2° veins (often obscure). Stems prostrate, rooting freely at nodes, reddish

Stems 4-angled, 2 opp faces shortly hairy (often densely so), the other 2 opp faces sparsely hairy or hairless. Lvs usu held horizontally, hairless or with sparse long septate hairs, scabrid-ciliate..........*Wild Thyme* **Thymus polytrichus** ssp **britannicus**

Stems hardly 4-angled, equally hairy all sides with short white weakly retrorse hairs. Lvs held upright, usu hairless, scabrid-ciliate. E Anglian Brecks ....*Breckland Thyme* **Thymus serpyllum**

Lvs not aromatic, not translucent-dotted (but *Saxifraga* has large pore at apex)

Lvs with revolute margins

Lvs (3)5-8 x 2-3mm, occ subopp, not decussate, oblong, obtuse, revolute, dark green and hairless above, paler and minutely hairy below, with stomata below only, with midrib sunken above, with obscure 2° veins. Petiole v short, not ciliate, with a pulvinus at base. Twigs ± prostrate, hairless, without interpetiolar ridge. Buds minute, adpressed, with 2 scales. Stony mtn hths. Scot..........*Trailing Azalea* **Loiseleuria procumbens**

Lvs 2-5mm, decussate, linear, *Erica*-like, strongly revolute, hairless above, minutely hairy below, with obscure veins, with stomata below only. Petiole v short, often ciliate, widely connate at base. Stems 1mm diam, prostrate, minutely papillate or hairy. Mat-forming subshrub. Saltmarshes, seacliffs (also hortal). Mostly S, E Eng..*Sea-heath* **Frankenia laevis***

Lvs 1-2mm, decussate, peg-like, strongly revolute (appearing trigonous), hairless or hairy above, with obscure veins, with stomata below only. Petiole absent but lvs with 2 short downward lobes at base (occ obscure). Shrub to 60cm. ❶. Hths, moors ....*Heather* **Calluna vulgaris**

Lvs with flat margins

Stipules absent

Lvs with 1(3) large lime-exuding pores at apex (often white), 4-5 x 2-2.5mm, oblong-linear, cuneate at base, sessile, often fascicled in lf axils, imbricate at least nr stem apices, ciliate with stiff white multiseriate (x100) hairs, 0-1-veined, with stomata both sides. Stems herbaceous but often mistaken for a trailing shrub, to 1mm diam, usu prostrate, sparsely hairy, reddish, round, with central stele. ❷. Mostly mtn rocks above 300m ....*Purple Saxifrage* **Saxifraga oppositifolia**

Lvs without pores at apex, 5-12(15) x 3-4mm, obovate to oblong, cuneate at base, ± sessile, ± connate, applanate when young, leathery, with minutely antrorsely crisped septate cilia, entire (or obscurely crenulate), usu with translucent midrib, with 2° veins obscure, with stomata both sides. Stems woody, 1-2mm diam, ± prostrate, minutely antrorsely crisped-hairy (esp above), round. Mtn rocks above 500m..........*Rock Speedwell* **Veronica fruticans**

Stipules small (best viewed underneath), silvery white, ovate-acuminate, ciliate

Lvs sparsely scabrid-ciliate when young (soon smooth), 3-7mm, elliptic-obovate, cuneate at base, with obscure veins, with stomata both sides. Stipules shortly ciliate. Stems prostrate, rarely woody at base, minutely hairy all round, with stele in TS. Ann to short-lived per ....*Smooth Rupturewort* **Herniaria glabra**

Lvs strongly ciliate, usu 2-6mm, often with 1 lf of opposite pr reduced or abortive (esp above) thus appearing alt, cuneate at base, with obscure veins, with stomata both sides. Stems prostrate, woody at base, with stele in TS. Per

Stems only usu hairy on upperside. Lvs broadly ovate to obovate. Cornwall, Alderney and Guernsey.....................................*Fringed Rupturewort* **Herniaria ciliolata** ssp **ciliolata**

Stems hairy all round. Lvs narrowly elliptic. Jersey ....*Fringed Rupturewort* (ssp) **Herniaria ciliolata** ssp **subciliata**

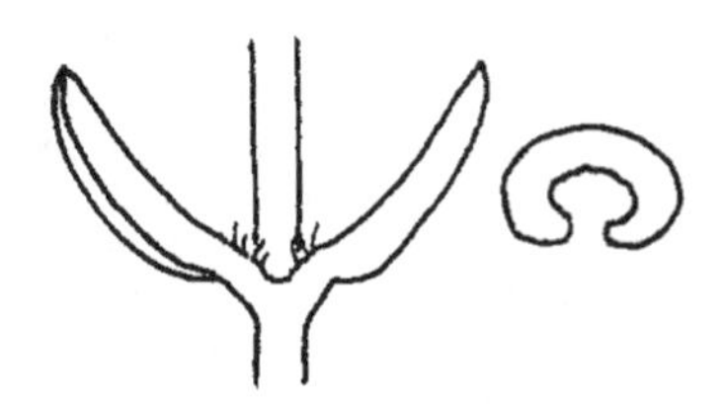

Frankenia laevis lvs & lf TS

Group LF – *Herb, occ woody at base. Latex present. Stems round.* ❶ *Lvs with black blotches*

■Sprawling or creeping woody evergreen per. Latex sparse, often confined to young shoots. Petiole with 1 pr extra-floral nectaries visible as pale green glands, with flat arc vb. Stems green, hardly branched, with interpetiolar ridge, with narrow outer ring of pale vascular tissue (stele) in TS. Lvs shiny dark green above, valvate when young, with c4 prs of anastomosing 2° veins slightly raised above (often opaque), with stomata below only

Lvs ciliate, hairless (veins occ hairy above)

Petiole 6-12mm. Lvs (2)4-7(9) x 2-6cm, ovate to lanc-ovate. Stems rooting at tips. PL 14 ....*Greater Periwinkle* **Vinca major***

Lvs not ciliate, hairless

Petiole 3-4mm. Lvs 2.5-4 x 2cm, lanc-elliptic. Stems rooting at some nodes. PL 9 ....*Lesser Periwinkle* **Vinca minor**

Petiole 7-10mm. Lvs 2.5-7 x 2-4.5cm, lanc to ovate. Stems rooting only at tips. VR alien ....*Intermediate Periwinkle* **Vinca difformis**

■Bi or ann. Latex abundant

Stems erect, usu solitary, to 120cm. Lvs decussate. Stipules absent. Bi. All yr

Lvs 4-20 x 0.5cm, occ subopp, linear, obtuse, often mucronate, rounded at base, sessile, often with recurved margins, dark glaucous-green above, paler with whitish midrib raised below, with stomata below only. Stems unbranched, with reddish buds below, glaucous-pruinose, purplish-reddish nr base, hairless, round (occ 2 shallow grooves), with stomata all around, hollow, with bluish latex..*Caper Spurge* **Euphorbia lathyris**

Stems ± prostrate, several. Lvs 2-ranked. Stipules present. Ann. Apr-Oct

Lvs usu with dark blotches above, with Kranz venation. ❶

Lvs to 15 x 5mm, oblong, obtuse to ± acute, shortly petiolate, thin, mid-green, paler below, occ hairless, occ with weak teeth, weakly 3-pli-veined, without 2° veins raised below, with stomata both sides. Stipules laciniate or dendritic, purplish. Stems often reddish, often hairless. Apr-Oct. R alien...*Spotted Spurge* **Euphorbia maculata**

Lvs without dark blotches

Lvs with Kranz venation, 3-10mm, ± oblong, obtuse to acute, rounded to cordate at base, ± sessile, thin to fleshy, dark green or reddish above, paler below, with long hairs (often denser below), usu weakly toothed. Stipules to 1.5mm, unequal, entire, soon falling. Stems hairy nr apices, often reddish. VR alien..........................*Thyme-leaved Spurge* **Euphorbia thymifolia**

Lvs without Kranz venation, 5-10(15)mm, ± oblong, obtuse or retuse, asymmetrically cordate at base (with large rounded auricle), shortly petiolate, fleshy, glaucous both sides with puplish veins, hairless, entire. Stipules to 1.5mm, subulate, entire. Stems hairless, reddish purple. Sand or shingle beaches, extinct, formerly SW Eng.............*Purple Spurge* **Euphorbia peplis**

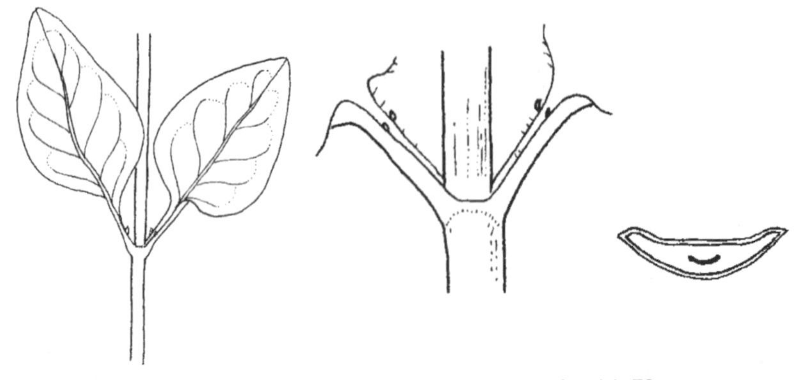

Vinca major lvs, extra-floral nectaries & petiole TS

Group LG – *Herb. Stipules (occ fused) or stipoid glands present.* Spergularia *and* Spergula *may appear whorled due to tufts (fascicles) of 2(6) lvs developing in lf axils. Stems solid (exc* Lythrum portula*).* ❶ *Lvs 3-veined*

▪Lvs linear or cylindrical. Stipules scarious, triangular-acuminate

Lvs ± cylindrical, fleshy, without long awn, with stomata all around. Usu damp coastal habs

Per, with woody rootstock

Stems densely glandular-hairy, often purplish. Lvs 10-20mm x 1.5-2mm, acute, sparsely glandular-hairy, ± flattened, often abruptly mucronate (esp on upper lvs). Stipules ± connate, ± silvery. Fl buds nodding. All yr. Coastal rocks, W Br ....*Rock Sea-spurrey* **Spergularia rupicola**

Stem usu hairless, often purplish. Lvs 10-25 x 1-2mm, obtuse to acute, with mucro, usu hairless. Stipules connate (fused to middle) when young, silvery. Fl buds held erect. Saltmarshes..........*Greater Sea-spurrey* **Spergularia media**

Ann (per)

Stem ± hairless to glandular-hairy. Lvs 10-25 x 1-2mm, ± acute, abruptly mucronate, ± hairless to glandular-hairy. Stipules connate (when young), silvery. Saline habs (inc roadsides). All yr..........*Lesser Sea-spurrey* **Spergularia marina**

Lvs linear, flattened, thin to ± fleshy (<0.5mm thick), with stomata both sides. Dry habs

At least some lvs with 0.2-0.4mm awn. Stipules ± connate, silvery, persistent

Lvs usu glandular-hairy, 4-25 x 1mm, often weakly channelled below. Stems hairy (often glandular) or hairless, hardly swollen at nodes. Ann or bi. All yr. Acidic sunny habs ....*Sand Spurrey* **Spergularia rubra**

Lvs sparsely glandular-hairy, 8-15 x 1mm, not channelled below. Stems hairy (often v glandular), hardly swollen at nodes. Ann. Ruderal, coastal, VR ....*Greek Sea-spurrey* **Spergularia bocconei**

All lvs without awn, (±) obtuse. Stipules free, occ pink, soon falling

Stems glandular-hairy, viscid. Lvs 1-3cm x 1mm, channelled below, green to grey-green, viscid glandular-hairy. Stems v swollen at or above nodes, appearing whorled, branched at lowest nodes. Ann. Often all yr. Acidic habs, usu arable.......*Corn Spurrey* **Spergula arvensis**

Stems ± glandular-hairy above, rarely viscid. Lvs 0.5-1.5(2)cm x 1mm, rarely channelled below, dark green, slightly fleshy, ± glandular-hairy. Stems hardly swollen at nodes, branched. Ann. Arable and nurseries, VR alien..........*Pearlwort Spurrey* **Spergula morisonii**

▪Lvs spathulate, flat, ovate to obovate. Stipules scarious to herbaceous, or gland-like

Stipules green, linear, on upper stem lvs only

Lvs basal or on stem, alt above, the basal dead at fl, to 3 x 1.5cm, elliptic-lanc, ± acute, sessile, 3-veined, with long simple hairs and stellate hairs esp below, with stomata both sides. Stems to 15cm, erect. Ann. ❶. Oct-Jul (occ spring germinating). Coastal, VR, W Br ....*Spotted Rock-rose* **Tuberaria guttata**

Stipules silvery

Stems erect or prostrate, angled

Lvs 8-13 x 3mm, obovate, narrowed to short petiole, the upper in whorls of 4 (usu unequal), with erose margins, with translucent midrib and obscure pinnate veins, with stomata both sides. Stipules lanc, acuminate, scarious. Stems to 25cm, 1mm diam, rough with low blunt papillae on angles. Ann. All yr (but frost-sensitive). Mostly SW Br. ....*Four-leaved Allseed* **Polycarpon tetraphyllum**

Stems prostrate, round

Stems minutely hairy all around, slender, green, with small ring of vascular tissue in TS. Lvs 3-7mm, elliptic-obovate, obtuse, ± sessile, hairless, sparsely scabrid-ciliate when young, with obscure veins. Stipules shortly ciliate, ovate. Ann to short-lived per. All yr. Disturbed sandy or gravelly habs..........*Smooth Rupturewort* **Herniaria glabra**

Stems hairless, thread-like, red, with central stele. Lvs 2-6mm, obovate, obtuse, shortly petiolate, hairless, not ciliate, occ with translucent midrib. Stipules entire, ovate. Ann. May-Oct. Winter-wet habs. Sch8..........*Coral-necklace* **Illecebrum verticillatum**

Stipules translucent, not silvery

Lvs often 4-whorled, to 8mm, spathulate, obtuse, without a visible hydathode, narrowed to short petiole, not connate, usu with obscure midrib, with stomata absent at least above. Stipules tiny, toothed. Stems rooting at nodes (many roots per node), round, with 6-10 hollows (cartwheel-like). Often submerged or fragments washed up

Fl stalk > fl bud length. Summer ann on mud, short-lived per underwater. Lakes, ponds, wet mud..*Six-stamened Waterwort* **Elatine hexandra***

Fl stalk < fl bud length (to absent). Ann. Ponds, small lakes, R ....*Eight-stamened Waterwort* **Elatine hydropiper**

Stipules gland- or bristle-like (or single root pustule each side of stem), not silvery

Sunny wet habs

Lvs 1-2cm, obovate-spathulate, obtuse, cuneate at base, applanate when young, often ± fleshy, with translucent midrib fading nr apex, mid-green, without hydathodes along margins, with ± opaque and often reddish 2° veins, hairless, with stomata absent or obscure above. Petioles short, not connate. Stipules (when present) minute, gland-like. Stems prostrate, rooting at nodes, reddish, without interpetiolar ridges, with 4 hollows in TS. Ann or short-lived per. All yr..*Water-purslane* **Lythrum portula***

Lvs 1.5-3(5)cm, ovate to broadly elliptic, shortly acuminate or acute, cuneate at base, obvolute when young, ± translucent, shiny dirty red-green, with sunken pale hydathodes along margins below, with 2° veins fading nr margins, with stomata both sides. Petioles short, not connate. Stipules (when present) minute, gland-like. Stems usu emergent, rooting at lower nodes, with weak interpetiolar ridges, with a green central stele surrounded by reddish aerenchyma. Short-lived per. Apr-Oct. VR, New Forest ....*Hampshire-purslane* **Ludwigia palustris**

Sunny dry habs

Lvs occ alt, to 2 x 0.8cm, oblong, obtuse, fleshy, with erose cartilaginous margins, often reddish, with translucent midrib, with Kranz venation, with stomata both sides. Petiole indistinct, 2.5mm. Stipules bristle-like, occ forming fringe, rarely absent. Stems not rooting at nodes, reddish, round, without interpetiolar ridges, snapping audibly, with vb's forming central stele in aerenchyma. Ann. VR alien.............................*Common Purslane* **Portulaca oleracea**

Shady habs (often damp)

Lvs 4-10cm, ovate, acuminate, truncate to ± cordate at base, thin, dull green above, paler and shiny below, ± hairless to minutely hairy, often ciliate, with 4-7 prs of 2° veins anastomosing nr margin. Petioles 2-4cm, connate at base with stipule-like extra-floral nectaries, hairy both sides, channelled. Stems 20-60cm, erect, ± sparsely glandular-hairy, round, with swollen lower nodes. Rhizomes slender, white, brittle........*Enchanter's-nightshade* **Circaea lutetiana***

Group LH – Hypericum. *Lvs all on stems, with black (oil) glands along margins and/or scattered translucent dots, (±) sessile, with orange terminal hydathode(s) (visible on upperside of lf apex), applanate when young, with v narrow erose cartilaginous margins, often with citrus or liquorice scent, with stomata below only (both sides in* H. olympicum*). Per (usu ann in* H. canadense*). May overwinter as small shoots produced in autumn. (*Anagallis arvensis *and* A. tenella *may key out here)*

■Stems hairy. Lvs without black glands on margins

Stems densely hairy with whitish septate hairs (underwater shoots hairless). Boggy pools

Lvs to 3cm, ± orb to broadly ovate, retuse with orange hydathode, ± clasping at base, with translucent dots all over surface, odourless, densely hairy both sides, 5(7)-pli-veined with anastomosing veins, not net-veined. Stems to 20cm, ascending, rooting at lower nodes, round. Fl buds with liquorice-scented reddish glandular hairs. Apr-Oct ....*Marsh St John's-wort* **Hypericum elodes**

Stems hairy with translucent unicellular hairs turning brown. Dry (often calc) habs

Lvs 2.5-5cm, ovate-elliptic, obtuse or retuse with obscure hydathode, with translucent dots all over surface, odourless, shortly hairy both sides, ciliate, not net-veined, with 2° veins mostly in proximal ½ and slightly sunken above and raised below. Stems to 100cm, erect, round. Fl buds with odourless glandular cilia....................................*Hairy St John's-wort* **Hypericum hirsutum**

■Stems hairless

Stems trailing. Lvs with black glands on margins both sides

Lvs to 10 x 4mm, oblong or obovate-lanc, with translucent dots abundant to absent, slightly scented, not or weakly net-veined, with c3 prs of 2° veins. Stems to 10cm tall, <1mm diam, reddish, round, with 2 obscure raised lines. All yr. Dry acidic habs ....*Trailing St John's-wort* **Hypericum humifusum**

Stems erect

Lvs with black glands on margins at least below

Lvs with translucent dots all over surface

Stems with 2 raised lines, not winged. Lvs odourless, without 3° veins clearly visible. Dry habs

Lvs 8-30 x 3-15mm, ovate to linear, obtuse (occ with tiny mucron), rounded at base, with black glands on margins below and occ above. Stems to 80cm, with few black glands on raised lines. Rhizomatous or root-budding ....*Perforate St John's-wort* **Hypericum perforatum***

Stems 4-angled, narrowly winged. Lvs scented, with 3° veins clearly visible. Damp habs

Lvs with flat margins, with 1-4 prs of veins mostly from base of midrib, to 4 x 2cm, oblong-ovate, obtuse, ± clasping at base, with black glands on margins both sides. Stems to 60cm, with black glands on wings. Stoloniferous ....*Square-stalked St John's-wort* **Hypericum tetrapterum***

Lvs with undulate margins, with 3(4) prs of veins in proximal ½, to 4 x 1(2)cm, oblong-ovate, obtuse, ± clasping at base, with black glands on margins both sides. Stems to 60cm, with black glands on wings. Stoloniferous.............*Wavy St John's-wort* **Hypericum undulatum***

Lvs with few or no translucent dots

Lvs linear (>4x as long as wide), hairless

Lvs 1-2(3)cm x 2mm, obtuse, ± clasping at base, with revolute margins, not net-veined, with 2° veins visible only. Stems to 40cm, round, often with 2 slightly raised lines, often reddish. Tufted. Acidic rocks, W Eng, NW Wales ....*Toadflax-leaved St John's-wort* **Hypericum linariifolium**

Lvs ovate to lanc, sparsely minutely hairy below

Stems to 60cm, round or v slightly 2-ridged. Lvs 1.5-3(7) x 0.5-1cm, ovate to lanc, obtuse, ± clasping at base, with black glands on margins at least below. Tufted. Dry calc habs ....*Pale St John's-wort* **Hypericum montanum**

Lvs ovate to lanc, hairless below

Stems ± 4-angled, not winged, to 60cm, with black glands on angles. Lvs 1.5-4 x 1.5-2cm, elliptic, obtuse, rounded at base, with black glands on margins both sides, odourless, (±) net-veined, with 2(3) prs of 2° veins in proximal ¼ slightly raised below. Rhizomatous ....*Imperforate St John's-wort* **Hypericum maculatum***

Stems round, usu with 2(4) raised lines, to 60cm. Lvs 3.4 x 2cm, ovate, obtuse, rounded at base, with black glands on margins both sides, odourless, hardly net-veined, with 1-2° veins raised below (the 3° veins hardly so). Rhizomatous. *H. maculatum* x *perforatum* ....*Des Etangs' St John's-wort* **Hypericum** x **desetangsii**

Lvs without black glands on margins

Stems ± round. Lvs with translucent dots absent from proximal part of lf

Lvs to 15(20) x 5-10mm, ovate, usu obtuse, cordate or ± clasping at base, dull green above, with orange glands along margins esp in distal ⅓, scented, not net-veined, with indistinct 2° veins. Stems to 50cm, often reddish. Tufted. All yr. Acidic habs ....*Slender St John's-wort* **Hypericum pulchrum**

Stems 2-ridged. Lvs with v obvious translucent dots all over surface

Lvs 8-15 x 4-8mm, elliptic-oblong, acute, ± fleshy, glaucous (overwintering lvs green), odourless, with indistinct 2° veins, with stomata both sides. Stems to 50cm, woody nr base, bushy, glaucous. VR alien............................*Olympic St John's-wort* **Hypericum olympicum**

Stems sharply 4-angled. Lvs with translucent dots all over surface

Lvs 10-15(20) x 2-2.5mm, elliptic to linear-oblanc, ± clasping at base, weakly scented, often purplish red, 1(3)-veined, all other veins invisible. Stems (3)12-20cm, 0.5mm wide, often purplish red. Ann (occ per). Damp acidic habs, VR Ire ....*Irish St John's-wort* **Hypericum canadense**

Hypericum maculatum
lf underside

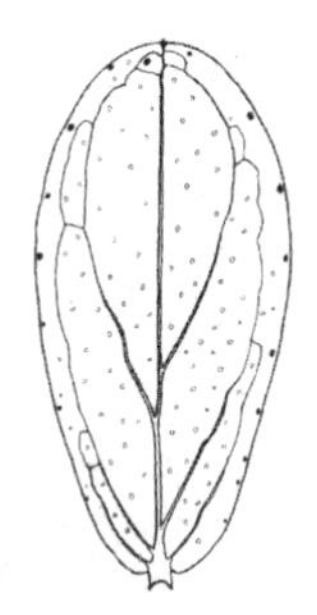

Hypericum perforatum
lf underside

Hypericum undulatum

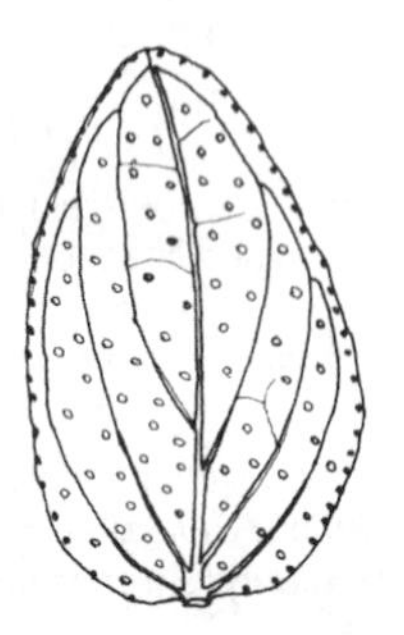

Hypericum tetrapterum
lf underside

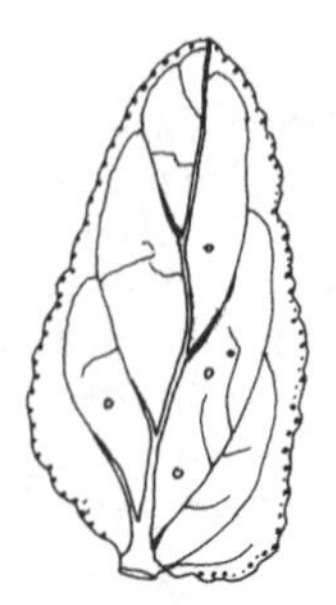

Hypericum undulatum
lf underside

Group LI – *Lvs with scattered orange glands (or confined to margins of recurved apex), sessile or petiolate, shortest nr base of stem (lowest soon dying and scale-like, basal rosette never formed), odourless, 2° veins anastomosing to form submarginal vein, with stomata below only.* ❶ *Lvs with 2° veins indistinct (distinct and translucent in other spp)*

■Stems prostrate, rooting at nodes, hairless. Lvs rolled when young. Stoloniferous. All yr ....*Creeping-Jenny* **Lysimachia nummularia***

■Stems erect, to 100cm, not rooting at nodes, usu with at least some septate hairs. Lvs revolute when young. Rhizomatous. May-Oct

Stem ridged, hairy. Dry habs

Lvs often whorled or alt, to 10 x 5cm, lanc to ovate, ± acute, apex recurved with 4mm orange gland, shortly petiolate, with orange dots below esp nr margins, hairy, ciliate. Petiole 2-3mm ....*Dotted Loosestrife* **Lysimachia punctata**

Stem hardly or not ridged (occ weakly angled), hairless to long-hairy

Dry or damp habs (often shady). Lvs without orange (or translucent) dots

Lvs occ whorled, 6-13 x 3-5cm, ovate (occ lanc), acuminate, ± cordate at base, hairless, scabrid-ciliate, with stomata below only. Petioles 10-20mm, connate at base, with ± long wavy cilia on cartilaginous channel margins. Stems square with rounded angles (occ double-channelled), reddish at base, hairless, solid. Often all yr. R alien ....*Fringed Loosestrife* **Lysimachia ciliata**

Wet habs (usu sunny)

Lvs with orange dots both sides

Lvs rarely whorled, 5-12 x 5cm, lanc to elliptic-lanc or ovate, acute, (±) sessile (petiole <1.5(4)mm), not connate at base, with orange dots along margins and scattered over surface both sides, hairy with short minutely glandular hairs above and longer nonglandular hairs below (occ ± hairless), ciliate. Stems with spreading long hairs and minute glandular hairs above, often hairless below..........................*Yellow Loosestrife* **Lysimachia vulgaris***

Lvs with orange dots above only

Lvs rarely whorled, (5)8-13 x 1-2.5(3)cm, oblong-lanc to lanc, acute to obtuse, ± clasping and connate at sessile base, undulate, with recurved margins, densely dotted with raised orange glands above only, occ with sparse glandular hairs on veins below, usu ciliate. Stems rooting at lower nodes, often purplish, with orange dots above, hairless, round, hollow. Plant resembles *Chamerion angustifolium* ....*Tufted Loosestrife* **Lysimachia thyrsiflora**

Lvs with elongate translucent dots both sides (dots eventually turning orange-black)

Lvs rarely whorled, 5-7 x 0.5-1cm, linear-lanc, narrowed to short or indistinct petiole (<5mm), not connate, with recurved margins, with sparse thin black patent minutely glandular hairs below, ciliate, with midrib raised below. Stems slightly angled, hairless, solid. ❶. VR alien, Lake Windermere...*Lake Loosestrife* **Lysimachia terrestris**

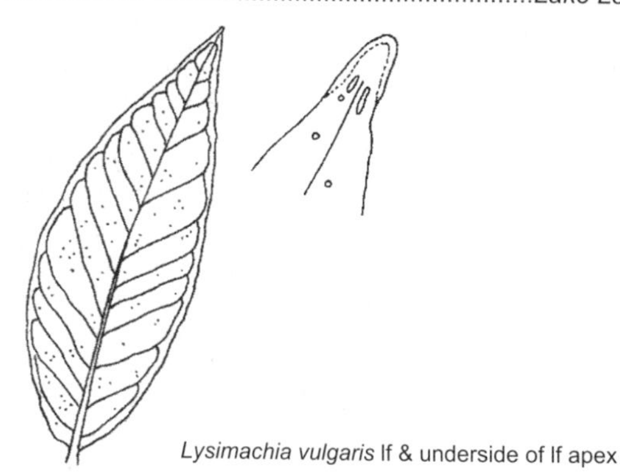

Lysimachia vulgaris lf & underside of lf apex

Group LJ – *Lvs <1.2mm wide, connate at base (exc some* Callitriche *spp and occ the upper lvs of* Montia fontana*), with stomata both sides. Stems round.* ❶ *Plant glaucous.* ❷ *Plant v viscid-hairy*

■Aquatic or wet habs, often submerged

Stems without a dark red ring below each node. Lvs usu notched at apex…(*Callitriche*) Go to **CAL**

Stems with a dark red ring below each node. Lvs not notched at apex

Per. Lvs 4-15(20) x 0.8-2mm, linear to ovate-lanc, acute, ± fleshy, usu with obscure midrib. Stems to 10(20)cm tall, decumbent to erect, rooting at nodes and mat-forming, with central stele occ reddish. All yr. Invasive alien..................*New Zealand Pigmyweed* **Crassula helmsii***

Ann. Lvs to 3(6) x 0.5mm, linear, (±) acute, fleshy, usu with obscure midrib. Stems to 5cm tall, decumbent, occ rooting at nodes, with central stele occ reddish. May-Sept. VR, Scot. Sch8 ....*Pigmyweed* **Crassula aquatica**

■Dry (to damp) habs, never submerged

Per. Basal lvs often forming a persistent rosette, often longer than stem lvs

Lvs obscurely 3-veined (best seen on persistent dead lvs), at least midrib raised below

Dry calc habs, N & W Br. Lvs 5-6(20) x 0.5mm, linear-subulate, ± acute to apiculate, straight to ± recurved, mid-green, hairless. Stems to 15cm tall, usu glandular-hairy at least above. Loosely mat-forming or forming weak cushions.................*Spring Sandwort* **Minuartia verna***

Calc mtn rocks, VR, Scot. Lvs 3-5 x 0.5mm, linear-subulate, ± acute, straight to curved, yellowish-green, hairless or sparsely glandular-hairy. Stems to 1.5(6)cm tall, glandular-hairy. Tufted. All yr..*Mountain Sandwort* **Minuartia rubella**

Acidic mtn rocks, VR, S Ire. Lvs 3-8 x 0.5mm, linear-subulate, ± acute to apiculate (mucro occ absent), recurved or curving upwards and to one side, mid-green, hairless. Stems to 5cm tall, glandular-hairy at least above. Densely tufted. All yr..*Recurved Sandwort* **Minuartia recurva***

Lvs 0(1)-veined

Plant forming a dense raised cushion to 3(8)cm tall

Lvs 6-12 x 1mm, linear, acute, not fleshy, obvolute to applanate when young, weakly channelled above, bright green, the lower reddish and bent sharply above ± colourless base, ± hairless, shortly stiffly scabrid-ciliate, obscurely 1-veined. Occ mat-forming with stolons. Mar-Oct (shoot apices overwinter). Usu calc mtn rocks ....*Moss Campion* **Silene acaulis**

Lvs 3(6) x 1mm, linear-subulate, obtuse to acute, ± fleshy, imbricate when young, flat or weakly channelled above, yellow-green, hairless, scabrid-ciliate, veinless. All yr. Usu mtn turf...*Cyphel* **Minuartia sedoides**

Lvs 2-6 x 0.5mm, linear-subulate, usu shortly apiculate, ± fleshy, occ purplish at base, hairless, with smooth margins, veinless. All yr. Calc mtns, VR ....*Snow Pearlwort* **Sagina nivalis**

Plant not forming a dense raised cushion

Lvs apiculate or with awn ≥0.2mm, acute

Plant often rooting readily at nodes, forming a large irreg mat

Fls with 4(5)stamens. Lvs 5-15 x 0.7mm, hairless or nonglandular ciliate, linear-subulate, with long awn 0.2-0.3mm, veinless. Stems to 20cm, occ patent-hairy. All yr ....*Procumbent Pearlwort* **Sagina procumbens**

Fls with 5-10 stamens. Lvs to 30mm, hairless. All yr. Mtns. *S. procumbens* x *saginoides* ....*Scottish Pearlwort* **Sagina** x **normaniana**

Plant not usu rooting at nodes, forming a neat circular mat. Fls with 10 stamens

Lvs forming a persistent rosette, glandular-ciliate, 3-12 x 0.5mm, linear-subulate, with 0.3(0.5)mm apiculus, dark green, some slightly channelled above and keeled below, veinless. Stems to 10cm, glandular-hairy, rarely hairless. All yr ....*Heath Pearlwort* **Sagina subulata***

Lvs not forming a persistent rosette, hairless, 10-20 x 0.5mm, linear-subulate, with apiculus <0.2mm, mid-green, not channelled or keeled either side, veinless. Stems to 10cm, hairless. All yr. Mtns....................................*Alpine Pearlwort* **Sagina saginoides**

Lvs not apiculate (may have slight apiculus) or awn <0.2mm, often obtuse

Lvs scabrid-ciliate at least nr base, glaucous (usu turning reddish)

Lvs to 6 x 1mm, with axillary lf-clusters, often without apiculus, often curved to 1 side, hairless, with obscure midrib. Stems to 20cm, ± woody below. ❶. All yr. E Anglian Brecks, Radnor. Sch8......................................*Perennial Knawel* **Scleranthus perennis**

Lvs not scabrid-ciliate, green

Lvs 4-11 x 0.5-1mm, linear-subulate, ± fleshy, clustered nr base and diminishing upwards to 1-2mm, forming lf clusters in upper lf axils giving 'knotted' appearance, hairless to ± glandular-hairy, veinless. Stems to 10cm, hairless or glandular-hairy. Plant usu forming compact turf, not isolated rosettes. All yr. Often slightly damp habs ....*Knotted Pearlwort* **Sagina nodosa***

Lvs 6-12 x 0.5mm, linear-subulate, not fleshy, clustered nr base and diminishing upwards, forming lf clusters in upper lf axils, often curved, mostly at base, hairless or sparsely glandular-hairy, veinless. Stems to 10cm, hairless or sparsely glandular-hairy. Plant laxly tufted, often reddish. Calc flushes, Teesdale. Sch8 ....*Teesdale Sandwort* **Minuartia stricta**

Ann. Basal lvs not forming a basal rosette (or soon dying), usu shorter than stem lvs

Lvs acute, not fleshy, minutely apiculate or awned (exc *Petrorhagia nanteuilii*)

Lvs with smooth margins, awnless

Lvs crowded nr base of stems, distant above (often giving 'knotted' appearance), 5-15 x 1mm, linear-subulate, straight or recurved, hairless. Stems to 20cm, erect, hairless, with swollen purplish nodes. Dry calc habs....................*Fine-leaved Sandwort* **Minuartia hybrida**

Lvs with scabrid-ciliate margins at least nr base

Lvs with a 0.3mm awn, 3-10 x 0.5-0.8mm, linear-subulate, often ciliate nr base, veinless. Stems to 15cm, erect, diffusely branched, often hairy. Oct-Jul ....*Annual Pearlwort* **Sagina apetala**

Lvs with a minute <0.1mm mucro, (5)10-18 x 0.5-0.6mm, linear-subulate, usu ciliate at least nr base, with midrib visible nr base. Stems to 20cm, decumbent to erect, branched, hairless or shortly crisped-hairy. Occ bi. Apr-Oct (all yr)............*Annual Knawel* **Scleranthus annuus**

Lvs without mucro, 10-20 x 1mm, linear-lanc, scabrid-ciliate, 0-1-veined, with midrib sunken above. Stems usu 1, usu to 10cm, erect, unbranched, hairless. ❶. May-Aug. Shingle, VR, S Eng, Channel Is. Sch8..*Childing Pink* **Petrorhagia nanteuilii**

Lvs blunt and/or fleshy, rarely apiculate or awned, with smooth margins

Lvs grooved above, with sunken midrib. ❶.......................*Upright Chickweed* **Moenchia erecta**

Lvs not grooved above, without sunken midrib

Lvs channelled below, viscid-hairy. ❷.................................*Corn Spurrey* **Spergula arvensis**

Lvs not channelled below, hairless (rarely ciliate in *Sagina maritima*)

Lvs 3-9 x 0.7(1)mm, ± cylindrical, (±) obtuse (occ apiculate), fleshy, occ in tiny basal rosette of 6-8 lvs, the lowest non-basal stem lvs longest, the uppermost short and ± flat above, all veinless. Stems to 15cm, erect, occ purplish (at least at nodes), occ sparsely hairy. Oct-Jul. Damp maritime silt and roadsides................*Sea Pearlwort* **Sagina maritima**

Lvs 4-15 x 1-4mm, narrowly spathulate to obovate, obtuse, with white hydathode at apex (often obscure), narrowed to indistinct petiole (rarely long), applanate when young, ± fleshy, with clearly visible cells (x20), often reddish, often with obscure midrib. Petiole broader at base with narrow hyaline margin. Stems usu <5cm, often reddish, with tiny central stele in TS..*Blinks* **Montia fontana** ssp **chondrosperma***

Lvs 1-2 x 0.5mm, ovate-oblong, ± acute, widely connate at base, forming concave cup around stem, often with lvs in axils, imbricate, thick, reddish, (0)1-veined. Stems 1-5cm, decumbent or ascending, reddish. May-Oct (all yr).........*Mossy Stonecrop* **Crassula tillaea**

Minuartia recurva lvs with lf apex & 3-veined dead lf

Minuartia verna lvs & lf apex

Sagina nodosa lf apex

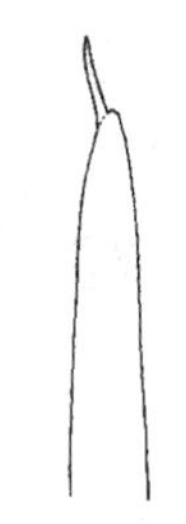

Sagina subulata lf apex

Group LK – *Plant hairless. Lvs in a basal rosette. Petiole >> lf*

▪Petiole triangular or laterally flattened, hollow, with 3(5) vb's

Basal lvs to 4cm, ovate-elliptic to ± broadly triangular, obtuse, rounded or truncate at base, obvolute when young, dark or bright green, scabrid-ciliate, entire or obscurely sinuate-toothed with white sunken hydathodes along margins above, with stomata both sides. Petiole to 7cm, channelled, with small hollow. Stolons long, round, solid. Per. Mar-Oct. Wet habs .... *Marsh Valerian* **Valeriana dioica***

▪Petiole ± flattened, solid, with 1 vb

Basal lvs with 3(5) ± parallel veins sunken above (not raised below), 1-5cm, ovate to rhombic, acuminate, with white apical hydathode turning dark, ± fleshy, shiny dark green above, paler below, with stomata obscure both sides. Stem lvs sessile but not perfoliate. Petiole to 8cm, weakly sheathing at base, channelled. Stems to 40cm, brittle. Ann to per. All yr (usu). Shady damp habs.. *Pink Purslane* **Claytonia sibirica**

Basal lvs with all veins obscure, 1-3cm, ovate to rhombic, ± obtuse, with white apical hydathode turning dark, ± fleshy, pale green, with stomata obscure both sides, the translucent rectangular cells give a crystalline appearance (esp below). Stem lvs perfoliate, forming a ± orb concave cup. Petiole to 5cm, weakly sheathing at base, not channelled. Stems to 30cm, brittle. Ann. Oct-Sept. Dry sunny sandy habs. PL 11.. *Springbeauty* **Claytonia perfoliata**

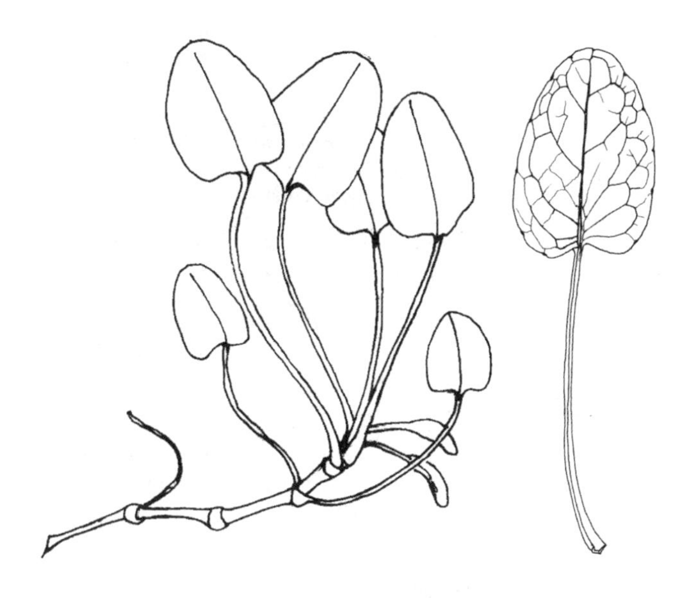

Valeriana dioica

Group LL – *Lvs clearly with 3(5) veins. (*Lathyrus aphaca *may key out here but usu has tendrils)*

■Plant with bitter taste, rhizomes short or absent. Lvs (1)3-5-veined, applanate when young ....(*Gentianaceae*) Go to **GENT**

■Plant tasteless, strongly rhizomatous. Lvs 3(5)-veined, rolled when young

Lvs 5-10cm, broadly ovate to elliptic, acute, with minutely crenate or scabrid-ciliate cartilaginous margins (often recurved), occ with pinnate veins but not net-veined, with stomata both sides. Petiole indistinct, with 3 flat vb's. Stem (and reduced lvs) purplish nr base, round. Apr-Oct ....*Soapwort* **Saponaria officinalis**

Group LM – *Plant glaucous. Lvs with scabrid margins, hairless (occ hairy both sides in* Silene vulgaris *and often hairy above in* Stellaria holostea*), connate at base (± so in* Stellaria holostea*). Stems erect (prostrate in* Silene uniflora*)*

■Lvs >5mm wide, oblong-lanc. Per

Lvs fleshy, all on stems, 0.5-4 x 1cm, ± oblong or oblong-spathulate, acute, obvolute when young, with serrate cartilaginous margins, waxy dull grey-green, without translucent dots, occ ciliate, with obscure 2° veins, with stomata both sides. Stems to 25cm, prostrate, ± woody at base, often purplish, usu hairless, with elastic stele. All yr. Maritime turf............*Sea Campion* **Silene uniflora***

Lvs thin, all on stems, to 6 x 1.5cm, obovate to oblong-lanc, apiculate, obvolute when young, with scabrid-ciliate to smooth margins, dull grey-green, often with translucent dots (HTL), with weakly translucent 2° veins (occ slightly raised both sides), with stomata below and often above. Stems to 80cm, erect, not woody, green, hairless or hairy, with elastic stele. Apr-Oct ....*Bladder Campion* **Silene vulgaris**

■Lvs <5mm wide, linear-lanc

Per, occ woody at base. Stems (±) 4-angled

Lvs with translucent midrib (2° veins obscure and opaque). Shortly rhizomatous

Lvs all on stems, 4-8cm x 5mm, acute, antrorsely scabrid-ciliate (with septate cilia retrorse proximally), with scabrid midrib raised below, with stomata both sides or below only. Stems to 60cm, with retrorse prickles on angles, brittle, with elastic stele. All yr ....*Greater Stitchwort* **Stellaria holostea**

Lvs usu with all veins obscure. Tufted

Lvs 4-6(9)cm x 3.5-4mm, acute, with crenulate-scabrid margins, with stomata both sides. Stems 20-50cm, usu branched above. Loosely tufted with false rosettes. All yr ....*Clove Pink* **Dianthus caryophyllus**

Lvs 3-5cm x 2-3mm, acute, scabrid-ciliate, with stomata both sides. Stems 15-30cm, usu branched above. Loosely tufted. All yr..*Pink* **Dianthus plumarius**

Lvs 2-3(6)cm x 1-2mm, ± acute, with remotely scabrid-ciliate margins (often retrorsely so), with stomata both sides. Stems 15-25cm, usu unbranched. Densely tufted. All yr. Sch8 ....*Cheddar Pink* **Dianthus gratianopolitanus**

Ann. Stems round

Stem(s) minutely glandular-hairy (but often hairless below each node)

Lvs 2-3cm x 2mm, acute, with stomata both sides. Stem usu 1, to 50cm, 1-2mm diam. Jun-Sept. VR alien..*Wilding Pink* **Petrorhagia dubia**

Stems hairless (occ nonglandular-hairy)

Lvs 2-3cm x 1-2mm, acute, scabrid-ciliate, with midrib raised below, with stomata both sides. Stems 1-several, 20-40(60)cm, 1mm diam, rarely densely hairy. Oct-Sept. R alien ....*Proliferous Pink* **Petrorhagia prolifera**

Lvs 1-2cm x 1mm, acute, scabrid-ciliate, with obscure veins but midrib or groove visible above, with stomata both sides. Stem usu 1, to 10cm, 1mm diam, always hairless. May-Aug. Coastal shingle, VR, S Eng, Channel Is. Sch8.................*Childing Pink* **Petrorhagia nanteuilii**

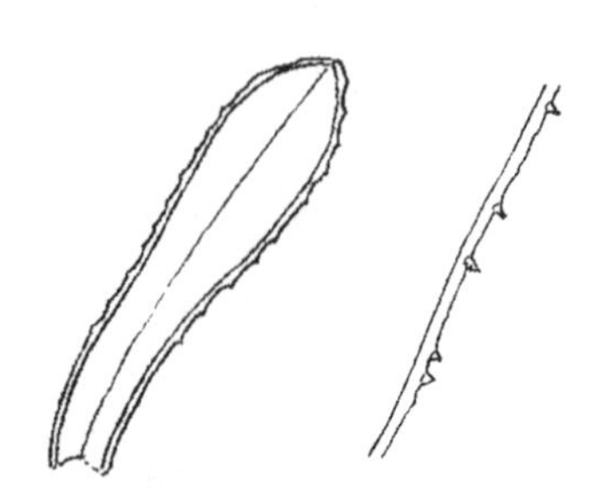

Silene uniflora & lf margin

Group LN – *Plant not glaucous, hairless. Lvs connate at base (exc* Valerianella*). (*Anagallis arvensis *may key out here)*

■Lvs distinctly pinnate-veined, with hydathodes along margins above.........(*Valerianella*) Go to **VAL**

■Lvs 1(5)-veined (2° veins often obscure or absent), without hydathodes along margins

Plant with bitter taste. Lvs decussate, usu sessile, (1)3-5-veined........(*Gentianaceae*) Go to **GENT**

Plant tasteless

Basal rosette dying at fl. Bi. Stems round or angled..................*Deptford Pink* **Dianthus armeria**

Basal rosette absent

Ann. Stems round or angled, erect

Lvs 5-8(12) x 2.5-3mm, the lower crowded at base and smaller than mid and upper (often held erect), oblong to obovate, ± obtuse, sessile, dark green, with minutely scabrid non-cartilaginous margins (often obscure), obscurely 3-veined proximally, with translucent midrib often fading before apex (and slightly raised below), with stomata both sides. Stems 5-8(25)cm, 0.5mm diam, branched, hairless, weakly angled. Mar-Jul. Dry habs ....*Fairy Flax* **Linum catharticum**

Lvs 10-25 x 5mm, oblong-ovate, the upper linear-lanc, obtuse to acute, ± sessile, pale green, with ± smooth non-cartilaginous margins, obscurely pinnate-veined, with translucent midrib, with stomata both sides. Stems 10-25cm, 1mm diam, branched, hairless, with 2 obscure raised lines (obscurely crenulate). Apr-Oct (all yr). Winter-wet habs. Sch8 ....*Grass-poly* **Lythrum hyssopifolium**

Per

Stems square, erect

Lvs 4-8 x 0.5cm, linear-lanc, acute, ± connate at base, hairy above or hairless, antrorsely scabrid-ciliate (cilia septate and retrorse proximally), with translucent scabrid midrib raised below, with obscure opaque 2° veins, with stomata both sides or below only. Stems with retrorse prickles on angles, brittle, with elastic stele. Shortly rhizomatous. All yr ....*Greater Stitchwort* **Stellaria holostea**

Stems round or angled, erect. Mtns

Lvs 3(6) x 1mm, crowded, imbricate, linear-subulate, obtuse to acute, flat or weakly channelled above, ± fleshy, yellow-green, hairless, scabrid-ciliate, veinless, with stomata both sides. Cushion-forming per to 3(8)cm tall. Usu mtn turf.....*Cyphel* **Minuartia sedoides**

Stems round or angled. Lowlands

Stems prostrate, rooting at nodes. Lvs broadly ovate

Lvs obtuse, with stomata both sides...............*Creeping-Jenny* **Lysimachia nummularia***

Lvs acute, with stomata below only..................*Yellow Pimpernel* **Lysimachia nemorum***

Stems erect, Lvs not ovate

Lvs with stomata below only (occ scattered above)

Lvs to 10 x 0.7-1.5cm, lanc to broadly elliptic, sessile or narrowed to indistinct petiole (occ purplish), obvolute when young, antrorsely septate-scabrid, yellow-green, with midrib raised below, with opaque netted veins slightly raised above. Stems 30-60cm, smooth, slightly angled. Bi or short-lived per. All yr (false rosette in winter) ....*Sweet-William* **Dianthus barbatus**

Lvs with stomata both sides

Lvs 1-2.5cm x 2-5mm, lower narrow oblanc and obtuse to ± acute, upper linear-lanc and acute, connate at base, valvate when young, with scabrid-ciliate margins (and midrib below), 1(3)-veined, with obscure pinnate veins. Stems to 45cm, scabrid-ciliate (septate and hair-like), with nodes ± swollen. Loosely tufted. All yr. Dry turf ....*Maiden Pink* **Dianthus deltoides**

Lvs 3-7(12)cm x 3-5mm, linear to spathulate, green, acute, scabrid-ciliate, 1-veined. Stems 25-65cm, simple or branched, hairless. Tufted. All yr. VR alien ....*Carthusian Pink* **Dianthus carthusianorum**

Group LO – *Lvs 0-1(3)-veined or pinnate-veined.* ❶ *Lvs dotted with red-black glands below.* ❷ *Stem with 4 hollows in TS.* ❸ *Stipules present. (*Linum catharticum *may key out here)*

■Stems 4-angled (often strongly so)

Stems (±) erect, with elastic stele. Lvs glaucous, (±) connate at base

Lvs 15-50 x 2-3mm, linear-lanc, acute, with recurved margins, with translucent midrib raised below, with 2° veins obscure or opaque, with stomata both sides. Stems 1mm wide, brittle, smooth, with slightly swollen nodes, with stomata all sides. Rhizomatous per. Calc marshes ....*Marsh Stitchwort* **Stellaria palustris***

Stems (±) prostrate, without elastic stele. Lvs green, not connate at base

Lvs sessile, occ 3(4)-whorled (esp bracts), 10-15mm, ovate, acute, applanate-reflexed when young, with erose-crenulate margins, rarely ± fleshy, obscurely 3-pli-veined (fading distally), with stomata both sides. Stems not rooting at nodes, green. Ann. Dry arable, ruderal. ❶. Often all yr..*Scarlet Pimpernel* **Anagallis arvensis***

Lvs narrowed to indistinct petiole, rarely 3-whorled, 10-20mm, obovate-spathulate, obtuse, applanate when young, with entire margins, often ± fleshy, with translucent midrib fading nr apex and 2° veins ± opaque (often reddish), with stomata below only or obscure above. Stems rooting at nodes, reddish. Ann or short-lived per. All yr. Winter-wet habs, rarely submerged for long. ❷...*Water-purslane* **Lythrum portula***

■Stems (±) round with 2 opposite grooves in TS (occ weak)

Stems freely rooting at nodes. Lvs not fleshy, not connate at base. All yr. Usu shady habs

Lvs acute, 2-4cm, ovate, applanate when young, often undulate, with crenate-serrate hyaline margins, without glands, with 3 prs of weakly anastomosing 2° veins, occ 3(5)-pli-veined, with stomata below only. Petiole to 3mm, flattened or channelled, with cusps at base, with 1 vb. Stem with 4 vb's (no stele)...*Yellow Pimpernel* **Lysimachia nemorum***

Lvs v obtuse, 1.5-3cm, broadly ovate to ± orb, rolled when young, not undulate, with entire margins, with orange glands both sides (esp nr margins) and along apex below, with translucent anastomosing 2° veins, with stomata both sides. Petiole to 6mm, channelled, with 1 flat arc vb. Stems with non-elastic stele...................*Creeping-Jenny* **Lysimachia nummularia***

Stems rarely rooting at nodes. Lvs often fleshy, rarely connate at base. Apr-Oct. Coastal

Lvs v fleshy, not pitted, 5-18 x 3-5mm, ovate, acute, with orange apical hydathode (best viewed end-on), sessile, strongly decussate, valvate when young, ± undulate, with erose cartilaginous margins, shiny bright green, with obscure veins, with midrib not or hardly raised below, with stomata both sides. Stems to 25cm, pale green, occ with swollen internodes, with elastic stele. Rhizomatous. Beaches...*Sea Sandwort* **Honckenya peploides**

Lvs ± not fleshy, pitted both sides, to 12 x 3mm, elliptic-oblong, acute to obtuse, ± sessile, decussate, valvate when young, not undulate, with v narrow erose hyaline margins, dull dark grey-green, occ with reddish speckles (glands), with obscure veins, with midrib not or hardly raised below, with stomata both sides. Stems to 25cm, reddish-speckled, occ with dark red horizontal line at nodes, with non-elastic stele. Rhizomatous. Saltmarshes......*Sea-milkwort* **Glaux maritima**

■Stems (±) round, not grooved

Lvs v fleshy and >15mm long

Woody-based evergreen per, rooting at nodes. Coastal, W Br

Lvs ± cylindrical and smooth, (1)2-4 x 0.5-1cm, sessile, connate at base, ± applanate when young, curved (like jelly-beans!), reddish-green, with stomata all around. Stems to 1m, round, with central stele...*Purple Dewplant* **Disphyma crassifolium**

Lvs strongly 3-angled (erose-serrulate on angles), 8-10 x 1-1.5cm, sessile, connate at base, applanate when young, slightly curved, green, with stomata all sides. Stems >>1m, angled, with central stele..*Hottentot-fig* **Carpobrotus edulis**

Ann, usu prostrate but not rooting at nodes. ❸. Ruderal

Lvs occ alt, to 2 x 0.8cm, oblong, obtuse, fleshy, with erose cartilaginous margins, often reddish, with translucent midrib, with Kranz venation, with stomata both sides. Petiole indistinct, 2.5mm. Stipules bristle-like, occ forming fringe, rarely absent. Stems reddish, round, without interpetiolar ridge, snapping audibly, with vb's forming central stele in aerenchyma. VR alien...*Common Purslane* **Portulaca oleracea**

Lvs not v fleshy, or <15mm long

Aquatic (often submerged), occ confined to wet turf or mud

Stems with a dark red horizontal ring below each node. Lvs not notched at apex, connate

Per. Lvs 4-15(20) x 0.8-2mm, linear to ovate-lanc, acute, ± fleshy, usu with obscure midrib. Stems to 10(20)cm tall, decumbent to erect, rooting at nodes and mat-forming, with central stele occ reddish. All yr. Invasive alien............. *New Zealand Pigmyweed* **Crassula helmsii***

Ann. Lvs to 3(6) x 0.5mm, linear, (±) acute, fleshy, usu with obscure midrib. Stems to 5cm tall, decumbent, occ rooting at nodes, with central stele occ reddish. May-Sept. VR, Scot. Sch8..*Pigmyweed* **Crassula aquatica**

Stems without a red ring below each node

Lvs notched at apex, usu connate..(*Callitriche*) Go to **CAL**

Lvs not notched, usu connate (at least lower lvs)

Lvs 4-15 x 1-4mm, narrowly spathulate to obovate, obtuse, with white hydathode at apex (often obscure), narrowed to indistinct petiole (rarely long), the upper free or ± connate, applanate when young, ± fleshy, with clearly visible cells (x20), often reddish, often with obscure midrib, with stomata both sides. Petiole broader at base with narrow hyaline margin. Stems usu <5cm, 0.5-1mm diam, often reddish, with tiny central stele. Occ per with stolons..*Blinks* (ssp) **Montia fontana** ssp **amporitana**

Lvs not notched, never connate

Stipules free, translucent, toothed ❸

Lvs occ 4-whorled, to 8mm, spathulate, obtuse, midrib usu obscure. Stems rooting at nodes (many roots per node), round, with 6-10 hollows in TS (cartwheel-like)

Fl stalk > fl bud length. Summer ann on mud, short-lived per underwater. Lakes, ponds, wet mud..*Six-stamened Waterwort* **Elatine hexandra***

Fl stalk < fl bud length (to absent). Ann. Ponds, small lakes, R ....*Eight-stamened Waterwort* **Elatine hydropiper**

Stipules (when present) minute, gland-like ❸

Lvs 1.5-3(5)cm, ovate to broadly elliptic, shortly acuminate or acute, cuneate at base, obvolute when young, ± translucent, shiny dirty red-green, with sunken pale hydathodes along margins below, with 2° veins fading nr margins, with stomata both sides. Petioles short, not connate at base. Stems usu emergent, to 20cm, rooting at lower nodes, with a central green stele surrounded by reddish aerenchyma. Short-lived per. Apr-Oct. VR, New Forest..*Hampshire-purslane* **Ludwigia palustris**

Stipules always absent

Lvs to 0.5cm, occ alt, ovate to ± orb, obtuse, applanate when young, with erose-crenulate margins, red-speckled, with red-black glands along margins below, with sweet disinfectant odour, obscurely 3-veined, with stomata both sides. Petioles short, not connate at base. Stems slender, rooting at nodes, without interpetiolar ridge. ❶. All yr. Short wet turf..*Bog Pimpernel* **Anagallis tenella***

Dry (occ winter-wet) habs

Lvs >20mm, with 2° veins visible. Per

Lvs all on stems, to 6 x 1.5cm, obovate to oblong-lanc, apiculate, obvolute when young, with scabrid-ciliate to smooth margins, thin, dull grey-green, often with translucent dots (HTL), occ hairy, with weakly translucent 2° veins (occ slightly raised both sides), with stomata below and often above. Stems to 80cm, erect, green, hairless or hairy, with elastic stele. Apr-Oct..*Bladder Campion* **Silene vulgaris**

Lvs basal and/or on stems, to 10cm, elliptic, cuneate and connate at base, rolled when young, with minutely crenate margins, shiny green to ± glaucous above, paler below, hairless (occ minute glandular hairs when young), with white sunken hydathodes visible along margins above, with midrib raised below, with sunken translucent 2° veins often originating from nr base, with stomata both sides. Petiole to 5cm, with 5 vb's. Stems to 100cm, brittle (snapping audibly), round to weakly angled, not swollen at nodes, with interpetiolar ridge, soon hollow. All yr. Old walls, ruderal. PL 9.........................*Red Valerian* **Centranthus ruber**

Lvs <20mm, with 2° veins not visible. Ann

Lvs 1-2mm, widely connate (forming cup around stem)

Lvs ovate-oblong, ± acute, imbricate, often with lvs in axils, thick, concave, reddish. Stems 1-5cm tall, decumbent or ascending, reddish. May-Oct (all yr) ....*Mossy Stonecrop* **Crassula tillaea**

Lvs 2-8mm, connate or free

Lvs alt above

Lvs 3-5 x 4mm, ovate, obtuse or apiculate, ± sessile, with erose margins, soon with border of blackish glands below, with obscure veins (midrib occ visible proximally). Stems 2-7cm, erect to decumbent, occ 3-ridged. May-Sept. Damp ground ....*Chaffweed* **Anagallis minima**

Lvs all opp

Stem lvs glaucous (or lvs all basal and dark green)

Basal lvs soon dying, 5-10(20)mm, linear-lanc, acute, ± sessile, connate at base, not glaucous, 1-veined, with stomata both sides (almost in parallel rows). Stem lvs sessile, ascending, rigid. Stems usu <5cm, erect, reddish at base. Dec-Jun ....*Upright Chickweed* **Moenchia erecta**

Stem lvs green

Upper lvs connate..........................(*Cicendia filiformis*, *Exaculum pusillum*) Go to **GENT**

Upper (or all) lvs usu free

Lvs 4-15 x 1-4mm, narrowly spathulate to obovate, obtuse, with white hydathode at apex (often obscure), narrowed to indistinct petiole (rarely long), the upper free or ± connate, applanate when young, ± fleshy, with clearly visible cells (x20), often reddish, often with obscure midrib, with stomata both sides. Petiole broader at base with narrow hyaline margin. Stems usu <5cm, 0.5-1mm diam, often reddish, with tiny central stele...*Blinks* **Montia fontana** ssp **chondrosperma***

Lvs 1.5-3 x 2mm, obovate to elliptic, acute to ± obtuse, with crenulate or erose hyaline margins, with translucent midrib, with stomata both sides. Stems 1-8cm, 0.5mm diam, ± prostrate to ascending, usu repeatedly forked....................*Allseed* **Radiola linoides**

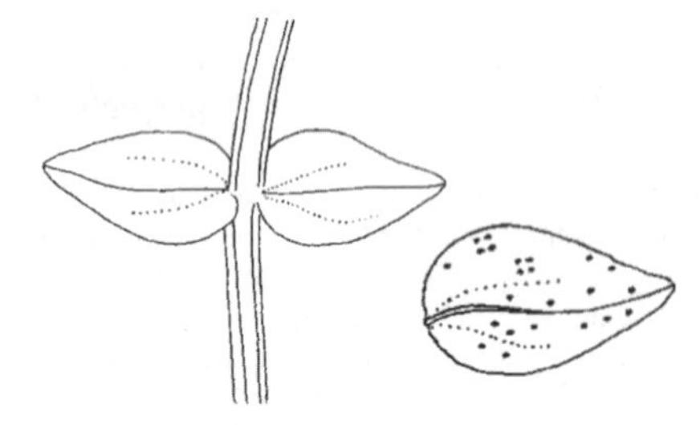

Anagallis arvensis lvs & lf underside

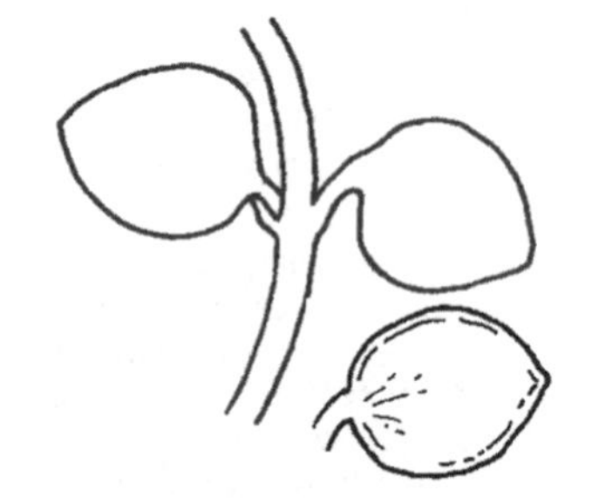

Anagallis tenella lvs & lf underside

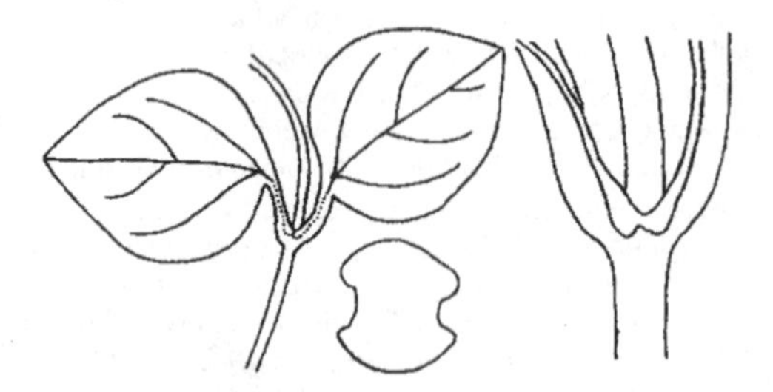

Lysimachia nemorum with stem TS & petiole cusps

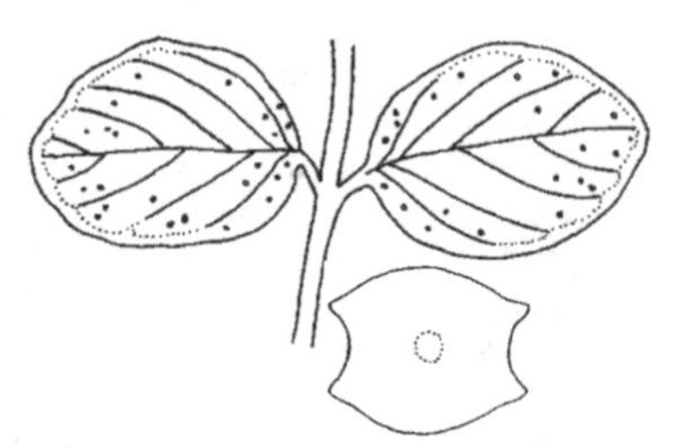

Lysimachia nummularia lvs & stem TS

Group LP – *Lvs with mealy, woolly, stellate or medifixed hairs*

▪Plant with sparse mealy bladder-like hairs when young. Ann

Lvs purplish below. Stems squarish, ridged...............(*Chenopodium polyspermum*) Go to **CHAT A**

▪Plant with stellate and long simple hairs (esp on lowerside of basal lvs). Ann

Lvs basal or on stem, alt above, the basal dead at fl, to 3 x 1.5cm, elliptic-lanc, ± acute, sessile, 3-veined, with stomata both sides. Stipules on upper stem lvs only, linear, green. Stems to 15cm, erect. Oct-Jul (occ spring germinating). Coastal, VR............*Spotted Rock-rose* **Tuberaria guttata**

▪Plant with medifixed hairs on lvs. Rhizomatous per

Lvs 1-2(3)cm, ovate or ovate-elliptic, acute or v shortly acuminate, ± sessile, green and medifixed-hairy above, glaucous and hairless below, sparsely ciliate, (3)5-veined, with stomata below only. Stems 6-20cm, hairless or adpressed-hairy, 4-angled with 2 sides grooved. Jun-Oct. Mtn moors. ....*Dwarf Cornel* **Cornus suecica**

▪Plant with dense woolly hairs (obscuring diagnostic features). Per (without rhizomes)

Plant erect or rosette-forming

Lvs to 20 x 5cm, oblanc to ovate-oblong, ± acute, narrowed to indistinct petiole, obvolute when young, occ undulate, occ with recurved margins, with midrib raised below only, with opaque indistinct 2° veins. Petiole long-ciliate at base, with 1 vb. Stems several, to 1m, becoming woody below, white-woolly, repeatedly forked, with swollen nodes. All yr ....*Rose Campion* **Lychnis coronaria**

Plant mat-forming

Lowlands. Lvs to 30 x 2.5-6mm, elliptic to linear-lanc, with slightly revolute margins, with dense short grey-white matted woolly hairs. Stems white-woolly, occ purplish above, often with axillary lf-clusters. All yr...*Snow-in-summer* **Cerastium tomentosum**

Mtns. Lvs 8-14 x 4-6mm, ± elliptic to ovate, with flat margins, with dense long (2-4mm) white wavy hairs. Stems with long and short wavy white hairs (occ glandular on infl stalks), not purplish, usu with axillary lf-clusters................................*Alpine Mouse-ear* **Cerastium alpinum**

LO

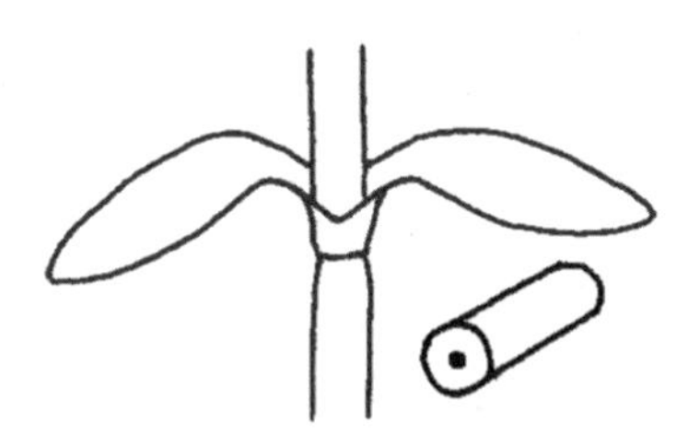

Crassula helmsii lvs & stem TS

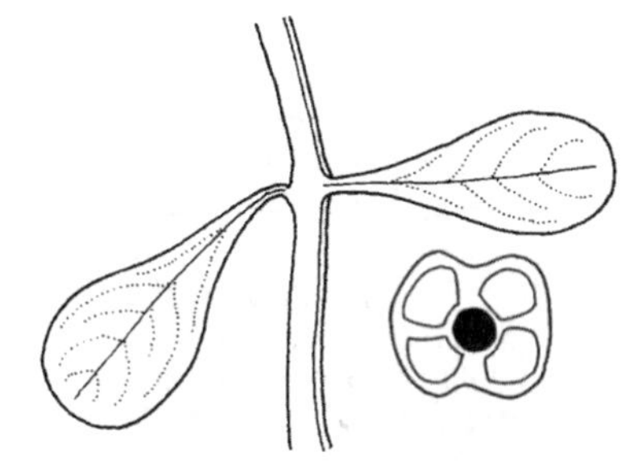

Lythrum portula lvs & stem TS

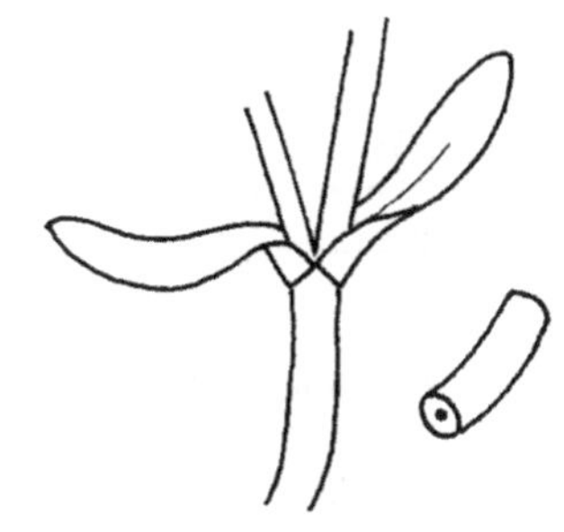

Montia fontana ssp *chondrosperma* lvs & stem TS

Group LQ – *Stems square, without elastic stele.* ❶ *Plant aromatic*

■Lvs with a distinct petiole and 2° veins visible

Stems rooting at lower nodes, deeply channelled on 2 sides......................(*Prunella*) Go to **PRUNE**

Stems not rooting at nodes, (±) erect, not channelled

Lvs of each pr usu unequal, 5-15 x 5-9cm, ovate, shortly acuminate, truncate to cordate at base, rolled when young, with narrow cartilaginous margins, dull dark green above, without translucent dots, shortly septate-hairy on midrib above and on 1-2° veins below, ciliate, weakly net-veined, with 2° veins anastomosing nr margins, with stomata both sides. Petiole to 5cm, with several vb's forming flat arc. Stems to 100cm, green, short-hairy esp in 2 wide lines, fetid (*Solanaceae*-like odour), obtusely 4-angled, with swollen nodes, solid, with scattered vb's. Per (ann), with tuberous roots. May-Oct. VR alien..*Marvel-of-Peru* **Mirabilis jalapa**

Lvs equal, 1.5-5 x 3cm, broadly ovate, obtuse, rounded at base, applanate when young, without cartilaginous margins, aromatic, shiny mid-green above, with translucent dots, sparsely septate-hairy both sides, ciliate, not net-veined, with 2° veins fading nr margins, with stomata usu below only. Petiole to 1cm, with 1(3) vb's. Stems 30-80cm, often purplish, hairy all sides, aromatic, 4-angled, without swollen nodes, solid. Per, without tuberous roots. ❶. Apr-Oct (young shoots may overwinter)..*Wild Marjoram* **Origanum vulgare***

■Lvs sessile or with indistinct petiole, usu with 2° veins obscure

Stems spreading-hairy (often minutely so), without elastic stele. Lvs not connate at base

Weakly aromatic ann. ❶..(*Valerianella*) Go to **VAL**

Odourless tufted per. Lvs 4-7(10) x 0.5-1.5cm, smaller nr base of stems, lanc to ovate, (±) acute, ± cordate or rounded at base, not connate at base, applanate when young, slightly undulate, shortly septate-hairy (to ± hairless) esp below, ciliate, not net-veined, with translucent 2° veins anastomosing nr margins, with 1-3° veins raised below, usu with stomata below only. Stems usu hairy, solid. Apr-Oct. Wet habs.................................*Purple-loosestrife* **Lythrum salicaria***

Stems minutely retrorsely hispid-hairy when young, without elastic stele. Lvs not connate at base

Lvs (1.5)3-10 x 0.4-1cm, linear-lanc, often with recurved scabrid-ciliate margins, with translucent midrib raised below, with indistinct anastomosing 2° veins, with stomata below only. Stems 20-40(60)cm, erect, often purplish, without interpetiolar ridge, occ obscurely 4-angled or with 2 shallow grooves. Hemiparasitic ann

Bracts usu entire. Lvs yellow-green to slightly purplish, often roughly hairy. Mtn woods, VR, Scot...*Small Cow-wheat* **Melampyrum sylvaticum**

Bracts usu with 1(2) prs of teeth nr base. Lvs dark green to purplish, with minute hispid antrorse hairs above, hairless below (occ minutely hispid on veins). Stems often with hairs on 2 opp sides...*Common Cow-wheat* **Melampyrum pratense**

Bracts with 1-4(6) prs of long teeth nr base. Lvs often purplish, minutely hispid-hairy (± hairless TNE). VR. Sch8...*Field Cow-wheat* **Melampyrum arvense**

Bracts with 6-9 prs of long teeth nr base. Lvs often purplish, minutely hispid-hairy. VR, S Eng ....*Crested Cow-wheat* **Melampyrum cristatum**

Stems hairless (but may be rough with prickles on angles), with elastic stele. Lvs ± connate at base

Lvs ciliate nr base

Stems rough with retrorse prickles on the angles (at least below) exc when young

Lvs green or glaucous, 4-8 x 0.5cm, linear-lanc, acute, hairy above or hairless, antrorsely scabrid-ciliate (cilia septate and retrorse proximally), with translucent scabrid midrib raised below (2° veins obscure and opaque), with stomata both sides or below only. Stems to 60cm, brittle. Shortly rhizomatous per. All yr................*Greater Stitchwort* **Stellaria holostea**

Stems smooth

Lvs slightly glaucous, 10-20 x 5mm, elliptic or oblanc to ovate, acute (with white or dark hydathode), valvate when young, hairless, with midrib slightly raised below, with stomata both sides. Stems decumbent and ascending, mat-forming, brittle. Rhizomatous per. All yr. Damp acidic habs...*Bog Stitchwort* **Stellaria uliginosa***

Lvs green, 10-40 x 5mm, linear-lanc to elliptic-ovate, acute (with white or dark hydathode), valvate when young, hairless, with midrib slightly raised below, with stomata both sides. Stems erect to mat-forming, brittle. Rhizomatous per. All yr. Usu dry habs ....*Lesser Stitchwort* **Stellaria graminea**

Lvs not ciliate nr base. Stems smooth

Lvs glaucous, 15-50 x 2-3mm, linear-lanc, acute, with recurved margins, with translucent midrib raised below, with 2° veins obscure or opaque, with stomata both sides. Stems 1mm wide, ± erect, brittle, with slightly swollen nodes, with stomata all sides. Rhizomatous per. Calc marshes..*Marsh Stitchwort* **Stellaria palustris***

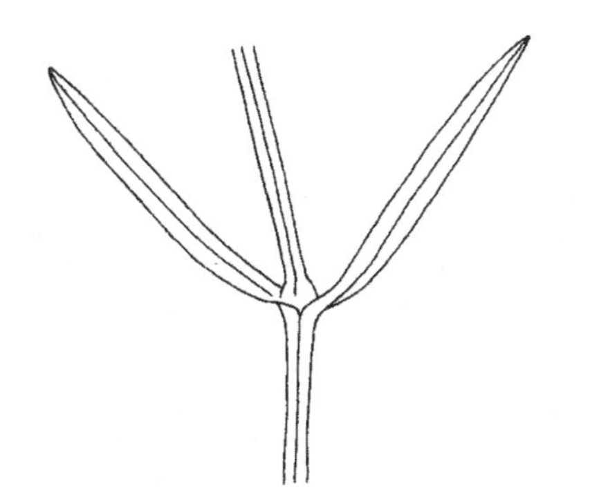

Stellaria palustris

Lythrum salicaria

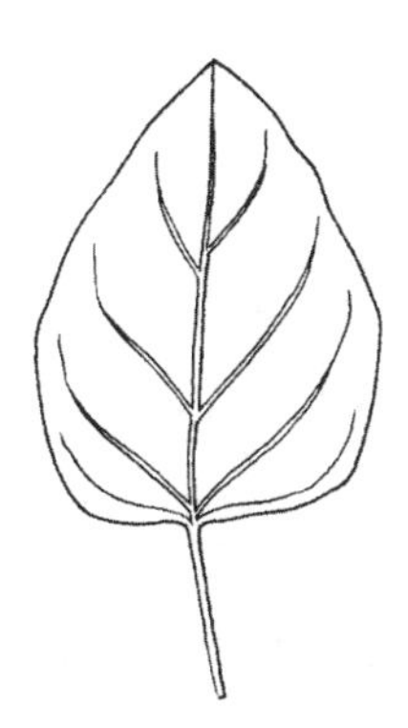

Origanum vulgare

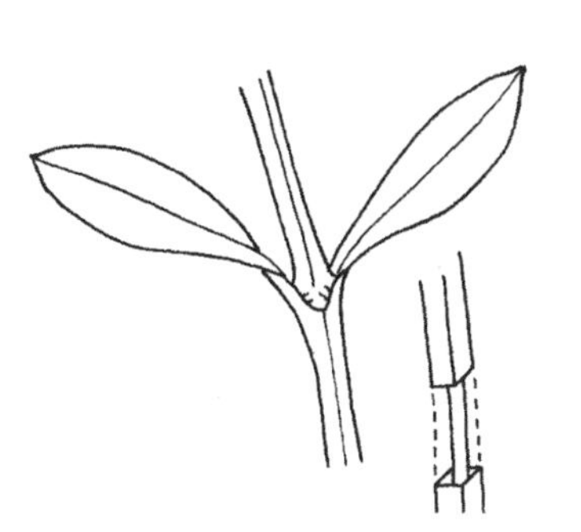

Stellaria uliginosa, with elastic stele & opaque veins

Group LR – *Stems with 1-2 lines of hairs, round. Lvs applanate when young and connate at base (free in* Epilobium brunnescens*)*

■Stem strongly rooted at nodes, with non-elastic stele. Hairs unicellular

Stems with 2 opp lines of minute hairs. Lvs 3-10mm, broadly ovate to ± orb, purplish below, entire to obscurely sinuate-toothed with 3 sunken hydathodes per side (the terminal one often indistinct), with 1(3) veins visible below, with stomata above only. Petiole 0.5-3mm. Per. All yr. Usu uplands..*New Zealand Willowherb* **Epilobium brunnescens***

■Stem not or weakly rooting at nodes (often straggling), with elastic stele. Hairs, when present, septate

Lvs 3(5)-pli-veined, with translucent dots

Lvs 6-25mm, ovate, acute to cuspidate, the lower stalked, the upper ± sessile, sparsely hairy, ciliate (cilia longest nr petiole). Stems with 2 dense lines of hairs and scattered hairs all around (hairs may be retrorsely curved), slightly swollen at nodes. Ann. Nov-Jul ....*Three-nerved Sandwort* **Moehringia trinervia**

Lvs pinnately-veined (usu obscurely so), without translucent dots

Stems mat-forming and/or rooting at nodes, ± woody at base. Per. Wet mtn rocks

Lvs 6-12 x 2-5mm, oblanc, obtuse, usu curving to one side, pale green or glaucous, with sparse 4-septate hairs both sides, with translucent midrib, with obscure 2° veins, with stomata both sides. Stems purplish, with 1 line of hairs. All yr ....*Starwort Mouse-ear* **Cerastium cerastoides**

Stems not or weakly mat-forming, rarely rooting at nodes, never woody at base. Ann. Usu lowlands

Lvs (10)20-30(70)mm, ovate, acuminate with apiculate white hydathode, ± cordate at base, the lower long-petiolate (to 2x lf length), the upper ± sessile (and larger), pale green, hairless, with opaque reticulations, with stomata both sides. Petiole sparsely ciliate, flattened. Stems 20-100cm tall, c1mm diam, ± prostrate to ascending, with hairs in 1 line. Nov-Jun. Damp shady habs..*Greater Chickweed* **Stellaria neglecta**

Lvs 6-25(30)mm, ovate, acute with apiculate white hydathode, the lower long-petiolate (to 2cm or equal to lf length), the upper ± sessile, pale green, sparsely hairy above or hairless both sides, without opaque reticulations, with stomata both sides. Petiole ciliate, flattened. Stems to 25(30)cm tall, 0.8-2mm diam, ± erect to straggling, with hairs in 1(2) lines. All yr (summer or winter ann)..*Common Chickweed* **Stellaria media**

Lvs 3-7mm, oblong or ovate, acute with apiculate white hydathode, the lower long-petiolate, the upper short-petiolate, pale yellow-green, hairless, with opaque reticulations, with stomata both sides. Petiole ciliate, flattened. Stems rarely >10cm tall, 0.4-0.7mm diam, weakly mat-forming, with hairs in 1 line. Oct-May. Dry well-drained soils ....*Lesser Chickweed* **Stellaria pallida**

Group LS – *Lvs usu long-ciliate nr base (hairs long, wavy, septate and longer than elsewhere on lf), usu in basal rosette (often dead at fl), usu spathulate or long-cuneate, usu obvolute when young, with stomata both sides (below only in* Silene dioica *and, rarely,* Lychnis flos-cuculi*). Petioles connate at base, with 1(3) vb's. Stems, if present, round, may pull apart at swollen nodes revealing stele. Hairs septate. (*Silene vulgaris, Valerianella *and* Arenaria *may key out here).* ❶ *Plant with prostrate stems (erect in other spp).* ❷ *Lf margins recurved*

■Bi or per. Lvs with 2° veins indistinct (but distinct and translucent in *Silene latifolia* ssp *alba*)

Basal lvs in persistent (or aerial) rosette on a short woody stock, narrowly spoon-shaped (lanc-spathulate). Sunny dry habs

Lvs (2)4-8cm, mucronate, narrowed to a long petiole, usu hairless (occ with dense short hairs), minutely ciliate (cilia often unequal). Stem lvs 0-1 pr. Stems to 60cm, not purplish, minutely retrorsely white-hairy below, viscid above. All yr. E Anglian Brecks...*Spanish Catchfly* **Silene otites**

Lvs 4-7cm, obtuse to acute, narrowed to a long petiole, sparsely hairy both sides, ciliate. Upper stem lvs narrowly lanc, ± sessile. Stems to 60cm, not purplish, viscid-hairy above. All yr ....*Nottingham Catchfly* **Silene nutans***

Lvs 4-6cm, ± acute to obtuse, narrowed to a long petiole, velvety with dense short hairs, ciliate. Upper stem lvs linear-lanc, sessile. Stems to 60cm, purplish at base, densely retrorsely hairy, viscid above. Plant more upright and tufted than *S. nutans*. All yr. VR alien ....*Italian Catchfly* **Silene italica**

Basal lvs often dying at fl (never in an aerial rosette on a woody stock), often oblanc

Shady habs

Lvs basal and/or on stems, 4-12 x 3-6cm, (ob)ovate, acute or acuminate, rounded to cuneate at base, narrowed into a long winged petiole, obvolute when young, often undulate, shiny green above, hairy at least above, with translucent to opaque 2° veins raised below, with 3° veins translucent, other veins visible as an opaque network. Upper stem lvs short-stalked to ± sessile. Petiole to 12cm, long-ciliate nr base, with 1(3) non-elastic vb's. Stems to 100cm, erect, sparsely hairy (usu weakly glandular-viscid above), hardly swollen at nodes, hollow. Bi or per. All yr. Shady habs, coastal cliffs, mtn scree. PL 11.............*Red Campion* **Silene dioica**

Lvs all on stem, 4-6cm, ovate, acuminate, ± cordate at base, valvate when young, usu undulate, light green, hairless or hairy either side, usu ciliate, with translucent or opaque 2° veins sunken above, with submarginal vein, other veins visible as an opaque network, with stomata absent or obscure above. Upper lvs ± clasping. Petiole (of lower lvs) 2-4cm, with long wavy hairs, channelled, with 1 elastic vb. Stems to 60cm, decumbent to ascending, weak, sparsely hairy, swollen above nodes, with strong elastic stele. Stoloniferous per. Mar-Oct. Damp woods, N & W Br..........*Wood Stitchwort* **Stellaria nemorum**

Sunny wet habs, occ in damp woods

Lvs mostly basal (often dead at fl but new rosettes soon developing), 4-10 x 0.6-1.5cm, oblanc, acute, narrowed to long indistinct petiole, hairless, long-ciliate at base, with midrib raised below, with opaque network of veins obscurely raised above. Stems to 75cm, erect, retrorsely hairy (occ sparsely so), slightly rough, angled, hollow. All yr........*Ragged-Robin* **Lychnis flos-cuculi***

Sunny dry habs

Plant usu of lowlands. Stems, if present, (±) erect

Basal lvs usu persisting at fl, 4-12 x 3cm, lanc, acute with purple apical hydathode, cuneate at base, often undulate, densely hairy both sides, ciliate, with translucent 2° veins raised below, other veins visible as an opaque network. Stem lvs similar to basal lvs, acuminate, sessile, occ 3-pli-veined. Petiole to 5cm, not winged, with long wavy cilia, with flat arc vb. Stems several, to 100cm, softly spreading-hairy, often glandular-viscid above. Short-lived per (occ ann or bi). All yr..........*White Campion* **Silene latifolia** ssp **alba**

Basal lvs dying by fl, 5-10 x 0.5-1.5cm, oblanc, obtuse, cuneate at base, not undulate, hairless or sparsely hairy on midrib below, long-ciliate nr base, scabrid-ciliate or smooth nr apex, with 1-3 translucent veins usu raised below (laterals opaque nr apex). Stem lvs 5-10cm x 3-8mm, linear-lanc, acute, often with antrorsely adpressed hairs both sides, 1-3 parallel veins (usu obscure). Stems 1-several, to 60cm, hairless or hairy for upper ½ of each internode, occ angled, hollow. Bi. All yr (usu). VR.......*Deptford Pink* **Dianthus armeria**

Plant usu of lowlands. Stems (±) prostrate. ❶. VR alien

Lvs all on stems, 0.6-2.5 x 0.3-1.4cm, ovate to oblanc, usu obtuse, narrowed to indistinct 5-7mm petiole, hairless, obscurely pinnate- or 3-pli-veined, with midrib raised below. Stems reddish, shortly scabrid-papillate, reddish. Per. All yr. VR alien ....*Rock Soapwort* **Saponaria ocymoides**

Plant of mtn/upland rocks. Stems (±) erect. VR

Lvs 3-12cm x 4-6mm, oblanc, acute or acuminate, narrowing to long petiole, often purplish, hairless, wavy long-ciliate proximally, with midrib visible but 2° veins indistinct. Stems to 45cm, purplish at least below nodes, hairless, v viscid beneath nodes, hollow. Per. All yr ....*Sticky Catchfly* **Lychnis viscaria**

Lvs 1.5-5cm x 5mm, oblong-lanc to linear, acute, sessile, with v minutely crenulate hyaline margins, hairless, v sparsely ciliate at base, with midrib raised below. Stems to 20cm, hairless. Per. All yr. Sch8..*Alpine Catchfly* **Lychnis alpina**

■Ann

Lvs 0-1-veined or obscurely pinnate

Lvs with cilia equal length along margins, margins revolute or recurved. ❷

Basal lvs in rosette, soon dying, to 4 x 0.5cm, oblanc, ± acute, narrowed to indistinct petiole, minutely softly hairy both sides, with midrib translucent only. Stem lvs linear, acuminate. Stem 1, usu 10-35cm, often branched at base, weakly retrorsely minutely hairy, densely glandular and viscid nr apex, swollen below purplish nodes, solid. Mar-Jul ....*Sand Catchfly* **Silene conica**

Lvs with cilia longest nr base, margins flat

Basal lvs usu absent, 5-15 x 0.5-2cm, linear-lanc, acute, narrowed to indistinct petiole, with adpressed and/or spreading white hairs both sides, with long (2-8mm) wavy cilia at least nr base, becoming adpressed-ciliate along lf, with midrib raised below, with ± parallel indistinct 2° veins slightly raised both sides. Stems 1-several, 30-100cm, with dense adpressed or spreading white hairs, hollow. (Oct) May-Sept. Arable, wildflower mixes ....*Corncockle* **Agrostemma githago**

Basal lvs dead at fl, 1-3(5) x 0.3-1.5cm, spathulate or oblanc, apiculate, narrowed to indistinct petiole, hairy both sides, with short cilia (longest at base), with midrib raised below, with obscure 2° veins. Stem lvs lanc to linear, sessile. Stem usu 1, to 40cm, hairy (glandular at least above), ± solid. Oct-Jul...*Small-flowered Catchfly* **Silene gallica**

Lvs 3-5-pli-veined at base, otherwise pinnately-veined

Lvs 5-10 x 5cm, obovate or ovate-lanc, narrowed to petiole-like base, 2° veins slightly raised below, with scattered hairs. Upper lvs narrowly oblong-lanc, v acute, sessile, with clasping petiole-like base. Stem usu 1, to 50cm, usu unbranched at base, glandular-viscid. May-Oct ....*Night-flowering Catchfly* **Silene noctiflora**

Lvs 3-pli-veined

Lvs 3-5cm, spathulate to ovate-lanc, petiolate, hairy. Upper lvs lanc, acute. Stems 1-several, 20-60cm, repeatedly forked, nonglandular-hairy. May-Oct............*Forked Catchfly* **Silene dichotoma**

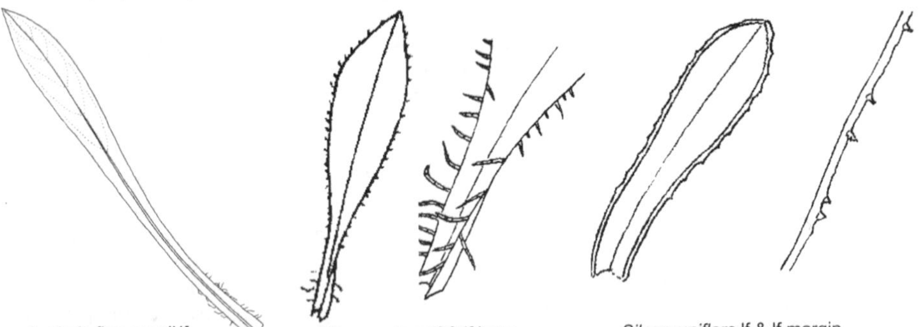

Lychnis flos-cuculi lf   *Silene nutans* lf & lf base   *Silene uniflora* lf & lf margin

Group LT – *Lvs with v long petiole (>> lf). Evergreen creeping per with stout rhizomes*

▪Lvs 2.5-10cm diam, reniform, wider than long, v obtuse, cordate at base, shiny green both sides, shortly hairy below and/or on veins above, opaquely net-veined, occ with 2° veins weakly translucent but not raised below, with stomata below only. Petiole with long wavy septate hairs when young (hairs with withering tips), channelled, with 3 vb's. Scale lvs 2(3), alt, ovate, with sessile red glands, ciliate. Stems prostrate, shortly hairy. Terminal shoot abortive or with fl. PL 13 .... *Asarabacca* **Asarum europaeum**

Group LU – *Lvs alt above. Stems solid, without swollen nodes or elastic stele, not connate (exc* Epilobium palustre*).* ❶ *Stems ridged*

▪Stems with patent septate hairs

Per (often short-lived). All yr

Lvs 3-5(7) x 1-2.5cm, ovate-lanc or linear-lanc, obtuse to acute, narrowed to indistinct petiole, often thick and snapping audibly, dark green above, paler below, glandular-ciliate at least when young, with midrib raised below, occ with obscure 2° veins, with stomata at least below. Stems 30-80cm, often woody at base, glandular-hairy above, hairless below, with stomata all around .... *Snapdragon* **Antirrhinum majus**

Ann. Apr-Oct

Lvs 30-50 x 2-4mm, linear-lanc to narrowly elliptic, ± acute, narrowed to indistinct petiole, thick, often with recurved margins, dark green above, sparsely glandular-hairy or glandular-ciliate, with 1 obscure tooth per side, obscurely 3-pli-veined (other pinnate veins visible), with stomata below and often above. Stems 20-50cm, simple or branched, hairy (usu glandular-hairy above) .... *Weasel's-snout* **Misopates orontium**

Lvs 10-25 x 2.5-4.5mm, linear to oblong-lanc, ± obtuse, narrowed to short petiole, ± thin, without recurved margins, dull dark grey-green to ± glaucous above, often ciliate, entire, obscurely 3-pli-veined, with midrib raised below and sunken above, with stomata both sides. Stems 8-25cm, branched, glandular-hairy........................ *Small Toadflax* **Chaenorhinum minus**

▪Stems with antrorse crisped unicellular hairs (rarely hairless). Per

Stem ridged ❶.. *Heath Milkwort* **Polygala serpyllifolia**

Stem round.. (*Epilobium palustre*) Go to **EPIL**

Group LV – *Lvs pinnate-veined (occ weakly so). Petioles connate at base (exc* Ludwigia x kentiana, Epilobium palustre*). (Exceptionally,* Saponaria ocymoides, Antirrhinum majus, Prunella *and* Valerianella *may key out here)*

■Plant with minute stipule-like extra-floral nectaries (often soon falling)

Lf veins often fading nr margins (eucamptodromus). Aquatic. R alien

Lvs 2-5cm, often oblanc, rolled when young, translucent, sparsely hairy. Stipules tiny, heart-shaped, brown, occ absent. Plant often greenish (reddish in *L. palustris*). *L. palustris* x *repens* ....*False Hampshire-purslane* **Ludwigia** x **kentiana**

Lf veins always anastomosing. Dry shady habs

Lvs 4-10cm, ovate, acuminate, truncate to slightly cordate at base, thin, dull green above, shiny pale green below, ± hairless to minutely hairy, often ciliate, with 4-7 prs of 2° veins anastomosing nr margins. Petiole 2-4cm, hairy both sides, channelled. Stems 20-60cm, ± sparsely glandular-hairy, round, with lower nodes swollen. Rhizomes slender, white, brittle ....*Enchanter's-nightshade* **Circaea lutetiana***

■Plant without extra-floral nectaries

Plant with spreading unicellular hairs (hairs often on minute swollen-bases). Basal rosette usu present

Lvs with narrow cartilaginous margins turning reddish. Petiole often purplish nr base

Basal lvs to 15(30)cm, elliptic, with apical hydathode occ purplish (others green and sunken along margin), obvolute when young, stiffly hairy above, often ciliate, (±) net-veined (esp below), with white midrib raised below (occ purple), with spiral fibres, with stomata both sides. Petiole with 3(5) vb's and spiral fibres. Stems to 100cm, roughly retrorsely hairy on lowest internode, antrorsely adpressed-hairy above. All yr....*Devil's-bit Scabious* **Succisa pratensis***

Lvs without cartilaginous margins. Petiole not purplish nr base

Basal lvs to 15(30)cm, elliptic to oblanc, hairy both sides (more hispid above), with pale green submarginal hydathodes visible above (the terminal visible below) and usu at least some traces of crenate teeth, the upper rosette lvs occ lyrate-pinnatifid, with midrib white above and green below, with stomata both sides. Petiole hairy, ciliate, often hollow, with 3(5) vb's and spiral fibres. All yr. PL 18..........*Field Scabious* **Knautia arvensis**

Basal lvs 5-15cm, obovate, sparsely or minutely hairy to hairless, the upper rosette lvs ± pinnately lobed, with long large end-lobe and free lateral lobes, with stomata both sides. Petiole hairless, ciliate, solid, with 3-5 vb's and spiral fibres. All yr. Calc turf ....*Small Scabious* **Scabiosa columbaria**

Plant with septate, crisped or glandular hairs. Basal rosette absent (exc *Silene latifolia* ssp *alba*)

Stem without elastic stele ....(*Epilobium palustre*) Go to **EPIL**

Stems with elastic stele

Petiole (of lower lvs) 2-4cm

Lvs 4-6cm, ovate, acuminate, ± cordate at base, valvate when young, usu undulate, light green, hairless or hairy either side, usu ciliate, with translucent or opaque 2° veins sunken above, with submarginal vein, other veins visible as an opaque network, with stomata absent or obscure above. Upper lvs ± clasping. Petiole with long wavy hairs, channelled, with 1 elastic vb. Stems to 60cm, decumbent to ascending, weak, sparsely hairy, round, swollen above nodes. Stoloniferous per. Mar-Oct. Damp woods, N & W Br ....*Wood Stitchwort* **Stellaria nemorum**

Petiole (of lower lvs) <1cm

At least some lvs basal. Short-lived per (occ ann or bi)

Basal lvs 4-12cm, lanc, acute with purple apical hydathode, narrowed to petiole, often undulate, densely hairy both sides, with translucent 2° veins raised below, other veins visible as an obscure opaque network. Upper lvs acuminate, sessile. Petiole to 5cm, not winged, with long wavy cilia, with flat arc vb. Stems to 100cm, swollen at nodes, softly spreading-hairy, often glandular-viscid above. All yr ....*White Campion* **Silene latifolia** ssp **alba**

All lvs on stems. Per

Stems clambering or sprawling, to 2m, much-branched (often at 90°), tough, with short retrorsely curved septate hairs. Lvs 5-7 x 3.5cm, ovate, acuminate, rounded at base, slightly undulate, yellow-green, sparsely hairy, ciliate, with weakly translucent 2° veins forming submarginal vein, opaquely net-veined, with stomata below only. Petiole c6mm, flattened, channelled, ciliate. Apr-Oct. Dry habs, R alien ....*Berry Catchfly* **Cucubalus baccifer**

Stems decumbent to erect, to 1m, branched, brittle, usu with glandular septate hairs above. Lvs 2-5 x 3cm, ovate, acuminate, ± cordate at base, usu undulate, mid-green, sparsely hairy to hairless, occ glandular-ciliate, with weakly translucent 2° veins forming submarginal vein, opaquely net-veined, with stomata both sides. Petiole short to ± absent. Apr-Oct (occ overwintering as prostrate stems). Damp habs.......*Water Chickweed* **Myosoton aquaticum**

Stems erect, to 0.8m, sparsely or not branched, hairless or hairy. Lvs to 6 x 1.5cm, obovate to oblong-lanc, apiculate, not or hardly undulate (margins scabrid-ciliate to smooth), dull grey-green, occ hairy, with weakly translucent 2° veins (occ slightly raised both sides), often with translucent dots (HTL), with stomata below and often above. Petiole ± absent. Apr-Oct. Dry habs..*Bladder Campion* **Silene vulgaris**

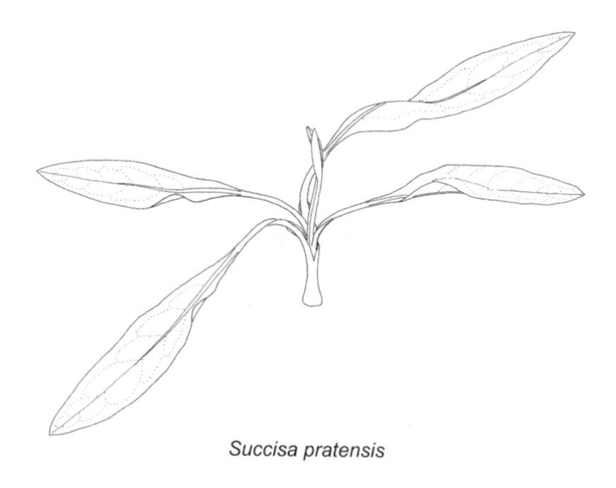

Succisa pratensis

Group LW – *Lvs (0)1-3(5)-veined. Stems (or lvs) with minute hairs (often retrorse), usu unicellular, rarely glandular.* ❶ *Plant (esp roots) aromatic*

■Lvs with 0-5 parallel veins (pinnate veins absent)

Lvs with 1(3) large lime-exuding pores at apex (often white). Stems with non-elastic stele. Hairs multiseriate (x100)

Lvs 4-5 x 2-2.5mm, oblong-linear, sessile, often in fascicles in lf axils, imbricate at least nr stem apices, ciliate with stiff white hairs, with stomata both sides. Petiole indistinct, ± connate at base. Stems to 1mm diam, usu prostrate, sparsely hairy, reddish, round. Mostly mtn rocks above 300m..*Purple Saxifrage* **Saxifraga oppositifolia**

Lvs without pores. Stems with elastic stele. Hairs unicellular (x10)

Stems with 2 dense lines of hairs and scattered hairs all around (hairs may be retrorsely curved)

Lvs 6-25mm, ovate, acute to cuspidate, the lower stalked, the upper ± sessile, all connate at base, applanate when young, translucent-dotted, sparsely hairy, ciliate (cilia longer nr petiole), 3(5)-pli-veined. Stems slightly swollen at nodes. Ann. Nov-Jul ....*Three-nerved Sandwort* **Moehringia trinervia**

Stems with retrorse stiff white hairs (equally distributed around stem, often like minute scabrid-cilia), or hairless. Lvs 3-8mm, ± acute, ± sessile, connate at base, occ translucent-dotted, scabrid-ciliate (cilia usu longer nr base), with 0-5 obscure veins, with stomata both sides

Ann, with few non-fl shoots

Widespread. Petals shorter than sepals. Lvs 3-8 x 2-4mm, broadly ovate to ovate-lanc, thick, greyish-green, with sparse white hairs either side, ± ciliate, 0-5-veined. Stems rarely with glandular hairs. Oct-Aug..............................*Thyme-leaved Sandwort* **Arenaria serpyllifolia**

Yorks (Eng). Petals longer than sepals. Lvs 4-6 x 1-2mm, narrowly obovate, thick, pale green, hairless, sparsely ciliate at least nr base, 0-1(3)-veined. Stems never with glandular hairs. Occ (bi). Apr-Oct. Sch8.................*English Sandwort* **Arenaria norvegica** ssp **anglica**

Per, with numerous branching non-fl shoots at base. Petals longer than sepals

NW Scot. Lvs 3-5 x 1-2mm, obovate, thick, dark green, hairless, sparsely ciliate at least nr base, 0-1-veined. Occ bi. All yr. Sch8...*Arctic Sandwort* **Arenaria norvegica** ssp **norvegica**

NW Ire. Lvs 3-7 x 1-2.5mm, oblong to lanc, thin, mid-green, hairless, usu ciliate only nr base, 0-1-veined (4-5(6)-veined when dry). All yr.......................*Fringed Sandwort* **Arenaria ciliata**

■Lvs with pinnate veins (often obscure)

Plant (esp roots) weakly aromatic. Hairs unicellular. Ann. ❶......................(*Valerianella*) Go to **VAL**

Plant odourless. Hairs septate. Per

Stems ± prostrate (occ ascending) and rooting at lower nodes

Lvs 5-20 x 15mm, ovate, obtuse, ± sessile, ± shiny light green above, hairless, ciliate, ± entire to weakly crenate nr apex, with few indistinct or obscure 2° veins. Stems 0.7-1.5mm diam, antrorsely crisped-hairy................................*Thyme-leaved Speedwell* **Veronica serpyllifolia**

Stems ± prostrate but not rooting at nodes

Lvs 6-25 x 3-14mm, ovate to oblanc, usu obtuse, narrowed to indistinct petiole, connate at base, obvolute when young, hairless, shortly septate scabrid-ciliate (cilia often longer nr base), mid-green, not scabrid-ciliate, obscurely pinnate-veined or weakly 3-pli-veined (2° veins weakly translucent to ± opaque), with midrib raised below, with stomata both sides. Stems reddish, shortly scabrid-papillate, pulls apart at swollen nodes revealing elastic stele. VR alien..*Rock Soapwort* **Saponaria ocymoides**

Lvs 10-25 x 2-5mm, the lower narrowly oblanc and obtuse to ± acute, upper linear-lanc and acute, all connate at base, valvate when young, dark bluish-green above, scabrid-ciliate, 1(3)-veined (pinnate veins obscure), the midrib scabrid below, with stomata both sides. Stems green, scabrid-papilllate to shortly septate-hairy, ± swollen at nodes. Loosely tufted. All yr. Dry turf..*Maiden Pink* **Dianthus deltoides**

Group LX – Cerastium. *Lvs 1-veined, with white hydathode below at apex, ± sessile at connate base, applanate-valvate when young, with faint submarginal vein, with midrib sunken above raised below, septate-hairy both sides (the hairs with a 'knee joint' at each septa), with stomata both sides. Stems with spreading septate hairs (often long), round, with elastic stele. (*Misopates orontium *may key out here)*

▪Hairs v dense, making plant appear greyish-white or woolly

Lowlands. Lvs with short grey-white matted woolly hairs ....*Snow-in-summer* **Cerastium tomentosum**

Mtns. Lvs with long (2-4mm) white wavy hairs....................*Alpine Mouse-ear* **Cerastium alpinum**

▪Hairs not making plant appear greyish or white

Per. Basal lvs often persisting to fl. All yr

Stems mostly with equal retrorse 0.2mm hairs. Rhizomatous

Lvs 10-20 x 3-6mm, linear-lanc, hardly narrowed to base, shortly softly hairy (± glandular nr infl), without long hairs. Stems often with axillary lf-clusters below, green to purplish, slightly swollen below nodes...*Field Mouse-ear* **Cerastium arvense**

Stems with unequal (often spreading) 0.5-1mm hairs. Rhizomatous or stoloniferous

Lvs 10-25 x 2-10mm, oblong to spathulate, often dark grey-green, sparsely hairy both sides (but hairier above). Stems branched from base, sparsely to densely hairy with few or no glandular hairs..*Common Mouse-ear* **Cerastium fontanum**

Stems with equal (straight) 0.5(1)mm hairs

Rhizomatous. Stems with or without short glandular hairs, with (3)8-septate whitish (to yellowish) hairs. Lvs to 12 x 6mm, narrowly elliptic, usu obtuse, mid-green, not glandular-hairy. Mtn rocks..*Arctic Mouse-ear* **Cerastium arcticum**

Tufted. Stems glandular-hairy (often also with short nonglandular hairs), with 4-6(8)-septate whitish (to yellowish) hairs. Lvs to 10 x 7mm, broadly obovate to ± orb, v obtuse, v dark green, turning reddish then black, glandular-hairy. Shetland...*Shetland Mouse-ear* **Cerastium nigrescens**

Ann (with slender taproot). Basal lvs dead by fl. Oct-Jun

Lvs hairy or densely so below, with long hairs protruding past apex

Lvs 20-30 x 5-10mm, elliptic, occ retuse, the lower lvs narrowed to a petiole-like base, pale yellow-green, hairy (occ glandular nr infl). Stems to 25cm, stout, green, densely hairy with ≥0.5mm hairs (glandular nr infl). Sunny habs..........*Sticky Mouse-ear* **Cerastium glomeratum**

Lvs 4-15 x 1.5-5(7)mm, lanc, acute, all narrowed to a petiole-like base, pale mid- to grey-green, sparsely long-hairy. Stems to 25cm, stout, simple, purplish below, shaggy with spreading-ascending nonglandular hairs. VR......*Grey Mouse-ear* **Cerastium brachypetalum**

Lvs hairless or sparsely hairy below, with short hairs barely protruding past apex (apex often purplish)

Bracts with broadly scarious margins (scarious apex ≥⅓ total length). Petals with veins not branched. Lvs 5-18 x 5mm, narrowly oblanc to elliptic-oblong, dark green. Stems often 1, usu <4cm tall, decumbent, branched, green, with spreading c0.5mm hairs (occ glandular above). Dry open sandy or calc habs...........................*Little Mouse-ear* **Cerastium semidecandrum***

Bracts with v narrow scarious margins. Petals with veins not branched. Lvs 5-18 x 5mm, oblanc to oblong-ovate, dark green, often purple-bordered. Stems often 1, usu <8cm tall, often ± erect, branched, green, with dense 0.2-0.5mm glandular hairs. Dry open sandy habs ....*Sea Mouse-ear* **Cerastium diffusum**

Bracts with v narrow scarious margins. Petals with branched veins. Lvs 5-15 x 3-6mm, elliptic or ovate-oblong, reddish. Stems often 1, 2-12cm tall, erect or ascending, branched, usu reddish, with abundant glandular and sparse nonglandular hairs. Bare calc slopes ....*Dwarf Mouse-ear* **Cerastium pumilum**

Cerastium semidecandrum lf apex (showing hydathode)

Key to Groups in Division M

(Lvs whorled or pseudowhorled)

- ■Small evergreen tree, shrub, or subshrub..**MA**
- ■Herb
  - Stipules or stipoid glands present..**MB**
  - Stipules absent
    - Stems square (or 4-furrowed), with whorls of 4-12 lvs..**MC**
    - Stems round or ridged
      - Stems with 1 whorl of lvs..**MD**
      - Stems with several whorls of lvs
        - Lvs with 0-1(3) parallel veins..**ME**
        - Lvs with pinnate veins or >3 parallel veins..**MF**

Group MA – *Small evergreen tree, shrub, or subshrub (woody at least at base).* ❶ *Lvs spine-tipped*

■Lvs pseudowhorled

Tall deciduous shrub to 2m. Twigs glandular-hairy..............*Yellow Azalea* **Rhododendron luteum**

Tall evergreen shrub to 3m. Twigs hairless. PL 9...........*Rhododendron* **Rhododendron ponticum**

Low evergreen shrub to 0.4m. Twigs minutely glandular-hairy when young, often reddish, weakly ridged or round. Lvs 4-6 x 0.8-1.2mm, linear, obtuse, strongly revolute (underside visible only as white line), hairless, glandular-ciliate when young, not net-veined, with stomata below only. Petiole v short. Buds minute. All yr. Mtn moorland, hths.................*Crowberry* **Empetrum nigrum***

■Lvs opp (decussate, appearing 4-whorled)

Lvs 1-2mm, peg-like, sessile, with 2 short downward lobes at base (occ obscure), ± imbricate, strongly revolute (appearing trigonous), hairless or hairy above, with obscure veins, with stomata below only. Shrub to 60cm. Hths, moors...*Heather* **Calluna vulgaris**

■Lvs in whorls of 3(4), usu spreading

Young twigs hairless. (A conifer)

Lvs always 3-whorled, (5)10-20mm, subulate, spine-tipped, sessile, glaucous above, green below, hairless, weakly keeled below. Small tree or shrub to 5m. ❶ ....*Common Juniper* **Juniperus communis***

Young twigs glandular-hairy

Lvs 1-3mm, ovate, weakly revolute, underside appearing whitish but midrib green, hairless, glandular-ciliate, with stomata below only. Twigs reddish. Dwarf shrub to 60cm ....*Dorset Heath* **Erica ciliaris**

Young twigs nonglandular-hairy

Lvs 1.2-2.2cm wide, elliptic or oblong

Lvs 3-whorled (occ opp), to 5cm, narrowly elliptic, ± obtuse but apiculate, revolute when young, with flat or slightly revolute margins at maturity, shiny green above, glaucous-pruinose below (at least when young), hairless or minutely hairy when young (hairs occ confined to midrib below), with short red glands below when young, with midrib raised below, with hardly translucent 2° veins often ± raised above. Petiole to 8mm, channelled. Young twigs minutely hairy, brittle, apricot-scented, ± round, without lenticels or interpetiolar ridges. Buds small, with 2 scales. Shrub to 1m..*Sheep-laurel* **Kalmia angustifolia**

Lvs 0.5mm wide, linear

Tall shrub, to 2(3)m. Lvs 5-9mm, acute, strongly revolute, with stomata below only. Twigs white-hairy

Hairs on young twigs dendritic or ciliate. Lvs rarely pseudowhorled..*Tree Heath* **Erica arborea**

Hairs on young twigs smooth. Lvs occ pseudowhorled....*Portuguese Heath* **Erica lusitanica**

Dwarf shrub, to 0.6m

Lvs 5(7)mm, strongly revolute, dark hairless green above, white-hairy below (often hidden by revolute margins). Buds minute, with 2-3 scales.....................*Bell Heather* **Erica cinerea**

■Lvs in whorls of 4-5(6)

Lvs hairy (often glandular)

Lvs erecto-patent esp nr twig apex, always 4-whorled, 2-4 x 0.5-1mm, linear, strongly revolute, grey-hairy, glandular-ciliate, with stomata below only. Twigs hairy when young (hairs often glandular). Damp hths...*Cross-leaved Heath* **Erica tetralix***

Lvs spreading horizontally from twigs, always 4-whorled, 2-5 x 2mm, oblong-lanc, weakly revolute, dark green and hairless above, white-hairy below, with long white cilia (usu gland-tipped), with stomata below only. Twigs ± hispid when young. Bogs, R, Ire ....*Mackay's Heath* **Erica mackaiana***

Lvs hairless

Lvs 6-12mm wide, ovate to elliptic-lanc, not revolute

Lvs 4-6-whorled, 1.5-6cm, tough, with strongly retrorsely scabrid cartilaginous margins, shiny dark green above, obscurely pinnate-veined, with raised midrib below (rough with retrorse prickles), with stomata below only. Stems clambering, woody nr base, hairless, sharply 4-angled, the angles rough with retrorse prickles..........................*Wild Madder* **Rubia peregrina**

Lvs to 1mm wide, linear, strongly revolute
Some lvs 5-whorled, 5-10mm, acute, bright green. Young twigs always hairless. Lizard, R hortal...*Cornish Heath* **Erica vagans**
All lvs 4-whorled, 5-6(8)mm, acute, dark green at maturity. Young twigs occ hairy
Fls Mar-Jun. Bogs, R, Ire. Hortal..*Irish Heath* **Erica erigena***
Fls Nov-Jun. Hortal. *E. carnea* x *erigena*...................*Darley Dale Heath* **Erica** x **darleyensis**

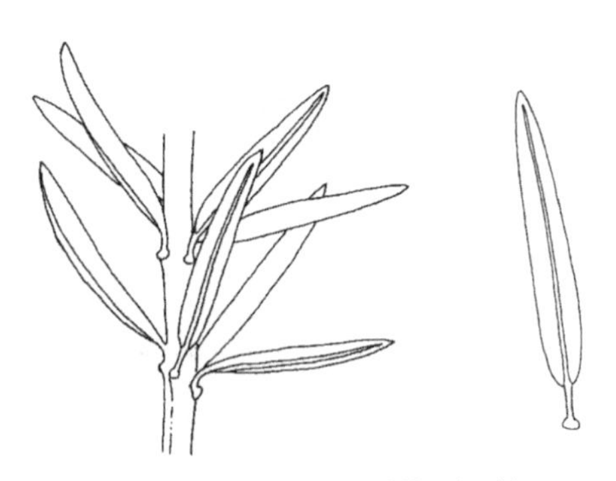

Erica erigena & lf underside

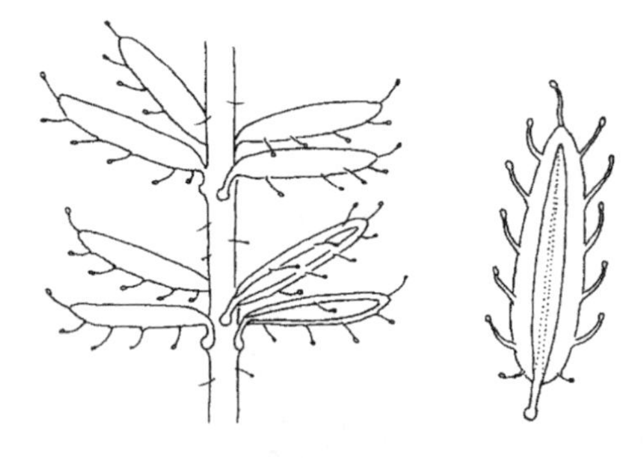

Erica mackaiana & lf underside

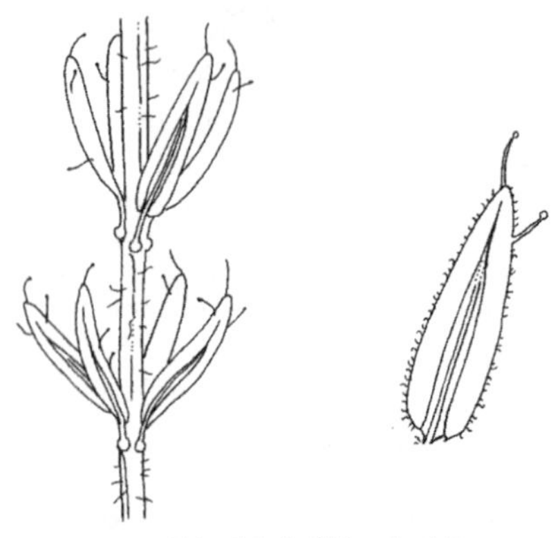

Erica tetralix & lf underside

Group MB – *Stipules or stipoid glands present. Hairless ann. (*Spergula arvensis *may key out here but is hairy and the lvs are in fascicles)*

- Stems hollow. Stipoid glands present..(*Impatiens glandulifera*) Go to **IMP**
- Stems solid. Stipules present

  Lvs 8-13 x 3mm, obovate, narrowed to short petiole, the upper in whorls of 4 (usu unequal), with erose margins, with translucent midrib and obscure pinnate veins, with stomata both sides. Stipules lanc, acuminate, scarious, silvery. Stems to 25cm, 1mm diam, erect to prostrate, rough with low blunt papillae on angles. Mostly SW Br....*Four-leaved Allseed* **Polycarpon tetraphyllum**

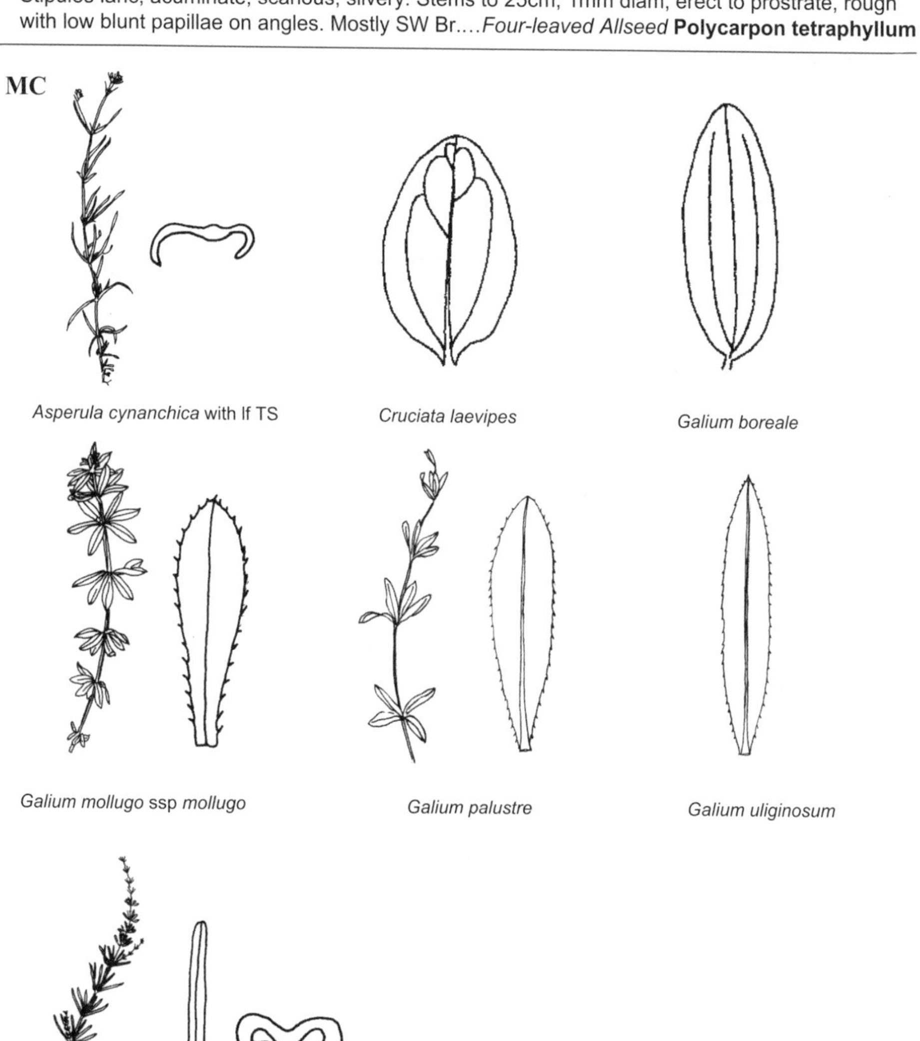

Asperula cynanchica with lf TS

Cruciata laevipes

Galium boreale

Galium mollugo ssp *mollugo*

Galium palustre

Galium uliginosum

Galium verum with lf & lf TS

Group MC – *Stems usu climbing or scrambling, with several to many whorls of 5-12 lvs, branching at whorls without axillary lvs, with hollow stele. Lvs often with prickly margins, with midrib raised below, often with colleters (minute papillae-like glands) at base in axils. Hairs, if present, unicellular.* ❶ *Hay-scented when bruised*

■Lvs clearly 3-veined or pinnate-veined, in whorls of 4

Lvs dark yellow-green, 12-25 x 4-10mm, elliptic-ovate, densely hairy, with smooth margins, usu pinnate-veined, with stomata below only. Stems hairy, 4-furrowed. Strongly rhizomatous. All yr ....*Crosswort* **Cruciata laevipes***

Lvs dark green (black when dry), 15-40 x 3-10mm, linear to elliptic-lanc, hairy or hairless, occ with antrorsely scabrid margins, usu 3-veined, with stomata both sides. Stems hairy at least below, 4-angled. Rhizomatous. N & W Br...*Northern Bedstraw* **Galium boreale***

■Lvs 1-veined (rarely obscurely pinnate-veined or upper lvs 3-veined), in whorls of 4-12

Lf margins with antrorse or patent prickles (or absent) at least nr midpoint. Stems often smooth

Lvs revolute and/or unequal in length, with few or no prickles on margins

Lvs 6-12-whorled, usu ± equal length, to 25 x 0.5-2mm, linear or obovate, usu mucronate, with revolute margins (winter lvs flat), dark green (black when dry), rough above with prickles (often tiny), hairy below, with midrib smooth, with stomata below only. Stems smooth, hairless to sparsely hairy, bluntly 4-angled, without stomata. All yr.....*Lady's Bedstraw* **Galium verum***

Lvs 4(6)-whorled, usu v unequal in length, to 10(20) x 0.8mm, linear-lanc, acute or mucronate, with flat or weakly revolute margins, mid-green, smooth above, hairless both sides, with midrib not scabrid, with stomata both sides. Stems often rough with short hairs (occ retrorse), 4-angled, with stomata all sides. All yr.........................*Squinancywort* **Asperula cynanchica***

Lvs not or rarely revolute, ± equal length

Damp sunny habs

Lvs 4-6-whorled, to 8 x 0.5-1mm, linear, ± acute to minutely apiculate, hairless, with stomata both sides. Stems 0.8mm diam, ± prostrate, smooth or occ slighly rough on angles. Rhizomatous. VR...*Slender Marsh-bedstraw* **Galium constrictum**

Dry open or shady habs

Stems smooth. Per. All yr

Lvs 6-15mm wide. ❶

Lvs 6-8-whorled, held patent, to 40mm, lanc or elliptic, ± cuspidate, with 25-60 antrorse prickles per side, hairless, not black on drying, occ with visible pinnate veins and sparse antrorse prickles on midrib below, usu with stomata below only. Stems erect, unbranched, hairless (exc below lf whorls), ending in flat-topped whorl of lvs. Rhizomatous. All yr..*Woodruff* **Galium odoratum**

Lvs <8mm wide

Lvs 6-whorled, 5-11 x 1-3mm, obovate-oblanc, acute, black on drying, hairless, with (0)3-7 prickles (often weakly curved) per side. Stems mat-forming, much-branched, 0.4-1mm diam, hairless, with nodes hardly swollen. Acidic gsld ....*Heath Bedstraw* **Galium saxatile**

Lvs 6-10-whorled, 5-30 x 2-7mm, oblong-oblanc, cuspidate or mucronate, green on drying, hairless, usu with 10-20 curved prickles per side. Stems clambering to ± prostrate, much-branched, 0.8-2mm diam, hairless or hairy, often with ± swollen nodes

Lvs upswept or patent..........................*Hedge Bedstraw* **Galium mollugo** ssp **mollugo***

Lvs downswept, ± adpressed to stem..*Hedge Bedstraw* (ssp) **Galium mollugo** ssp **erectum**

Stems rough

Per, often mat-forming (with short rhizomes). All yr

Lvs 6-9-whorled, 1-3cm x 4mm, linear-lanc, acute, with stomata below only. Stems to 80cm, 2mm diam, sprawling. VR alien...............*Caucasian Crosswort* **Phuopsis stylosa**

Ann. Mar-Oct

Lvs usu 4-5-whorled below (soon withering), 5-6-whorled above, usu spreading, 7-12(18) x 2.5-3.5mm, obovate to elliptic, acute to cuspidate, with cartilaginous margins (occ weakly revolute), not black on drying, hispid-hairy above, occ 3-pli-veined at least proximally, with midrib scabrid (or hairy) below, with stomata sparse above. Stems to 1mm diam, prostrate, rough with retrorse or spreading stiff hairs ....*Field Madder* **Sherardia arvensis**

Lvs 6-10-whorled, spreading then reflexed, 3-5 x 1-1.5mm, linear-oblong, shortly mucronate, without cartilaginous margins, black on drying, hairless, always 1-veined, often with midrib scabrid below, often with stomata below only. Stems to 0.8mm diam, ascending, rough with retrorse prickles, hairless.........*Wall Bedstraw* **Galium parisiense**

Lf margins with retrorse prickles in proximal ⅓ (occ antrorse in distal ½), not revolute

Stems smooth, without prickles

Loosely tufted erect per, with few or no non-fl shoots. Lvs 5-9-whorled, often held erect, 7-10(16) x 1.5mm, oblanc-linear, shortly mucronate, usu hairless, with a few curved prickles at least distally (rarely all antrorse), usu with stomata below only. Stems 0.4-0.8mm diam, hairless, often with purplish nodes. All yr. Calc gsld, S Eng....*Slender Bedstraw* **Galium pumilum**

Mat-forming per, with abundant non-fl shoots much-branched at base. Lvs 5-8-whorled, usu patent, 3-10(14) x 1mm (longer up stem), oblanc-linear, with long apiculus, with curved prickles (often sparse), with stomata below only (occ v sparse above). Stems 0.4-0.8mm diam, usu hairless, without purplish nodes. All yr. Calc gsld, mostly N Br ....*Limestone Bedstraw* **Galium sterneri**

Stems rough, with prickles

Lvs with cartilaginous margins. Stems woody nr base

Lvs 4-6-whorled, 1.5-6cm x 6-12mm, ovate to elliptic-lanc, tough, with strongly retrorsely scabrid cartilaginous margins, shiny dark green above, obscurely pinnate-veined, with retrorse prickles on midrib below, with stomata below only. Stems clambering, sharply 4-angled, the angles rough with retrorse prickles, hairless.........*Wild Madder* **Rubia peregrina**

Lvs without cartilaginous margins. Stems herbaceous

Per. Damp or marshy habs

Lvs obtuse to ± acute (occ with short mucro when young), 6-8 whorled, 5-15(20) x 1-3mm, linear-oblanc, hairless, with weak prickles often straight or v sparse, with stomata below only (or sparse above). Stems 0.8-1.6(2)mm diam, erect, ± rough (rarely smooth), often purplish, blackening on drying. Plant hardly rough to touch. All yr (wintergreen as low mat) ....*Common Marsh-bedstraw* **Galium palustre***

Lvs with long (≥0.5mm) apiculus, 6(8)-whorled, 7-20 x 1-3mm, linear-oblanc, hairless, with strong curved prickles, usu with stomata below only. Stems 0.6-1mm diam, erect, v rough, often purplish, not blackening on drying. Plant v rough to touch. Apr-Oct (all yr) ....*Fen Bedstraw* **Galium uliginosum***

Ann. Dry habs, often ruderal or arable

Frs >3mm, with bristles. Lvs 6-8-whorled, 1-6cm x 2-8mm, linear-oblanc to broadly oblanc-elliptic, long-awned to 1.5mm, occ with slightly recurved margins, bristly hairy above, with prickles often strongly curved (occ antrorse but always retrorse proximally), with stomata below only. Stems (1)2-3mm diam, hairless or hairy, retrorsely scabrid on angles, occ purplish below nodes. Oct-Jul...*Cleavers* **Galium aparine**

Frs >3mm, papillate. Lvs 6-8-whorled, 2-3cm x 1-4mm, linear-lanc, mucronate-cuspidate, hairless and smooth above, with v stout strongly hooked broad-based prickles on margins and midrib below, with stomata below only. Stems 1-3mm diam, with prickles, hairless. VR ....*Corn Cleavers* **Galium tricornutum**

Frs <3mm, smooth. Lvs 6-7-whorled, 2-3cm x 1-4mm, linear-oblanc, abrubtly acuminate with long awn to 1.5mm, occ with stiff hairs or prickles above, with strongly hooked prickles on margins, with stomata below only. Stems 1-2(3)mm diam, v rough with strongly hooked prickles. VR. Jun-Nov...*False Cleavers* **Galium spurium**

Group MD – *Stems round, with 1 whorl of lvs nr top only. Lvs with stomata below only. Rhizomatous*

■Lvs 4(5)-whorled, equal, with 3(5) parallel veins and pinnate veins

Lvs 5-12 x 2-5cm, obovate to ± orb, shortly acuminate, cuneate at base, ± sessile, dull green above, shiny green below, with erose to papillate hyaline margins. Stems to 40cm, unbranched, hairless, odorous. Apr-Aug. Calc woods..*Herb-Paris* **Paris quadrifolia**

■Lvs 5-7-whorled, unequal, with all veins pinnate

Lvs 1-4(8) x 1cm, 5(6)-whorled, obovate to obovate-lanc, acute or obtuse, cuneate at base, (±) sessile, with erose-crenulate hyaline margins, shiny light green, entire or finely serrulate nr apex, not net-veined, with 5 prs of anastomosing 2° veins. Stems to 20cm, rarely with small lower lvs, rarely branched, reddish, with reddish papillae just below lf whorl (x20). May-Oct. Pine woods, moors, N Br..*Chickweed-wintergreen* **Trientalis europaea**

Lvs 6-8(10) x 2-2.5cm, (6)7-whorled, elliptic-lanc, acute, cuneate at base, (±) sessile, with erose-crenulate hyaline margins, light green, minutely serrate-crenate along length, not net-veined, with 6-9 prs of anastomosing 2° veins. Stems to 20cm, with 1-2 adpressed scale-like lvs below and 2 bract lvs just below lf whorl, rarely branched, with reddish papillae just below lf whorl (x20). May-Oct. VR alien, S Eng...*Starflower* **Trientalis borealis**

Group ME – *Stems round, with several whorls of lvs (or branches). Lvs with 0-1(3) parallel veins. (*Spergula arvensis*,* Carum verticillatum *and* Asparagus *may key out here in error)*

■Aquatic, usu with emergent stems..*Mare's-tail* **Hippuris vulgaris***

■Terrestrial

Per. Lvs 5-whorled (often alt or pseudowhorled above), to 4cm, linear-lanc, ± acute, sessile, often with recurved margins, with midrib hardly translucent. Stems to 1m, ± erect.

Stems to 6mm diam, usu purplish, pruinose at least near base, never glandular nr infl. Lvs glaucous or dark bluish-green, the lower often purplish, 4-6mm wide, with stomata both sides or below only. Fls purple. Tufted..*Purple Toadflax* **Linaria purpurea**

Stems to 2mm diam, green to glaucous, occ pruinose, occ glandular nr infl. Lvs pale glaucous-green, 1.5-5mm wide, with stomata both sides. Fls yellow. Patch-forming, occ extensively, by stems arising from root buds. All yr.....................................*Common Toadflax* **Linaria vulgaris**

Stems to 1.5mm diam, green, occ purplish, rarely pruinose, often glandular nr infl. Lvs glaucous-pruinose, rarely purplish, to 2.5mm wide, with stomata below only. Fls pale lilac. Patch-forming, not v extensively, by stems arising from root buds *Pale Toadflax* **Linaria repens**

Ann. VR alien, SW Br

Lvs 3-whorled below (occ pseudowhorled), 3-10 x 1.5mm, lanc or oblanc, ± acute, narrowed to short petiole, with flat (or revolute) margins, light green, viscid-hairy, without veins visible. Stems to 15cm, often ascending-erect, much-branched, viscid-hairy. Sand dunes ....*Sand Toadflax* **Linaria arenaria**

Lvs 3-5-whorled, 1-3cm x 2.5mm, linear, ± acute, narrowed to short petiole, with flat margins, light bluish-green, ± hairless, (0)1-veined. Stems 5-20cm, decumbent to ascending, much-branched, glandular nr infl..*Prostrate Toadflax* **Linaria supina**

Group MF – *Lvs all on erect stems, not connate at base (exc* Lysimachia ciliata*), with pinnate veins or several (>3) parallel veins visible, with stomata below only. Stems solid. Hairs septate (absent in* Lilium martagon*)*

■Lvs (4)5-11-whorled below, alt above (occ single lf below), not revolute when young. Plant with bulb

Lvs 8-14 x 3-6cm, obovate-lanc, narrowed to base, shiny green at least below, with minutely crenate-papillate margins, with parallel 2° veins raised below and sunken above, with pinnate cross-veins, with stomata below only. Petiole 0-1cm (often indistinct). Stems round, often yellowish at extreme base, with scattered vb's. Apr-Aug................*Martagon Lily* **Lilium martagon**

■Lvs mostly 3-4-whorled (occ in prs), revolute when young. Plant with purplish rhizomes

Lvs usu in prs, rarely 3-whorled, without orange dots. R alien

Lvs 6-13 x 3-5cm, ovate (occ lanc), acuminate, ± cordate at base, hairless, finely scabrid-ciliate, with stomata below only. Petioles 1-2cm, connate at base, with ± long wavy cilia on cartilaginous margins, channelled. Stems to 80cm, square with rounded angles (occ double-channelled), reddish at base, hairless. Often all yr. Dry or damp habs (often shady) ....*Fringed Loosestrife* **Lysimachia ciliata**

Lvs 3-4-whorled, with orange dots at least nr margins below

Petiole <1.5(4)mm (lvs ± sessile). Lvs hairy esp below (occ sparsely so). Stems round to slightly ridged, often ± hairless

Lvs 3-4-whorled, 5-12 x 5cm, lanc to ovate, acute, with short (2mm) orange apex, with orange dots along margins and scattered over surface both sides, hairy with short minutely glandular hairs above and longer nonglandular hairs below (occ ± hairless), ciliate. Stems usu with spreading long hairs and minute glandular hairs above, often hairless below, with reduced lvs below (occ reddish). May-Oct. Wet habs (often sunny) ....*Yellow Loosestrife* **Lysimachia vulgaris***

Petiole 4-15mm (lvs shortly petiolate). Lvs hairy both sides. Stems ridged, densely nonglandular-hairy

Lvs 4-whorled, occ alt, to 10 x 5cm, lanc to ovate, ± acute, with orange gland along recurved apex for 4mm, with orange dots below esp nr margins, nonglandular-hairy, ciliate. Stems reddish at extreme base, often with purplish nodes. May-Oct. Dry habs ....*Dotted Loosestrife* **Lysimachia punctata**

Key to Groups in Division N

(Lvs toothed, alt)

■Tree, shrub or subshrub
Branches usu spiny or thorny....................NA
Branches not spiny or thorny (lvs may be spiny)
Dwarf shrub, <20cm tall....................NB
Shrub or tree, >30cm tall
Lvs white-hairy or with silvery scales below....................NC
Lvs not white-hairy and without scales
Lvs evergreen....................ND
Lvs deciduous
Lvs with 2° veins straight (parallel) and usu ending in teeth....................NE
Lvs with 2° veins curving towards margins, usu forking several times or anastomosing before reaching margins....................NF
■Herb (occ woody at base)
Stipules, or scale-like reduced lvs, present at base of petiole....................NG
Stipules absent, or represented by glands or inflated sheathing base
Latex abundant. Stems with lf scars (abscission scars) nr base (*Euphorbia*)....................NH
Latex usu sparse. Stems without lf scars
Lvs with at least some hairs forked or hooked....................NI
Lvs hairless or with all hairs simple
Lf teeth with distinct white hydathodes (*Campanulaceae*)....................NJ
Lf teeth without distinct white hydathodes....................NK
Latex absent
Lvs with at least some hairs forked or stellate....................NL
Lvs with dendritic hairs (*Verbascum*)....................NM
Lvs with white or yellowish meal (like flour) below, or with bladder-like hairs....................NN
Lvs with cottony or woolly hairs (occ v sparse)....................NO
Lvs with hairs all simple (may be glandular), or hairless. (If hairs hooked Go to **KM**; if hairs sharply hispid Go to **KN**)
Lvs with 3-9 parallel veins (usu raised below) (*Plantago*)....................NP
Lvs with pinnate or palmate veins (occ obscure) or midrib visible only
Petiole developing 1-2 hollows (may have translucent channel and sheathing base) ....(*Ranunculaceae*) Go to **RAN**
Petiole solid or absent (*Veronica* may key out here but lower lvs are opp)
Lvs with strongly net-veined with translucent veins....................NQ
Lvs not strongly net-veined (veins may be opaque, beware young lvs)
Lvs with at least some glandular hairs (teeth may have submarginal hydathodes), or viscid ....NR
Lvs without glandular hairs (occ sessile glands on lf surface), hairy or hairless, never viscid
Lvs fleshy (or waxy) or ± so. Often coastal....................NS
Lvs not fleshy or waxy
Lf surfaces with at least some hairs. Lvs usu pinnate-veined
Hairs septate....................NT
Hairs unicellular....................NU
Lf surfaces (±) hairless (margin may be ciliate) AND/OR lvs 1-veined....................NV

Group NA – *Tree or shrub. Branches usu spiny or thorny. Lvs net-veined (exc some* Berberis *spp and* Ribes uva-crispa*), with stomata below only (occ above in* Pyracantha coccinea*). Petiole channelled. Twigs (±) round. Buds with several scales.* ❶ *Twigs with odourless yellow inner bark.* ❷ *Twigs with odorous orange inner bark*

■Spine(s) at stem nodes (may be 3-partite)

Stipules large, lf-like

Twigs minutely hairy

Lvs 2.5-7 x 3cm, elliptic-lanc to ovate, cuneate at base, folded when young, occ undulate, dark green above, paler below, occ reddish, serrate, with anastomosing 2° veins not or hardly raised below. Petiole 0.5-1.5cm. Twigs with 2cm spine at each node. Buds occ collateral, small, acute. Deciduous shrub to 1m. Hortal. *(C.* x *superba* (*C. japonica* x *speciosa*) is perhaps the commonest taxon but is not separable from *C. japonica*) ....*Japanese Quince* **Chaenomeles japonica**

Twigs hairless

Lvs 4-8 x 4cm, ovate to obovate, cuneate at base, ± rolled when young, occ undulate, dark green above, paler below, serrate, with anastomosing 2° veins not or hardly raised below. Petiole 0.5-1.5cm. Twigs usu with 2cm spine at each node. Buds occ collateral, small, acute. Deciduous shrub to 2m. Hortal................................*Chinese Quince* **Chaenomeles speciosa**

Stipules absent

Lvs with spine-tipped teeth. Deciduous or evergreen shrub. ❶...................(*Berberis*) Go to **BER**

Lvs without spine-tipped teeth, 2-5cm, as wide or wider than long, 3-5 lobed, broadly cuneate to ± cordate at base, usu shortly unicellular hairy, up to 7 veins from base raised below. Petiole with few short glandular hairs and nonglandular hairs, long-ciliate at base, with 3 vb's. Twigs whitish, hairless to minutely hairy or bristly, much-branched, with 1-3 spines at each node, with short shoots. Buds to 10mm, spindle-shaped, acute, white, turning brown. Deciduous shrub to 1m. PL 15...*Gooseberry* **Ribes uva-crispa**

■Thorns often at end of branches, never 3-partite. Stipules present at least when young

Petiole <10mm (<⅓ lf length)

Evergreen. Lvs ± leathery

Lvs 3-5 x 1.5-2cm, lanc to oblanc, acute to obtuse, cuneate at base, shiny above, dull below, hairless or sparsely hairy esp below, crenate-serrate with (0)9-31 teeth per side (each tooth with fragile claw-like dark gland), without 2° veins strongly raised or sunken either side. Petiole 5-12mm, hairy. Twigs densely hairy to hairless, with 6-10mm reddish or green thorns. Buds small, occ collateral. Many other hybrid *Pyracantha* cultivars are planted as ornamentals...*Firethorn* **Pyracantha coccinea**

Deciduous. Lvs not leathery

Twigs with odorous orange inner bark ❷

Lvs often subopp, 3-6 x 2-3(5)cm, ovate, obtuse (occ cuspidate), dull green, with minute translucent dots (HTL), usu hairy both sides, serrate with (20)40-45 teeth per side (each with claw-like gland), with 2-3 prs of 2° veins converging nr apex. Petiole 5-20mm. Twigs often branching at 90° (often subopp), dark, minutely hairy when young, with short shoots. Buds adpressed, 2-6mm, dark brown, hairless, with 5 scales. PL 12 ....*Buckthorn* **Rhamnus cathartica**

Twigs with inner bark not odorous or orange

Lvs 2-4cm. Suckering shrub, often forming dense thickets

Lvs oblong-obovate to elliptic-lanc, acute or obtuse, cuneate at base, ± rolled when young (occ obvolute), often dull dark green above, hairy to hairless, serrate with 20-40 teeth per side (lowest occ gland-tipped). Petiole 2-10mm. Stipules linear, glandular-toothed. Twigs branching at 45-90°, dull blackish, minutely hairy when young, with thorns often minutely hairy. Buds ovoid, occ along thorns or collateral.......................*Blackthorn* **Prunus spinosa**

Lvs >4cm. Small tree, not suckering. (*Mespilus germanica* may key out here)

Lvs to 12 x 6cm, ovate, rounded to ± cordate at base, dull pale green above, ± hairless above, patent-hairy below, 2-serrate, occ shallowly lobed. Petiole to 15mm. Buds to 6mm, reddish..*Scarlet Thorn* **Crataegus mollis**

Lvs to 8 x 5cm, ovate, cuneate at base, shiny dark green above, paler and dull below, hairless above, hairy on 2° veins below, 1(2)-serrate, with 7 prs of 2° veins (all ending in teeth). Petiole to 10(15)mm. Buds to 6mm, reddish ....*Broad-leaved Cockspurthorn* **Crataegus persimilis**

Petiole >10mm. Deciduous tree. Lvs not leathery

Lvs <8cm, each tooth with a fragile claw-like gland on distal side

Lvs with reddish glands along midrib above (often v sparse)

Lvs 3-5cm, ovate, acuminate or cuspidate, broadly cuneate or rounded at base, sparsely hairy on veins when young (soon hairless), crenate-serrate, with 1-3° veins raised. Petiole ≥ lf, reddish at base, soon hairless, with 1 arc vb. Stipules soon falling. Twigs with short shoots (each with 2-3 lvs). Buds ciliate....................................*Crab Apple* **Malus sylvestris***

Lvs without reddish glands along midrib above

Lvs 3-8cm, elliptic to ± orb, obtuse to cuspidate, rounded or ± cordate at base, involute when young, ± shiny green above, turning black, hairless or hairy below, occ woolly when young (esp below), ciliate, ± entire to crenate-serrate with c60 teeth per side, occ ± entire, with midrib raised below. Petiole 2-8cm, occ >> lf. Stipules linear, soon falling. Twigs hairless or sparsely hairy, with short shoots. Buds 5mm, sharply acute, ciliate, with c7 scales. Bark in square plates

Twigs often without thorns. Buds purple-black, hairy to hairless. Frs >5cm, usu pear-shaped, soft and sweet at maturity. Hortal..*Pear* **Pyrus communis**

Twigs with thorns on lower branches. Buds dark chestnut-brown, hairless. Frs 1.5-4cm, globose to obovoid, hard and sour even at maturity. R.................*Wild Pear* **Pyrus pyraster**

Lvs 1-4cm, ovate, acute, rounded to cordate at base, involute when young, hairless or hairy below, not ciliate, crenate-serrulate with 20-120 teeth, with midrib raised below. Petiole (1.5)2-4cm, >½ lf (often longer). Stipules linear, soon falling. Twigs hairless, with short shoots, often without thorns. Buds 5mm, ovoid, purple-brown, hairless, sparsely ciliate, with erose scales. Frs 0.8-2cm, globose to obovoid, hard and sour even at maturity. Devon, Cornwall. Sch8..*Plymouth Pear* **Pyrus cordata**

Group NB – *Dwarf or low shrub <15(60)cm tall. Lvs with stomata below only (exc some* Salix*). Twigs round. Often moors, hths or uplands. (Dwarfed* Betula nana *may key out here)*

■Lvs strongly net-veined

Evergreen. Buds without scales. Lvs white-woolly below

Lvs 0.5-2 x 1cm, oblong, obtuse, cordate at base, revolute when young, with margins recurved at maturity, dark green and soon hairless above, with white-dendritic stout brown hairs on midrib below (occ on petiole), deeply crenate (-dentate) with 6-8 teeth per side, with 2° veins sunken above. Petiole 4-10mm. Stipules scarious, brownish, adnate to petiole for ½ length, with long subulate apex, hairy. Stems much-branched, twisted, rooting at nodes. Usu calc rocks. PL 11..*Mountain Avens* **Dryas octopetala**

Evergreen. Buds with 5 scales. Lvs hairless, 5-12 x 7cm, broadly ovate to ± orb. PL 10, 23.
....*Shallon* **Gaultheria shallon**

Deciduous

Buds with 1 scale. Lvs white-woolly to hairless, rugose or not.......................(*Salix*) Go to **SAL C**

Buds with ≥2 scales. Lvs (±) hairless

Lvs rugose

Lvs 1-2.5cm, obovate, obtuse to ± acute, cuneate at base, shiny bright or dark green above, paler below, turning reddish in autumn, usu with long cilia at least nr base, serrate with c15 teeth per side. Petiole short. Twigs hairless. Buds tiny, with c3 scales
....*Arctic Bearberry* **Arctostaphylos alpinus**

Lvs not rugose

Lvs 1-3cm, ovate, acute, finely serrate with 15-30 teeth per side, each tooth with stalked orange gland on distal side. Twigs green, ± zig-zagging, strongly ridged, hairless, with stomata. Buds adpressed, to 4mm, green, hairless, with 2 scales
....*Billberry* **Vaccinium myrtillus***

■Lvs not or hardly net-veined

Deciduous..(*Salix*) Go to **SAL C**

Evergreen

Lvs with revolute margins

Lvs to 18 x 8mm, elliptic to obovate, slightly notched, with terminal hydathode, rounded to cuneate at base, leathery, shiny dark green above, paler and gland-dotted below, hairless, with thick cartilaginous margins occ remotely crenulate by hydathode-teeth, with ± obscure 2° veins. Petiole to 3mm. Twigs minutely hairy when young......*Cowberry* **Vaccinium vitis-idaea**

Lvs with flat margins

Lvs c5, often clustered nr stem apices, 20-50 x 20mm, ovate-lanc, acute, cuneate at base, with cartilaginous margins, shiny dark green above, paler below, hairless, serrate with 12-15 bristle-tipped teeth per side. Petiole to 6mm, reddish. Twigs hairless, green, with occ stomata. Buds 2-3mm, obtuse, with reddish-green scales. Rhizomatous or stoloniferous shrub, to 15cm tall. R alien..*Checkerberry* **Gaultheria procumbens**

Lvs many, all along stems, 5-10 x 1-1.5mm, linear, obtuse, cuneate at base, with cartilaginous margins, dark green above, paler below, often with sessile glands above, with hairy midrib below, serrulate or glandular-denticulate, with midrib sunken both sides. Petiole to 0.5mm. Twigs with sparse glands when young, brown, without stomata. Buds <1mm, obtuse, with waxy scales. Low *Erica*-like domed shrub, to 20cm tall. Mtns, VR, Perth. Sch8
....*Blue Heath* **Phyllodoce caerulea**

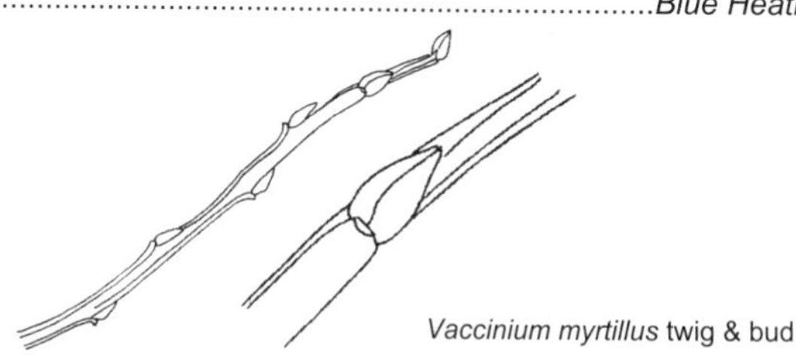

Vaccinium myrtillus twig & bud

Group NC – *Tree or shrub. Lvs white-hairy or with silvery peltate scales below, with obscure stomata below only. (*Malus domestica *and* Aronia arbutifolia *may key out here)*

■Plant with peltate scales. Evergreen

Lvs to 10 x 6cm, broadly elliptic, acute, rounded at base, occ undulate, thick, shiny dark green with sparse silvery fimbriate scales above (leaving pits when rubbed off), silvery below with scales, crenulate-undulate and sinuate-toothed, not or obscurely net-veined, with 5-8 prs of 2° veins not raised below. Petiole to 2cm, not channelled, with 1 vb. Twigs angled, with reddish-brown scales. Buds with 2 scales. PL 10.............*Broad-leaved Oleaster* **Elaeagnus macrophylla**

■Plant with stellate hairs

Lvs deciduous, net-veined..(*Tilia* 'Petiolaris', *T. tomentosa*) Go to **TIL**

Lvs evergreen, net-veined or not

Trunk with corky bark..*Cork Oak* **Quercus suber**

Trunk with bark not corky..*Evergreen Oak* **Quercus ilex**

■Plant without stellate hairs or peltate scales

Deciduous tree

Young twigs white-woolly. Petiole >2cm...(*Populus*) Go to **POP A**

Young twigs not or hardly white-woolly. Petiole ≤2cm...................................(*Sorbus*) Go to **SOR**

Evergreen shrub

Young twigs white-felted or -woolly

Lvs 3-8 x 4cm, ovate, obtuse, shiny green above with scattered minute white papillae (looking like stomata), white-felted below, entire to obscurely sinuate-toothed with hydathodes, net-veined, with 2° veins not raised or sunken either side. Petiole to 2cm, channelled, with 5 vb's. Buds without scales...*Shrub Ragwort* **Brachyglottis** x **jubar***

Young twigs brown and adpressed-hairy

Lvs 5-8 x 4cm, ovate-oblong, rounded at base, often with cartilaginous margins slightly recurved, dark green with sparse adpressed hairs above, white adpressed-hairy below, deeply toothed with upturned spine-like teeth, net-veined. Petiole to 2.5cm. Twigs brown, angled...*New Zealand Holly* **Olearia macrodonta**

Brachyglottis x *jubar* lf & petiole TS

Group ND – *Evergreen tree or shrub. Lvs tough or leathery, with cartilaginous margins, with midrib raised below, with stomata below only. Twigs round unless otherwise stated. (*Prunus serotina *is deciduous but may appear evergreen and key out here in error)*

■Twigs (and lvs) stellate-hairy, often with terminal bud cluster, with linear stipules (soon falling)

Lvs 2-6(9) x 1-3cm, acute, usu with undulate or recurved margins, shiny dark green and soon hairless above, grey or glaucous with stellate hairs below (often dense), usu net-veined, with 2° veins anastomosing or ending in teeth. Petiole to 15mm, often swollen at base. Twigs densely stellate-hairy when young, soon ± hairless. Buds ovoid, obtuse, brown, stellate-woolly

Bark corky, pale brown. Lvs shallowly lobed with 5-6 spine-tipped teeth per side. Buds 2-3mm. R planted..*Cork Oak* **Quercus suber**

Bark not corky, dark brown. Lvs with 0-8 spiny teeth per side (on young or vigorous shoots), usu entire on mature branches. Buds 1-3mm................................*Evergreen Oak* **Quercus ilex**

■Twigs hairy or hairless (never stellate-hairy), without terminal bud cluster or obvious stipules

Lvs spiny (at least at apex)

Twigs green. Lvs 3-10cm. Stipules vestigial, minute, dark. Tree >1.5m, not suckering

Lvs ovate to elliptic or oblong, spine-tipped, usu undulate, shiny dark green above, dull pale green below, strongly spiny sinuate-dentate Twigs hairless or minutely hairy. Buds often purplish, with 2 minute lf-like sharp scales...*Holly* **Ilex aquifolium**

Lvs ovate to broadly so, spine-tipped, often undulate, shiny dark green above, dull pale green below, occ variegated, with 0-7 spiny teeth. Twigs minutely hairy when young. Buds often purplish, with 2 minute lf-like sharp scales. *I. aquifolium* x *perado* ....*Highclere Holly* **Ilex** x **altaclerensis**

Twigs reddish. Lvs 1-2cm. Stipules absent. Shrub <1.5m, suckering

Lvs spirally arranged, to 6mm wide, ovate to elliptic-lanc, spine-tipped to 1.2mm, shiny dark green above, dull pale green below, remotely serrate with fragile gland-tipped teeth, with indistinct anastomosing 2° veins. Petiole to 2mm. Twigs usu minutely hairy (often gland-tipped), often rough with longer swollen-based bristles...*Prickly Heath* **Gaultheria mucronata**

Lvs not spiny

Young twigs glandular-hairy or reddish bristly hairy (or lvs glandular-ciliate when young)

Lvs sweetly aromatic, with ± sessile glands below

Lvs 2-7cm, obovate, 1-2-serrate, acute, cuneate at base, rolled when young, shiny dark green above, paler below, hairless, ± net-veined. Petiole 0-5mm. Twigs with long-glandular and short simple hairs (glandular hairs stout, red-tipped and often minutely hairy themselves)...*Escallonia* **Escallonia macrantha**

Lvs odourless, without glands below

Tall shrub or tree to 5m. Lvs 4-10 x 3.5cm, elliptic-oblong or elliptic-obovate, cuneate at base, plastic-feel, shiny dark green above, often discoloured with purple-black spots, paler or glaucous below, hairless, occ with long glandular cilia when young, serrate with c25 irreg teeth per side, usu net-veined, with anastomosing 2° veins. Petiole 6-12mm, not channelled. Twigs green to reddish, usu with bristly reddish glandular multiseriate hairs, weakly angled. Buds 1.5-2mm, acute, usu red, with 8-16 scales. Bark reddish-brown, rough, often flaking in strips. Pl 10...*Strawberry-tree* **Arbutus unedo**

Low suckering shrub to 1.5m. Lvs 5-12 x 7cm, broadly ovate to ± orb, acute with long terminal hydathode (often recurved), rounded to cordate at base, often dark green above, paler below, hairless, serrate with 45-75 teeth per side, each tooth with a long fragile bristle, net-veined, with anastomosing 2° veins. Petiole 5-7mm, with 1 vb. Twigs green to reddish, zig-zagging, bristly glandular-hairy when young, flattened, pith-filled, with irritant sap. Buds often well above lf axils, 6mm, acute, green, ciliate, with 5 scales. Pl 10, 23 ....*Shallon* **Gaultheria shallon**

Young twigs nonglandular-hairy

Lvs 2-5 x 1-2.5cm...*Mediterranean Buckthorn* **Rhamnus alaternus**

Lvs 7-20 x 5cm

Lvs ovate-lanc, acuminate, shiny dark green above, dull and paler below, hairless (TNE), with short adpressed orange- or red-tipped glandular hairs below (glands not swollen, x20), serrate with c30 long (0.5mm) cartilaginous teeth per side, not net-veined. Petiole to 1cm, often reddish, minutely hairy, channelled. Twigs arching, green, occ with reddish glands. Buds often above lf axils, 3mm, reddish-brown, not ciliate. Hortal, VR alien ....*Dog-hobble* **Leucothoe fontanesiana**

Young twigs hairless

Petiole to 15mm, green. Lvs 5-20cm, oblong-obovate, acuminate, folded when young, almond-scented when crushed, shiny bright green above, hairless, weakly serrate with up to 15 teeth per side (occ ± entire). Stipules minute, soon falling. Twigs green when young. Buds 5mm, green. PL 10...*Cherry Laurel* **Prunus laurocerasus**

Petiole 15-25mm, red. Lvs 8-12cm, oblong-ovate, ± acuminate, folded when young, odourless, shiny dark bluish-green above, paler below, hairless, serrate with 20-35 teeth per side. Stipules to 10mm, subulate, soon falling. Twigs reddish when young. Buds 5mm, reddish. PL 10..*Portugal Laurel* **Prunus lusitanica**

NE

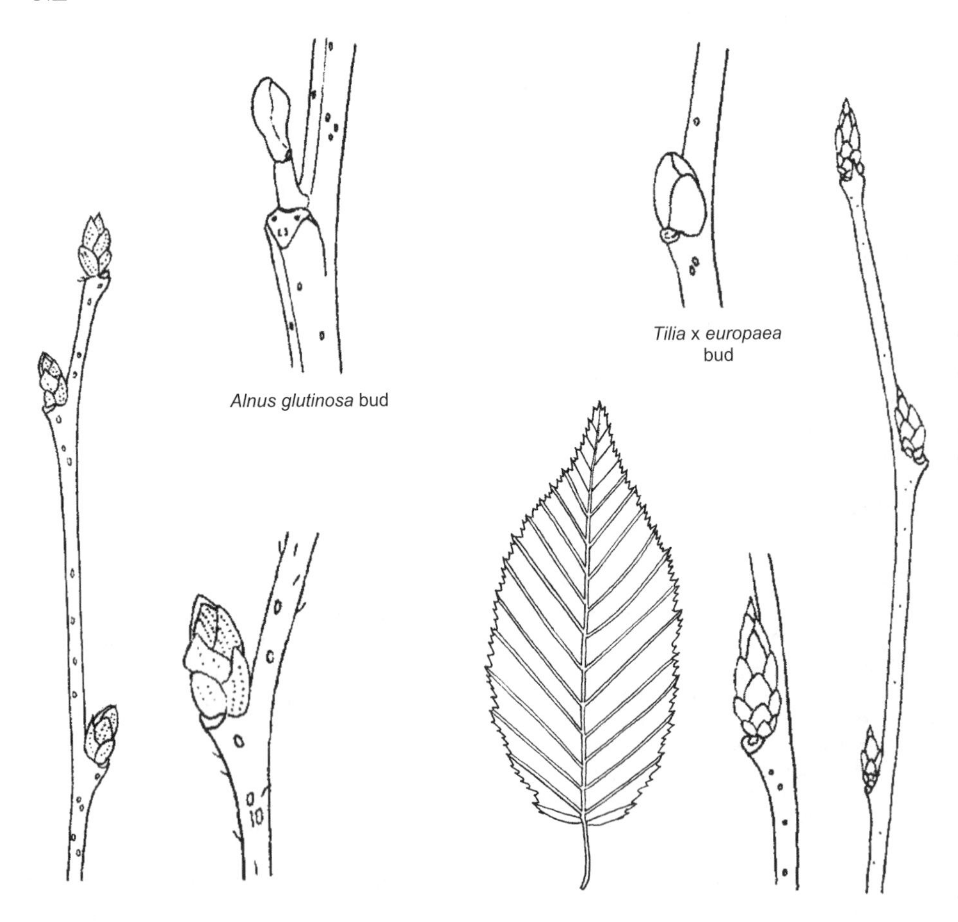

Ostrya carpinifolia twig & bud

Carpinus betulus lf , bud & twig

Group NE – *Deciduous tree or shrub. Lvs with 2° veins straight and usu ending in a tooth, net-veined, with stomata below only. Stipules present, soon falling. Twigs sympodial (monopodial in* Fagus*), round unless stated.* ❶ *Twigs green*

■Buds on 2-3mm stalk, with 2 scales

Lvs with at least some stellate hairs. Wood pale, not changing colour when cut

Lvs 8-12 x 4-7cm, ovate-elliptic, folded (-plicate) when young, undulate, shiny at least above, shallowly crenate-lobed, 3-pli-veined from base, sparsely floccose below esp in vein axils. Petiole 5mm, swollen below at base, often with sparse stellate or floccose hairs. Stipules to 8 x 2mm, hairy. Twigs zig-zagging, with lenticels. Buds to 6mm, flattened, dark brown, scurfy or floccose with stellate hairs. PL 20..*Persian Ironwood* **Parrotia persica**

Lvs without stellate hairs. Wood turning orange when cut

Lvs truncate or retuse at apex, with 4-7 prs of 2° veins, viscid when young

Twigs ± hairless and viscid when young. Lvs 3-9cm, obovate, rounded or cuneate at base, folded-plicate when young, with obscurely thickened or recurved margins, dark green, with minute orange glands both sides, hairless (exc for tufts in vein axils below), 1(2)-dentate or lobed, with 4-7 prs of 2° veins. Petiole 1-3cm. Twigs ridged when young, with short shoots, with raised orange lenticels. Buds 6-10mm, obtuse, purplish, usu resin-coated, with large outer scale ± hiding inner scales, with raised orange glands. Bark dark brown, soon fissured. PL 12..*Alder* **Alnus glutinosa***

Lvs acute to acuminate, with 7-15 prs of 2° veins, not viscid when young (may have resin granules)

Twigs shortly grey-hairy, greenish and weakly angled when young, soon brown and round with orange lenticels. Lvs 8-12 x 5-7cm, ovate, usu rounded at base, folded-plicate when young, with obscurely recurved margins, dull green above, grey-green or glaucous with orange-black glands below, sparsely hairy on veins above, hairy below (but later on veins only), 2-serrate, occ lobed, with 8-12 prs of 2° veins. Petiole 1-2(3)cm. Stipules to 2cm, oblong-lanc. Buds 8mm, curved, purplish, resin-coated, densely grey-hairy. Bark deep grey to greenish, smooth, with small flaky vertical lenticels. PL 23.................*Grey Alder* **Alnus incana**

Twigs hairless, greenish, with orange lenticels. Lvs to 8-12 x 5-7cm, ovate-lanc, rounded at base, with minutely recurved margins, dark green above, slightly greyish below, hairless (exc midrib) above, with sparse rusty hairs on veins below, 2-serrate, with (10)12-13(15) prs of 2° veins. Petiole to 0.3cm. Buds to 10mm, reddish, usu resin-coated, hairless. Bark grey, smooth, developing rectangular plates..*Red Alder* **Alnus rubra**

■Buds sessile, with >2 scales

Lvs with aristate teeth, with 2° veins ending in a tooth (no teeth between adjacent vein endings)

Lvs 10-25cm, oblong-lanc, acute or acuminate, broadly cuneate to ± cordate at base, ± folded and viscid when young, with cartilaginous margins (rare in deciduous trees), dark green and hairless above, hairy below when young, serrate-dentate, with 13-25 prs of 2° veins. Petiole (0.5)1-2(3)cm. Stipules 7-10 x 2mm, linear-lanc. Twigs stout, ± straight, olive-brown, usu hairless, with prominent lenticels. Buds often asymmetric, with (2)3 (often retuse) scales. PL 21, 22..*Sweet Chestnut* **Castanea sativa**

Lvs without aristate teeth, teeth between adjacent 2° vein endings

Twigs usu with red-tipped glandular hairs

Shrub (multi-stemmed), suckering. Lvs 5-12cm, ± orb, cuspidate, ± cordate at base, folded when young, hairy both sides at least when young, 2-serrate, occ lobed (lobes cut to <⅓ way), with 5-8 prs of 2° veins. Petiole to 2cm. Stipules ovate, occ persistent. Twigs zig-zagging. Lf buds 4-6mm, ovoid, flattened, green with brown margins, ciliate, with 7-8 scales. Fl buds globose. PL 12, 20..*Hazel* **Corylus avellana**

Tree (single trunk), not suckering. Lvs 6-17cm, broadly ovate to ± obovate, acuminate, ± cordate at base, folded when young, hairy above, occ woolly below, with 7-8(10) prs of 2° veins. Petiole 0.7-4cm. Stipules lanc. Twigs ± straight. Lf buds ovoid, hairy.....*Turkish Hazel* **Corylus colurna**

Twigs hairy or hairless, without glandular hairs

Lvs with resin glands and/or 2-7 prs of 2° veins..(*Betula*) Go to **BET**

Lvs without resin glands, or >7 prs of 2° veins

Lvs often rough above, ± folded when young, often strongly asymmetrical at base. Twigs often corky-ridged, zig-zagging. Buds usu blackish, spreading.................(*Ulmus*) Go to **ULM**

Lvs rarely rough above, ± plicate when young. Twigs never corky-ridged, often straight. Buds brown or green

Lvs with obvious hair-tufts in vein axils below. Tree

Buds asymmetric, with 3 scales (1 smaller). Twigs strongly zig-zagging. Lvs cordate at base (often asymmetrically so)...(*Tilia*) Go to **TIL**

Buds symmetric, >3 scales (all equal). Twigs slightly zig-zagging

Buds adpressed or incurved, 5-10mm, pale brown, not striate, with >7 scales. Lvs 11 x 6cm, ovate-oblong, acute or acuminate, equal to unequal at cordate or rounded base, hairless exc for long adpressed hairs on veins below, 2-serrate, with 7-15 prs of unbranched 2° veins. Petiole 5-15mm, often adpressed-hairy. Twigs dark brown-blackish, minutely hairy in 1st yr, often with sparse longer ± adpressed hairs to 1.5mm persisting until 2nd yr, with lenticels not prominent. Trunk fluted, with smooth bark. PL 12..*Hornbeam* **Carpinus betulus***

Buds spreading from twig, to 5mm, green, faintly striate, with 7 scales. Lvs v like *Carpinus betulus* but sparsely adpressed-hairy above and with 14-18 prs of 2° veins (some often branched). VR planted........................*Hop-hornbeam* **Ostrya carpinifolia***

Lvs without obviously longer hair-tufts in vein axils below

Shrub. Buds green

Lvs to 7 x 2.5cm, ovate, acuminate, rounded at base, bright green, sparsely hairy above, shortly hairy below, 2-serrate, with 2° veins strongly sunken above and raised below. Petiole 6mm, adpressed-hairy, channelled. Stipules small, triangular to linear, red-brown. Twigs green, hairless, brittle, ridged when young, pith-filled, with stomata. Buds often collateral, 3.5mm, ± globose, ciliate. ❶. PL 14...........*Kerria* **Kerria japonica**

Tree. Buds brown or reddish

Lvs with 6-9 prs of 2° veins usu ending at tooth apex, 4-9cm, ovate-elliptic, often asymmetric at base, soon hairless exc for silky buff hairs on veins below, long-ciliate, entire to sinuate-toothed. Petiole to 1cm. Stipules to 18mm, linear. Twigs with sparse long hairs when young. Buds spreading or terminal, to 2cm, reddish-brown, sparsely hairy, ciliate. PL 12, 21...*Beech* **Fagus sylvatica***

Lvs with (7)9-11 prs of 2° veins ending at tooth apex, 3-8cm, oblong, usu asymmetric at base, ± hairless (or shortly hairy both sides), coarsely 2-serrate. Petiole to 1cm, shortly hairy. Stipules 2mm, ovate-linear. Twigs shortly hairy in 1st yr, soon hairless. Buds adpressed, 3-4mm, ovoid, obtuse, pale brown, long-ciliate. Early leafing (Feb) ....*Roble* **Nothofagus obliqua**

Lvs with 10-14 prs of 2° veins ending indefinitely but not at base of sinus, 4-7cm, ovate-oblong, usu asymmetric at base, hairy on veins below, finely serrate. Stipules 2mm, ovate-linear. *N. alpina* x *obliqua*..................*Hybrid Roble* **Nothofagus** x **dodecaphleps**

Lvs with 14-23 prs of 2° veins (± deeply sunken) appearing to end at sinus, 4-10 x 4cm, ovate-oblong, ± symmetric at base, often hairy both sides or sparsely hairy on veins below, finely serrate. Petiole 7mm, hairy. Stipules 2mm, ovate-linear. Twigs hairy, with rough warts. Buds spreading, 4-13mm, reddish, hairless, not ciliate ....*Rauli* **Nothofagus alpina**

Group NF – *Tree or shrub. Lvs with 2° veins curving towards margins, usu forking several times or anastomosing before reaching margins, net-veined (exc* Baccharis halimifolia*), with stomata below only (both sides in* Baccharis halimifolia *and some* Salix *and* Populus*). Deciduous (occ semi-evergreen in* Baccharis halimifolia*). (*Rhamnus cathartica *may key out here).* ❶ *Lvs aromatic.* ❷ *Lvs 3-pli-veined.* ❸ *Twigs and petioles with white latex*

■Buds on 5mm stalk, naked, scurfy. Petiole often ≥½ lf length on some lvs

Lvs 5-8cm, broadly ovate, shortly acuminate, cordate at base, rolled when young, ± shiny dark green above, often orange-tinged when young, paler with scattered raised reddish glands below, with tufts of white-orange hairs in vein axils below, shallowly serrate with 40-55 teeth per side. Petiole 2-3.5cm. Stipules papery, orange. Twigs dark brown, angled when young, with low orange lenticels. Buds 5mm, globose, pale green, speckled red-brown. Bark ± smooth but blistered ....*Italian Alder* **Alnus cordata***

■Buds sessile, with 1 scale. Petiole <½ lf........................(*Salix*) Go to **SAL**

■Buds sessile, with 3 scales (buds asymmetric). Petiole often ≥½ lf length on some lvs..(*Tilia*) Go to **TIL**

■Buds sessile, usu with >3 scales (buds symmetric)

Petiole ≥½ lf length on some lvs. Stipules linear, soon falling

Petiole often laterally flattened. Twigs without short shoots..........................(*Populus*) Go to **POP**

Petiole not laterally flattened but channelled above. Twigs usu with short shoots (woody spurs, with 2-6 lvs)

Lvs silky-hairy or glaucous below, without glands on teeth. Twigs never thorny ....*Juneberry* **Amelanchier lamarckii**

Lvs woolly or densely hairy at least below when young, with fragile claw-like glands on distal side of teeth. Twigs rarely thorny

Lvs to 8 x 5cm, ovate, obtuse to acuminate, weakly involute when young, with reddish glands along midrib above, usu serrate with 40-60 teeth per side. Petiole turning reddish, densely hairy even at maturity. Twigs woolly when young, soon hairy or sparsely so. Buds 2-5mm, adpressed, hairy to hairless, ciliate, with obtuse scales.......*Apple* **Malus domestica**

Lvs sparsely hairy on veins below when young, soon hairless. Twigs often thorny

Lvs without reddish glands along midrib above, with 2° veins hardly raised below ....(*Pyrus*) Go to **NA**

Lvs with reddish glands along midrib above (often v sparse), with 1-3° veins raised below

Lvs 3-5 x 4cm, ovate, acuminate or cuspidate, broadly cuneate or rounded at base, dull above, crenate-serrate with 25 teeth per side, with fragile claw-like glands on distal side of teeth. Petiole reddish at base, soon hairless, with arc vb. Twigs hairless or sparsely hairy when young. Buds 2-5mm, (±) hairless, ciliate.......................*Crab Apple* **Malus sylvestris***

Petiole <½ lf length

Petiole often with glands (extra-floral nectaries) at apex. Twigs often with short shoots. Buds often clustered at twig apex. Stipules soon falling.......................................(*Prunus*) Go to **PRU**

Petiole without glands at apex. Twigs with or without short shoots. Buds usu solitary at twig apex

Lvs toothed only in distal ½. Stipules absent

Lvs with sessile glands, sweetly aromatic. ❶

Lvs 2-6cm, oblanc, cuneate at base, dark green, with yellow sessile glands both sides, with short adpressed hairs above when young, ± hairy below, toothed in distal ½, with c15 prs anastomosing 2° veins, with midrib raised below. Young twigs brown, soon purplish, hairy. Buds 1.5-2mm, ovoid, obtuse, orange- or reddish-brown, ciliate. Fl buds all nr twig apex, acute, the male 5-8mm, the female to 3mm. Suckering. Bogs.........*Bog-myrtle* **Myrica gale**

Lvs 4-10cm, obovate or oblong, cuneate at base, shiny pale green, with yellow/orange sessile glands mostly below (scent weaker but sweeter than *M. gale*), hairy both sides (hairless below TNE), toothed nr apex only, with 4-5 prs of 2° veins. Young twigs green, hairy. Buds 1.5-2mm, ovoid, obtuse, purple-brown with green-centred scales. Suckering. VR alien..*Bayberry* **Myrica pensylvanica**

Lvs without sessile glands, odourless

Lvs pinnate-veined. Often suckering..(*Spiraea*) Go to **SPI**

Lvs 3-pli-veined, to 7 x 3cm, obovate, narrowed to a short or indistinct petiole, often salt-encrusted, obscurely pitted esp below, with (0)3-5 teeth per side, with midrib ± only raised below, with 3-4 prs of 2° veins, with stomata both sides. Twigs green, hairless, ridged when young. Buds small, with scales. Not suckering. ❷. Coastal, VR alien ....*Tree Groundsel* **Baccharis halimifolia**

Lvs regularly toothed

Stipules absent. Lvs revolute when young

Lvs pseudowhorled, 8-17 x 3-5cm, oblong-lanc, mucronate or with short aristate point, cuneate at base, occ slightly undulate, aromatic, yellow-green both sides, hairy above and at least on main veins below, glandular-ciliate, often entire, with midrib and 2° veins raised below. Petiole 0.8-1.5cm, channelled. Twigs bristly glandular-hairy when young, ± round or angled. Lateral (lf) buds 1mm, with 2-3 ciliate scales. Terminal (fl) buds to 20mm ....*Yellow Azalea* **Rhododendron luteum**

Stipules absent. Lvs rolled when young

Buds 1-2(3)mm, pale brown, with scales not acuminate..........................(*Spiraea*) Go to **SPI**

Buds 5-6mm, reddish, with v acuminate scales. Lvs 3-9 x 4cm, ovate or elliptic, mucronate, with weakly recurved margins, leathery, green and soon hairless above, paler and hairy below, reddish in autumn, ± entire to minutely toothed, ± net-veined. Petiole 2-3mm, reddish. Twigs densely hairy to hairless, round, occ green with raised patterning when young. VR alien...*Blueberry* **Vaccinium corymbosum**

Stipules present (often falling early leaving scars). Lvs folded or rolled when young

Twigs hairy

Twigs and petioles with white latex ❸

Lvs 8-15 x 15cm, broadly ovate, ± acuminate, cordate at base, folded when young, shiny dark green and often rough above with adpressed hispid hairs on slightly swollen bases, less hispid below, deeply serrate-dentate, rarely lobed, with 5-7 prs of 2° veins. Petiole to 3cm, soon red, sparsely hairy, round, with 7 vb's in ring. Twigs softly hairy, green, later maroon-brown with lenticels, the inner bark weakly odorous. Buds to 8mm, ovoid, sharply acute, green or chestnut-brown, hairless, ciliate, with 5 scales ....*Black Mulberry* **Morus nigra**

Twigs and petioles without latex

Lvs 5-14 x 3-6cm, oblong-ovate or elliptic-oblong, acute, cuneate to ± cordate at base, without glands along midrib above, softly short-hairy both sides, shallowly serrate (often entire nr base). Petiole to 5mm. Stipules small, lf-like, green, glandular-serrate, persistent. Twigs often thorny, densely hairy, round. Buds 3-4mm, sharply acute, purplish, ciliate. Shrub or tree to 9m.................................*Medlar* **Mespilus germanica**

Lvs 5-8 x 4cm, narrowly obovate to elliptic, abruptly acuminate or cuspidate, cuneate to rounded at base, with red-black glands along midrib above, densely hairy to ± woolly below, with 5-7 prs of 2° veins. Petiole to 7mm. Stipules small, green, toothed, soon falling. Twigs not thorny, densely adpressed-hairy to sparsely woolly when young. Buds to 6(8)mm, acute, reddish-brown, ± hairless. Suckering shrub to 3m ....*Red Chokeberry* **Aronia arbutifolia**

Twigs hairless

Lvs with red-black glands along midrib above, 3-7 x 5cm, obovate, acute, cuneate to rounded at base, rolled when young, shiny dark green and hairless above (midrib may be strigose), paler and sparsely hairy below, serrate, with 5 prs of curved 2° veins (veins turning reddish). Petiole 6mm, reddish, hardly channelled. Stipules 8mm, linear, glandular-fimbriate. Twigs olive-brown. Buds adpressed, to 8mm, flattened, acute, shiny dark red, with 5 acute glandular-ciliate scales (occ with two notches at tip). Suckering shrub to 1.5m...*Black Chokeberry* **Aronia melanocarpa**

Lvs without red-black glands along midrib above, 5-7(9) x 3-4cm, obovate to oblong, acute to shortly acuminate, rounded to cordate at base, folded and purplish when young, dull green and hairless above, ± glaucous and white silky-hairy below, soon hairless, later turning crimson, serrate with 17-36 teeth per side (without claw-like glands), with 7-10 prs of 2º veins. Petiole 1-3.5cm, yellowish, with U-shaped vb. Stipules partly adnate to petiole, 5mm, linear, entire. Twigs wth orange lenticels. Buds ± adpressed, to 10mm, spindle-shaped, acute, purplish, hairless, ciliate. Shrub or small tree ....*Juneberry* **Amelanchier lamarckii**

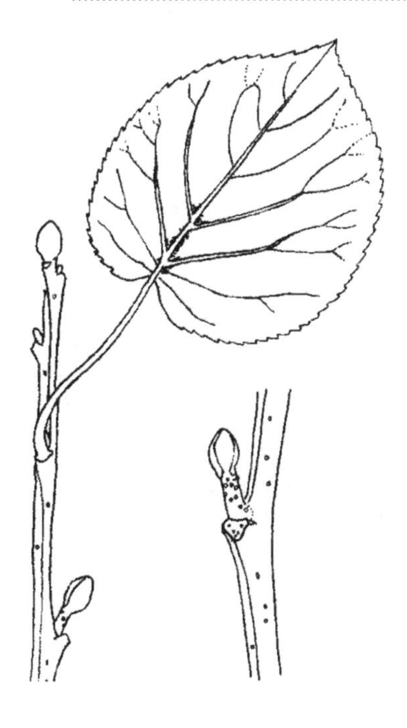

Alnus cordata

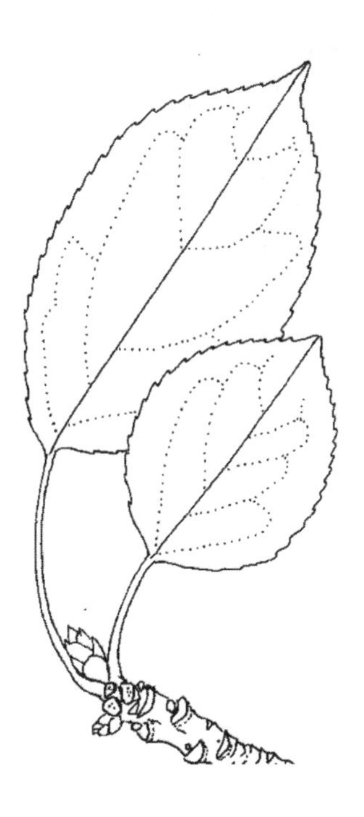

Malus sylvestris

Group NG – *Herb. Stipules, or scale-like reduced lvs, present. (*Caltha palustris *and* Bergenia *spp may key out here due to their inflated sheaths.* Moneses uniflora *may key out here due to its alt stipule-like scale lvs but the typical lvs are opp)*

■Stems >50cm, climbing or sprawling

Stems sprawling, smooth. R alien..*German-ivy* **Delairea odorata**

Stems clockwise-climbing, rough, striate. Lvs usu alt, to 10 x 9cm, ovate (nettle-like), acuminate, cordate at base, v rough above, with sessile glands esp below, deeply dentate-serrate, net-veined, with stomata below only. Petiole to 3cm, channelled, solid, with several vb's in centre. Stipules 1cm, lanc or oblong, entire. PL 17...............................(male plant) *Hop* **Humulus lupulus**

■Stems <50cm, not climbing or sprawling

Stipules fused into ochreae. Lvs revolute when young....(*Polygonaceae* esp *Rumex*) Go to **DOCK**

Stipules free, not fused into ochreae. Lvs involute when young............................(*Viola*) Go to **VIO**

Stipules present as minute gland-like extra-floral nectaries. Lvs rolled when young

Lvs 4-8 x 1-1.2cm, oblanc (occ ovate to broadly elliptic), narrowed to petiole, dull green above, patent-hairy both sides (obscurely septate), antrorsely ciliate, with few obscure teeth, with 6-11 prs of translucent 2° veins anastomosing or fading nr margins, with stomata both sides. Stipules tiny, heart-shaped, brown, occ absent. Stems to 80cm, green, occ reddish, long-hairy, round, ridged. Stoloniferous. Aquatic habs. R alien..........*Water-primrose* **Ludwigia grandiflora**

Stipule-like alt dark brown scales present on caudex (short stem between roots and lvs). Lvs rolled when young

Lvs basal, in raised (aerial) rosette, leathery, obscurely crenate with hydathode-teeth, with translucent anastomosing 2° veins slightly raised above (± not raised below). Petiole often triangular, with 1 vb. Stipule-like scales 5-9(15) x 2-5mm, alt. Shortly rhizomatous per

Lvs strongly net-veined

Lvs in lax rosette, to 3cm, ovate to elliptic-ovate, acute, slightly undulate, dull light green above, finely serrate or crenate with 11-15 teeth per side. Petiole 5-15mm, << lf. All yr. N Br ....*Serrated Wintergreen* **Orthilia secunda**

Lvs not or weakly net-veined

Lvs mostly with >16 teeth per side, with stomata nr veins above. Basal rosettes often in clusters of 2(3)

Lvs to 4.2cm, orb to ovate, obtuse to mucronate, rounded to truncate at base, dull light green above, with (12)14-24(28) teeth per side. Petiole 0.5-4cm, usu < lf ....*Common Wintergreen* **Pyrola minor**

Lvs mostly with <16 teeth per side, with stomata below only. Basal rosette solitary

Lvs dull dark green above, with 9-15(20) teeth per side, to 3cm, orb (to ovate-orb), obtuse, rounded to ± cuneate at base. Petiole 1.5-4.5cm, ≥ lf. N Br ....*Intermediate Wintergreen* **Pyrola media**

Lvs shiny dark or light green above, with (10)14-16(22) teeth per side, to 3.5cm, orb or ovate, obtuse, occ ± cordate at base. Petiole 2.5-8cm, > lf. Dune slacks, chalk-pits ....*Round-leaved Wintergreen* **Pyrola rotundifolia**

Group NH – Euphorbia. *Latex abundant. Lvs all on stems, (±) sessile, spirally arranged, rolled when young (but may appear folded), usu with translucent midrib raised below and 2° veins obscure, with stomata below only (both sides in* E. helioscopia*). Stipules absent. Stems (±) erect, round, often lfless below leaving raised lf scars*

■Per, with v short rhizomes. Stems not scaly at base

Lvs slightly glaucous and sparsely hairy below, hairless above, 5-10 x 0.6-1.2cm, oblong to oblanc-oblong, obtuse, often serrate in distal ½. Stems 30-60cm, reddish, patent-hairy with hairs to 1mm, with bluish latex. Apr-Oct..*Balkan Spurge* **Euphorbia oblongata**

■Per, with short rhizomes. Stems scaly at base

Lvs green with dense patent hairs below to 1mm, (±) hairless above, 3-5 x 1.5cm, oblong, obtuse, often serrate nr apex. Stems 30-60cm, sparsely hairy with hairs to 0.5mm, with bluish latex. Apr-Oct..*Sweet Spurge* **Euphorbia dulcis**

■Ann. Stems never scaly at base

Lvs v obtuse, serrulate in distal ½ (occ more), cuneate at base

Stem lvs 2-3 (with whorl of 5 lvs nr stem apex), 1-3cm, obovate, hairless. Stem usu 1, to 40cm, often sparsely patent-hairy, often with slightly bluish latex. Apr-Oct (all yr) ....*Sun Spurge* **Euphorbia helioscopia**

Lvs (±) acute (occ obtuse), serrulate exc nr base, often cordate at base

Lvs hairless, 2-13 x 0.7(1)cm, obovate-lanc to oblong-lanc, serrulate. Stems to 70cm, reddish, hairless exc for minute sparse hairs on lf scars, with white latex. Shady habs. Apr-Sept ....*Upright Spurge* **Euphorbia serrulata**

Lvs sparsely hairy, 1-5 x 1(1.5)cm, obovate-lanc to oblong-lanc, serrulate. Stems to 70cm, green, hairless or hairy, with white latex. Sunny habs. Apr-Oct ....*Broad-leaved Spurge* **Euphorbia platyphyllos**

Group NI – *Latex usu sparse. Stems without lf scars. Lvs with at least some hairs forked or hooked (unicellular), sinuate-toothed to ± lobed (occ ± entire), with stomata both sides*

▪Lvs with hooked hairs (adhering to woolly clothing). Taprooted. Latex white. Fl buds erect when young

Lvs with some hairs on white swollen bases

Basal lvs usu dead at fl, to 20cm, elliptic to oblanc, with forked (occ trifid) minutely hooked hairs and simple hispid hairs, with midrib not reddish, with translucent 2° veins. Stem lvs clasping. Stems to 80cm, with long simple hairs and short hooked hairs, with cartilaginous ridges, solid. Bi (per). All yr..*Bristly Oxtongue* **Picris echioides**

Lvs without hairs on white swollen bases

Basal lvs dead at fl, to 20cm, linear-lanc, with forked hooked hairs, with midrib reddish above, with indistinct 2° veins. Stem lvs clasping. Stems to 100cm, with long simple hairs and short hooked hairs, with cartilaginous ridges, solid. Bi (per). All yr ....*Hawkweed Oxtongue* **Picris hieracioides**

▪Lvs with forked (or trifid) hairs. Shortly rhizomatous. Latex bluish-white. Fl buds nodding when young

Lf margin with dense hairs (5 hairs per mm). Lvs to 20cm, oblanc, narrowed to petiole-like base, held ± erect, hairy (hairs often trifid), the hairs rarely red-based on midrib above, distantly sinuate-toothed to runcinate-lobed, with indistinct 2° veins. Petiole with (3)5(7) vb's. Scape to 35cm, hairy, twisted, with large hollow. Per. Basic gsld. PL 18 ....*Rough Hawkbit* **Leontodon hispidus**

Lf margin with ± sparse hairs (1-3 hairs per mm). Lvs to 20cm, oblanc, narrowed to a petiole-like base, ± flat on ground, hairy (hairs mostly forked), the hairs often red-based on midrib above, distantly sinuate-toothed to runcinate-lobed, with indistinct 2° veins. Petiole with 3 vb's. Scape to 20cm, hairless (exc at base), ± straight, with small hollow. Bi or per. Often acidic gsld ....*Lesser Hawkbit* **Leontodon saxatilis**

Group NJ – *Latex usu sparse. Lvs hairless or with simple white unicellular hairs, scabrid-ciliate, regularly toothed (occ obscurely so), each tooth tipped with a white hydathode, the veins occ raised above. Petiole channelled, with 1 vb. Stems angled, often with internodal ridges, usu solid.*
➊ *Latex yellow or green*

■Basal or lower lvs sessile or narrowed to indistinct petiole

Lf hydathodes submarginal and extended into long papillae (may shrivel mid-summer)

Lvs 1-5 x 0.4cm, linear-oblong to linear-oblanc, narrowed to short petiole, undulate, with cartilaginous margins ± patently minutely scabrid and sparsely ciliate, crenate to ± entire, with translucent midrib raised below only, with stomata both sides. Stems 5-40cm. Woody stock with persistent lf bases. Bi. All yr. Sunny acidic habs....................*Sheep's-bit* **Jasione montana**

Lf hydathodes not extended, usu confined to tooth apex

Ann

Lvs to 8cm, spathulate to oblong-ovate, ± clasping on stems, usu undulate, minutely white hispid-hairy to hairless, ± antrorsely scabrid-ciliate, obscurely toothed by submarginal hydathodes (turning reddish), with translucent midrib often purplish below, with obscure 2° veins and opaque network of minor veins, with stomata obscure above. Stems to 30cm, ridged, retrorsely white-scabrid-ciliate. Arable

Fls 4-10mm diam. Widespread...............................*Venus's-looking-glass* **Legousia hybrida**

Fls 15-20mm diam. VR, N Hants ....*Large Venus's-looking-glass* **Legousia speculum-veneris**

Lvs 1-2cm, oblong, narrowed to indistinct petiole, ± sessile and slightly decurrent down stems, pale-green, with acrid odour, sparsely hairy, scabrid-ciliate, with 1-5 teeth per side, with obscure 2° veins, occ with stomata below. Stems to 30cm, sparsely hairy to hairless, ± round with 2 ridges, with bitter latex. Hortal, casual............................*Garden Lobelia* **Lobelia erinus**

Bi or per

Lvs (or stems) with at least some hairs

Lvs 3-7 x 1.5-2.5cm, ovate to lanc, ± sessile, mid-green, sparsely hairy (to hairless) above, hairy below on veins, 2-serrate with 20-25 teeth per side, with stomata below only. Stems to 80cm, strongly ridged, hairy. Tufted per. Apr-Oct........*Milky Bellflower* **Campanula lactiflora**

Lvs 3-5 x 1cm, obovate-oblong, narrowed to a short or indistinct petiole, greyish-green, retrorsely scabrid-hairy above and on margins, ± hairless below, crenate with 13-20 teeth per side, with stomata below only. Stems to 50cm, angled, hairless. Bi. All yr ....*Spreading Bellflower* **Campanula patula**

Lvs (and stems) hairless

Lvs often basal, 5-12 x 0.6-1.3cm, linear-lanc to narrowly so, acute, narrowed to indistinct petiole, folded when young, dark green above, with minutely patently scabrid margins, weakly crenate-serrate with 10-25 teeth per side, with midrib raised below, with obscure 2° veins, with stomata below only. Petiole retrorsely scabrid. Stems 30-70cm, ridged. Tufted per. All yr. Dry habs, hortal.......................*Peach-leaved Bellflower* **Campanula persicifolia**

Lvs all on stems, 3-8 x 2.5cm, obovate-oblong, acute, sessile, slightly decurrent and running down stem as v narrow wings, weakly rolled when young, mid-green, minutely ciliate-papillate, weakly serrate, with white midrib raised below, with weakly translucent 2° veins, with stomata both sides. Stems 20-50cm, ridged, square to angled, the v bitter latex with weak acrid odour. Rhizomatous per. Mar-Oct. Acidic damp habs ....*Heath Lobelia* **Lobelia urens**

■Basal or lower lvs with long or distinct petiole (upper stem lvs may be sessile)

Plant sprawling or with above ground rhizomes. Lvs weakly palmate-veined

Damp acidic habs. All yr....................................*Ivy-leaved Bellflower* **Wahlenbergia hederacea**

Dry walls, hortal. All yr

Lvs 3-5cm diam, ovate to ± orb, cordate at base, light green above, bristly hairy both sides, scabrid-ciliate with long cilia to 0.6mm, 2-serrate or 2-dentate with 10-30 teeth per side, with stomata below only. Petiole to 15cm, retrorsely hairy. Fl buds hairy ....*Trailing Bellflower* **Campanula poscharskyana**

Lvs 1-2.5cm diam, ovate to ± orb or reniform, cordate at base, dark green above, sparsely hairy above (often minutely so), scabrid-ciliate with short cilia to 0.1mm, deeply 2-dentate with 6-10 teeth per side, with stomata below only. Petiole to 10cm, often retrorsely hairy. Fl buds hairless..*Adria Bellflower* **Campanula portenschlagiana**

Dry gsld, dunes. All yr

Basal lvs 0.4-1.5cm diam, ovate to ± orb, usu cordate at base, occ purplish below, usu hairless, scabrid-ciliate, entire or serrate-crenate with 0-9 hydathodes, with stomata both sides, with 2° veins slightly raised above and flat below. Stem lvs 15-25 x 1-3mm, linear-lanc, acute, petiolate or narrowed to indistinct petiole, hairless, scabrid-ciliate, occ entire. Petiole >> If on basal lvs, shortly hairy to hairless, with retrorsely scabrid-ciliate channel. Stems to 30cm, creeping to erect, minutely retrorsely hispid below, hairless above, weakly angled, with stomata. All yr...*Harebell* **Campanula rotundifolia***

Plant tufted or with underground rhizomes. Lvs pinnate-veined (± palmate-veined in *C. rhomboidalis*)

Lvs hairless (may be ciliate). Stems hairless

Acidic habs. Basal lvs 3-8 x 2.5-4cm, ovate, obtuse, cordate at base, with minutely scabrid-ciliate (to bluntly so) margins, not ciliate, 1-2-serrate or crenate, net-veined, often with stomata along veins above. Upper stem lvs lanc, sessile. Petiole 4-15cm, hairless, with latex turning yellowish. Stems 30-80cm. Taproot ± swollen. ❶. Mar-Aug. Sussex. Sch8 ....*Spiked Rampion* **Phyteuma spicatum**

Calc turf. Basal lvs 2-4 x 1.5-2cm, ovate-lanc, rounded to ± cordate at base, with minutely scabrid-ciliate (to bluntly so) margins, usu ciliate, crenate-serrate with 8-9 teeth per side, net-veined, with stomata both sides or below only. Lower stem lvs (occ mistaken for basal lvs) cuneate at base. Upper stem lvs linear-lanc, sessile. Petiole 3cm, hairless, often narrowly winged, with white latex. Stems 5-40cm..........*Round-headed Rampion* **Phyteuma orbiculare**

Walls, limestone cliffs, VR alien. Basal lvs 2-5 x 1.5-2cm, orb-ovate, cordate at base, strongly undulate and ± channelled, with ± papillate or antrorsely scabrid-ciliate margins, with c17 teeth per side, net-veined, with midrib raised below, with stomata both sides. Petiole to 8cm ....*Oxford Rampion* **Phyteuma scheuchzeri**

Lvs hairy. Stems hairy or hairless

Plant rhizomatous. Basal lvs 1-several (absent at fr)

Basal lvs long-petiolate, 5-12 x 7cm, triangular-ovate, ± acute, often cordate at base, deeply serrate, roughly hairy, ciliate, net-veined, with 2° veins slightly raised below, with stomata below only. Stem lvs narrower, ± acuminate, the lower petiolate, the upper sessile. Petiole to 30cm. Stems 20-60cm, tough, hairless, weakly ridged. Rhizomes long, ± thick. Roots tuberous (carrot-like). Apr-Oct.....................*Creeping Bellflower* **Campanula rapunculoides**

Plant tufted, often with ± swollen vertical taproot. Basal lvs, if present, usu several in rosette

Lvs >5cm

Basal and lower lvs cordate at base

Basal lvs 6-10cm, ovate to triangular, acuminate (with long terminal tooth), petiolate, ± hispid both sides, 2-serrate-dentate, net-veined. Stem lvs ± truncate at base, shortly petiolate. Petiole to 10cm. Stems 50-100cm, sharply angled and ridged with narrow wings, with creamy green latex. ❶. Apr-Oct ....*Nettle-leaved Bellflower* **Campanula trachelium**

Basal and lower lvs rounded to cuneate at base

Stems 50-120cm. Lvs all on stems

Lvs 6-10(20)cm, ovate or broadly lanc, acuminate, usu slightly decurrent along petiole, 1-2 serrate, shortly hairy both sides. Upper lvs ± sessile. Petiole to 10cm, winged. Stems hairy above, obscurely and bluntly angled. Apr-Oct ....*Giant Bellflower* **Campanula latifolia**

Stems to 60cm. Lvs in basal rosette and on stems

Per (bi), with ± swollen taproot. Lvs 6-8 x 2.5-3cm, linear-elliptic to ovate-oblong, obtuse, cuneate at base, thin, light green, hairy or sparsely so, with minutely scabrid margins, weakly crenate-serrate, with stomata both sides or below only. Stem lvs 3-7 x 0.5-0.8cm, oblanc to linear-lanc, sessile, undulate. Petiole to 8cm. Stems often hispid-hairy below, hairless or hairs mostly on 4-angles above. Oct-Aug ....*Rampion Bellflower* **Campanula rapunculus**

Bi, with taproot not swollen. Lvs 6-11cm, ovate-oblong, obtuse, cuneate at base, narrowed to long or short petiole, dull pale yellow-green, hairless, long-ciliate, crenate-dentate or serrate. Stem lvs to 7cm, lanc, sessile. Petiole to 6cm. Stems hairy. All yr ....*Canterbury-bells* **Campanula medium**

Lvs <5cm. Per

Basal lvs usu 2-4cm, oblong, ovate or linear, obtuse to ± acute, often cordate at base, densely short-hairy, crenate with >30 shallow teeth per side, net-veined. Stem lvs ± clasping or the lower petiolate. Petiole 2-4cm, retrorsely adpressed-hairy. Stems 3-20(50)cm, retrorsely adpressed-hairy, round, solid, with white latex. Mar-Dec. Chalk turf ....*Clustered Bellflower* **Campanula glomerata***

Basal lvs to 2.5cm, orb, obtuse, cordate at base, shortly hispid-hairy above, (±) hairless below, with unequal cilia, crenate or crenate-serrate with up to 10 teeth per side, not net-veined, with stomata below only. Stem lvs shortly petiolate (1-10mm), ovate, to 3 x 1.5cm, shallowly serrate with up to 7 teeth per side. Petiole to 5cm, hairless but channel margins usu ciliate. Stems to 30cm, sparsely hairy but hairless at base, 3-ridged esp above, hollow, with creamy latex. ➊. All yr. VR alien......*Broad-leaved Harebell* **Campanula rhomboidalis**

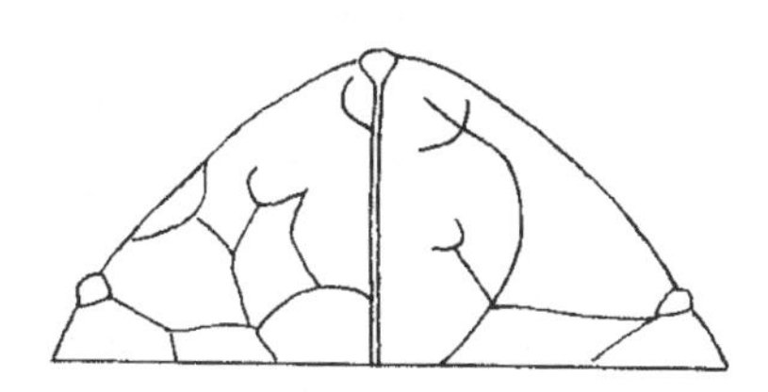

Campanula rotundifolia lf apex

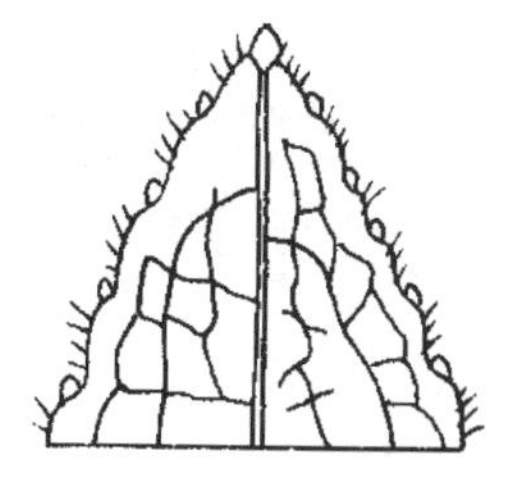

Campanula glomerata lf apex

Group NK – Asteraceae. *Latex usu sparse. Lf teeth without white hydathodes. Scabrid and woolly hairs are unicellular, the simple hairs are septate. (*Sonchus, Arctium *and* Centaurea montana *may key out here)*

■Lvs with some long scabrid hairs (*Hieracium* may key out here)

Lvs white-woolly below with stellate hairs, the lowest rosette lvs may be ± hairless in winter, rolled when young, 1-6 x 0.8-1.5cm, dark green above, obscurely toothed. Per. All yr

Lvs with short (<6mm) scabrid hairs above (occ sparse or confined to margin). Stolons long, slender..*Mouse-ear-hawkweed* **Pilosella officinarum**

Lvs with long (6-10mm) scabrid hairs above. Stolons short, thick. Usu calc sea-cliffs, S Eng ....*Shaggy Mouse-ear-hawkweed* **Pilosella peleteriana**

Lvs not white-woolly below

Rhizomatous (occ stoloniferous). Basal lvs persisting at fl, 5-15(25)cm, oblanc, apiculate, narrowed to a winged petiole (occ indistinct), rolled when young, with bristly scabrid hairs to 6mm (occ with purple swollen bases), with sparse tiny stellate hairs below, sinuate-toothed, with 2° veins weakly visible, with opaque reticulations, with stomata both sides. Stems occ purplish at extreme base, with stellate hairs and yellow-tipped black glandular hairs that have swollen black bases, hollow..*Fox-and-cubs* **Pilosella aurantiaca**

Stoloniferous, stolons often purplish, with swollen-based hairs. Basal lvs persisting at fl, 5-15cm, oblanc, apiculate, narrowed to a winged petiole (occ indistinct), rolled when young, with bristly scabrid hairs 4-6(10)mm, sinuate-toothed, with 2° veins weakly visible, with opaque reticulations, with stomata both sides. Stems reddish at base, with swollen-based hairs, hollow. R native (Shetland) or alien.......................*Shetland Mouse-ear-hawkweed* **Pilosella flagellaris**

■Lvs hairless, or with smooth hairs

Lvs net-veined, usu with margins and/or midrib below with weak spines (often soft to touch)

Lvs usu with midrib reddish above

Basal lvs 15-35 x 4-10cm, obovate, narrowed to petiole, dull green above, ± glaucous or pruinose below, hairless, toothed to pinnately lobed, with stomata both sides. Stem lvs spirally arranged, sessile, clasping and cordate at base, with ± adpressed rounded auricles, ± glaucous. Petiole with 3(7) vb's around small hollow. Stem usu 1, to 2.5m, tough, purplish esp nr base, round, with latex drying orange after 3 minutes (tasteless but with odour of opium-poppy). Ann or bi. All yr..*Great Lettuce* **Lactuca virosa***

Lvs usu with midrib not reddish

Basal lvs 7-15(30) x 1-3(8)cm, linear-oblanc, petiolate, revolute when young, dull grey-green, with sparse white moniliform hairs when young, with midrib raised below, with stomata both sides. Stem lvs sessile, clasping and often sagittate. Petiole with 5(7) vb's. Stem usu 1, to 1.5m, bristly-hairy, round. Ann (bi). Often all yr. Ruderal.......*Prickly Lettuce* **Lactuca serriola***

Basal lvs 4-7 x 0.5-2(5)cm, hairless, ± entire to narrowly acutely lobed (occ runcinate) with long slender terminal lobe. Stem lvs usu held erect and in 1-plane, occ lobed and clasping, sagittate or not, the uppermost linear-lanc and usu entire. Stem usu 1, to 1m, whitish, hairless or bristly below, round. Ann. May-Oct. Shingle and sea-walls, SE Eng. Sch8 ....*Least Lettuce* **Lactuca saligna**

Lvs not net-veined, without spines on margins or midrib

Petiole hollow

Rhizomatous per. Lvs basal and/or on stems, the basal and lower narrowed to a short winged petiole, thin, (±) hairless (hairs occ in lf axils), sinuate-toothed or with runcinate teeth, with stomata below only or obscure above. Stem lvs sessile, clasping with rounded non-adpressed long-pointed auricles. Petiole pinkish-red at base, with 7 vb's. Stems 30-90cm, 3-4mm diam, erect, often reddish below, hairless, slightly ridged esp above, hollow. Mar-Oct. Wet habs, N Br..*Marsh Hawk's-beard* **Crepis paludosa**

Taprooted per. Lvs all basal, shiny bright green above, with midrib often reddish nr base, (±) hairless, with stomata both sides. PL 18........................*Dandelion* **Taraxacum officinale** agg*

Petiole solid

Latex bluish-white (young lvs of other *Crepis* may key out here)

Ann or bi, with taproot

Lvs 5-15cm, ± oblanc, thin, hairless to hairy (occ glandular-hairy in N Br), with midrib sparsely hairy below, toothed to lobed, the terminal lobe usu broader than lf, the lateral lobes more triangular. Often all yr. PL 18..............*Smooth Hawk's-beard* **Crepis capillaris***

Per, with short rhizomes

Lvs 5-15(20)cm, narrowly to broadly oblanc, obtuse to acute, narrowed to a winged petiole, hairy or hairless, ciliate, with reddish midrib, deeply narrowly pinnately lobed to ± sinuate-toothed, the terminal lobe much larger than laterals but usu ± same width as rest of lf, the lateral lobes often linear-oblong and directed out at 90°. Stem lvs absent or reduced to bracts. Petiole with 3-5 vb's. Stems to 60cm, usu hairless, ridged, solid. All yr ....*Autumn Hawkbit* **Leontodon autumnalis***

Lvs 5-10cm, oblanc to ovate, obtuse, narrowed to a winged petiole (occ long), hairless or shortly bristly, with midrib not reddish, entire to sinuate-toothed with 3-12 distant hydathodes, with stomata both sides. Stem lvs to 5(10)cm, oblong, ± acute, sessile, ± clasping with short ± rounded auricles. Petiole with 3 vb's. Stems to 60cm, hairless or hairy, with several shallow cartilaginous ridges, hollow. Apr-Oct (all yr). R, N Eng, Scot ....*Northern Hawk's-beard* **Crepis mollis**

Latex white, turning brown (after 1-2 minutes)

Lvs dark green, never purple-spotted, 4-25cm, broadly oblong-lanc, narrowed to indistinct petiole, usu roughly hairy (occ with reddish-based hairs on midrib), with upturned teeth. Petiole with 3 vb's. Per. All yr...*Cat's-ear* **Hypochaeris radicata***

Lvs dark green with purple spots or streaks above, often with midrib reddish at least nr base, 4-15(30)cm, obovate-oblong, narrowed to indistinct petiole, usu roughly hairy, ± sinuate-toothed (occ pinnately lobed). Petiole with 5-7 vb's. Per. All yr ....*Spotted Cat's-ear* **Hypochaeris maculata**

Lvs usu waxy pale green, often reddish, 1-3cm, oblanc, narrowed to indistinct petiole, (±) hairless, ciliate, sinuate-toothed to -pinnatifid. Petiole with (1)3. Ann. Nov-May ....*Smooth Cat's-ear* **Hypochaeris glabra**

Latex white, not changing colour

Per, with taproot. Lvs basal and on stems, roughly hairy to ± hairless, runcinate-lobed, (±) ciliate, often with reddish midrib. Stem lvs ± clasping at base, sessile, ± glaucous, the upper lanc, entire or distantly toothed, with clasping auricles. Petiole short, retrorsely hairy, with 3-7 vb's. Stems 30-120cm tall, stiffly branched, tough, shallowly grooved, hollow ....*Chicory* **Cichorium intybus***

Ann. Lvs all basal, spathulate or oblanc, narrowed to short petiole, hairless or ± hairy, ciliate, coarsely toothed. Arable. Extinct....................................*Lamb's Succory* **Arnoseris minima**

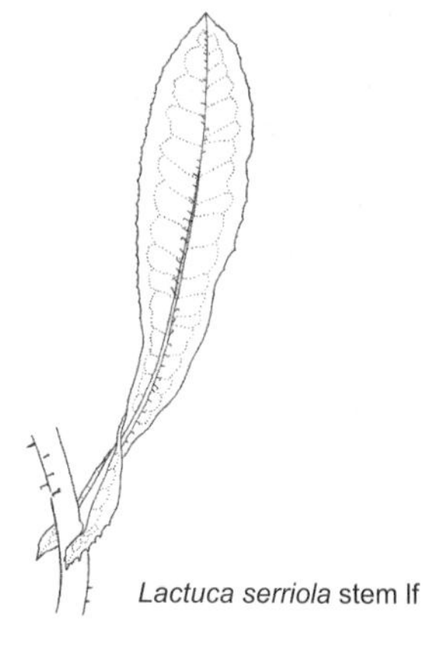

Lactuca serriola stem lf

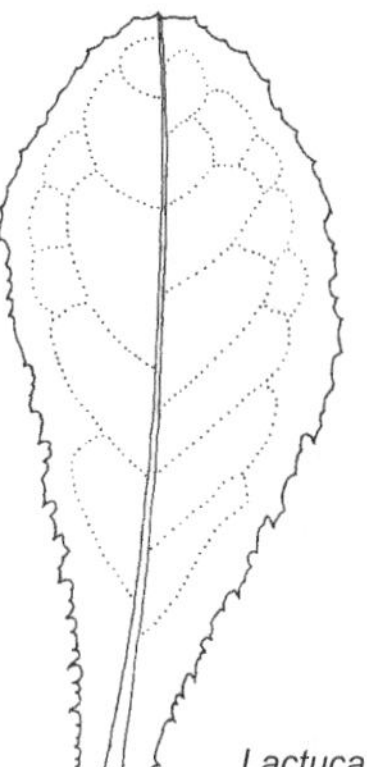

Lactuca virosa basal lf

Group NL – Brassicaceae *(and* Verbascum lychnitis*). Lvs with at least some hairs forked or stellate (hairs unicellular), often with opaque or weakly translucent veins, occ with opaque reticulations, often cress-scented when crushed, with stomata both sides. Petiole with 1(3) vb's unless otherwise stated. Stem(s) (±) round*

■Basal lvs mostly >15cm

Lvs densely white-hairy with stellate hairs

Basal lvs to 30cm, ovate, ± entire. Stem usu 1, to 150cm, stout, hairy. Bi. Calc habs ....*White Mullein* **Verbascum lychnitis**

Basal (and stem) lvs to 15(25)cm, lanc-oblanc, the lower sinuate or pinnately lobed. Stems several, to 60cm, stout, hairy, with occ glandular hairs. Per. Sand dunes, cliffs, R ....*Sea Stock* **Matthiola sinuata**

Lvs not white-hairy

Basal (and lower) lvs 15-50 x 15cm, elliptic, acute, cuneate at base, with mostly forked hairs both sides (occ sparse and with midrib ± retrorsely hispid below), often toothed and/or lobed, with 0-2 prs of ± free lfts nr base. Petiole to 10cm, purplish at base, retrorsely hairy, channelled (often slightly winged), with strong cress odour, solid or with small hollow, with 5-7 vb's. Stems 30-100cm, purplish at base, retrorsely hairy esp at base (sparse above), occ hairless or with sparse glandular hairs, ± round below, ridged above. Per (bi). Mar-Nov. R alien ....*Warty-cabbage* **Bunias orientalis**

Basal (and lower) lvs 10-25 x 8cm, ovate-lanc, acute, rounded to cuneate at base, roughly hairy with simple and short forked hairs, toothed with pale hydathodes. Petiole to 10cm, with 1 pr of minute extra-floral nectary glands at base, channelled, ± winged, with weak cress odour, solid, with 1-7 vb's. Stems 40-90cm, purplish at base, with short simple/forked or stellate hairs, occ retrorsely simple-hairy below, round. Bi (per). All yr.............*Dame's-violet* **Hesperis matronalis**

■Basal lvs <15cm (inc petiole)

Plant usu in rocky mtn habs, often <20cm tall

Lf teeth obscure, usu absent from some lvs

Basal lvs to 10 x 2mm, oblong-lanc, cuneate to sessile base, mid-green, mostly simple- or stellate-hairy, with forked hairs esp on margins, entire or with 1-3 teeth per side. Stem lvs 0-3. Stems to 20cm, stellate-hairy. Per, with 1-2 rosettes on stock ....*Rock Whitlowgrass* **Draba norvegica**

Basal lvs to 15 x 5mm, oblong-lanc, narrowed to petiole or sessile, grey-green, densely hairy (occ sparse) with sessile stellate hairs below and simple or short forked hairs above (occ sessile stellate) above and on margins, entire or with (1)2(4) acute teeth per side. Stem lvs >3, lanc to narrowly ovate, rounded or slightly clasping at base, densely stellate-hairy, ciliate, coarsely toothed, occ entire. Stems 20-60cm, densely hairy with all hair types (variable). Bi. All yr. Usu basic rocks..*Hoary Whitlowgrass* **Draba incana**

Lf teeth (or lobes) clearly present

Mat-forming per. Petiole of basal lvs << lf, broad. Basal lvs 3-6cm, obovate-oblong, narrowed to short broad petiole (± sessile), sparsely stellate-hairy with a few forked and simple hairs, coarsely dentate with 5-7 teeth per side. Stem lvs with clasping rounded auricles ± as long as broad. Stems sparsely stellate-hairy, with a few forked and simple hairs. All yr. VR, Skye. Sch8...*Alpine Rock-cress* **Arabis alpina**

Tufted per. Petiole of basal lvs < lf, narrow. Basal lvs 1.5-3.5cm, lyrate-pinnatifid, long-petiolate, hairless or ± hispid with simple and stellate hairs. Stem lvs much reduced, spathulate, ± hairless, ± entire. Stems hispid below with spreading simple hairs, ± hairless and pruinose above. All yr..*Northern Rock-cress* **Arabis petraea**

Plant usu in lowlands (may climb mtns)

Basal (or lower stem) lvs densely hairy with mostly stellate hairs

Mat-forming

Stem lvs sessile or broadly petiolate. Basal lvs to 3 x 1.5cm, spathulate to rhombic, cuneate at base, greyish, stellate-hairy with 3-5-rayed hairs both sides (occ simple and forked hairs), usu with 1-3 deep teeth per side. All yr. Hortal...........................*Aubretia* **Aubrieta deltoidea**

Stem lvs clasping with ± acute auricles. Basal lvs to 5 x 2cm, oblanc or obovate, ± clasping at cuneate base, grey-green or whitish, with dense short stellate hairs both sides, with 2-4 teeth each side. All yr. Hortal..*Garden Arabis* **Arabis caucasica**

Not mat-forming

Lvs shortly grey- to white-woolly both sides. Per

Basal lvs persistent at fl, 5-12 x 1-2cm, oblanc, narrowed to petiole, ± obvolute when young, with dense adpressed 6-rayed stellate hairs, with (0)4-6 sinuate teeth per side. Stem lvs to 3 x 0.8cm. Petiole with flat arc vb. Stems to 45cm, slender, ± woody at base. All yr ....*Golden Alison* **Alyssum saxatile**

Lvs not woolly. Bi (all yr)

Basal lvs to 8(10)cm, oblanc, broadly petiolate, grey-green, with adpressed stellate hairs above, with stellate and simple adpressed hairs below, entire to sinuate-toothed. Stem lvs to 0.5-2 x 0.6cm, broadly linear-oblong, sessile, entire. Stems 20-80cm, purplish, with adpressed stellate hairs...*Hoary Alison* **Berteroa incana**

Basal lvs to 8cm, oblanc to spathulate, narrowed to indistinct petiole, dark green, with stalked stellate hairs both sides (mostly 3-rayed), weakly denticulate to coarsely sinuate-toothed or lobed. Stem lvs clasping with sagittate auricles, glaucous-pruinose, ± hairless, remotely denticulate. Stems 30-90cm, purplish below, pruinose above, sparsely hairy below with soft (mostly simple) hairs.....................................*Tower Mustard* **Arabis glabra**

Basal lvs to 8cm, ovate-elliptic, narrowed to indistinct petiole, grey-green, densely softly hairy with v short-stalked stellate hairs (mostly 4-rayed), toothed to lobed at least proximally. Stem lvs clasping at base with rounded auricles. Stems 30-90cm ....*Tower Cress* **Arabis turrita**

Basal lvs densely hairy to sparsely so, with mostly forked or simple hairs

Lvs with hairs medifixed (some occ stellate)

Basal lvs dying before fl, 1-5 x 0.4cm, lanc-elliptic, acute to obtuse, shortly petiolate, entire or obscurely sinuate-toothed (rarely lobed). Stems to 100cm, with scattered 2(3)-rayed medifixed hairs. Ann (occ overwintering)...........*Treacle-mustard* **Erysimum cheiranthoides**

Lvs with stellate, forked or simple hairs

Per, with 1 or more compact basal rosettes on short woody stock

Basal lvs to 2.5cm, oblanc, narrowed to indistinct petiole, shiny dark green above, often purplish below, with forked hairs and few simple hairs (mostly along margin), sinuate-lobed with 4-6 rounded lobes per side (length increasing along lf), with water-cress taste. Stem lvs 1-3, tiny, oblong, sessile with ± clasping base, less lobed than rosette lvs. Stems 3-20(30)cm, usu hairless. All yr. VR, Bristol. Sch8..............*Bristol Rock-cress* **Arabis scabra**

Ann (bi in *Arabis hirsuta*), with single basal rosette

Stem lvs clasping

Basal lvs 5-15cm

Basal lvs often in a persistent rosette, oblanc, weakly cress-scented, with some stellate hairs (often sparse above), ± entire or deeply to shallowly pinnately lobed. Stem lvs clasping with ± acute auricles at base, with longer simple hairs. Stems to 20(30)cm tall, with mostly stellate hairs. Plant sparsely to densely hairy, with (3)5-rayed stellate or simple (occ septate) hairs. All yr (usu).......*Shepherd's-purse* **Capsella bursa-pastoris**

Basal lvs rarely persisting to fl, spathulate, broadly petiolate, sparsely to densely hairy, sparsely toothed. Stem lvs often with small clasping auricles. Stems to 100cm, with forked and stellate hairs (and few simple hairs). VR alien ....*Gold-of-pleasure* **Camelina sativa**

Basal lvs to 5cm, in a persistent rosette

Basal lvs to 2.5cm wide, oblong-elliptic to oblanc, narrowed to indistinct petiole, shiny dark green, occ purplish below, hairy (occ densely so) both sides with forked hairs (occ with simple or 3-rayed stellate hairs above), with (0)4-6 teeth per side. Stem lvs often held erect, clasping with short rounded auricles, often hairless, long-ciliate, distantly sinuate-dentate (occ entire). Petiole often ciliate with simple hairs. Stems 20-50(90)cm, usu densely stiffly simple-hairy below, (±) hairless above. All yr .... *Hairy Rock-cress* **Arabis hirsuta**

Basal lvs to 1.5cm wide, obovate to oblanc, narrowed to indistinct petiole, dull grey-green, with forked or stellate hairs above, with mostly stellate hairs below, entire to dentate with up to 7 teeth per side (each tooth with a purple-black hydathode). Stem lvs broadly ovate, obtuse (with v broad terminal tooth), ± clasping at base, with 10 distinct teeth per side. Stems 8-30cm, densely stellate-hairy at least below. Nov-Jul .... *Wall Whitlowgrass* **Draba muralis**

Stem lvs not clasping

Basal lvs 5 x 1.5cm, elliptic or spathulate, ± obtuse, narrowed to petiole, weakly cress-scented, grey-green, with simple/forked/trifid or 3-rayed stellate hairs, with 0-7 teeth per side. Stem lvs few, rounded at base, sessile, ± hairless or with forked hairs below and on margins, ± entire. Petiole often purplish. Stems 5-35cm, hairy below with mostly simple patent hairs, ± hairless above. Nov-Jul............ *Thale Cress* **Arabidopsis thaliana**

Stem lvs absent

Basal lvs 0.5-2 x 0.2cm, usu narrowly spathulate, acute, green, turning reddish, with few forked/stellate or simple hairs, with 0-1 teeth per side. Petiole often as long as lf. Stems 1-10cm. Oct-Apr.. *Common Whitlowgrass* **Erophila verna** agg

Group NM – Verbascum. *Lvs with dendritic hairs, often woolly at least when young, crenate or dentate (± entire in* V. speciosum*), net-veined, with stomata both sides. Petiole with involute vb. Tall bi (per in* V. phoeniceum, V. pyramidatum *and bi to per in* V. chaixii*). All yr*

■Basal lvs with long distinct channelled petiole

Basal lvs cordate at base

Basal lvs to 20cm, ovate, often acuminate, ± rugose, hairy esp below with 0.5-1mm dendritic to forked hairs, crenate, with midrib often purple-tinged nr base. *Dark Mullein* **Verbascum nigrum**

Basal lvs rounded (to broadly cuneate) at base

Petiole (of basal lvs) 5-25cm. Basal lvs 20-30 x 4-12cm, oblong-ovate, sparsely hairy above, densely hairy below. Stem lvs 10-18 x 4-5cm, oblong-lanc ....*Nettle-leaved Mullein* **Verbascum chaixii**

Petiole to 3cm. Basal lvs 4-10 x 6cm, ovate to oblong, hairless to hairy. Stem lvs small, sessile ....*Purple Mullein* **Verbascum phoeniceum**

■Basal lvs sessile or with short (or indistinct) petiole

Lvs (and stems) with at least some glandular hairs

Basal lvs 5-20cm, elliptic, cuneate at base, shortly petiolate, not rugose, sparsely short-hairy both sides, occ glandular-hairy, often with forked or weakly dendritic hairs, dentate with c18 teeth per side. Stems with decurrent ridges. Fls yellow, 1-6 per node ....*Twiggy Mullein* **Verbascum virgatum**

Basal (and stem) lvs 5-20cm, elliptic-lanc, truncate to cuneate at base, shortly petiolate, rugose, shiny green, with sessile glands above, occ dendritic-hairy, crenate-dentate or lobed. Stems without decurrent ridges. Fls pinkish or yellow, 1 per node...*Moth Mullein* **Verbascum blattaria**

Lvs (and stems) without glandular hairs

Lf hairs v short, ≤0.5mm

Lvs (±) entire, sessile

Basal lvs to 40 x 10cm, lanc, usu 3-5x longer than wide, with strongly branched dendritic hairs (occ yellow and appearing stellate from above). Petiole winged ....*Hungarian Mullein* **Verbascum speciosum**

Lvs distinctly crenate, ± sessile or narrowed to short petiole

Basal lvs 10-30 x 16cm, broadly ovate to oblong, grey-woolly when young, with minute stalked dendritic hairs (to 0.5mm), crenate. Stems round, weakly striate ....*Hoary Mullein* **Verbascum pulverulentum**

Basal lvs 10-30 x 10cm, ovate-elliptic, ± sessile, white-woolly, becoming green, with ± sessile 0.2mm hairs (slightly longer below), deeply crenate. Stem lvs ± clasping, less hairy than basal lvs. Stems weakly angled, often glandular nr infl ....*Caucasian Mullein* **Verbascum pyramidatum**

Basal lvs 10-30 x 6cm, ovate or oblong to lanc, whitish-woolly at least below when young, with minute sessile ± dendritic hairs (appearing stellate), shallowly crenate. Stems angled ....*White Mullein* **Verbascum lychnitis**

Lf hairs longer, >0.5mm

Stem lvs strongly decurrent down stem

Basal lvs 15-45cm, broadly ovate, (±) acute, narrowed to a 2-3cm winged petiole, densely greyish- or white-woolly both sides, with (0.5)1-5mm dendritic hairs (5-7-whorled). Stem lvs sessile, usu ± decurrent to lf below. Stems winged. Stigma swollen at apex ....*Great Mullein* **Verbascum thapsus**

Basal lvs similar to *V. thapsus*. Stem lvs sessile, decurrent to lf below. Stems strongly winged. Stigma spathulate. VR alien........*Dense-flowered Mullein* **Verbascum densiflorum**

Stem lvs not or weakly decurrent

Basal lvs to 15 x 6cm, broadly ovate, yellowish-woolly, with hairs to 2mm. Petiole to 5(6)cm ....*Orange Mullein* **Verbascum phlomoides**

Basal lvs to 15 x 6cm, broadly ovate, white-woolly esp below, with hairs matted and adpressed above. Petiole to 1cm. VR alien......*Broussa Mullein* **Verbascum bombyciferum**

Group NN – *Lvs with white or yellowish meal at least below when young, or with bladder-like hairs*

■Lvs all on stems, applanate when young

Lvs (or stems) with bladder-like hairs....................................(*Chenopodium*, *Atriplex*) Go to **CHAT**

■Lvs all basal, revolute when young (often with recurved margins at maturity)

Lvs with yellow sessile glands below, 10-30cm, rugose...*Candelabra Primrose* **Primula helodoxa**

Lvs white- or yellowish-mealy below

Lvs 10-30cm, rugose

Lvs to 8(10)cm wide, obovate to oblanc, narrowed to petiole, dentate. VR alien ....*Mealy Cowslip* **Primula pulverulenta**

Lvs <6cm, not rugose

Lvs 2-6 x 0.5-1.5cm, oblanc-spathulate to obovate, obtuse, gradually narrowed to petiole (not purplish at base), whitish below (and above when young), minutely sinuate-denticulate. Per. Apr-Oct. N Eng..*Bird's-eye Primrose* **Primula farinosa**

Lvs 2-3 x 0.8-2cm, obovate to oblong-ovate, obtuse to ± acute, abruptly narrowed to short petiole (purplish at extreme base), yellowish-white below, minutely crenulate. Per (bi). All yr. N'most Scot...*Scottish Primrose* **Primula scotica**

Group NO – Asteraceae. *Lvs with some woolly or cottony hairs (occ v sparse), the simple hairs are septate*

■Basal (or lower stem) lvs with distinct petiole (often long)

Basal lvs cordate at base

Petiole channelled above, ridged below, rarely with sparse latex. Tufted monocarpic per

Basal lvs 2-several, to 40 x 30cm, ovate, rolled when young, often undulate, rugose, septate-hairy above, cobwebby below at least when young, usu crenate-denticulate, with prominent green hydathodes turning purple, net-veined, with stomata both sides. Petiole hollow, with 7(10) vb's. Apr-Sept..*Burdock* **Arctium minus** agg

Petiole round, without latex. Rhizomatous per

Basal lvs 2(3) per shoot, 10-20cm diam, ± orb, revolute when young, with white cobwebby hairs above when young, septate-hairy below on veins (minutely gland-tipped, the tips soon withering), sinuate-denticulate with purple hydathodes, net-veined, with stomata below only. Petiole reddish, sheathing at base, glandular septate-hairy, solid or with small hollow, soon turning yellow-brown, with 10-15 scattered vb's. Sept-Jul. PL 12, 23 ....*Winter Heliotrope* **Petasites fragrans***

Basal lvs not cordate. Petiole channelled

Basal lvs green below, 5-20 x 5cm, ovate-lanc to lanc, acute, petiolate, rolled when young, septate-hairy both sides, sinuate-toothed (occ pinnately lobed) with protruding purple hydathodes (occ obscure), with midrib often purplish nr base, with stomata both sides. Stem lvs sessile, entire to pinnately lobed. Petiole to 15cm, winged, with 3(5)vb's. Stems usu septate-hairy and cottony, grooved. Tufted per. Feb-Nov ....*Common Knapweed* **Centaurea nigra**

Basal lvs white-cottony below, 5-11 x 4cm, ovate-lanc, acute, petiolate, rolled when young, soon hairless above, sharply sinuate-toothed with hydathodes (to ± entire), with stomata both sides. Stem lvs sessile. Petiole to 5cm, purplish, cottony, with 3(5) vb's. Stems ± cottony, angled to grooved. Shortly stoloniferous per. Apr-Oct. Usu mtns............*Alpine Saw-wort* **Saussurea alpina**

■Basal lvs (if present) with indistinct or short petiole (to sessile)

Sand dunes or shingle, SE Ire

Plant aromatic, white-woolly. Lvs all on stems, spirally arranged, to 2 x 1cm, oblong-lanc, sessile, weakly toothed nr apex. Rhizomatous. All yr..............*Cottonweed* **Otanthus maritimus**

Damp or wet habs

Stem lvs 7-15cm, sharply serrate, sessile. Basal lvs narrowed to short petiole-like base

Stem lvs 1(2)cm wide, narrowly lanc, acute, ± clasping at base, ± cottony below, with anastomosing 2° veins ± forming submarginal vein. Stems 0.8-2m, 5-10mm diam, sparsely cottony, round, v shallowly ridged, hollow. Tufted per. Apr-Oct. Fen ditches, VR, Cambs. Sch8..*Fen Ragwort* **Senecio paludosus**

Stem lvs <6cm, obscurely toothed with hydathodes, sessile. Basal lvs absent

Rhizomatous per. Lvs to 5(8) x 1-2.5cm, oblong-lanc, ± acute, clasping with auricles at base, revolute when young, with recurved margins at maturity, undulate, rugose, without glands, weakly aromatic, hairy both sides, with stomata both sides. Stems to 50cm, not purplish, hairy, cottony, round. Apr-Oct.................................*Common Fleabane* **Pulicaria dysenterica**

Ann. Lvs to 4 x 0.5-1.5cm, elliptic-lanc to oblanc, ± clasping, not or hardly cordate at base, undulate, with yellow sessile glands below, fetid, sparsely hairy both sides, with stomata both sides. Stems to 40cm, purplish, slightly glandular-hairy, round. May-Oct. VR, New Forest. Sch8..*Small Fleabane* **Pulicaria vulgaris**

Dry calc turf (rarely on sea-cliffs in N Wales)

Basal lvs persistent at fl, ± adpressed to ground, 2-5(10) x 1.5-3(5)cm, broadly ovate to ± orb, abruptly narrowed to short winged petiole, leathery, hairy both sides, with cottony hairs esp when young, entire to remotely dentate, with indistinct 2° veins. Stems 7-30cm, cottony when young. Bi or per. Apr-Oct (all yr)....................................*Field Fleawort* **Tephroseris integrifolia**

Ruderal, hortal or arable

Lvs revolute when young, with recurved margins at maturity

Basal lvs absent. Stem lvs 4-7(12)cm x 5mm, linear, weakly clasping-auriculate at base, sessile, ± entire to (rarely) pinnately lobed. Stems 30-60cm, ± woody at base, ± hairless, angled. Tufted per. Apr-Oct. Ruderal................*Narrow-leaved Ragwort* **Senecio inaequidens**

Lvs rolled when young, with ± flat margins when mature

Plant viscid, weakly citrus-scented, with glandular hairs

Lvs 7-12 x 1.5-2.5cm, oblong, with apiculate hydathode at apex, sessile, clasping, bright green, with prominent hydathode-teeth along margins, often weakly 3-pli-veined at base, with midrib raised below, with indistinct opaque pinnate 2° veins not sunken above, with stomata both sides. Stems to 60cm, ridged, pith-filled. Ann to per. Usu all yr ....*Pot Marigold* **Calendula officinalis**

Lvs 3-7 x 1-2cm, oblong-lanc, with apiculate hydathode at apex, sessile, clasping, grey-green, with prominent hydathode-teeth along margins (soon black), often weakly 3-pli-veined at base, with midrib raised below, with indistinct opaque pinnate 2° veins not sunken above, with stomata both sides. Stems to 30cm, ridged, pith-filled. Ann. Usu all yr. VR alien ....*Field Marigold* **Calendula arvensis**

Plant not viscid, odourless, without glandular hairs

Per, with rhizomes (and occ stolons). Basal lvs 10-15 x 2-4cm, broadly oblanc, with long recurved terminal hydathode at apex, narrowed to indistinct petiole, cottony to woolly (at least when young), usu with additional septate hairs both sides, later hairless, entire but with distant hydathodes (often obscured by hairs), not or hardly net-veined, with stomata both sides. Upper lvs sessile. Petiole strongly winged, decurrent, with 3(5) vb's, often with sparse orange sap. Stems 30-80cm, usu unbranched, cottony, broadly winged. All yr ....*Perennial Cornflower* **Centaurea montana**

Ann (bi). Basal lvs ± absent or similar to *C. nigra*. Lower stem lvs 5-20 x 0.2-4cm, usu lyrate-pinnatifid with narrow distant lobes, rarely oblanc, acute with hydathode at apex, narrowed to distinct petiole, greyish with persistent cottony hairs, entire but with distant hydathodes, not or hardly net-veined, with stomata both sides. Upper lvs 7-15 x 0.4-1cm, linear-lanc, narrowed to ± sessile base, with obscure hydathodes, occ lobed, 3-veined. Petiole winged, with 3(5) vb's. Stems 20-90cm, branched, sparsely cottony, grooved or angled, with cartilaginous ridges. Apr-Sept (all yr).....................................*Cornflower* **Centaurea cyanus**

Group NP – Plantago. *Lvs all basal, spirally arranged in rosette(s), septate-hairy or hairless, always with silky hairs at extreme base, with 3-9 parallel veins raised below, with stomata both sides*

■Lvs acute or apiculate. Fl stalk ribbed. Shortly rhizomatous per, with 1-3 rosettes

Lvs 5-30cm, linear-lanc (to elliptic), sessile and narrowing gradually to base, often concave, not or hardly undulate, hairy to ± hairless, with long slender (>1 x 0.05mm) hairs, with 3-5 translucent main veins slightly sunken above. Petiole not purplish at base, with long white silky hairs in rosette centre. PL 9..*Ribwort Plantain* **Plantago lanceolata***

■Lvs (±) obtuse. Fl stalk round

Shortly rhizomatous per, with 1-3 rosettes. Lvs ± sessile or with short petiole. All yr

Lvs 5-8cm, ovate, hairy, with medium stout (<1 x 0.2mm) hairs, thick, convex, undulate, with 3-5(7) opaque main veins (HTL), without cross-veins. Petiole often purplish at base, with brown silky hairs in rosette centre..*Hoary Plantain* **Plantago media***

Tufted per or ann, with single rosette. Lvs with distinct petiole. Mar-Nov

Lvs 7-15cm, elliptic to ovate, often undulate, ± thin, often with shortish stout (to 0.6 x 0.1mm) hairs, with (3)5-9 translucent main veins. Petiole not purplish at base, with white (occ brown) silky hairs in rosette centre

Lvs ± hairless, cuneate to ± cordate at base, usu dark green, 5-9-veined. Per. Usu dry habs ....*Greater Plantain* **Plantago major** ssp **major**

Lvs minutely hairy, cuneate at base, usu pale green, 3-5-veined. Usu ann. Damp or saline habs..*Greater Plantain* (ssp) **Plantago major** ssp **intermedia**

Group NQ – *Lvs strongly net-veined with translucent veins. (If hairs sharply hispid and unicellular (*Trachystemon, Borago officinalis*)* <u>*Go to*</u> ***KN***. *If plant has revolute lvs when young and the remains of ochreae (*Rumex*)* <u>*Go to*</u> ***DOCK****)*

■At least some lvs basal (usu in a rosette). Per or bi, tufted or rhizomatous

Lvs revolute when young, with margins recurved at maturity, odourless, to 20cm, ovate-oblong to obovate-spathulate, often rugose above, crenate with prominent green hydathodes, with stomata both sides. Petiole reddish at base, with 1(3) vb's. Hairs septate. All yr (occ summer dormant)

Lvs hairy above, suddenly narrowed into petiole

Lvs shortly hairy both sides (minutely glandular). Petiole < lf. PL 11......*Cowslip* **Primula veris***

Lvs hairy to sparsely so above, hairy below. Petiole often ≥ lf. Woods, East Anglia ....*Oxlip* **Primula elatior**

Lvs hairless above (exc midrib), gradually narrowed into petiole

Lvs hairy below. PL 11..........*Primrose* **Primula vulgaris***

Lvs with sessile yellow glands below. VR alien...........*Candelabra Primrose* **Primula helodoxa**

Lvs not revolute when young

Lvs (weakly) aromatic when crushed. Hairs septate

Basal lvs 25-60 x 20cm, ovate-elliptic, cuneate to rounded at base, involute when young, rugose, dull yellow-green above, hairy above (soon hairless), softly hairy (± woolly) below, crenate, each tooth with a prominent purplish hydathode, with 2-3° veins raised below, with stomata both sides. Stem lvs short-petiolate or sessile, the upper cordate and clasping. Petiole to 30cm, winged nr apex, with c10 vb's around margin. Stems to 150cm, tough, ± woolly, round. Rootstock tuberous. Apr-Oct..........*Elecampane* **Inula helenium**

Basal lvs to 15(25) x 5cm, obovate or oblanc, cuneate at base, rolled when young, often crinkled, dull yellow-green above, the midrib and margins often purplish, sparsely hairy both sides or on midrib below only, ciliate, serrate with (3)12 teeth per side, with 2° veins hardly raised below, with stomata both sides. Stem lvs sessile, slightly clasping stem, occ entire. Petiole to 15cm, winged, with 3(5)vb's. Stems to 70cm, ± brittle, hairless or hairy, ± round. Apr-Oct..........*Goldenrod* **Solidago virgaurea**

Basal lvs to 12 x 4cm, elliptic-lanc, cuneate at base, rolled when young, rugose, dark yellow-green above, densely softly hairy both sides, crenate, each tooth with a prominent purplish hydathode, with 2-3° veins raised below, with stomata both sides. Petiole to 5cm, winged, with 1(3) vb's. Stems to 125cm, tough, densely hairy, ± round or weakly angled. All yr. Calc habs..........*Ploughman's-spikenard* **Inula conyzae**

Lvs weakly garlic or cress/mustard scented when crushed. Hairs, if present, unicellular (often confined to petiole). Petiole channelled, without sheathing base (but ± dilated)

Basal lvs cordate at base

Lvs weakly garlic-scented when crushed. Petiole with 3(5) vb's

Basal lvs to 8cm diam, reniform to ovate, the basal sinus open, rolled when young, weakly rugose, ± hairless, ciliate, crenate, each tooth retuse with a sunken hydathode, with distinct 4° veins, with stomata both sides. Stem lvs triangular-ovate, acuminate, cordate. Petiole often >10cm, purplish nr base, sparsely long-ciliate (esp nr base). Stems purplish at base, hairy below, hairless and pruinose above. Bi (ann). All yr. PL 13, 23 ....*Garlic Mustard* **Alliaria petiolata**

Basal lvs to 12cm diam, reniform to ovate, the basal sinus closed, rolled when young, not rugose but thicker than *A. petiolata*, hairless, usu ciliate, crenate, each tooth with a sunken black-purple hydathode (± not retuse), with obscure 4° veins, with stomata both sides. Rhizomatous per. All yr. VR alien..............*Caucasian Penny-cress* **Thlaspi macrophyllum**

Lvs weakly cress-scented when crushed. Petiole with 3(7) vb's

Upper lvs sessile. Bi. All yr. PL 13..........*Honesty* **Lunaria annua**

Upper lvs petiolate. Rhizomatous per. Mar-Oct. VR alien ....*Perennial Honesty* **Lunaria rediviva**

Basal lvs not cordate at base, mustard-scented when crushed. Petiole with >7 vb's

Basal lvs hairless, often >30cm, ovate to oblong, rounded at base, crenate-serrate with retuse hydathodes involute when young. Stem lvs linear to elliptic, sinuate-toothed, the lower often ± pinnately lobed. Petiole to 30cm, slightly sheathing at base, laterally flattened, acutely V-channelled, with 17 vb's around margin. Stems 50-120cm, purplish at base, hairless, strongly ridged. Tufted, with swollen roots tasting of horse-radish (or strong mustard). All yr..*Horse-radish* **Armoracia rusticana***

Basal lvs shortly white-hairy both sides, to 30cm, ovate, slightly decurrent down petiole, toothed (or pinnately lobed with large terminal and 2 smaller lateral lobes), the veins slightly raised both sides, with stomata both sides. Stem lvs hairless, (±) entire, acute. Petiole to 15cm, slightly sheathing at base, v shallowly channelled, with 7-11 vb's around margin. Stems 50-120cm, hairless. Rhizomatous. Apr-Oct (all yr)......*Dittander* **Lepidium latifolium***

Lvs odourless when crushed

Lvs (±) palmately veined

Lvs basal or on stems, 20-45 x 20-40cm, broadly triangular or ovate, acuminate, cordate at base, rarely with cottony hairs, scabrid-ciliate, deeply dentate, occ lobed, palmately veined, with veins raised below, with stomata both sides or below only. Petiole v long, ± sheathing at base, often purplish, hairless or with cottony hairs, round, with many scattered vb's. Stem lvs to 12cm. Stems to 150cm, purplish, hairless, round. Per. *Leopardplant* **Ligularia dentata**

Lvs basal, to 8 x 6cm, ovate, obtuse, cordate at base, hairless, crenate-serrate (each tooth mucronate), usu net-veined, with anastomosing 7-9 ± palmate veins slightly raised both sides, with stomata both sides. Upper leaves sessile, 3-5-lobed. Petiole to 13cm, not sheathing at base, with wide shallow channel above, striate below, flat, with 10-12 oblong hollows in TS, with 10-12 vb's in centre of partitions. Stems to 90cm, whitish to purple, hairless, round. Per. All yr..*Blue Eryngo* **Eryngium planum**

Lvs pinnately veined

Lvs with sheathing base...............................(*Caltha palustris*, *Ranunculus ficaria*) Go to **RAN**

Lvs without sheathing base

Lvs with glandular or forked hairs (often sparse sessile glands above)

Lvs in basal rosette or on stem, 5-20cm, oblong, the lower shortly petiolate, the upper sessile and pinnately lobed, ± rugose, occ dendritic-hairy, occ minute forked and glandular hairs below, shortly glandular-hairy both sides, crenate-serrate or dentate, occ lobed, not v strongly net-veined, with whitish 1-4° veins raised below, with stomata both sides. Petiole to 2cm, with flat arc or involute vb. Stem usu 1, 30-120cm, sparsely glandular above, angled. Infl glandular. Bi.................*Moth Mullein* **Verbascum blattaria**

Lvs with unicellular hairs (occ hairless)

Lvs often undulate or rugose, with some whitish 2° veins fading nr margin. Petiole with a flat arc to involute vb. Taproot long..(*Oenothera*) Go to **OENO**

Lvs with septate hairs

Basal lvs 10-30cm, ovate to lanc, rugose, hairy to densely so both sides, crenate, with stomata both sides. Petiole often purplish nr base, winged, with flat arc vb. Bi. All yr. Acidic habs...*Foxglove* **Digitalis purpurea**

Basal lvs 5-20cm, ovate-lanc to lanc, not rugose, hairy both sides, sinuate-toothed (occ pinnately lobed) with protruding purple hydathodes (occ obscure), with stomata both sides. Stem lvs sessile, entire to pinnately lobed. Petiole often purplish nr base, winged, with 3(5)vb's. Tufted per. Feb-Nov........................*Common Knapweed* **Centaurea nigra**

■All lvs on stem(s) (never in a rosette). Ann

Stems translucent, v hollow ....(*Impatiens*) Go to **IMP**

Stems opaque, pith-filled or with small hollow

Lvs elliptic-lanc to oblanc, pinnate-veined. Sch8...................*Small Fleabane* **Pulicaria vulgaris**

Lvs broadly ovate, pinnate-veined...(*Nicandra physalodes*) Go to **SOL**

Lvs broadly ovate, strongly 3-veined at base................................*Sunflower* **Helianthus annuus**

■All lvs on stems (rarely in a rosette in wintergreen plants). Rhizomatous per

Lvs revolute when young (margins often remaining recurved at maturity)

Lvs 10-40 x 3-6cm, elliptic to lanc, acute, cuneate and hardly clasping at sessile base, not undulate, with cartilaginous margins (occ scabrid-ciliate), dark green above, paler below, serrate or dentate with hydathode-tipped teeth, occ with all veins opaque, usu with stomata below only. Stems to 2m, occ purplish at base, hairless below, hairy above (occ glandular-hairy), ridged. Apr-Oct. R alien................................... *Broad-leaved Ragwort* **Senecio fluviatilis**

Lvs 5-15 x 2cm, elliptic-lanc, acute with dark scale at apex (and double hydathode), ± sessile, often ± undulate, mid-green above, becoming reddish (at least midrib and margin), glaucous below, with small or obscure distant hydathode-teeth, with midrib raised below, with anastomosing 2° veins forming a wavy submarginal vein, with stomata below only. Stems to 1.5m, often reddish at base, rarely with minute white adpressed hairs above, ± round. Apr-Oct .... *Rosebay Willowherb* **Chamerion angustifolium***

Lvs rolled when young

Lvs 1-veined at base, without glandular vein swellings

Lvs >2.5cm wide, broadly ovate

Lvs opp in unequal-sized prs (at least above) on same side of stem, broadly cuneate to ± cordate at base, sparsely hairy both sides, entire to coarsely dentate. Stems with swollen nodes above, fetid... (*Physalis alkekengi*) Go to **SOL**

Lvs usu <2.5cm wide, lanc to elliptic-oblong

Lvs without 2° veins raised below... (*Aster*) Go to **AST**

Lvs with 2° veins raised below, 2-6 x 0.5-2.5cm, held erecto-patent, acute with long terminal hydathode, cordate at clasping base, rolled when young, often folded at maturity, dull green above, paler below, hairless or ± hairy on veins below, ciliate, entire or remotely denticulate with hydathodes (often blackish). Stems 25-75cm, ± hairless, ridged. Strongly rhizomatous, occ rooting at lowest nodes. VR, Ire.. *Irish Fleabane* **Inula salicina**

Lvs weakly 3-veined from base, with minute glandular vein swellings (weakly citrus-aromatic)

Lvs hairy at least below on veins, 5-16 x 1.5-3cm, lanc, acuminate to ± aristate, cuneate at base, ± sessile, with (3)10-14 ± equal teeth per side, with stomata below only. Stems to 1.5m, 3-5mm diam, green, purplish at base, sparsely septate-hairy above, often hairless below, nodding in bud, round, solid. Rhizomes purplish... *Canadian Goldenrod* **Solidago canadensis**

Lvs hairless but antrorsely ciliate, 5-16 x 2-4cm, elliptic, acuminate to ± aristate, cuneate at base, ± sessile, with 15-18 often v unequal teeth per side, with stomata below only. Stems to 2m, 5-8mm diam, often pruinose, purplish at base, hairless or sparsely minutely septate-hairy above (hairs crisped or glandular), hairless below, nodding in bud, round, solid. Rhizomes purplish... *Early Goldenrod* **Solidago gigantea**

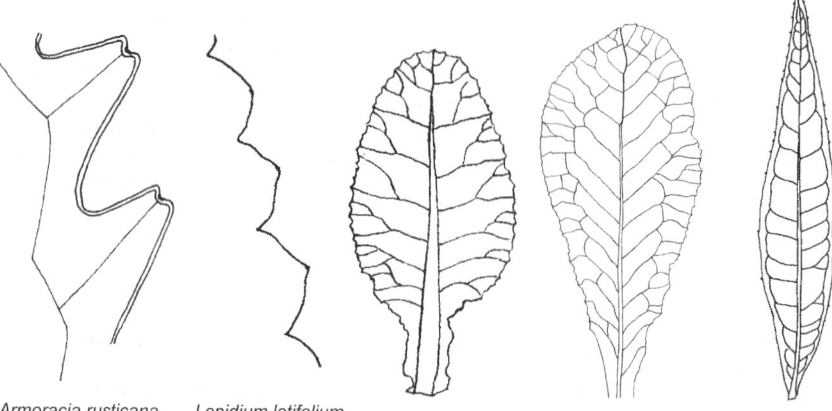

Armoracia rusticana lf teeth | *Lepidium latifolium* lf teeth | *Primula veris* | *Primula vulgaris* | *Chamerion angustifolium*

Group NR – *Lvs not net-veined, with at least some glandular-septate hairs (occ viscid)*

■Lf teeth with submarginal pore above (HTL) and cartilaginous margins. Lf veins obscure (opaque to weakly translucent), hardly raised or sunken either side. Petiole with at least some septate (occ glandular) hairs, with 1(3) vb's. Per. All yr

Petiole long, distinct. Lvs with stomata below only

Lvs cordate, orb, with narrow cartilaginous margins. Petiole 1-2mm diam, (±) round

Lvs 3-6cm, thick, dark to bright green, often purplish below, bristly, dentate with 7-9 teeth per side, terminal tooth longer than laterals (usu much wider). Petiole >> lf, bristly (hairs often minutely gland-tipped). Stems glandular-hairy. Hairs multiseriate (x100). Rosette-forming. VR alien..*Round-leaved Saxifrage* **Saxifraga rotundifolia**

Lvs 1.5-3cm, thin, bright green, bristly, dentate with (6)11-13 teeth per side, terminal tooth shorter than laterals (usu slightly wider). Petiole >> lf, bristly (hairs often minutely reddish gland-tipped). Stems glandular-hairy. Hairs multiseriate. Stoloniferous ....*Kidney Saxifrage* **Saxifraga hirsuta***

Lvs cuneate to ± truncate, spathulate, with wide cartilaginous margins. Petiole 2-4mm wide, flattened

Lvs crenate or crenate-serrate with 4-10 obtuse teeth per side, terminal tooth shorter and broader than laterals, 2-3(5)cm, hairless, dark green, thick. Petiole usu < lf, densely long-ciliate. Stems glandular-hairy, reddish. Hairs multiseriate on margins. Cushion forming ....*Pyrenean Saxifrage* **Saxifraga umbrosa**

Lvs dentate with 9-12 teeth per side, with terminal tooth ± equal to laterals, 2-3(5)cm, thick, dark green, glandular-hairy to hairless. Petiole often > lf, with wavy cilia (occ glandular). Stems sparsely glandular-hairy. Hairs multiseriate. Stoloniferous. *S. spathularis* x *umbrosa*. PL 22..*Londonpride* **Saxifraga** x **urbium**

Lvs dentate with 4-7 acute teeth per side, with terminal tooth as long or longer than laterals, 2-3(5)cm, thick, dark green, hairless. Petiole >> lf, sparsely ciliate nr base. Stems glandular-hairy. Hairs multiseriate. Stoloniferous. Ire. PL 11 ....*St Patrick's-cabbage* **Saxifraga spathularis***

Petiole short or indistinct. Lvs with stomata both sides. Mtns (*Saxifraga aizoides* may key out here in error since the scabrid-cilia of the lf margin resemble teeth)

Lvs serrate-dentate with 1-3 distal teeth per side, 0.5-3cm, obovate-spathulate or oblanc, acute or obtuse, ± thick, often purplish below, with sparse long (minutely glandular) hairs above, usu hairless below, ciliate, obscurely 3-pli-veined. Petiole ± indistinct. Stems with scattered minutely glandular hairs. Hairs multiseriate. Stoloniferous, with 1-several basal rosettes, often elongating into lfy shoots..*Starry Saxifrage* **Saxifraga stellaris**

Lvs crenate-serrate with 4-5 teeth per side, 1-2cm, orb or obovate-spathulate, obtuse, ± thick, with margins often upturned, usu purplish below, hairless above, sparsely glandular-hairy below, minutely ciliate. Petiole to 1.5cm, broad, with long wavy cilia nr base. Stems on short thick stock, with purple glandular hairs. Hairs uniseriate. Rhizomatous, with 1-few basal rosettes..*Alpine Saxifrage* **Saxifraga nivalis**

■Lf teeth without submarginal pore but often with narrow cartilaginous margins

Basal lvs 2-5, 1.5-3.5cm, orb, cordate at base, yellow-green, with sparse stout patent minutely glandular hairs both sides, crenate with 9-11 shallow slightly retuse truncate lobes, each lobe with a protruding hydathode, with obscure veins. Petiole to 9cm, hairy, channelled, with 1 vb. Stems ± prostrate, weak, with elastic stele. Non-fl stems (stolons) lfless. All yr ....*Alternate-leaved Golden-saxifrage* **Chrysosplenium alternifolium**

■Lf teeth without submarginal pore or cartilaginous margins

Stems trailing or prostrate. Ann

Plant densely glandular-hairy. Lvs 2.5-6cm, ovate to ± orb, obtuse, rounded or ± cordate at base, shortly petiolate, grey-green, hairy both sides, occ slightly toothed, 3(5)-pli-veined from base, with stomata both sides. Stems usu 20-50cm, c1mm diam, branched from base, with patent long nonglandular and short glandular hairs, round. May-Oct ....*Round-leaved Fluellen* **Kickxia spuria**

Plant (sparsely) hairy, hardly glandular. Lvs 1.5-3 x 1-2cm, ovate, ± obtuse to acute, hastate at base (at least upper and middle lvs), shortly petiolate, grey-green, hairy both sides, occ with 4 teeth per side, with few anastomosing veins (occ fading nr margins), with stomata both sides. Stems usu 20-50cm, c1mm diam, branched from base, with unequal spreading to patent nonglandular septate hairs to 2mm (the longest hairs often widely spaced) and short glandular hairs, round. May-Oct..*Sharp-leaved Fluellen* **Kickxia elatine***

Stems (±) erect

Ann. Basal lvs usu absent. (*Euphrasia* spp may key out here)

Lvs with distinct petiole, weakly fetid..(*Solanum nigrum* ssp *schultesii*, *S. physalifolium*) Go to **SOL**

Lvs narrowed to indistinct petiole, with menthol odour

Lvs to 6 x 1.2cm, oblanc, reflexed, ± acute, hairy both sides but with short glandular hairs below, with prominent green hydathodes, with stomata both sides. Stems to 60cm, patent-hairy (wispy with a thick septate base) and with short glandular hairs, round. Jun-Oct. VR alien..*Stinking Fleabane* **Dittrichia graveolens**

Lvs sessile, clasping, weakly citrus-scented

Lvs 7-12 x 1.5-2.5cm, oblong, with apiculate hydathode at apex, bright green, with prominent hydathode-teeth along margins, often weakly 3-pli-veined at base, with midrib raised below, with indistinct opaque pinnate 2° veins not sunken above, with stomata both sides. Stem to 60cm, viscid-hairy, ridged.........................*Pot Marigold* **Calendula officinalis**

Lvs 3-7 x 1-2cm, oblong-lanc, with apiculate hydathode at apex, grey-green, with prominent hydathode-teeth along margins (soon turning black), often weakly 3-pli-veined at base, with midrib raised below, with indistinct opaque pinnate 2° veins not sunken above, with stomata both sides. Stems to 30cm, viscid-hairy, ridged. VR alien..*Field Marigold* **Calendula arvensis**

Bi or per. Basal lvs often present or in an aerial rosette

Plant strongly odorous

Plant resin-scented, viscid. Lvs on stems, to 11 x 2cm, lanc, acute, clasping at base, glandular-hairy both sides, with hydathode-teeth, with indistinct veins, with stomata both sides. Stems to 100cm, erect, woody, glandular-hairy...*Woody Fleabane* **Dittrichia viscosa**

Plant paraffin-scented. Lvs on stems, to 12cm, usu lanc, entire to deeply dentate. Stems to 100cm, glandular-hairy. Ann to short-lived per....*Mexican-tea* **Chenopodium ambrosioides**

Plant fetid. Basal lvs 15-20cm, oblong-ovate, irreg lobed to ± entire, usu glandular-hairy (at least on stem) esp on veins both sides, with c6 prs whitish anastomosing 2° veins raised below, with stomata both sides. Stem lvs smaller, clasping. Petiole with flat arc vb and 2 rib bundles. Stems to 80cm, ± woody at base, glandular-hairy above, round. Bi (ann). All yr ....*Henbane* **Hyoscyamus niger**

Plant odourless or ± so

Basal lvs cordate at base. Plant never with aerial rosettes. Rhizomatous. Oct-Aug. ....*Leopard's-bane* **Doronicum pardalianches**

Basal lvs not cordate. Plant often with aerial rosettes. Tufted. All yr

Lvs 3-8 x 1-3cm, obovate-spathulate, ± obtuse with apical hydathode below, narrowed to petiole, rolled when young, dull grey-green, with v tiny glandular hairs both sides (appearing hairless TNE), ciliate, usu with 1-4 teeth (or purple hydathodes) per side nr apex, with opaque or weakly translucent veins raised above and often fading nr margins, with stomata both sides. Stem lvs sessile, entire. Petiole flat, often purplish at base, often with recurved wings, with 1 vb. Stems to 50cm, often purplish esp at base, ridged, with long septate hairs and tiny glandular ones. Tufted per with woody rootstock. All yr ....*Seaside Daisy* **Erigeron glaucus**

Lvs to 2 x 0.4cm, obovate to spathulate, narrowed to petiole, hairless or septate-hairy at least above (usu glandular or minutely so), ciliate, crenate or dentate with c4 teeth per side, usu 1-veined, with 2° veins obscure. Stem lvs occ opp, sessile. Stems 5-15cm, hairy, round..*Fairy Foxglove* **Erinus alpinus**

For illustrations see NT

Group NS – *Lvs fleshy (or waxy) or ± so. Often coastal. (Stem lvs of* Isatis tinctoria *may appear waxy but are never fleshy)*

■Lvs confined to stems, hairless (but often ciliate in *Inula crithmoides*)

Lvs aromatic, teeth without submarginal hydathode below, all veins obscure

Lvs to 50 x 6mm, linear, usu 3-toothed at apex, sessile, weakly channelled above, green, often septate-ciliate esp nr base, with midrib sunken below. Stems 15-90cm. Per. Mar-Oct. Sea-cliffs, saltmarshes...*Golden-samphire* **Inula crithmoides**

Lvs odourless, teeth with submarginal hydathode below (often obscure, HTL)

Lvs with translucent midrib, not falling. Ann

Lvs to 10(15)cm, narrowed to petiole-like base, obovate or oblanc to deeply pinnate-lobed, with midrib and obscure 2° veins raised both sides, with stomata both sides. Petiole with c15 vb's around margin. Apr-Oct..*Sea Rocket* **Cakile maritima**

Lvs with all veins obscure, the lower falling leaving scars on stem. Per

All lvs 1-4cm. Rhizomes often visible above ground, without tuberous roots

Lvs obovate-oblong, dull glaucous grey-green, with 3-6 dentate teeth per side distally, with obscure veins, with stomata both sides. Apr-Oct. Mtns, sea cliffs, N Br ....*Roseroot* **Sedum rosea**

At least some lvs >4cm. Rhizomes below ground, with tuberous roots

Lvs 2-10cm, ovate-oblong, pale ± waxy grey-green, with stomata both sides. Stems to 60cm, 4-8mm diam, erect, round, without stomata. Apr-Oct

Fls fertile, with stamens not protruding. Lvs shallowly dentate-serrate with 5-10 teeth per side at least in distal ½ (occ ± entire)..*Orpine* **Sedum telephium**

Fls fertile, with stamens protruding. Lvs shallowly dentate-serrate with 5-10 teeth per side at least in distal ½ (occ ± entire)...............................*Butterfly Stonecrop* **Sedum spectabile**

Fls sterile, with stamens ± absent. Lvs deeply serrate. *S. spectabile* x *telephium* ....*Autumn Stonecrop* **Sedum 'Herbstfreude'**

■Lvs mostly or all basal

All veins obscure. Lvs hairless exc at extreme base......................*Sea Plantain* **Plantago maritima**

At least midrib visible. Lvs hairless (often hairy in *Plantago coronopus*)

Petiole short or indistinct, with 3 vb's. Lvs 3-10cm.....*Buck's-horn Plantain* **Plantago coronopus**

Petiole long, distinct

Petiole with ≥10 vb's. Lvs 20-50cm...*Sea-kale* **Crambe maritima**

Petiole with 3-7 vb's. Lvs 4-40cm..................................*Sea Beet* **Beta vulgaris** ssp **maritima***

Petiole with 1 vb. Lvs <4cm

Basal lvs cuneate. Bi or per. All yr. Saltmarshes

Lvs obovate-ovate, entire or with 1(4) tooth per side. Stem lvs petiolate, or sessile and clasping..*English Scurvygrass* **Cochlearia anglica**

Basal lvs cordate or rounded

Ann. Oct-Jun. Dry saline habs (inc roadsides)

Lvs to 1.5cm, orb to ivy-shaped. Stem lvs sessile, often clasping ....*Danish Scurvygrass* **Cochlearia danica**

Per (bi). All yr

Saltmarshes. Lvs ovate-orb, to 5cm, entire or 1(4) teeth per side. Stem lvs sessile, often clasping...*Common Scurvygrass* **Cochlearia officinalis**

Mtns. Lvs orb (occ ivy-shaped), to 2cm, entire or 1(4) teeth per side. Stem lvs sessile, often clasping...*Mountain Scurvygrass* **Cochlearia micacea**

Group NT – *Lvs septate-hairy, usu with stomata both sides. (*Veronica *and* Euphrasia *spp may key out here)*

■Lvs with translucent 2° veins sunken above

Basal lvs to 8 x 8cm, ovate, cordate at base, long-petiolate, thin, the hairs often glandular or with withered tips (hairs v short when mature), odourless, sinuate-toothed, opaquely net-veined. Stem lvs clasping with auricles at base (at least the upper). Petiole hairy, channelled, often hollow, with 5 vb's. Stems erect, hairy, round, ridged, hollow. Rhizomatous per ....*Leopard's-bane* **Doronicum pardalianches**

Basal lvs absent. Stem lvs broadly ovate (usu as wide as long), hairy (occ glandular) to hairless, odorous, irreg toothed. Stems much-branched with wide spreading outline, tough, solid. Ann to per....................(*Solanum*) Go to **SOL**

■Lvs with opaque or weakly translucent 2° veins flat or raised above (often not or hardly raised below)

Lvs with <6 teeth per side (rarely more)

Basal lvs 1-3(5)cm (upper stem lvs entire), often aromatic when crushed. Per

Lvs obovate, 3-lobed, adpressed-hairy above, ciliate; lobes to 2 x 1cm. Stems to 50cm, sparsely hairy, ridged or angled. Often on walls...*Mexican Fleabane* **Erigeron karvinskianus**

Basal lvs 2-5cm (stem lvs never present), odourless. Shortly rhizomatous per

Lvs obovate-spathulate, obtuse, abruptly narrowed to petiole, sparsely hairy, crenate with 4-5(7) teeth per side, with midrib raised below, with veins often weakly translucent. Petiole short, purplish at base, winged, ciliate, with 1 vb. Stems 3-12cm, round. All yr. PL 11 ....*Daisy* **Bellis perennis**

Basal lvs >5cm (stem lvs often smaller), often aromatic when crushed. Ann (bi)

Lvs roughly adpressed-hairy (± strigose) above

Basal lvs with long widely spaced cilia to 10(15)mm, obovate-lanc, dull pale yellow-green both sides, ± toothed to entire. Stem lvs 1-5 x 0.2-1cm, narrowly lanc or linear, entire to weakly toothed. Petiole indistinct, with 3 vb's. Stems (3)50-100cm, rough with swollen-based hairs, ridged. Oct-Aug....................*Canadian Fleabane* **Conyza canadensis**

Basal lvs with short 0.2mm cilia, obovate-lanc, often shiny dark green above, with 4-5 teeth per side (more on stem lvs). Stem lvs 2-8 x 0.5-3cm, oblanc, deeply toothed. Petiole indistinct, with 3 vb's. Stems to 150cm, rough with swollen-based hairs, ± ridged. Oct-Aug ....*Bilbao's Fleabane* **Conyza bilbaoana**

Lvs softly adpressed- to patent-hairy above

Basal lvs with short cilia to 0.3mm, slightly grey-green, with 0-7 teeth per side (not as deeply toothed as *C. bilbaoana*). Stem lvs 5-toothed, occ with minute sessile glands when young. Stems to 150cm, hairy, ridged. All yr....................*Guernsey Fleabane* **Conyza sumatrensis***

Basal lvs not ciliate. Stems to 100cm. All yr....................*Argentine Fleabane* **Conyza bonariensis**

Lvs with >6 teeth per side

Lvs gland-dotted both sides. Rhizomatous, often strongly so

Lvs usu all on stems, 5-12 x 1.5-4(5)cm, oblong to lanc, occ ± clasping at sessile base, pale green, occ hairless, deeply serrate, rarely pinnately lobed at base. Stem usu 1, to 150cm, hairy. Fls Oct....................*Autumn Oxeye* **Leucanthemella serotina**

Lvs not gland-dotted

Lvs in basal rosettes on non-fl plants. Shortly rhizomatous

Basal and lower stem lvs to 5cm, ovate to obovate-spathulate, cuneate at base, petiolate, rolled when young, occ ± fleshy, dark green above, sparsely hairy (occ unicellular), with 5-10 teeth per side, occ weakly 3-pli-veined at base, with midrib raised below. Stem lvs to 5(8)cm, oblong, ± clasping and pinnately fringed at sessile base, toothed to pinnately lobed. Petiole to 5cm, pinnately fringed (at least on upper lvs) at dilated base, often winged, not channelled, with 3 vb's. Stems to 70cm, purplish, hairy (hair tips soon withering), 5-angled. All yr. PL 18....................*Oxeye Daisy* **Leucanthemum vulgare***

Basal and lower stem lvs to 10cm, elliptic-oblong, cuneate at base, petiolate, rolled when young, shiny dark green above, sparsely minutely hairy to hairless above, hairy below, with 17-38 shallow teeth per side (occ ± entire), with midrib raised below. Stem lvs 10-18 x 5cm, broadly oblong, often clasping at sessile base, weakly 3-pli-veined. Petiole to 7cm, reddish, ± clasping, channelled, with 3(5) vb's. Stems to 150cm, rarely purplish, sparsely hairy, slightly ridged to round. *L. lacustre* x *maximum*. All yr......*Shasta Daisy* **Leucanthemum** x **superbum**

Lvs all on stem. Strongly rhizomatous (occ rooting at lowest nodes)

Lvs 2-6 x 0.5-2.5cm, held erecto-patent, acute with long terminal hydathode, cordate at clasping base, rolled when young, often folded at maturity, dull green above, paler below, hairless or ± hairy on veins below, ciliate, entire or remotely denticulate with hydathodes (often blackish), with 2° veins raised below. Stems 25-75cm, ± hairless, ridged. VR, Ire ....*Irish Fleabane* **Inula salicina**

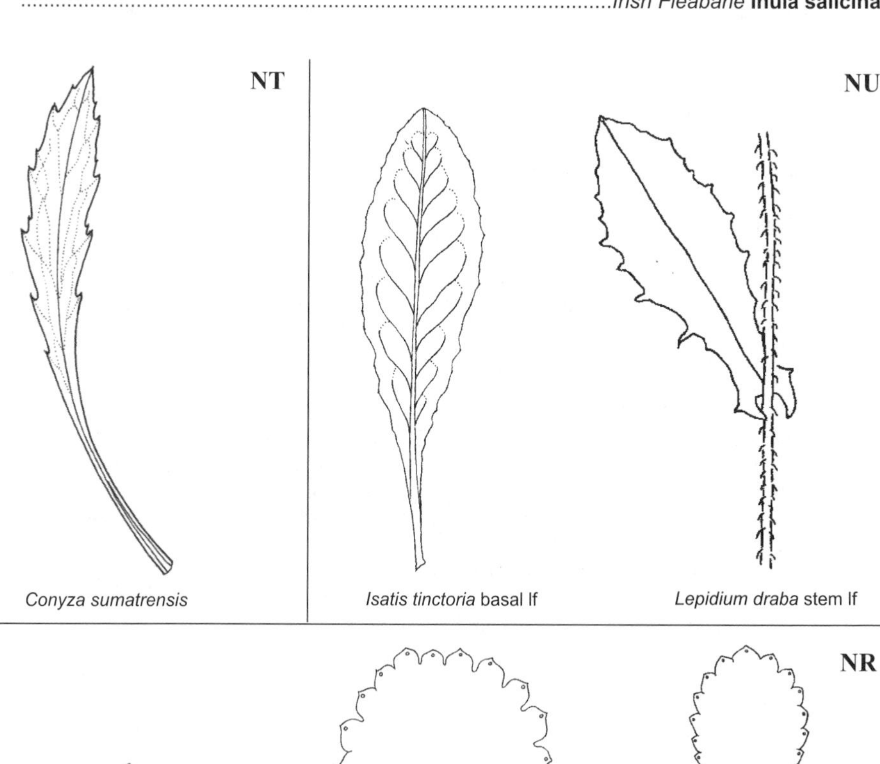

Conyza sumatrensis

Isatis tinctoria basal lf

Lepidium draba stem lf

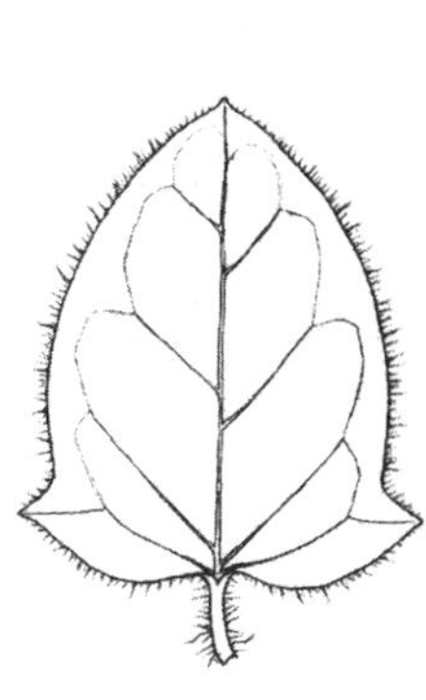

Kickxia elatine

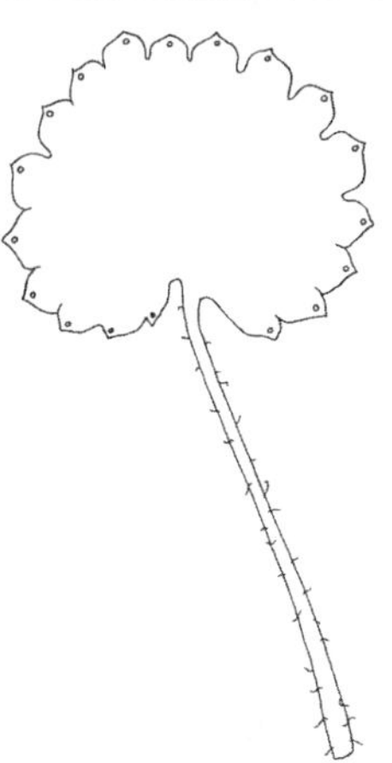

Saxifraga hirsuta

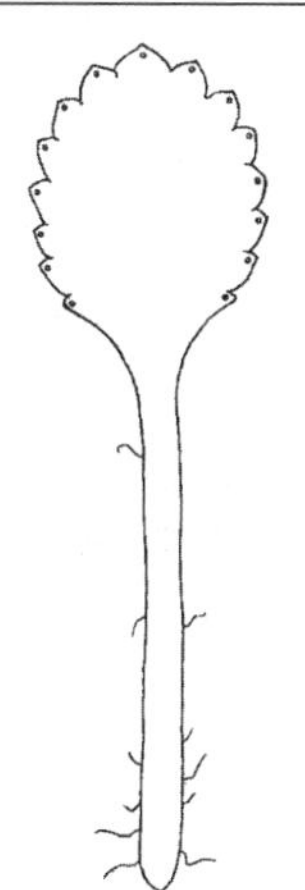

Saxifraga spathularis

Group NU – *Lvs with unicellular hairs, usu with stomata both sides*

■Lvs sharpy hispid..(*Boraginaceae*) Go to **KN**

■Lvs softly hairy (*Lepidium latifolium* may occ key out here)

Lvs odourless

Wet habs. Lvs with sheathing base, adpressed-hairy. Petiole often hollow ....(*Ranunculus flammula, R. lingua*) Go to **RAN**

Dry habs. Lvs without sheathing base, hairless or hairy. Petiole solid...(*Oenothera*) Go to **OENO**

Lvs with cress odour (often weak). Stems (and lower petioles) purplish at base

At least basal lvs with ± adpressed retrorse white hairs above. Rhizomatous

Stems to 60cm, with curved white retrorse 0.2-0.3mm hairs, (±) round below, ridged above. Basal lvs usu dead by fl, obovate, grey-green, softly hairy to ± hairless, ≤10 sinuate hydathode-teeth per side, with opaque or weakly translucent 2° veins. Stem lvs to 6 x 3cm, ovate-oblong, ± acute, with acute auricles at base, toothed to ± entire. Petiole to 6cm, retrorsely hairy, channelled, with 3(5) vb's. Jan-Jul (all yr)........*Hoary Cress* **Lepidium draba***

All lvs with patent or spreading white hairs above. Tufted or with short rhizomes

Stems glaucous or ± so

Basal lvs not persisting to fl, to 30 x 6cm, lanc, mid-green, hairy both sides, often obscurely sinuate-toothed with black hydathodes, occ lobed. Stem lvs 7-15 x 1.5-4cm, sagittate at clasping base, sessile, glaucous. Petiole slightly winged, not channelled, with 7(9) vb's. Stems 60-120cm, (±) hairless. Bi or per. All yr.....................................*Woad* **Isatis tinctoria***

Basal (and lower) lvs occ persisting to fl, (2)8-20 x (1)2-7cm, dark green, usu hairless above, sparsely bristly below, sparsely ciliate, irreg toothed to pinnately lobed with a large terminal lobe and c3 prs of small laterals. Upper lvs 2-5 x 0.5-1.5cm, ovate to linear. Petiole with W-shaped groove above, with 5 vb's. Stems 20-70(110cm), with stiff hairs (occ retrorse) at least below, ± hairless above. Ann or bi. Nov-Aug (all yr) ....*Bastard Cabbage* **Rapistrum rugosum**

Stems not glaucous

Basal lvs dead by fl (resprouting after fr). Stem lvs with auricles

Stems with ± patent 0.3-0.5mm hairs, branched below middle, to 50cm. Basal and lower stem lvs to 3cm, oblong to lanc, obtuse, cuneate at base, grey-green, softly hairy to hairless, entire to sinuate-toothed, often lyrate, with weakly translucent 2° veins. Stem lvs sagittate, clasping the stem with long narrow acute auricles, distantly sinuate-toothed. Petiole to 3cm, sparsely hairy or hairless, densely ciliate, channelled, with 1(3) vb's. Tufted per (bi), with long taproot. All yr....................*Smith's Pepperwort* **Lepidium heterophyllum**

Stems with slightly retrorse <0.3mm hairs, branched above middle, to 40cm. Basal and lower stem lvs to 5cm, oblong to lanc, obtuse, cuneate at base, green to grey-green, densely softly shortly hairy (hairless TNE), entire or with small distant teeth, often lyrate, with weakly translucent 2° veins. Stem lvs sessile, clasping with long narrow acute auricles. Petiole to 3cm, usu densely hairy, channelled, with 1(3) vb's. Ann (bi) ....*Field Pepperwort* **Lepidium campestre**

Basal lvs never present (occ in false rosette). Stem lvs without auricles

Lvs often held erect, 8-12 x 2.5-4cm, elliptic, mid-green, shortly hairy both sides, sinuate-toothed, opaquely net-veined, with 2° veins translucent. Petiole (0)1-6cm, channelled, with 7 vb's. Stems to 50cm, densely retrorsely hairy with 1.5mm hairs, (±) round, striate. Shortly rhizomatous per. VR alien............................*Perennial Rocket* **Sisymbrium strictissimum**

Group NV – *Lvs (±) hairless (lf margin may have cilia or stems occ papillate).* ❶ *Lvs 1-veined with translucent midrib visible only*

▪Petiole sheathing with ± membranous non-auriculate base. Lvs with white hydathodes along margins above (turning dark). Wet habs.......................(*Ranunculus lingua*, *R. flammula*) Go to **RAN**

▪Petiole sheathing with herbaceous green auricle-like base

Lvs to 35 x 25cm, cordate to rounded at base, slightly decurrent along petiole, thick, with cartilaginous margins, gland-pitted both sides, shiny dark green above, paler below, denticulate with long hydathodes, with ± opaque 2° veins raised below and slightly sunken above, with stomata both sides. Petiole to 25cm, round, channelled nr lf, pith-filled, with scattered curved (C-shaped) or linear vb's and spiral fibres. Rhizomes 1-3cm diam, scaly. All yr. Hortal

Lvs toothed without bristles, ovate, flat, reddish in winter.....*Elephant-ears* **Bergenia crassifolia**

Lvs bristle-toothed, ovate, flat, green in winter. The commonest taxon. *B. ciliata* x *crassifolia* ....*Hybrid Elephant-ears* **Bergenia** x **schmidtii***

Lvs bristle-toothed, ovate, flat, reddish in winter (frost-sensitive) ....*Hairy Elephant-ears* **Bergenia ciliata**

Lvs not bristle-toothed, obovate, bullate with wavy margins, green to reddish in winter ....*Heart-leaved Elephant-ears* **Bergenia cordifolia**

▪Petiole not (or hardly) sheathing at base

Plant with cress, mustard or garlic odour. (*Lepidium latifolium* may rarely key out here)

Stems hollow. Per

Stem lvs ± clasping and often auriculate, 3-7 x 2cm, lanc to broadly lanc, (±) hairless, shallowly to deeply toothed. Lower lvs broadly ovate to narrowly elliptic-oblong, narrowed to petiole, entire to sharply dentate (occ lobed). Basal lvs lobed, not persisting. Stems to 1.2m, often rooting at nodes, hairless. Tufted or with stolons. Mar-Oct (overwinters as a tiny rosette). Wet habs...*Great Yellow-cress* **Rorippa amphibia**

Stem lvs clasping with obtuse auricles, 3-7 x 1.5cm, lanc to narrowly elliptic, (±) hairless, with 10-15 weak teeth per side, with 1-3° veins raised both sides, minutely hairy when young. Lower lvs ovate to oblong, narrowed to petiole, dentate. Basal lvs rarely lobed, not persisting. Stems to 1m, never rooting at nodes, hairless or minutely hairy. Strongly rhizomatous. Mar-Oct (often overwinters as a tiny rosette). R alien......*Austrian Yellow-cress* **Rorippa austriaca**

Stems solid or absent. Ann (bi)

Stem lvs ± sagittate or clasping at base

Stem lvs glaucous, the upper ± hairless

Lower lvs with at least some stalked stellate hairs (mostly 3-rayed). All yr ....*Tower Mustard* **Arabis glabra**

Lower lvs with simple hairs or hairless. All yr.....................................*Woad* **Isatis tinctoria***

All lvs green, hairless

Stems hairy below, hairless above. Plant with garlic odour when bruised

Stem lvs with acute auricles. Stems grooved.........*Garlic Penny-cress* **Thlaspi alliaceum**

Stems hairless. Plant with mustard or cress odour when bruised

Basal lvs rarely persisting to fl, to 6 x 2cm, oblanc to obovate, cuneate at base, mid-green, entire, opaquely net-veined, with stomata both sides. Stem lvs many, oblong or lanc, usu obtuse, clasping with short acute auricles, with c4 irreg teeth per side. Petiole indistinct. Stems grooved (strongly angled or with cartilaginous ridges). Stems 1-several, 10-60cm. Often all yr ....*Field Penny-cress* **Thlaspi arvense**

Basal lvs occ persisting to fl, to 5 x 2cm, (ob)ovate, cuneate at base, grey-green, entire or toothed, opaquely net-veined, with stomata both sides. Stem lvs few, elliptic-lanc, often acute, clasping with rounded auricles, weakly toothed. Petiole to 2cm, often purplish. Stems 1-several, 3-12cm, round to ridged. Oct-Jun (all yr). Sch8 ....*Perfoliate Penny-cress* **Thlaspi perfoliatum**

Stem lvs not sagittate or clasping. Lvs not glaucous

Basal lvs usu present. Rhizomatous per. All yr

Lvs 9-15 x 6cm, ovate, obtuse, petiolate, thickish, green, hairless, deeply toothed to lobed, with ovate to hastate terminal lobe and 0-4 prs lateral lobes, the upper lvs lanc, entire. Petiole to 3cm, channelled. Stems to 75cm, occ sparsely hairy below, round. VR alien ....*Russian Mustard* **Sisymbrium volgense**

Basal lvs usu absent. Ann. Apr-Oct

Lvs with 1-2 lobe-like teeth per side nr apex, to 2 x 0.5cm, spathulate to linear-lanc, ± acute, narrowed to base, ± hairless, white-ciliate, with obscure veins, with v bitter taste, with stomata both sides. Stems to 30cm, shortly hairy (often looking like papillae), weakly angled. Bare calc habs...*Wild Candytuft* **Iberis amara**

Lvs ± entire, to 8 x 1cm, linear-lanc, ± acute, narrowed to base, hairless, rarely ciliate, usu with obscure veins, tasteless, with stomata both sides. Stems to 40cm, hairless, weakly angled. Apr-Oct. Hortal..*Garden Candytuft* **Iberis umbellata**

Plant with meaty or fetid odour when crushed

Basal lvs absent. Stem lvs usu broadly ovate..(*Solanaceae*) Go to **SOL**

Basal lvs absent. Stem lvs linear, 4-7(12)cm x 5mm, weakly clasping at base, revolute, slightly purplish at base, sparse septate hairs in axils, weakly toothed to entire (rarely) pinnatifid, (±) 1-veined, with midrib raised below, with stomata both sides. Stems 30-60cm, ± woody at base, ± hairless, angled. Tufted per. ❶. Apr-Oct..........*Narrow-leaved Ragwort* **Senecio inaequidens**

Plant aromatic when crushed (or lvs with sessile glands)

Basal lvs in aerial rosette, 3-8cm, obovate-spathulate, ± obtuse with apical hydathode below, narrowed to petiole, rolled when young, dull grey-green, with v tiny glandular hairs both sides (appearing hairless TNE), ciliate, usu with 1-4 teeth (or purple hydathodes) per side nr apex, with opaque or weakly translucent veins raised above and often fading nr margins, with stomata both sides. Stem lvs sessile, entire. Petiole flat, often purplish at base, often with recurved wings, with 1 vb. Stems to 50cm, often purplish esp at base, ridged, with long septate hairs and tiny glandular ones. Tufted per with woody rootstock. All yr.............*Seaside Daisy* **Erigeron glaucus**

Plant odourless

Basal lvs usu absent

Stems hollow, brittle, hairless. Ann..(*Impatiens*) Go to **IMP**

Stems solid, tough, hairy above. Rhizomatous per

Lvs 10-40 x 3-6cm, elliptic to lanc, acute, narrowed to sessile base, weakly clasping, with cartilaginous margins (occ scabrid-ciliate), dark green above, serrate or dentate with hydathode-tipped teeth, with opaque veins and reticulations (occ translucent), with stomata both sides. Stems to 200cm, occ glandular-hairy above, ridged. R alien ....*Broad-leaved Ragwort* **Senecio fluviatilis**

Lvs 1.5-8 x 0.6cm, linear-lanc, acute, sessile, weakly clasping, with cartilaginous margins, hairless or sparsely septate-hairy (may appear unicellular), serrulate (teeth almost bristle-tipped), 1-veined, with midrib raised below, with stomata both sides. Stems 20-60cm, angled. ❶. Apr-Oct..*Sneezewort* **Achillea ptarmica**

Basal lvs usu present

Lvs (±) orb

Lvs peltate, emerging along rhizome. Wet habs. PL 24 ....*Marsh Pennywort* **Hydrocotyle vulgaris**

Lvs cordate at base, 1-several from corm, reniform to ± orb, with cartilaginous margins, ± fleshy, dark green usu marbled with white above, purplish below, toothed with hydathodes, with stomata below only. Petiole long, v shortly papillate at least nr lf, round, with 1 vb. Fls in spring. Oct-May..*Eastern Sowbread* **Cyclamen coum***

Lvs not orb

Lvs with translucent midrib only ❶..*Fairy Foxglove* **Erinus alpinus**

Lvs with opaque or weakly translucent 2° veins flat or raised above (often ± not raised below)

Lvs with <6 teeth per side (rarely more), basal, to 5cm, obovate-spathulate. PL 11
..........Daisy **Bellis perennis**

Lvs with >6 teeth per side

Basal and lower stem lvs to 10cm, elliptic-oblong
..........*Shasta Daisy* **Leucanthemum** x **superbum**

Basal and lower stem lvs to 5cm, ovate to obovate-spathulate. PL 18
..........*Oxeye Daisy* **Leucanthemum vulgare***

Lvs with translucent 2° veins sunken above and raised below

Lvs with some 2° veins fading nr margins, often with unicellular cilia
..........(*Oenothera*) Go to **OENO**

Lvs without 2° veins fading nr margins, with septate cilia nr base, 12-20 x 2-2.5cm, lanc to oblong, rolled when young, dull green above, hairless (rarely with occ hair below), serrate with 8-20 teeth per side, rarely ± net-veined, with stomata below and often above nr veins. Stem lvs sessile. Petiole indistinct, with flat arc vb. Stems to 1m, hairless to sparsely hairy, (±) round. Per (bi). All yr. VR alien..........*Straw Foxglove* **Digitalis lutea**

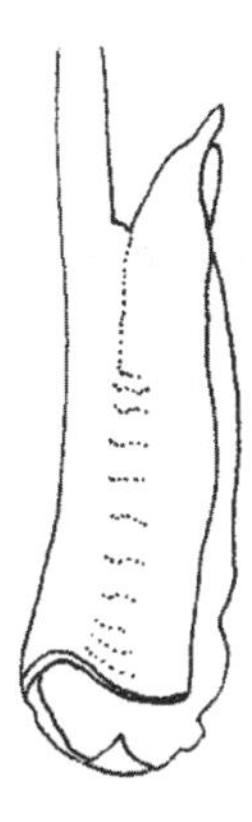

Bergenia x *schmidtii* petiole base

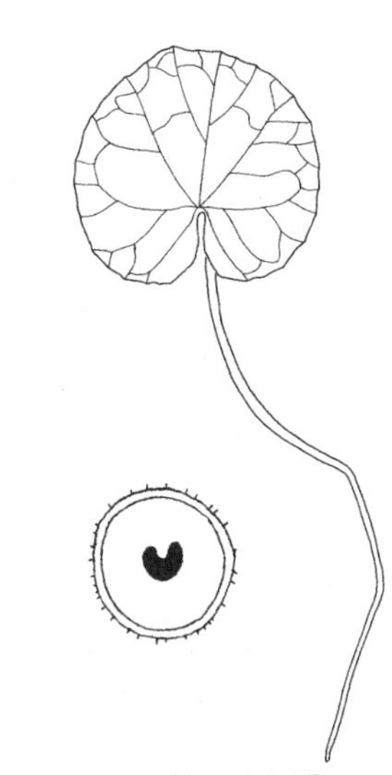

Cyclamen coum with petiole TS

Key to Groups in Division O

(Lvs toothed, opp or whorled)

■Shrub (occ dwarf or trailing)
  Twigs with interpetiolar ridge. Lvs exactly opp (or whorled)..**OA**
  Twigs without interpetiolar ridge. Lvs often subopp..**OB**
■Herb
  Stipules present...**OC**
  Stipules absent or reduced to glands
    Lvs with ± spiny midrib below (*Dipsacus*)...**OD**
    Lvs without spiny midrib below
      Lvs with dendritic hairs...**OE**
      Lvs with mealy or bladder-like hairs (esp below, or in lf axils)
      ....(*Chenopodium*, *Atriplex*) Go to **CHAT**
      Lvs with simple, woolly or glandular hairs, or hairless
        Stems square, or absent
          Lvs strongly mint-scented..**OF**
          Lvs pleasantly aromatic (may be lemon- or orange-scented), AND/OR white-woolly to densely white-hairy..**OG**
          Lvs fetid or ± odourless
            Lvs all on stem, with petiole 0-6mm...**OH**
            Lvs in basal rosette, OR on stem with petiole >6mm (at least on lower lvs)
              Lvs hairless (may have ± sessile glands), not ciliate.................................**OI**
              Lvs hairy (hairs occ sparse or lvs ciliate only)
                Petiole with 2(4) vb's (central vb usu absent; always an even number)..........**OJ**
                Petiole with 1 or 3(7) vb's (central vb usu present; always an odd number)
                  Lvs cordate (to truncate) at base..**OK**
                  Lvs cuneate or narrowed to petiole (rarely rounded at base).......................**OL**
        Stems round (may be ridged but not 4-angled), always present
          Stems aromatic with black or purple resin canals..**OM**
          Stems not aromatic, without resin canals
            Lvs or petioles hairy. Stems hairy or hairless
              Stems ± prostrate, often rooting at least at lower nodes ....**ON**
              Stems ± erect, never rooting at nodes..**OO**
            Lvs and petioles hairless (but may be papillate or ciliate). Stems hairy or hairless
              Plant in v wet (or mtn) habs...**OP**
              Plant in dry habs...**OQ**

Group OA – *Shrub or creeping subshrub. Twigs with interpetiolar ridge. Lvs always exactly opp, (±) net-veined (exc* Fuchsia magellanica*), with stomata below only.* ❶ *Lvs yellow-blotched.* ❷ *Lvs 3-pli-veined.* ❸ *Stems prostrate with lvs <1.5cm*

■Twigs with dendritic hairs. Buds naked. Evergreen..................*Jerusalem Sage* **Phlomis fruticosa**

■Twigs with stellate hairs (white-woolly when young, occ with glandular hairs). Buds naked

Evergreen (or ± so). Stipules interpetiolar, growing with age. Lvs 10-25 x 6cm, ovate-lanc to lanc, acuminate, soon dark green and (±) hairless above. Petiole v short

Lvs rounded to cuneate at base, not or weakly aromatic, white-felted below, serrate with 5-100 teeth per side. Twigs ± angled when young..............................*Butterfly-bush* **Buddleja davidii***

Lvs cuneate at base, sweetly aromatic, white- or yellow-felted below, crenate with 60-100 teeth per side. Twigs 6-ridged when young. R hortal....................*Orange-ball-tree* **Buddleja globosa**

Evergreen (or ± so). Stipules absent

Lvs 8-25 x 7cm, ovate to obovate, obtuse, cuneate to cordate at base, rugose, shiny dark green above, stellate-woolly below, serrate to entire. Petiole 1-3cm, round. Twigs scurfy hairy with stellate hairs when young. Buds adpressed, stellate-hairy ....*Leatherleaf Viburnum* **Viburnum rhytidophyllum**

Deciduous. Stipules absent

Lvs 5-10cm, ovate to obovate, usu acute, ± involute when young, cordate at base, rugose, sparsely stellate-woolly below, serrate with 40-45 teeth per side (each with a long acute green hydathode), with c10 prs of 2° veins forking 1-2x before margins (and ending in teeth). Petiole 1-3cm. Twigs scurfy hairy with stellate hairs when young, angled, pith-filled. Buds adpressed, naked, pale yellow-green, stellate-hairy. PL 12, 23................*Wayfaring-tree* **Viburnum lantana**

■Twigs with stellate scales (occ sparse) when young, never woolly. Buds with scales. Deciduous

Lvs 4-8 x 4cm, ovate, rounded to cordate at base, folded when young, rough with 5-rayed stellate scales above, the scales 7-10-rayed below, with 13-30 long hydathode-tipped teeth often upcurved, net-veined, with anastomosing 2° veins, with spiral fibres. Petiole 2-6mm, channelled. Twigs angled when young, later round with flaky bark. Buds adpressed, 2mm, brown, with several persistent scales..*Deutzia* **Deutzia scabra**

■Twigs with simple hairs or hairless. Buds with scales (naked in terminal buds of *Hydrangea macrophylla*)

Evergreen. Lvs 8-20cm, hairless

Lvs shiny dark green with yellow blotches above, dull and paler below, lanc to obovate, folded when young, not rugose, with flat margins, with 0-5 hydathode-tipped teeth per side, not net-veined, with anastomosing 2° veins hardly raised below. Petiole 1-3cm, green, occ purplish. Twigs green, adpressed-hairy at nodes, strongly fetid, round. Fl buds terminal, large, with several scales. ❶...*Spotted-laurel* **Aucuba japonica**

Lvs shiny dark green above, paler below, elliptic, weakly rugose, with recurved margins, remotely denticulate (to ± entire), net-veined and strongly 3-pli-veined, with 3 main veins sunken above and raised below. Petiole 1-2cm, reddish. Twigs green, minutely hairy when young, odourless, round. Lf buds 5-7mm, with 2 scales. ❷. Widely planted ....*David's Viburnum* **Viburnum davidii**

Evergreen. Lvs 0.5-1.5cm, sparsely unicellular-hairy. ❸

Lvs broadly ovate to orb, valvate-applanate when young, often with sparse ± sessile glands, crenate-serrate with 1-3 teeth per side, with anastomosing 2° veins. Petiole to 3mm, ciliate. Stems creeping, usu rooting at nodes, 1mm diam, reddish, hairy, round. Buds with 2 scales. Pine woods, rock ledges, Scot..*Twinflower* **Linnaea borealis**

Deciduous

Buds hidden within base of petiole, with 2 ± valvate scales. Lvs ± acuminate, applanate when young, with spiral fibres. Twigs with persistent bud scales, brittle, weakly angled, dark to warm brown, without lenticels, pith-filled. Hairs septate. (Many garden hybrids occur)

Lvs 4-12 x 7cm, ovate, (±) hairless, occ hair-tufts in vein axils below, with few unequal teeth (some large) in distal ½, with 2° veins mostly in proximal ½. Petiole to 10mm. Twigs with peeling bark..*Mock-orange* **Philadelphus coronarius**

Lvs 4-12 x 7cm, ovate, hairy esp below (silky-hairy below when young), denticulate with 5-12 teeth per side, with 2° veins mostly in proximal ½. Petiole 4-5mm. Twigs with bark hardly peeling. ?*P. coronarius* x *microphyllus* x *pubescens* ....*Hairy Mock-orange* **Philadelphus** x **virginalis**

Lvs to 4 x 1.3cm, ovate to lanc, hairless below exc for adpressed hairs on midrib and occ 2° veins, with (0)1-2 teeth per side, 3-5-pli-veined. Petiole 3mm ....*Littleleaf Mock-orange* **Philadelphus microphyllus**

Buds exposed

Young twigs green, hairless

Stems v hollow, pruinose when young, ± arching, round, with stomata, without lenticels. Lvs 5-18 x 6cm, ± ovate, acuminate, truncate to rounded at base, involute when young, with sparse short hairs on veins both sides, ciliate, shallowly toothed (occ entire), with c8 prs of (often purplish) anastomosing 2° veins, with spiral fibres. Petiole to 15mm, often reddish. Buds to 6mm, with 2-4 acuminate scales (often persistent on twig) ....*Himalayan Honeysuckle* **Leycesteria formosa**

Stems pith-filled, not pruinose, not arching, round, without stomata, with flat horizontal lenticels (often purplish). Lvs to 16 x 11cm, broadly ovate, abruptly cuneate at base, obvolute when young, hairless, not ciliate, serrate, with c9 prs of whitish anastomosing 2° veins, with spiral fibres. Petiole 15-30mm, green. Buds 3-6mm, with several scales, the terminal bud to 2cm and naked..........*Hydrangea* **Hydrangea macrophyllla**

Young twigs brown or reddish, minutely hairy

Stipules or extra-floral nectaries present. Twigs without persistent bud scales

Lvs occ 3-whorled, 2-8cm, ovate-oblong, acuminate (with 3 hydathodes at apex), dark green to purplish, with reddish dots (HTL), sparsely hairy to hairless, often ciliate, denticulate with 7-16 reddish hydathode-teeth per side, with reddish 2° veins fading nr margins. Petiole 3-15mm, usu reddish. Twigs reddish when young, round, with flaking bark in 2nd yr..........*Fuchsia* **Fuchsia magellanica***

Stipules and stipular nectaries absent. Twigs with persistent bud scales. Hortal

Lvs serrate with 30-50 short hydathode-tipped teeth per side, 5-10 x 2.5-5cm, ovate to elliptic-oblong, acuminate, involute when young, dark green above (occ variegated), paler below, ± undulate, hairy esp below (soon hairless above), sparsely ciliate, with 5-7 prs of anastomosing 2° veins, with spiral fibres. Petiole 2-6mm. Twigs sparsely hairy in 2 opp lines (on ridges) in 1st yr. Buds adpressed, 4mm, brown, with 4 scales ....*Weigelia* **Weigela florida**

Lvs serrate with up to 10 long hydathode-tipped teeth per side, 2.5-7.5 x 2-5cm, ovate, acuminate, dark green above, paler below, flat, sparsely rough hairy above, ciliate, with 3-4 prs of anastomosing 2° veins, with spiral fibres. Petiole 3-5mm. Twigs minutely hairy all round (soon hairless), pale or reddish-brown, brittle, with peeling bark. Buds spreading, 2-4mm, brown, with 4-6 scales..........*Beautybush* **Kolkwitzia amabilis**

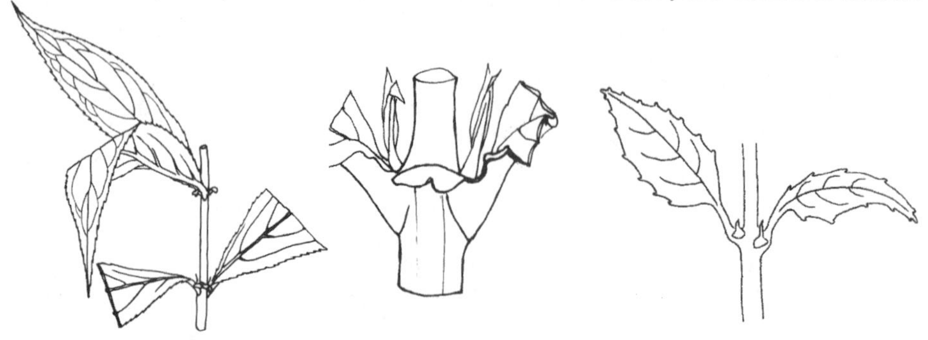

Buddleja davidii & interpetiolar stipules

Fuchsia magellanica

Group OB – *Shrub. Twigs without interpetiolar ridge. Lvs often subopp. Petiole channelled, with 1 vb. (*Fuchsia magellanica *may key out here).* ➊ *Branches thorny, with wood turning orange when cut*

■Young twigs green, with stomata (and tiny round whitish lenticels)

Lvs evergreen, 2° veins never raised below, with stomata both sides (often obscure above)

Lvs 2-9 x 5cm, elliptic to obovate, rolled when young, shiny dark green and often variegated above, dull pale green below, hairless, with 16-30 serrate teeth per side (each with a fragile claw-like gland), not net-veined, with 3-6 prs of anastomosing 2° veins (occ indistinct). Stipules 0.5-2mm, present as minute brown scales or linear glands. Petiole to 15mm. Twigs round, with circular stele. Buds to 2cm, spindle-shaped, with green scales fringed with reddish-brown gland-tipped cilia. PL 9..*Evergreen Spindle* **Euonymus japonicus**

Lvs deciduous, 2° veins raised below, with stomata below only

Lvs with minute translucent dots (x20, HTL), 7-16cm, elliptic to obovate, acuminate, hairless, serrate with c120 glandular-ciliate teeth per side, net-veined. Petiole to 5mm. Stipules minute, soon falling. Twigs hairless, round, soon 4-angled with lenticels, with dumbbell-shaped stele. Buds adpressed, to 15mm, spindle-shaped, purple-brown, hairless, with several opp scales, without free or reflexed tips. VR alien......................*Large-leaved Spindle* **Euonymus latifolius**

Lvs without minute dots, 3-13cm, ovate-lanc to elliptic, acuminate, often turning reddish in autumn, hairless, serrate with 38-72 teeth per side (each with a claw-like gland), net-veined. Petiole to 5mm. Stipules minute, soon falling. Twigs hairless, round, soon 4-angled with lenticels, with round stele. Buds adpressed, to 5mm, ovoid, green, hairless, with several opp scales, with long free or reflexed brown tips. PL 9......................*Spindle* **Euonymus europaeus**

■Young twigs usu brown, without stomata (but may have obvious lenticels)

Twigs square

Lvs evergreen, to 7 x 3cm, ovate to ovate-lanc, cuneate or rounded at base, pitted above, shiny green above, paler below, hairless, serrate, not net-veined, with stomata both sides. Petiole to 25mm, often indistinct, with tiny stipule-like auricles at base. Twigs occ with minute reddish glandular hairs when young, with lenticels, brittle, solid, weakly odorous. Buds naked ....*Cape Figwort* **Phygelius capensis**

Lvs deciduous, to 8 x 4cm, ovate to ovate-lanc, cuneate at base, pitted both sides, dull dark green above, paler below, hairless (occ minutely hairy above), serrate at least distally, not net-veined, with stomata below only. Petiole to 15mm, without auricles. Twigs hairless, yellow- to olive-brown, becoming round, with lenticels, occ with chambered pith. Buds often clustered, yellow-brown, hairless, with opp scales. *F. suspensa* x *viridissima* ....*Forsythia* **Forsythia x intermedia**

Twigs round (occ ridged when young)

Lvs evergreen, leathery, with cartilaginous margins

Lvs 2-5 x 1-2.5cm, ovate to obovate, obtuse but apiculate, cuneate at base, shiny dark green above (occ variegated), paler below, hairless, serrate or with small spiny teeth, with 3-6 prs of anastomosing 2° veins (occ raised above), not net-veined, with stomata below only. Petiole 3-10mm, minutely hairy. Twigs minutely hairy when young, with weakly odorous lime-green inner bark. Buds with scales, 1-4mm................*Mediterranean Buckthorn* **Rhamnus alaternus**

Lvs 4-7 x 1-3cm, elliptic-lanc, acute to obtuse, rounded at base, shiny dark green above, paler and pitted below, hairless, with plastic-feel, crenate-serrate with (6)10-16 teeth per side, occ ± net-veined, with stomata below only. Petiole to 7mm, minutely hairy. Twigs minutely hairy when young, odourless. Buds naked, 2-3mm, yellow-green, ± acute, hairy ....*Mock Privet* **Phillyrea latifolia**

Lvs deciduous, thin, without cartilaginous margins

Twigs without inner bark orange and odorous (but v bitter-tasting)..(*Salix purpurea*) Go to **SAL**

Twigs with orange odorous inner bark, 90° to main stem, occ with short shoots, dark black-brown, usu minutely hairy when young. Lvs 3-6 x 2-3cm, ovate to ± elliptic, obtuse (occ cuspidate), dull green, with minute translucent dots (x20), usu hairy both sides, serrate with (20)40-45 teeth per side (each with claw-like hydathode), net-veined, with 2-3(5) prs of 2° veins curving to apex, with stomata below only. Petiole 5-20mm. Buds adpressed, 2-6mm, dark brown, with 5 scales. ➊. Calc habs. PL 12.....................*Buckthorn* **Rhamnus cathartica**

Group OC – *Herb. Stipules present. Petiole channelled. Stems erect. Hairs unicellular. (*Euphorbia thymifolia *and* E. maculata *may rarely key out here but have prostrate stems and latex.* Moneses uniflora *often has alt stipule-like brown scales at base of petioles)*

■Stipules >5mm. Plant with stinging hairs (occ sparse or absent) with elongate swollen bases

Per, patch-forming, with long rhizomes and rooting at nodes. Stipules 2 prs at each node, to 8mm, lanc to ovate-lanc, entire, soon falling. Stems to 150cm, ± square, usu unbranched, purplish, hollow. Lvs 4-15 x 2-6cm, rounded to cordate at base, acuminate, folded when young, coarsely 1(2)-serrate, net-veined, 3-7-pli-veined, with stomata below only. Petiole longest on mid-stem lvs, with 5(7) vb's. Apr-Oct (all yr)

Lvs with long stinging hairs, ovate (occ long and narrow). Stems usu sparsely bristly with stinging hairs. PL 22..*Common Nettle* **Urtica dioica** ssp **dioica**

Lvs with dense short stingless hairs, lanc (usu long and narrow). Stems densely white-hairy (hairs to 1mm), with stinging hairs sparse or absent. Damp habs ....*Fen Nettle* **Urtica dioica** ssp **galeopsifolia**

Ann, with taproot. May-Oct

Stipules 1 pr at each node (2 fused prs), >2mm, ovate, entire. Lvs pale green. Stems to 100cm. VR alien..*Mediterranean Nettle* **Urtica membranacea**

Stipules 2 prs at each node, c1mm, ovate-lanc, toothed or lobed. Lvs dark green, 2-8 x 1.2-4cm, ovate to elliptic, acute, cuneate at base, translucent-dotted, deeply toothed (more deeply cut than other *Urtica* spp). Petiole of lower lvs longer than lf. Stems to 30(60)cm, often branched, furrowed...*Small Nettle* **Urtica urens**

■Stipules to 2mm (occ absent from lowest lvs). Plant without stinging hairs

Lvs involute when young. Stipules triangular, green

Rhizomatous per. Stems unbranched, 10-50cm, hairy to ± hairless, ± round or with 2 ridges, swollen above nodes, fetid. Lvs 3-8cm, elliptic-ovate, minutely hairy both sides, ciliate, crenate-serrate with 10-40 pale hydathode-tipped teeth per side, net-veined, with whitish midrib and all veins translucent, with stomata below only. Petiole 3-15mm, with 1 pr minute extra-floral nectaries nr lf, with 3 vb's and spiral fibres. Apr-Oct (all yr) ....*Dog's Mercury* **Mercurialis perennis***

Ann. Stems branched, 10-50cm, hairy to ± hairless, 4-grooved to ± round, swollen above nodes, fetid. Lvs 1.5-5cm, ovate to elliptic-lanc, hairless, ciliate, crenate-serrate with (4)8-18 pale hydathode-tipped teeth per side, hardly net-veined, with midrib and 2° veins white and opaque, with stomata below only. Petiole 2-20mm, with 1 pr minute extra-floral nectaries nr lf, with 3 vb's and spiral fibres. May-Oct..*Annual Mercury* **Mercurialis annua***

Lvs not involute when young. Stipule-like extra-floral nectaries minute, soon falling

Petiole crisped-hairy both sides. Stolons absent. Apr-Oct ....*Enchanter's-nightshade* **Circaea lutetiana***

Petiole crisped-hairy above or on channel margins. Stolons often present. Apr-Oct. *C. alpina* x *lutetiana*..*Upland Enchanter's-nightshade* **Circaea** x **intermedia**

Petiole hairless. Stolons usu present. May-Oct. Uplands ....*Alpine Enchanter's-nightshade* **Circaea alpina**

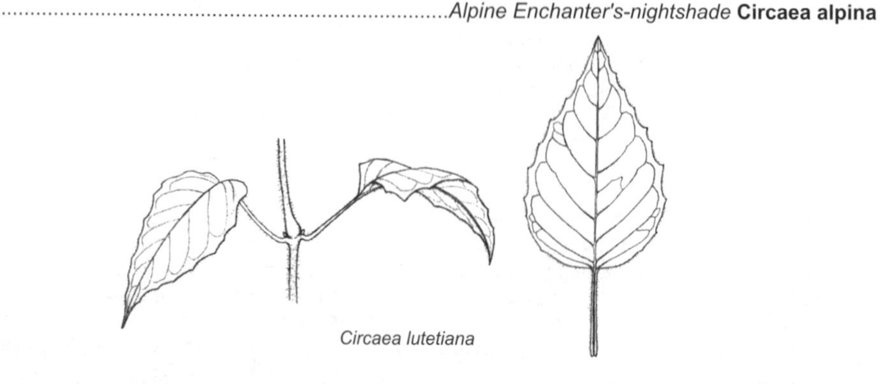

Circaea lutetiana

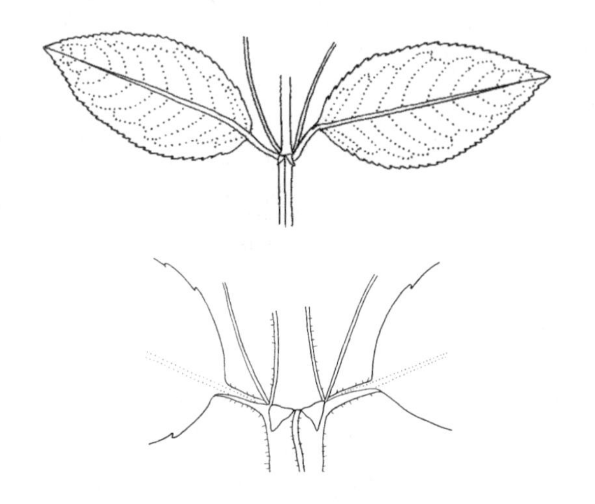

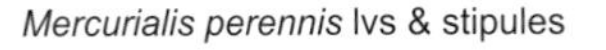

Mercurialis perennis lvs & stipules

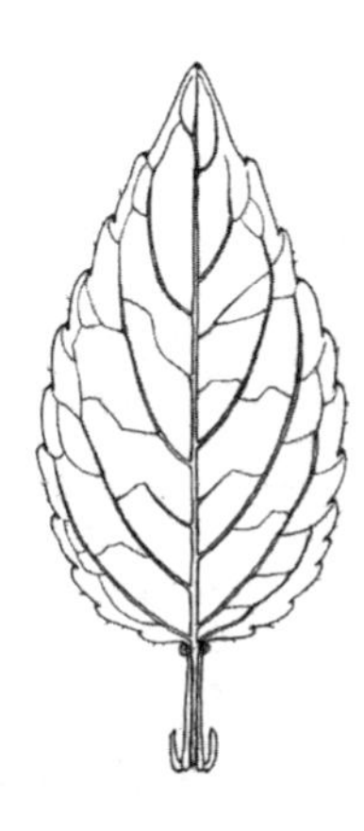

Mercurialis annua

Group OD – Dipsacus. *Lvs usu with midrib ± spiny (actually stiff bristles) and raised below (often forming acute keel). Hairs unicellular (rarely septate on the lowerside of lvs of* D. pilosus*). Bi*

■Basal lvs with swollen-based stiff hairs both sides (otherwise hairless or with glandular hairs below)

Basal lvs 5-40cm, oblong to elliptic-oblanc, obtuse, shortly petiolate, bullate, crenate-dentate with submarginal hydathodes along margins above, net-veined, with 2-3° veins raised below, with spines on 2° veins below, with stomata both sides. Stem lvs narrowly lanc, connate at base, forming a water-collecting cup. Petiole often hollow, with 5 vb's. Stems 50-200cm, hairless but with prickles on angles. All yr..*Wild Teasel* **Dipsacus fullonum**

■Basal lvs with stiff hairs both sides (hairs swollen-based on lowerside only)

Basal lvs 4-9 x 2-4.5cm, ovate, acute or shortly acuminate, narrowed to long petiole, occ with free lobes, involute when young, dark green, crenate with hydathodes not submarginal, net-veined with 2(3)° veins raised below, with spines on midrib below, with stomata both sides. Stem lvs to 17 x 7cm, ovate to narrowly elliptic, connate at base but not forming a water-collecting cup, simple or with 1 pr of ± free small or unequal lobes at base, entire to crenate. Petiole to 17cm on basal lvs, to 6cm on stem lvs, sparsely bristly or spiny, slightly winged, channelled, with 7 vb's around hollow. Stems 30-120cm, angled, furrowed, with sparse weak prickles on angles, hollow. All yr..*Small Teasel* **Dipsacus pilosus**

■Basal lvs hairless above, soon stiffly hairy on veins below

Basal lvs 15-40cm, cuneate at base, long-ciliate, crenate to laciniate-lobed esp proximally, net-veined, 2-3(4)° veins raised below, with stomata both sides. Stem lvs connate at base, forming water-collecting cup. Petiole winged, with 5(7) vb's. Stems 50-200cm. All yr ....*Cut-leaved Teasel* **Dipsacus laciniatus**

Group OE – *Lf surface with dendritic hairs. Plant also with septate hairs. Per*

Basal lvs 6-20 x 15cm, ovate, cordate at base, dendritic-hairy both sides but ± whitish woolly below, crenate-serrate, net-veined, with 2(3)° veins raised below. Stem lvs petiolate, the uppermost rounded or cuneate. Petiole to 30cm, slightly woolly, channelled, with 1(3) vb's (the central vb involute). Stems to 100cm, woolly and with stout septate hairs below branches, square, pith-filled. All yr...*Turkish Sage* **Phlomis russeliana**

Group OF – *Stems square or absent. Lvs strongly mint-scented (pitted with sessile glands at least below), applanate when young, usu hairy, with stomata below only. Petiole (when present) channelled, with flat arc vb. Stems pith-filled, soon hollow. Hairs septate. Rhizomatous (exc* Agastache*). Apr-Oct (overwintering as short shoots with reduced lvs). The range of variation in* Mentha *is ± continuous*

■Stems prostrate, rooting at most nodes

Lvs 1-2cm, elliptic-ovate, cuneate at base, often strongly channelled, yellow-green, gland-pitted both sides (glands often translucent), sickly-scented, hairless to minutely hairy, obscurely crenate-serrate with 1-4(6) teeth per side. Petiole 3-4mm. Stems 5-20cm tall, minutely hairy. All yr. Sch8..........*Pennyroyal* **Mentha pulegium**

Lvs to 0.5cm, ± orb, cordate at base, dull green, gland-pitted below, pungently scented, sparsely long hairy, crenate to entire. Petiole c2mm. Stems <1cm tall, minutely ± retrorsely hairy. All yr ....*Corsican Mint* **Mentha requienii**

■Stems usu erect, occ rooting at lower nodes

Petiole 0-3(5)mm. Lvs rugose, strongly net-veined, with branched hairs below

Lvs usu ± orb to oblong-lanc, with teeth not or hardly recurved

Lvs 2-6 x 1.5-4cm, broadly rounded at base, v hairy below (but less so than *M. suaveolens*). Petiole 1-5mm. R alien. *M. longifolia* x *suaveolens*.....*False Apple-mint* **Mentha** x **rotundifolia**

Lvs usu ± orb to oblong-ovate, with recurved teeth

Lvs 2-6 x 1.5-4cm, occ lanc, rounded at base, spearmint-scented, hairy above, often woolly below, crenate-serrate with c15 teeth per side (some often recurved), strongly net-veined. Petiole 1-3mm. Stems densely hairy. Hortal. *M. spicata* x *suaveolens* ....*Apple-mint* **Mentha** x **villosa**

Lvs 2-5 x 1.5-3cm, ± cordate at base, sickly scented, hairy above, grey- or white-woolly below, often variegated with pale margins, crenate-serrate. Petiole to 5mm. Stems sparsely to densely hairy. R..........*Round-leaved Mint* **Mentha suaveolens**

Petiole 0-3(5)mm. Lvs not rugose (2° veins sunken only), weakly net-veined, with hairs all simple or absent, lanc to broadly ovate, without recurved teeth (exc *M. spicata* 'Moroccan')

Lvs deeply serrate with 6-12 acuminate teeth per side (teeth revolute when young), 3-10 x 1.5-3cm, lanc to broadly ovate, ± acuminate, broadly cuneate to rounded at base, rarely rugose, spearmint-scented, hairless to densely hairy, rarely long-hairy. Petiole 0-1mm (2-3mm in long-hairy form). Stems often purplish, densely hairy to hairless. The cultivar 'Moroccan' may have recurved teeth..........*Spear Mint* **Mentha spicata**

Lvs serrate with c6 teeth per side, 4-8 x 4cm, ovate-lanc, obtuse to shortly acuminate, rounded to cuneate at base, usu spearmint-scented, sparsely hairy (to ± hairless). Petiole 1-4mm. *M. arvensis* x *spicata*..........*Bushy Mint* **Mentha** x **gracilis**

Lvs serrate with 8-15 ± patent acuminate teeth per side, 3-10 x 1.5-3cm, lanc-elliptic, acute, rounded to ± cuneate at base, rarely rugose, hairy. Petiole 0-1mm. *M. longifolia* x *spicata* ....*Sharp-toothed Mint* **Mentha** x **villosonervata**

Lvs serrate with 10-20 teeth per side, 5-10 x 2.5cm, lanc-elliptic, acute, rounded at base, fetid, softly grey-hairy. Petiole 0-1mm. Stems densely long-hairy. VR alien ....*Horse Mint* **Mentha longifolia**

Petiole >3mm. Lvs not rugose, not net-veined, with hairs all simple or absent, ovate-orb to elliptic, without recurved teeth

Damp habs. Rhizomes often emerging as stolons

Plant often purplish. Lvs 2-6 x 1.5-4cm, ovate, obtuse or acute, rounded at base, with sessile yellow glands when young, densely hairy to hairless, serrate or crenate-serrate with (5)7-15 teeth per side. Petiole (2)4-20mm. Stems often with 2 concave and 2 convex sides, with retrorsely crisped hairs (esp on angles). Infl lfy..........*Water Mint* **Mentha aquatica**

Plant usu green. Lvs 2-6 x 1.5-4cm, usu ovate to elliptic, acute or obtuse, cuneate to ± cordate at base, with sickly pine-disinfectant scent, hairy both sides (often densely so), serrate with 5-14 teeth per side. Petiole to 15mm. Infl lfy. *M. aquatica* x *arvensis* ....*Whorled Mint* **Mentha** x **verticillata**

Dry habs. Rhizomes never emerging as stolons

Lvs without translucent dots (HTL)

Petiole 2-3cm

Lvs 4-9cm, ovate, pitted below, minutely softly hairy both sides, crenate-serrate with 7-12 teeth per side. Petiole and stems minutely retrorsely hairy. VR alien ....*Korean Mint* **Agastache rugosa**

Petiole <2cm

Lvs shallowly crenate or serrate with 7-9 teeth per side. Plant never purplish

Lvs 2-6 x 1-2cm, ovate to elliptic, usu obtuse, cuneate to rounded at base, with sweet fruity scent, ± hairy both sides. Petiole 4-15mm. Stems ± hairy ....*Corn Mint* **Mentha arvensis**

Lvs serrate with (8)11-25 teeth per side. Plant often purplish

Lvs >2x longer than wide, 3-9 x 1.5-4cm, ovate to oblong-lanc, cuneate to rounded at base, peppermint-scented, hairless to sparsely hairy. Petiole (2)5-11mm. Stems to 90cm, sparsely hairy to hairless (but often with abundant sessile glands). *M. aquatica* x *spicata* ....*Peppermint* **Mentha** x **piperita**

Lvs <2x as long as wide, 3-6 x 1.5-3.5cm, ovate, rounded to broadly cuneate, often spearmint-scented, hairless to sparsely hairy. Petiole <5mm. Stems to 150cm, ± hairless. *M. aquatica* x *arvensis* x *spicata*..*Tall Mint* **Mentha** x **smithiana**

Lvs with translucent dots (HTL)

Lvs (3)4-7 x 3.5cm, ovate, usu obtuse, broadly cuneate to rounded at base, serrate with 5-10 teeth per side (teeth slightly recurved when young), hairy both sides, ± net-veined. Petiole (5)15-20mm. Stems with spreading (or slightly retrorse) hairs. Isle of Wight. Sch8 ....*Wood Calamint* **Clinopodium menthifolium**

Lvs 1.5-4 x 3.5cm, ovate to orb-ovate, obtuse, broadly cuneate to rounded at base, with sessile glands below, hairy both sides, ciliate, shallowly crenate-serrate with (3)5-8 teeth per side, the uppermost lvs ± entire. Petiole 8-10mm. Stems with slightly retrorse to spreading 0.5mm hairs....................................*Common Calamint* **Clinopodium ascendens**

Lvs 1.5-3 x 2.5cm, ovate, obtuse, usu broadly cuneate at base, densely greyish-hairy with long and short hairs above (more long hairs below), shallowly crenate or crenate-serrate with 4-5 teeth per side. Petiole to 5mm. Stems greyish with long (0.7-1mm) soft spreading hairs...*Lesser Calamint* **Clinopodium calamintha**

OG

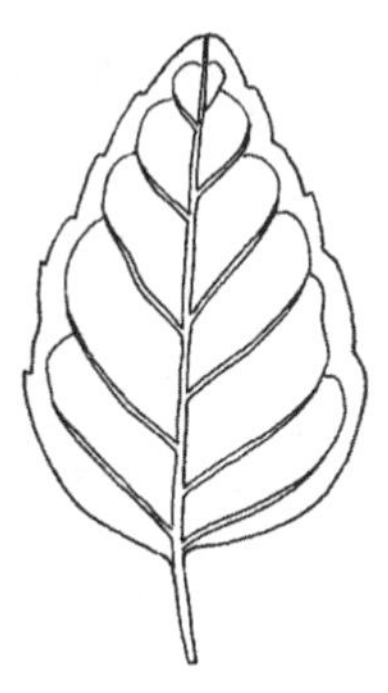

Clinopodium vulgare

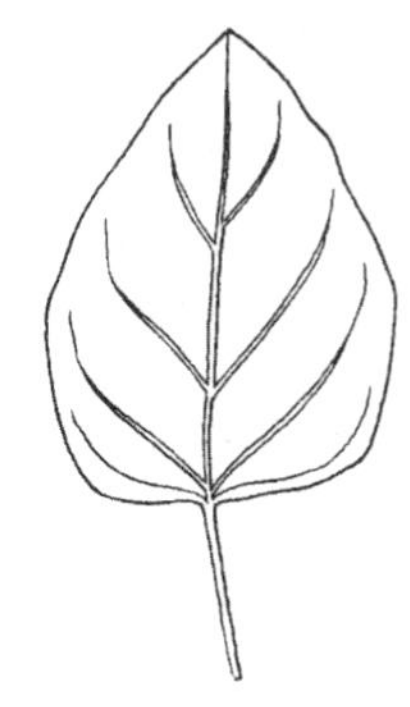

Origanum vulgare

Group OG – *Lvs pleasantly aromatic (occ lemon/orange-scented) or fetid AND/OR white-woolly to densely white-hairy, (±) net-veined (exc* Origanum vulgare, Clinopodium acinos*). Petiole channelled. Stems (±) square, erect, pith-filled or hollow. Hairs septate. Per (often ann in* Clinopodium acinos*). Mar-Oct (all yr in* Stachys byzantina*, others overwintering as short shoots with reduced lvs). (*Teucrium *spp may key out here). ❶ Petiole with 2 vb's (all others have 1 flat arc vb and 2 tiny rib bundles)*

■Lvs whitish with white hairs at least below, strongly scented. Stems white-hairy
 Lvs cuneate at base
  Lvs 5-8cm, oblong, obvolute when young, weakly fetid, densely white-woolly, with 50-80 teeth per side (obscured by hairs). Petiole to 2cm. Stems to 50cm, white-woolly. Stoloniferous. All yr ....*Lamb's-ear* **Stachys byzantina**
 Lvs usu cordate at base
  Petiole with 1(3) vb's
   Lvs 3-7 x 3-5cm, ovate, ± obtuse (the terminal tooth wider than long) but mucronate with hydathode, sessile, flat, with orange glands above (not pitted HTL), with coconut-curry odour, shortly densely grey-hairy esp below, serrate with 10-12 teeth per side, with stomata both sides. Petiole to 2cm. Stems 50-100cm, densely short-hairy with recurved hairs to 0.4mm. Tufted..........*Cat-mint* **Nepeta cataria**
   Lvs to 4cm, ovate, applanate-obvolute to weakly revolute when young, usu slightly channelled, ± rugose, with chewing-gum odour, white short-hairy both sides, deeply serrate with 8-11 teeth per side, with 2(3)° veins raised below. Petiole to 0.5cm. Stems 30-70cm. Rhizomatous. *N. nepetella* x *racemosa*..........*Garden Cat-mint* **Nepeta** x **faassenii**
  Petiole with 2 vb's ❶
   Lvs 1.5-4.5cm, ovate-orb, the lower long-petiolate, the upper shortly petiolate, all obtuse, weakly involute when young, strongly rugose, aromatic or absinth-scented, with dense short white hairs, crenate. Stems 15-60cm, stout, white-woolly..........*White Horehound* **Marrubium vulgare**

■Lvs not whitish-hairy. Stems hairy
 Lvs with 2° veins fading towards margins (eucamptodromus), with 0-2(3) obscure teeth per side
  Lvs 1.5-5 x 3cm, ovate, obtuse, rounded at base, ± applanate when young, with flat margins, hairy and pitted with glands both sides, ciliate, with few 2° veins, with stomata both sides or below only. Petiole to 2cm, spreading-hairy. Stems 30-80cm, retrorsely to spreading-hairy ....*Wild Marjoram* **Origanum vulgare***
  Lvs 0.5-1.5 x to 0.5cm, ovate to elliptic, often acute, cuneate at base, ± applanate when young, with slightly recurved margins, weakly aromatic, not pitted, hairy both sides (occ hairless), ciliate, with few 2° veins, with stomata usu below only. Petiole to 0.2cm, antrorsely hairy. Stems to 10cm, retrorsely hairy. Per or ann. All yr..........*Basil Thyme* **Clinopodium acinos**
 Lf veins not fading towards margins (anastomosing), usu with >3 distinct teeth per side
  Lvs strongly lemon-scented
   Lvs 3-7cm, ovate, obtuse (terminal tooth as wide as long), cordate at base, ± rugose, often purplish below when young, sparsely hairy to hairless, ciliate, crenate-serrate with 7-16 teeth per side (recurved when young), with stomata below only. Petiole to 4cm, hairy or hairless. Stems 30-60cm, hairless or hairy..........*Balm* **Melissa officinalis**
  Lvs strongly mock-orange scented
   Lvs 5-9cm, ovate, acute (the terminal tooth often longer than wide), ± valvate when young (with teeth slightly revolute), hairy, crenate with 16-18 teeth per side. Petiole hairy. Stems to 60cm, with long patent (rarely retrorse) hairs. Rhizomatous ....*Bastard Balm* **Melittis melissophyllum**
  Lvs weakly aromatic
   Lvs 2-4(5)cm, ovate, rounded to ± cuneate at base, weakly revolute when young (applanate initially), with obscure glands, hairy (hairs often minutely swollen-based), shallowly crenate-serrate with ≤10 teeth per side (often obscure), with stomata below only. Petiole to 0.6cm, spreading-hairy. Stems to 60cm, with 0.5-0.7mm retrorse or spreading hairs. Rhizomatous ....*Wild Basil* **Clinopodium vulgare***

Group OH – *Lvs fetid or odourless, all on stems, with petiole 0-6mm. Petiole (when present) with 1 flat arc vb (and often 2 tiny rib bundles).* ❶ *Hairs unicellular (septate in all other taxa)*

■Lvs viscid with dense glandular hairs

Lvs alt above, 1.5-4cm, oblong to lanc, (±) acute, not connate at base, glandular-hairy both sides, coarsely serrate, weakly 3-5-pli-veined, with 2-3° veins ending in a sinus, with 2° veins raised below, with whitish bullate or 'crazy paving' pattern below. Stem usu 1, to 50cm, viscid glandular-hairy (hairs yellow-tipped when young), odourless, occ round. Ann. Jun-Sept .... *Yellow Bartsia* **Parentucellia viscosa**

■Lvs not viscid (but may have glandular hairs)

Lvs with 2° veins ending in a sinus, sessile, turning black on drying. Ann, usu with 1 stem

Stems often shortly streaked with black, without interpetiolar ridge

Lvs 2-4(6)cm x (2)5-8(12)mm wide, oblong to linear, ± cordate at base, not connate, applanate when young, with recurved scabrid-ciliate margins, crenate-dentate with (0)7-12 teeth per side, shortly hispid or adpressed-hairy to hairless above, weakly net-veined, with 2° veins (occ blue when young) raised below, often with bullate or 'crazy paving' pattern below, with stomata both sides (often obscure above). Stem to 50cm, sparsely hairy to hairless, solid. Apr-Sept.. *Yellow-rattle* **Rhinanthus minor** agg*

Lvs 4-8cm x 4-15mm, lanc or linear-lanc, with recurved scabrid margins, often strongly crenate-serrate. Stem 30-60cm, hairless, solid. VR. Sch8 .... *Greater Yellow-rattle* **Rhinanthus angustifolius**

Stems not streaked with black, without interpetiolar ridge

Lvs 1.2-4cm, lanc to linear-lanc (with long terminal tooth), not connate at base, dentate with 1-5 teeth per side, ± antrorsely hairy, ciliate, with stomata both sides. Stem to 50cm, often purple-tinged, retrorsely unicellular-hairy, solid. Apr-Sept.........*Red Bartsia* **Odontites vernus**

Lvs without 2° veins ending in sinus (anastomosing 3° veins may end in sinus), occ shortly petiolate or broadly winged to base, not turning black on drying

Ann. Lvs with hydathodes along margins of upperside. ❶. Usu dry habs ....(*Valerianella*) Go to **VAL**

Per. Lvs without hydathodes along margins of upperside

Damp to wet habs

Lvs net-veined, not rugose

Lvs 5-10 x 1.5-3cm, oblong-lanc or linear-lanc (3-5x long as wide), acute (with long terminal tooth), rounded to ± cordate at base, applanate when young, shortly roughly hairy both sides (hairs usu longer on midrib below), weakly fetid, crenate-serrate with 16-35 teeth per side, usu with stomata below only. Petiole 0-5(7)mm (<10% total lf), long-hairy. Stems to 100cm, with purplish nodes, the angles with long retrorse hairs on rough swollen bases, the faces often minutely hairy, hollow. Rhizomatous, with elongated tubers. Apr-Oct .... *Marsh Woundwort* **Stachys palustris**

Lvs 2-7 x 1-2cm, oblong-lanc, ± obtuse, cordate at base, applanate when young, often with recurved margins at maturity, occ purplish below (esp lower lvs), shortly hairy both sides (esp below), odourless, shallowly crenate with 6-10(23) teeth per side, with stomata below only. Petiole 2-8mm, short-hairy. Stems 20-50cm, lfless below, branched above, with sparse retrorse crisped hairs on angles, hollow. Shortly rhizomatous. Apr-Oct .... *Skullcap* **Scutellaria galericulata**

Lvs not net-veined, rugose

Lvs 1-2(7) x 0.5-1cm, ovate, ± acute, cordate at sessile base (rarely petiolate), with teeth revolute when young and margins recurved at maturity, strongly sage- or onion-scented, sparsely hairy to woolly, crenate-serrate with 6-9 teeth per side, with stomata below only. Stems 5-50cm, rooting at lowest nodes, branched, sparsely hairy to woolly. All yr (overwinters as stolons with reduced lvs). VR. Sch8 .... *Water Germander* **Teucrium scordium**

Lvs 1-2.5 x 0.6-1.5cm, ovate, obtuse (to ± acute), rounded at sessile base, hairless to glandular-hairy above, odourless, crisped-hairy below, crenate-serrate with 6-8(14) teeth per side, with 2° veins raised below (often blue when young), with stomata below only. Stems 5-20cm, not rooting at nodes, usu unbranched, retrorsely white-hairy, often glandular-hairy above. Shortly rhizomatous. May-Sept. Uplands, mtns .... *Alpine Bartsia* **Bartsia alpina**

Lvs not net-veined, not rugose

Lvs 1-3 x 1cm, lanc, sparsely adpressed-hairy both sides when young, ciliate, with (0)1-4 teeth per side at base, with anastomosing 2° veins raised below, with stomata both sides. Petiole 1.5-4mm, channelled. Stems 10-20cm, with sparse crisped hairs. Shortly rhizomatous.. *Lesser Skullcap* **Scutellaria minor***

Dry habs

Lf teeth clearly revolute when young. Lvs weakly aromatic, 1-3cm, ovate, sparsely hairy both sides, with sunken or sessile glands at least below, with 4-6 teeth per side, not or rarely net-veined, with stomata below only. Petiole 3-5(7)mm. Hairs septate. All yr

Stems 20-40cm, densely white-hairy (occ weakly retrorse). Lvs shiny dark green above, shallowly dentate-serrate. Fls sterile. Hortal. *T. chamaedrys* x *lucidum* .... *Hedge Germander* **Teucrium** x **lucidrys**

Stems to 20cm, weakly retrorsely hairy (often densely so). Lvs dull or shiny mid-green above, deeply serrate-crenate or lobed. Fls fertile *Wall Germander* **Teucrium chamaedrys**

Lf teeth never revolute when young. Lvs odourless. Hairs unicellular ❶

Lvs net-veined, all opp, 4-10 x 1.5cm, oblong-lanc, acute, narrowed to petiole, ± folded when young, rugose, shortly hispid both sides (occ glandular), ciliate, serrate with 10-20 teeth per side, with stomata obscure both sides. Petioles often indistinct, to 1cm, v broadly winged, clasping but rarely connate at base. Stems 30-100cm, roughly hairy (hairs on swollen-bases), often glandular, hollow. Apr-Oct (all yr) .... *Argentinian Vervain* **Verbena bonariensis**

Lvs not net-veined, the upper alt...............(*Epilobium tetragonum*, *E. obscurum*) Go to **EPIL**

Rhinanthus minor agg

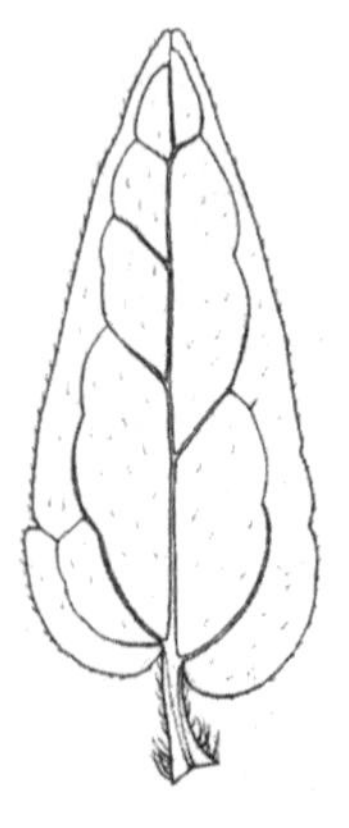

Scutellaria minor

Group OI – *Lvs in basal rosette, OR on stem with petiole >6mm (at least on lower lvs). Lvs hairless (occ with ± sessile glands). Per. (Basal lvs of* Dipsacus laciniatus *may occ key out here)*

■Lvs with 5-7 ± parallel veins (and weaker pinnate veins) ....(*Mimulus*) Go to **MIM**

■Lvs with pinnate veins

Lvs net-veined. Petiole deeply V-shaped with keel below, with 1(3) flat arc vb's

Basal rosette absent. Petiole ± unwinged. Stems unwinged but strongly 4-angled

Lvs 6-13cm, ovate to elliptic, acute, rounded to cordate at base (often asymmetrically decurrent), without translucent dots, fetid, with ± sessile glands when young, rarely hairy both sides, 1-2-serrate with 25-60 teeth per side, usu with stomata below only. Stems hairless to hairy. Rhizomes short, tuberous, with swollen nodes. *Common Figwort* **Scrophularia nodosa**

Basal rosette present. Petiole winged. Stems 4-winged

Lvs 6-12cm, elliptic to ovate, usu obtuse, deeply cordate to rounded at base, with or without translucent dots (HTL), weakly odorous (with sessile glands when young), occ sparse glandular hairs below, crenate with 25-32 blunt hydathode-tipped teeth per side (occ ± serrate esp on upper lvs), with stomata both sides. Petiole often with 2 lobes nr lf. Stems winged, hollow. Tufted. All yr.....................................*Water Figwort* **Scrophularia auriculata***

Lvs 6-15cm, elliptic to ovate (to lanc), ± acute to obtuse (v broad triangular terminal lobe), ± cordate to cuneate at base, with translucent dots (HTL), fetid, with ± sessile glands when young, serrate-crenate with 40-50 hydathode-tipped teeth per side, with stomata below only. Petiole without basal lobes. Stems strongly winged, hollow. Rhizomatous. All yr ....*Green Figwort* **Scrophularia umbrosa**

Lvs not net-veined. Petiole channelled (lvs often sessile in *Epilobium*)

Stipule-like scales absent ....(*Epilobium*) Go to **EPIL**

Stipule-like alt dark brown scales present on short stem between roots and lvs. Lvs in basal rosette, 1-2 x 1.5cm, ovate, obscurely crenate with 2-12 hydathode-teeth per side, with 2° veins raised above. Petiole 2-10mm....................*One-flowered Wintergreen* **Moneses uniflora**

OJ

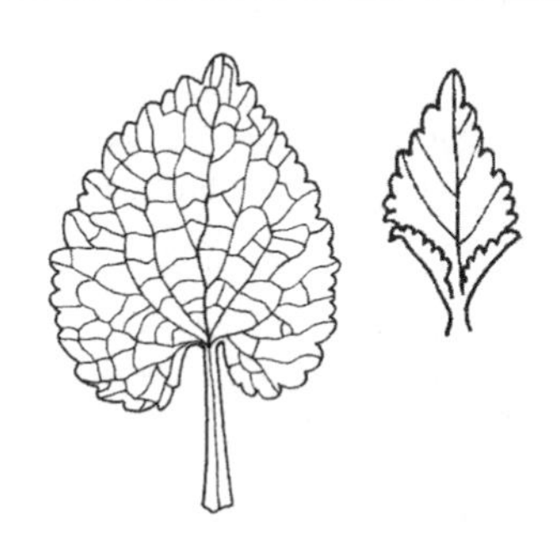

Ballota nigra & vernation of young lf

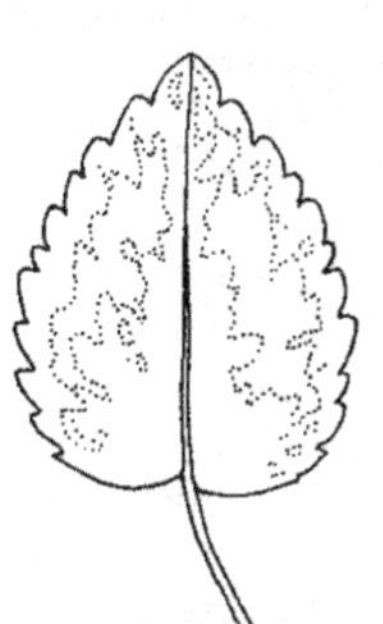

Lamiastrum galeobdolon ssp *argentatum*

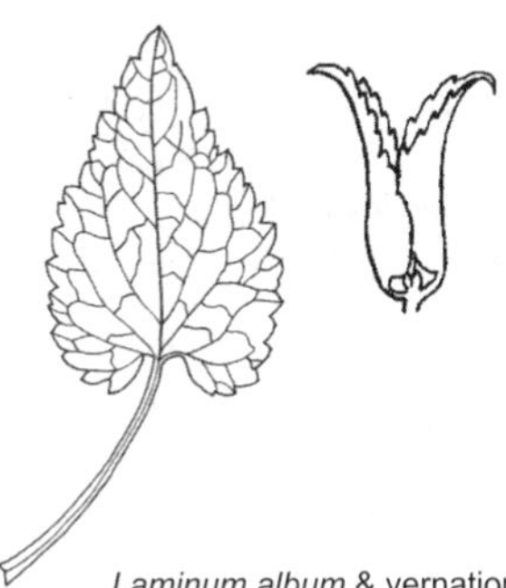

Laminum album & vernation of young lvs

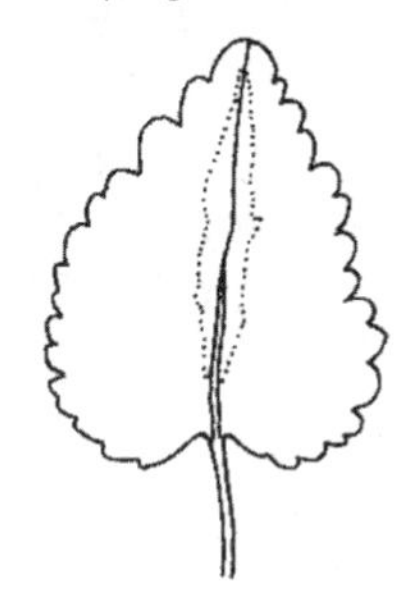

Lamium maculatum

Group OJ – *Petiole with 2(4) vb's (always an even number, rarely also with a tiny central vb). Lvs net-veined (exc young lvs of* Galeopsis*). Stems pith-filled, soon hollow. Hairs septate*

▪Plant with basal rosette. Lvs rugose, weakly sage-scented. Stems with long retrorse hairs. Per (bi)

Basal lvs 7-15cm, ovate to ovate-oblong, obtuse, cordate at base, hairless or minutely hairy above, sparsely long hairy on veins below, crenate or 2-dentate with c30 blunt teeth per side, occ shallowly lobed, with stomata both sides. Petiole to 15cm, hairy, flat, with 1-3 vb's. Stems to 90cm, glandular-hairy above. Mar-Oct (all yr). Sch8................................*Meadow Clary* **Salvia pratensis**

Basal lvs 4-12cm, oblong-ovate to slightly fiddle-shaped, obtuse, cordate to rounded at base, hairy both sides (often minutely so), ciliate, crenate-serrate to lobed, with stomata both sides. Petiole to 8cm, hairy, flat, narrowly winged, with 1-5 vb's. Stems to 80cm, glandular-hairy above. All yr ....*Wild Clary* **Salvia verbenaca**

▪Plant without basal rosette (false rosette may be present)

Per, often sparsely branched

Lvs with obvious whitish blotch along midrib

Lvs 2-5cm, ovate, usu acute, cordate at base, ± applanate when young, mid-green, hairy, strongly paraffin-scented, crenate-serrate, with stomata both sides. Petiole 1-2cm. Stems retrorsely hairy. Stoloniferous (occ rhizomatous). All yr ....*Spotted Dead-nettle* **Lamium maculatum***

Lvs with obvious large white blotches either side of midrib

Lvs 3-7cm, ovate, obtuse (v broad terminal tooth) to acuminate, truncate at base, incurved and applanate when young, ± shiny dark green above, often purplish below, sparsely hairy, weakly paraffin-scented, crenate-serrate with 10-14 teeth per side, with stomata below only. Petiole 1-3cm. Stems usu rampant, usu retrorsely hairy (esp on angles). Stoloniferous. All yr ....*Variegated Yellow Archangel* **Lamiastrum galeobdolon** ssp **argentatum***

Lvs with white blotches indistinct or absent (may have some whitish speckling)

Lvs with 20-30 teeth per side, without whitish speckling

Lvs to 13 x 7cm, ovate, cordate at base, sparsely minutely crisped-hairy on veins below, ciliate, serrate-crenate, with stomata below only. Petiole to 6cm, minutely antrorsely hairy, with the 2 vb's occ merging into single U-shape. Stems to 1m, minutely antrorsely crisped-hairy on angles, occ glandular-hairy above. Tufted. All yr (overwintering as ± basal lvs) ....*Somerset Skullcap* **Scutellaria altissima**

Lvs usu with 10-18 teeth per side, occ with whitish speckling

Lvs obtuse (with short terminal tooth wider than long)

Lvs 2-5cm, ovate, usu cordate at base, involute when young, hairy (often greyish-hairy below), fetid (rarely paraffin-scented), crenate, with stomata below only. Stems usu densely softly retrorsely hairy, with hairs often >1mm. Shortly rhizomatous. All yr ....*Black Horehound* **Ballota nigra** ssp **meridionalis***

Lvs acute to acuminate (with long terminal tooth 2x longer than wide)

Rhizomatous. Lvs occ white-speckled, 3-7cm, ovate, acuminate, cordate at base, weakly folded when young, slightly rugose, hairy, fetid, acutely crenate-serrate, with stomata below only. Stems often purplish esp at upper nodes, retrorsely hairy, with hairs <1mm. All yr. Pl 13...*White Dead-nettle* **Lamium album***

Stoloniferous. Lvs occ indistinctly white-blotched either side of midrib, 3-5(7)cm, ovate, truncate at base, incurved and applanate when young, ± shiny dark green above, often purplish below, sparsely hairy, weakly paraffin-scented to odourless, crenate-serrate to serrate, with stomata below only. Stems retrorsely (to adpressed-) hairy or sparsely so. All yr

Stolons long................*Yellow Archangel* (ssp) **Lamiastrum galeobdolon** ssp **montanum**

Stolons short. VR or overlooked ....*Yellow Archangel* **Lamiastrum galeobdolon** ssp **galeobdolon**

Ann, often much-branched

Lvs with 2° veins ending in a sinus

Stems hispid (sharply bristly), swollen below nodes, 10-100cm. Lvs hispid both sides (soft when young), 2.5-10cm, ovate to ovate-lanc, acuminate, cuneate at base, weakly folded when young, crenate-serrate with 5-13 teeth per side, with 7-9 prs of 2° veins per side, with stomata below only. Petiole to 3cm. Apr-Oct

Stems strongly swollen below nodes, slightly retrorsely hispid with yellow-tipped glandular hairs below nodes (esp nr infl)..................*Large-flowered Hemp-nettle* **Galeopsis speciosa**

Stems slightly swollen below nodes, with retrorse hairs, occ red-tipped glandular hairs below nodes..*Common Hemp-nettle* **Galeopsis tetrahit** agg*

Stems not hispid (hairy to hairless), not swollen below nodes, usu <50cm

Lvs adpressed-hairy all over surface below

Lvs 2-5 x 1.5cm, ovate-lanc to lanc, cuneate at base, adpressed-hairy above, serrate with 4-7 teeth per side. Stems minutely retrorsely hairy. May-Oct. VR alien ....*Downy Hemp-nettle* **Galeopsis segetum**

Lvs adpressed-hairy only on veins below

Lvs 2-8 x 1-3cm, ovate to ovate-lanc, acute to obtuse, cuneate at base, serrate with 3-8 teeth per side. Petiole to 3cm. Stems with soft adpressed hairs. May-Oct. VR alien ....*Broad-leaved Hemp-nettle* **Galeopsis ladanum**

Lvs 1.5-8 x 0.8cm, linear-lanc to oblong-lanc, acute, cuneate at base, serrate with 1-4 teeth per side, adpressed-hairy above (occ sparsely so). Petiole to 1cm. Stems ± hairless or with short retrorse crisped hairs. Jun-Oct.............*Red Hemp-nettle* **Galeopsis angustifolia**

Lvs with 2° veins ending in a tooth apex. Stems not hispid, without swollen nodes, to 30cm. Lvs patent-hairy to ± hairless, fetid. Bracts often larger than lvs. Jan-Oct (all yr)

All lvs shortly petiolate (uppermost bracts sessile)

Bracts with shallow teeth <2mm

Lvs usu <5cm, ovate, obtuse (terminal tooth ± wider than long), often cordate at base, ± applanate (with teeth slightly revolute) when young, rugose, usu hairier above, ± crenate with 4-12 blunt teeth per side, with stomata below only. Stems v sparsely hairy to ± hairless...*Red Dead-nettle* **Lamium purpureum**

Bracts mostly with deep teeth >2mm

Lvs 0.8-4cm, ovate, obtuse, cordate to truncate at base but ± decurrent down petiole (at least upper lvs), incised-dentate. Plant slender, less hairy ....*Cut-leaved Dead-nettle* **Lamium hybridum**

Upper lvs (and bracts) sessile, the lower shortly petiolate (1-5cm)

Calyx ± patent-hairy. Lvs 0.7-2.5cm, orb to ovate-orb (lower often wider than long), obtuse, truncate to (±) cordate at base, the lower shortly petiolate, crenate with 5 teeth per side, shortly hairy..*Henbit Dead-nettle* **Lamium amplexicaule**

Calyx ± adpressed-hairy. Lvs 1-4cm, broadly ovate, obtuse, truncate to (±) cordate at base, the lower shortly petiolate, crenate with 5 teeth per side. Plant more robust. N Br ....*Northern Dead-nettle* **Lamium confertum**

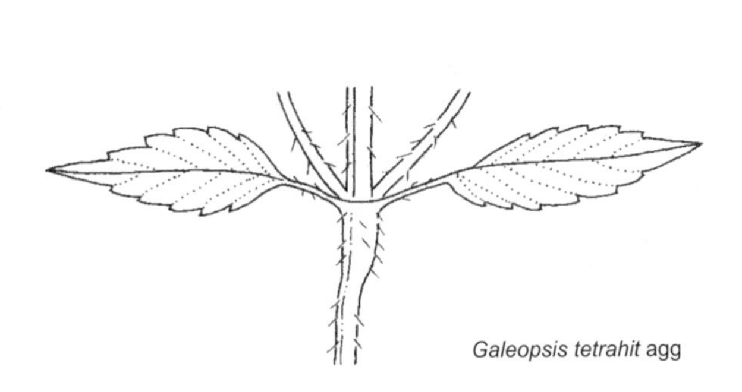

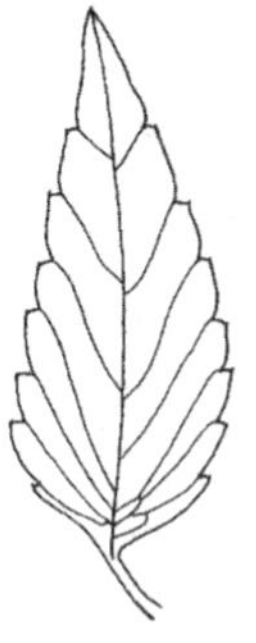

Galeopsis tetrahit agg

Group OK – *Lvs hairy (occ glandular), cordate (to truncate) at base, (±) net-veined or with at least 3° veins translucent (exc* Prunella, Scutellaria x hybrida*). Petiole channelled, with 1 or 3(7) vb's.* ❶ *Petiole hollow*

■Stems rooting at least at lower nodes. All yr

Lvs ovate, odourless, dull green above, with stomata both sides...............(*Prunella*) Go to **PRUNE**

Lvs reniform, fetid, dull dark green, with stomata below only

Lvs to 4cm, obtuse with broad truncate tooth, occ rounded at base, rolled when young, pitted with glands below, hairy to ± hairless, crenate with 6-10 teeth (each with a sunken pale hydathode visible end-on). Petiole to 10cm, retrorsely hairy, with 1(3) vb's. Stems usu retrorsely hairy, the interpetiolar ridge fringed with long hairs, soon hollow. PL 13 ....*Ground-ivy* **Glechoma hederacea**

■Stems not rooting at nodes, or absent

Lf teeth revolute when young. Lvs septate-hairy

Lvs crenate to 2-dentate-serrate with up to 60 hydathode-tipped teeth per side, (4)5-10 x 6cm, ovate, obtuse to acute, hardly rugose, not pitted either side, dark green above, greyish-hairy (with short 0.3-0.5mm hairs), fetid, with stomata below only. Petiole to 3cm, hairy, with 1(3) vb's. Stems to 100cm, with short spreading hairs, hollow. Tufted per, often with basal and stem lvs. All yr...*Balm-leaved Figwort* **Scrophularia scorodonia**

Lvs crenate with 20-40 teeth per side, to 5 x 3cm, ovate-oblong, ± obtuse, rugose, not pitted either side, dark green above, shortly hairy both sides, sage-scented, with stomata below only. Petiole 0.5-3cm, hairy, with 1(3) vb's. Stems to 50cm, spreading to ± retrorsely hairy (often densely so), hollow. Rhizomatous per, with lvs mostly on stem. All yr ....*Wood Sage* **Teucrium scorodonia***

Lvs crenate with 7-14 obtuse teeth per side, 3-7 x 2.5cm, oblong or ovate-oblong, not rugose, obtuse, pitted esp below, often dull dark green above, usu hairy esp below, odourless, with stomata both sides. Petiole to 7cm, usu hairy, with 1(3) vb's. Stems to 60cm, usu retrorsely hairy, solid. Tufted per, with basal lf rosette (usu persisting at fl). All yr ....*Betony* **Stachys officinalis***

Lf teeth not revolute when young

Lvs unicellular-hairy

Plant with weak cress odour. Basal lvs not in a true rosette. Upper lvs alt

Upper lvs sessile. Lower lvs to 15 x 8cm, broadly ovate, ± acuminate, rolled when young, grey-green above, densely hairy, crenate-dentate with blackish hydathodes, often with 5 veins from base, with stomata both sides. Petiole to 10cm, with stiff retrorse hairs, with 3(7) vb's. Stems to 100cm, sparsely stiffly hairy, round but square at base. Bi. All yr. PL 13 ....*Honesty* **Lunaria annua**

Upper lvs petiolate. Lower lvs to 18 x 13cm, broadly ovate, ± acuminate, involute when young, grey- or mid-green, hairy, dentate with hydathodes, with 5 veins from base, usu with stomata below only. Petiole to 8cm, often with stiff retrorse hairs, with (1)3-5(7) vb's. Stems to 140cm, retrorsely hairy (densely so nr base), round. Per, usu rhizomatous. Mar-Oct ....*Perennial Honesty* **Lunaria rediviva**

Plant weakly aromatic to odourless. Basal lvs in a true basal rosette. All lvs opp

Basal lvs to 15 x 15cm, ovate to ± orb, with sparse long hairs on veins above, densely minutely hairy on veins below, dentate, with stomata above esp along veins. Stem lvs similar but occ with 1-2 prs lateral lfts. Petiole often reddish and dilated or ± sheathing at base, sparsely hairy, hollow, with 7-9 vb's. Stems to 120cm. Tufted per. ❶. All yr ....*Pyrenean Valerian* **Valeriana pyrenaica**

Lvs septate-hairy (may be glandular)

Lvs (or stems) with some glandular hairs (occ sparse) AND/OR strongly fetid

Petiole shallowly channelled (U-shaped). Basal rosette never present

Lvs 4-9cm, ovate, acuminate, cordate at base, rolled when young, softly hairy, fetid, crenate-serrate with 9-20 teeth per side, with stomata below only. Petiole 4-12cm, usu densely hairy, with 1(3) vb's. Stems to 80cm, often turning purple, with weakly retrorse or spreading hairs, usu soon hollow. Rhizomatous. Apr-Oct ....*Hedge Woundwort* **Stachys sylvatica***

Petiole shallowly channelled (U-shaped). Basal rosette often present

Lvs (4)7-12(16)cm, oblong-ovate, ± obtuse, cordate at base, obvolute when young, softly hairy, weakly fetid, dentate-serrate with 35-45 teeth per side, with stomata below only. Petiole 3-12cm, hairy, with 1(3) vb's. Stems to 80cm, often glandular-hairy above. Tufted per. Apr-Oct. SW Eng, Wales. Sch8.........................*Limestone Woundwort* **Stachys alpina**

Petiole deeply channelled (V-shaped). Basal rosette present, often dead by fl

Lvs with abundant translucent dots (HTL), to 5 x 4cm, ± ovate, obtuse, ± truncate to cordate at base, dark green, minutely glandular-hairy both sides, fetid, obtusely 1-2-serrate with 25-45 teat-like hydathode-teeth per side (only 10-20 main teeth), with stomata both sides or below only. Petiole to 5cm, with 1 flat arc vb and 2 tiny rib bundles. Stems to 100cm, minutely glandular-hairy, 4-ridged, hollow or pith-filled. Often all yr ....*Italian Figwort* **Scrophularia scopolii***

Lvs with v sparse or no translucent dots (HTL), 4-15 x 10cm, broadly ovate, ± obtuse, cordate at base, light green, softly minutely glandular-hairy both sides (slightly viscid), fetid, deeply obtusely 2-serrate with 30-50 teat-like hydathode-teeth per side (10-35 main teeth), with stomata below only. Petiole to 17cm, glandular-hairy, with 1 flat arc vb and 2 tiny rib bundles. Stems to 100cm, strongly 4-ridged, glandular-hairy, hollow. Bi (per). All yr ....*Yellow Figwort* **Scrophularia vernalis**

Lvs without glandular hairs (may be weakly fetid)

Lvs net-veined or ± so (at least 3° veins visible)

Ann

Lvs 1.5-4 x 2.5cm, ovate, obtuse (terminal tooth often truncate), usu cordate at base, hairy, weakly or not odorous, crenate-serrate with 9-11 teeth per side. Petiole to 1.5cm. Stems to 25cm, roughly hairy esp on angles, hollow. Often all yr ....*Field Woundwort* **Stachys arvensis**

Per (bi)

Lvs in basal rosette and on stems, 5-12 x 4cm, oblong-lanc, acute or obtuse, usu cordate at base, rugose, ± thick, grey-green, silky-hairy (with dense hairs to 3mm), weakly odorous, crenate-serrate with 60-80 teeth per side, with stomata both sides but sparse above. Petiole to 70mm. Stems to 80cm, with long silky hairs. Tufted bi or short-lived per. All yr. Sch8..*Downy Woundwort* **Stachys germanica**

Lvs all on stems, 2-7 x 1-2cm, oblong-lanc, ± obtuse, cordate at base, applanate when young, not rugose but often with recurved margins at maturity, thin, occ purplish below (esp lower lvs), shortly hairy both sides (esp below), odourless, shallowly crenate with 6-10(23) teeth per side, with stomata below only. Petiole 2-8mm, hairy. Stems 20-50cm, lfless below, with sparse retrorse crisped hairs on angles, hollow. Rhizomatous per. Apr-Oct..*Skullcap* **Scutellaria galericulata**

Lvs not or hardly net-veined. Rhizomatous per

Lvs 1-3.5cm, all petiolate, with (4)6-13 teeth per side (usu nr base), with ± adpressed hairs above, v sparsely hairy on veins below, with stomata both sides. Petiole 1-8mm. Stems to 30cm. Fls rare. *S. galericulata* x *minor*......................*Hybrid Skullcap* **Scutellaria** x **hybrida**

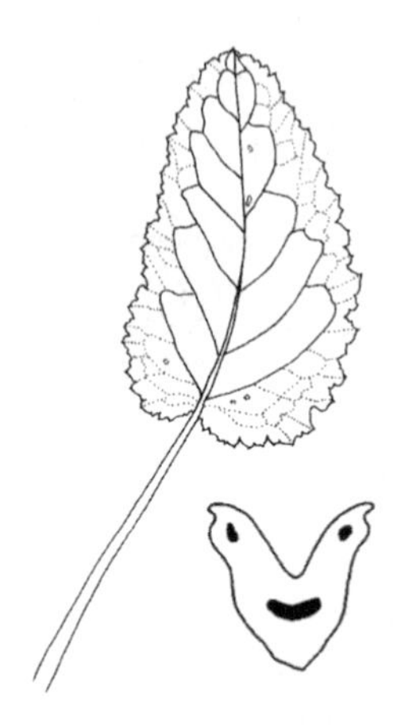

Scrophularia scopolii & petiole TS

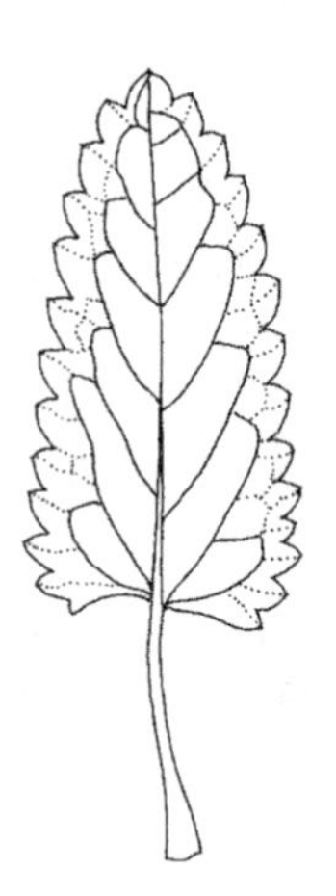

Stachys officinalis

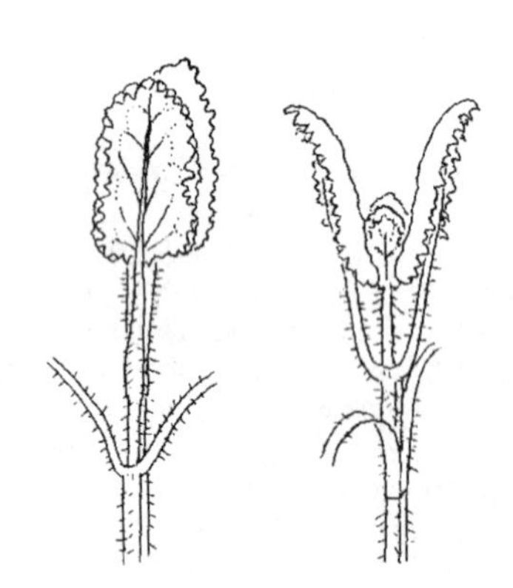

Teucrium scorodonia

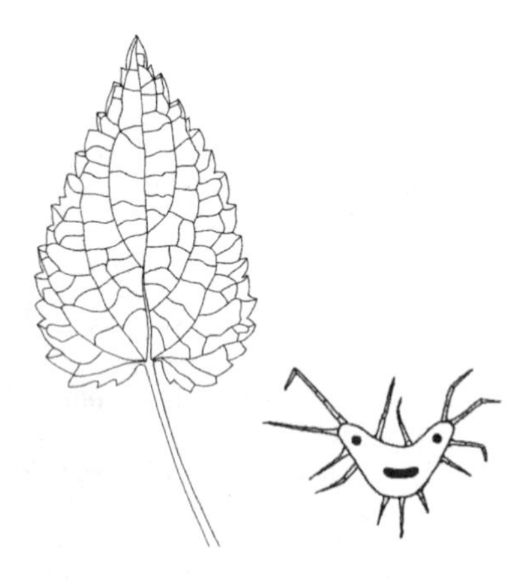

Stachys sylvatica & petiole TS

Group OL – *Lvs hairy (occ glandular), cuneate or narrowed to petiole (rarely rounded at base), odourless (but weakly aromatic in* Valerianella*). Petiole with 1 or 3(7) vb's. (*Epilobium tetragonum, Teucrium chamaedrys *and* T. x lucidrys *may key out here)*

■Lvs unicellular-hairy, the hairs often on minutely swollen bases

Petiole with 3(5) vb's, with spiral fibres (and in lf veins). Lvs ± without 2° veins raised below

Basal lvs with narrow cartilaginous margins turning reddish, to 15(30)cm, elliptic, acute with apical hydathode occ purplish, obvolute when young, stiffly hairy above, often ciliate, obscurely toothed with sunken hydathodes, rarely deeply toothed to lobed, (±) net-veined (esp below), with white midrib raised below (occ purple), with stomata both sides. Petioles to 15cm, ± connate and often purplish at base. All yr......................*Devil's-bit Scabious* **Succisa pratensis***

Basal lvs without cartilaginous margins, to 15(30)cm, elliptic to oblanc, acute, rolled when young, hairy both sides (more hispid above), ciliate, with pale green submarginal hydathodes visible above (but the terminal visible below) and at least some traces of crenate teeth, the upper rosette lvs occ lyrate-pinnatifid, net-veined or not, with midrib white above and green below, with stomata both sides. Petioles to 15cm, connate but not purplish at base, often hollow. All yr. PL 18..*Field Scabious* **Knautia arvensis**

Petiole with 1 vb, without spiral fibres

Ann. Lvs with indistinct teeth, weakly aromatic, without 2-3° veins raised below ....(*Valerianella*) Go to **VAL**

Per. Lvs with distinct teeth, odourless, with 2-3° veins raised below

Basal lvs dead at fl, 2-8(10) x 1-5cm, ovate to oblong, narrowed to petiole, dark green, sparsely hairy both sides, ciliate with swollen-based antrorse cilia, dentate to deeply lobed, ± net-veined, with stomata both sides. Stems 1-several, 30-140cm, tough, rough with swollen-based hairs, deeply furrowed on 2 opp sides, pith-filled. All yr.....*Vervain* **Verbena officinalis***

■Lvs septate-hairy (may be glandular), hairs without swollen bases (exc *Prunella*)

Ann. Lvs net-veined or ± so (3° veins visible), weakly fetid

Lvs 2-6cm, oblong, obtuse, cuneate at base, shallowly crenate. Petiole short. Stems 10-30cm, much-branched, hairy (occ glandular). VR alien........*Annual Yellow-woundwort* **Stachys annua**

Per. Lvs not or rarely net-veined, odourless

Lvs minutely pitted both sides, gradually narrowed to petiole, obvolute when young

Basal lvs (2)4-8 x 4cm, obovate-oblong, obtuse, long-petiolate, shiny dark green (to purplish) above, sparsely hairy, obscurely crenate to ± entire (each hydathode sunken in a tooth sinus), with 4-6 prs of anastomosing 2° veins slightly raised below, with stomata both sides. Stem lvs obovate, ± sessile. Petiole distinct from lf, channelled, winged, with 1(3) vb's. Stem 1, 10-30cm, unbranched, usu ± hairy in 2 rows, square. Stolons often developing (Apr-Oct). All yr ....*Bugle* **Ajuga reptans**

Basal lvs to 4(8) x 2cm, obovate to oblong, obtuse, shortly petiolate, light green, with long shaggy hairs or ± hairless, crenate to ± entire (with ± obtuse hydathodes), with c4 prs of anastomosing 2° veins slightly raised below, with stomata both sides. Stem lvs usu ovate, shortly petiolate to sessile. Petiole indistinct from lf, not channelled, winged, with 1 vb. Stem 1, 10-20(30)cm, unbranched, hairy all sides, ± round. Stolons absent but thin rhizomes present. All yr. Scot, Ire. Sch8..*Pyramidal Bugle* **Ajuga pyramidalis**

Lvs not or hardly pitted, truncate to cuneate (occ slightly winged) at base, applanate when young..(*Prunella*) Go to **PRUNE**

Group OM – *Stems erect, round, with longitudinal black resin canals, aromatic when broken. Hairs septate. Ann. Jun-Oct*

▪Lvs all sessile, revolute when young. Dry habs

Lvs to 11 x 3cm, lanc, ± clasping and ± connate at base, rough with sparse short adpressed hairs both sides, serrate, becoming net-veined, with 2° veins raised below, with stomata both sides. Stem 1, to 80(150)cm, purple or purple-spotted, hairy (hairs often antrorsely crisped), soon furrowed and hollow. Ruderal, birdseed alien..*Niger* **Guizotia abyssinica***

▪Lvs petiolate at least below, folded or obvolute when young. Wet habs

Lvs 4-15 x 1-1.5cm, acuminate, cuneate at base, sparsely hairy to ± hairless both sides, shortly (scabrid-) ciliate, serrate (with teat-like hydathodes), ± opaquely net-veined, with 2° veins often opaque and hardly raised below, with stomata below only. Petiole long-ciliate, with 3 vb's. Stems to 80cm, hairless to hairy (esp above), the interpetiolar ridge with long cilia, with spiral fibres and submarginal vb's

Lvs only petiolate below, toothed but never lobed. Fl heads nodding in bud

Frs (achenes) smooth, brown. Lvs lanc with 5-15 coarse teeth per side. Petioles (±) connate at base. Stems hairy at least above..............................*Nodding Bur-marigold* **Bidens cernua***

Lvs all petiolate, at least some lobed. Fl heads always erect

Frs (achenes) ± smooth (with few or no tubercles), brown. Lvs lanc-elliptic, the lower usu pinnatisect, the terminal lobe deeply serrate with 9-11 teeth per side ....*Trifid Bur-marigold* **Bidens tripartita**

Frs (achenes) densely tubercled, ± black. Lvs lanc to ovate-lanc, sharply irreg toothed, unlobed or pinnate with a large terminal lobe and a pr of stalked lateral lobes (small plants may have partial lobes like *B. tripartita*)......................................*Beggarticks* **Bidens frondosa**

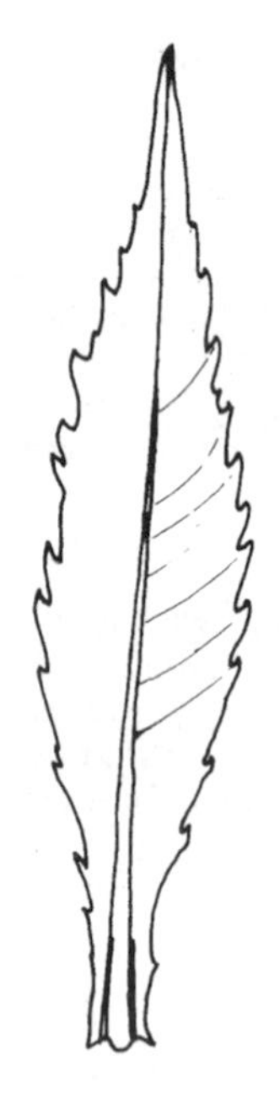

Bidens cernua

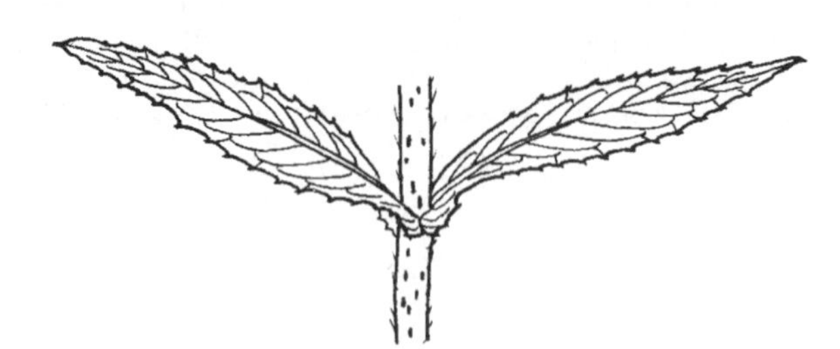

Guizotia abyssinica

Group ON – *Stems round, prostrate to decumbent, often rooting at nodes. Lvs (or petiole) hairy. Petiole with 1 vb. Hairs septate. (*Ajuga reptans *may key out here)*

■Plant viscid-hairy

Dry habs. Lvs to 5 x 5cm, broadly ovate to ± orb, cordate at base, grey-green, occ purplish below, crenate to shallowly lobed, ± palmately veined, not net-veined, with veins 2° raised below, with stomata below only. Petiole 1.5-3cm, channelled. Stems prostrate, to 8cm tall. All yr ....*Trailing Snapdragon* **Asarina procumbens**

Wet habs. Lvs 1-4(6)cm, ovate or elliptic, rounded at base, mid-green, distantly toothed, ± parallel-veined, rarely net-veined, with 2° veins raised below, with stomata below only. Petiole 0-1.5cm, flat or channelled. Stems decumbent, to 40cm tall. Apr-Oct.................*Musk* **Mimulus moschatus**

■Plant not viscid-hairy (hairs may be glandular)

Lvs with submarginal hydathodes (in centre of weakly retuse lobes) and narrow cartilaginous margins

Lvs 0.5-2cm, orb, obtuse, truncate or broadly cuneate at base, sparsely (often adpressed) glandular-hairy to hairless, glandular-ciliate, with 0-7 shallow obtuse teeth per side, with obscure veins, with stomata both sides. Petiole 0.5-2cm, ciliate, channelled. Stems hairy, occ channelled on opp sides, with central stele. All yr ....*Opposite-leaved Golden-saxifrage* **Chrysosplenium oppositifolium**

Lvs with hydathodes along margins or not visible, without cartilaginous margin

Stems usu hairless

Lvs 2-4cm x 4-6mm, held erecto-patent, linear-lanc to lanc, acute, ± clasping at sessile base, applanate when young, with slightly recurved margins, stiff, often purplish esp below, hairless, rarely hairy or glandular-hairy, remotely denticulate with (0)2-6 hydathodes per side esp distally, with translucent midrib sunken above and raised below, with 2° veins indistinct or absent, with stomata both sides. Stems to 15cm tall, 1-3mm diam, creeping, then ascending, often purplish, occ hairy or glandular-hairy, usu snapping audibly, with aerenchyma around stele. All yr. Wet habs...*Marsh Speedwell* **Veronica scutellata**

Stems with 2 opp lines of long white hairs (often antrorsely curved), hairless between (occ hairy on young shoots)

Lvs 1-2.5cm, ovate, ± obtuse, dull green above, hairy esp on veins below, ciliate, crenate-serrate with 4-7 teeth per side, pinnate-veined (and 3-5-pli-veined), with stomata below only. Petiole 0-5(10)mm. Per. All yr...........................*Germander Speedwell* **Veronica chamaedrys**

Stems hairy all round (rarely in 2-3 lines)

Lvs with (4)5-13 teeth per side

Ann, often rooting at lower nodes. Lvs alt above, with 2° veins clearly translucent

Lvs 7-9-pli-veined, 2-3cm, ovate, obtuse, ± cordate to rounded at base, light green, hairy both sides (mainly on veins below), crenate-serrate usu with 6-8 teeth per side, with stomata below only. Petiole to 6mm. Stems to 3mm diam, with sparse spreading hairs, with elastic stele. Oct-Jul. VR alien..............*Crested Field-speedwell* **Veronica crista-galli***

Lvs 3-5-pli-veined to pinnate-veined, 1-3cm, ovate, ± acute to obtuse, truncate at base, dull light green (occ purplish below), hairy esp below and on margins, crenate-serrate with 5-7 teeth per side, with stomata below only. Petiole to 6mm. Stems to 1.2mm diam, with spreading or antrorsely crisped hairs (hairs occ in 2-3 lines), with stele not elastic. All yr (late winter ann)...*Common Field-speedwell* **Veronica persica**

Per, always rooting at least at lower nodes. Lvs all opp, with 2° veins ± obscure or indistinct

Petiole ≤0.6cm

Lvs crenate-serrate with 5-12 teeth per side, with stomata below only, 1-2(3) x 1cm, ovate, ± acute, applanate when young, flat at maturity, dull dark green above, hairy both sides, with 3-4 prs of obscure anastomosing 2° veins. Stems with spreading hairs. All yr ....*Heath Speedwell* **Veronica officinalis***

Petiole 0.6-1.5cm

Lvs deeply crenate-serrate with 4-5 teeth per side, with stomata below only, (1)2-3 x 1.6cm, ovate to ovate-orb (<2x longer than wide), ± obtuse (terminal tooth as wide or wider than long), applanate when young, flat at maturity, light green (occ purplish below), hairy, with 3-4 prs of indistinct anastomosing 2° veins. Upper lvs similar to lower lvs. Petiole 0.5-1.5cm. Stems to 10cm tall, with spreading hairs (occ minutely glandular). All yr. Humus-rich woods..*Wood Speedwell* **Veronica montana**

Lvs crenate-serrate with 7-13(20) teeth per side, with stomata both sides, 1.5-3 x 1cm, oblong-lanc to narrowly so (>2x longer than wide), ± obtuse, applanate when young, slightly folded at maturity, dull dark green above, hairy, with 3-4 prs of distinct anastomosing 2° veins. Upper lvs ± sessile, linear-lanc, ± entire. Petiole 0.8-1.5cm. Stems to 30(50)cm tall, with spreading hairs (occ glandular). All yr. Calc habs, VR. Sch8 ....*Spiked Speedwell* **Veronica spicata**

Lvs with 2-4 teeth per side

At least lower lvs as wide as long. Lvs alt above

Stems rooting at most nodes. Per

Lvs 4-10mm, reniform or orb, v obtuse (terminal tooth much wider than long) to v minutely retuse, with sparse stout hairs to 0.5mm (mostly on veins below), crenate with 3-4 teeth per side, pinnate- or 3-5-pli-veined, with stomata below only. Petiole to 5mm. Stems 0.8mm diam, minutely retrorsely hairy with hairs ≤0.1mm. All yr ....*Slender Speedwell* **Veronica filiformis**

Stems rooting only at lower nodes (if at all). Ann

Lower lvs wider than long, 5-15mm, ovate, obtuse, cordate to truncate at base, with 4 obtuse teeth per side, dull dark green above, occ purplish below, hairy, 3-5-pli-veined (midrib strongest) to pinnate-veined, with stomata below only. Petiole to 4mm. Stems 1-2mm diam, spreading-hairy, with stele not elastic. All yr ....*Grey Field-speedwell* **Veronica polita**

All lvs wider than long, to 12mm, reniform, ± obtuse (with v broad terminal tooth), truncate to ± cordate at base, light or yellow-green, sparsely hairy both sides, ciliate, with 2-3 deep ± acute teeth per side, ± 3(5)-pli-veined (often obscure), with stomata both sides. Petiole to 10mm. Stems 1-2mm diam, spreading-hairy or sparsely so (hairs often in (2)3 lines), with elastic stele. (Oct) Mar-Jun ....*Ivy-leaved Speedwell* **Veronica hederifolia**

All lvs longer than wide

Ann, not or rarely rooting at nodes. Lvs alt above, 5-15mm, ovate, hardly pitted above, light green, v sparsely hairy above, crenate-serrate with 3-4 teeth per side, with stomata below only. Petiole to 2mm. Stems minutely patent- to ± antrorsely hairy, occ with sparse longer hairs. Mar-Sept...*Green Field-speedwell* **Veronica agrestis**

Per, rooting at most nodes. Lvs all opp, 0.5-1.5 x 0.5-1cm, ovate-orb, pitted above, dull green above, minutely glandular-hairy to hairless, ± entire to weakly crenulate, with stomata both sides. Petiole to 2mm. Stems usu hairless below, glandular-hairy above. All yr. VR alien...*Corsican Speedwell* **Veronica repens**

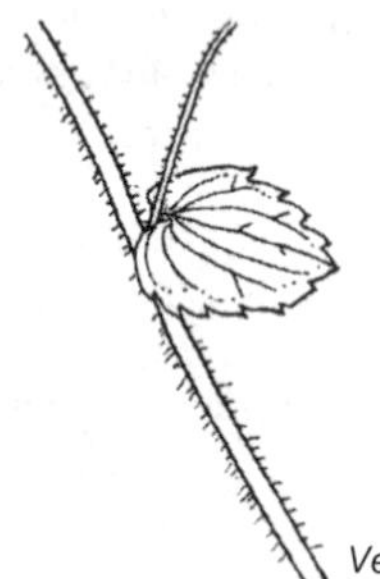

Veronica crista-galli upper stem lf

Veronica officinalis

Group OO – *Stems (±) round and erect, never rooting at nodes. Lvs (or petiole) hairy. (Exceptionally,* Ajuga pyramidalis *may key out here).* ❶ *Plant viscid-hairy*

■Per (exc *Lunaria annua* which is usu bi)

Lvs v roughly hairy, with septate hairs (often with swollen-bases). (*Succisa pratensis* may key out here but has unicellular hairs)

Lvs to 18 x 6(10)cm, oblong-lanc, acute, cuneate but abruptly narrowed to petiole-like base, rolled when young, serrate with 14-18 teeth per side, net-veined, with 3 main veins nr base raised below. Petiole indistinct, to 3cm, narrowly winged, long-ciliate at base. Stems to 200cm, roughly hispid, with interpetiolar ridge, pith-filled, with vb's around margin. Rhizomatous. Apr-Oct. *H. pauciflorus* x *tuberosus*............................*Perennial Sunflower* **Helianthus** x **laetiflorus***

Lvs not roughly hairy, with septate hairs

Dry habs. Lvs rarely 4-whorled, 3-12 x 2-3cm, narrowly ovate to lanc, acute or acuminate, cuneate to ± cordate at base, softly ± crisped-hairy both sides to hairless, 1-2-serrate with 30-35 teeth per side, with stomata both sides. Petioles 1-2cm, connate at base, channelled. Stems to 120cm tall, 2-3mm diam, woody nr base, hairless to minutely hairy (slightly retrorse). Apr-Oct...*Garden Speedwell* **Veronica longifolia**

Wet habs. Lvs always opp, 2-4 x 0.4-0.6cm, linear-lanc to lanc, acute, ± clasping at sessile base, hairless, rarely hairy or glandular-hairy, remotely denticulate with (0)2-6 hydathodes per side esp distally, with stomata both sides. Stems to 15cm tall, 1-3mm diam, creeping, then ascending to erect, usu hairless, occ hairy or glandular-hairy. All yr............*Marsh Speedwell* **Veronica scutellata**

Lvs not roughly hairy, with unicellular hairs

Extra-floral nectaries present at petiole base, minute, soon falling

Petioles crisped-hairy both sides, 2-4cm, connate, channelled, with flat arc vb and spiral fibres. Lvs 4-10cm, ovate, acuminate, truncate to ± cordate at base, dull green above, paler and shiny below, ± hairless to minutely hairy, often crisped-ciliate, not net-veined, with 4-7 prs of 2° veins anastomosing nr margins, with stomata below only. Stems 20-60cm, hairy with retrorse crisped hairs (often minutely glandular) to hairless, with swollen lower nodes. Stolons absent. Rhizomes slender, white, with scale lvs. Apr-Oct ....*Enchanter's-nightshade* **Circaea lutetiana***

Petiole crisped-hairy above or on channel margins, (±) hairless below, to 3cm, connate, channelled, with flat arc vb and spiral fibres. Lvs 3-8cm, ovate, acuminate, cordate at base, bright green, hairless or with adpressed hairs above (occ sparse hairs on veins below), not net-veined, with 4-7 prs of 2° veins anastomosing nr margins, with stomata below only. Stems 10-45cm, sparsely (glandular-) hairy. Stolons often present. Rhizomes slender, often pinkish, with scale lvs. Apr-Oct. *C. alpina* x *lutetiana* ....*Upland Enchanter's-nightshade* **Circaea** x **intermedia**

Extra-floral nectaries absent

Lvs odourless, ± net-veined, ± without 2° veins raised below. All lvs opp

Basal lvs with narrow cartilaginous margins turning reddish, to 15(30)cm, elliptic, acute with apical hydathode occ purplish, obvolute when young, stiffly hairy above, often ciliate, obscurely toothed with sunken hydathodes, rarely deeply toothed to lobed, (±) net-veined (esp below), with white midrib raised below (occ purple), with spiral fibres, with stomata both sides. Petioles to 15cm, ± connate and often purplish at base, with 3(5) vb's and spiral fibres. Stems to 100cm, roughly retrorsely hairy at lowest internode, becoming antrorsely adpressed above. All yr...*Devil's-bit Scabious* **Succisa pratensis***

Basal lvs without cartilaginous margins, to 15(30)cm, elliptic to oblanc, acute, rolled when young, hairy both sides (more hispid above), ciliate, with pale green submarginal hydathodes visible above (but the terminal one visible below) and at least some traces of crenate teeth, the upper rosette lvs occ lyrate-pinnatifid, net-veined or not, with midrib white above and green below, with sparse spiral fibres, with stomata both sides. Petioles to 15cm, connate but not purplish at base, often hollow, with 3(5) vb's and spiral fibres. Stems to 100cm, roughly hairy at least below with stiff retrorse hairs. All yr. PL 18 ....*Field Scabious* **Knautia arvensis**

Lvs odourless, not net-veined, ± with or without 2° veins raised below. Upper lvs usu alt ....(*Epilobium*) Go to **EPIL**

Lvs with weak cress odour, ± net-veined, with 2° veins raised below. Upper lvs alt

Upper lvs sessile. Lower lvs to 15 x 8cm, broadly ovate, ± acuminate, rolled when young, grey-green above, densely hairy, crenate-dentate with blackish hydathodes, ± net-veined, often with 5 veins from base, with stomata both sides. Petiole to 10cm, with stiff retrorse hairs, channelled, with 3(7) vb's. Stems to 100cm, sparsely stiffly hairy, usu square at base. Bi. All yr. PL 13..........*Honesty* **Lunaria annua**

Upper lvs petiolate. Lower lvs to 18 x 13cm, broadly ovate, ± acuminate, involute when young, grey- or mid-green, hairy, dentate with hydathode-teeth, ± net-veined, with 5 veins from base, usu with stomata below only. Petiole to 8cm, often with stiff retrorse hairs, channelled, with (1)3-5(7) vb's. Stems to 140cm, retrorsely hairy (densely so nr base). Per, often rhizomatous. Mar-Oct..........*Perennial Honesty* **Lunaria rediviva**

■Ann. Hairs septate. (*Epilobium* spp may appear ann but has unicellular hairs)

Stems viscid-hairy at least above. ❶..........*Yellow Bartsia* **Parentucellia viscosa**

Stems glandular-hairy at least above (but not viscid to touch)

Ephemeral ann, dead by May. E Anglian Brecks

Lvs deeply serrate, 0.5-1.2cm, ovate, shortly petiolate ....*Breckland Speedwell* **Veronica praecox**

Lvs lobed or with lobe-like teeth

Lvs usu digitately 3-7-lobed with spathulate or oblong lobes, 0.5-1cm, petiolate (upper lvs sessile), glandular-hairy. Stem minutely hairy with scattered long red-tipped glandular hairs. Oct-May. Sch8..........*Fingered Speedwell* **Veronica triphyllos**

Lvs pinnatifid with 3-7 lobes to >½ way, 0.5-1.2cm, petiolate (mid and upper lvs sessile), shortly nonglandular-hairy. Stems with white nonglandular hairs below, glandular-hairy above..........*Spring Speedwell* **Veronica verna**

Persistent ann, all yr

Lf veins ending in sinus..........(*Euphrasia*) Go to **EUPH**

Lf veins ± ending in teeth

Lvs to 1.5cm, ovate, all longer than wide, the lowest petiolate, the upper sessile and elliptic-lanc, hairy, all crenate-serrate with (1)2-5 teeth per side, 3-pli-veined, with stomata both sides (obscure above). Stems 1-8(20)cm, 1mm diam, simple or branched at base, with long spreading hairs all round and v short antrorsely crisped hairs (often in 2 lines), often glandular-hairy above. Dec-Jul..........*Wall Speedwell* **Veronica arvensis**

Lvs to 1cm, ovate, obscurely crenate to ± entire. Stems glandular-hairy (at least above), 5-15cm tall, hairy, greenish or purplish. VR alien..........*French Speedwell* **Veronica acinifolia**

Stems nonglandular-hairy to hairless.

Lvs alt above

Lvs 8-20cm, broadly ovate, shortly hispid-hairy, dentate, net-veined, pinnate-veined but 3-pli-veined at base and raised below, with stomata both sides. Petiole channelled, with 3(5) vb's. Stem to 3m, erect, pithy, with vb's around margin. Jun-Oct..........*Sunflower* **Helianthus annuus**

Lvs to 1(1.5)cm..........*Wall Speedwell* **Veronica arvensis**

Lvs all opp, 2-7cm, ovate, acute to acuminate, cuneate, soon limp, sparsely hairy, ± ciliate, serrate with 6-10 dark hydathode-tipped teeth, 3-pli-veined at base, otherwise pinnate-veined, with weakly translucent 2° veins, other veins visible as an opaque network, with stomata below only. Petiole 0.3-3cm, channelled, with 3(5) vb's. Stems 10-25cm, often ± translucent, ± hairless or with long spreading and minute crisped hairs (often minutely glandular), striate (HTL), with abundant sap. Jun-Oct

Fls with 4(5) ray florets, the ligules 1mm; receptacle bracts usu trifid; pappus scales never awned..........*Shaggy-soldier* **Galinsoga quadriradiata**

Fls with (4)5 ray florets, the ligules 1-1.5mm; receptacle bracts ± entire (rarely trifid); pappus scales often narrowed into an awn..........*Gallant-soldier* **Galinsoga parviflora**

Group OP – *Stems round, with stele not elastic (if present). Lvs and petioles hairless. Plant in v wet (or mtn) habs.* ❶ *Stems antrorsely crisped-hairy.* ❷ *Lvs net-veined.* ❸ *Lvs strongly 3-5-pli-veined*

■Lvs mostly <1.5cm. Stems solid, rooting at least at lower nodes. Per

Stems minutely hairy in 2 lines

Lvs 0.3-1cm, broadly ovate to ± orb, purplish below, entire to obscurely sinuate-toothed, with 0-3 sunken hydathodes per side (the terminal one often indistinct), with 1(3) veins visible below, with stomata above only. Petiole 0.5-3mm. Stems rooting strongly at nodes. All yr. Usu uplands ....*New Zealand Willowherb* **Epilobium brunnescens***

Stems minutely antrorsely crisped-hairy all round ❶

Lvs 0.5-2cm, ovate, obtuse, rounded at base, ± sessile, obvolute when young, ± shiny above, weakly pitted both sides, rarely hairy, occ ciliate, ± entire to weakly crenulate with 4-6 pale hydathodes, weakly 3-pli- or pinnate-veined with few 2° veins (occ obscure) slightly raised below, with stomata both sides. Stems 0.7-1.5mm diam. All yr ....*Thyme-leaved Speedwell* **Veronica serpyllifolia**

Stems sparsely hairy all round (occ hairless below)

Lvs 1-1.5 x 1.3cm, ovate, cuneate at base, serrate with 3-6 teeth per side, with stomata both sides. Petiole to 2.5mm. Mtns..*Alpine Speedwell* **Veronica alpina**

Stems hairless

Lvs 0.3-0.8 x 0.3-0.5cm, ovate to ± orb, ± acute, ± sessile, connate or not at base, applanate when young, without cartilaginous margins, with 6 minute reddish gland-tipped teeth, with veins invisible when fresh, with stomata both sides. Stems slender, brittle, ridged. Mar-Oct. Boggy peat, R, W Galway (Ire)..*Creeping Raspwort* **Haloragis micrantha***

■Lvs mostly >1.5cm

Stems hollow or becoming hollow

Stems translucent, not rooting at nodes, hairless. Ann. Often ❷................(*Impatiens*) Go to **IMP**

Stems opaque, rooting at lowest nodes

Lvs 5-7-pli-veined. Stems hairless below, occ glandular-hairy above. Per. ❸. (*Mimulus*) Go to **MIM**

Lvs pinnate-veined, 5-12 x 1-2cm, ± connate at base, slightly pitted above, with obscure 2° veins, with stomata both sides. Stems occ hairy below, hairless above, 10-40cm tall, green or purplish, weakly 3-ridged, with aerenchyma around doughnut-like stele. Ann (per). (The following two spp form a vigorous sterile hybrid to 90cm tall, *V.* x *lackschewitzii*, which may replace the parents)

Fls usu blue. Lvs rarely petiolate below, narrowly ovate to lanc, with many deep teeth or obscurely crenate with 6-15 teeth (or hydathodes) per side. Upper lvs lanc, acute, ± clasping at base, remotely serrulate......*Blue Water-speedwell* **Veronica anagallis-aquatica**

Fls usu pink. Lvs sessile, linear to linear-lanc, with few shallow teeth. Upper lvs similar to lower lvs...*Pink Water-speedwell* **Veronica catenata**

Stems solid

Rhizomatous

Lvs linear..(*Epilobium palustre*) Go to **EPIL**

Lvs ovate, (1)2-6(8)cm, ± acuminate, cordate at base, thin, ± translucent and shiny both sides, with crisped cilia, deeply dentate with (3)5-6 teeth per side, with stomata below only. Petiole 1cm, channelled, with 1 pr of stipule-like glands at base. Stems 5-10(30)cm, usu hairless. Stolons usu present. Rhizomes with spindle-shaped tubers. May-Oct. Uplands ....*Alpine Enchanter's-nightshade* **Circaea alpina**

Stoloniferous or rooting at nodes

Petiole long (to 7cm), distinct

Basal lvs to 4cm, ovate-elliptic to ± broadly triangular, obtuse, rounded or truncate at base, obvolute when young, dark or bright green, scabrid-ciliate, entire or obscurely sinuate-toothed with white sunken hydathodes along margins above, with stomata both sides. Petiole channelled, with 3(5) vb's around small hollow. Stolons long, round, solid. Per. ❷. Mar-Oct. Wet habs...*Marsh Valerian* **Valeriana dioica***

Petiole short or indistinct

Extra-floral nectaries usu present at petiole base, tiny, gland-like, black

Lvs 1.5-3(5)cm, ovate to broadly elliptic, shortly acuminate or acute, cuneate at base, obvolute when young, ± translucent, shiny dirty red-green, with sunken pale hydathodes along margins below, with 2° veins fading nr margins, with stomata both sides. Petioles short, not connate. Stems usu emergent, to 20cm, rooting at lower nodes, with a weak interpetiolar ridge, with a green central stele surrounded by reddish aerenchyma. Short-lived per. Apr-Oct. VR, New Forest........................*Hampshire-purslane* **Ludwigia palustris**

Extra-floral nectaries absent

Lvs alt above...(*Epilobium*) Go to **EPIL**

Lvs all opp, 2.5-6 x 3cm, usu held patent, ovate to oblong, obtuse, rounded at base, applanate when young, often thick, minutely pitted both sides, shallowly crenate-serrate with 9-20 teeth per side (with ± sunken tiny hydathodes turning purple), usu with 2-4 prs of 2° veins raised below, with stomata both sides. Petioles 0-10mm, connate at base. Stems 20-60cm, 3-6mm diam, prostrate but soon ascending, occ pruinose, occ hairy below, with doughnut-like stele. All yr...*Brooklime* **Veronica beccabunga**

Lvs all opp, 2-4 x 0.4-0.6cm, held erecto-patent, linear-lanc to lanc, acute, ± clasping at sessile base, applanate when young, with slightly recurved margins at maturity, stiff, not pitted, remotely denticulate with (0)2-6 hydathodes per side esp distally, with translucent midrib sunken above and raised below, with 2° veins indistinct or absent, with stomata both sides. Stems to 15cm, 1-3mm diam, creeping but ascending to erect, often purplish, occ hairy or glandular-hairy, usu snapping audibly, with aerenchyma around stele. All yr. Wet habs...*Marsh Speedwell* **Veronica scutellata**

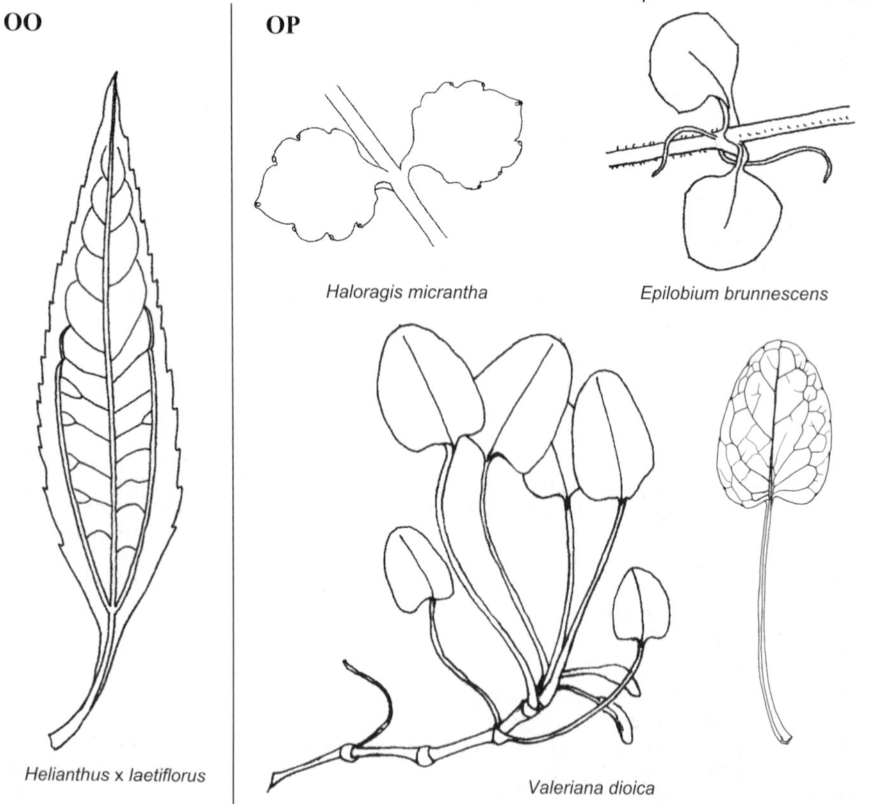

Helianthus x *laetiflorus*

Haloragis micrantha

Epilobium brunnescens

Valeriana dioica

Group OQ – *Stems round. Lvs and petioles hairless. Plant in dry habs*

■Stems creeping and rooting at nodes. Per

Lvs ± fleshy (dropping leaving scars on stems), each tooth with a submarginal hydathode below (HTL)

Lvs cuneate at base. Stems with reddish tissue surrounding green central stele

Stems papillate, often reddish. Lvs 1.5-3 x 1-2cm, obovate, obtuse, ± applanate when young, with long-papillate margins (at least nr base), crenate-serrate with 4-6 teeth per side, with obscure veins, with obscure stomata both sides. Petioles indistinct, not connate. All yr (occ reduced to buds in winter)..*Caucasian-stonecrop* **Sedum spurium***

Stems not papillate, reddish. Lvs 1.5 x 0.5-1.3cm, obovate, obtuse, ± applanate when young, with margins not papillate (occ crenulate), crenate-serrate with 5-6 v shallow teeth per side, with obscure veins, with stomata both sides. Petioles indistinct, not connate. All yr (occ reduced to buds in winter).........................*Lesser Caucasian-stonecrop* **Sedum stoloniferum**

Lvs rounded at base. Stems with scattered vb's, not papillate

Lvs 2-4 x 3cm, ovate-oblong, v obtuse to truncate, grey-green to reddish, applanate when young, crenate (-serrate) with 8-14 teeth per side, with obscure veins, with stomata both sides. Petioles 3-6mm, not connate. All yr...........*Lamb's-tail* **Chiastophyllum oppositifolium**

Lvs not fleshy, teeth without submarginal hydathodes below

Stems with 2 opp lines of minute hairs. Lvs 3-10(14) x 5mm, ovate, purplish below, with 4-6 acutely dentate teeth per side, with indistinct pinnate 2º veins, with stomata above only. Petiole 0.5(1)mm. All yr..*Rockery Willowherb* **Epilobium pedunculare***

Stems minutely antrorsely crisped-hairy all round. Lvs 5-20 x 15mm, ovate, ± shiny above, occ ciliate, ± entire to weakly crenulate with 4-6 pale hydathodes, weakly 3-pli- or pinnate-veined with few 2º veins (occ obscure), with stomata both sides. Petiole 0-1mm. All yr. Occ in wet habs..*Thyme-leaved Speedwell* **Veronica serpyllifolia**

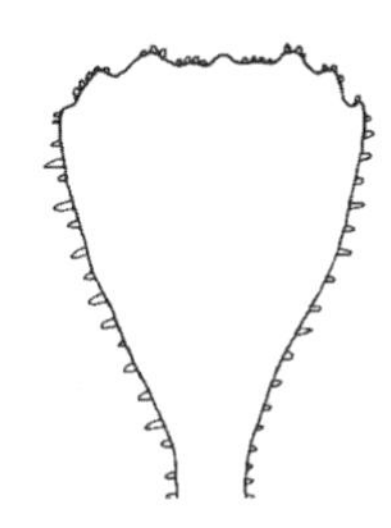

Sedum spurium lf upperside

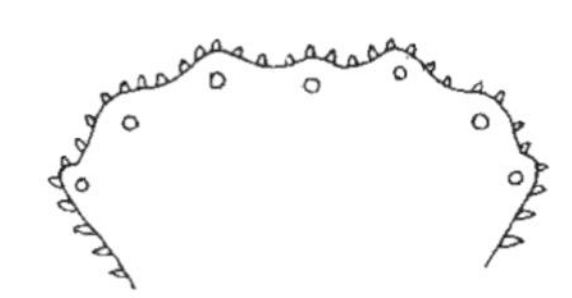

Sedum spurium underside of lf apex

Epilobium pedunculare

■Stems ± erect, not rooting at nodes

Per

Lvs fleshy, the upper alt

Lvs 2-10cm, ovate-oblong, pale ± waxy grey-green, shallowly dentate-serrate with 5-10 teeth per side at least in distal ½ (occ ± entire), with obscure veins, with stomata both sides. Stems to 60cm, round, with lf scars. Rhizomes short with tuberous roots. Apr-Oct ....*Orpine* **Sedum telephium**

Lvs thin, the upper alt..(*Epilobium*) Go to **EPIL**

Lvs ± thin, all opp

Lvs to 10cm, elliptic, cuneate and connate at base, with minutely crenate hyaline margins, shiny green to ± glaucous above, with white sunken hydathodes visible along margins above, with midrib raised below, with sunken translucent 2° veins often originating from nr base, with stomata both sides. Petiole to 5cm, with 5 vb's. Stems to 1m, brittle (snapping audibly), round to weakly angled, not swollen at nodes, with interpetiolar ridge, soon hollow. All yr. Old walls, ruderal. PL 9..*Red Valerian* **Centranthus ruber**

Ann. Lvs not fleshy, the upper alt

Stems hollow. Lvs (±) net-veined..(*Impatiens*) Go to **IMP**

Stems solid. Lvs with 2° veins translucent ....(*Euphrasia*) Go to **EUPH**

Stems solid. Lvs with midrib translucent only

Lvs 1-2.5 x 0.3-0.8cm, oblanc, usu cuneate at base, thickish, dull above (occ shiny below), v obscurely pitted above, crenate-serrate with 3-4 teeth per side, with midrib sunken above, with stomata both sides. Upper lvs entire. Petiole 0-1cm. Stems to 12(25)cm tall, 1-2mm diam, occ glandular-hairy. Apr-Oct. R alien.............*American Speedwell* **Veronica peregrina**

Key to Groups in Division P

(*Lvs lobed, alt*)

- ■Tree, with single trunk, always unarmed....................................**PA**
- ■Shrub, multi-stemmed (occ trailing or creeping), OR tree with spines/thorns....................**PB**
- ■Climbing herb or woody per, occ with tendrils or adventitious roots....................**PC**
- ■Non-climbing herb (occ woody or trailing)
  - Plant with latex (occ sparse)
    - Hairs forked....................................**PD**
    - Hairs simple and scabrid (*Hieracium*, *Papaver*)....................................**PE**
    - Hairs simple and smooth, or absent
      - Lf midrib ± spiny below AND/OR lf margins prickly or with bristle-tipped teeth....................**PF**
      - Lf midrib or lf margins never spiny or prickly....................................**PG**
  - Plant without latex (rarely present in roots of *Cirsium*)
    - Lvs spiny (occ weakly so)....................(Thistles & thistle-likes) Go to **THIS**
    - Lvs not spiny
      - Stipules (or ochreae) present
        - Petiole with (1)3 vb's....................................**PH**
        - Petiole with ≥3 vb's
          - Hairs stellate or swollen-based. Petiole often with 6 vb's in a ring (*Malvaceae*)....................**PI**
          - Hairs (if present) not stellate or swollen-based....................................**PJ**
      - Stipules absent but stipoid glands nr base of lf (*Reseda*)....................................**PK**
      - Stipules and stipoid glands absent (may have lf lobes nr lf axil)
        - Plant with branched or stellate hairs....................................**PL**
        - Plant with bladder-like hairs ....(*Chenopodium*, *Atriplex*) Go to **CHAT**
        - Plant with cottony or woolly hairs (often confined to rosette centre, lf axils or underside of lvs)
          - Lvs orb, palmately lobed, with cordate base....................................**PM**
          - Lvs not orb, pinnately lobed, without cordate base
            - Plant aromatic, often strongly so (but never with glandular hairs)....................**PN**
            - Plant (±) odourless (occ weakly aromatic with glandular hairs)....................**PO**
        - Plant with simple hairs (occ glandular), or hairless
          - Lvs palmately or digitately lobed (often ± orb in outline and/or with cordate base)
            - Stems creeping and rooting at nodes....................................**PP**
            - Stems not creeping or rooting at nodes
              - Lvs hairless....................................**PQ**
              - Lvs hairy....................................**PR**
          - Lvs pinnately or irregularly lobed (not orb, without cordate base)
            - Lvs hairless (may be ciliate)
              - Lvs fetid....................................**PS**
              - Lvs not fetid (may be aromatic or have cress/mustard odour)
                - Lvs and stems glaucous-pruinose, and/or lvs fleshy....................................**PT**
                - Lvs and stems green, lvs not or slightly fleshy....................................**PU**
            - Lvs hairy
              - Lvs (or stems) with cress or mustard odour when crushed. Hairs unicellular
                - Basal rosette present....................................**PV**
                - Basal rosette never formed....................................**PW**
              - Lvs without cress odour when crushed (may have other odour). Hairs usu septate....**PX**

Group PA – *Tree, never spiny or thorny. (*Alnus *spp may rarely key out here but has lvs toothed not lobed).* ❶ *Buds hidden within petiole base.* ❷ *Twigs and petioles with white latex*

■Lvs white- to grey-woolly below, or velvety to touch, at least when young

Lvs stellate-hairy both sides. All buds absent. Wintergreen tree-like herb .... *Tree-mallow* **Lavatera arborea***

Lvs stellate-hairy below only. Terminal buds clustered at apex of twigs. Evergreen

Bark corky.. *Cork Oak* **Quercus suber**

Bark not corky.. *Evergreen Oak* **Quercus ilex**

Lvs with hairs not stellate. Deciduous

Lvs pinnately veined. Petiole channelled..(*Sorbus*) Go to **SOR**

Lvs palmately veined. Petiole laterally flattened..(*Populus*) Go to **POP A**

Lvs palmately veined. Petiole round (not channelled)

Lvs 12-25cm diam, 3-5-lobed, truncate to cordate at base, floccose-woolly when young, soon hairless or with dendritic hairs confined to veins below, net-veined, with 3 main palmate veins. Petiole 3-10cm, v swollen at base (base hollow and forming a hood over the bud), usu dendritic-hairy. Stipules fused, ochrea-like, soon falling, leaving scars. ❶

Lvs lobed to ½ way. *P. occidentalis* x *orientalis*............*London Plane* **Platanus** x **hispanica***

Lvs lobed >⅔ way.. *Oriental Plane* **Platanus orientalis**

■Lvs not woolly or velvety. Deciduous

Terminal buds clustered at twig apices; scales 5-ranked. Branches wide, twisting, often falling

Terminal bud cluster with persistent long (to 25mm) narrow stipules (occ obscuring bud cluster)

Lvs 5-15cm, ± oblong, usu cuneate at base, greyish stellate-hairy both sides when young, soon dull or shiny with sparse minute hispid hairs and stellate hairs above, with 5-8 prs unequal ovate-triangular lobes per side, with intercalary veins; lobes acute to ± obtuse, often tipped with protruding hydathodes. Petiole 0.5-2.5cm. Twigs stellate-hairy when young. Buds to 4mm, with <20 reddish-brown scales, with grey stellate hairs... *Turkey Oak* **Quercus cerris**

Terminal bud cluster usu without long or persistent stipules

Petiole 0-1cm

Lvs 5-12cm, obovate-oblong to oblanc, usu cordate with reflexed auricles at base, dull dark green above, hairless or with simple hairs below, with 3-5 deep irreg lobes per side (often with smaller lobes), usu with intercalary veins. Twigs hairless, angled. Buds 4-9mm, ovoid, usu with <20 scales, chestnut-brown, hairless, ciliate. PL 21, 22 .... *Pedunculate Oak* **Quercus robur**

Petiole 1-2.5cm (shorter on young or epicormic lvs)

Lvs 5-12cm, elliptic, usu cuneate at base, with auricles absent or obscure, ± shiny dark hairless green above, with adpressed stellate hairs below, with long simple hairs at midrib base and along 2° veins below (and in vein axils), with 5-7 shallow regular lobes per side, without intercalary veins. Twigs hairless, angled. Buds 3-8mm, ovoid, usu with >20 scales, pale grey-brown, hairless, ciliate.. *Sessile Oak* **Quercus petraea**

Petiole (2)2.5-5cm

Lvs 12-22cm, oblong, truncate to cuneate at base, yellow when young, dark green above, turning red in autumn, hairless (exc for orange stellate hair-tufts in vein axils below), with 4-5 lobes per side, further toothed with aristate teeth, with 2° veins not sunken above, with intercalary veins. Petiole occ reddish. Buds in clusters of (1)4, 3-7mm, usu with <20 scales, pale brown, usu hairless, ciliate. Twigs hairless

Lvs dull above, with lobes ≤½ way to midrib, with inconspicuous hair-tufts in vein axils below. Buds ovoid .... *Red Oak* **Quercus rubra**

Lvs shiny above, with lobes ≥½ way to midrib, with obvious hair-tufts in vein axils below. Buds spherical.. *Scarlet Oak* **Quercus coccinea**

Terminal buds solitary at twig apices (absent in *Trachycarpus*); scales not 5-ranked. Branches not wide or twisting, rarely falling

Buds hidden within petiole base ❶

Lvs lobed to ½ way. *P. occidentalis* x *orientalis*..............*London Plane* **Platanus** x **hispanica***

Lvs lobed >⅔ way..*Oriental Plane* **Platanus orientalis**

Buds absent. Evergreen. Lvs in terminal crown. Trunk covered with black fibres and petiole bases

Lvs to 90cm diam, ± orb, fan-shaped, divided ± from the base into 30-40 lobes; lobes 3cm wide, occ sharply acute, folded when young, occ with scabrid margins, shiny, with translucent parallel veins, the mid-vein raised below only, with stomata in lines (more abundant below). Petiole as long as lf, flattened, with scabrid margins. Buds absent. Trunk 3-6m. Hortal ....*Chusan Palm* **Trachycarpus fortunei**

Buds exposed. Deciduous. Lvs along twigs (and occ on short shoots). Trunk covered in bark

Twigs and petioles with white latex ❷

Lvs 7-20cm, palmately 3-5-lobed with obovate lobes (terminal lobe long), weakly odorous, rough with minute hispid hairs above, densely minutely hairy below, net-veined, with stomata below only. Petiole sparsely minutely hairy, round, with many vb's in ring. Stipules 2cm, long-acute, almost fused at base, reddish, soon falling, the scars ± encircling twig. Twigs sparsely minutely hairy, soon hairless, round, odorous, pith-filled; lf scars with many traces in ellipse. Terminal buds >10mm, conical, acute, with 1 infolded striate scale, green, hairless, not ciliate. Lateral buds often collateral, to 6mm, globose, with several scales. PL 15..*Fig* **Ficus carica**

Twigs and petioles without latex

Stipules 2-5cm, soon falling, leaving scars encircling twig. Lvs to 12 x 12cm, uniquely shaped, vaguely resembling the outline of a tulip flower, often with long apiculus, shiny light green above, ± glaucous below, hairless exc for sparse hairs on midrib below. Petiole to 10cm, swollen at base, green, hairless. Twigs hairless, round. Buds to 15mm, oblong, flattened, green, hairless, with 2 scales (stipules) containing young lvs curled over next pr of stipules. PL 20..*Tulip-tree* **Liriodendron tulipifera**

Stipules to 1.5cm, soon falling, leaving indistinct scars. Lvs to 10cm, ± ovate, truncate at base, green both sides, sparsely hairy below, deeply lobed, net-veined, with 4-6 prs of 2° veins; lobes acuminate, 1-2-serrate, the lowest pr cut ½ way to midrib and spreading at 90° to midrib. Petiole 1.5-5cm, reddish at base. Twigs sparsely woolly when young, round, with lenticels. Buds 5-6mm, globose, green, hairless, ciliate ....*Wild Service-tree* **Sorbus torminalis**

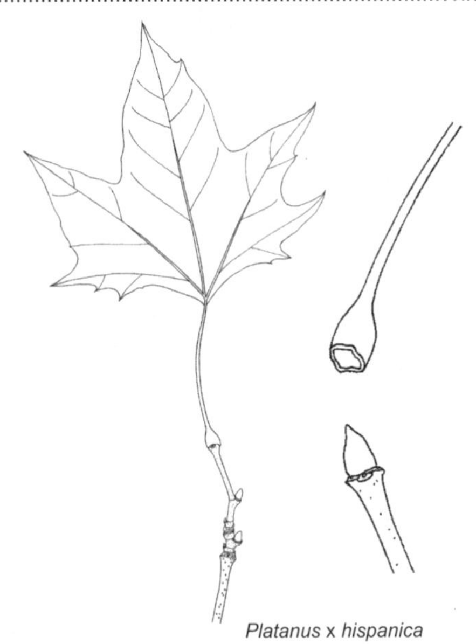

Platanus x *hispanica*

Group PB – *Shrub (spiny or not) or tree with thorns.* ❶ *Twigs and petioles with white latex*

■Deciduous tree armed with thorny branches. Short shoots present as woody spurs

Lvs 1.5-5cm, shiny dark green above, with veins running to the sinuses as well as to the lobe apices. Stipules, if present, lf-like. Twigs reddish-brown, zig-zagging, round, with curved thorns to 1.5cm. Buds 1-2, 3mm, ovoid, reddish-brown, with erose ciliate scales

Lvs deeply (3)5(7)-lobed, ovate to obovate, hairless or sparsely hairy on midrib below, occ with hair-tufts in vein axils below; lowest pr of lobes cut >½ way to midrib, usu acute, longer than wide, often with 2° veins curving downwards. Twigs hairless. PL 19, 20 ....*Hawthorn* **Crataegus monogyna**

Lvs shallowly 3-lobed (to ± simple), obovate, hairless or with scattered hairs on veins both sides, without hair-tufts in vein axils; lowest pr of lobes cut <½ way to midrib, obtuse, wider than long, usu with lowest pr of 2° veins straight or curving upwards. Twigs occ with a few long hairs when young. Woods, usu on heavy soils.......................*Midland Hawthorn* **Crataegus laevigata**

■Deciduous shrub armed with 1 or 3 spines at each stem node. Short shoots present as woody spurs

Lvs 2-5cm, as wide or wider than long, broadly cuneate to ± cordate at base, 3-5-lobed, hairless or with short unicellular hairs both sides, with up to 7 veins from base, net-veined, with 2-3° veins raised below. Petiole to 2.5cm, with few short glandular hairs and long gland-tipped dendritic hairs to densely minutely hairy, with 3 vb's. Twigs whitish, hairless to minutely hairy or bristly. Buds to 10mm, spindle-shaped, acute, white, turning brown. Much-branched shrub to 1m. PL 15 ....*Gooseberry* **Ribes uva-crispa**

■Unarmed (sub)shrub (may have bristly hairs). Short shoots absent

Lvs white-woolly or grey-velvety on one or both sides, hairs may be stellate. Evergreen

Shrub, >1m tall

Lvs 5-12cm, orb to lanc, 3-5-lobed with long end-lobe, plicate when young, grey-green, velvety with adpressed simple or stellate hairs above, with 6-8-rayed stellate hairs below, net-veined, with stomata both sides; lobes crenate-serrate or dentate. Petiole to 4cm, stellate-hairy, channelled, solid, with 4-6 vb's in a ring. Stipules to 8mm, ovate to lanc, soon falling. Twigs purplish, hairless or stellate-hairy (with swollen-based hairs), round. *L. olbia* x *thuringiaca*..*Garden Tree-mallow* **Lavatera** x **clementii**

Low shrub, <0.6m tall

Stems ± erect, not rooting at nodes. Lvs strongly aromatic. Stipules absent

Lvs to 1cm, pinnately lobed, sessile, white-woolly; lobes alt, appearing 4-ranked, 1mm, ovoid, obtuse, fleshy....................................*Lavender-cotton* **Santolina chamaecyparissus**

Stems ± prostrate, rooting at nodes. Lvs odourless. Stipules present

Lvs 4-12 x 3-8cm, ovate to oblong, shallowly 3-7 lobed, plicate when young, with ± recurved margins at maturity, with translucent dots (HTL), shiny dark green and hairless above (exc for sparse hairs on main veins), whitish-woolly below, with long brown bristles on raised 1-2° veins, with stomata below only. Petiole with dense brown bristly (spine-like) hairs. Stipules to 8-20 x 8mm, ovate, toothed. Stems trailing, with brown bristly gland-tipped hairs (hairs soon breaking off), round...*Chinese Bramble* **Rubus tricolor**

Lvs 0.5-2 x 1cm, oblong, obtuse, cordate at base, revolute when young, with ± recurved margins at maturity, dark green and soon hairless above, densely woolly below, with white-dendritic thick brown hairs on midrib below, deeply crenate (-dentate) with 6-8 teeth per side, net-veined. Petiole 4-10mm, occ with white-dendritic stout brown hairs. Stipules scarious, brownish, adnate to petiole for ½ length, with long subulate apex, hairy. Stems twisted. Usu calc rocks. PL 11..*Mountain Avens* **Dryas octopetala**

Lvs not woolly or velvety, hairs never stellate

Evergreen. Stipules absent. Lvs to 25cm diam, palmately 7(9)-lobed, leathery, shiny dark green above, dull and paler below, hairless, with stomata below only. Petiole to 30cm, round. Lf scars ½ encircling stem. Hortal...*Fatsia* **Fatsia japonica**

Deciduous

Stipules present

Stems usu bi (canes dying at end of 2nd yr). Rhizomatous

Lvs 10-25cm diam, palmately 5-lobed, with sparse minute translucent dots, hairy both sides, hairs occ glandular nr petiole, with stomata below only. Petiole to 15cm, with red gland-tipped hairs and short nonglandular hairs all along, round, pith-filled, with obscure vb's. Stipules to 10 x 2mm, oblong-lanc, green. Stems to 2m tall, with sparse long glandular hairs and dense minute nonglandular hairs, round when young ....*Purple-flowered Raspberry* **Rubus odoratus**

Lvs 5-15cm diam, palmately 5-lobed, without translucent dots, hairy both sides, hairs often glandular on veins below or nr petiole, with stomata below only. Petiole to 12cm, with red gland-tipped hairs nr lf, round, pith-filled, with obscure vb's. Stipules to 10 x 2mm, oblong-lanc, green. Stems to 3m tall, with sparse short hairs, weakly angled when young ....*Thimbleberry* **Rubus parviflorus**

Stems per (twigs), growing for several yrs. Not rhizomatous (but may sucker)

Lvs 5-12cm, shallowly pinnately lobed. PL 12, 20............................*Hazel* **Corylus avellana**

Lvs 7-20cm, palmately 3-5-lobed with obovate lobes (terminal lobe long), weakly odorous, rough with minute hispid hairs above, densely minutely hairy below, net-veined, with stomata below only. Petiole minutely sparsely hairy, round, with many vb's in ring. Stipules 2cm, long-acute, almost fused at base, reddish, soon falling, with scars ± encircling twig. Twigs sparsely minutely hairy, soon hairless, round, odorous, pith-filled; lf scars with many traces in ellipse. Terminal buds >10mm, conical, acute, with 1 infolded striate scale, green, hairless, not ciliate. Lateral buds often collateral, to 6mm, globose, with several scales. ➊. PL 15..*Fig* **Ficus carica**

Stipules absent

Lvs rolled when young, not cordate at base. Petiole to 0.5cm, not dilated at base, with 1 vb, without spiral fibres...(*Spiraea*) Go to **SPI**

Lvs plicate-folded when young, usu cordate at base. Petiole 1-8cm, dilated at base, with 3 curved vb's (often ± merging) and spiral fibres

Lvs (and buds) odorous (with glands), net-veined

Lvs 5-12cm wide, wider than long, acutely 3-5-lobed, cordate at base, blackcurrant-scented, shiny or dull green above (often yellow-mottled), hairy only on veins below, with shiny yellow sessile glands below. Petiole 4-6cm, minutely hairy, occ with sessile glands, with few stout dendritic hairs nr base. Twigs green, shortly hairy, with sessile glands. Buds clustered at each node of each yrs growth, to 6mm, conical, dark brown to yellow-green, sparsely hairy to hairless............................*Black Currant* **Ribes nigrum**

Lvs 4-10cm wide, usu wider than long, obtusely 3-5-lobed, cordate at base, with fruity odour, dull green both sides, with sparse yellow short-stalked or ± sessile glands below, with stomata below only. Petiole to 3cm, with abundant short stout glandular hairs and long stout dendritic hairs at base. Twigs reddish, minutely crisped-hairy, with sessile glands, reddish. Buds not clustered, to 15mm, spindle-shaped, green to reddish, with several apiculate scales.....................................*Flowering Currant* **Ribes sanguineum***

Lvs odourless (glands, if present, scentless)

Lvs net-veined

Lvs hairless or veins sparsely hairy below, to 8cm wide, wider than long, obtusely (3)5-lobed, cordate at base, dull dark green above, ciliate, with stomata below only. Petiole 4-8cm, ciliate with few long nonglandular simple hairs nr dilated base, shortly crisped-hairy at least nr lf, with few red short-stalked or ± sessile glands. Buds clustered, 5-7mm, purple-brown..*Red Currant* **Ribes rubrum**

Lvs softly hairy both sides, to 8cm wide, wider than long, obtusely 5-lobed, truncate to shallowly cordate at base, dull grey-green above, ciliate, with stomata below only. Petiole to 4cm, shortly crisped-hairy, with stout dendritic hairs and few short-stalked or ± sessile red glands at base. Buds not clustered, to 5mm, whitish. Limestone districts, R, N Br..*Downy Currant* **Ribes spicatum**

Lvs not net-veined

Lvs 3-5cm wide, usu as wide as long, 3(5)-lobed (lobes acutely dentate), truncate to ± cordate at base, shiny green esp below, with sparse minutely glandular hairs at least above, with 3 main veins sunken above but hardly or not raised below, with stomata below only. Petiole to 1.5cm, hardly dilated at base, hairless or with sparse red gland-tipped hairs. Twigs hairless, pale or grey-brown, with short shoots. Buds adpressed, 4-6mm, white (but each scale with a dark brown apiculus), hairless, often glandular-ciliate...*Mountain Currant* **Ribes alpinum***

Lvs 3-8cm wide, wider than long, 3-5-lobed (lobes hardly toothed), truncate to ± cordate at base, (±) shiny green above, hairless above, with yellow sessile glands below esp when young, with stomata both sides, with 1-2° veins raised above. Petiole 1.5-4cm, minutely white-hairy, with v sparse dendritic glandular hairs at expanded base. Twigs densely minutely white-hairy, without short shoots. Buds usu spreading, 5-7mm, brown to purple-brown, minutely hairy, not glandular-ciliate. VR alien ....*Buffalo Currant* **Ribes odoratum**

Ribes alpinum *Ribes sanguineum*

Group PC – *Climbing or clambering herbaceous ann or per (occ woody), often with tendrils or adventitious roots.* ❶ *Lvs eccentrically peltate*

■Tendrils present

Stipules present (at least when young). Per

Petiole with 1-2 prs of red disc-like glands (like 'golf tees'). Stipules persistent, auricle-like

Lvs to 12cm diam, palmately 5-7-lobed; lobes 6-12 x 2.5cm, ovate-lanc, mucronate, folded when young, shiny above, dull below, hairless, entire, net-veined, with stomata below only. Tendrils coiled, unbranched. Stipules with excurrent midrib. Stems angled. Often all yr. Hortal .... *Passion-flower* **Passiflora caerulea***

Petiole without glands. Stipules soon falling (but scars visible), not auricle-like

Tendrils usu forked, stout, lf-opposed, absent from every 3rd node. Lvs orb, cordate at base, palmately 5-7-lobed with toothed lobes, ± hairless above, net-veined, with 5 main veins raised below, with stomata below only. Petiole to 10cm, round or narrowly channelled, with spiral fibres. Stipules ovate. Stems stout, woody at least below, solid; lf scars with several traces. Terminal bud absent; lateral buds with 2 scales. Apr-Oct

Lvs sparsely hairy to woolly below, 6-20cm diam. Stems soon hairless, round to angled, occ ribbed when young.. *Grape-vine* **Vitis vinifera**

Lvs brown-felted below, 10-30cm diam. Stems with long woolly hairs, ribbed .... *Crimson-glory-vine* **Vitis coignetiae**

Tendrils 5-9-branched, with adhesive red pads, lf-opposed, usu present at every node. Lvs 3-17 x 2.5-15cm, ovate, cordate at base, palmately 3-lobed (occ unlobed or with 3 lfts), bronze when young, hairless or sparsely hairy on veins below; lobes with 5 irreg forward-curved lobes per side, net-veined, with stomata below only. Petiole swollen at base, laterally flattened, or round with narrow groove above. Stems grey, swollen at nodes, angled when young, developing lenticels. Buds often collateral, with 2-3 scales. Apr-Oct .... *Boston-ivy* **Parthenocissus tricuspidata**

Stipules always absent

Tendrils branched, lf-opposed

Tendrils sharply bristly-hairy. Lvs 10-35cm diam, broadly ovate, cordate at base, palmately 5-7-lobed (occ shallowly so), hispid-hairy (some hairs with swollen-bases), toothed with green hydathodes, net-veined, with 2-3° veins raised below, with stomata below only. Petiole 9-25cm, stout, with unequal hispid septate hairs (some on swollen bases), channelled. Stems clambering, hispid, striate, hollow. Ann. Apr-Sept.............................. *Marrow* **Cucurbita pepo**

Tendrils unbranched, lf-opposed

Tendrils spirally-coiled, clockwise. Lvs 5-8 x 5-10cm, palmately 5-lobed, cordate with wide square sinus at base, dark green above, shortly roughly septate-hairy; lobes sinuate-dentate with protruding green hydathodes, net-veined, with 2° veins raised below, with stomata below only. Petiole 2-5cm, rough with swollen-based hairs, round, solid, with 7-10 vb's and abundant ± viscid sap. Stems climbing to 4m, rough with swollen-based hairs, occ with minute glandular hairs, angled, solid. Tuberous per. PL 15......... *White Bryony* **Bryonia dioica**

Tendrils not coiled. Lvs 5-15 x 5-15cm, reniform, 5-angled or shallowly 5-lobed, yellow-green, hispid to softly unicellular-hairy; lobes toothed with green hydathodes, net-veined, with 2° veins raised below, with stomata below only. Petiole 2-10cm, hispid. Stems clambering to 1(2)m, hispid, ± round, soon with irreg hollow. Ann. Apr-Sept................. *Melon* **Cucumis melo**

■Tendrils absent

Adventitious roots present. Per.. (*Hedera*) <u>Go to</u> **IVY**

Adventitious roots absent. Per

Lvs ivy-shaped.. *German-ivy* **Delairea odorata**

Lvs ovate with 2 lobes nr base.. (*Solanum dulcamara*) <u>Go to</u> **SOL**

Adventitious roots absent. Ann. Lvs eccentrically peltate ❶

Lvs to 14cm diam, with slightly angular margins, limp, fetid, dull pale green and hairless above, paler with minute white septate hairs below, with 9 palmate main veins ending in slight indentations along margins, with anastomosing 2° veins, with stomata nr veins above and scattered below. Petiole hairless, round, pith-filled to hollow, with c10 vb's around margin. Stipule-like glands 2-3 prs per node. Stems clambering, often reddish-streaked, round. Apr-Oct. PL 24..*Nasturtium* **Tropaeolum majus***

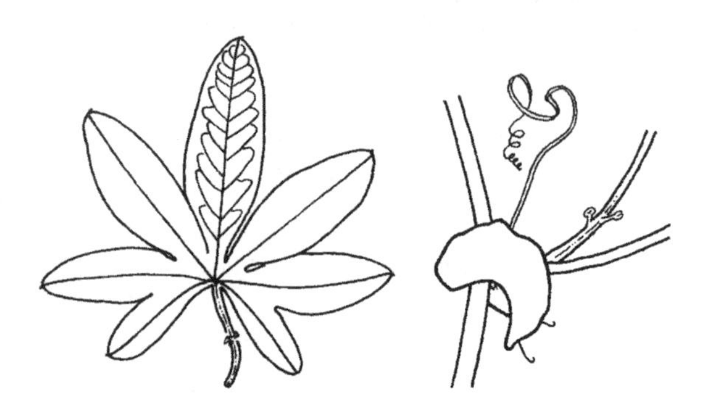

Passiflora caerulea lf & stipules

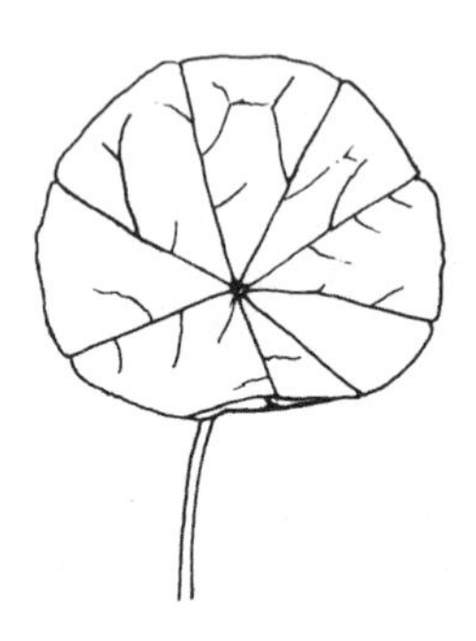

Tropaeolum majus

Group PD – *Latex present. Hairs forked (or trifid) at apex*

▪Hairs with hooked tips (adhering to woolly clothing) ....*Hawkweed Oxtongue* **Picris hieracioides**

▪Hairs without hooked tips. Lvs to 20cm, oblanc, narrowed to petiole-like base, distantly sinuate-toothed to runcinate-lobed, with indistinct 2° veins. Shortly rhizomatous. Latex bluish-white. Fl buds nodding when young (often visible in centre of rosette long before fl). All yr

Lf margin with dense hairs (5 per mm). Lvs held ± erect, hairy (hairs often trifid), the hairs rarely red-based on midrib above. Petiole with (3)5(7) vb's. Scape hairy, twisted, with large hollow. Per. Basic gsld. PL 18..*Rough Hawkbit* **Leontodon hispidus**

Lf margin with ± sparse hairs (1-3 per mm). Lvs ± flat on ground, hairy (hairs mostly forked), the hairs often red-based on midrib above. Petiole with 3 vb's. Scape hairless (exc at base), not or hardly twisted, with small hollow. Per (bi). Often acidic gsld..*Lesser Hawkbit* **Leontodon saxatilis**

Group PE – Hieracium, Papaver. *Latex present. Hairs scabrid, unicellular (but additional septate hairs in many* Hieracium *spp). Lvs often ± net-veined.* ❶ *Latex occ yellow*

■Per. Latex white or bluish, never drying red or orange-brown. Lvs pinnately lobed. Plant usu with additional septate, stellate, woolly, or glandular hairs.....(not keyed out) *Hawkweeds* **Hieracium** spp

■Per. Latex white, drying orange-brown. Lvs >10cm, usu 1-2-pinnately lobed. Stems with wavy spreading white hairs at base (may be adpressed distally)

Lvs usu 10-20cm, dull grey-green, densely antrorsely adpressed white-hairy both sides, with stomata below only or scattered along veins above; lobes with a short (0.5-0.8mm) terminal bristle. Petiole flat or U-channelled, weakly ridged below, with ≥5 vb's. Stems slender, hairs usu <4mm. All yr...*Atlas Poppy* **Papaver atlanticum**

Lvs usu 20-35cm, shiny dark green, with long spreading and v minute hairs (x20) both sides, net-veined; lobes with a long (to 1.5mm) terminal bristle. Petiole broad, flat or channelled, strongly 8-10-ridged below, with 8-10 vb's. Stems v stout (10mm diam), hairs mostly >4mm. All yr

Bracts absent. Fl buds nodding. Petals usu unspotted ....*Bractless Oriental-poppy* **Papaver orientale**

Bracts 0-5 below fl, usu entire. Fl buds nodding. Petals with dark spot nr base ....*Oriental-poppy* **Papaver pseudoorientale**

Bracts 1-5 below fl, usu lobed. Fl buds erect. Petals with dark spot nr base ....*Bracteate Oriental-poppy* **Papaver bracteatum**

■Ann. Latex white, drying red. Lvs <15cm, usu 1-2-pinnately lobed

Stems with stiff patent purple-tinged hairs at base

Lvs pinnately lobed to pinnate, with terminal lobe usu much larger than laterals; lobes with narrowly acute and bristle-pointed ± deeply toothed segments; terminal bristle <1mm. Cotyledons 3-8x as long as wide. Ovary hairless. Fr capsules globose, hairless. Apr-Oct (all yr). Arable, ruderal. PL 18...*Common Poppy* **Papaver rhoeas**

Stems with antrorsely adpressed white hairs at base

Lvs 1-2-pinnately lobed, dark green, stiffly hairy with spreading or patent hairs, with terminal lobe ± same size as laterals; ultimate segments v narrow (1mm wide), abruptly acute. Cotyledons >8x as long as wide. Ovary bristly. Fr capsules elongate, bristly. Apr-Aug. Arable ....*Prickly Poppy* **Papaver argemone**

Stems with wavy spreading white hairs at base (may be adpressed distally)

Lvs ± patent-hairy with soft long white spreading hairs (often sparse esp above), greyish-green, 1-2-pinnately lobed; ultimate segments broad, abruptly acute, with long 1(2)mm terminal bristle. Cotyledons 3-8x as long as wide. Ovary hairless. Fr capsules elongate, hairless. ❶. Apr-Oct (all yr). Arable, ruderal. PL 18...*Long-headed Poppy* **Papaver dubium**

Lvs ± adpressed-hairy with short stiff hairs (less hairy than *P. argemone*), often soon hairless, dark green, pinnately lobed; ultimate segments narrow (1.5mm wide), ± gradually acute, with terminal bristle to 1mm. Cotyledons >8x as long as wide. Ovary bristly. Fr capsules globose, bristly. Arable...*Rough Poppy* **Papaver hybridum**

Group PF – Asteraceae, Astrantia. *Latex white, at least initially. Hairs smooth or absent. Lf midrib ± spiny below and/or lf margins prickly-toothed (often not prickly to touch).* ❶ *Petiole with sheathing base*

■Lf midrib ± spiny below and lf margin prickly-toothed (not to touch)

Lvs usu with reddish midrib raised below, net-veined. Taprooted

Basal lvs 15-35 x 10cm, obovate, narrowed to petiole, dull green above, ± glaucous or pruinose below, hairless, toothed to pinnately lobed, with stomata both sides. Stem lvs spirally arranged, sessile, clasping and cordate at base with ± adpressed rounded auricles, ± glaucous. Petiole with 3(7) vb's around small hollow. Stem usu 1, to 2.5m, tough, purplish esp nr base, round, usu solid, with tasteless latex smelling of opium-poppy and drying orange after 3min. Ann or bi. All yr...*Great Lettuce* **Lactuca virosa***

Lvs with whitish midrib raised below, net-veined

Basal lvs 7-15(30) x 1-3(8)cm, acute, ± glaucous, hairless, unlobed to sinuate, with stomata both sides. Stem lvs often 2-ranked, usu held erect, clasping with auricles, occ runcinate-lobed with acute terminal lobe and a few distant prs of narrowish acute lateral lobes backcurved distally, the uppermost often entire and clasping with acute auricles. Stem usu 1, to 1.5m, rarely red nr base, often spiny, round, solid, with tasteless latex smelling of lettuce. Ann (bi). Often all yr. Ruderal..*Prickly Lettuce* **Lactuca serriola***

Basal and stem lvs 4-7 x 0.5-2(5)cm. Ann. May-Oct. Shingle and sea-walls, SE Eng. Sch8 ....*Least Lettuce* **Lactuca saligna**

■Lf midrib smooth below but lf margin weakly spiny or with bristle-tipped teeth (often not to touch)

Rhizomatous. Stems glandular-hairy above, rarely branched at base. Apr-Oct

Stems bluntly angled, unwinged, 0.6-1.2m, hairless below, with red-based stout glandular hairs above, hollow. Basal lvs toothed to rucinate-lobed with short triangular-oblong lobes, narrowed to winged petiole, green to ± glaucous above, glaucous below, net-veined, with stomata both sides or below only. Stem lvs less divided, sessile, with cordate-clasping base with rounded deeply-cut adpressed auricles. Petiole soon hollow, with acrid latex drying orange. PL 18 ....*Perennial Sow-thistle* **Sonchus arvensis***

Stems sharply (2)4-winged, 0.9-3m, hairless below, with red-based stout glandular hairs above, hollow. Basal lvs with a long acute sagittate sessile base, pinnatifid with a few distant lateral lobes and a larger lanc acute terminal lobe, ± glaucous at least below, net-veined, with stomata both sides or below only. Stem lvs less lobed, with acute (to obtuse) long narrow adpressed auricles. Petiole hollow, with acrid latex drying orange. Damp habs, occ saline, R ....*Marsh Sow thistle* **Sonchus palustris***

Tufted (with taproot). Stems never glandular-hairy, often branched at base. Oct-Sept

Lvs orb, net-veined. Per. Mar-Oct

Basal lvs 6-17cm diam, divided ¾ to base into 3-7 lobes; lobes ± dull above, coarsely serrate with bristle-tipped teeth (1mm bristles), with stomata usu below only. Petiole long, laterally flattened, with narrow shallow channel above, with 12 vb's around margin of large hollow. Stems ridged, hollow. ❶..*Astrantia* **Astrantia major***

Lvs oblanc to obovate, (±) net-veined. Ann (bi). Nov-Oct

Stem lvs with rounded adpressed auricles, shiny dark green above, glaucous below, usu crisped and spiny to touch, runcinate-lobed, the terminal usu larger than laterals. Seedling lvs dull darker glaucous-green, with stomata both sides. Petiole with 7(9) vb's, latex quickly turning dirty orange, with ± bitter taste. Stems 20-80cm, pruinose, hairless, 5-ridged below, round above, hollow..*Prickly Sow-thistle* **Sonchus asper**

Stem lvs with obtuse to acute spreading auricles, ± dull glaucous both sides, not spiny to touch, with white moniliform hairs when young, soon hairless, runcinate-lobed, the terminal lobe usu wider than uppermost pr of laterals. Seedling lvs soon developing large rounded terminal lobe, with stomata below only. Petiole with 5 vb's, latex turning orange >3min, with bitter taste. Stems 20-80cm, pruinose, hairless, 5-angled to round, hollow ....*Smooth Sow-thistle* **Sonchus oleraceus**

PF

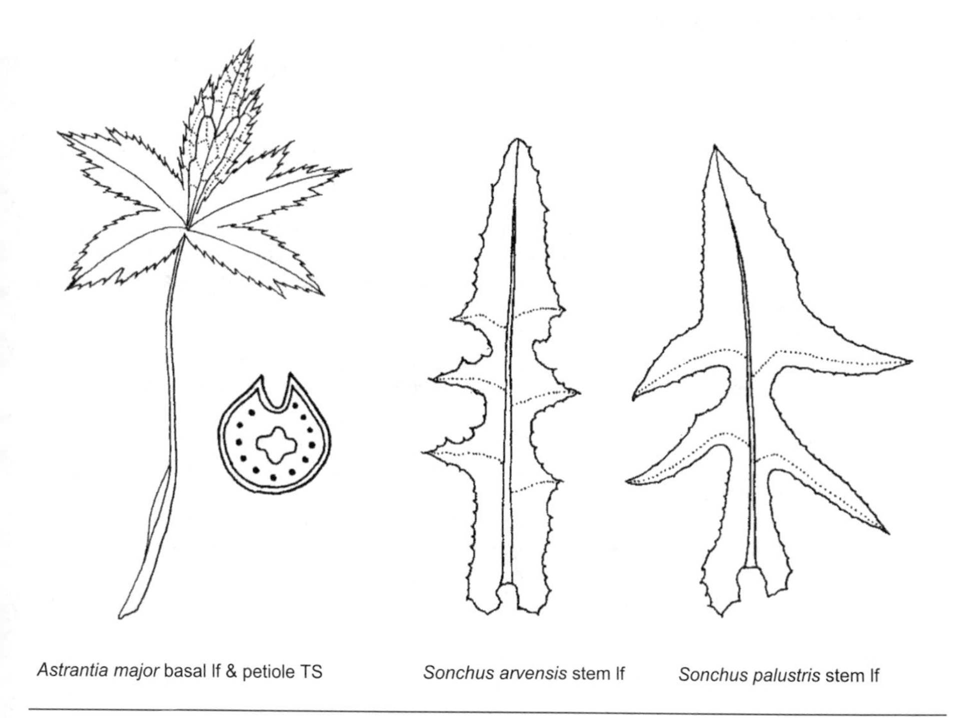

Astrantia major basal lf & petiole TS

Sonchus arvensis stem lf

Sonchus palustris stem lf

PG

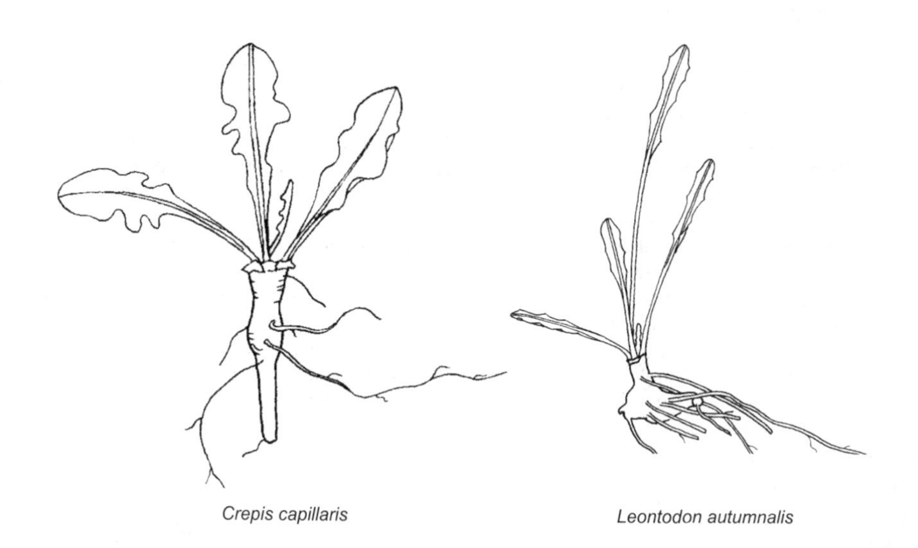

Crepis capillaris

Leontodon autumnalis

Group PG – Asteraceae, Chelidonium. *Latex present, often sparse. Hairs smooth or absent. Lf midrib and margins never spiny. Hairs septate (unicellular in* Wahlenbergia*).* ❶ *Lvs fetid or almond-scented*

■Lvs with v large wide terminal lobe and much smaller pr(s) of lateral lobes, not dandelion-like

Lvs hairy, occ sparsely so

Rhizomatous, basal lvs often emerging singly

Basal lvs to 20cm, green above, ± glaucous below, sparsely hairy on veins below, ciliate, ± lyrate, with large cordate terminal lobe, usu with 1 pr of lateral lobes decurrent into a winged stalk with a broad cordate-clasping base, sinuate, with 35-40 hydathode-teeth per side, with stomata below only. Petiole to 20cm, usu sparsely hairy, winged, often laterally flattened, channelled, with 7 vb's around hollow, latex turning off-white or brown. Stems 60-150cm, often hairy or glandular below. Mar-Oct....*Common Blue-sow-thistle* **Cicerbita macrophylla** ssp **uralensis**

Taprooted, several basal lvs in rosette

Latex yellow, abundant

Lvs 15-35cm, dull grey-green above, ± glaucous below, ± pinnate with 4-7 prs ovate to oblong lfts, ± hairless or with scattered white hairs (hairs obscurely septate), with stomata below only; terminal lobe often 3-lobed; lateral lobes usu with a lobe at base, crenate. Petiole brittle, with sparse long wavy septate hairs, hollow. Stems to 90cm. All yr ....*Greater Celandine* **Chelidonium majus**

Latex orange, sparse, usu confined to petiole (occ absent)

Lvs 10-25cm, often shiny green above, not glaucous below, 1-2-pinnately lobed (first lvs entire or toothed at base), septate-hairy, with purple hydathodes, with stomata both sides; uppermost lvs ± entire. Petiole not brittle, with 5(7) vb's. Stems 30-90, erect, usu branched above, ± hairy, grooved. Per. Feb-Oct. PL 18.........*Greater Knapweed* **Centaurea scabiosa**

Latex white, often sparse

Basal lvs to 15cm, rarely hairless, petiolate, usu lyrate-pinnatifid, with terminal lobe often ± cordate and 1(3) prs small lateral lobes, all sinuate-toothed, often with stomata both sides. Upper lvs shortly petiolate to sessile. Petiole solid, with 3(5) vb's. Stems to 100cm, branched above, hairy at least below, weakly ridged, hollow

Ann. Upper lvs distinctly toothed. Often all yr. PL 19 ....*Nipplewort* **Lapsana communis** ssp **communis**

Per. Upper lvs ± entire. All yr. VR.....*Nipplewort* (ssp) **Lapsana communis** ssp **intermedia**

Lvs (±) hairless. Tufted or shortly rhizomatous

Lvs glaucous or purplish below. Lowland, rarely to 500m or on open limestone

Basal lvs with 1-4 prs lateral lobes, with sharply angular cordate terminal lobe (occ hastately 3-lobed), petiolate, thin, dull dark green above, sinuate-toothed with long purple hydathodes, ± net-veined, with 2° veins raised below, usu with stomata below only. Stem lvs sessile, with auricle-like clasping base. Petiole winged, occ with hairs at base, with 5 vb's around hollow. Stem 1, 25-100cm, branched above, hairless, round, pith-filled or with small hollow. Per. Sept-Jul...*Wall Lettuce* **Mycelis muralis**

Lvs ± glaucous below, never purple

Mtns >700m, VR. Basal lvs 8-25cm, narrowed to winged petiole, dark green above, with broadly acuminate-triangular terminal lobe and 2(3) prs of small lateral lobes. Stem lvs widened to cordate-clasping base. Stems 50-200cm, bristly hairy below, usu with dense reddish glandular hairs above, furrowed, hollow. Shortly rhizomatous per. Sch8 ....*Alpine Blue-sow-thistle* **Cicerbita alpina**

Lowland, VR alien. Basal lvs 5-60cm, narrowed to winged petiole, mid-green above, with broadly triangular terminal lobe and 3-6 prs of small lateral lobes. Stem lvs clasping at base. Stems 60-130cm, usu hairless, furrowed, hollow. Tufted per.....*Hairless Blue-sow-thistle* **Cicerbita plumieri**

■Lvs without large terminal lobe, often dandelion-like (with backward-pointing lobes)

Petiole hollow

Lvs (±) hairless, often shiny green above, the lateral lobes often triangular, midrib often reddish nr base. Per. All yr. PL 18...*Dandelion* **Taraxacum officinale** agg*

Petiole solid

Latex bluish (occ yellow-white), not turning brown

Rootstock (rhizome) horizontal. Per

Lvs 5-15(20)cm, narrowly to broadly oblanc, obtuse to acute, narrowed to a winged petiole, hairy or hairless (hairs occ purplish nr base), ciliate, with reddish midrib, deeply narrowly pinnately lobed to ± sinuate-toothed, the terminal lobe much larger than laterals but usu ± same width as rest of lf, the lateral lobes often linear-oblong and at 90° to midrib, with stomata both sides. Stem lvs absent or reduced to bracts. Petiole with 3-5 vb's. Stems to 60cm, usu hairless, ridged, solid. Fl buds tapered into stalk. All yr ....*Autumn Hawkbit* **Leontodon autumnalis***

Rootstock (taproot) vertical. Ann to per

Stem lvs with sagittate base and acute auricles

Stems 1(7), with whitish hairs (rarely hairless), 20-90cm, usu unbranched below, (±) hairy, reddish at base, weakly ribbed, hollow. Lvs 5-15 x 0.8-5cm, ± oblanc, thin, hairless to hairy (occ glandular-hairy in N Br), with midrib sparsely hairy below, toothed to lobed, the terminal lobe usu broader than lf, the lateral lobes more triangular, with 2° veins indistinct, with stomata both sides. Petiole usu with 3-5 vb's. Fl buds flask-shaped. Ann or bi. Often all yr. PL 18..*Smooth Hawk's-beard* **Crepis capillaris***

Stem 1, with yellowish bristles, 8-70cm, stout, often reddish, roughly hairy. Lvs 5-30 x 1-8cm, roughly hairy, ± bristly along midrib, lobed (occ just toothed), the terminal lobe large. Petiole usu with 5(7) vb's. Ann (bi). R alien (mostly in grass-seed) ....*Bristly Hawk's-beard* **Crepis setosa**

Stems with whitish bristles (often short), usu 1, 25-110cm, often reddish below, densely roughly hairy, strongly ribbed. Lvs 10-20 x 2-4cm, oblanc, roughly hairy (hairs often short and yellowish), slightly rough. Petiole usu with 5(7) vb's. Bi (ann). VR alien (?extinct) ....*French Hawk's-beard* **Crepis nicaeensis**

Stem lvs without sagittate base

Lvs fetid or with strong almond odour ❶

Lvs oblanc, the lobes often linear and serrate, densely hairy both sides. Stem lvs few, small, lanc, sessile, ± clasping, toothed or lobed nr base. Stems 1(5), 10-60cm, hairy, slightly furrowed. Fl heads nodding in bud. Ann (bi). Oct-Jul. VR, Kent. Sch8 ....*Stinking Hawk's-beard* **Crepis foetida**

Lvs odourless

Petiole often purplish nr base, with bitter bluish-white latex, with 5(7) vb's. Lvs 10-35 x 2-8cm, oblanc, densely hairy to (±) hairless, ciliate, with red-based hairs at least on midrib above, with large terminal lobe (occ toothed only), with sharply toothed lobes. Stem lvs clasping. Stems 1-several, 15-80(120)cm, purplish below, roughly hairy below or hairless (occ with black bristles), ridged. Fl buds to 10mm. Bi (occ ann or per). All yr ....*Beaked Hawk's-beard* **Crepis vesicaria**

Petiole rarely purplish nr base, with tasteless yellow-white latex, with 5(7) vb's. Lvs 15-30 x 1-8cm, oblanc, with triangular terminal lobe, with hairs not red-based, with stomata both sides. Stem lvs with forward-pointing auricles, roughly hairy, sessile, ± not clasping. Stem usu 1, 30-120cm, stout (to 11mm diam), often purplish below, hairy esp below, ridged, v hollow. Bi. All yr...*Rough Hawk's-beard* **Crepis biennis**

Latex white, turning brown (after 1-2 mins)

Lvs dark green, never purple-spotted, 4-25cm, broadly oblong-lanc, narrowed to indistinct petiole, usu roughly hairy (occ with reddish-based hairs on midrib), with upturned teeth. Petiole with 3 vb's. Per. All yr..*Cat's-ear* **Hypochaeris radicata***

Lvs dark green with purple spots or streaks above, often with midrib reddish at least nr base, 4-15(30)cm, obovate-oblong, narrowed to indistinct petiole, usu roughly hairy, ± sinuate-toothed, occ pinnately lobed. Petiole with 5-7 vb's. Per. All yr ....*Spotted Cat's-ear* **Hypochaeris maculata**

Lvs usu waxy pale green, often reddish, 1-3cm, oblanc, narrowed to indistinct petiole, (±) hairless, ciliate, sinuate-toothed to -pinnatifid. Petiole with (1)3 vb's. Ann. Nov-May .... *Smooth Cat's-ear* **Hypochaeris glabra**

Latex white, not turning brown

Stems ± prostrate, rooting at nodes

Lvs occ opp (or ± so) nr stem apices, 5-25mm diam, broadly ovate to orb-reniform, angled or shallowly lobed, ivy-shaped, ± cordate at base, petiolate, thin, pale shiny green, sparsely bristly-hairy above or below, acutely toothed, with protruding white hydathodes. Stems 0.5mm diam, with elastic stele and v sparse latex .... *Ivy-leaved Bellflower* **Wahlenbergia hederacea**

Stems (if present) ± erect, never rooting at nodes

Per, with short rhizome. Stems often several. Wet habs, N Br .... *Marsh Hawk's-beard* **Crepis paludosa**

Per, with taproot. Stems often several

Petiole often long, hairy at base, channelled, with 3-5(7) vb's. Mar-Oct. PL 18 .... *Welsh Poppy* **Meconopsis cambrica**

Petiole short, retrorsely hairy, not channelled, with 3-7 vb's. Basal lvs bristly hairy to ± hairless, runcinate-lobed, ciliate or sparsely so, often with reddish midrib. Stem lvs ± clasping at base, sessile, ± glaucous, the upper lanc, entire or distantly toothed and clasping with auricles. Stems 30-120cm, stiffly branched, tough, shallowly grooved, hollow. All yr.. *Chicory* **Cichorium intybus***

Ann (occ bi in *Lactuca sativa*), with taproot. Stem usu 1

Basal lvs persistent, ovate to lanc, rounded at base, hairless, pale green to purple, weak lettuce odour, toothed to runcinate-lobed (occ entire), with stomata both sides. Stem lvs ovate to orb, sessile, clasping at cordate base, with acute or obtuse auricles. Stem to 1m, usu whitish, round. Taproot with abundant lateral roots....... *Garden Lettuce* **Lactuca sativa**

Basal (soon dying) and lower stem lvs lettuce-like, waxy, narrowed to short winged stalk, wavy, with opium odour, usu shallowly lobed, with white veins, with stomata below only. Mid-stem lvs sessile, clasping, occ unlobed. Stem to 1m, often with sparse patent stiff bristles above (non-scabrid unlike other *Papaver* spp), solid. Taproot with few lateral roots. Often all yr

Lvs glaucous, hairless.................... *Opium-poppy* **Papaver somniferum** ssp **somniferum**

Lvs weakly or not glaucous, with sparse stiff hairs on veins below .... *Small-flowered Opium-poppy* **Papaver somniferum** ssp **setigerum**

Cichorium intybus petiole TS

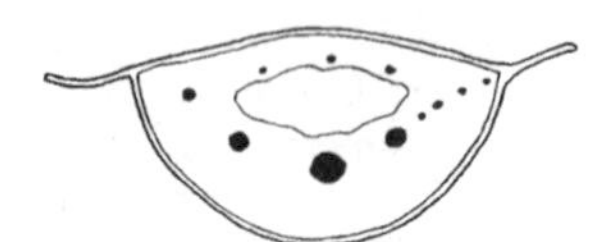

Taraxacum officinale agg petiole TS

Hypochaeris radicata petiole TS

Group PH – *Stipules (or ochreae in* Ranunculus aquatilis *agg) present. Petiole with (1)3 vb's. Hairs unicellular*

■Plant with bristly glandular hairs (occ sparse, slightly swollen-based). Lvs opaquely net-veined

Lvs mostly basal, broadly ovate to ± orb, cordate at base, palmately lobed, the lobes toothed with bristle-tipped green hydathodes, with main veins ending in teeth, with stomata below only. Petiole 5-30cm, not sheathing at base, with long swollen-based hispid glandular hairs (tips soon withering) and short glandular hairs, round, with 3 vb's and spiral fibres. Stipules partly adnate to petiole, laciniate on stem lvs. Stems to 70cm, roughly hairy, round. Shortly rhizomatous per. All yr

Lvs with a bud at junction with petiole (producing plantlets in moist conditions), 4-10cm diam, hairy above and on veins below, with 5-7 shallow lobes....*Pick-a-back-plant* **Tolmiea menziesii**

Lvs without a bud or plantlets at junction with petiole, (4)7-12cm diam, broadly ovate to orb, often purplish below, with long white hispid hairs and short glandular hairs both sides, with (7)9-11 shallow lobes..*Fringecups* **Tellima grandiflora**

■Plant without glandular hairs. Lvs with some translucent veins

Aquatic or wet habs. Lvs 0.5-3(4)cm diam. Ann to per ....(*Ranunculus aquatilis* agg*) Go to **RAN-BAT**

Dry habs (stem lvs of *Geum* spp may key out here but the basal lvs are 1-pinnate)

Lvs >3cm diam

Trailing shrub. Lvs shortly white-woolly below, ovate, shallowly 3-7-lobed. All yr ....*Chinese Bramble* **Rubus tricolor**

Tufted per. Lvs hairy or hairless, orb, 7-11-lobed (often shallowly so). Herb. Apr-Nov ....(*Alchemilla*) Go to **ALC**

Lvs 0.2-1.5cm. Ann. Oct-Aug

Lvs 3-lobed, fan-shaped, shortly stalked; lobes divided into 3(5) oblong lobes, each with reddish apical hydathode, with long thin hairs, sparsely ciliate, not net-veined, with stomata below only. Petiole weakly channelled, with 1(3) vb's. Stipules ± auricle-like when young, soon toothed, ciliate. Stems <5cm

Lvs truncate at base, greyish-green. Stems stout, vigorous, with short internodes. Basic to ± acidic habs..*Parsley-piert* **Aphanes arvensis**

Lvs cuneate at base, mid-green. Stems slender, weak, with long internodes. Acidic habs ....*Slender Parsley-piert* **Aphanes australis**

Group Pl – Malvaceae. *Stipules usu herbaceous and persistent, entire (unless otherwise stated), ± parallel-veined. Lvs broadly ovate to orb, cordate at base (occ truncate in* Althaea officinalis*), plicate when young, palmately lobed, crenate (without the reddish hydathodes of* Rosaceae*), (±) net-veined, usu with at least 2° veins raised below, with stomata both sides. Petiole with ≥3 vb's (often 6 in a ring). Hairs stellate or swollen-based (unicellular)*

■Basal rosette present (at least in 1st yr). Stems usu erect

Petiole channelled, solid, with 3 vb's (seedlings of *Malva neglecta* may key out here)

Basal lvs 2.5-8cm wide, reniform, long-petiolate, (5)7-lobed, shiny green above, hairless or sparsely hairy above, sparsely hairy on veins below (occ swollen-based), ciliate, crenate. Stem lvs cut to 9/10 way to base, with 3-7 primary divisions deeply lobed into ± linear lobes, successively short-petiolate and more deeply divided. Petiole with spreading to retrorse swollen-based hairs (occ sparse). Stipules long, linear-lanc, hairy, green nr apex, toothed. Stems 1-several, occ woody at base, often purple-spotted, with spreading to retrorse simple hairs (bases swollen, purple), round or weakly angled, solid. Per. All yr ....*Musk-mallow* **Malva moschata**

Basal lvs similar to *M. moschata* but dull above, hairy both sides with simple and V-shaped forked hairs above and with sessile 3-6-rayed stellate hairs below. Stem lvs also similar to *M. moschata* but less divided (cut to 6/7 way to base). Petiole stellate-hairy. Stem 1, branched above, rough due to swollen-based hairs (also on petiole and lvs). Per. All yr ....*Greater Musk-mallow* **Malva alcea**

Petiole channelled, usu solid, with (3)4 vb's

Basal (and lower) lvs 2-4cm wide, reniform, long-petiolate, grey-green, ± 5-lobed, crenate, with obtuse to ± acute lobes, with simple swollen-based hairs to 1.5mm above (occ tiny sessile forked or stellate hairs), with mostly simple and a few sessile 2-rayed stellate hairs below. Upper lvs short-petiolate, deeply 3-5-lobed and crenate. Petiole to 6cm. Stems several, 8-60cm, ascending, hispid. Ann (bi). Oct-Jul. Sch8........................*Rough Marsh-mallow* **Althaea hirsuta**

Petiole weakly channelled, usu solid, with 6 vb's

Basal lvs 5-10cm wide, orb, dark green, usu purple in centre, sparsely to densely hairy with mostly simple hairs above, usu with V-shaped hairs below (slightly swollen-based), with v shallow crenate lobes. Stem lvs lobed to ⅓ way into 5-7 crenate lobes. Petiole rough with swollen-based simple or stellate hairs. Stipules hairless, ciliate, greenish. Stems 1-several, 20-80cm, with simple hairs. Per. All yr. PL 15...........................*Common Mallow* **Malva sylvestris***

Basal lvs 5-12cm wide, orb, grey-green, rarely purplish in centre, velvety hairy with short (0.5mm) sessile 4-7-rayed stellate hairs above and 2-3-rayed stellate hairs below, with 5 triangular-acute toothed lobes. Petiole occ round, with 4-7-rayed stellate hairs. Stems 1, 30-150cm, stellate-hairy. Ann or bi. Often all yr. VR.............*Smaller Tree-mallow* **Lavatera cretica**

Petiole not channelled, soon hollow, with 10 vb's

Lvs 10-30cm wide, shallowly 5-7-lobed or angled, rugose, sparsely ± roughly hairy with simple or V-shaped hairs above, densely sessile stellate-hairy (often on swollen bases) below, crenate-dentate. Petiole ± round. Hairs simple to 3-rayed stellate from swollen base. Stipules herbaceous, purplish, 3-dentate. Stems to 300cm, stout, simple-hairy. Per or bi ....*Hollyhock* **Alcea rosea**

■Basal rosette absent (present in seedlings)

Stem(s) usu 1, erect. Lvs velvety both sides. Petiole with 6 vb's. Per

Stem to 300cm, woody at base, trunk-like, unbranched, round. Lvs 7-25cm wide, orb, 3-5(7)-lobed, with dense simple or stellate hairs above and sessile 2-7-rayed stellate hairs below (slightly swollen-based), crenate, with 2-3° veins slightly raised below. Petiole 5-30cm, with 5-rayed stellate hairs, slightly channelled above, occ laterally flattened, soon hollow. Monocarpic. All yr ....*Tree-mallow* **Lavatera arborea***

Stem 60-120cm, not woody or trunk-like, simple or branched, round to weakly angled. Lvs 3-11cm wide, roundish ovate-triangular, slightly 3-5-lobed, irreg toothed, with short sessile long-rayed stellate hairs both sides, with 2° veins raised below. Petiole 1-4cm, stellate-hairy, round, flattened to v weakly channelled above, often with small hollow. Not monocarpic. May-Oct ....*Marsh-mallow* **Althaea officinalis**

Stems usu several, ± prostrate. Lvs roughly hairy above. Petiole with 3-4 vb's. Ann

Basal lvs (1.5)3-7cm wide, roundish reniform, dark bluish- or grey-green above, often with tiny purplish centre to petiole join (but not on lf), usu with stellate, V-shaped and/or simple hairs above, with stellate (and forked) hairs adpressed below (like *M. sylvestris*), with 5-7-shallow acutely crenate lobes. Upper lvs lobed to ¼. Petiole more strongly channelled than *M. sylvestris*, with ± equal hairs (not spreading-hairy), with swollen-based short stellate hairs. Stems with swollen-based stellate hairs. Apr-Oct..........................*Dwarf Mallow* **Malva neglecta**

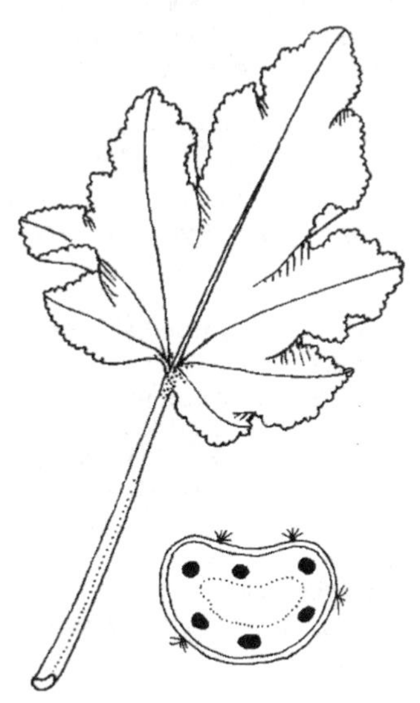

Lavatera arborea lf & petiole TS

Malva sylvestris lf & petiole TS

Group PJ – *Stipules (or ochreae) present. Petiole with ≥3 vb's (often obscure). Hairs not stellate or swollen-based.* ❶ *Lvs peltate*

▪Stems trailing and rooting at nodes

Stipule-like auricles (pseudostipules) present, enlarging as lf matures, ciliate. Lvs all on stems

Plant with ragwort odour. Lvs 3-10cm diam, ± orb to broadly ovate, ivy-like, usu with 5-7 acute angles or triangular lobes, cordate at base, often thick or rubbery, light green, hairless, ciliate, opaquely net-veined, with indistinct palmate veins, with stomata obscure both sides. Petiole usu > lf, round to weakly channelled, solid, with 5 vb's and abundant clear sap. Stems round, solid, with 8 vb's in outer ring. Woody-based branched trailing or clambering twiner. All yr. R alien, SW Br...*German-ivy* **Delairea odorata**

Stipules membranous, closed (ochrea-like), v thin, not adnate to petiole. Lvs all basal

Petiole hairless

Lvs floating or emergent, to 7cm diam, orb, 3-7-lobed, shiny dark green above, paler below, with minute stomata both sides. Petiole round, with aerenchyma and 4-5 obscure vb's. Stems often floating, without stomata, snapping audibly, each node with 20-40 roots. All yr. Invasive aquatic..*Floating Pennywort* **Hydrocotyle ranunculoides**

Petiole with strongly retrorse hairs

Lvs 0.5-2cm diam, orb, cordate at base, crenate. VR alien ....*New Zealand Pennywort* **Hydrocotyle novae-zeelandiae**

Petiole usu with scattered patent (or weakly retrorse) hairs in distal ⅓

Lvs hairless, 0.8-5cm diam, orb, peltate, crenate with 7 shallow lobes, occ net-veined (veins not raised below), net-veined, with stomata both sides. Petiole 1-25cm. ❶. May-Oct (overwinters as tiny lvs). Damp to wet habs. PL 24....*Marsh Pennywort* **Hydrocotyle vulgaris**

Lvs with stiff adpressed hairs (esp above), 0.5-2cm diam, orb, cordate at base, with 5-7 toothed lobes cut ½ way to base, with reddish dots along margins and veins above, net-veined, with stomata both sides. Petiole 1-5cm, with sparse weakly retrorse hairs in distal ⅓. All yr (but frost-sensitive). Dry to damp habs, VR alien, Ire ....*Hairy Pennywort* **Hydrocotyle moschata**

▪Stems, if present, not rooting at nodes

Plant hairless

Ochreae present on all lvs..(*Oxyria digyna*, *Rheum*) Go to **DOCK**

Ochreae absent from basal lvs but present on stem lvs...................(*Caltha palustris*) Go to **RAN**

Plant with at least some hairs

Stipules herbaceous, green or reddish...(*Geranium*) Go to **GER**

Stipules scarious, whitish (occ with green or red veins). Uplands

Basal lvs 4-7cm diam, ± orb, cordate at base, palmately 5-7-lobed, rugose, sparsely glandular-hairy on veins above, sparsely hairy below (occ glandular), net-veined, with stomata below only; lobes dentate, with reddish hydathodes. Petiole 1-4(7)cm, sparsely hairy, channelled, with 3(5) vb's and spiral fibres. Stipules to 3mm, ovate. Stems to 20cm, round. Rhizomatous per. Apr-Oct...*Cloudberry* **Rubus chamaemorus**

Stipules membranous, whitish (occ with green or red veins). Coasts, S & W Br

Basal lvs 0.5-2cm, ovate, rounded at base, with white stout hairs above, minutely ciliate, pinnately lobed to ½(⅔) way, with 1 pr of small lobes at base of large terminal lobe (often held in vertical plane). Petiole to 1.5cm, with stout white hairs (often antrorsely crisped), slightly flattened, with 4 vb's. Stipules ± obtuse. Stems to 10cm, round. Ann (bi). Oct-Jul (all yr)..*Sea Stork's-bill* **Erodium maritimum**

Group PK – Reseda. *Stipules absent but lvs with stipoid glands nr base, 0.5mm, reddish, v easily missed or occ absent. Lvs* with main vein of lobes raised only below, *with stomata both sides. Stems 1-several, rough with low papillae, ridged, soon hollow. Plant hairless. All yr*

■Basal lvs unlobed, oblong-spathulate. Stem lvs with 1 pr lobes

Stems 20-40cm, ascending. Ann or bi. Arable, VR alien.........*Corn Mignonette* **Reseda phyteuma**

■All lvs pinnately lobed with 1-2 prs lobes

Stem lvs to 10cm, with papillate margins. Stems 30-75cm, ascending to erect. Per ....*Wild Mignonette* **Reseda lutea***

■All lvs pinnately lobed with 5-8 prs lobes

Stem lvs to 10cm, often dark bluish-green, with entire undulate margins; lobes to 2.5cm, linear-oblong, unequal, decurrent. Stems 30-90cm, erect. Ann or per with long taproot. R alien ....*White Mignonette* **Reseda alba**

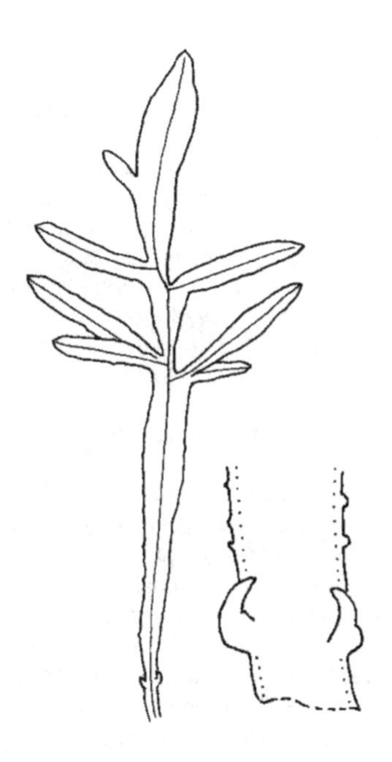

Reseda lutea (showing stipoid glands)

Group PL – Brassicaceae. *Plant with at least some branched or stellate hairs. Lvs with stomata both sides. Stems erect, solid*

■Lvs >20cm

Basal (and lower) lvs to 50 x 15cm, elliptic, acute, cuneate at base, with mostly forked hairs both sides (occ sparse and with midrib ± retrorsely hispid below), often toothed and/or lobed, with 0-2 prs of ± free lfts nr base. Petiole to 10cm, purplish at base, retrorsely hairy, channelled (often slightly winged), with strong cress odour, solid or with small hollow, with 5-7 vb's. Stems 30-100cm, purplish at base, retrorsely hairy esp at base (sparse above), occ hairless or with sparse glandular hairs, ± round below, ridged above. Per (bi). Mar-Nov. R alien ....*Warty-cabbage* **Bunias orientalis**

■Lvs <20cm

Lvs densely grey-hairy (to ± woolly)

Bi (per), with persistent rosettes at end of stolons. Lvs to 15(25)cm, sinuate or pinnately lobed, lanc-oblanc, cuneate at base, grey-white woolly with stellate hairs, with midrib raised below, with obscure 2° veins. Petiole indistinct, with 1 vb. Stems 40-80cm, stout, round, stellate-hairy, glandular nr infl. All yr. Littoral sand dunes & cliffs, R.....................*Sea Stock* **Matthiola sinuata**

Ann, with rosette soon dying and weak taproot. Lvs to 8cm, 2-3-pinnately lobed; lobes usu 1-2mm wide, linear-lanc, acute, stellate-hairy, with main vein visible only. Stems to 100cm, 1.5-2mm diam, purplish at base, greyish with stellate hairs. May-Sept. Sandy arable habs ....*Flixweed* **Descurainia sophia**

Lvs green, not densely grey-hairy

Ann (bi), with sessile stellate hairs mostly 5-rayed (occ 3-rayed or simple)

Basal lvs in a persistent rosette, 5-15cm, oblanc in outline, petiolate, pinnately lobed to ± entire, weakly cress-scented, shiny green at least above, sparsely to densely hairy with at least some stellate hairs (often sparse above), with weakly translucent veins. Stem lvs clasping with ± acute auricles at base, sparsely to densely hairy, with longer simple hairs. Petiole with 1(3) vb's. Stems to 20(30)cm, mostly stellate-hairy. Usu all yr ....*Shepherd's-purse* **Capsella bursa-pastoris**

Ann (bi), with sessile stellate hairs mostly 3-rayed

Basal lvs in a rosette, not persisting, to 3cm, petiolate, pinnately lobed. Stem lvs many, sessile. Stems hairless or with sparse stellate hairs. Oct-May ....*Hutchinsia* **Hornungia petraea**

Bi or per, with stalked hairs mostly forked or 3-rayed. (Basal lvs of *Erysimum cheiranthoides* may key out here)

Per, with 1 or more compact rosettes on short woody stock. VR, Bristol

Basal lvs to 2.5cm, oblanc, narrowed to indistinct petiole, shiny dark green above, often purplish below, with forked hairs and few simple hairs (mostly along margin), sinuate-lobed with 4-6 rounded lobes per side (lengths increasing along lf), with water-cress taste. Stem lvs 1-3, tiny, oblong, sessile and ± clasping, less lobed than rosette lvs. Stems 3-20(30)cm, usu hairless. All yr. Sch8..*Bristol Rock-cress* **Arabis scabra**

Per, with 1 rosette only (but tufted with short rhizomes). Rocky mtn habs

Basal lvs 1.5-3.5cm, lyrate-pinnatifid (occ with separate lfts), long-petiolate, shiny dark green above, cress odour (and taste), hairless or sparsely hairy with at least some forked or stellate hairs above or on margins (simple hairs v sparse or absent), all veins ± obscure, with stomata both sides. Stem lvs much reduced, spathulate to elliptic-lanc, sessile, ± hairless, ± entire. Petiole of basal lvs < lf, narrow, with 1 vb. Stems 8-20cm, with spreading stiff simple hairs below, ± hairless and pruinose above. All yr ....*Northern Rock-cress* **Arabis petraea**

Bi, with 1 rosette only. Lowland sandy habs

Basal lvs to 8cm, pinnately lobed, with stalked 3-rayed hairs both sides. Petiole with 1(3) vb's. Stem to 100cm, purplish nr base. Bi................................*Tower Mustard* **Arabis glabra**

Group PM – Asteraceae. *Plant with cottony or woolly hairs (often confined to underside of lvs). Lvs ± orb and palmately lobed (toothed only in* Petasites fragrans*), all basal, usu emerging in prs (occ 3's) from rhizome (but 5-7 in* Tussilago*), cordate at base, revolute and cobwebby or white-felted above when young, often with additional simple septate hairs both sides. Petiole long, usu sheathing at base.* ❶ *Lvs <4cm diam*

■Petiole with c9 vb's in an arc, solid, hardly sheathing at base. Lvs shallowly lobed with angular margin, not net-veined, with stomata both sides

Lvs 4-20cm diam, with 5-12 shallow acute lobes toothed by distant blackish-purple hydathodes, white-felted both sides, soon ± hairless above (± hairless below in autumn), the lowest pr of 2° veins not bordering sinus. Petiole channelled, with faint odour when cut. Apr-Oct ....*Colt's-foot* **Tussilago farfara***

■Petiole with 5 vb's around hollow, hardly sheathing at base. Lvs sinuate-dentate, not net-veined, with stomata below only. ❶

Lvs to 4cm diam, orb to reniform, shiny dark green above, paler and occ purplish below, with crisped and glandular hairs above when young, cottony below with additional septate hairs on veins, shallowly. Petiole to 9cm, hairy. All yr. Mtns, VR, Scot (Clova). Sch8 ....*Purple Colt's-foot* **Homogyne alpina**

■Petiole with ≥10 randomly scattered vb's, solid or hollow, sheathing at base. Lvs denticulate, net-veined, with stomata below only

Petiole channelled. Lvs with basal lobes usu overlapping (sinus bordered by lowest pr of veins)

Petiole with 7-10(11) ridges below, hollow (occ <<⅓ diam), with strong odour when cut, turning orange-brown, with >30 scattered vb's and spiral fibres. Lvs 10-100cm diam, dull green above, greyish- or whitish-cobwebby below, obtusely denticulate but with larger distant teeth where 2° veins end at margin. Mar-Oct..*Butterbur* **Petasites hybridus**

Petiole with 12-14 ridges below, hollow (c⅓ diam), weakly odorous, soon turning purplish when cut, with >30 scattered vb's and spiral fibres. Lvs 10-100cm diam, dull green above, greyish- or whitish-cobwebby below, sharply but irreg dentate with 2.5mm green hydathode-teeth. Mar-Oct ....*Giant Butterbur* **Petasites japonicus**

Petiole round. Basal lobes usu spreading (sinus rarely bordered by lowest pr of veins in *P. albus*)

Lvs 15-40cm diam, white-woolly below, dentate with acute prominent teeth, the intervening spaces sharply denticulate with hydathode-teeth. Petiole solid, the odour weak or absent. Mar-Oct...*White Butterbur* **Petasites albus**

Lvs 10-20cm diam, hairy below on veins at least when young (hairs minutely gland-tipped, the tips soon withering), regularly sinuate-denticulate with purple hydathodes. Petiole solid or with small hollow, the odour weak or absent, with 10-15 scattered vb's, soon turning yellow-brown when cut. Sept-Jul. Pl 12, 23..*Winter Heliotrope* **Petasites fragrans***

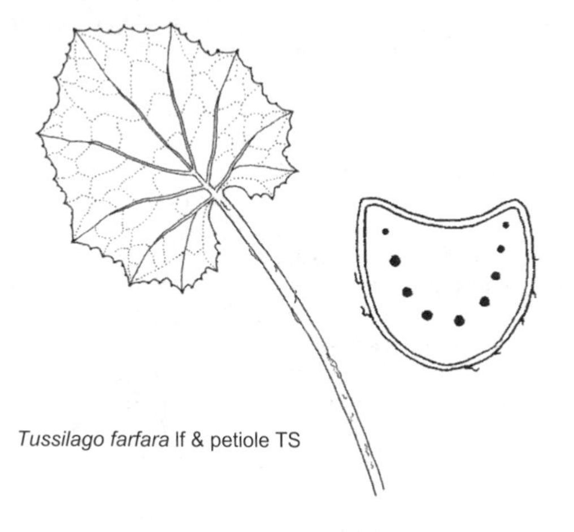

Tussilago farfara lf & petiole TS

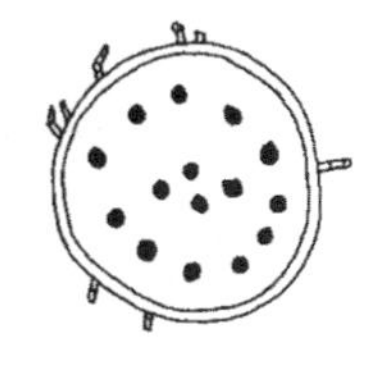

Petasites fragrans petiole TS

Group PN – Asteraceae. *Per (occ woody at base) with at least some unicellular cottony or woolly hairs (occ sparse), additional septate hairs may also be present. Lvs pinnately lobed, not cordate at base, aromatic (but not with glandular hairs), not net-veined. Stem lvs usu with tiny pinna-like lobes at base. (*Senecio sylvaticus *may key out here).* ❶ *Mtns*

▪Plant absinth-scented, always with cottony or silky hairs

Lvs densely white-woolly both sides, not revolute when young

Lvs 2-5cm, 2-3-pinnately lobed, the lower stalked and auriculate, the upper pinnate and sessile, the uppermost pinnately lobed or entire; ultimate lobes often opp, c1mm wide, linear, obtuse. Stems 20-50cm, branched, weakly ridged. Tufted per. Apr-Oct. Upper saltmarshes ....*Sea Wormwood* **Artemisia maritima**

Lvs white silky-hairy both sides but esp below, not revolute when young

Tall tufted per. Basal (and lower) lvs 3-pinnate, the mid-stem lvs 2-pinnate, the uppermost unlobed or 1-pinnate; ultimate lobes to 4 x 2-3mm, lanc or linear-oblong, usu obtuse. Petiole without pinnae at base, channelled, with 3 vb's. Stems 30-100cm, ± woody below, white-hairy, angled. Ruderal...*Wormwood* **Artemisia absinthium**

Dwarf tufted or shortly rhizomatous per. Basal lvs 2cm, petiolate, ± palmately lobed into (3)5 deeply toothed lobes. Stem lvs ± sessile. Stems 3-10cm, hairy. ❶. Mar-Oct. Mtns ....*Norwegian Mugwort* **Artemisia norvegica**

Lvs white-cottony below, soon (±) hairless, revolute when young (margins weakly revolute at maturity)

Strongly aromatic rhizomatous per with overwintering rosettes. Lvs bright green above, sparsely cobwebby to hairless above, with short crisped hairs below, pinnately lobed, with stomata below only. Stem lvs with lobes at clasping petiole base; terminal lobe of lower stem lvs 30-50 x 5-6mm, linear-lanc to linear, acute or apiculate. Petiole channelled, with 1(3) vb's. Stems 60-120cm, usu purple, crisped-hairy, grooved, pith-filled. Fls Nov-Dec. All yr ....*Chinese Mugwort* **Artemisia verlotiorum**

Weakly aromatic tufted per, without overwintering rosettes. Lvs dark green above, white-cottony below (whiter than *A. verlotiorum*), the basal ± lyrate-pinnatifid or 3-foliate, with stomata below only. Stem lvs sessile, clasping, 2-pinnately lobed (uppermost pinnate), with remote tiny lobes nr base; terminal lobe of lower stem lvs 7-30 x 3-6mm, short lanc to oblong, acute or apiculate. Petiole channelled, with 1(3) vb's. Stems 60-120cm, usu purple, sparsely cottony to hairless, grooved, pith-filled. Mar-Oct. PL 19...*Mugwort* **Artemisia vulgaris**

▪Plant tansy-scented, often without cottony or silky hairs

Rhizomatous. Lvs strongly gland-dotted, with tiny intercalary lobes. Stems 30-120cm

Lvs to 25cm, the lower petiolate, the upper sessile and ± clasping, all hairless or sparsely hairy, pinnately lobed with up to 18 prs of alt linear-oblong lobes (lobes entire to deeply toothed), with cartilaginous hydathode-teeth. Petiole channelled, angled below, with 5 vb's. Stems usu purplish, sparsely hairy to hairless, angled. Mar-Oct. PL 18..............*Tansy* **Tanacetum vulgare**

Tufted. Lvs not or weakly gland-dotted, without tiny intercalary lobes

Stems to 120cm

Lvs 10-18cm, with 5-8 prs of ± free lfts along winged rachis and tiny lobes at base, pale grey-green esp below, softly ± adpressed cottony-hairy both sides, with stomata below only; lobes 3-5mm wide. Stems hairy, round but weakly angled. R alien ....*Rayed Tansy* **Tanacetum macrophyllum**

Stems 20-60cm

Lvs 3-7cm, 1-3-pinnately lobed, with smaller lobes nr base, not gland-dotted, grey-green, adpressed-hairy with silky long hairs esp below, occ ± hairless or cottony, ciliate, with obscure veins; lobes with bristle-tipped teeth. Petiole with 3 vb's. Stems usu branched, tough, purplish, ridged, cottony and often with swollen-based septate hairs (hair bases elongate and turning purple-black). All yr..*Yellow Chamomile* **Anthemis tinctoria**

Lvs 1-12 x 0.5-6cm, 1-2-pinnately lobed, weakly gland-dotted below, grey-cottony; lobes 2mm wide, ± acute. Stems woody below, white-cottony, often with short glandular hairs, angled. Hortal, R alien...................................*Sicilian Chamomile* **Anthemis punctata** ssp **cupaniana**

Group PO – Asteraceae. *Plant (±) odourless (occ weakly aromatic with glandular hairs) but the stems/petioles of all* Senecio *spp are weakly fetid when broken. Lvs with cottony hairs (often sparse or confined to lf axils or rosette centre), often with additional septate hairs. Stem lvs with tiny lobes at base. Stems usu purplish nr base, often tough and angled, solid (but soon hollow in* Senecio vulgaris *and* S. vernalis*)*

■Lvs persistently grey- or white-woolly or felted

Rhizomatous per. Lvs 3-7cm, densely white-felted both sides, 2-pinnate or 2-pinnately lobed; lobes 5-10mm wide, obtuse. Petiole with 1(3) vb's. Stems 30-60cm, densely white-felted, angled. Mar-Oct. VR alien...*Hoary Mugwort* **Artemisia stelleriana**

Tufted per. Lvs 8-20cm, shortly grey-woolly, all petiolate, ± involute when young; lobes 5-10mm wide, obtuse. Petiole flat above, hardly angled, with 5 vb's. Stems 40-80cm, ± woody at base, white-woolly, slightly ridged or angled. All yr...................*Ragwort Knapweed* **Centaurea cineraria**

Tufted per. Lvs 3-15cm, shortly white-woolly, petiolate (exc uppermost), revolute when young; lobes 4-6mm wide, obtuse. Petiole channelled above, 5-angled, with 5 vb's. Stems 25-100cm, woody at least below, white-woolly, angled. All yr (but frost-sensitive) ....*Shrub Ragwort* **Senecio cineraria**

■Lvs not woolly (or only when young)

Plant with glandular septate hairs (tips soon withering), often viscid to touch. Ann

Basal rosette lvs to 10cm, shortly petiolate, grey-green, weakly resin-scented or aromatic, densely hairy with slightly viscid obscurely glandular hairs, irreg pinnately lobed with toothed lobes, with stomata both sides. Upper stem lvs sessile or clasping with auricles. Petiole green, with cottony hairs at base (in lf axils), with 3(5) vb's. Stems to 70cm, ± straight, with ascending branches, occ purplish below, ± glandular-hairy. Fl buds oblong. Oct-Jul ....*Heath Groundsel* **Senecio sylvaticus**

Basal rosette lvs absent. Lower lvs to 5cm, shortly petiolate, yellowish-green, strongly fetid, viscid glandular-hairy, pinnately lobed with ± equal toothed or lobed lobes, the lobes with slightly recurved margins, with stomata both sides. Upper stem lvs sessile, weakly clasping without auricles. Petiole green, with cottony hairs in axils. Stems to 60cm, ± wavy, usu with many spreading branches, purplish below, densely viscid-hairy. Fl buds oblong. Apr-Oct ....*Sticky Groundsel* **Senecio viscosus***

Plant with nonglandular septate hairs, never viscid to touch

Rhizomatous per. Lower lvs with stomata below only

Lower lvs 5-12cm, obovate-lanc, petiolate, deeply lobed, with ± recurved margins, whitish-grey cottony when young, soon dull green and with long septate hairs both sides, occ cobwebby esp below; lobes c5 per side, oblong, each with 0-6 irreg teeth, with white-tipped hydathodes turning purplish. Mid-stem lvs petiolate, with a small narrow acute terminal lobe and long linear-oblong lateral lobes, the lobes entire or sparsely toothed. Upper lvs sessile, clasping. Petiole with 3(5) vb's. Stems to 100cm, not purplish when young, sparsely cottony below. Mar-Oct...*Hoary Ragwort* **Senecio erucifolius***

Tufted bi or per. All lvs with stomata both sides

Lvs ± clasping stem (at least the middle and upper lvs), revolute when young

Basal lvs in rosette, usu dead at fl

Basal lvs often lyrate-pinnatifid with a large ovate blunt terminal lobe and 0-6 prs of much smaller oblong lobes, to 15cm, dark or grey-green, crisped, usu without recurved margins, often with cobwebby hairs below and long hairs both sides, sinuate-toothed or lobed. Petiole often purple, often hairy, with 5 vb's. Stem(s) to 120cm. Per (bi). All yr. Dry habs ....*Common Ragwort* **Senecio jacobaea***

Basal lvs often lyrate-pinnatifid with a large ovate-oblong terminal lobe and 1(3) prs small lobes at base, to 15cm, dark or yellowish-green, often puplish below, with slightly wavy (but not crisped) margins, hairless, weakly crenate. Petiole purplish at base, with v sparse cobwebby hairs confined to extreme base in rosette centre or lf axils, with 5 vb's. Stem(s) to 80cm, purplish nr base, sparsely hairy. Bi (per). All yr. Wet habs ....*Marsh Ragwort* **Senecio aquaticus**

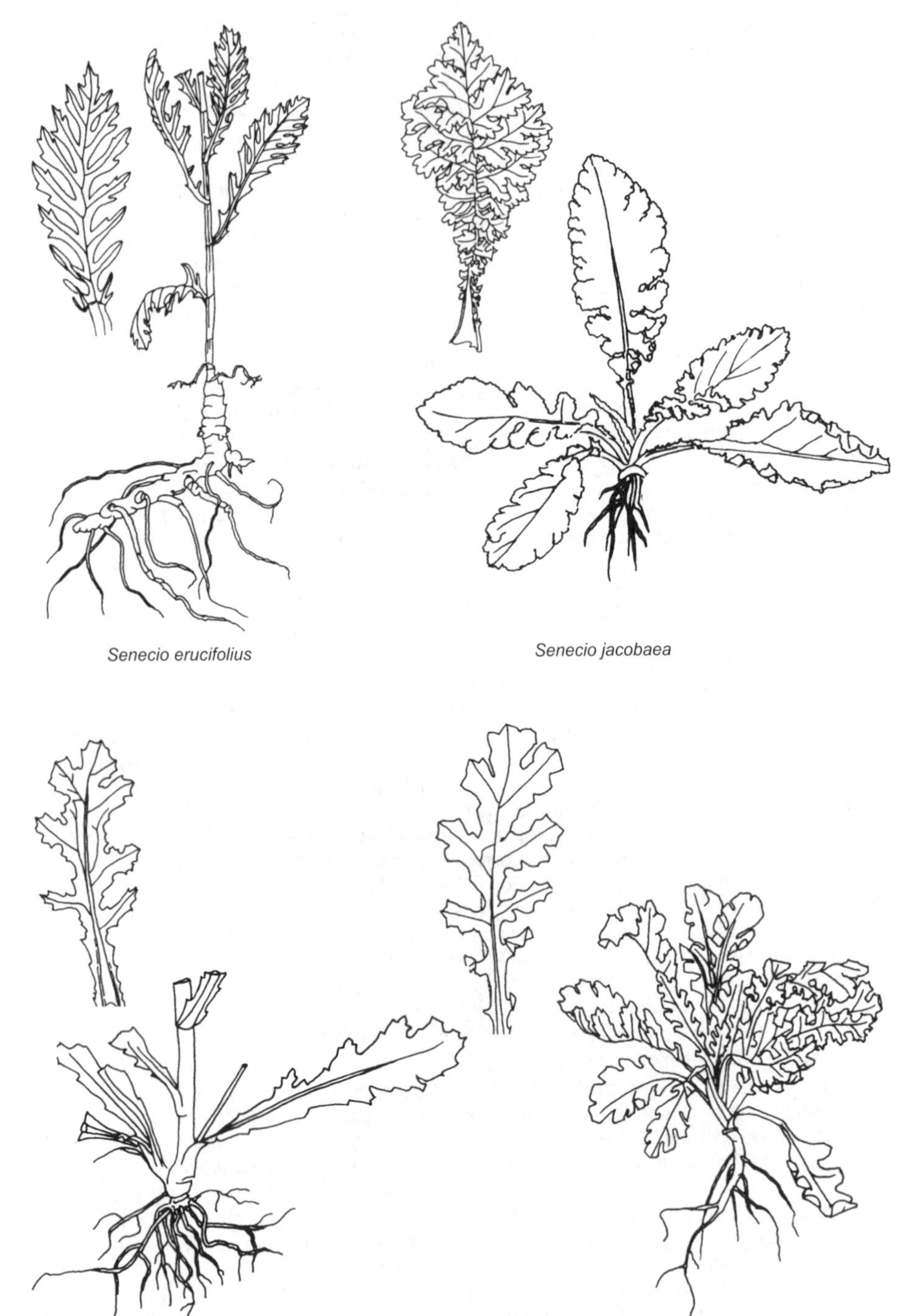

Senecio erucifolius

Senecio jacobaea

Senecio vulgaris

Senecio viscosus

Basal lvs absent. Lvs revolute when young

Lower lvs held upright, with ± clasping winged petiole, tough, ± shiny bright green above, ± hairless, usu deeply lobed, with 2-7 prs of distant oblong lobes (toothed with purple hydathodes), with recurved margins, with midrib and 2° veins raised below. Upper lvs sessile, clasping. Petiole sparsely cottony, hairy above esp nr base, with 3 vb's. Stems to 60cm, ± woody at base, wavy, branched, ± hairless. Fl buds ± square. Short-lived per (ann). All yr..*Oxford Ragwort* **Senecio squalidus**

Lvs not clasping (occ decurrent down stems), rolled when young

Basal lvs 5-20cm, petiolate, septate-hairy, rarely lobed, lobes with protruding purple hydathodes (occ obscure), with midrib often purplish nr base. Stem lvs sessile, entire to pinnately lobed, occ with pinnae at base. Petiole to 15cm, winged, channelled, with 3(5) vb's. Stems 20-80cm, usu septate-hairy and cottony, grooved. Feb-Nov ....*Common Knapweed* **Centaurea nigra**

Basal and lower lvs 10-25cm, often shiny green above, 1-2-pinnately lobed, septate-hairy, with purple hydathodes; uppermost lvs ± entire. Petiole with 5(7) vb's, often with orange sap. Stems 30-90cm, ± hairy, grooved. Feb-Oct. PL 18 ....*Greater Knapweed* **Centaurea scabiosa**

Ann (rarely bi in *Centaurea solstitialis* and *C. cyanus*). All lvs with stomata both sides

Lvs revolute when young

Lower lvs 3-20(30)cm, lobed, narrowed to indistinct petiole, occ ± fleshy, cottony to ± hairless; lobes distant, oblong-linear, obtuse, with v slightly recurved margins, dentate, with protruding hydathodes turning purple, often opaquely net-veined. Mid and upper lvs ± clasping with auricles. Petiole 1-3cm, winged, with 3(5) vb's. Stems to 30(40)cm, purplish, cottony, ± ridged, soon hollow. Jan-Sept

Fl buds oblong. Plant sparsely to densely cottony. PL 18........*Groundsel* **Senecio vulgaris***

Fl buds wider than long. Plant densely cottony at least when young. VR alien ....*Eastern Groundsel* **Senecio vernalis**

Lower lvs 3-12cm, deeply lobed, petiolate, cottony (esp below) when young, soon ± hairless; lobes distant, oblong-linear, obtuse, dentate, occ lobulate. Mid and upper lvs ± clasping, with auricles. Petiole 1-3cm, with cottony hairs in lf axils. Stems to 50cm, occ purplish, sparsely cottony to hairless, ridged. Fl buds square. Occ per. Plant intermediate between *S. squalidus* and *S. vulgaris* but more robust than either. Often all yr. R, N Wales, Scot ....*Welsh Groundsel* **Senecio cambrensis**

Lvs rolled when young

Basal lvs 6-15cm, deeply lyrate-pinnatifid with distant narrow toothed or entire lobes, petiolate, with sparse cottony hairs both sides, the teeth with v acute cartilaginous tips. Stem lvs lanc, sessile, decurrent into stem wings, ± entire. Stems 20-90cm, stiff, simple or branched, winged, cottony, ridged. VR alien..........*Yellow Star-thistle* **Centaurea solstitialis**

Basal lvs ± absent (usu unlobed if present). Lower stem lvs 5-20 x 0.2-4cm, usu lyrate-pinnatifid with narrow distant lobes, petiolate, greyish with persistent cottony hairs, the teeth without cartilaginous tips. Upper lvs linear-lanc, ± sessile, occ lobed. Stems 20-90cm, branched, sparsely cottony, grooved or angled, with cartilaginous ridges. Apr-Sept (all yr). Arable, wildflower mixes...*Cornflower* **Centaurea cyanus**

Group PP – *Lvs palmately or digitately lobed, all on stems (or basal in* Chrysoplenium alternifolium*). Stems prostrate, rooting at nodes*

■Lvs glandular-hairy (hairs may be sparse, or have withered tips), often ciliate

Wet habs. Basal lvs 2-5, 1.5-3.5cm diam, orb, cordate at base, yellow-green, with sparse patent minutely glandular hairs both sides, crenate with 9-11 shallow slightly retuse truncate lobes, each lobe with a protruding hydathode, with obscure veins. Petiole to 9cm, hairy, channelled, with 1 vb. Non-fl stems (stolons) lfless, with elastic stele. All yr ....*Alternate-leaved Golden-saxifrage* **Chrysosplenium alternifolium**

Dry habs, hortal. Lvs mostly alt, 1-9cm diam, ivy-shaped, cordate at base, 7-lobed, thick, ± dull green above, densely and shortly glandular-hairy both sides, without veins raised below. Petiole (and stems) mostly densely nonglandular-hairy. All yr.............*Italian Toadflax* **Cymbalaria pallida**

■Lvs nonglandular-hairy and/or ciliate

Lvs divided into 3-5 finger-like lobes (occ unlobed on sterile shoots)

Lf lobes widely spreading (c75°), 6-10 x 0.5-1.5mm, elliptic-linear, acute or acuminate with aristate point to 0.5mm and an indistinct pale apical pore, sessile or narrowed to a petiole, ± hairless, sparsely ciliate esp nr base with long wavy nonglandular hairs, with obscure veins, with stomata below only. Stems usu with axillary lfy bulbils, hairless or sparsely wavy-hairy (occ glandular), round. Hairs uniseriate. All yr................*Mossy Saxifrage* **Saxifraga hypnoides**

Lvs palmately 5(7)-lobed with crenate lobes

Lvs peltate, all alt. PL 24..*Marsh Pennywort* **Hydrocotyle vulgaris**

Lvs not peltate

Lvs opp below, alt above..*Slender Speedwell* **Veronica filiformis**

Lvs all alt, 0.5-2cm diam, reniform, pale green, septate-hairy at least above, crenate with 7 broad lobes, the lobes v slightly retuse, often with indistinct palmate veins (ending in an obscure hydathode at lobe apices), with stomata both sides. Petiole 0.5-2cm, shortly patent-hairy, round, hardly channelled, with 1 vb. Stems patent-hairy, with central stele. All yr. W Br ....*Cornish Moneywort* **Sibthorpia europaea**

■Lvs hairless, not ciliate

Aquatic or wet habs...(*Ranunculus aquatilis* agg) Go to **RAN-BAT**

Dry walls, shingle beaches. PL 16....................................*Ivy-leaved Toadflax* **Cymbalaria muralis**

Group PQ – *Lvs palmately lobed, hairless (but may be ciliate). Petiole occ hairy. Stems (if present) usu ± erect, never rooting at nodes.* (Astrantia major *may key out here if latex overlooked).* ❶ *Lvs peltate*

■Lvs with bristle-tipped (± spinescent) teeth

Basal lvs 2-8cm diam, dull above, shiny below, with 3-5(7) lobes often ± divided to base; lobes deeply serrate with bristle-tipped teeth to 0.3mm, each tooth with a white sunken hydathode, net-veined, with veins raised above, with stomata below only. Petiole 4-25cm, purplish at sheathing base, channelled, with 3 vb's. All yr. Woods...*Sanicle* **Sanicula europaea**

■Lvs without bristle-tipped teeth

Plant with a corm. Lvs all basal, cordate at base, usu dark green with silver-white marbling above, purplish below, 5-9-angled, crenate, not net-veined, with stomata below only. Petiole long, round, with 3 vb's. (*Cyclamen coum* would key out here but lvs are toothed only and has 1 vb in the petiole)

Petiole densely minutely papillate esp nr lf. Lvs 3-8(14)cm, ovate, with minute scattered papillae above, without protruding hydathodes. Fls in autumn. Sept-May ....*Sowbread* **Cyclamen hederifolium***

Petiole with sparse ± sessile red glands nr lf. Lvs 3-10cm, ovate-orb, hairless above, with protruding hydathodes. Fls in early spring. Oct-May. R alien ....*Spring Sowbread* **Cyclamen repandum**

Plant tuberous, rhizomatous, or tufted

Lvs emerging singly after fl. Basal lf 1, ± orb, deeply (3)5-7-lobed, toothed, with white hydathodes (best viewed end-on), with stomata below only. Stem lvs (bracts) in 1 whorl, 3-foliate, short-stalked to sessile

Plant with at least some hairs (occ sparse and confined to lf margin or petiole)

Rhizomes long, not tuber-like. Fls (and buds on rhizome) white. Basal lvs often ciliate, not purplish below. Petiole usu hairy. Fr stalk drooping. Mar-Jul. Pl 16 ....*Wood Anemone* **Anemone nemorosa**

Rhizomes short, tuber-like. Fls (and buds on rhizome) usu blue. Basal lvs occ purplish below. Petiole hairy to hairless. Fr stalk drooping. Mar-Jul ....*Balkan Anemone* **Anemone blanda**

Plant totally hairless

Plant with long rhizomes. Basal lvs occ absent. Fls (and buds on rhizome) yellow. Mar-Jun ....*Yellow Anemone* **Anemone ranunculoides**

Plant with small round tubers. Basal lvs 1(2), to 11cm diam, orb, palmately 3-5-lobed, the lobes further divided into lobes, with veins indistinct. Petiole to 15cm, round, hollow. Stem lvs (bracts) forming whorl around infl. Fls yellow. Jan-May. Pl 16 ....*Winter Aconite* **Eranthis hyemalis**

Lvs not emerging singly, always present before fl

Petiole with inflated sheathing base, or with ochreae

Lvs rolled when young..(*Caltha palustris*) Go to **RAN**

Lvs revolute when young..(*Rheum palmatum*) Go to **DOCK**

Petiole sheathing base not inflated

Dry to damp habs..(*Ranunculus*) Go to **RAN**

Aquatic or wet habs (lvs usu floating, additional submerged thread-like lvs often present) ....*Water-crowfoots* Go to **RAN-BAT**

Petiole not sheathing at base

Lvs fleshy, peltate. ❶

Petiole hairless, with 4-5 purple vb's. Lvs mostly basal, 1.5-7cm diam, orb, with dimple in centre, crenate with v broad teeth, with obscure veins, with stomata both sides. Stem(s) usu 1, 3-40cm. Tuberous per. Oct-Jun.................................*Navelwort* **Umbilicus rupestris**

Lvs fleshy, usu cordate at base

Lvs to 1.5cm, orb to ivy-shaped. Stem lvs sessile, often clasping. Ann, with fibrous roots. Dry saline habs (inc roadsides)................................*Danish Scurvygrass* **Cochlearia danica**

Lvs thin, cuneate at base

Lvs all on stems. Stems 50-100cm, fetid. Shortly rhizomatous per, often with a pr of swollen taproots..*Monk's-hood* **Aconitum napellus**

Lvs all basal. Stems to 12cm, odourless. Rhizomatous per ....*Moschatel* **Adoxa moschatellina**

Cyclamen hederifolium lf & petiole TS

Group PR – *Lvs palmately or digitately lobed, hairy. Stems (if present) usu ± erect (trailing in* Ecballium elaterium*), never rooting at nodes.* ❶ *Lvs peltate*

■Lvs >15cm (usu much larger), orb, cordate at base. Rhizomatous per. Damp habs

Lvs to 200cm diam, palmately 5-9-lobed, rugose, dull above, weakly soap-scented, hispid with short hairs and spiny projections both sides, denticulate with acute hydathodes, net-veined (veins often red), with stomata below only. Petiole to 150cm, with stout herbaceous spiny projections, round, with many scattered vb's and spiral fibres. Apr-Nov.........*Giant-rhubarb* **Gunnera tinctoria**

Lvs 15-60cm diam, palmately 7(11)-lobed, not rugose, shiny above, odourless, hairless above, with sparse swollen-based hairs below, serrate, net-veined, with stomata both sides. Petiole to 100cm, rough with projections, round, with many scattered vb's and spiral fibres. ❶ ....*Indian-rhubarb* **Darmera peltata**

■Lvs (3)4-16(20)cm

Basal lvs several, usu in a rosette. Petiole sheathing at base

Lvs 8-15cm diam. *A. hupehensis* x *vitifolia*...................*Japanese Anemone* **Anemone** x **hybrida**

Lvs usu <8cm diam....................(*Ranunculus*) Go to **RAN**

Basal lf 1. Petiole not sheathing at base

Rhizomes long, not tuber-like. Fls (and buds on rhizome) white

Basal lvs with adpressed to patent hairs both sides, often ciliate, not purplish below. Petiole usu hairy. Fr stalk drooping. Mar-Jul. PL 16...................*Wood Anemone* **Anemone nemorosa**

Rhizomes short, tuber-like. Fls (and buds on rhizome) usu blue

Basal lvs adpressed-hairy both sides, rarely purplish below. Petiole usu hairy. Fr stalk erect Mar-Jul....................*Blue Anemone* **Anemone apennina**

Basal lvs adpressed-hairy above, soon (±) hairless, occ purplish below. Petiole hairy to hairless. Fr stalk drooping. Mar-Jul....................*Balkan Anemone* **Anemone blanda**

Basal lvs absent. Petiole not or weakly sheathing at base

Stems 50-100cm, erect, fetid. Shortly rhizomatous per, often with a pr of swollen taproots

Lower lvs 5-10cm diam, divided to base, petiolate, minutely hairy to hairless, with weakly crenulate-erose narrow hyaline margins, with stomata below only. Mid-stem lvs to 15cm diam, short-petiolate, 3-5-lobed; lobes deeply divided into linear acute lobes. Petiole to 10cm, weakly sheathing at base, hairless, channelled, hollow or pith-filled, with 4 vb's. Stems minutely hairy or hairless, round, pith-filled or hollow. Mar-Oct ....*Monk's-hood* **Aconitum napellus**

Stems 20-100cm, (±) erect, often aromatic. Ann, with fibrous roots

Lvs 4-16cm, broadly ovate, shallowly to deeply palmately 3-7-lobed (occ pinnately lobed), cordate (rarely cuneate) at base, yellow-green above, often strongly aromatic, with short white stout hairs, occ with sessile yellow glands, entire to ± deeply serrate-dentate, net-veined, with 1-2° veins raised below and slightly raised above, with stomata both sides. Petiole to 5cm, weakly channelled, with several scattered vb's and spiral fibres. Stems occ glandular-hairy. Infl glandular. May-Oct....................*Rough Cocklebur* **Xanthium strumarium**

Stems to 50cm, trailing, odourless. Per, with tuberous roots

Lvs 8-20cm, broadly ovate, shallowly lobed, obtuse, yellow-green (greyish when young), hispid above (occ with glandular hairs), septate-hairy below, net-veined, with stomata both sides. Petiole long, hairy, rough with scattered projections, ± round, not channelled, with c11 vb's around margin. Stems striate, pith-filled. R casual, occ persisting ....*Squirting Cucumber* **Ecballium elaterium**

■Lvs <3cm

Lf teeth without submarginal pores

Basal lvs without auricles at sheathing base. Per to ann....................(*Ranunculus*) Go to **RAN**

Basal lvs with auricles at sheathing base. Ann

Lvs truncate at base, greyish-green. Stems <5cm, stout, vigorous, with short internodes. Basic to ± acidic gsld....................*Parsley-piert* **Aphanes arvensis**

Lvs cuneate at base, mid-green. Stems <5cm, slender, weak, with long internodes. Acidic gsld....................*Slender Parsley-piert* **Aphanes australis**

Lf teeth with submarginal hydathode (pore) above (occ obscure). Petiole not sheathing at base, with 1 vb. Hairs uniseriate

Lvs cordate at base, reniform, with septate hairs (often gland-tipped), with raised cells giving a crystalline appearance, with cartilaginous margins and obscure veins. Petiole long-ciliate

Lowland

Ann (bi). Lvs ± all on stem, 1-2.5cm diam, pale yellow-green, minutely glandular-hairy, with 5(9) shallow lobes. Petiole < lf. Stems to 15cm, minutely glandular-hairy. Apr-Sept (all yr) ....*Celandine Saxifrage* **Saxifraga cymbalaria**

Per. Lvs mostly basal (usu dead by fl), 0.5-3cm diam, dark green (occ reddish), with sparse wavy (mostly nonglandular) hairs, crenate-lobulate with ≥5 lobes, with stomata v sparse or absent above. Petiole >> lf, with wavy hairs (often glandular). Stems 15-30cm, with dense wavy hairs, round, with bulbils in axils of lowest lvs or on short rhizome. Oct-May ....*Meadow Saxifrage* **Saxifraga granulata**

Mtns. Per

Basal lvs forming rosette, (0.5)1-1.8cm diam, whitish below, with long minutely glandular hairs v sparse or absent, with (3)5 shallow obtuse lobes cut ¼ way to base, with stomata scattered above and abundant below. Petiole 1-3cm, not toothed at base. Stem 1, 3-15cm, with many red bulbils in bract axils. Apr-Oct (all yr). Dry calc rocks above 920m. Sch8 ....*Drooping Saxifrage* **Saxifraga cernua**

Basal lvs scarcely forming rosette, 0.4-1.5cm diam, pale green below, ± hairless, with 3-7 obtuse to ± acute lobes, with stomata below only. Petiole 0.7-5cm, with a tooth each side of the dilated base. Stems 1-5, 2-8cm, without bulbils. Apr-Oct (all yr). Wet rocks above 795m ....*Highland Saxifrage* **Saxifraga rivularis**

Lvs not cordate at base, deeply 3-5(7) digitately lobed (finger-like) with broad limb-like petiole

Per, usu cushion-forming with persistent dead lvs at base. All yr. (*Saxifraga hypnoides* roots at the nodes but would otherwise key out here)

Lvs 3-5(7)-lobed, to 15mm, narrowed to indistinct petiole, often reddish; lobes 3-10(12) x 1.5mm, hairless or with sparse long hairs (occ gland-tipped). Petiole with long wavy septate nonglandular cilia. Stems 6-15cm. Plant in ± loose cushion. R, W Ire ....*Irish Saxifrage* **Saxifraga rosacea***

Lvs (0)3(5)-lobed, to 12mm, narrowed to indistinct petiole, pale green; lobes 1-2(10) x 1(2)mm, shortly glandular-hairy to hairless both sides; upper stem lvs entire. Petiole with short patent unicellular cilia (glandular or not). Stems 4-6cm. Plant in dense cushion. Mtns, VR. Sch8 ....*Tufted Saxifrage* **Saxifraga cespitosa**

Ann. Nov-May

Stem lvs 3-5-lobed, reddish or yellow-green, with red-tipped glandular hairs, with obscure veins. Basal lvs soon dying, to 1cm, obovate-spathulate, cuneate at base, entire. Petiole with 3(5) obscure vb's. Stems to 10cm, red, ± glandular-hairy ....*Rue-leaved Saxifrage* **Saxifraga tridactylites**

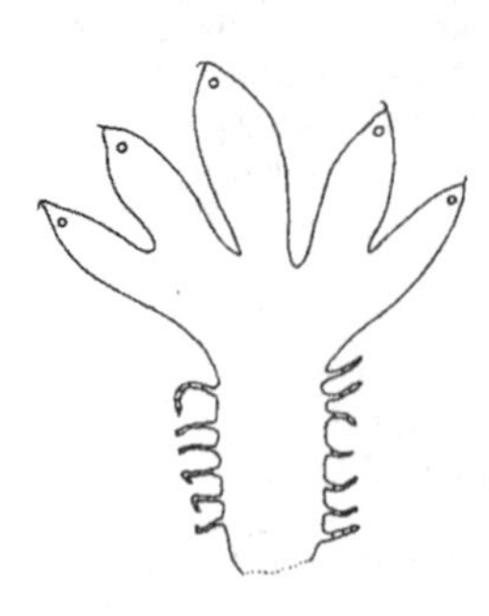

Saxifraga rosacea

Group PS – Brassicaceae. *Lvs pinnately lobed, (±) hairless (but may be ciliate), fetid, not net-veined, with stomata both sides. Stems ± round. Hairs, if present, unicellular*

■Plant low-growing with ± prostrate stems. Basal lvs dying at fl

Basal lvs to 10cm, sparsely ciliate at base, pinnately lobed (to 2-pinnate) with 3-5(7) prs of lobes and long narrow terminal lobe, each lobe with a fragile bristle at apex and indistinct veins. Stem lvs with narrow ± entire lobes. Petiole ± flattened, not channelled, with 1(3) vb's. Stems usu <8cm, hairy, solid. Ann or bi. All yr................................*Lesser Swine-cress* **Coronopus didymus**

■Plant tall with ± erect stems. Basal lvs absent or soon dying

Lowest lvs sinuate-toothed only, becoming pinnately lobed up stem with 1-3 prs of lobes

Ann. Lvs to 10cm, obovate, dull greyish-green, hairless (occ with stiff hairs on midrib above), with sparse stiff cilia, pinnately lobed with 1-2(3) ± free prs of lfts, with stomata both sides. Petiole to 7cm, with sparse stiff hairs. Stems to 100cm, purple at base, with bristly hairs spreading in all directions, round, soon hollow...........................*Garden Rocket* **Eruca vesicaria**

All lvs pinnately lobed

Per. Basal lvs absent. Stem lvs to 12cm, elliptic-spathulate, narrowed to long winged petiole, slightly fleshy, with minutely crenulate hyaline margins, ± waxy glaucous grey-green, sparsely ciliate, with whitish opaque midrib and 2° veins raised below; lobes ± linear, ≥3x as long as wide, entire or with few deep teeth, the teeth often with a fragile bristle to 0.5mm; terminal lobe longer but hardly wider than laterals. Petiole with 1(5) vb's. Stems to 80cm, ± glaucous, purplish at base, tough, hairless, solid. Petals 8-15mm. Apr-Oct
....*Perennial Wall-rocket* **Diplotaxis tenuifolia**

Ann (per). Basal lvs soon dying, to 10cm, elliptic-spathulate, narrowed to long winged petiole, with minutely crenulate hyaline margins, greyish-green, sparsely ciliate, occ toothed only (unlobed), with whitish opaque midrib and 2° veins raised below; lobes triangular, to 2x as long as wide, entire or with a few distant hydathode-teeth, the teeth often with a fragile bristle to 0.5mm; terminal lobe longer but hardly wider than laterals. Petiole with 3(5) vb's. Stems to 50cm, rarely glaucous, purplish at base, tough, usu sparsely hairy below, solid. Petals 4-9mm. Apr-Oct..*Annual Wall-rocket* **Diplotaxis muralis**

Group PT – Brassicaceae, Plantago coronopus. *Lvs pinnately lobed, hairless (occ ciliate), with stomata both sides. Lvs and stems glaucous-pruinose, and/or lvs fleshy*

■Lvs v fleshy, >0.5mm thick, snapping audibly. Coastal

Lvs 20-50cm, strongly glaucous or with waxy bloom. Per

Basal lvs ovate, 1-4mm thick, purplish esp when young, with faint radish odour, strongly sinuate-undulate, minutely denticulate. Petiole with ≥10 scattered vb's. Rhizomes short, thick. Mar-Oct. Beaches and maritime rocks..*Sea-kale* **Crambe maritima**

Basal lvs broadly oblong, 0.5mm thick, often purplish, with strong cabbage odour, sinuate-undulate, usu lyrate-pinnatifid with 2 small prs of basal lobes and v large terminal lobe, toothed with apiculate hydathodes, with veins hardly raised below. Upper lvs entire, sessile or partly clasping. Petiole dilated and ± clasping at base, with 10 vb's (1 vb lower than the others). Tufted, with thick woody stock (with lf scars and ann rings). All yr. Sea cliffs, or garden escape ....*Cabbage* **Brassica oleracea**

Lvs 3-10cm, not glaucous

Bi or per. Lvs all basal, to 12cm, narrowed to indistinct petiole, pitted both sides, odourless, with long silky hairs at extreme base, usu with white septate cilia, pinnately lobed to entire, often with c3 teeth per side in distal ½ (with purplish hydathodes), with translucent midrib not sunken above. Petiole with 3 vb's. All yr...............................*Buck's-horn Plantain* **Plantago coronopus**

Ann. Lvs all on stem, to 10(15)cm, oblanc, narrowed to indistinct petiole (the upper sessile), not pitted, strongly cress-scented, deeply pinnately lobed to ± entire, often with translucent midrib, with 2° veins obscure. Petiole with 5 vb's. May-Oct (all yr). Sandy or shingle beaches ....*Sea Rocket* **Cakile maritima**

■Lvs not v fleshy, usu <0.5mm thick, rarely snapping audibly

Stem lvs clasping with rounded auricles

Basal (and stem) lvs v glaucous

Basal lvs to 30cm, soon dying, rarely sparsely hispid, with (0)1-4(6) prs lobes. Stem lvs pruinose, hairless, unlobed. Petiole with 7(9) vb's. Stems to 130cm, pruinose or glaucous above, round. Ann or bi (per), occ with swollen taproot. All yr................*Rape* **Brassica napus**

Basal lvs usu green but stem lvs glaucous (at least underneath)

Basal lvs to 30cm, soon dying, sparsely hispid esp on veins below, sparsely ciliate, lyrate-pinnatifid, with c3-4 prs surcurrent lobes. Stem lvs sessile, hairless to sparsely hispid on veins below, the uppermost toothed to entire. Petiole with 3(7) vb's. Stems to 100cm, often purplish at least nr base, pruinose to ± glaucous above, not or hardly ridged. Ann (bi), occ with swollen taproot..*Turnip* **Brassica rapa**

Stem lvs not clasping, without auricles

Basal lvs in false rosette, soon dying, petiolate, mid-green, occ sparsely hairy, ciliate, lyrate with toothed ovate to linear-lanc lobes. Stem lvs sessile, 1-2-pinnately lobed, often glaucous. Stem 1, 20-40cm, pruinose, hairless, round. Ann. Apr-Oct...............*Garden Cress* **Lepidium sativum**

Group PU – *Mostly* Brassicaceae. *Lvs pinnately lobed, hairless (but occ ciliate), not or slightly fleshy. Lvs and stems green (occ reddish), never glaucous. (*Senecio aquaticus *and* Leucanthemum *spp may key out here.* Reseda *spp key out here if stipoid glands overlooked)*

■Petiole (occ indistinct) sheathing at base

Lvs not clasping at base..(*Ranunculus*) Go to **RAN**

Lvs clasping at base, mostly basal, 2-5cm, sessile or with an indistinct petiole, slightly fleshy, weakly aromatic (tansy-scented), rarely hairy, deeply toothed or irreg 1-2(3)-pinnately lobed (esp distally) with long narrow lobes, rarely entire, teeth tipped with a minute fragile bristle, with weakly translucent midrib (other veins obscure), with stomata both sides. Stem usu 1, 5-20cm, erect to prostrate, branched, round, hairless. Ann (per). Damp, esp saline, habs ....*Buttonweed* **Cotula coronopifolia**

■Petiole not sheathing at base (may be dilated)

Basal lvs with large broadly ovate or ± orb terminal lobe, entire or sinuate-toothed, shiny, with stomata both sides. Stem lvs ± clasping with auricles at base. Petiole solid, with 1(3) vb's. Stems 10-90cm, v tough, often purplish at base, ± hairless, ridged or furrowed, solid. Usu bi. All yr. (*Rorippa palustris* may key out here)

Basal lvs with 1-3 prs of small lateral lobes and ovate-oblong non-cordate terminal lobe longer than rest of lf, the terminal pr shorter than width of terminal lobe, yellow-green. Stem lvs with 0-2 prs lobes, becoming entire nr infl. Fl buds hairy ....*Small-flowered Winter-cress* **Barbarea stricta**

Basal lvs with 1-2(3) prs of oblong lateral lobes and orb (often) cordate terminal lobe, the terminal pr ± equal to width of terminal lobe, retuse, dark green, with water-cress taste. Stem lvs with (0)4-6 prs lobes, the uppermost shallowly lobed or toothed. Fl buds hairless ....*Winter-cress* **Barbarea vulgaris**

Basal lvs with 3-5(6) prs of lateral lobes, the terminal pr > width of ovate-cordate terminal lobe, retuse, dark green, with v mild taste. Stem lvs all deeply pinnately lobed with long narrow lateral lobes and a ± larger terminal lobe, the uppermost pinnately lobed with >2 prs of lobes. Fl buds hairless...*Medium-flowered Winter-cress* **Barbarea intermedia**

Basal lvs with 6-10 prs of lateral lobes, the terminal pr ≥ width of the ovate-cordate terminal lobe, entire or sinuate-toothed, dark green, with strong water-cress taste. Stem lvs all deeply pinnately lobed with 5-8 prs of long narrow lobes, the uppermost pinnately lobed with >2 prs of lobes. Fl buds hairless..*American Winter-cress* **Barbarea verna**

Basal lvs, if present, without large round terminal lobe

Per, with stout rootstock or rhizomes (occ slender)

Lvs with cress odour when crushed

Rhizomes slender, white. Stems solid. Damp to dry habs

Basal lvs soon dying, to 10(15)cm. Lower lvs petiolate, not or hardly clasping, usu without auricles, often shiny, pinnate to deeply pinnately lobed with 3-4 prs of small oblong or lanc acute lobes, ciliate, usu toothed or lobed, with the midrib and 2° veins crenulate below, with stomata both sides. Upper lvs sessile, usu pinnately lobed with narrow lobes (occ ± entire). Petiole not channelled, with 3 vb's. Stems to 50cm, purple at base, hairless or minutely hispid below, angled or ridged. Feb-Oct (all yr) ....*Creeping Yellow-cress* **Rorippa sylvestris**

Rhizomes stout, brown. Stems solid. Ruderal, VR alien

Basal (and lower) lvs persistent, to 9-15 x 6cm, ovate, obtuse, petiolate, thickish, hairless, with 0-4 prs of lateral lobes and an ovate to hastate terminal lobe, irreg deeply toothed. Upper lvs lanc, entire. Petiole to 3cm, channelled. Stems to 75cm, hairless (occ sparsely hairy below), round. All yr.....................................*Russian Mustard* **Sisymbrium volgense**

Rhizomes absent. Stems hollow. Wet habs

Basal (and lowest) lvs soon dying, to 25cm, pinnately lobed, with 1-3(7) prs of usu acute lobes nr base (rarely along length), dentate. Stem lvs broadly ovate to narrowly elliptic-oblong, cuneate, occ auriculate and clasping at base (auricles occ ciliate), bright or yellowish green, sinuate to lobed, with stomata both sides. Upper stem lvs shortly petiolate to sessile, often sparsely hairy both sides. Petiole channelled, with 1(3) vb's. Stems to 120cm, often rooting at lowest nodes, purplish at base, round below, ridged and hairy above..*Great Yellow-cress* **Rorippa amphibia**

Lvs with meaty odour when crushed..(*Solanaceae*) Go to **SOL**

Lvs odourless when crushed

Lvs 30-100cm, occ with basal pr of small lobes....................*Bear's-breech* **Acanthus mollis**

Lvs 5-20cm..*Saw-wort* **Serratula tinctoria***

Ann (occ bi in *Lepidium ruderale*), with slender taproot or fibrous roots

Stems prostrate. Lvs with cress odour when crushed

Basal lvs dying at fl, to 10(20)cm, ± fleshy, dull grey- or bluish-green, sparsely ciliate at base, pinnately lobed, each lobe with short lobes esp on distal side. Upper stem lvs narrower, ± entire. Petiole ± flattened, unchannelled, with 3 vb's. Stems to 25cm. Mar-Oct ....*Swine-cress* **Coronopus squamatus**

Basal lvs usu persisting, to 4cm, often ± fleshy and snappable, shiny dark green, with 2-4 prs of free lobes, with stomata both sides. Stems to 15cm, angled, solid. Apr-Sept ....*Northern Yellow-cress* **Rorippa islandica**

Stems erect or ± so

Lvs odourless when crushed. Basal lvs soon dying..........*Love-in-a-mist* **Nigella damascena**

Lvs with meaty odour when crushed. Basal lvs absent........(*Solanum laciniatum*) Go to **SOL**

Lvs with cress odour when crushed. Basal lvs usu dying at fl

Damp habs (occ ruderal)

Basal (and lower) lvs (4)6-15cm, petiolate, green, lyrate-pinnatifid with an ovate lobed rounded terminal lobe and 1-6 prs of lobes (occ with intercalary lfts), the lobes narrow and irreg sinuate-toothed. Upper stem lvs short-petiolate or sessile, with ciliate auricles. Petiole with 1(3) vb's. Stems 5-20cm, hairless or sparsely hairy below, angled, usu hollow. Mar-Oct (all yr). Usu damp habs...............*Marsh Yellow-cress* **Rorippa palustris**

Basal lvs to 4cm, often ± fleshy and snappable, shiny dark green, with 2-4 prs of free lobes, with stomata both sides. Stems to 15cm, angled, solid. Apr-Sept ....*Northern Yellow-cress* **Rorippa islandica**

Dry habs

Basal lvs 2-7cm, petiolate, often reddish-brown, pinnately divided into 4-5 prs of narrow lobes (the lobes themselves lobed), midrib visible. Lower lvs pinnately lobed into narrow entire acute lobes. Mid and upper lvs sessile, without auricles, entire. Petiole flattened, with 3 vb's. Stem(s) 10-25cm, 1.5mm diam, branched above, purplish at base, often with minute spreading hairs, round, solid, with stomata. All yr. Ruderal, often nr coast ....*Narrow-leaved Pepperwort* **Lepidium ruderale**

Basal lvs to 3cm, petiolate, pinnately lobed, with small elliptic lobes. Stem lvs numerous, sessile. Stems 2-10cm, hairless or with sparse stellate hairs (mostly 3-rayed). Oct-May. Dry calc habs, R...*Hutchinsia* **Hornungia petraea**

Group PV – Brassicaceae. *Lvs pinnately lobed, unicellular-hairy, with cress, mustard or cabbage odour when crushed, with stomata both sides. Stems pith-filled. Basal rosette present (occ dead by fl). (*Bunias orientalis *may key out here)*

■Basal and/or stem lvs ± glaucous at least below, with ± free lobes

Lf margin with any cilia confined to teeth or lobe apices (cilia soon falling)

Basal lvs persisting, to 15cm, dark glaucous-green above, ± fleshy, usu sparsely hairy on rachis only, with 4-5 prs lobes (occ reflexed); lobes deeply toothed or lobed. Stem lvs 0-2. Petiole sparsely hairy, channelled, with 3 vb's. Stems to 50cm, purplish at base, pruinose, (±) hairless, round. Bi. All yr. Dunes, W Br..............*Isle of Man Cabbage* **Coincya monensis** ssp **monensis**

Basal (and lower) lvs often persisting, to 10cm, green to ± glaucous above, not fleshy, sparsely hispid both sides, with 3-5 prs of lobes (occ reflexed); lobes deeply toothed or lobed. Petiole retrorsely hairy, channelled, with 3 vb's. Stems to 50cm, purplish at base, ± glaucous above, ± hispid with scattered spreading or retrorse hairs below, ± hairless above, round. Bi (ann). All yr. Ruderal...*Wallflower Cabbage* **Coincya monensis** ssp **cheiranthos**

Lf margin without cilia confined to teeth (cilia may be present or absent)

Basal (and stem) lvs v glaucous

Basal lvs soon dying, to 30cm, rarely sparsely hispid, with (0)1-4(6) prs of lobes. Stem lvs clasping with rounded auricles, pruinose, hairless, unlobed. Petiole with 7(9) vb's. Stems to 130cm, pruinose or glaucous above, round. Ann or bi (per), occ with swollen taproot. All yr ....*Rape* **Brassica napus**

Basal lvs usu green but stem lvs glaucous at least beneath

Basal lvs soon dying, to 30cm, sparsely hispid esp on veins below, sparsely ciliate, lyrate-pinnatifid, with c3-4 prs of surcurrent lobes. Stem lvs clasping with rounded auricles, hairless to sparsely hispid on veins below, the uppermost toothed to entire. Petiole with 3(7) vb's. Stems to 100cm, often purplish at least nr base, pruinose to ± glaucous above, ± round. Ann (bi), occ with swollen taproot..*Turnip* **Brassica rapa**

Basal lvs in persistent rosette, to 45cm, v hairy above with ascending hairs, lyrate-pinnatifid, with 0-2 prs small of lateral lobes (often runcinate) and large obtuse terminal lobe. Stem lvs not clasping, with 1-5 prs lobes, the uppermost glaucous, linear, entire. Petiole with 7(9) vb's. Stems to 130cm, woody at base, purplish nr base, pruinose above, retrorsely or adpressed-hairy. Bi (per). All yr. Lundy Is. Sch8....................................*Lundy Cabbage* **Coincya wrightii**

■Basal and stem lvs green, with ± free lobes

Basal lvs with a truncate (occ round) terminal lobe, usu persisting to fl

Basal lvs 5-15cm, with (1)3-5 prs of surcurrent lobes, ± toothed, sparsely ± bristly-hairy. Stem lvs with a long hastate terminal lobe and 1-3 prs of lateral lobes, the upper lvs often entire. Petiole often purplish, hairy, with (3)5 vb's. Stems to 100cm, purplish at base, occ pruinose, usu bristly with retrorse hairs (hairs often swollen-based), weakly ridged or striate. Ann or bi. Often all yr...*Hedge Mustard* **Sisymbrium officinale**

Basal lvs with a hastate terminal lobe, rarely persisting (usu dead by fl)

Basal lvs 5-15cm, grey-green, with (1)2-4 prs of broadly triangular reflexed lobes, shortly softly patent-hairy both sides. Upper lvs with a linear-oblanc terminal lobe and 0-1(2) prs of narrow lateral lobes. Petiole purplish nr base, channelled, solid, with 9 vb's. Stems to 80cm, retrorsely hairy, round. Ovary (and young frs) with sparse patent hairs nr base. Ann. Often all yr. PL 19 ....*Eastern Rocket* **Sisymbrium orientale**

Basal lvs 5-15cm, bright green, with 2-6 prs of narrow lobes (lobes reflexed when young), hairless or sparsely hairy on flanges (densely hairy in Ire), ciliate. Upper lvs with a linear-oblanc terminal lobe and 0-3 prs of narrow lateral lobes. Petiole not purplish, channelled, hollow, with 7(9) vb's. Stems to 60(80)cm, hairless or with short adpressed hairs, round. Ovary (and young frs) hairless. Ann...*London-rocket* **Sisymbrium irio**

Basal lvs with a large or broad terminal lobe, persisting to fl

Basal lvs to 60cm, bright green, lyrate-pinnatifid, with 1-8(11) prs of lobes often overlapping (occ alt in size, or basal reflexed and v small), ± sharply hispid with stout unequal 0.2-1mm hairs, toothed with prominent green hydathodes. Petiole channelled, with 5(7) vb's. Stems to 100cm, pruinose or glaucous above, with spreading or retrorse hispid hairs (esp below). All yr

Bi or per. Coastal....................................*Sea Radish* **Raphanus raphanistrum** ssp **maritimus**
Ann..*Wild Radish* **Raphanus raphanistrum** ssp **raphanistrum**
Basal lvs to 35cm, grey-green, lyrate-pinnatifid, with 1-5(9) prs of lateral lobes never overlapping, shortly hairy above with hairs <0.5mm. Petiole channelled, with 5(7) vb's. Stems to 120cm, glaucous above, retrorsely shortly hispid at least nr base. Per (ann). Oct-Sept ....*Hoary Mustard* **Hirschfeldia incana***
Basal lvs with a large terminal lobe; occ persisting (usu dead by fl)
Basal (and lower) lvs dark green, usu hairless above, sparsely bristly below, sparsely ciliate, with c3prs of small laterals, irreg toothed. Petiole with W-shaped groove above, with 5 vb's. Stems to 110cm, ± glaucous, with stiff hairs (occ retrorse) at least below. Ann or bi. Often all yr ....*Bastard Cabbage* **Rapistrum rugosum**
Basal lvs with narrow terminal lobe; not persisting (dead by fl)
Aquatic. Per...*Great Yellow-cress* **Rorippa amphibia**
Dry habs. Ann. Basal lvs to 25cm, roughly hairy, rucinate-pinnatifid with 6-8 narrow triangular distantly toothed lobes. Stem lvs pinnately lobed with narrow distantly toothed or hastate lobes, the uppermost sessile and deeply lobed with linear or thread-like entire lobes; lobes ≤1mm wide, acute, v slightly undulate, with 2° veins obscure. Stems 20-60cm, to 2mm diam, hairy below, usu hairless and pruinose above, round.................*Tall Rocket* **Sisymbrium altissimum**
■Basal and stem lvs green, without free lobes (often deeply lobed)
Lvs 7-50cm
Basal (and lower) lvs to 50cm, elliptic, mid-green, often with 0-2 prs of ± free lfts nr base. Petiole channelled, with 5 or 7 vb's. Stems 30-100cm, retrorsely hairy esp at base (sparse above), occ hairless or with sparse glandular hairs. Per (bi). Mar-Nov. R alien ....*Warty-cabbage* **Bunias orientalis**
Basal lvs to 30 x 6cm, lanc, mid-green, hairy both sides, occ lobed nr base. Petiole not channelled, with 7(9) vb's. Stems 60-120cm, (±) hairless. Bi or per. All yr *Woad* **Isatis tinctoria***
Lvs 2-5cm
Stems with ± patent 0.3-0.5mm hairs, branched below middle, to 50cm. Basal (and lower stem) lvs to 3cm, oblong to lanc, grey-green. Petiole channelled, with 1(3) vb's. Per (bi). All yr ....*Smith's Pepperwort* **Lepidium heterophyllum**
Stems with slightly retrorse <0.3mm hairs, branched only above middle, to 40cm. Basal (and lower stem) lvs to 5cm, oblong to lanc, mid- to grey-green. Petiole channelled, with 1(3) vb's. Ann (bi)...*Field Pepperwort* **Lepidium campestre**
Lvs to 2(5)cm
Lvs all basal, up to 50 in a rosette, ± fleshy, often purplish, sparsely hairy above, ± hairless below, lyrate-pinnatifid with a few short round lateral lobes and a broader often 3-lobed terminal lobe, all veins obscure. Petiole > lf, winged at base. Stems 2-10cm. Ann. Nov-Jul ....*Shepherd's Cress* **Teesdalia nudicaulis**

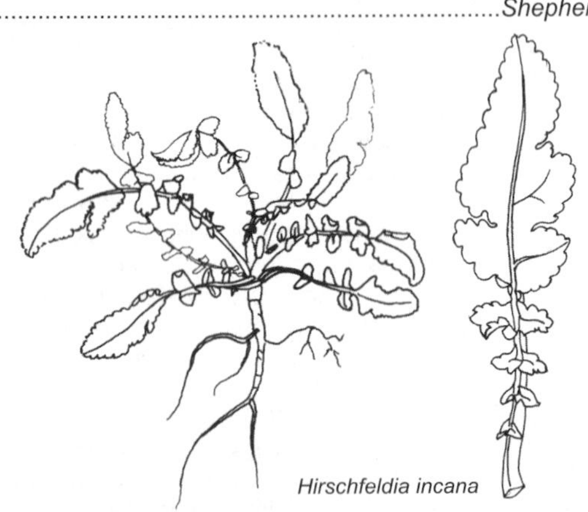

Hirschfeldia incana

Group PW – Brassicaceae. *Basal rosette never formed (basal or lower lvs may form a weak false rosette). Lvs pinnately lobed, unicellular-hairy, with cress, mustard or cabbage odour when crushed, with stomata both sides. Ann (bi)*

■Lvs v glaucous, with (0)1-4(6) prs of lobes..*Rape* **Brassica napus**

■Lvs green

Lower lvs usu with 3-8 prs of lobes. Ann (bi)

Basal lf rarely present. Lower lvs to 25cm, lyrate with patent or forward-pointing lobes (not free), occ 1-2 prs of free lobes, hairy. Mid-stem lvs deeply pinnately lobed with ± distant oblong lobes, the basal lobes not clasping stem, sparsely hairy both sides (often ± hairless above). Petiole with 3(5) vb's. Stems 20-60cm, with short white stiff retrorse or adpressed hairs below, pith-filled. Mar-Oct..*Hairy Rocket* **Erucastrum gallicum**

Lower lvs with 1-3 prs of lobes. Ann

Lvs to 2cm

Basal lvs usu absent. Lvs spathulate to linear-lanc, ± acute, narrowed to base, ± hairless, white-ciliate, with 1-2 lobe-like teeth per side nr apex, with all veins obscure, with v bitter taste, with stomata both sides. Stems 5-30cm, shortly hairy (often looking like papillae), weakly angled, solid. Apr-Oct. Bare calc habs..*Wild Candytuft* **Iberis amara**

Lvs 5-20cm

False rosette lvs with (0)1-2 prs of ± free lfts

Lvs usu stiffly hairy both sides (occ hairless), toothed, with midrib hollow at maturity. Petiole with 3(5) vb's. Stems 20-120cm, occ pruinose above, hairless or bristly retrorsely hairy below, ridged esp above, pith-filled. Often all yr...........................*Charlock* **Sinapis arvensis**

False rosette lvs with 3 prs of free lfts

Lvs hairless or with a few v sparse bristles, ± undulate, lyrate-pinnatifid or pinnate, with terminal lobe larger than laterals. First lvs deeply lobed (unlobed in *Brassica rapa*, *Sinapis arvensis*, etc). Petiole with 5(7) vb's. Stems 20-100cm, purplish at base, not pruinose above, usu with stiff weakly retrorse (± patent) hairs, hollow.................*White Mustard* **Sinapis alba**

False rosette lvs with 1-2(3) prs lfts

Lvs hispid to hairless, v thin, lyrate-pinnatifid with a large terminal lobe not divided to midrib (unlike *Sinapis alba*); lobes not decurrent; upper stem lvs with 0-1 pr of lobes, the uppermost lvs lanc or narrowly elliptic. Petiole with 7 vb's around hollow. Stems 40-100cm, glaucous-pruinose above, usu with sparse patent to slightly retrorse hispid hairs below, round, hollow. Mar-Oct..*Black Mustard* **Brassica nigra**

Group PX – *Lvs pinnately lobed, hairy, without cress odour when crushed (may have other odour). Plant usu with septate hairs (but unicellular in* Artemisia campestris, Consolida ajacis *and some* Solanaceae*). (*Conyza *spp key out here if lvs appear lobed)*

■Plant with glandular hairs and/or strong-smelling

Per or bi (rarely ann). Plant strong-smelling

Basal lvs 15-20cm, fetid, oblong-ovate, irreg lobed to ± entire, softly nonglandular-hairy esp on veins both sides, with c6 prs of whitish anastomosing 2° veins raised above, with stomata both sides. Stem lvs oblong, clasping at base, glandular-hairy. Petiole with flat arc vb and 2 rib bundles. Stems to 80cm, stout, ± woody at base, glandular-hairy above, round. Bi (ann). All yr ....*Henbane* **Hyoscyamus niger**

Basal/lower lvs 3-8cm, tansy-scented, pinnate with 1-2(5) prs of deep lobes, yellow-green, with spreading or adpressed hairs both sides or ± hairless, minutely ciliate, with main veins raised both sides; lobes 2.5-3cm, the ultimate lobes c6 x 6mm, obtuse but apiculate with hydathodes. Petiole dilated at slightly sheathing base, hairy, ± triangular, not channelled, with flat arc vb and 2 rib bundles. Stems 20-60cm, sparsely hairy, round but ridged. Per. All yr. PL 19 ....*Feverfew* **Tanacetum parthenium**

Bi. Plant odourless

Lvs to 10cm, oblong, the basal/lower shortly petiolate, the upper sessile and pinnately lobed, ± rugose, shortly glandular-hairy or hairless but with sparse minute sessile glands, occ with dendritic or forked hairs, crenate-serrate or dentate, occ lobed, weakly net-veined, with whitish 1-4° veins raised below, with stomata both sides. Petiole to 2cm, with flat arc or involute vb. Stem usu 1, 30-120cm, erect, sparsely glandular above, angled. Infl glandular ....*Moth Mullein* **Verbascum blattaria**

Ann. Plant strong-smelling or odourless

Lvs deeply divided into narrow-linear lobes. Basal lvs long-petiolate or absent

Stem lvs 1-5cm, ± orb, deeply divided into numerous narrowly linear acute lobes (often appearing palmate), hairy. Stem usu 1, to 80cm, erect, ± glandular-hairy above, the hairs crisped with swollen yellow glandular bases..........*Larkspur* **Consolida ajacis**

Lvs not deeply divided into narrow lobes. Basal lvs absent

Stems decumbent to ascending. Lvs weakly fetid when crushed........(*Solanaceae*) Go to **SOL**

Stems (±) erect

Lvs 4-16cm, broadly ovate, cordate to cuneate at base, yellow-green above, often strongly aromatic, with short stout white hairs, occ with sessile yellow glands, net-veined, with 1-2° veins raised below and slightly so above, with stomata both sides. Petiole to 5cm, weakly channelled, with several scattered vb's and spiral fibres. Stems 20-100cm, occ glandular-hairy. Infl glandular. May-Oct..........*Rough Cocklebur* **Xanthium strumarium**

Lvs to 5cm, elliptic-ovate, cuneate at base, deeply 3-5-lobed (often oak-like), weakly musky-aromatic, with short stout hairs, usu with sessile glands. Petiole to 3cm, channelled. Stems 10-80cm, reddish at base, with thick short hairs, usu glandular nr lfy inf ....*Schrader's Goosefoot* **Chenopodium schraderianum**

■Plant with nonglandular hairs, never strong-smelling

Lf teeth bristle-tipped

Lvs 20-60cm. Petiole to 25cm, ± round, hairless, with 1 large vb and 2 rib bundles

Lvs shiny dark green above, occ weakly marbled, paler below, sparsely hairy on veins both sides, shortly ciliate, teeth tipped with 0.5mm spine-like bristles, not net-veined, with stomata below only. Per. Mar-Oct. R alien, hortal..........*Spiny Bear's-breech* **Acanthus spinosus**

Lvs 5-20cm

Petiole channelled, ± hairless, with 3(7) vb's. Basal and lower lvs 5-20cm, rolled when young, dark green (with purplish teeth), hairless or sparsely hairy (tips soon withering), ciliate, toothed to deeply lyrate-pinnatifid with narrow lateral lobes and a larger ± narrow elliptic terminal lobe, net-veined, with anastomosing 2° veins occ slightly raised below, with stomata both sides. Stems to 70cm, hairless, tough, grooved. Per. Apr-Oct ....*Saw-wort* **Serratula tinctoria***

Petiole not channelled, sparsely hairy, with 5 vb's. Basal and lower lvs 5-10cm, sparsely hairy (tips soon withering), with entire to deeply toothed lobes. Stem lvs sessile to ± clasping, toothed to entire, not net-veined, with stomata both sides. Stems to 100cm, ± hairless or sparsely hairy, weakly grooved, with pale cartilaginous ridges. Fl buds (actually phyllaries) spiny. Bi, with stout taproot. All yr. VR alien..................*Red Star-thistle* **Centaurea calcitrapa**

Lvs 1-3(5)cm. Often on walls..................................*Mexican Fleabane* **Erigeron karvinskianus**

Lf teeth not bristle-tipped (lvs may be ciliate)

Lvs glaucous

Shingle beaches. Bi. Basal lvs to 30cm, petiolate, fleshy, unwettable, with stout white hairs, ± lyrate, with 7-9 lobes increasing in size along lf, with stomata both sides; lobes further lobed or coarsely toothed. Upper lvs sessile, ± clasping. Petiole with reddish dilated base, flat above, with 7-10 vb's. Stems to 90cm, hairless. Taproot orange. All yr ....*Yellow Horned-poppy* **Glaucium flavum**

Arable. Ann. Basal lvs absent. Lowest lvs 2-10 x 2cm, narrowed to indistinct winged petiole, often ± fleshy, hairless exc for sparse short hairs on midrib both sides, toothed nr base, lobed distally, with green hydathodes often upturned, with obvious midrib and opaque 2-3° veins often raised above, with stomata both sides. Mid-stem lvs sessile, ± clasping at base. Upper lvs toothed only. Stem usu 1, 20-60cm, usu branched, usu hairless, round, striate, pith-filled or hollow. Mar-Oct (all yr)..*Corn Marigold* **Glebionis segetum**

Lvs not glaucous

Woody-based per

Basal and lower lvs 4-12cm, 2-3-pinnate, long-petiolate, clasping at base. Upper lvs shortly petiolate, less divided, the uppermost lvs sessile, linear, entire. Lf lobes 0.5-1mm wide, linear, acute or mucronate, sparsely silky-hairy both sides, soon hairless. Stems to 60cm, tough, branched, reddish-brown, ± hairless, ± round. All yr. E Anglian Brecks. Sch8 ....*Field Wormwood* **Artemisia campestris** ssp **campestris**

Herbaceous per (or bi in *Plantago coronopus*)

Damp hths

Basal lvs to 3cm, revolute when young, with strongly recurved margins at maturity, limp, with weak radish or cucumber/dill scent, often reddish, with deeply dentate lobes to 4mm, with abundant sap, with veins ± obscure, with stomata below only. Petiole 5-sided, weakly channelled, with (1)3 vb's. Stems >1, decumbent, densely branched, often sparsely hairy esp in lf axils, brittle, often purple, ± grooved or 4-5 angled, solid. Rootstock stout. Often all yr. Damp hths...*Lousewort* **Pedicularis sylvatica**

Dry habs (occ saltmarshes in *Plantago coronopus*)

Basal lvs 30-100cm, shiny dark green above, ± hairless above, shortly hairy on veins below, ciliate, lobed, occ with basal pr of small lobes, with hydathode-teeth, with stomata below only. Petiole to 50cm, stout, minutely or sparsely hairy, round, slightly triangular at base, with 1 round vb...*Bear's-breech* **Acanthus mollis**

Basal (and lower stem) lvs 10-25cm, often shiny green above, hairy both sides, 1-2-pinnately lobed (first lvs entire or toothed at base), with purple hydathodes, with stomata both sides; uppermost lvs ± entire. Petiole with 5(7) vb's, often with sparse orange sap. Stems 30-90cm, erect, usu branched above, ± hairy, grooved. Feb-Oct. PL 18 ....*Greater Knapweed* **Centaurea scabiosa**

Basal lvs 3-12cm, mid-green and pitted both sides, hairless to hairy, with long silky hairs at extreme base, usu with white septate cilia, lobed to entire, with c3 teeth per side in distal ½ (with purplish hydathodes), with translucent midrib visible below (not sunken above). Petiole indistinct, with 3 vb's. Bi to per. All yr.....*Buck's-horn Plantain* **Plantago coronopus**

Basal lvs to 5cm, dark green above, occ ± fleshy, sparsely hairy both sides, with 5-10 teeth per side. Stem lvs to 5(8)cm, oblong, ± clasping and pinnately fringed at sessile base, toothed to pinnately lobed. Petiole to 5cm, pinnately fringed (at least on upper lvs) at dilated base, often winged, not channelled, with 3 vb's. Stems to 70cm, purplish, hairy (tips soon withering), 5-angled. All yr. PL 18...................*Oxeye Daisy* **Leucanthemum vulgare***

Ann (occ bi in *Pedicularis palustris*)

Lvs with weakly fetid odour..(*Solanaceae*) Go to **SOL**

Lvs with weak radish or cucumber/dill scent

Basal lvs to 2-4(8)cm, revolute when young, with strongly recurved margins at maturity, limp, with weak radish or cucumber/dill scent, often reddish, with deeply dentate lobes to 15mm, with abundant sap, with veins ± obscure, with stomata below only. Petiole sparsely hairy, 5-sided, weakly channelled, with (1)3 vb's. Stem 1, erect, loosely branched, often sparsely hairy esp in lf axils or stem grooves, brittle, often purple, ± grooved or 4-5 angled, solid. Taprooted or with fibrous roots. Apr-Oct. Bogs, fens ....*Marsh Lousewort* **Pedicularis palustris**

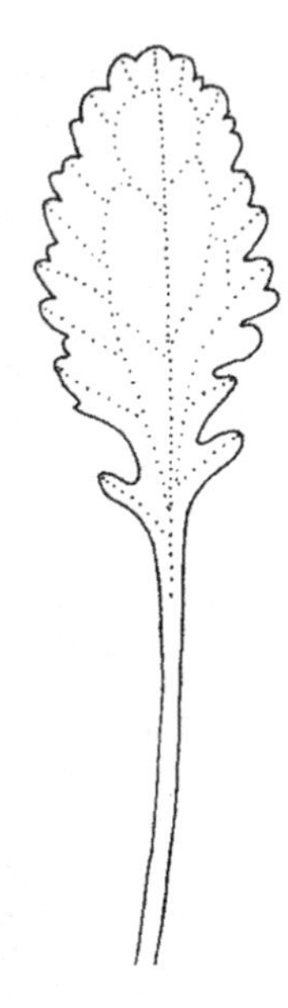

Leucanthemum vulgare

Serratula tinctoria

Key to Groups in Division Q

(Lvs lobed, opp)

■Tree or shrub.......... **QA**
■Woody climber.......... **QB**
■Herb, never climbing
  Stems rooting at least at lower nodes.......... **QC**
  Stems not rooting at nodes, usu erect or absent
    Stipules present..........(*Geranium*) Go to **GER**
    Stipules absent
      Lvs strongly aromatic or fetid when crushed.......... **QD**
      Lvs not or weakly odorous when crushed (stems may be aromatic when broken)
        Lvs with midrib bristly below (*Dipsacus*) .... **QE**
        Lvs without midrib bristly below
          Lvs strongly 5-9-pli-veined..........(*Mimulus*) Go to **MIM**
          Lvs pinnate-veined
            Lvs (and stems) hairless, exc for sessile glands, never ciliate (*Scrophularia auriculata*) .... **QF**
            Lvs (or stems) with at least some hairs or cilia (occ confined to lf base).......... **QG**

Group QA – *Deciduous tree or shrub. Lvs opp, palmately lobed, plicate when young (folded in* Viburnum*), net-veined, with 2° veins raised below, with stomata below only. Twigs sympodial; lf scars narrow and with 3 traces (exc* Paulownia *and* Catalpa*). Buds with opp scales.* ❶ *Buds superposed.* ❷ *Buds often collateral*

■Lvs mostly 20-40cm. Petiole round, without latex

Petiole (and twigs) minutely glandular-hairy. Lvs occ 3-whorled, broadly ovate to ± orb, shallowly 3-lobed to entire, acute, cordate at base, softly septate-hairy both sides (hairs minutely gland-tipped), with small cup-shaped glands above at petiole join. Twigs stout, glandular-hairy when young, without interpetiolar ridge, with chambered pith; lf scars circular. Buds glandular-hairy. ❶ ....*Foxglove-tree* **Paulownia tomentosa**

Petiole minutely nonglandular-hairy to hairless, round. Lvs mostly 3-whorled (occ opp), broadly ovate to ± orb, shallowly 3-lobed to entire, acute, rounded to ± cordate at base, hairless above (exc on veins), softly unicellular- or obscurely septate-hairy below, weakly odorous, with minute spherical glands above at petiole join. Twigs stout, ± hairless, without interpetiolar ridge, with solid pith; lf scars circular, with many traces in a ring........*Indian Bean-tree* **Catalpa bignonioides**

■Lvs mostly <20cm

Petiole channelled, with 1(3) prs disc-like glands (extra-floral nectaries) nr apex. Stipules to 1cm, subulate. Buds with 1 pr of hairless scales

Lvs 5-8cm, 3-lobed to 4/5 way, truncate at base, dull or shiny green above, later reddish, hairless above, usu minutely hairy below, ciliate, irreg dentate. Petiole 1-3cm. Twigs greyish or reddish, hairless, angled when young, pith-filled. Buds to 6mm, flattened-globose, adpressed or spreading, greenish, hairless. Shrub. Pl 19........................*Guelder-rose* **Viburnum opulus***

Petiole round, without glands, swollen at base. Stipules absent. Buds with several prs of ciliate scales

Petiole with white latex

Large tree. Lvs 5-13cm, 5(7)-lobed to ⅓ way, cordate at base

Lvs bright green above, hairless (exc in vein axils below); lobes triangular-acuminate, sinuate-dentate with few large acuminate teeth. Petiole 5-10(20)cm, often red, hairless or minutely hairy at base. Twigs olive-brown. Buds 4-10mm, reddish. Pl 15 ....*Norway Maple* **Acer platanoides**

Lvs dark green above, hairless (exc in vein axils below); lobes v acuminate (caudate), entire. Petiole 5-10cm, rarely red, hairless. Twigs pruinose when young. Buds 5-10mm, green ....*Cappadocian Maple* **Acer cappadocicum**

Small tree. Lvs 3-5cm, (3)5-lobed to ½ way, cordate at base

Lvs dark green above, hairy at least on veins above when young, hairy esp on veins below; lobes ovate, obtuse, entire or with shallow teeth, occ 3-lobed. Petiole 1.5-8cm, sparsely hairy. Twigs stripy, brown and striate with green lenticels, occ with corky wings. Buds to 5mm, reddish to brown or green, adpressed, hairless or sparsely hairy. Pl 15 ....*Field Maple* **Acer campestre**

Petiole without latex (but with abundant clear sap in *Acer saccharum*). Large tree

Lvs glaucous silver-white below, with minute adpressed hairs. Twigs fetid

Lvs to 15cm, deeply 5-lobed >½ way, dark green and hairless above, cordate at base; lobes acute-acuminate, sharply toothed. Petiole 5-12cm, green to pinkish. Twigs warm brown to reddish. Buds to 10mm, green or reddish. ❷......................*Silver Maple* **Acer saccharinum**

Lvs green or weakly glaucous below, hairless or with patent hairs below. Twigs odourless

Lvs 10-22cm, 5-lobed to ⅓ way, dull green above, shiny green below, hairless both sides (exc in vein axils below), occ papillate along veins above. Petiole to 12cm, green. Twigs green, red-spotted. Buds to 6mm, dark brown......................*Sugar Maple* **Acer saccharum**

Lvs 7-16cm, 5-lobed to ½ way, bronze when young, soon ± dull dark green above and ± glaucous below, hairless above (but occ short papillae on veins), soon hairless below or with buff hairs in vein axils (var *villosum* is densely hairy below); lobes acute, irreg crenate-serrate to lobed. Petiole 10-20cm, usu red. Twigs grey-brown. Buds 5-10mm, green. Pl 15, 21..........*Sycamore* **Acer pseudoplatanus**

Group QB – *Deciduous woody climber*

■Plant with anvil-shaped hairs

Lvs 10-15cm, broadly ovate, ± cordate at base, plicate when young, deeply 3-5-lobed, roughly hairy above, with yellow disc-shaped sessile glands below, ciliate; lobes acuminate, irreg dentate, with green hydathodes. Petiole long, encircling stem at base, channelled. Stipules large, deeply bifid, with several veins. Stems twining clockwise, round. PL 17..................*Hop* **Humulus lupulus**

Group QC – *Per herb. Stems rooting at nodes, <15cm tall. Per. All yr.* ❶ *Plant viscid glandular-hairy.* ❷ *Lvs fetid*

■Stems square

Lvs cordate at base, pitted below. ❷. All yr. PL 13......................*Ground-ivy* **Glechoma hederacea**

Lvs truncate or rounded at base, not or weakly pitted below. All yr............(*Prunella*) Go to **PRUNE**

■Stems round. Lvs usu cordate at base

Lvs orb

Lvs alt above, opp below..*Slender Speedwell* **Veronica filiformis**

Lvs all opp

Lvs 2-5cm, cordate at base, viscid glandular-hairy, crenate to shallowly lobed, not net-veined, with 2° veins raised below, with stomata below only. Petiole 1.5-3cm, channelled. ❶. All yr. Dry habs..*Trailing Snapdragon* **Asarina procumbens**

Lvs 0.5-2cm, truncate or broadly cuneate at base, sparsely (often ± adpressed) glandular-hairy to hairless, with 0-7 shallow obtuse teeth per side, with obscure veins not raised below, with stomata both sides. Petiole 0.5-2cm, channelled. All yr. Damp habs ....*Opposite-leaved Golden-saxifrage* **Chrysosplenium oppositifolium**

Lvs ivy-shaped, mostly opp

Lvs hairless, 1-4cm, usu (3)5(7)-lobed, thick, shiny green above, occ purplish below, with protruding hydathodes, with stomata obscure both sides; lobes rounded to triangular. Petiole to 2cm, channelled. Stems 1mm diam, rooting sparingly at lowest nodes, often purplish, often with elastic stele. Dry walls, shingle beaches. PL 16 ....*Ivy-leaved Toadflax* **Cymbalaria muralis**

Lvs densely short glandular-hairy both sides, 1-9cm, 7-lobed, thick, ± dull green above, with protruding hydathodes, with stomata obscure both sides. Petiole (and stems) mostly densely nonglandular-hairy. All yr. Dry walls, R hortal......................*Italian Toadflax* **Cymbalaria pallida**

QA

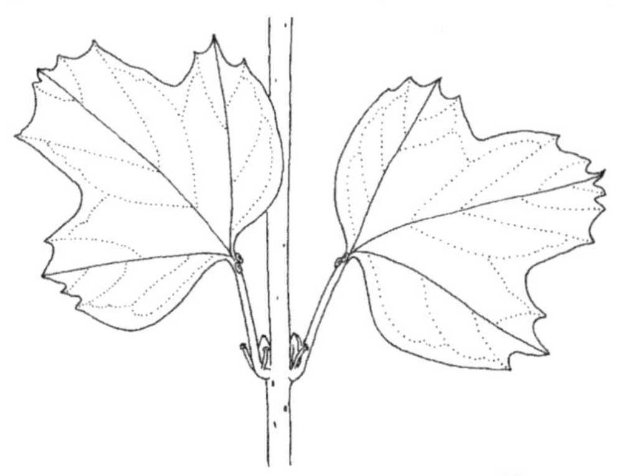

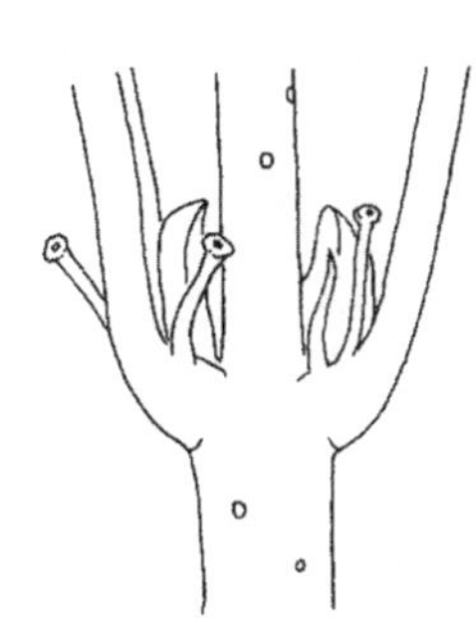

Viburnum opulus

Group QD – *Herbs. Stems erect. Lvs strongly aromatic or fetid. Petioles not connate*

■Lvs fetid, rugose. Shortly rhizomatous per

Basal lvs 3-6cm, ovate-orb, cordate at base, petiolate, palmately 5-7-lobed (rarely with 3 ± free lfts), dark green above, softly septate-hairy, with stomata below only. Stem lvs 6-12cm, cuneate (to sessile) base, 3-lobed with lanc toothed lobes. Petiole to 7cm, purplish, densely shortly retrorsely hairy, channelled, occ hollow, with 2 (or 4) vb's. Stems 60-120cm, branched, sparsely hairy, square, hollow. All yr... *Motherwort* **Leonurus cardiaca**

■Lvs pleasantly aromatic, not rugose. Ann (bi)

Lvs hairless (occ septate hairs in lf axils). Stems round to angled

Stem lvs 2-10cm, pinnately lobed, with additional tiny lobes at base, dark blue-green above, with large sunken semi-translucent yellow glands, with stomata both sides; lobes and teeth usu bristle-tipped. Stems to 40cm, solid. Petiole channelled, solid, with 5 vb's. Hortal .... *French Marigold* **Tagetes patula**

Lvs hairy. Stems square. Sch8

Stem lvs 2-4cm, pine- or citrus-scented, softly nonglandular-hairy, divided into 3 linear lobes, each with 2-7 obtuse lobes (or deep teeth) distally. Basal lvs toothed, soon dying. Stems 5-20cm, hairy. May-Sept... *Ground-pine* **Ajuga chamaepitys**

Stem lvs 1-2.5cm, pineapple-scented, softly glandular-hairy, with white sessile glands below, pinnately lobed; lobes 1-2mm wide, oblong, obtuse, with recurved margins, often lobed, with translucent main veins raised below. Stems 5-25cm, with glandular and nonglandular hairs. Apr-Oct... *Cut-leaved Germander* **Teucrium botrys**

Group QE – Dipsacus. *Stems erect, tall. Lvs with sparse stiff spine-like bristles on midrib below (occ absent from basal lvs), midrib raised below (often forming acute keel), odourless. Hairs unicellular (rarely septate on the lowerside of* D. pilosus *lvs). Bi. All yr*

■Basal lvs with stiff hairs both sides (hairs swollen-based on lowerside only)

Basal lvs 4-9 x 2-5cm, ovate, acute or shortly acuminate, cuneate at base, occ with free lobes, involute when young, dark green, crenate with hydathodes not submarginal, net-veined with 2(3)° veins raised below, with stomata both sides. Stem lvs to 17 x 7cm, ovate to narrowly elliptic, connate at base but not forming a water-collecting cup, simple or with 1 pr of ± free small or unequal lobes at base, entire to crenate. Petiole to 17cm on basal lvs (to 6cm on stem lvs), sparsely bristly or spiny, slightly winged, channelled, with 7 vb's around hollow. Stems 30-120cm, angled, furrowed, with sparse weak prickles on angles, hollow..... *Small Teasel* **Dipsacus pilosus**

■Basal lvs hairless above, soon stiffly hairy on veins below

Basal lvs 15-40cm, cuneate at base, long-ciliate, crenate to laciniate-lobed esp proximally, net-veined, 2-3(4)° veins raised below, with stomata both sides. Stem lvs connate at base, forming water-collecting cup. Petiole winged, with 5(7) vb's. Stems to 200cm .... *Cut-leaved Teasel* **Dipsacus laciniatus**

Group QF – *Stems erect. Lvs (and stems) hairless exc for sessile glands, not ciliate, weakly fetid to odourless. Tufted per*

▪Basal lvs 6-12cm, elliptic to ovate, usu obtuse, deeply cordate to rounded at base, with or without translucent dots (HTL), with sessile glands when young, occ sparse glandular hairs below, crenate with 25-32 hydathode-tipped teeth per side (occ ± serrate esp on upper lvs), net-veined, with stomata both sides. Petiole winged, often with 2 lobes nr lf, deeply V-shaped with keel below, with 1(3) flat arc vb's. Stems 4-winged, hollow. All yr..................*Water Figwort* **Scrophularia auriculata***

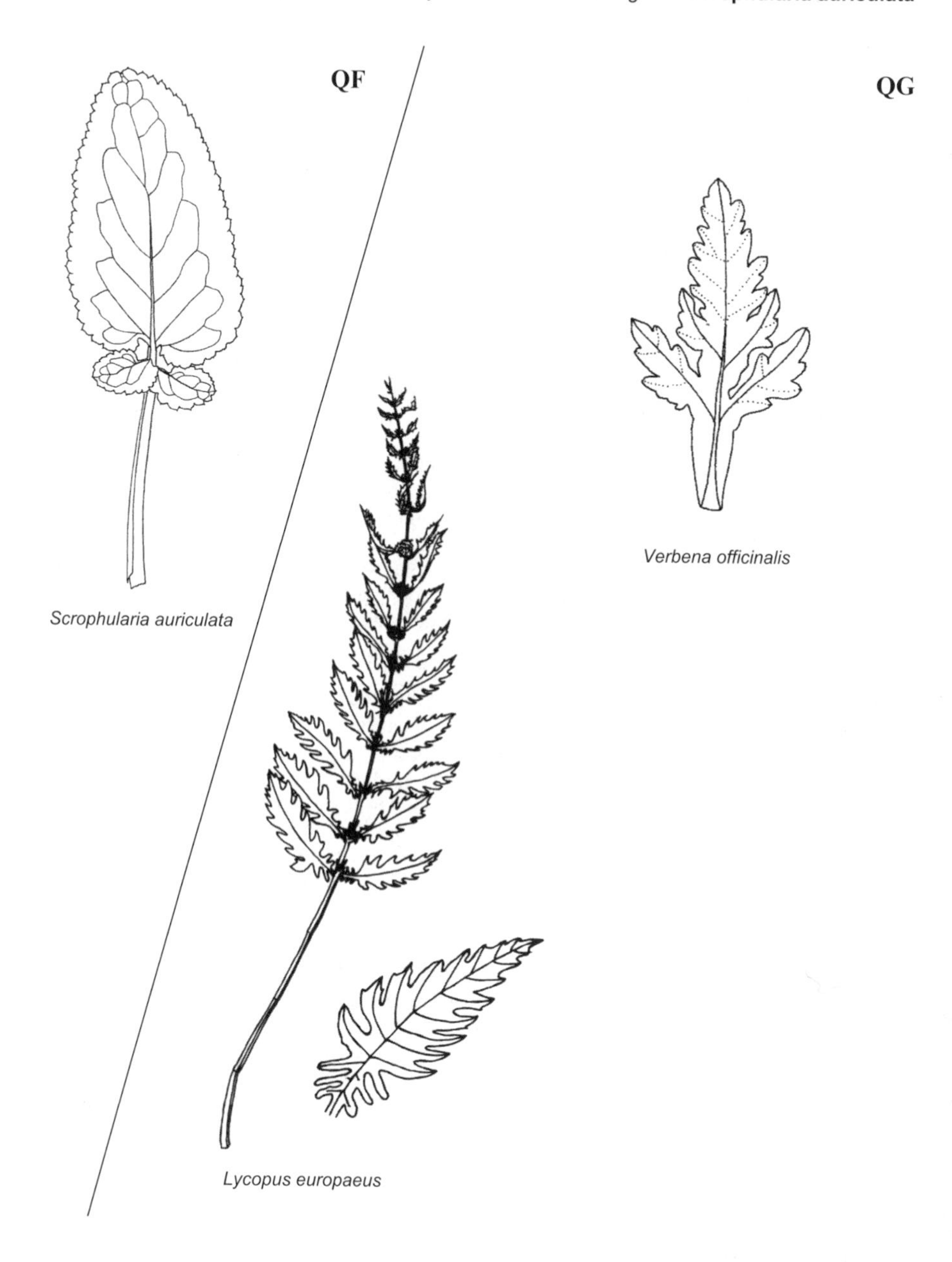

Verbena officinalis

Scrophularia auriculata

Lycopus europaeus

Group QG – *Stems erect. Lvs (or stems) with at least some hairs or cilia, not or weakly aromatic. (*Melampyrum arvense*,* Leonurus cardiaca *and* Teucrium chamaedrys *may key out here)*

■Lvs basal and on stems, rugose, usu cordate at base, weakly sage-scented, toothed to pinnately lobed, petiolate, septate-hairy, ciliate, net-veined, with stomata both sides. Stem lvs more deeply lobed, shortly petiolate to sessile. Stems 30-80cm, hairy (often retrorsely so), square, soon hollow

Basal (and stem) lvs with 1 pr of small ± free lobes nr base

Lvs 5-15cm, broadly ovate, hairy both sides, irreg crenate-dentate with 20-25 teeth per side. Petiole weakly retrorsely hairy, slightly channelled, with flat arc vb (and 2 rib bundles). Often all yr..*Whorled Clary* **Salvia verticillata**

Basal lvs without small lobes nr base

Basal lvs 7-15cm, oblong to ovate, crenate or 2-dentate, rarely lobed, hairless or minutely hairy above, sparsely long-hairy on veins below. Petiole hairy, flat, with 1-3 vb's. Stems glandular-hairy above. Mar-Oct (all yr). Sch8..*Meadow Clary* **Salvia pratensis**

Basal lvs 4-12cm, oblong-ovate or slightly fiddle-shaped, crenate-serrate, usu lobed, hairy both sides (often minutely so). Petiole hairy, flat, with 1-5 vb's. Stems glandular-hairy above. All yr ....*Wild Clary* **Salvia verbenaca**

■Lvs usu all on stems, not rugose, not cordate at base, not sage-scented (may have other odour)

Stems square or ridged, always present

Stems pith-filled. Dry habs

Lvs 2-8(10)cm, cuneate at base, dark green, sparsely unicellular-hairy both sides, ciliate with swollen-based antrorse cilia, deeply lobed, with stomata both sides. Petiole with 1 vb. Stems 1-several, 30-140cm, tough, rough with swollen-based hairs, deeply furrowed on 2 opp sides. Tufted per. All yr..*Vervain* **Verbena officinalis***

Stems hollow. Wet habs

Lvs with some septate hairs at least on veins below (hairs occ unicellular), to 10cm, ovate-lanc or elliptic, acute, sessile, usu deeply lobed near base (lobes occ reflexed), otherwise deeply serrate, with stomata both sides. Stems to 60(90)cm, 4-angled, with crisped retrorse or antrorse hairs. Rhizomatous. Apr-Oct..............................*Gypsywort* **Lycopus europaeus***

Lvs hairless, to 8cm, pinnately lobed, with oblanc-lanc terminal lobe to 3cm and tiny lateral lobes at base, sessile, with white sunken hydathodes along upper margin, scabrid-ciliate (occ confined to base), net-veined, with stomata both sides. Stems to 30cm, ridged, hairless. Stoloniferous..*Marsh Valerian* **Valeriana dioica***

Stems round (to weakly angled) or absent, solid

Stems aromatic when broken (and often lvs when crushed). Hairs septate

Stem with purple-black resin canals. Lvs 5-15cm, lanc-elliptic, acuminate, petiolate, sparsely hairy to hairless, shortly ciliate, (0)3(5)-partite, deeply serrate with 9-11 hydathode-tipped teeth per side of main lobe, opaquely net-veined (Kranz venation), with 2° veins ± raised both sides and ending in teeth apices, with stomata both sides. Petioles often ± connate at base, long-ciliate nr base, with 3 vb's and spiral fibres. Stems to 80cm, hairless to ± hairy, with long-ciliate interpetiolar ridge, with spiral fibres around submarginal vb's. Ann

All lvs (0)3-lobed..*Trifid Bur-marigold* **Bidens tripartita**

Some lvs often 5-lobed. VR alien..*Beggarticks* **Bidens frondosa**

Stem without purple-black resin canals

Lvs with colourless sessile glands below, and minutely translucent gland-dotted. Per

Lvs 5-10cm, 3(5)-lobed, petiolate (the upper ± sessile), sparsely hairy, ciliate, net-veined, with 2° veins raised below, with stomata below only; lobes elliptic to oblanc, serrate with hydathode-tipped teeth. Petioles to 1cm, connate at base, flat, with 3 vb's. Stems to 150cm, purplish, antrorsely crisped-hairy, striate. Mar-Oct ....*Hemp-agrimony* **Eupatorium cannabinum**

Lvs without glands (occ translucent gland-dotted in *Ambrosia artemisiifolia*)

Per. Lvs occ alt above, to 20cm, hairy both sides, deeply ternately (occ pinnately) lobed, the lowest often entire, with midrib raised below, with stomata both sides. Petiole to 15cm, channelled, with 7(9) vb's. Stems 30-100cm, branched above, hairy or sparsely so (hairs with withering tips), slightly angled. Apr-Oct ....*Large-flowered Tickseed* **Coreopsis grandiflora**

Ann. Lvs alt above, to 10(15)cm, dark green above, ± grey-green below, sparsely hairy to hairless above, shortly hairy below, ciliate, 1-2-pinnately lobed, the lobes narrow, acuminate and toothed with tiny pale hydathodes. Petiole to 3cm, with 3 vb's. Stems 25-135cm, branched above, ± densely shortly adpressed-hairy and with purple-based hairs, bluntly ridged to round. May-Oct.....................................*Ragweed* **Ambrosia artemisiifolia**

Stems not aromatic when broken. Hairs unicellular

Stems hollow. Lvs ± fetid esp when dried

Stems round. Basal lvs with 0-5 prs of lobes. All yr. VR alien, Wales ....*Turkey Valerian* **Valeriana phu**

Stems ridged. Basal lvs usu with 4 prs of lobes. Mar-Oct. Marshes or calc gsld

Basal (and lower) lvs rarely 3-whorled, petiolate; terminal lobe 3-lobed, the lateral lobes lanc, short-stalked, distantly irreg toothed (occ entire) with sunken hydathodes, all usu hairless above, minutely hairy below, ciliate, net-veined, with 2° veins raised below, with stomata below only. Upper lvs ± sessile, often 2-pinnate. Petiole sheathing at base on basal lvs, often ciliate at base, channelled, occ laterally flattened, often hollow, with spiral fibres. Stems to 150cm, hairy or retrorsely so below, ± hairless above ....*Common Valerian* **Valeriana officinalis**

Stems solid, ± round. Lvs odourless

Basal lvs to 15(30)cm, elliptic to oblanc, hairy both sides (more hispid above), with pale green submarginal hydathodes visible above (the terminal visible below) and usu at least some traces of crenate teeth, the upper rosette lvs occ lyrate-pinnatifid, with midrib white above and green below, with stomata both sides. Stem lvs usu deeply pinnatifid with elliptic terminal lobe and linear-oblong lateral lobes. Petiole hairy, ciliate, often hollow, with 3(5) vb's and spiral fibres. Stems to 100cm, roughly hairy at least below with stiff retrorse hairs. All yr. Usu calc gsld. PL 18..*Field Scabious* **Knautia arvensis**

Basal lvs 5-15cm, obovate, sparsely or minutely hairy to hairless, the upper rosette lvs ± pinnately lobed, with long large end-lobe and free lateral lobes, with stomata both sides. Petiole hairless, ciliate, solid, with 3-5 vb's and spiral fibres. Stems to 70cm. All yr. Calc gsld ....*Small Scabious* **Scabiosa columbaria**

Key to Groups in Division R

(Lvs with 3-17 lfts (usu palmately arranged))

■Lfts entire (occ notched at apex), with veins anastomosing before margins
Lvs with 3 lfts (3-foliate)
Tree or shrub..**RA**
Herb
Stipules present (may appear lft-like in *Lotus* and *Tetragonolobus*)..**RB**
Stipules absent or reduced to sheathing base
Lfts usu notched at apex. Dry habs (*Oxalis*)...**RC**
Lfts not notched at apex. Bogs (*Menyanthes*)...**RD**
Lvs with 4-17 lfts...**RE**
■Lfts toothed (at least distally) or lobed, or with veins hitting margins
Stems with prickles...**RF**
Stems without prickles
Tree or non-climbing shrub. Lvs opp or alt...**RG**
Woody climber (or scrambling shrub). Lvs opp or alt...**RH**
Herb (occ woody at base). Lvs alt
Stipules absent
Basal lvs arising singly, or appearing after fl (*Anemone*, *Eranthis*, *Menyanthes*)..................**RI**
Basal lvs in rosette, or stem lvs present
Lvs hairy at least below..**RJ**
Lvs hairless (occ with ± sessile glands when young)...**RK**
Stipules present (occ lft-like)
Lfts deeply lobed..(*Geranium*) Go to **GER A**
Lfts toothed (often deep), occ entire in *Trifolium* but with veins hitting margins
Lfts 3, not net-veined (3° veins not visible), without 2° veins raised below (*Fabaceae*)
Lvs with viscid glandular hairs...**RL**
Lvs with nonglandular hairs or hairless
Terminal lft with 1.5-7mm stalk (stalk much longer than that of sessile lateral lfts)......**RM**
Terminal lft with 0-1mm stalk (all lfts sessile)...**RN**
Lfts 3, net-veined (3° veins visible), usu with 2° veins raised below (*Rosaceae*)...............**RO**
Lfts (3)4-9, net-veined (3° veins visible), usu with 2° veins raised below (*Rosaceae*, *Cannabis*)...**RP**

Group RA – *Tree or shrub. Lfts 3, entire. Hairs unicellular*

■Unarmed deciduous tree

Lfts to 7 x 3.5cm, ± elliptic, mucronate, folded when young, hairless above, net-veined, with midrib raised below, with stomata below only. Petiole to 8cm. Stipules linear. Twigs silky-hairy when young, with short shoots. Buds to 5mm, ovoid, green, silky-hairy, with several scales

Lfts adpressed-hairy below with dense short (0.1-0.3mm) hairs and sparse longer (0.4-0.7mm) hairs. Petiole adpressed-hairy..*Laburnum* **Laburnum anagyroides**

Lfts with sparse long (0.4-0.7mm) adpressed hairs below (esp on veins), hairless above. Petiole sparsely adpressed-hairy. *L. anagyroides* x *alpinum*.....*Hybrid Laburnum* **Laburnum** x **watereri**

Lfts hairless or with sparse long (0.4-0.7mm) adpressed hairs on veins below. Petiole hairless ....*Scottish Laburnum* **Laburnum alpinum**

■Spiny shrub, 1-2.5m. Lvs 3-foliate on young or vigorous shoots, reduced to scales at maturity ....(*Ulex*) Go to **EA**

■Unarmed shrub

Evergreen. Lvs (sub)opp

Lfts 3(4), to 8 x 3cm, ovate-oblong, obtuse, cuneate at base, shiny dark green above, paler below, odorous, with unequal translucent glandular dots (HTL), hairless, rarely toothed, not net-veined, without veins raised below, with stomata below only. Petiole to 4(8)cm, channelled, with 1 arc vb. Twigs minutely hairy to hairless, ± round, without interpetiolar ridge ....*Mexican Orange* **Choisya ternata**

Deciduous. Lvs alt

Lvs all 3-foliate. Twigs persistently hairy

Stipules 1-2mm, densely hairy. Stems with 7-9 ridges. Petiole <5mm. Lfts 15 x 7mm, adpressed-hairy esp below. R alien....................*Montpellier Broom* **Genista monspessulana**

Stipules absent. Stems round or obscurely ridged. Petiole >10mm. Lfts (6)10-30 x (2)5-10(16)mm, oblong to linear, dark green above, paler below, adpressed-hairy when young, soon hairless above. VR alien...*Black Broom* **Cytisus nigricans**

Lvs 3-foliate below, becoming simple towards stem apices. Twigs soon hairless. Stipules absent

Twigs usu with 5 deep ridges. Petiole 1-4mm, on 3-foliate lvs only (simple lvs sessile)

Lfts 6-20 x 3-5mm, narrowly elliptic to obovate, acute, with slight mucro, patent-hairy above, hairless or with adpressed hairs below, with midrib raised below, with few indistinct 2° veins, with stomata both sides. Twigs green, sparsely long-hairy when young, with stomata ....*Broom* **Cytisus scoparius**

Twigs with c8 deep ridges. Petiole <2mm. VR alien..............*White Broom* **Cytisus multiflorus**

Twigs with 10-15 shallow ridges. Petiole 2-4mm. R alien

Lfts 6-9 x 2.5-4mm, narrowly elliptic-obovate, often retuse, dark green, soon hairless above, adpressed-hairy below. Twigs green, sparsely hairy at least in 1st yr ....*Hairy-fruited Broom* **Cytisus striatus**

Group RB – *Herb. Lfts 3, entire. Petiole channelled. Stipules obvious or lf-like (actually lfts). Hairs unicellular.* ❶ *Stipels present (stipule-like outgrowths below each lft)*

■'Stipules' lft-like (true stipules minute, <0.2mm, brown, soon falling)

Terminal lft usu >3x longer than wide, always <5mm wide

Lfts 8-14 x 1.2-2.5mm, linear-lanc to oblanc, hairless or occ with sparse adpressed hairs, with obscure veins, with stomata both sides; lateral lfts subopp. Petiole 3mm. Stems solid. Per. Damp clayey habs..*Narrow-leaved Bird's-foot-trefoil* **Lotus tenuis**

Lfts 4-10 x 2.5-4mm, narrowly obovate, hairy, with obscure veins, with stomata both sides; lateral lfts often subopp. Stems solid. Ann. Dry, often sandy, habs ....*Slender Bird's-foot-trefoil* **Lotus angustissimus**

Terminal lft <3x longer than wide, usu >5mm wide

Lfts thin, with 2° veins visible and translucent (HTL) or raised above. Stems hollow

Lfts 12-20(25) x 10(15)mm, obovate, ± glaucous below, densely hairy to ± hairless, usu with stomata both sides; lateral lfts often subopp. Petiole to 1cm. Stems weakly erect, usu with sparse long hairs. Mar-Oct (all yr). Per, often with stolons. Damp habs ....*Greater Bird's-foot-trefoil* **Lotus pedunculatus***

Lfts thick, with 2° veins obscure or opaque (HTL). Stems solid

Lfts >17mm, with translucent dots (HTL)

Per. Lfts rhombic to obovate, asymmetric each side of midrib, acute, sparsely hairy, with few obscure 2° veins, with stomata obscure both sides; lateral lfts opp. Stems densely hairy. Apr-Oct. Calc turf..*Dragon's-teeth* **Tetragonolobus maritimus**

Lfts <15mm, without translucent dots

Per. Lfts 3-15 x 4-9mm, lanc to broadly ovate, often ± glaucous, hairless to sparsely hairy, with stomata both sides; lateral lfts sessile, opp. Petiole to 5mm. Stems round nr base, square above. All yr (lvs often much reduced in winter)

Stems ± prostrate, hairless to sparsely short-hairy, solid ....*Common Bird's-foot-trefoil* **Lotus corniculatus** var **corniculatus***

Stems often weakly erect, long-hairy, hollow. Alien, often sown ....*Common Bird's-foot-trefoil* (var) **Lotus corniculatus** var **sativus**

Ann. Lfts 4-7(13) x 3(6)mm, ovate to obovate, greyish-green, densely hairy, with stomata both sides; lateral lfts shortly stalked, often subopp. Stems ± prostrate, densely hairy with long spreading hairs to 2mm, round. Oct-Jun........*Hairy Bird's-foot-trefoil* **Lotus subbiflorus**

■Stipules not lft-like, adnate to petiole for much of their length............................(*Trifolium*) Go to **RN**

■Stipules gland-like. ❶

Ann, with taproot or fibrous roots. Stems to 3m, occ climbing......*French Bean* **Phaseolus vulgaris**

Per with tuberous rootstock. Stems to 5m, climbing

Lvs pinnately 3-foliate, with rachis 4-6cm. Lfts to 15 x 13cm, broadly ovate, acuminate, broadly cuneate at base, folded when young, dull green and minutely hispid above, sparsely minutely hairy below, sparsely ciliate, net-veined, pinnate-veined but 3-pli-veined at base, with stomata both sides or below only; lateral lfts stalked to 8mm. Stipels to 6 x 3mm, ovate to lanc, green. Petiole swollen at base, tough, soon hairless, with vb's around small hollow. Stems minutely hairy, ribbed, ± solid..*Runner Bean* **Phaseolus coccinea**

Lotus corniculatus var *corniculatus*

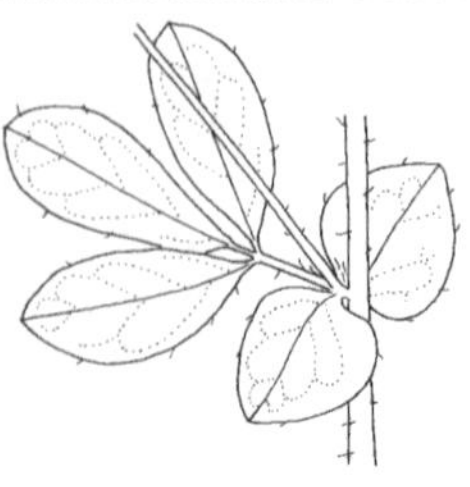

Lotus pedunculatus

Group RC – Oxalis. *Herb. Stipules absent, minute or reduced to an adnate sheathing base. Lfts typically obcordate or obovate, often notched at apex, without hydathodes along margins, folded when young (drooping, with 'sleep' movements), with stomata below only. Petiole round, with 1 vb. Hairs unicellular.* ➊ *Lvs whorled on branched aerial stems*

■Lfts with raised orange (occ pale or pinkish) dots below

Lfts (±) hairless, not ciliate. Lvs whorled on branched aerial stems ➊

Lfts 5-15mm, pale green, with sparse pale orange dots nr margins below nr apex. Petiole 2-6cm. Bulbils present at or above ground. Tuberous. Nov-Jul...........*Pale Pink-sorrel* **Oxalis incarnata**

Lfts usu hairy and/or ciliate. Lvs in a rosette

Lvs on a short stock, originating from a bulb with many bulbils. Lfts with dots scattered below

Lfts to 35mm, adpressed-hairy esp below. Petiole to 20cm, antrorsely hairy. SW Eng ....*Bermuda-buttercup* **Oxalis pes-caprae**

Lvs emerging from taproot. Lfts with dots mostly confined to margins below

Lfts densely adpressed-hairy both sides, 25-40mm. Petiole to 15cm, sparsely adpressed-hairy to hairless. Taproot without bulbils. Oct-Jul.................................*Pink-sorrel* **Oxalis articulata***

Lfts sparsely adpressed-hairy to hairless either side, 25-40mm. Petiole 5-15cm, sparsely hairy. Taproot surrounded by bulbils. Mar-Oct....................*Large-flowered Pink-sorrel* **Oxalis debilis**

■Lfts without raised dots below (occ with scattered translucent dots)

Lvs alt along stems (often prostrate and rooting at the nodes). Petiole with 'knee-joint' at base

Lfts 2-5mm wide, rarely (or faintly) purplish below, hairless or with scattered adpressed hairs, ciliate. Petiole 0.5-4cm. Stems prostrate, rooting at nodes, reddish. Per (ann). Apr-Oct ....*Least Yellow-sorrel* **Oxalis exilis**

Lfts usu 6-25mm wide, often purplish at least below, sparsely hairy to ± hairless, ciliate, occ with translucent dots. Petiole 1-12cm, with antrorse or spreading hairs. Stems often densely hairy. All yr. (The following 3 spp may best be lumped into *O. corniculata* agg)

Stems prostrate, rooting at nodes. Fl stalks patent or reflexed in fr. Usu per ....*Procumbent Yellow-sorrel* **Oxalis corniculata***

Stems usu ± erect, not or rarely rooting at nodes. Fl stalks patent or reflexed in fr. Usu ann ....*Sussex Yellow-sorrel* **Oxalis dillenii**

Stems usu ± erect, not or rarely rooting at nodes. Fl stalks erect in fr. Usu ann ....*Upright Yellow-sorrel* **Oxalis stricta**

Lvs whorled on branched aerial stems (with bulbils). Petiole without 'knee-joint'. ➊

Lfts 20-45mm, usu much wider than long, hairless, ciliate. Bulbils on stolons to 2cm ....*Garden Pink-sorrel* **Oxalis latifolia**

Lvs in a rosette (may be at the end of a woody stock or rhizome). Petiole without 'knee-joint'

Lvs tufted at the end of slender rhizomes (covered with swollen lf scars). Lfts thin, 12-20mm x 20(25)mm, wider than long, often purplish below, with translucent dots (HTL), with sparse long hairs, ciliate, with midrib visible. Petiole 5-15cm, reddish nr base. All yr. Humus-rich woods. PL 17..*Wood-sorrel* **Oxalis acetosella**

Lvs tufted from a short stock, with swollen taproot. Lfts fleshy, obovate, rounded to shallowly notched at apex, with crystalline appearance below, hairless, with midrib usu obscure. VR alien, SW Eng...*Fleshy Yellow-sorrel* **Oxalis megalorrhiza**

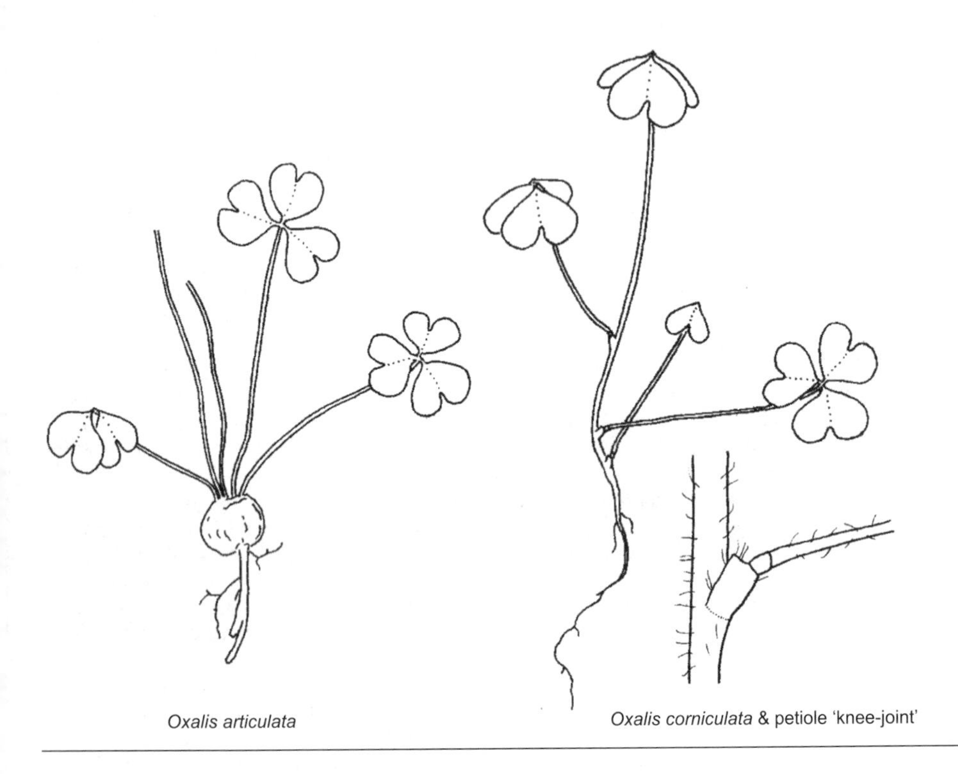

Oxalis articulata *Oxalis corniculata* & petiole 'knee-joint'

Group RD – *Herb. Stipules absent (but petiole with an adnate sheathing base). Lvs alt. Lfts rounded at apex, rolled when young, with c10 white hydathodes along margins above. Petiole with 6 vb's. Bogs*

■Lfts 3-9cm, obovate to ovate-elliptic, thick, hairless, pinnately veined, often net-veined, with stomata both sides. Petiole 7-20cm, often with prominent rounded auricles at long sheathing base, round, with elongate stomata, with aerenchyma in TS. Rhizomes with adventitious roots. Mar-Oct ....*Bogbean* **Menyanthes trifoliata***

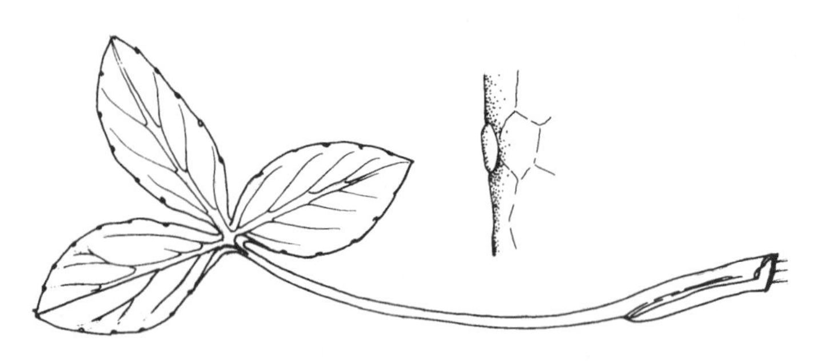

Menyanthes trifoliata lf & lf margin

Group RE – *Lvs with 4-17 lfts (digitately arranged exc* Potentilla*). Lfts folded when young (revolute in* Potentilla*). Hairs unicellular.* Lupinus *spp all have hollow petioles with partly adnate stipules*

■Shrub <200cm. Stipules present

Lfts 5, net-veined

Lfts ± palmately arranged with v short rachis, 1-3cm, oblong-lanc, acute, with recurved margins, silky-hairy both sides, with sunken veins, with stomata below only. Petiole 5-10mm, with antrorse adpressed hairs. Stipules to 1cm, lanc, scarious, persisting for 1-3 yrs. Twigs hairy when young, with reddish-brown bark flaking in later yrs......................*Shrubby Cinquefoil* **Potentilla fruticosa**

Lfts (5)6-10(12), with midrib visible only

Lfts to 4 x 1cm, oblanc, mucronate, channelled at maturity, hairless above, silky-hairy below, with stomata both sides. Petiole silky-hairy. Stipules adnate for ½ length, with 8mm linear-lanc free part, silky-hairy. Twigs reddish, hairy, round. All yr (but frost-sensitive). Usu coastal, often on sand dunes..*Tree Lupin* **Lupinus arboreus**

■Herb, dying down to nr (or below) ground level in winter

Stipules absent. Plant <30cm tall

Lfts 4, 2-5cm, narrowly obovate to oblanc, sparsely hairy, ciliate. Taproot surrounded by bulbils on underground stolons to 6cm. R alien....................*Four-leaved Pink-sorrel* **Oxalis tetraphylla**

Lfts 5-10, 2-5cm, narrowly obovate to oblanc, hairless, not ciliate. Taproot surrounded by sessile bulbils. VR alien...*Ten-leaved Pink-sorrel* **Oxalis decaphylla**

Stipules present (occ absent or reduced to sheathing base on basal lvs). Plant usu >50cm tall

Petiole with long antrorsely adpressed or shaggy hairs

Lfts 6-9, 2.5-5 x 1.5cm, elliptic, mucronate, cuneate at base, adpressed-hairy esp below, with midrib raised below, with c4 prs of 2° veins. Petiole ≥2x lfts. Stipules 10-14mm, subulate. Stems adpressed-hairy (occ shaggy). Per. River shingle, R alien, Scot ....*Nootka Lupin* **Lupinus nootkatensis**

Petiole with sparse short or adpressed hairs

Lfts 1.5-2.5cm wide. Per (bi)

Stems unbranched. Lfts 10-17, 4-15cm, oblanc. Petiole ≥2x lfts. Usu all yr ....*Garden Lupin* **Lupinus polyphyllus**

Stems usu branched

Lfts 9-15, 6-12cm, oblanc, apiculate, hairless to adpressed-hairy below, with long antrorse cilia, with 2° veins hardly raised below, with stomata below only or nr midrib above. Petiole with spreading long or short hairs (occ sparse). Stipules to 2cm, linear, acute. Stems with spreading hairs, weakly angled. Ruderal and river shingle. *L. arboreus* x *polyphyllus* ....*Russell Lupin* **Lupinus** x **regalis**

Lfts usu 9, 6-12cm, oblanc, densely hairy both sides with hairs slightly antrorse, with stomata both sides. Stems densely hairy. River shingle, VR alien, Scot. *L. nootkatensis* x *polyphyllus*..**Lupinus** x **pseudopolyphyllus**

Lfts <1.5cm wide. Ann. Mar-Oct. VR alien or crop relic

Stem(s) much-branched, with sparse short antrorse crisped hairs. Lfts 5-9, to 4 x 0.5cm, linear, with all veins obscure, with stomata both sides ....*Narrow-leaved Lupin* **Lupinus angustifolius**

Stem(s) sparsely branched, hairy. Lfts 5-9, to 6 x 1.5cm, oblanc, with all veins obscure, with stomata both sides..*White Lupin* **Lupinus albus**

Group RF – *Stems with prickles (occ absent in* R. saxatilis*). Lvs folded-plicate when young. Lfts pinnately arranged (palmately arranged in some spp of* Rubus fruticosus *agg), with terminal rachis >> lateral rachis, all with hydathode-tipped teeth, net-veined, with stomata below only. Petiole with 3 vb's and spiral fibres. Stipules adnate to base of petiole. Hairs unicellular*

▪Erect deciduous shrub

Stems with dense reddish gland-tipped bristles. Often suckering

Lfts densely white-hairy below, dull pale green and sparsely hairy above, 3(5), to 11 x 10cm, ovate, without translucent dots, irreg 2-serrate, with sparse weak patent prickles on veins. Petiole to 7cm. Stems bi, to 2(3)m, round, with sparse weak patent prickles ....*Japanese Wineberry* **Rubus phoenicolasius**

Stems without gland-tipped bristles. Suckering

Lfts green and ± hairless both sides, 3, to 13cm, obovate to ovate, without translucent dots, irreg toothed or lobed. Petiole to 5cm. Stipules long, linear or subulate. Stems per, to 3m, rooting at tips, green, round, with scattered narrow prickles and sparse hairs when young ....*Salmonberry* **Rubus spectabilis**

Lfts densely white-woolly below, green and minutely hairy above, 3-5(7), 5-12cm, ovate or ovate-lanc, acuminate, without translucent dots, 2-dentate with mucronate teeth. Petiole 2-7cm, minutely hairy, round. Stipules to 1cm, thread-like. Stems bi, to 1.5m, cane-like, ± pruinose, round, with slender weak straight prickles and minute hairs................*Raspberry* **Rubus idaeus**

▪Trailing or sprawling shrub

Lvs with (3)5(7) lfts, at least some wintergreen (not true of all *R. fruticosus* agg)

Stems bi to per, tangled and sprawling, usu rooting at tips, rarely pruinose, 5-angled at least when young. Lfts 4-12cm, ovate to oblong-lanc, often with minute translucent dots (HTL), hairy at least below, 2-serrate. Petiole round. Stipules to 15mm, thread-like to lanc ....*Bramble* **Rubus fruticosus** agg

Lvs with 3 lfts (pinnately arranged), deciduous

Stems ± prostrate, pruinose, hairless, with sparse weak curved prickles, round. Terminal lft to 5cm diam, ovate to rhombic, acute or acuminate, usu cordate at base with 1-2(3)cm stalk, weakly rugose, usu with minute translucent dots (HTL), usu hairy both sides, 2-dentate or shallowly lobed, occ 3-lobed, with acute purple hydathodes. Petiole weakly prickly, sparsely hairy, channelled. Stipules lanc to ovate, herbaceous, pinnate-veined ....*Dewberry* **Rubus caesius**

Stems arching, not pruinose, minutely hairy, with strong ± straight prickles (and 4mm slender purplish prickles), round. Terminal lft 5-10cm, ovate, acute, cordate at base, weakly rugose, without translucent dots, hairy both sides, 3-lobed, with midrib (and 2° veins) sparsely spiny below. Petiole spiny, weakly channelled. Stipules to 12mm, thread-like. R hortal ....*Loganberry* **Rubus loganobaccus**

▪Trailing or sprawling herb

Stems often purplish, not pruinose, hairy (occ glandular so), with weak prickles or unarmed. Lvs rarely simple. Lfts 3-9cm, ovate to obovate, ± acute, broadly cuneate at base, bright or mid-green above, paler below, hairy both sides or ± hairless above, irreg 1(2)-serrate-dentate; terminal lft with 5-20mm stalk; lateral lfts ± sessile, often with shallow rounded lobe on lower margin. Petiole 2-7cm, armed as stem, channelled. Stipules 5mm, ovate to linear-lanc, green, entire. Apr-Oct. Usu upland..*Stone Bramble* **Rubus saxatilis**

Group RG – *Deciduous tree or non-climbing shrub. Lfts toothed at least distally*

■Lvs alt. Stipules to 1cm, linear, entire or weakly toothed

Lfts 3, pinnately arranged, to 2 x 1.5cm, obovate, mucronate, the terminal with channelled stalk to 5mm, hairless above, densely adpressed-hairy below, toothed distally, with stomata both sides. Petiole to 2cm, channelled. Twigs white silky-hairy, angled. Shrub to 1.5m. VR alien ......*Tree Medick* **Medicago arborea**

■Lvs opp. Stipules absent

Lfts 3 or 5, pinnately arranged

Tree, 3-20m. Twigs with interpetiolar ridge, often pruinose when young

Lfts 3(5), to 14 x 7cm, ovate-elliptic, acute, pale (yellow-) green, hairless above (exc midrib), sparsely hairy on veins and in vein axils below, ± toothed (mostly in distal ½). Petiole to 10cm, swollen at base, round. Twigs hairless, round, developing lenticels. Buds to 5(10)mm, minutely white-hairy......*Ashleaf Maple* **Acer negundo**

Shrub, to 3m. Twigs without interpetiolar ridge, never pruinose. Lfts 3 (or with simple 3-partite lvs)......*Forsythia* **Forsythia x intermedia**

Lfts 5-7(9), palmately arranged. Large tree, to 32m

Lfts 8-20(30) x 10cm, acuminate, cuneate at base, hairless above, sparsely orange-woolly below when young, irreg crenate-serrate, net-veined, with stomata below only. Petiole 15-25cm, round. Lf scars semi-cylindrical, with 5(7) traces. Buds to 25mm, hairless, occ ciliate

Twigs monopodial, ending in a terminal bud

Buds dark reddish-brown, viscid. Lfts obovate, sessile, dull dark green above, green below, soon hairless both sides. PL 21......*Horse-chestnut* **Aesculus hippocastanum**

Twigs sympodial, ending in a pr of buds (at least on weaker shoots)

Buds green with reddish scale margins, not or hardly viscid. Lfts broadly obovate, hardly stalked, slightly crinkly, shiny dark green above, green below, occ with white hairs in vein axils below, often with midrib reddish at least nr base. Petiole often reddish ......*Red Horse-chestnut* **Aesculus carnea**

Buds green to pinkish-red, slightly viscid. Lfts ± narrowly elliptic-obovate, stalked to 1cm, not crinkly, dull dark green above, grey-green below, hairless both sides, with midrib rarely reddish. Petiole rarely reddish. Fls 1 month later than other 2 spp ......*Indian Horse-chestnut* **Aesculus indica**

RH

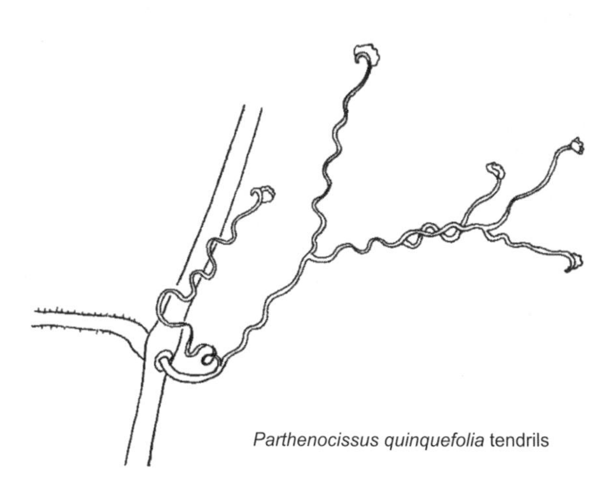

Parthenocissus quinquefolia tendrils

Group RH – *Woody climber (or scrambling shrub). Deciduous*

■Lvs opp. Tendrils absent. Stipules absent. (*Clematis vitalba* may rarely key out here)

Scrambling shrub

Lfts 3, to 3 x 1.5cm, dark green above, paler below, with recurved serrate cartilaginous margins (occ entire), with few indistinct 2° veins hardly raised below, with stomata below only. Petiole with sparse short hispid hairs, channelled. Twigs green, 4-ridged, with stomata, pith-filled. Buds 2-4mm, adpressed, green, with 5 scales. PL 17............*Winter Jasmine* **Jasminum nudiflorum**

Woody climber

Lfts 3, to 10cm, elliptic-ovate, often dark green, sparsely hairy to hairless above, hairy on veins below, ciliate, toothed distally, ± net-veined, with stomata below only; terminal lft with channelled stalk to 10mm; lateral lfts stalked to 2mm. Petiole 4-10cm, hairy, channelled, pith-filled, with 7 vb's in ring. Stems purplish, striate (not ridged), with interpetiolar ridge. Buds to 7mm, often purplish, silky-hairy, with 5 scales...............*Himalayan Clematis* **Clematis montana**

Lfts (3)5, to 4cm, ovate, light green, ± hairless, coarsely toothed, net-veined, with stomata below only. Petiole to 10cm, semi-cylindrical, with 11 vb's in ring. VR alien ....*Chocolate-vine* **Akebia quinata**

■Lvs usu alt (occ opp above). Tendrils lf-opposed, branched. Stipules present

Lfts 5, palmately arranged, 9-13 x 4.5-6cm, ovate, shortly stalked, bronze when young, later dull green above, sparsely septate-hairy or hairless both sides, with 5-10 teeth per side (each with long hydathode), net-veined, with stomata below only. Petiole to 8cm, swollen at base, sparsely hairy, laterally flattened or round with narrow channel above, with 12 vb's in outer ring. Stipules to 10mm, ovate-lanc, soon falling. Stems swollen at nodes, with lenticels; lf scars round, with 12 traces in ring. Buds to 7mm, reddish, with several scales

Tendrils 2-5-branched, with indistinct or no sucker pads at tips ....*False Virginia-creeper* **Parthenocissus inserta**

Tendrils 5-12-branched, usu with obvious red sucker pads at tips ....*Virginia-creeper* **Parthenocissus quinquefolia***

Group RI – *Herb. Stipules absent. Basal lvs arising singly (or appearing after fl). Per*

■Lfts 3, unlobed. Bogs..*Bogbean* **Menyanthes trifoliata***

■Lfts 3(5), deeply cut into serrate lobes. *Anemone* have lvs ± orb in outline, with white or purple hydathodes (best viewed end-on), with 2° veins weakly or not raised below, with stomata below only. Stem lvs (bracts) in 1 whorl, 3-foliate, short-stalked to sessile. Petiole to 8cm, not sheathing at base (but may be dilated), channelled, with 3 vb's

Plant with at least some hairs (occ sparse)

Rhizomes long, not tuber-like. Fls (and buds on rhizome) white

Basal lvs with adpressed to patent hairs both sides (occ hairless), often ciliate, not purplish below. Petiole usu hairy. Fr stalk drooping. Mar-Jul. PL 16 ....*Wood Anemone* **Anemone nemorosa**

Rhizomes short, tuber-like. Fls (and buds on rhizome) usu blue

Basal lvs adpressed-hairy both sides, rarely purplish below. Petiole usu hairy. Fr stalk erect. Mar-Jul..*Blue Anemone* **Anemone apennina**

Basal lvs adpressed-hairy above, soon (±) hairless, occ purplish below. Petiole hairy to hairless. Fr stalk drooping. Mar-Jul..................................*Balkan Anemone* **Anemone blanda**

Plant totally hairless

Plant with long rhizomes. Basal lvs occ absent. Fls (and buds on rhizome) yellow. Mar-Jun ....*Yellow Anemone* **Anemone ranunculoides**

Plant with small round tubers. Basal lvs 1(2), to 11cm diam, orb, palmately 3-5-lobed, the lobes further divided into lobes, with veins indistinct, with stomata below only. Petiole to 15cm, round, hollow. Stem lvs (bracts) forming whorl around infl. Jan-May. PL 16 ....*Winter Aconite* **Eranthis hyemalis**

Group RJ – *Herb. Stipules absent. Basal lvs in rosette, or stem lvs present. Lvs hairy at least below*

■Petiole sheathing at base (on basal lvs at least). Hairs unicellular

Plant fetid

Basal lvs to 30cm, long-petiolate, 2-ternate or with 2 prs of 2-pinnate divisions; terminal lft ovate, often 3-lobed, acute, with channelled stalk, dark green above, paler below, hairless above (or with minute curved hairs on veins), sparsely long-hairy on veins below, minutely ciliate (hairs curved), incised-serrate, with pale hydathodes visible above, with 2(3)° veins raised below, with stomata below only. Stem lvs 1-4, much smaller. Petiole weakly channelled to triangular, with large irreg hollow, with >20 vb's (turning brown) around margin. Stems to 60cm, minutely hairy, ± round, solid to hollow. Mar-Oct. Limestone habs, N Eng.................*Baneberry* **Actaea spicata**

Plant odourless

Petiole channelled (occ weakly so)

Wet habs. Rhizomatous..*Marsh Cinquefoil* **Potentilla palustris***

Dry to damp habs. Tufted

Basal lvs mostly <10cm diam...(*Ranunculus*) Go to **RAN**

Basal lvs mostly >10cm diam. Lfts 3, palmately or pinnately arranged, 7-12 x 9cm, dark green above, often purplish below, densely hairy to ± hairless, deeply 3-7-lobed, serrate with white hydathodes, net-veined, with veins raised below, with stomata below only. Petiole long, hairy, round, weakly channelled, pith-filled, with scattered vb's. Hortal. *A. hupehensis* x *vitifolia*...*Japanese Anemone* **Anemone** x **hybrida**

Petiole round

Basal lvs long-petiolate, (1)2-ternate; lfts 2 prs, 3-6cm, ovate, glaucous below, often hairless above, softly hairy below, irreg 3-lobed and crenate with white hydathodes (visible end-on or above) in retuse apices, ± net-veined, with veins not or hardly raised below, with stomata below only. Petiole often purplish, usu softly hairy, with sheathing base, with 10-12 vb's around margin. Stems to 100cm, hairy, round. Mar-Oct

Fls with strongly hooked spurs. Petiole with hairy sheathing base ....*Columbine* **Aquilegia vulgaris**

Fls with ± straight spurs (or hooked at end). Petiole with hairless sheathing base. Hortal, alien...*Garden Columbine* **Aquilegia vulgaris** cv

■Petiole not sheathing at base

Hairs unicellular. Stems fetid when broken. Lvs alt.....................*Monk's-hood* **Aconitum napellus**

Hairs septate. Stems aromatic when broken. Lvs opp

Stem with purple-black resin canals. Lvs without glands or translucent dots. Ann

Lvs 5-15cm, lanc-elliptic, acuminate, petiolate, sparsely hairy to hairless both sides, shortly ciliate, deeply serrate with 9-11 hydathode-tipped teeth per side of main lobe, opaquely net-veined (Kranz venation), with 2° veins ± raised both sides and ending in teeth apices, with stomata both sides. Petioles often ± connate at base, long-ciliate nr base, with 3 vb's and spiral fibres. Stems to 80cm, hairless to ± hairy, with long-ciliate interpetiolar ridge

All lvs with (1)3 lfts..*Trifid Bur-marigold* **Bidens tripartita**

Some lvs often with 5 lfts. VR alien..*Beggarticks* **Bidens frondosa**

Stem without purple-black resin canals. Lvs with colourless sessile glands below, and minute translucent dots. Per

Lvs 5-10cm, 3(5)-lobed, petiolate (the upper ± sessile), sparsely hairy both sides, ciliate, net-veined, with 2° veins raised below, with stomata below only; lobes elliptic to oblanc, serrate with hydathode-tipped teeth. Petioles to 1cm, connate at base, flat, with 3 vb's. Stems to 150cm, purplish, antrorsely crisped-hairy, round to weakly angled, striate. Mar-Oct ....*Hemp-agrimony* **Eupatorium cannabinum**

Group RK – *Herb. Stipules absent. Basal lvs in rosette, or stem lvs present. Lvs hairless (*Helleborus *have ± sessile glands when young).* ❶ *Upper lvs perfoliate*

▪Petiole hollow (occ solid when young), sheathing at base

Petiole with abundant greenish-cream latex (turning brown)

Lvs 1-2-ternate; lfts 4-10cm, broadly ovate, pine- or celery-scented, with narrow cartilaginous margins (entire or remotely scabrid), irreg lobed, with stomata below and occ scattered along veins above; lobes serrate with shortly aristate teeth. Petiole usu reddish at base, with minute papillae-like hairs nr lf axils, round, shallowly ridged, with vb's scattered around 2 vertical hollows. Stems 30-100cm, ridged, hollow. Tufted per. Mar-Oct .... *Masterwort* **Peucedanum ostruthium**

Petiole with white latex (occ sparse)

Lvs 1-3-ternate, the basal soon dying, the upper sessile; lfts to 4 x 3cm, ovate to rhombic, often asymmetric at base, stalked, shiny dark green both sides, crenate-serrate, with white protruding hydathodes (unlike other *Apiaceae*), net-veined, the veins sunken above but with raised cartilaginous ridges, with stomata below only. Petiole often auriculate and ciliate at base, with minute hairs in lf axils, laterally flattened, with (6)11 vb's around hollow. Stems to 150cm, furrowed, solid, later hollow, celery-scented. Tufted per. Nov-Jul. Often nr coast .... *Alexanders* **Smyrnium olusatrum***

Lvs 1-2-ternate, the basal soon dying, the upper perfoliate; lfts 2-6 x 5cm, ovate, often asymmetric at base, stalked, light green, with stomata both sides but often confined to vein margins above (the upper lvs usu with stomata below only). Petiole with ciliate sheathing base, with dense tuft of minute bristles in lf axils, channelled to flat above, with 7 vb's, odourless. Stems to 120cm, round below, with 2 ciliate cartilaginous ridges above, pith-filled or hollow. Tufted bi or per, with swollen taproot. ❶. Feb-Aug. VR alien .... *Perfoliate Alexanders* **Smyrnium perfoliatum**

Petiole without latex, without latex

Lvs 1-2-ternate; lfts 2-5 x 3cm, ovate to rhombic, shiny bright green, the terminal with channelled stalk, the laterals sessile, serrate distally with cartilaginous teeth apices occ purple, occ lobed, with stomata below only. Petiole purplish at auriculate base, laterally flattened, not channelled, striate, with sweet celery odour. Stems to 50cm, stout, often purplish nr base, hairless, ± round, striate. Tufted per. Apr-Oct. Coastal, Scot .... *Scots Lovage* **Ligusticum scoticum**

▪Petiole solid

Petiole sheathing at base

Lvs 1-2-ternate; lfts 5-30cm, linear-lanc to linear, ± curved, glaucous, with strongly serrate (± spinescent) cartilaginous margins, with midrib raised below, with anastomosing 2° veins and with ± parallel lateral veins each side of midrib, with stomata both sides. Petiole to 15cm, with long non-auriculate sheathing base, channelled at least nr base, with 5 vb's. Rhizomes swollen. Mar-Oct (all yr). R alien .... *Longleaf* **Falcaria vulgaris**

Petiole not sheathing at base (may be dilated)

Lfts >5cm, tough

Basal lvs 2, arising with fls. Mar-Oct

Lfts 7(11), to 10 x 3cm, narrowly elliptic, acute, rolled when young, faintly fetid, dark green above, paler below, serrate (exc nr base), net-veined, with midrib and 2-3° veins raised below, with stomata below only. Stem lvs (bracts) smaller, sessile. Petiole to 30cm, channelled, with 7-11 vb's in U-shape. Stems to 40cm, hairless or sparsely hairy above. Humus-rich calc woods........................ *Green Hellebore* **Helleborus viridis** ssp **occidentalis**

Basal lvs absent. All yr

Lfts (3)5-9(11), 7-11 x 0.7-1.5cm, narrowly lanc, acute, rolled when young, fetid, dark bluish-green above, paler below, occ reddish at petiole join, serrate at least distally, not net-veined, with midrib raised below, with stomata below only. Petiole usu channelled, with 3-9 vb's. Stems to 80cm, with lf scars ≥½ diam. Calc woods or garden escape .... *Stinking Hellebore* **Helleborus foetidus**

Lfts 3, to 15 x 7cm, elliptic-ovate, acute, rolled when young, not fetid, shiny dark green above, paler below, deeply serrate, reddish at petiole join, net-veined, with midrib raised below, with stomata below only. Petiole channelled, with several vb's. Stems to 120cm, with lf scars ≥½ diam. VR alien................................*Corsican Hellebore* **Helleborus argutifolius**

Lfts to 2cm, thin to translucent

Lfts obovate, apiculate, with minutely crenate-serrate hyaline margins (x20), dull light green above, with veins ± obscure, with stomata below only. Petiole not dilated at base, triangular, channelled, brittle, without sap, with 1(3) vb's and spiral fibres. Rhizomes long, whitish, with persistent lf scars. Mar-Jun..*Moschatel* **Adoxa moschatellina**

Group RL – *Herb with viscid glandular septate hairs. Lvs pinnately 3-foliate, with lfts folded when young, serrate, with obscure hydathodes, with midrib raised below, not net-veined, with ± opaque 2° veins, with stomata both sides. Stipules adnate to petiole for >½ length, clasping the stem, serrate. Stems round*

■Per. Stems 10-60cm. Stipules 3-15mm

Stems ± prostrate, tough but hardly woody, usu green, rarely spiny (var *horrida*), ± hairy all round

Terminal lft 10-20 x 6-15mm, obovate, obtuse to emarginate, with terminal tooth usu ≤ laterals, with 14-18 teeth per side. Petiole 3-5mm. Rhizomatous and stoloniferous. Mar-Oct ....*Common Restharrow* **Ononis repens**

Stems ± erect, woody, usu reddish, rarely spineless (var *mitis*), hairy mostly in 1(2) lines

Terminal lft 4-12 x 4-6mm, narrowly elliptic, usu acute, with terminal tooth > laterals, with 10-12 teeth per side. Spines to 2(3)cm, straw coloured. Tufted. Mar-Oct. Gsld on clay ....*Spiny Restharrow* **Ononis spinosa**

■Ann. Stem(s) 4-8cm, erect. Stipules 1-4mm. VR

Terminal lft 3-5mm, obovate, cuneate, serrate at apex, with terminal tooth equal to laterals, densely glandular-hairy. Petiole 2mm. Stem(s) glandular-hairy. Mar-Aug. Coastal cliffs, Devon, S Wales. Dunes, Channel Is. Sch8..*Small Restharrow* **Ononis reclinata**

Group RM – *Lvs pinnately 3-foliate, with lfts folded when young, not net-veined; terminal lft with 1.5-7mm stalk.* ❶ *Plant usu with black-blotched lfts and septate hairs (hairs unicellular in other taxa).* ❷ *Lvs only pinnately 3-foliate above lowest few nodes*

■Stipules deeply laciniate. Ann. Plant odourless, usu <30cm tall

Stipules adnate to petiole only nr base

Lfts (±) hairless, to 10(25)mm, obovate to obcordate, truncate at apex, serrate nr apex, with 6-9 prs of 2° veins; terminal lft with 3(4)mm stalk. Stems hairless. Oct-Jun ....*Toothed Medick* **Medicago polymorpha**

Lfts with ± spreading hairs both sides, to 10mm, obtruncate or rhombic, mucronate, dentate nr apex, with c8 prs of 2° veins; terminal lft with 3(4)mm stalk. Stems hairy. VR alien ....*Strong-spined Medick* **Medicago truncatula**

Stipules adnate to petiole for ≥½ way. Lfts hairless above

Lfts 5-12 x 5-13mm, obcordate to obovate, without darker flecks, sparsely adpressed-hairy below, with recurved mucro, with (3)4(6) teeth per side, 4-5 veins per side (occ forking); terminal lft with stalk ≥4x lateral lft stalks. Stems densely white adpressed-hairy when v young, soon ± hairless. Plant soon hairless. VR alien............*Early Medick* **Medicago praecox***

Lfts 8-11 x 4-6mm, obcordate, with darker flecks, adpressed-hairy below. Plant sparsely hairy. VR alien............*Tattered Medick* **Medicago laciniata**

■Stipules shallowly toothed to entire

Plant hay-scented, with sparse adpressed hairs when young, soon hairless. Stipules usu bristle-like, 3-8mm. Lfts apiculate, thick, hairless above, soon hairless below, deeply serrate (often to nr base), with obscure or opaque veins, with stomata below only. Stems weakly angled, solid or hollow

Lfts to 12 x 6mm; terminal lft with stalk 1-2.5x lateral lft stalks. Stipules scarious, whitish, dilated and toothed nr base. Stems usu <50cm tall. Ann. Ruderal, R alien ....*Small Melilot* **Melilotus indicus**

Lfts 15-30mm; terminal lft with stalk 3.5-7x lateral lft stalks. Stipules herbaceous, green, entire, stiff. Stems 60-120cm tall. Bi (per)

Lfts pale green when young, contrasting with dull grey-green older lfts, to 15mm wide, rhombo-ovate (upper oblanc). Petiole to 8mm. Fls white. Frs hairless. Ruderal ....*White Melilot* **Melilotus albus***

Lfts all mid-green. Fls yellow

Lfts to 25mm wide, obovate to ovate, apiculate, those of upper lvs oblong-elliptic. Frs hairless. Ruderal............*Ribbed Melilot* **Melilotus officinalis**

Lfts to 8mm wide, oblong or obovate, apiculate, those of upper lvs ± parallel-sided. Frs with adpressed hairs. Usu open calc habs............*Tall Melilot* **Melilotus altissimus**

Plant odourless, often hairy. Stipules never bristle-like

Stems 30-90cm tall, ± erect. Per

Lfts hairy to ± hairless, toothed mostly in distal ½, usu with 6-11 opaque veins per side, with stomata both sides; terminal lft with stalk to 6x lateral lft stalks. Petiole to 11mm. Stipules adnate for ≤ length, often almost clasping at base, with free part to 1cm, ovate-lanc, acuminate. Stems hairy to hairless, often with small hollow. All yr

Tufted. Lfts to 30 x 12mm, ovate, often mucronate. Stems ± square (to ± round). Fls purple ....*Lucerne* **Medicago sativa** ssp **sativa***

Shortly rhizomatous. Lfts to 12(20) x 5-8mm, narrowly elliptic, mucronate. Stems ± round. Fls yellow............*Sickle Medick* **Medicago sativa** ssp **falcata**

Stems usu <30cm tall, often ± prostrate. Ann (rarely per)

Lfts 3-6mm, densely softly or silky-hairy both sides

Lfts obovate, often emarginate and apiculate, entire or serrate at apex, with c6 veins per side, with stomata both sides; terminal lft stalked to 3mm. Stipules ovate, acute or acuminate, entire to ± entire............*Bur Medick* **Medicago minima**

Lfts >6mm, or if 3-6mm then sparsely hairy to hairless above

Lfts usu with a large black blotch above. ❶

Lfts 7-25mm, obovate or obcordate, minutely apiculate, hairless above, sparsely adpressed-hairy on veins below, serrate nr apex, with 4-6 translucent veins per side, with stomata occ both sides. Petiole with translucent channel, with 3 vb's. Stipules ovate, acute to acuminate, toothed (to ± entire). Stems square........*Spotted Medick* **Medicago arabica***

Lfts without a black blotch (or small and inconspicuous if present)

Plant hairless (but stipules may be ciliate)

Lfts 3-10mm, obovate to obcordate, truncate, not mucronate, cuneate at base, occ with black (or pale) blotch above and obscure red dots esp nr margins, sharply serrate at least nr apex, with 3-5 prs of 2° veins, with stomata below only. Petiole to 4cm. Stipules adnate to petiole for >½ length, to 15mm, lanc, acuminate. Sept-Jul ....*Bird's-foot Clover* **Trifolium ornithopodioides**

Plant usu with some hairs (often v sparse or confined to young stems and lvs)

Lfts 6-15mm wide

Lfts (3)6-20mm, obovate, apiculate, (silky-) hairy to hairless both sides, serrate nr apex, with translucent veins, with stomata both sides. Petiole with 1(3) vb's. Stipules adnate to petiole for ½ length, acuminate with aristate apex, clasping at base, weakly toothed or entire. Stems usu ± prostrate, hairy to hairless (v rarely glandular-hairy). All yr ....*Black Medick* **Medicago lupulina***

Lfts to 5mm wide. ❷

Stems with dense spreading or adpressed hairs. Lfts 6-10mm, obovate to obcordate, often without tiny mucro, cuneate at base, hairless above, occ sparsely adpressed-hairy on midrib below, with up to 11 teeth per side in distal ½, with translucent veins mostly unbranched, with stomata both sides; terminal lft with stalk >1.5mm. Petiole to 17mm, hairy. Stipules adnate to petiole for <½ length, clasping at base, with free part 5mm, ovate, acute, ± entire...*Hop Trefoil* **Trifolium campestre**

Stems sparsely hairy to hairless. Lfts 3-8(11)mm, obcordate or obovate, often without tiny mucro, cuneate at base, occ with red dots above, hairless or with scattered long hairs (occ adpressed) below, with translucent veins mostly unbranched, with stomata below and often above; terminal lft with 0.5-1mm stalk. Petiole 2-4mm, sparsely hairy to hairless. Stipules adnate to petiole for ≥½ length, broadly ovate, acuminate ....*Lesser Trefoil* **Trifolium dubium**

Medicago arabica stipules

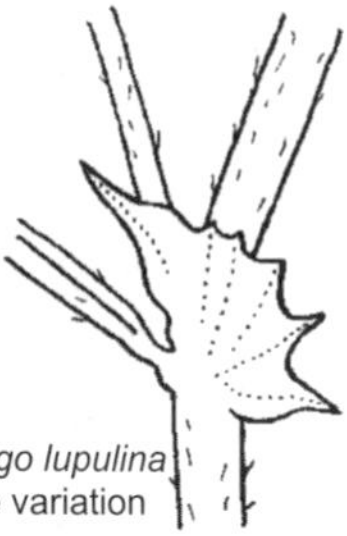

Medicago lupulina stipule variation

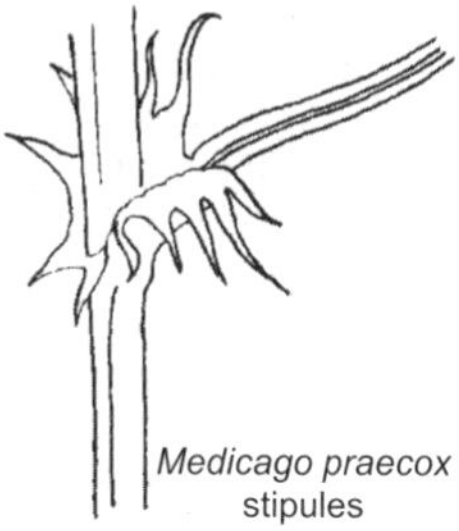

Medicago praecox stipules

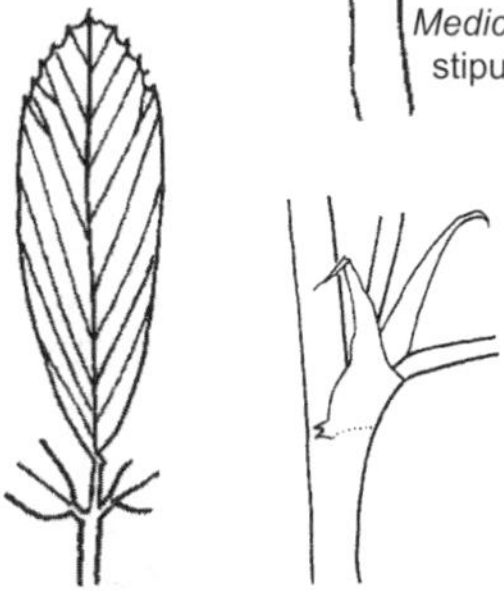

Medicago sativa ssp *sativa* lft & stipules

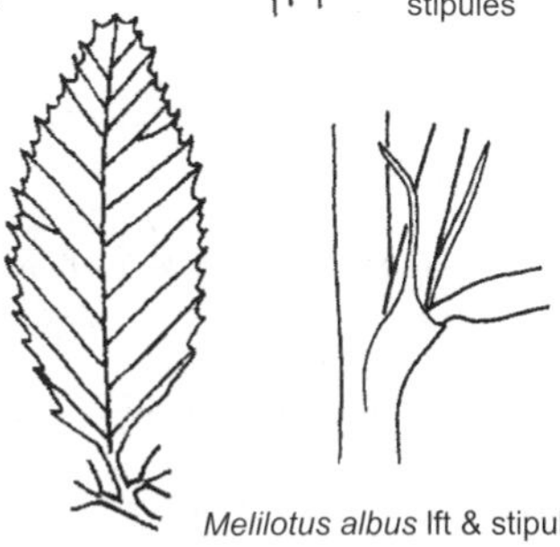

Melilotus albus lft & stipules

Group RN – Trifolium. *Lvs pinnately 3-foliate, with lfts folded when young, not net-veined, with stomata below (and often above); terminal lft with 0-1mm stalk. Stems round. Hairs unicellular.* ❶ *Lfts usu white-blotched.* ❷ *Lfts often black-blotched.* ❸ *Lfts with veins thickened nr margins*

■Terminal lft usu >2.5x as long as wide (beware lfts of first lvs, which may be wider)

Lfts hairless. Ann

Lfts 5-15 x 5mm, narrowly elliptic, shiny green below (often with reddish veins), with v narrow hyaline margins, sharply serrate with >20 teeth (and veins) per side ending in stalked glands. Petiole 5-10mm. Stipules broadly ovate, hyaline, whitish, with gland-tipped teeth. Oct-Jun. VR, Lizard..*Upright Clover* **Trifolium strictum**

Lfts hairy at least below

Rhizomatous per

Lfts hairless above, adpressed-hairy below, 20-60 x 7-20mm, ovate to elliptic, ± obtuse, often with faint whitish blotch above, entire to weakly toothed, with recurving 2° veins. Petiole to 8cm, usu ± adpressed-hairy, channelled. Stipules with subulate free part to 2cm, occ red-veined. Stems 10-30cm tall, usu ± adpressed-hairy. ❶. Mar-Oct (all yr) ....*Zigzag Clover* **Trifolium medium***

Lfts adpressed-hairy both sides, 30-60 x 20mm, lanc, truncate or emarginate, entire. Petiole spreading-hairy, channelled. Stipules with linear free part to 3cm (occ bifid), green-veined. All yr. VR alien..*Hungarian Clover* **Trifolium pannonicum**

Ann

Lfts 5-15(25) x 2-6mm, narrowly obovate-oblong, minutely mucronate, grey-green, softly ± adpressed- to spreading-hairy, often slightly toothed at apex. Petiole 2-30mm, spreading-hairy. Stems adpressed-hairy (occ spreading-hairy). Stipules bristle-tipped to 8mm, hairy, reddish-veined. Sandy or rocky habs. Oct-Aug...............*Hare's-foot Clover* **Trifolium arvense**

Lfts 10-25 x 5(9)mm, narrowly obovate, often minutely mucronate, mid-green, (±) adpressed-hairy both sides, ± entire, with 12 prs of 2° veins. Petiole 1-10cm, ± hairy. Stipules with long linear-lanc free part to 12mm, not reddish-veined. Stems spreading- or adpressed-hairy. Oct-Aug. Grassy habs nr sea, R, S Eng.....................................*Sea Clover* **Trifolium squamosum**

■Terminal lft always <2.5x as long as wide

Lfts hairy above and/or with (±) entire margins

Per. Lvs >12mm

Stipules soon red-veined, with 3mm bristle-tipped apex. Lfts 15-30 x 13mm, ovate, occ minutely mucronate, often with white chevron above, hairy at least below (often adpressed), obscurely crenate or entire. Petiole to 20cm, spreading-hairy to hairless. Stems with spreading hairs. All yr...*Red Clover* **Trifolium pratense***

Stipules green-veined, with long thread-like apex to 12mm. Lfts 15-30 x 11mm, oblong, obtuse to retuse, with dense ± adpressed or spreading hairs, entire. Petiole to 10cm, spreading-hairy. Stems with spreading hairs. All yr. E Eng ....*Sulphur Clover* **Trifolium ochroleucon**

Ann. Lvs mostly >12mm

Lfts (5)12-20(35)mm, obovate, entire. Petiole to 15cm. Stipules ovate, ± acute to obtuse, often dentate, green-veined. Stems 20-50cm, with spreading hairs. Alien, crop-plant relic ....*Crimson Clover* **Trifolium incarnatum** ssp **incarnatum**

Ann. Lvs mostly <12mm

Lfts and stems with spreading hairs

Lfts 5-12mm, broadly obcordate, not mucronate, often black-blotched, entire to obscurely toothed. Petiole usu with spreading hairs. Stipules broadly ovate, acute (but not aristate), herbaceous, not reddish-veined. Stems hairy. ❷ ....*Subterranean Clover* **Trifolium subterraneum**

Lfts 5-12mm, obovate to obdeltoid, often with minute mucro, toothed nr apex. Petiole with spreading hairs. Stipules broadly ovate, toothed, ± obtuse, with green apex, with purplish-green veins. Stems densely hairy. Coastal, VR.................*Starry Clover* **Trifolium stellatum**

Lfts and stems with adpressed hairs

Lfts truncate to acute at apex

Lfts 4-10 x 5mm, obovate, often mucronate, occ with dark blotch above, adpressed-hairy esp below, weakly toothed in distal ½, with 2° veins thickened and recurved at margins (obscure nr midrib). Petiole to 1.2cm, often purple-blotched. Stipules with a bristle-like tip to 0.5mm, often purple-veined. Stems reddish-spotted. ❷❸ ....*Rough Clover* **Trifolium scabrum***

Lfts 5-15mm, obovate, mucronate, adpressed-hairy esp below, weakly toothed to ± entire. Petiole to 4cm, occ ± spreading-hairy. Stipules with a 2-5mm subulate apex, often reddish or reddish-veined. Stems not spotted..............................*Knotted Clover* **Trifolium striatum**

Lfts retuse at apex

Lfts 5-12(35)mm, obcordate to ± orb, minutely mucronate, antrorsely hairy both sides. Stipules ovate or oblong, ± acute, entire or often dentate, green-veined. Petiole to 5(8)cm, adpressed-hairy. Stems 5-15cm. Sea cliffs, SW Eng ....*Long-headed Clover* **Trifolium incarnatum** ssp **molinerii**

Lfts hairless above, toothed at least distally

Per, with creeping stolons rooting at nodes, patch-forming. Lvs often >12mm

Lfts dull bluish-green below

Lfts 8-20mm, ovate or obcordate, oblong, slightly retuse, often minutely mucronate, dark green above, often with white chevron, with sparse long hairs on midrib below, serrate with 20-30 teeth per side (deepest in proximal ¼ of lft), with 2° veins branching repeatedly and slightly recurving nr margins. Petiole to 10cm, sparsely long-hairy or hairless, channelled or round, with 5 vb's. Stipules 15mm, with free part 6-10mm, acuminate, entire, never red. Often ❶. Apr-Oct. Often coastal..............................*Strawberry Clover* **Trifolium fragiferum**

Lfts ± shiny mid-green below

Lfts 10-30mm, obovate or obcordate, not mucronate, green above, usu with white chevron, (±) hairless, sharply serrate (often to nr base) with 5-8 teeth per side, with 7 ± unbranched 2° veins per side. Petiole usu hairless, channelled. Stipules oblong, with 1mm subulate apex, occ red-veined. ❶. All yr. PL 17..................................*White Clover* **Trifolium repens**

Lfts shiny dark green below

Lfts 5-10mm, obcordate to ± orb, often retuse, not mucronate, thick, dull ± glaucous-green above, occ with white chevron, hairless, often with bristle-tipped teeth, with 2° veins more (or only) translucent nr margins. Petiole sparsely hairy. Stipules with aristate apex to 1.5mm, purple-red. Stems hairless. Occ ❶. All yr. Coastal, SW Eng, S Wales, E Ire ....*Western Clover* **Trifolium occidentale**

Ann (occ per in *T. hybridum*), without stolons

Lfts mostly >12mm

Lfts (10)15-25(35) x 9-20mm, (±) hairless, often with midrib sparsely adpressed-hairy nr base below, often strongly serrate with cartilaginous teeth (excurrent veins), with 2° veins branched 1-3x, with stomata below only. Petiole to 10cm, hairless. Stipules with acuminate free part to 25mm, entire or toothed. Stems several, ascending, occ sparsely adpressed-hairy. Often all yr...*Alsike Clover* **Trifolium hybridum**

Lfts 8-25 x 6-9mm, narrowly obovate, hairy esp nr apex, toothed distally, with 8 prs of 2° veins (usu unbranched). Petiole 0.2-1cm, hairy to hairless. Stipules linear-oblong, acuminate, hairy nr apex. Stems 1-several, erect. VR alien..*Large Trefoil* **Trifolium aureum**

Lfts usu <12mm

Lfts (and/or stems) usu with some hairs (young plants of *T. dubium* may key out here)

Lfts 2.5-5 x 3mm, obcordate or obovate, occ with red dots above, occ hairs on midrib below, ± crenate at least in distal ½, with 3-6 veins per side. Stipules oblong to ovate, often ciliate. Stems usu hairless. Oct-Jul...................................*Slender Trefoil* **Trifolium micranthum**

Lfts 5-10(15)mm, obovate, occ mucronate, hairless above, sparsely adpressed-hairy below. Petiole to 5mm. Stipules oblong, with ciliate subulate apex, occ reddish. Stems ± hairy. Cornwall, VR..*Twin-headed Clover* **Trifolium bocconei**

Lfts (and stems) always hairless (young plants of *T. ornithopodiodes* may key out here)

Fls usu not clustered at ground level, pink, the corolla > calyx. Lfts 4-12mm, obovate, unnotched but minutely mucronate, bright green, sharply serrate with 7-10 prs of 2° veins, with stomata both sides. Petiole to 2cm. Stipules lanc to ovate, with long aristate point ....*Clustered Clover* **Trifolium glomeratum**

Fls often tightly clustered at ground level, white, the corolla < calyx. Lfts 5-10mm, obovate, occ emarginate, deep green and usu with dark chevron above, sharply serrate. Petiole to 2cm. Stipules ovate, acuminate, whitish when old. ❷ ....*Suffocated Clover* **Trifolium suffocatum**

Trifolium medium *Trifolium pratense* *Trifolium scabrum*

Group RO – *Lfts always 3, cuneate at base, net-veined, with 2° veins raised below and sunken above, the teeth tipped with reddish hydathodes. Petiole often hollow, with 1 or 3 vb's and spiral fibres. Hairs unicellular. (*Potentilla palustris *may key out here).* ❶ *Stipules lft-like*

■Lfts with terminal tooth usu shorter than its neighbours

Lfts dull dark bluish-green and hairy above, pale glaucous below, folded when young

Petiole with spreading hairs, channelled. Lfts 0.5-3.5cm, broadly ovate, ± truncate at apex, weakly adpressed-hairy above, more densely ± spreading-hairy below, adpressed-ciliate, crenate-dentate with (3)4-7 teeth per side, with stomata below only; terminal lft short-stalked to 2.5mm, the lateral lfts sessile. Stipules partly adnate to petiole, ovate-lanc, green to papery brown. Shortly stoloniferous..*Barren Strawberry* **Potentilla sterilis**

Lfts (±) shiny dark bluish-green and usu hairless above, paler below, folded-plicate when young

Petiole with sparse long ± adpressed hairs, weakly or not channelled. Lfts 5-8cm, orb to ovate, the terminal rounded at base, all stalked, ± adpressed-hairy below, deeply serrate-dentate with c6 teeth per side; terminal lft with stalk to 8mm, the laterals to 4mm. Stipules papery, long-acute. Stoloniferous. All yr. *F. chiloensis* x *virginiana*.. *Garden Strawberry* **Fragaria** x **ananassa**

■Lfts with terminal tooth ≥ its neighbours (may be narrower)

Lfts with >4 teeth per side

Lfts adpressed-hairy below. Stoloniferous

Terminal lft with a short stalk <1mm, the lateral lfts ± sessile

Lfts usu shiny yellowish-green and ± adpressed-hairy above, pale glaucous below, ciliate, 1-6cm, obovate to oblong, folded-plicate when young, serrate-dentate with 8-13 teeth per side. Petiole spreading or retrorsely hairy, weakly channelled. Stipules ovate-lanc, herbaceous or papery, entire. All yr. PL 17, 22...................*Wild Strawberry* **Fragaria vesca***

Lfts dull green and adpressed-hairy both sides, ciliate, to 3cm, ovate, folded-plicate when young, serrate-dentate with 7-9 teeth per side. Petiole antrorsely adpressed-hairy, weakly channelled. Stipules ovate-lanc, herbaceous or papery, entire. All yr. VR alien ....*Dry Strawberry* **Waldsteinia ternata**

Terminal lft with a stalk to 3.5mm, the lateral lfts with a 1-3mm stalk

Lfts to 5 x 3cm, ovate to elliptic, v weakly folded-plicate when young, ± shiny dark green and adpressed-hairy above, pale green below, deeply crenate (occ 2-serrate) with 9-13 teeth per side, with stomata below only. Petiole with antrorsely adpressed hairs, channelled, soon hollow. Stipules of basal lvs linear-lanc, herbaceous, entire, those on stem lvs smaller, oblong-ovate to linear and toothed. All yr...... *Yellow-flowered Strawberry* **Duchesnea indica**

Lfts spreading-hairy below

Stoloniferous

Lfts 5-7 x 5cm, ovate, ± shiny yellow-green and spreading-hairy above, weakly glaucous and densely hairy below, with 10-12 deep teeth per side, with stomata below only; terminal lft with a c2mm stalk, the laterals lfts with a 1-2mm stalk. Petiole densely spreading-hairy (occ retrorse). All yr...*Hautbois Strawberry* **Fragaria moschata**

Tufted (stem lvs of *Geum* may key out here but the basal lvs are 1-pinnate)

Lfts 1-7 x 1.5cm, obovate to elliptic, mid-green, spreading-hairy above (hairs often swollen-based), occ with scattered sessile glands, (sparsely) hairy below, deeply serrate or serrate-dentate with 6(8) teeth per side, weakly net-veined, with stomata both sides. Upper lvs ± sessile, rarely with 5 lfts. Petiole with hairs to 2.5mm (often swollen-based). Stipules c20mm, with free part 8mm, lanc, long-adnate, those of upper lvs larger, ovate and occ toothed nr base. All yr...*Ternate-leaved Cinquefoil* **Potentilla norvegica**

Lfts with ≤4 teeth per side

Lvs all basal, petiolate. Stipules entire

Lfts 0.5-1.5cm, ± orb, mid-green, with sparse adpressed hairs only on midrib below, ciliate, usu with 2-3 teeth per side, ± net-veined, with stomata both sides. Petiole sparsely adpressed-hairy, not channelled. Stipules long-adnate, ciliate. Aug-Mar ....(winter rosette) *Tormentil* **Potentilla erecta***

Lfts 0.5-1.5(2)cm, obovate, folded when young, bluish-green, occ purplish below, with sparse adpressed hairs both sides, ciliate, 3-dentate at truncate apex with terminal tooth narrower (but not shorter) than laterals, with stomata both sides. Stipules ovate, adnate to petiole for ½ length, with free part to 5mm, ± lanc, acute, green. Compact tufted per with persistent lf bases. Mar-Oct. Mtn turf..*Sibbaldia* **Sibbaldia procumbens**

Lvs mostly on stems, ± sessile. Stipules lft-like, large ❶

Lfts 1-2cm, with adpressed long hairs at least on veins below (often above), ciliate, incised-serrate with (2)3-4(5) teeth per side, with terminal tooth occ shorter than laterals, with stomata below only; terminal lft with stalk 0-2mm. Stems ± prostrate, not rooting at nodes, with adpressed or spreading hairs...*Tormentil* **Potentilla erecta***

Lvs mostly on stems, petiolate. Stipules entire (to deeply toothed), often large

Stem lvs with 3 lfts, each with 2-4(5) teeth per side, entire nr base. Basal lvs mostly with 5 lfts, rarely persisting; lfts 0.5-2 x 0.8cm, narrowly obovate, adpressed-hairy below, deeply dentate with 4-6 teeth per side. Petiole 1-3(10)cm. Stems ± prostrate, rooting at nodes, often branching. All yr...*Trailing Tormentil* **Potentilla anglica**

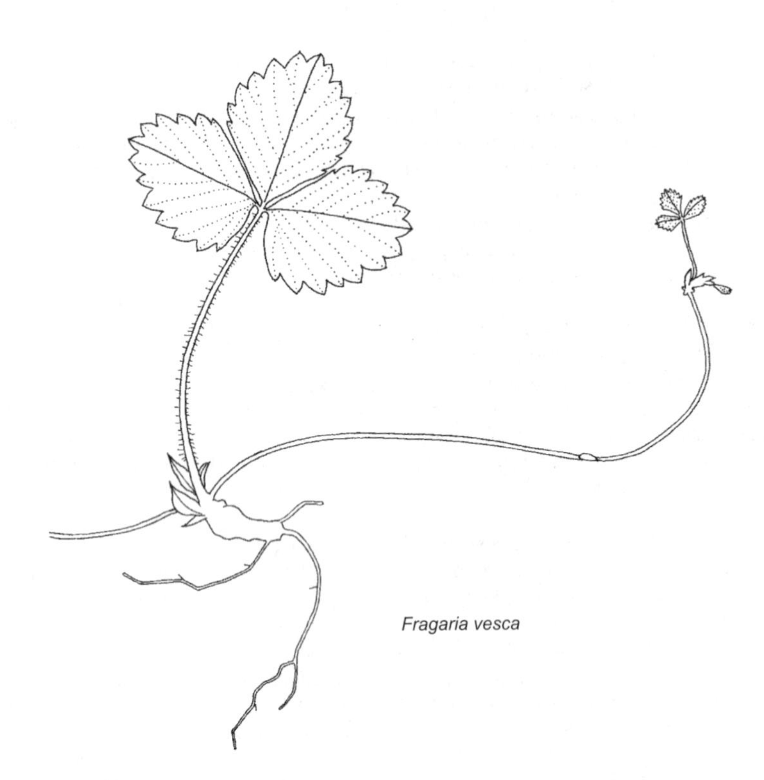

Fragaria vesca

Group RP – *Lfts 3-5, cuneate at base, the teeth usu tipped with reddish hydathodes (often absent in* Cannabis*), net-veined, with 2° veins raised below and sunken above, with stomata both sides (unless stated). Petiole channelled. Stipules partly adnate to petiole (free in* Cannabis*). Hairs unicellular*

■Lvs with a mixture of 3-5 lfts (lower lvs usu with 4 lfts). Per. Lfts not gland-dotted

Basal lvs not persisting. Stipules toothed. Stems rarely rooting at nodes. *P. anglica* x *erecta* ....**Potentilla** x **suberecta**

Basal lvs persisting, with (4)5 lfts; lfts 1-3 x 1.2cm, narrowly obovate, adpressed-hairy both sides, with 4-5 teeth per side, terminal tooth may be narrower but not shorter. Stem lvs with 3(5) lfts, each with 3 teeth. Petiole 1-2cm, rarely shorter than the shortest lft, often much longer. Stipules entire. Stems prostrate, rooting at nodes, rarely branched. *P. erecta* x *reptans* or *P. anglica* x *reptans*. All yr...*Hybrid Cinquefoil* **Potentilla** x **mixta**

Basal lvs rarely persisting, mostly with 5 lfts; lfts 0.5-2 x 0.8cm, narrowly obovate, adpressed-hairy below, deeply dentate with 4-6 teeth per side. Stem lvs with 3 lfts, each with 2-4(5) teeth per side but entire nr base. Petiole 1-3(10)cm. Stems ± prostrate, rooting at nodes, often branching. All yr ....*Trailing Tormentil* **Potentilla anglica**

■Lvs with 5 or more lfts (occ fewer on upper lvs)

Ann. Lfts gland-dotted

Lfts (3)5-7(9), 7-15 x 2cm, lanc, acuminate, dull grey-green above, paler below, resin-scented or odourless, rough above with v short hairs arising from glands, with short patent white hairs below, occ viscid-hairy, with 14 teeth per side, with 2° veins raised below and curving to tooth apices, with stomata below only. Petiole to 3cm, minutely antrorsely adpressed-hairy, with 1 involute vb. Stipules short, subulate (abortive lfy shoots may appear in lf axils). Stem usu 1, to 2.5m, erect, weakly or not branched. Apr-Oct...*Hemp* **Cannabis sativa**

Per. Lfts not gland-dotted

Lfts silvery silky-hairy below, hairless above, 5-9, with stomata absent above ....(*Alchemilla*) <u>Go to</u> **ALC A**

Lfts shortly white-woolly below, adpressed-hairy above

Lfts (3)5, to 3cm, with v narrowly recurved margins, deeply crenate-serrate with 5-6 teeth per side. Petiole 2-8cm, with 1(3) vb's and spiral fibres. Stipules with free linear herbaceous part to 1cm. All yr...*Hoary Cinquefoil* **Potentilla argentea**

Lfts not white-hairy below, hairy or hairless above

Lfts with >5 teeth per side

Lfts 4-10cm

Lfts 5-7, lanc to oblanc, densely spreading- or adpressed-hairy, serrate-dentate with 6-11(17) teeth per side. Petiole to 7cm, with spreading hairs, usu solid. Stipules with free part 0.9-2cm, the lowest almost thread-like, the upper broader, all entire or toothed (esp on stem) with long narrow terminal lobe. Stems with spreading hairs. All yr ....*Sulphur Cinquefoil* **Potentilla recta**

Lfts mostly <4cm

Lfts usu 5, 0.5-5 x 2cm, oblong-obovate, hairless to sparsely adpressed-hairy, ciliate, dentate with 5-10 reddish hydathode-tipped teeth per side. Petiole to 18cm, sparsely adpressed-hairy, with 3 vb's. Stipules with free part (0)2-8mm, linear to broadly lanc, entire. Stems rooting at nodes in summer. All yr................*Creeping Cinquefoil* **Potentilla reptans**

Lfts (3)5, 1-4 x 2cm, obovate, softly long-hairy, occ densely greyish long-hairy below, ciliate, deeply serrate-dentate to incised with 6 teeth per side. Mid and upper lvs with 3 lfts. Stipules to 10mm, linear, the upper usu toothed. Stems never rooting at nodes. All yr ....*Russian Cinquefoil* **Potentilla intermedia**

Lfts with <5 teeth per side

Lfts 0.5-2cm

Lvs mostly on stems, ± sessile; lfts 5 (lowest pr are actually stipules), with long adpressed hairs at least on veins below (and often above), ciliate, deeply serrate with (2)3-4(5) teeth per side, with terminal tooth occ shorter than laterals, with stomata below only; terminal lft with a 0-2mm stalk. Stems with adpressed or spreading hairs..*Tormentil* **Potentilla erecta***

Lfts mostly 2-4cm

Sepals clearly visible between separated petals. Lfts 5, 1.5-3.5 x 0.5-2cm, obovate, dark green and hairless above, paler and hairy on at least midrib below, ciliate, deeply serrate with 2-4(5) teeth per side in distal ½, with terminal tooth usu smaller than laterals, weakly net-veined, with 2° veins not raised below. Petiole adpressed-hairy, with 1(3) vb's. Stipules of stem lvs linear-lanc, acute, entire. Stems prostrate, mat-forming, hairy. All yr. Calc habs, usu lowland..*Spring Cinquefoil* **Potentilla neumanniana**

Sepals not visible between close or overlapping petals. Lfts 5, 1-3(4) x 0.5-2cm, obovate, dark green above, paler below, hairless or hairy above, ± adpressed-hairy on veins below, deeply dentate with 2-4 teeth per side, with terminal tooth hardly smaller than laterals, net-veined, with 2° veins weakly raised below. Petiole sparsely hairy, with 1(3) vb's. Stipules with free part 3 x 1mm, lanc, obtuse, herbaceous, with scarious margins, adpressed-hairy, entire. Stems ascending, only forming mats when grazed. All yr. Calc mtns ....*Alpine Cinquefoil* **Potentilla crantzii**

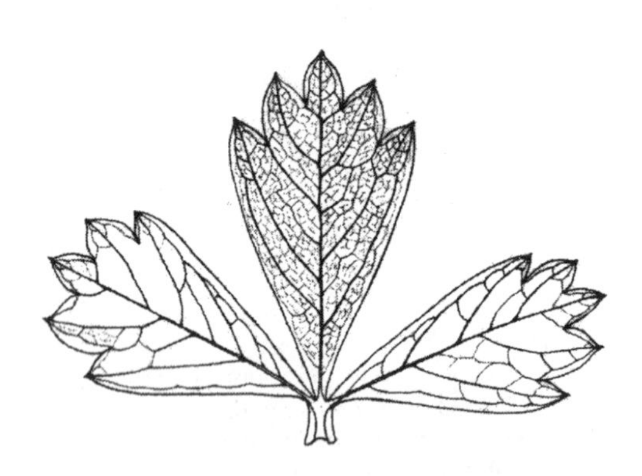

Potentilla erecta

Key to Groups in Division S

(Lvs 1-pinnate)

■Stem or petiole with prickles.......... **SA**
■Stem and petioles without prickles (lvs or stipules may be spiny)
 Tree or non-climbing shrub >10cm tall
  Lfts with entire margin.......... **SB**
  Lfts with toothed margins (rarely with single tooth)
   Lvs opp or whorled.......... **SC**
   Lvs alt.......... **SD**
 Woody climber, or scrambling shrub (lvs opp).......... **SE**
 Herbaceous plant or creeping dwarf shrub <10cm tall
  Lfts with entire margins
   Lvs whorled. Aquatic (*Myriophyllum aquaticum*).......... **SF**
   Lvs alt
    Terminal lft present; tendrils absent (or reduced to subulate points).......... **SG**
    Terminal lft absent; tendrils often present
     Lfts (0)1 pr.......... **SH**
     Lfts 2-16 prs .... **SI**
  Lfts with toothed margins (occ ± entire or lobed)
   Lvs opp (rarely 3-whorled).......... **SJ**
   Lvs alt
    Intercalary lfts present (i.e. lfts alternating large and small).......... **SK**
    Intercalary lfts absent (i.e. lfts same size or increasing along lf)
     Plant strongly tansy-scented.......... **SL**
     Plant not tansy-scented (may have cucumber, cress, celery or parsnip scent when)
      Stipules present.......... **SM**
      Stipules absent
       Petiole with sheathing base
        Lfts hairy above.......... **SN**
        Lfts hairless (or with sparse scattered hairs) above.......... **SO**
       Petiole without sheathing base.......... **SP**

Group SA – Rosa, Rubus. *Stems and/or petioles with prickles (suckers more prickly than typical stems). Lfts usu regularly toothed, folded when young (plicate in* Rubus*), short-stalked to 1mm, net-veined, with stomata below only. Stipules present, partly adnate to petiole. Stems round. Buds to 3mm, usu reddish, with 5 scales. For* Rosa, *frs are generally required for identification (many fertile hybrids also occur).* ❶ *Low shrub <50cm, with blackish hips (frs).* ❷ *Semi-evergreen.* ❸ *Prickles often opp on stem at base of lf*

■Stems pruinose, or lvs white/glaucous below

Lvs pale green both sides

Stems bi, whitish-pruinose, often with straight broad-based prickles, the older stems unarmed. Lfts 3-4 prs, opp, 4-7 x 2-3cm, ovate, sessile, minutely hairy above, hairy below, 1(2)-serrate, each tooth with a protruding hydathode. Terminal lft to 10cm, acuminate. Rachis hairy and with hooked spines. Suckering............................... *White-stemmed Bramble* **Rubus cockburnianus**

Lvs white or glaucous below

Stems bi, cane-like, ± pruinose, green, round, with slender weak straight prickles and minute hairs. Lfts 2(3) prs, 5-12cm, ovate or ovate-lanc, acuminate, without translucent dots (unlike *Rubus fruticosus* agg), green and minutely hairy above, densely white-woolly below, irreg dentate with mucronate teeth. Terminal lft with stalk usu much longer than those of lateral lfts. Petiole 2-7cm, minutely hairy. Stipules to 1cm, thread-like. Suckering........................ *Raspberry* **Rubus idaeus**

Stems per, not cane-like, pruinose, often reddish, round, with few slender prickles (abundant on suckers). Lfts glaucous blue-grey or purplish below, hairless, serrate. Stipules v narrow, sparsely gland-fringed. Stems ± wavy, hairless. Suckering... *Red-leaved Rose* **Rosa ferruginea**

■Stems woolly, not pruinose. Lvs not white/glaucous. Non-climbing shrub, to 2m tall

Lfts 2-4(9)prs, 2-5 x 3cm, ± elliptic to ovate, acute, rugose, occ with recurved margins, shiny dark or yellow green above, hairless above, hairy to woolly below (and with ± sessile glands), crenate-serrate, with stout prickles on midrib below. Petiole 5-13cm, densely hairy, with unequal prickles. Stipules 1.5-2.5cm, hairy, shallowly toothed, with red gland-tipped cilia. Stems v spiny, with spines v unequal (larger spines hairy at least nr the base), densely hairy, often with stout glandular hairs. Rhizomatous, occ suckering................................... *Japanese Rose* **Rosa rugosa**

■Stems not woolly or pruinose. Lvs not white/glaucous

Low shrub <50cm, suckering ❶

Lfts 3-5 prs, 0.5-1.5cm, ± orb to ovate, obtuse, with purple margins, usu hairless, serrate, with midrib raised below. Stipules with narrow free part ± divergent, lf-like, toothed with stalked glands. Stems with dense unequal stiff slender prickles. Mostly coastal .... *Burnet Rose* **Rosa spinosissima**

Climbing or trailing shrub >50cm, with arching stems, not suckering

Lfts with glands below

Lfts 1.5-3.5cm, with strongly aromatic glands below, hairy below, glandular-multiserrate. Stems with strong broad-based prickles, without acicles (unlike *R. rubiginosa*). Fl stalks 1-2cm, glandular. Sepals reflexed, soon falling. Stylar orifice small .... *Small-flowered Sweet-briar* **Rosa micrantha**

Lfts 3-6cm, with weakly resin-scented (or odourless) glands below, hairy to densely so both sides, 2-serrate. Stems green, with strong slender prickles and acicles. Fl stalks 2-3.5cm, densely glandular. Sepals spreading after fl, persistent. Stylar orifice small .... *Harsh Downy-rose* **Rosa tomentosa**

Lfts without glands (occ on petiole or rachis)

Lfts hairless or with hairs below (occ sparse or confined to midrib)

Lvs semi-evergreen ❷

Lfts 3-4 prs, 2-6 x 1-3cm, broadly obovate, occ acuminate, ± thin, shiny dark green above, paler below, hairless or sparsely hairy on midrib below, serrate with hydathode-tipped teeth. Petiole (and rachis) hairy, usu sparsely glandular, with pricklets. Stipules 10-15 x 1.5mm, v narrow, glandular-laciniate. Stems to 3m, prostrate and occ rooting at the nodes, or climbing and hanging, hairless at maturity, the prickles sparse, small, slender, curved and occ reddish.. *Memorial Rose* **Rosa luciae**

Lvs deciduous

Lfts bluish-green at least below, often reddish

Lfts 1-2-serrate. Stems often reddish, weakly prickly with small broad-based strongly curved prickles. Stipules wide, often reddish. Fl stalks 0.5-1.5cm, smooth, often hidden by large lfy bracts. Sepals spreading-erect. Stylar orifice ⅓ disc diameter. Stigmas form hairy dome

Lfts hairless...*Glaucous Dog-rose* **Rosa caesia** ssp **caesia**

Lfts hairy at least on veins below................*Hairy Dog-rose* **Rosa caesia** ssp **vosagiaca**

Lfts mid- to grey-green above, rarely reddish

Lfts usu hairy below

Stems with v broad-based curved prickles, usu hairless. Lfts 2-3 prs, 3-6 x 1.5-2(3)cm, ovate-lanc, dark green above, paler below, serrate. Petiole (and rachis) usu hairy, glandular, with pricklets. Stipules 15-25 x 2-5mm, sparsely gland-toothed ....*Short-styled Field-rose* **Rosa stylosa**

Stems unarmed or with small broad-based curved prickles, usu hairless. Lfts 2-4 prs, 1-5 x 0.8-2.8cm, ovate to obovate, dull mid-green above, paler below, shallowly crenate-serrate. Petiole (and rachis) with short glands and pricklets. Stipules to 15 x 1mm, glandular-pinnate. ❷.................................*Many-flowered Rose* **Rosa multiflora**

Lfts hairless, or with sparse hairs confined to midrib (occ 2° veins) below

Fl stalks 1.5-2.5cm, smooth. Sepals 2-5cm, pinnately lobed, reflexed. Styles hardly protruding. Lfts 2-3 prs, 1.5-4cm, ovate to obovate or elliptic, acute, bright to dark green above, usu 2-serrate with c17 teeth per side. Petiole often reddish. Stems ± erect, ± self-supporting, stout, ± straight, often reddish, hairless; stem prickles narrowed from broad 5mm base (± equal spine length), strongly curved or hooked ....*Dog-rose* **Rosa canina**

Fl stalks 2-5cm, with stalked glands. Sepals to 1cm, ± entire, spreading. Styles strongly protruding. Lfts 2-3 prs, 1-2.5cm, ovate to elliptic, acute, grey-green above, serrate with 7-15 teeth per side. Petiole usu green. Stems weakly trailing or climbing and hanging down, slender, zig-zagging, green, hairless; stem prickles slender, occ abruptly narrowed from narrow base (much shorter than spine length), usu ± hooked. Shady habs..*Field-rose* **Rosa arvensis**

Non-climbing shrub >50cm

Lfts with abundant colourless viscid apple-scented stalked glands below

Lfts 2-3 prs, 1-2cm, dark green, hairy on veins (esp below), glandular-ciliate. Stipules hairless, glandular-ciliate. Stems erect; stem prickles usu unequal, broad-based, hooked ....*Sweet-briar* **Rosa rubiginosa**

Lfts with abundant small reddish resin-scented ± sessile glands below (occ confined to 2° veins), OR glands absent

Lfts strongly rugose, sparsely hairy both sides

Lfts 3-4 prs, 3-6cm, bright green, shallowly crenate-serrate. Petiole (and rachis) densely hairy, often glandular. Stems hairy or glandular-hairy; stem prickles v unequal, slender, slightly downcurved, yellowish, often sparsely hairy...............*Dutch Rose* **Rosa 'Hollandica'**

Lfts never rugose, usu hairless

Lfts 2(3) prs, to 9 x 6cm, shiny bluish-green above, often bronze when young, occ hairy on veins below. Stipules (1)3-4cm, with free part to 0.5cm, v fimbriate toothed. Stems with v broad-based prickles, often with glandular acicles......*Red Rose (of Lancaster)* **Rosa gallica**

Lfts not or weakly rugose, at least sparsely hairy on veins below

Lfts cuneate at base

Lfts narrow, cuneate at base, densely glandular below. Stems erect but ± wavy; stem prickles ± equal, curved. Sepals reflexed, soon falling. Stylar orifice small. Hips occ slightly urn-shaped (narrowed below the disc)..................*Small-leaved Sweet-briar* **Rosa agrestis**

Lfts rounded at base

Lfts 2-3.5cm, broadly ovate, not rugose, dark green, often with odourless glands below, hairy at least on veins below, finely 2-serrate with reddish hydathode-tipped teeth. Petiole (and rachis) hairy, often with glands. Stipules to 15 x 4mm, glandular-ciliate, with narrow acute auricles at base. Stems 1-2m, arching; stem prickles broad-based, abruptly contracted to a long fine point, strongly curved. Fl stalks 0.5-1cm, without glandular hairs. Sepals 1.5-2cm, 2-pinnate, strongly reflexed after fl, soon falling. Hips 1-1.5cm. Stigmas forming a small globose head. Stylar orifice <<⅓ disc diameter. Not suckering ....*Round-leaved Dog-rose* **Rosa obtusifolia**

Lfts 2-4cm, often elliptic, not rugose, grey-green, often densely glandular below and strongly resin-scented when young, usu densely hairy below, irreg glandular-serrate. Petiole (and rachis) glandular-hairy. Stipules 15-20 x 7-12mm, glandular-ciliate, without auricles. Stems to 1.5(2)m, erect, straight; stem prickles slender, straight. Fl stalks 0.5-1cm, glandular. Sepals 2-2.5cm, ± entire, with lft-like apices, erect after fl, persistent. Hips 1.5-3cm. Stigmas forming a large hairy dome. Stylar orifice ½ disc diameter. Suckering, with ± compact habit..*Soft Downy-rose* **Rosa mollis**

Lfts 2.5-4cm, ovate to broadly elliptic, ± rugose, bluish-green above, with or without glands below, densely hairy both sides (occ only on the veins below), glandular-multiserrate. Petiole (and rachis) densely hairy, with sparse glandular hairs. Stipules 15-20 x 8-10mm, gland-fringed, without auricles. Stems 1-2m, slender, erect, ± glaucous or reddish, often zig-zagging below or wavy nr base and apex; stem prickles usu slender, arcuate-acuminate or declining, with weak bases. Fl stalks 1-1.5cm, glandular. Sepals 1.5-2cm, pinnately lobed, erect after fl, not v persistent. Hips 1.5-2.5cm. Stigmas forming a large hairy dome. Stylar orifice ⅓ disc diameter. Not suckering ....*Sherard's Downy-rose* **Rosa sherardii**

Group SB – *Tree or shrub. Lvs with terminal lft. Lfts with entire margins. Stipules present (absent in* Juglans regia*).*

■Tree to 30m. Buds naked. Deciduous

Lfts 5-9 prs, 2-5 x 1-3cm, oblong, often retuse or with mucro, with 1-3mm stalk, each with a 1mm extra-floral nectary (or stipel) at base (soon falling), folded when young, dull dark green above (yellowish in cultivar 'Frisia'), grey-green below, with minute sparse adpressed hairs when young, net-veined, with 2° veins flat above, with stomata below only. Petiole 0.5-4cm, swollen at base and hiding bud. Stipules 4mm, often spiny when mature. Twigs reddish-brown, zig-zagging, angled, developing lenticels, hard-wooded, soon hairless. Buds 3-4 together, minute, naked. Usu suckering..*False-acacia* **Robinia pseudoacacia**

Lfts usu 2-4 prs, 8-23 x 4-10cm, obovate to elliptic, acute or acuminate, ± sessile, folded when young, dull dark yellow-green, with fruity scent, rarely toothed (on vigorous shoots), hairy when young, soon hairless exc for vein axils below with c10 prs of 2° veins slightly raised above, with stomata below only. Petiole glandular-hairy, swollen at base. Stipules absent. Twigs glandular-hairy when young, soon hairless, greenish-grey, with chambered pith; lf scars Y-shaped, with 3 large traces each consisting of 3 small traces. Buds to 7mm, broadly ovoid, squat, blackish, hairless but terminal bud greyish and furry. Not suckering............................*Walnut* **Juglans regia**

■Shrub to 2.5m. Buds with scales. (The strongly aromatic *Santolina chamaecyparissus* may key out here but has naked buds)

Evergreen. Lfts ± waxy-looking, hairless

Lfts 3-4 prs, to 1.5 x 1cm, obdeltoid, mucronate, with c4 prs of obscure opaque veins, with stomata both sides. Stipules <1mm, brown, scarious. VR alien ....*Shrubby Scorpion-vetch* **Coronilla valentina**

Deciduous. Lfts not waxy-looking, usu hairy

Lfts 2 prs (appearing palmately arranged at first glance), 1-3cm, oblong-lanc, acute, with revolute margins, silky-hairy both sides, net-veined, with sunken veins above, with stomata below only. Petiole 5-10mm, antrorsely adpressed-hairy. Stipules to 10mm, lanc, scarious, persisting for 1-3 yrs. Twigs hairy when young, with reddish-brown bark flaking in later yrs. Buds 2-3mm, reddish-brown, naked.............................*Shrubby Cinquefoil* **Potentilla fruticosa**

Lfts 3(4) prs, to 2.5 x 1.5cm, obovate, slightly retuse to truncate at apex, with tiny mucro, with 0-1mm stalk, dull green above, ± glaucous below, hairless or sparsely adpressed-hairy below, with indistinct opaque 2° veins, with stomata below only. Stipules 1mm, green. Twigs green, soon hairless, ridged. Buds 2-4, green, with 1 pr of scales. VR alien ....*Scorpion Senna* **Hippocrepis emerus**

Lfts 4-11 prs, to 2 x 1.5cm, ovate, slightly retuse or truncate at apex with tiny mucro, wth 1-3mm stalk, folded when young, dull green above, silky-hairy below, ± net-veined, with midrib raised below, with stomata occ above. Petiole to 15mm, antrorsely adpressed-hairy, slightly channelled. Stipules 4mm, ovate-lanc, brown-green, acute, entire. Twigs pale green, purplish above, adpressed-hairy (x20), angled. Buds 2-4mm, superposed, the upper developing into a shoot, with several scales...*Bladder-senna* **Colutea arborescens**

Group SC – *Deciduous tree or shrub. Lvs and lfts opp. Lfts with toothed margins, net-veined, with stomata below only. Hairs unicellular. (Toothed lvs of* Juglans regia *will key out here).* ❶ *Lvs usu 3-whorled at twig apex*

■Large tree. Lfts folded when young. Stipules absent. Twigs solid (hard-wooded)

Twigs with interpetiolar ridge, often pruinose, sympodially branched. Buds with 2-3 prs scales

Buds minutely white-hairy. Lfts 1-3 prs, to 14 x 7cm, ovate-elliptic, acute, pale (yellow-) green, hairless above (exc midrib), sparsely hairy on veins and in vein axils below, ± toothed (mostly in distal ½). Petiole to 10cm, round, swollen at base. Twigs round, developing lenticels, otherwise smooth, hairless..*Ashleaf Maple* **Acer negundo**

Twigs without interpetiolar ridge or raised lenticels, greyish, monopodial; lf scars ± round, with shallow arc of 7 traces. Buds with 3 prs scales. Petiole swollen at base

Buds black, scurfy. Lfts 4-6 prs, 6-9 x 2-3cm, oblong, mid-green, hairless or white-hairy along midrib nr base below, serrate with 16-30 teeth. Rachis ± not wavy. Twigs flattened below nodes. Pl 21..*Ash* **Fraxinus excelsior***

Buds brown, scurfy. Lfts 3-5 prs, 6-8 x 1-1.4(2.5)cm, lanc, dark green above, occ purplish, hairless or white-hairy along 1 side of midrib nr base below, serrate with 12-18 teeth per side. Rachis slightly wavy. Twigs not flattened below nodes. ❶
....*Narrow-leaved Ash* **Fraxinus angustifolia**

Buds purplish-grey, minutely hairy. Lfts 3(4) prs, 5-9 x 3-3.7cm, ovate, pale green above, slightly crinkly, orange- or white-hairy along both sides of midrib nr base below, serrate with 25-40 teeth per side. Rachis wavy. Twigs ± flattened below nodes
....*Manna Ash* **Fraxinus ornus**

■Small tree or shrub. Lfts involute when young. Stipules present. Twigs pith-filled (soft-wooded), with interpetiolar ridge (occ obscure)

Buds green, with 1 pr scales. Twigs with white pith, odourless

Lfts 1-2(3) prs, 5-10 x 5cm, ovate, acuminate, occ with long apiculus (excurrent midrib), hairless, occ with midrib sparsely hairy nr base below, 1(2)-serrate with 40-55 teeth per side. Stipels present, glandular or herbaceous. Petiole often ridged, not or hardly channelled. Stipules 1(3) prs, glandular or herbaceous, usu persistent. Twigs green, occ pruinose, round, smooth, developing low lenticels..*Bladdernut* **Staphylea pinnata**

Buds reddish, with >2 prs scales (persisting at base of shoot)

Twigs with white pith and fetid gravy odour, usu green, ridged with rough lesions, becoming fluted; lf scars with 3 traces. Lfts 2-7 prs, 3-9cm, ovate to elliptic (rarely orb or dissected), acuminate, sparsely hairy on veins both sides, ciliate, serrate with 12-40 teeth per side. Stipels often present, similar to stipules. Petiole 3-4cm, often hairy, deeply channelled above, with cartilaginous wings, with spiral fibres. Stipules (extra-floral nectaries) 1.5-3(10) x 0.3(1)mm, stalk-like, often absent from lower lvs. Buds 2-9mm, reddish, hairless. Mar-Oct
....*Elder* **Sambucus nigra**

Twigs with orange-brown pith, often with faint blackcurrant odour, green, sparsely hairy or hairless, developing lesions; lf scars with traces. Lfts 2(3) prs, 6-15 x 2.5-7cm, ovate to elliptic, acuminate, shortly hairy or hairless below, sparsely ciliate, serrate with 20-30 teeth per side. Stipels absent. Petiole to 6cm, round. Stipules 1mm, gland-like. Buds 5-12mm, reddish-green, hairless..*Red-berried Elder* **Sambucus racemosa**

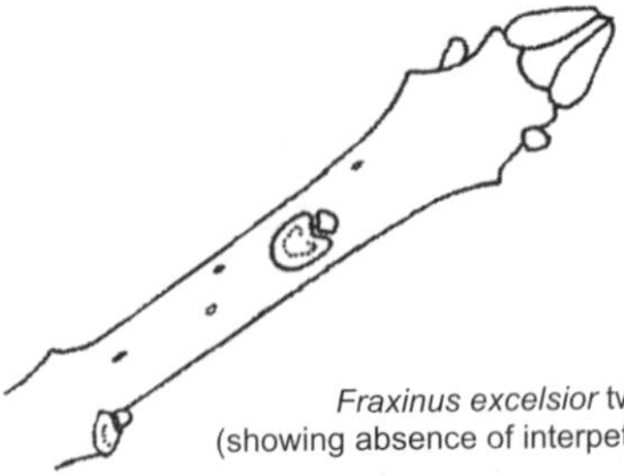

Fraxinus excelsior twig
(showing absence of interpetiolar ridge)

Group SD – *Tree or shrub. Lvs alt. Lfts with toothed margins, (±) net-veined, with stomata below only.* ❶ *Twigs odorous (smelling of mice).* ❷ *Lvs evergreen with spiny lfts (all other spp deciduous and not spiny)*

▪Twigs with white latex

Lfts usu 5-7 prs, 5-12 x 2-3cm, oblong-lanc, acute or acuminate, the terminal lft stalked, the laterals ± sessile, all folded when young, dark green to reddish above, paler below, ± hairless above (glandular-hairy on rachis), softly hairy below, 1(2)-serrate, with c25 prs of 2° veins ending in teeth. Petiole glandular-hairy. Stipules absent. Twigs stout, densely brown-hairy. Buds 2-4mm, naked, hidden by petiole base. Suckering.................................*Stag's-horn Sumach* **Rhus typhina**

▪Twigs without latex

Large tree

Buds naked. Twigs with chambered pith, odorous

Lfts 7-9 prs, to 12 x 5cm, elliptic-ovate, acuminate, asymmetric at ± sessile base, folded when young, yellow-green, hairless above, glandular-hairy below, serrate with c50 teeth per side. Petiole glandular-hairy, dorsally flattened. Twigs glandular-hairy when young, grey; lf scars with 3 U-shaped traces. Buds 5-6mm, sessile, pale scurfy-hairy. Not suckering. VR alien ....*Black Walnut* **Juglans nigra**

Lfts (1)3-6 prs, to 13 x 5cm, oblong, clasping at sessile base, rolled-folded when young, brown-hairy when young, soon hairless exc for sparse stellate hairs in vein axils below, with up to 52 teeth per side. Petiole hairless, brown-dotted on sides. Twigs grey, round, with low lenticels; lf scars with 3 U-shaped traces. Buds to 20mm, stalked, reddish-hairy, superposed, the upper one developing into shoot. Suckering..*Caucasian Wingnut* **Pterocarya fraxinifolia***

Buds with scales. Twigs without chambered pith

Twigs fetid (smelling of mice), pith-filled (softwood). Stipules absent

Lfts 5-10 prs, to 15 x 4cm, ovate-oblong, acuminate, folded when young, shortly hairy at least below when young, with 1-4 teeth nr base (each with submarginal extra-floral nectary below), with veins anastomosing at margins; terminal lft usu present (occ reduced), often 3-lobed. Petiole 7-12cm, swollen at base, minutely hairy, not channelled. Twigs stout, yellow-green to brown, minutely hairy when young; lf scars U-shaped, with (6)8-14 traces. Buds all lateral, to 4(6)mm, squat, flattened, with 2-4 scales. Bark smooth. Suckering. ❶ ....*Tree-of-heaven* **Ailanthus altissima**

Twigs odourless, solid (hardwood). Stipules present. (For partially pinnate *Sorbus* ssp Go to **SOR**)

Buds densely hairy to woolly (esp on margins)

Lfts 5-7 prs, 3-6 x 2cm, oblong, (±) acute, dark green above (occ red-tinged), grey-green below, hairy when young, soon hairless, 1(2)-serrate with 13-20 teeth per side. Petiole 2-4cm, long-hairy when young, narrowly channelled, with spiral fibres. Stipules toothed, unlobed. Twigs greyish, smooth, with long hairs when young. Terminal bud to 8mm, the lateral buds adpressed, all conical, dark brown or purplish, with 2-5 scales, opening simultaneously with fl buds. Usu acidic soils................................*Rowan* **Sorbus aucuparia**

Buds hairless, viscid. VR

Lfts (4)6-8 prs, 3-6 x 1.5cm, oblong-lanc, acute to acuminate, often pale or yellow-green above, grey-green, ± woolly when young, usu shortly hairy above, serrate (at least in distal ½) with 13 teeth per side. Petiole 2-4cm, ± grey-woolly, narrowly channelled, with spiral fibres. Stipules lobed, toothed, viscid. Terminal bud to 15mm, the lateral buds ± spreading, all ovoid, shiny green or pale brown, with 4-9 scales, opening after fl buds. Usu basic soils ....*Service-tree* **Sorbus domestica**

Buds with red-brown or white hairs ± confined to apex. VR, planted

Lfts 7-10 prs, 3-6 x 1.5-2cm, oblong-lanc, acute, dark green above, paler below, not papillate. Twigs reddish-brown. Buds to 12-16mm, ovoid to conical, purplish, with a tuft of red-brown hairs, the lateral buds with keeled scales. Bark grey or reddish-grey ....*Kashmir Rowan* **Sorbus cashmiriana**

Lfts 4-8 prs, 3-5 x 1-1.8cm, narrowly obovate, acute, bluish-green above, paler and papillate below. Twigs red. Buds to 12mm, conic-ovoid, reddish, with a tuft of red-brown or white hairs. Bark grey-brown to purplish-brown. *Hubei Rowan* **Sorbus pseudohupehensis**

Shrub (often tall)

Evergreen. Lfts with spiny margins. Stipules to 4mm, adnate to petiole, thread-like, herbaceous

Lfts 2-4 prs, 3-10 x 2-5cm, ovate-elliptic to oblong, acute, (±) sessile, dark green above, hairless, sinuate-dentate with spiny tips ≥1mm, with anastomosing 2° veins. Petiole dilated at base, green, hairless, round to weakly channelled. Twigs occ corky, with yellow inner bark. Buds 2-4mm, with c6 scales. ❷

Lvs shiny above. Widely naturalised. PL 23....................*Oregon-grape* **Mahonia aquifolium**

Lvs dull above. Widely planted, rarely naturalised. *M. japonica* x *lomariifolia* ....*Lily-of-the-valley-bush* **Mahonia** x **media**

Deciduous. Lfts 2-serrate. Stipules to 15mm, free, oblong-ovate, herbaceous

Lfts (±) opp, 5-10 prs, lanc, acuminate, ± sessile, folded (-plicate) when young. Petiole weakly laterally flattened. Buds all lateral, with c4 scales. Suckering

Lfts 5-12 x 1.8-2.5cm, (±) hairless above, stellate-hairy below when young, with 12-16 prs of 2° veins. Shrub to 2.5m...*Sorbaria* **Sorbaria sorbifolia**

Lfts 5-12 x 1.5-2cm, hairless above, sparsely simple-hairy below (esp in vein axils) at least when young (occ stellate-hairy), with 15-23 prs of 2° veins. Shrub to 6m ....*Chinese Sorbaria* **Sorbaria kirilowii**

Lfts 5-9 x 1.5cm, hairless above, v sparsely simple-hairy along veins below, with c25 prs of 2° veins. Shrub to 6m...*Himalayan Sorbaria* **Sorbaria tomentosa**

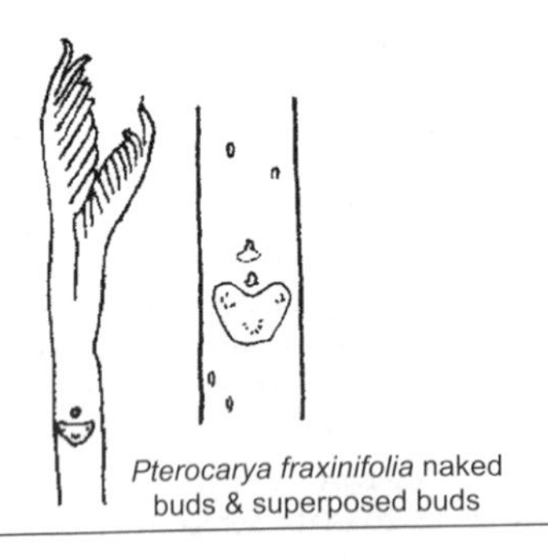

Pterocarya fraxinifolia naked buds & superposed buds

SE

Clematis vitalba stem TS

Group SE – *Deciduous woody climber or scrambling shrub to 10m. Lvs opp, with stomata below only*

■Climbing with spiralling petiolules. Stems with solid pith

Stems with 6 ridges. Lvs 1-pinnate. Lfts (3)5, 3-10cm, narrowly ovate, acute to acuminate, rounded to ± cordate at base, folded when young, silky-hairy when young, soon sparsely hairy, entire to lobed, the teeth tipped with purple hydathodes, net-veined. Petiole channelled. Buds to 5mm, silky-hairy, with several prs of opp scales. Mar-Oct............*Traveller's-joy* **Clematis vitalba***

Stems with 14 shallow ridges. Lvs 1-2-pinnate. Lfts 5-11, ± hairless, entire or deeply lobed, net-veined. Petiole sparsely hairy, channelled...............................*Virgin's-bower* **Clematis flammula**

■Scrambling or weakly climbing without spiralling petiolules. Stems with chambered pith

Stems round to 6-8-ridged, without interpetiolar ridge, green, usu hairless. Lvs 2-4 prs, to 11cm; lateral lfts to 4cm, curved, acute, shortly petiolate, often shortly hairy above and on veins below, scabrid-ciliate, occ with raised glands above, pitted below, not net-veined; terminal lft to 5cm, acuminate, often lobed. Petiole green, clasping at base with small cartilaginous auricles and minute red gland, hairless or minutely hairy along channel, with flat arc vb. Buds 3-5mm, ± acute, green or reddish, hairy, with persistent scales..................*Summer Jasmine* **Jasminum officinale**

Group SF – *Lvs whorled. Aquatic*

■Lvs 5-6-whorled

Stems emergent, unbranched, usu 3-4mm diam, with c25 hollows around central stele. Lvs 4-6 whorled, with 4-15 lobes per side, feathery, pale blue-green, with abundant sessile glands. All yr. Usu eutrophic water. Invasive alien.............................*Parrot's-feather* **Myriophyllum aquaticum***

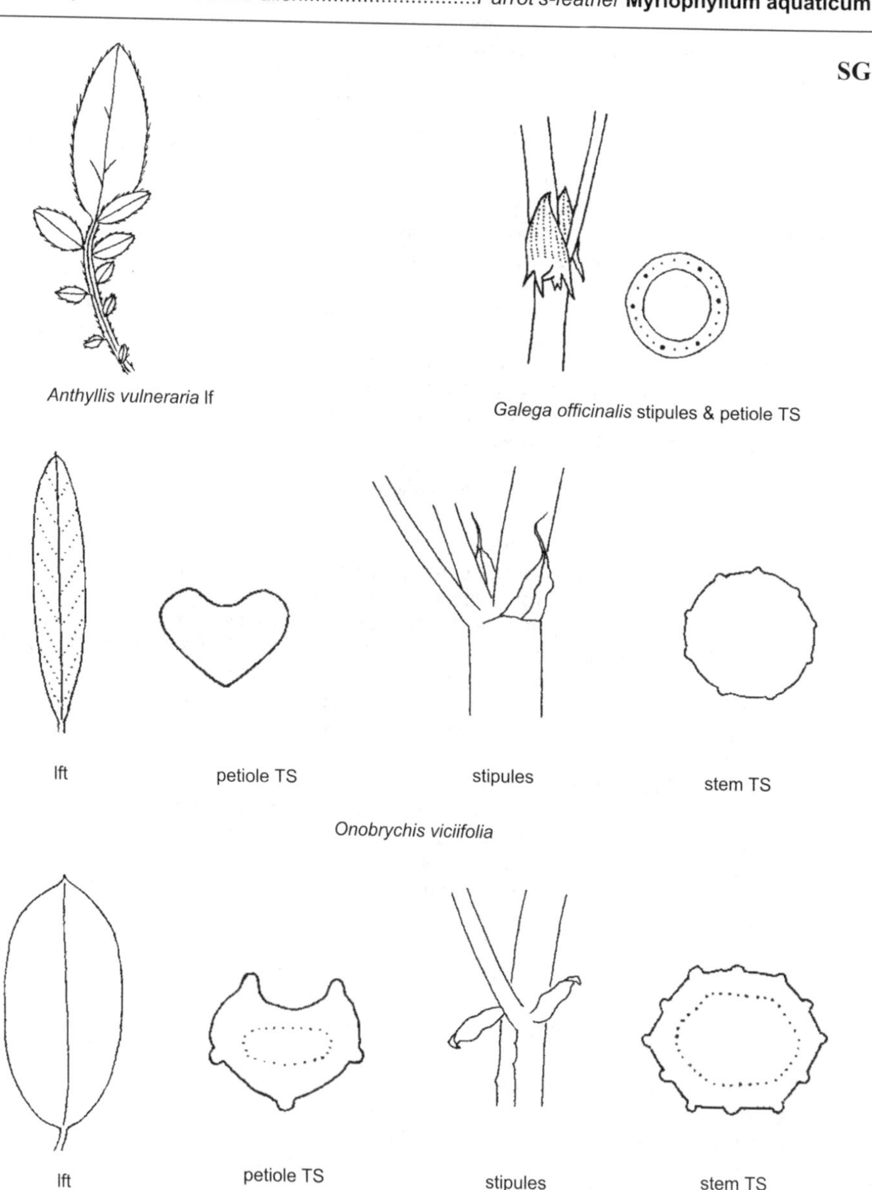

Anthyllis vulneraria lf

Galega officinalis stipules & petiole TS

Onobrychis viciifolia

Securigera varia

Group SG – *Herb. Lvs alt. Lfts entire, folded when young; terminal lft present; tendrils absent or reduced to subulate points.* ❶ *Hairs septate (unicellular in all other spp).* ❷ *Petiole sheathing at base and terminal lft much longer than laterals*

■Stipules (and sheathing base) absent, occ present as tiny brown dots in *Lotus*

Lfts 2 prs..(*Lotus*) Go to **RB**

Lfts 6-12 prs, 2-4 x 1.2cm, acute to acuminate, ± sessile, fetid, hairless or sparsely hairy esp below, not net-veined, with stomata below only. Petiole long on basal lvs, slightly winged, brittle, with flat arc vb. Stems to 90cm, hairless below, ± glandular-hairy above, angled, hollow. Roots strongly fetid. Per. ❶. All yr (but overwintering as a small rosette)

Lfts >3x as long as wide, ± lanc....................................*Jacob's-ladder* **Polemonium caeruleum**

Lvs ≤3x as long as wide, ± ovate..................*Garden Jacob's-ladder* **Polemonium caeruleum** cv

■Stipules (or sheathing base) present, often v small

Lfts usu densely hairy both sides. R, Scot

Lfts 6-15 prs, 1-1.5 x 0.5cm, linear-oblong to oblong-lanc, sessile, patent-hairy above, adpressed-hairy below, with 2° veins obscure, with stomata both sides. Petiole with long spreading or adpressed hairs, with 1 vb. Stipules connate, adnate to petiole for ½ way, with 7mm free part, membranous, adpressed-hairy. Stems white-hairy. Per ....*Yellow Oxytropis* **Oxytropis campestris**

Lfts 9-14 prs, 0.5-1.2 x 0.3cm, elliptic, ± acute, ± sessile, spreading- (occ adpressed-) hairy both sides (hairs to 2mm), with stomata both sides (but obscured by hairs below). Petiole long, flattened, channelled, with 1(3) vb's. Stipules connate, adnate to petiole for <½ way, with 15mm free part, persistent. Stems with long spreading white hairs and short antrorse black hairs above. Per. Mar-Oct..*Purple Oxytropis* **Oxytropis halleri**

Lfts sparsely hairy to hairless above (may be hairy below)

Petiole hollow

Stipules usu >1cm. Petiole channelled above

Stipules (0.7)1-2cm, broadly ovate, sagittate, entire. Lvs circinate when young (but lfts folded). Lfts 7-9 prs, 3-5 x 1.7cm, oblong or ovate-oblong, usu retuse with mucro at apex, ± sessile, hairless or sparsely hairy below, obscurely pinnate-veined, with stomata both sides. Stems hollow. Per. All yr...*Goat's-rue* **Galega officinalis***

Stipules 1(2)cm, ovate, not sagittate, occ toothed. Lvs not circinate when young. Lfts 3-7prs, 1.5-2(4) x 1.5cm, oblong-elliptic, obtuse or mucronate, ± sessile, hairless or antrorsely hairy below, obscurely pinnate-veined, with stomata below only. Stems solid. Per. Mar-Oct ....*Wild Liquorice* **Astragalus glycyphyllos**

Stipules 0.5cm. Petiole 5-ridged, ± round

Lfts 7-12 prs, 0.6-2 x 0.5-1cm, oblong-elliptic, obtuse to retuse with long mucro, the lowest short stalked to 1mm, hairless, with obscure veins, with stomata both sides. Petiole occ with 1 pr lfts at extreme base. Stipules ± oblong, not sagittate. Stems sprawling to 120cm, strongly 11-13-ridged, hollow. Rhizomatous per. Mar-Oct....*Crown Vetch* **Securigera varia***

Petiole solid. Stipules <1cm

Lfts >1cm long. Per

Lfts with translucent netted veins. Tendrils reduced to 1.5-4mm subulate point ....*Wood Bitter-vetch* **Vicia orobus**

Lfts without translucent netted veins. Tendrils reduced to 6mm subulate point ....*Bitter-vetch* **Lathyrus linifolius**

Lfts without translucent netted veins. Tendrils (subulate points) absent

Lfts 1-3(5) prs, occ long terminal lft only, the terminal to 6 x 2cm, elliptic, the laterals 0.3-2.5cm, ovate to elliptic, all thick, often hairless above, silky-hairy below when young, ciliate, with few ± obscure veins, with stomata both sides. Petiole sheathing at base. Stipules minute, brown (do not confuse with small lfts at lf base), usu absent from basal lvs. ❷. All yr. Often calc turf...*Kidney Vetch* **Anthyllis vulneraria***

Lfts (6)8-14 prs, occ 1 remote pr nr base, 1-3.5 x 0.7-1cm, ovate to linear-oblong, obtuse with mucro, thin, dark blue-green, with pale dots (HTL), sparsely hairy to hairless both sides, with c8 prs of obscure pinnate veins, with stomata both sides. Petiole channelled, occ hollow. Stipules to 6mm, ovate-acuminate to lanc, scarious, those on stem with a bristle-like point to 2mm. Stems occ reddish at base, striate to weakly ridged. Mar-Oct. Usu calc habs...*Sainfoin* **Onobrychis viciifolia***

Lfts <1cm long

Ann, often with lowest pair lfts at base of rachis and distant from the others

Lfts with sparse hairs above, hairy below, ciliate, 4-12 prs, 2-4mm, elliptic-orb to linear-oblong, not retuse or mucronate, with obscure veins, with stomata both sides. Lvs to 6cm. Stipules with tiny linear free part to 1.5mm. Oct-Jul. Well-drained soils, often sandy habs ....*Bird's-foot* **Ornithopus perpusillus**

Lfts hairless above, hairy below, 2-13 prs, 3-5mm, linear-lanc to narrow obovate. Lvs 1-2.5cm. Stipules 1-2mm. Sandy habs, Scilly Is, Channel Is ....*Orange Bird's-foot* **Ornithopus pinnatus**

Per, never with lowest pr lfts at base of rachis and distant from the others

Lfts 2-5(8) prs, 3-8 x 2-3mm, obovate to oblong, usu retuse with tiny mucro, rarely acute, ± fleshy, usu translucent-dotted HTL, dull green and hairless above, sparsely adpressed-hairy or hairless below, adpressed-ciliate, with obscure 2° veins, with stomata both sides. Petiole channelled. Stipules 4mm, triangular-ovate. Stems woody at base, ridged, often hairless. All yr. Calc turf..*Horseshoe Vetch* **Hippocrepis comosa**

Lfts 6-13 prs, 5-8(12) x 2-3(4)mm, ovate, dull green above, (±) obtuse, not mucronate or retuse, not translucent-dotted, with sparse stiff white hairs above and sparse adpressed stout hairs below, ciliate, with obscure 2° veins, with stomata both sides (fewer below). Petiole channelled, with 3 vb's. Stipules 5mm, connate at base, hairless, ciliate. Stems with ± adpressd antrorse hairs (often sparse) and sparse black hairs on nodes. Mar-Oct (all yr). Calc turf...*Purple Milk-vetch* **Astragalus danicus**

Lfts 6-9 prs, 2-5 x 2-3mm, ovate, obtuse, occ minutely apiculate, shortly stalked, ± hairless above, adpressed-hairy below, with 3 prs of 2° veins. Stipules 5-12mm, ovate, obtuse. Mtn turf, VR, Scot..*Alpine Milk-vetch* **Astragalus alpinus**

SH

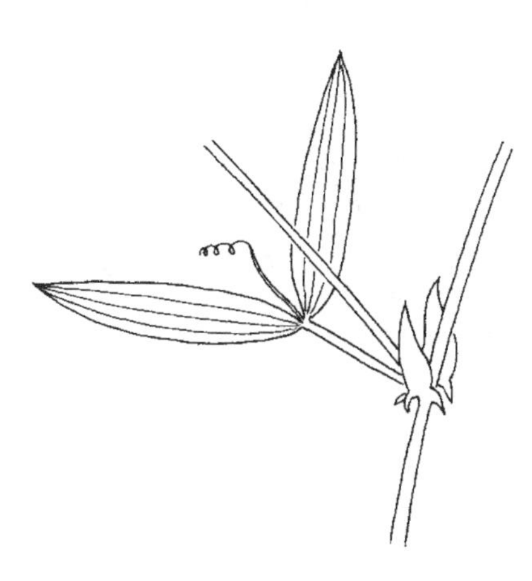

Lathyrus pratensis

Group SH – Fabaceae. *Herb. Lvs without terminal lft but with tendrils (tendrils branched unless otherwise stated). Lfts (0)1 pr, rolled when young and ± parallel-veined (± not anastomosing), with stomata both sides. Hairs unicellular*

■Lfts absent, replaced by lft-like stipules

Stipules to 2(5) x 1.5(4)cm, ovate, hastate, applanate when young, hairless, parallel-veined, without veins raised either side. Tendrils unbranched. Stems 1mm wide, unbranched, ± square, solid. Ann. Apr-Sept...*Yellow Vetchling* **Lathyrus aphaca**

■Lfts 1 pr

Stems with 2 broad wings >1mm. Stipules semi-sagittate (with 1 single basal lobe)

Lfts with strongly undulate or crisped margins. Ann

Lfts 6-10 x 4-9cm, ovate-oblong, mucronate, usu glaucous, hairy or sparsely so, opaquely net-veined. Petiole to 4cm, winged. Stipules to 3cm, as wide as stem ....*Sweet Pea* **Lathyrus odoratus**

Lfts with flat margins

Ann

Lfts 1.5-6(8) x 1.6cm, linear-oblong, mucronate, slightly glaucous, hairless. Stipules to 1.5(1.8)cm x 2mm, c½ as wide as stem, subulate. Stems 0.5cm wide. VR alien ....*Hairy Vetchling* **Lathyrus hirsutus**

Per

Lfts 9-13 x 2-4cm, lanc-oblong, mucronate, green, hairless, not net-veined, with 3-4 opaque parallel veins per side. Petiole to 2 x 0.8cm, winged. Stipules to 7 x 2cm, <½ as wide as stem, lanc. Stems 1cm wide. VR alien.......*Norfolk Everlasting-pea* **Lathyrus heterophyllus**

Lfts 5-10 x 2-4cm, ovate-oblong, mucronate, usu ± glaucous and pruinose, hairless, ± net-veined, with 3-4 translucent parallel veins per side. Petiole to 6 x 1.5cm, winged. Stipules to 7 x 2cm, >½ as wide as stem, ovate. Stems 0.8-1.5cm wide. Apr-Nov ....*Broad-leaved Everlasting-pea* **Lathyrus latifolius**

Lfts 4-8 x 2cm, linear-lanc, acute, green or glaucous, hairless, not net-veined, with 1-2 opaque parallel veins per side. Petiole to 4 x 0.8cm, winged. Stipules 1-3 x 0.2cm, <<½ as wide as stem, narrowly lanc. Stems to 1cm wide. Apr-Nov ....*Narrow-leaved Everlasting-pea* **Lathyrus sylvestris**

Stems with 2(4) ridges (or narrow wings <0.5mm). Stipules semi-sagittate or sagittate. Per

Plant with at least some hairs

Lfts ± parallel-veined 1-3.5 x 1cm, elliptic-lanc, acute, sessile, not undulate, ± glaucous, sparsely to densely short-hairy both sides or below only. Tendrils often unbranched. Petiole to 2cm. Stipules 1-3 x 1cm, sagittate (but lobes unequal), entire. Stems square, sparsely hairy, solid. Apr-Oct...*Meadow Vetchling* **Lathyrus pratensis***

Lfts with anastomosing pinnate veins, 3-5 x 3cm, broadly ovate, mucronate, shortly stalked (to 2mm), usu undulate and glaucous, sparsely hairy esp below. Petiole to 3cm. Stipules 3-10 x 0.5-1.5mm, semi-sagittate, occ toothed. Stems ridged, sparsely hairy, usu solid ....*Two-flowered Everlasting-pea* **Lathyrus grandiflorus**

Plant hairless

Lfts 15-50 x 1(1.5)cm, obovate to linear-lanc, obtuse to ± acute, mucronate, with weakly crenulated hyaline margins, with weakly translucent pinnate anastomosing veins. Petiole to 15mm. Stipules 12mm, ≥ petiole width, semi-sagittate. Stems ridged. Apr-Oct ....*Tuberous Pea* **Lathyrus tuberosus**

Group SI – Fabaceae. *Herb. Lvs wihout terminal lft, tendrils present or not. Lfts 2-15 prs, with stomata both sides.* Vicia *and* Arachis *lfts are folded when young and pinnately veined.* Lathyrus *lfts are usu rolled when young and ± parallel-veined (± not anastomosing).* Pisum *lfts are rolled when young and pinnately veined. Hairs unicellular.* ❶ *Lfts widest below middle.* ❷ *Stems strongly winged*

■Tendrils absent or reduced to a mucro (often long) (beware *Vicia sativa* and *V. lathyroides*)

Lfts (6)7-15 prs, with translucent (±) netted veins. Per

Lfts 1-1.5(2)cm, elliptic, mucronate, sparsely adpressed-hairy below, occ ciliate, with stomata above only; terminal lft occ present. Tendrils reduced to 1.5-4mm subulate point. Stipules to 13mm, semi-sagittate, slightly toothed nr base or ± entire. Stems not winged, sparsely hairy. Mar-Oct..*Wood Bitter-vetch* **Vicia orobus**

Lfts 4 prs, with translucent (±) netted veins. Per

Lfts to 2.5 x 1.5cm, elliptic, obtuse, mucronate, dark green above, blackening when dried, with stomata below only. Tendrils reduced to 5mm subulate point. Stipules to 8mm, lanc, entire. Stems 2(4)-ridged, hairless. VR alien..*Black Pea* **Lathyrus niger**

Lfts 1-4 prs, without netted veins

Rhizomatous tuberous per. Stems narrowly winged

Lfts 2(4) prs, to 4.5 x 1.1cm, narrowly lanc to elliptic, acute or obtuse, mucronate, dark blue-green to glaucous, sparsely hairy or hairless below, with 2-3 parallel veins per side, with stomata both sides; terminal lft occ present. Tendrils reduced to 6mm subulate point. Stipules 7mm, linear-lanc, semi-sagittate, entire to weakly toothed. All yr *Bitter-vetch* **Lathyrus linifolius**

Ann. Stems unwinged

Lfts 2 prs, 1-6 x 1-3cm, elliptic to obovate, obtuse to retuse, sparsely long-hairy to ± hairless. VR casual...*Peanut* **Arachis hypogaea**

Lfts (1)2-4 prs, 5-7(10) x (1)3-4cm, ovate to elliptic, obtuse, mucronate, hairless, with erose cartilaginous margins, with stomata both sides. Tendrils reduced to 15mm subulate point. Stipules 10-20mm, ovate, with purple spot (extra-floral nectary), semi-sagittate, toothed. Stems 60-120cm, erect, hairless, square, hollow. (Oct) Mar-Sept.........*Broad Bean* **Vicia faba**

■Tendrils present, often branched

Plant strongly glaucous, hairless

Per. Lfts 3-4 prs, 2.5-4(5) x 1-3cm, elliptic, obtuse, mucronate. Tendrils branched. Stipules 10-25mm, ovate, hastate, entire. Stems to 20cm tall, prostrate, occ purplish below, 4-angled, solid. Apr-Oct. Shingle beaches...*Sea Pea* **Lathyrus japonicus**

Ann. Lfts (1)2(3)prs, 4-7 x 5cm, ovate, truncate, minutely mucronate, hairless, with stomata both sides. Tendrils v branched. Stipules to 80 x 50mm, larger than lfts, ovate, rounded at base, toothed esp nr base. Stems 30-150cm tall, erect, hairless, ± round or weakly angled, hollow. Apr-Oct...*Garden Pea* **Pisum sativum**

Plant green (lfts occ glaucous below in *L. palustris*), hairy to hairless

Stipules laciniate or v deeply toothed (the upper occ entire in *V. hirsuta*)

Lfts 2(3) prs

Ann. Lfts 2-5.5 x 2.3cm, elliptic or ovate, mucronate, hairless, sparsely ciliate, with erose hyaline margins. Terminal rachis (after end-pr of lfts but before tendrils) to 3cm. Tendrils branched. Stipules 10-20mm, ovate-acuminate, sparsely ciliate. Stems angled, ridged to narrowly winged, hollow. Apr-Sept.....................................*Bithynian Vetch* **Vicia bithynica**

Lfts >3prs

Per. Lfts (5)6-9(12) prs, occ 1 pr at base of rachis, 5-15(20) x 5-7mm, oblong-elliptic (occ widest below middle), obtuse, mucronate, hairless, not ciliate, with c5 prs of obscure 2° veins, with stomata both sides or above only. Tendrils branched. Stipules unequal, crescent-shaped, mucronate, laciniate at base. Stems with v sparse hairs, ridged, ± square. Occ ❶. Apr-Sept...*Wood Vetch* **Vicia sylvatica**

Ann. Lfts 4-10 prs, occ 1 pr at base of rachis, 5-13 x 1.5-4mm, linear-oblong, truncate to emarginate, mucronate, sparsely adpressed-hairy on midrib below, with 2° veins ± opaque and obscure, with stomata above and v sparse below. Tendrils branched. Stems sparsely hairy to hairless, angled, ± square. Oct-Jul....................................*Hairy Tare* **Vicia hirsuta***

SI

Stipules entire to toothed (may be semi-sagittate)

Stipules with a dark (or green) extra-floral nectary (always green in *V. lathyroides*, usu green and obscure in other spp when young)

Lfts widest below middle. Per. ❶

Lfts 5-9 prs, 1-3 x 2cm, truncate, mucronate, dark green above, minutely hairy, ciliate, with 6-8 prs of indistinct 2° veins, with stomata below only. Tendrils simple or branched. Stipules toothed or entire. Stems hairless or sparsely hairy, ridged, hollow. All yr. Shady habs........*Bush Vetch* **Vicia sepium***

Lfts widest at or above middle. Ann

Lfts 2(4) prs, usu opp. Tendrils unbranched, occ reduced to a mucro or absent on upper lvs

Lfts 4-10mm, the 1st pr remote from stem by 1 lft length, linear-oblong or obovate, usu emarginate with mucro usu ≤ sinus, dull dark green above, paler below, hairy both sides. Stipules narrowly lanc, semi-sagittate, entire, with green obscure extra-floral nectary, Seeds papillate. Oct-Jul. Sand dunes........*Spring Vetch* **Vicia lathyroides**

Lfts 3-16 prs, alt. Tendrils usu branched

Lfts (2)3-6(9) prs, 6-15(25)mm, linear to obovate, truncate with acute mucro > sinus, hairy to sparsely so, densely ciliate. Stipules 4-8mm, semi-sagittate, toothed or entire. Stems hairy to hairless. Seeds smooth (unlike *V. lathyroides*). Oct-Aug

Lfts 4-10mm wide........*Common Vetch* **Vicia sativa** ssp **segetalis***

Lfts 1-4mm wide. Often sandy habs........*Narrow-leaved Vetch* **Vicia sativa** ssp **nigra***

Lfts 3-7(10) prs, 6-12(25) x 5mm, linear-oblong to elliptic, usu obtuse with tiny mucro visible below (< sinus), hairless to ± hairy, remotely long-ciliate. Stipules 1-2mm, triangular, semi-sagittate. Stems (±) hairless. Oct-Jul........*Yellow-vetch* **Vicia lutea***

Lfts 4-16 prs, 8-18mm, linear-oblong, emarginate with mucro > sinus. Stipules 2-4mm, lanc, entire. VR alien........*Hungarian Vetch* **Vicia pannonica**

Stipules without an extra-floral nectary

Lvs mostly with 6-15 prs lfts

Lfts densely adpressed-hairy at least below (hairs occ spreading above), the lowest pr ± at base of lf, 10-25 x 2-5mm, oblong-lanc to linear-lanc, acute or mucronate, grey-green, with stomata above only (or sparse nr midrib below). Tendrils branched. Petiole with adpressed hairs. Stipules to 10mm, semi-sagittate, usu entire, sparsely hairy. Stems shortly crisped-hairy, ridged. Per........*Tufted Vetch* **Vicia cracca**

Lfts sparsely spreading-hairy both sides, the lowest pr often ± at base of lf, 5-25 x 2-8mm, oblong, mucronate, pale or dark green, with stomata above and sparse below. Tendrils branched. Petiole with sparse long spreading hairs. Stipules to 8mm, often semi-sagittate, usu entire, often v hairy. Stems sparsely long-woolly or shortly crisped-hairy, ridged. Ann........*Fodder Vetch* **Vicia villosa**

Lvs with 2-6 prs lfts

Per. Stems distinctly winged, (±) hairless. Lfts 2-5 prs, 3-6cm, narrowly lanc, acute or obtuse, mucronate, dark green above, often glaucous below, with 2-3 prs of parallel veins per side, with stomata along midrib above. Tendrils usu branched. Stipules 10-20mm, lanc, semi-sagittate, entire. ❷. Mar-Oct. Fens, R........*Marsh Pea* **Lathyrus palustris**

Ann. Stems not (or v narrowly) winged

Lfts (3)4-6 prs, usu alt, 10-20 x 2-3mm, linear-oblong, acute, mucronate, hairless above, sparsely adpressed-hairy below, with stomata above but v sparse or absent below. Tendrils mostly unbranched. Stipules to 5mm, occ semi-sagittate. Stems often sparsely hairy when young, unwinged........*Smooth Tare* **Vicia tetrasperma***

Lfts 2-5 prs, often opp, 15-20 x 1.5-2mm, linear-lanc, acuminate or v acute, light pale green, patent-hairy above, sparsely adpressed-hairy below, with stomata above but v sparse below. Tendrils unbranched. Stipules 5(7)mm, semi-sagittate, entire. Stems sparsely adpressed-hairy, v narrowly winged........*Slender Tare* **Vicia parviflora**

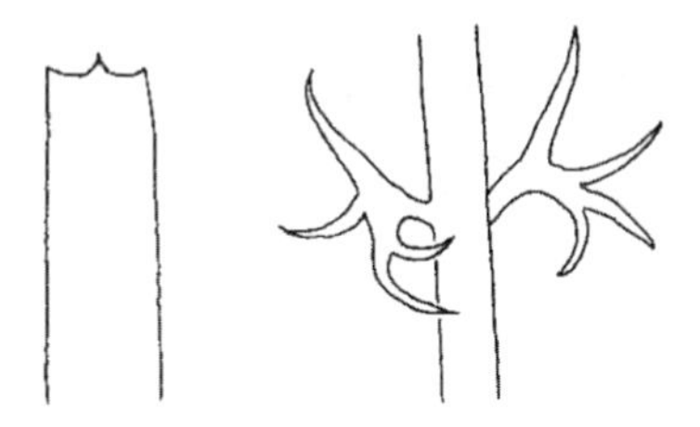

Vicia hirsuta lft apex & stipules

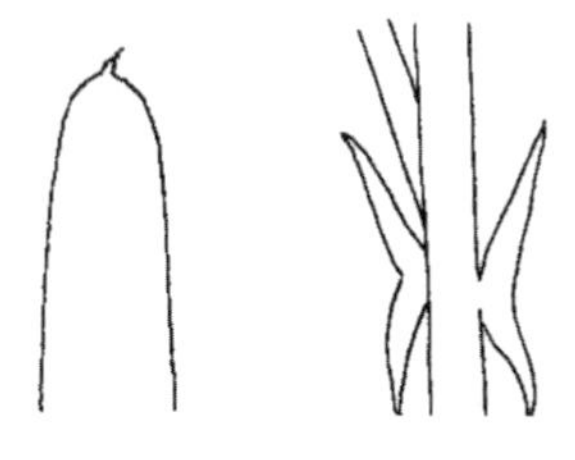

Vicia tetrasperma lft apex & stipules

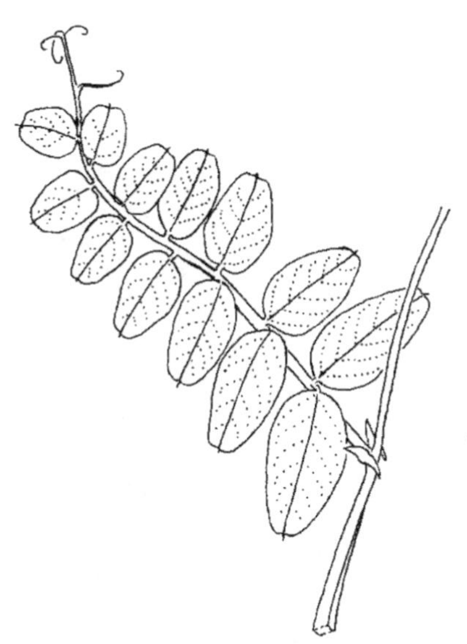

Vicia sepium

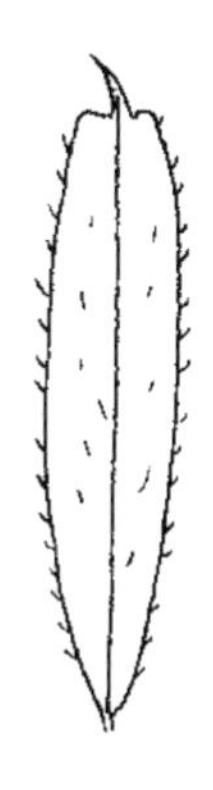

Vicia sativa ssp *segetalis* lft

Vicia sativa ssp *nigra* lft

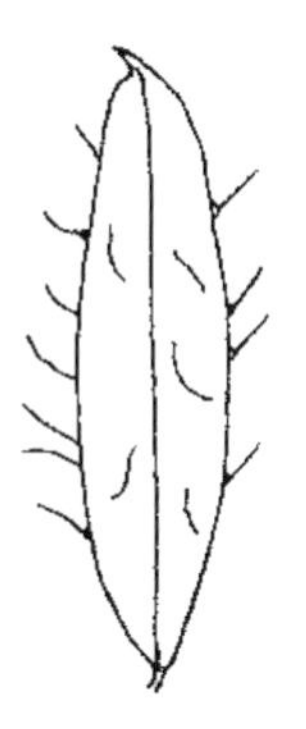

Vicia lutea lft

Group SJ – *Herb. Lvs opp (rarely 3-whorled). Hairs unicellular*

■Stipules present. Lvs always opp

Lvs all petiolate. Lfts (2)3-8 prs, 6-8 x 2-2.5cm, oblong or oblong-lanc, acuminate, asymmetric at base, involute when young, fetid (gravy odour), hairy below on veins (and above on midrib), serrate with 25-55 hooked teeth per side, each with a reddish hydathode, with stomata below only. Stipels often present, subulate, gland-like. Petiole 1-3.5cm, channelled with cartilaginous ridges. Stipules ± ovate, lft-like. Stems to 120cm, furrowed with cartilaginous ridges. Rhizomatous shrub-like herb..*Dwarf Elder* **Sambucus ebulus***

■Stipules absent. Lvs rarely 3-whorled

Lvs petiolate below, the upper ± sessile, odorous (esp when dried). Lfts usu 4 prs, 2-8 x 0.3-4cm, lanc, hairy or hairless either side, ciliate, toothed to entire, ± net-veined, with 2° veins raised below, with stomata below only; terminal lft often 3-lobed. Upper lvs often 2-pinnate. Petiole with sheathing base on basal lvs (absent from stem lvs), hairy to hairless, occ laterally flattened, channelled, hollow or solid, with several vb's and spiral fibres. Stems to 150cm, retrorsely hairy below, ± hairless above, ridged, hollow. Mar-Oct (all yr). Marshes or calc gsld ....*Common Valerian* **Valeriana officinalis**

SK

Sambucus ebulus

Potentilla anserina with a pr of intercalary lfts

Group SK – *Herb. Lvs alt, with adjacent lfts v unequal, alternately large and small (intercalary lfts). Hairs unicellular (septate in* Lycopersicon esculentum*).* (Acaena novae-zelandiae *and* Potentilla rupestris *may rarely key out here).* ❶ *Lvs with antiseptic odour.* ❷ *Lfts silky-hairy below.* ❸ *Lvs tansy-scented*

■Basal lvs with terminal lft much larger than lateral lfts and often 3-lobed

Plant with antiseptic odour. Terminal lft 1.5-8cm, ovate, usu 3-lobed

Lfts 2-5 main prs, 1.5-8cm, larger along rachis, ovate, acute, dark green and hairless above, hairy or shortly white-woolly below, 2-serrate, net-veined, with stomata below only; intercalary lfts 1-4mm, 2-5 prs between main lfts. Petiole reddish, ± hairless, channelled, with 3 vb's. Stipules adnate to petiole or stem for ≥½ length, scarious or herbaceous, toothed. ❶. Mar-Oct. Damp or wet habs..*Meadowsweet* **Filipendula ulmaria**

Plant odourless. Terminal lft 3-10cm, ± orb. Stipules usu absent from basal lvs (but present on ± 3-foliate stem lvs)

Terminal lft usu cuneate at base, ± as wide as long. All yr

Basal lvs with (1)2-4 prs lfts; lfts 5-20mm, dark green, hairless or ± adpressed-hairy, ciliate, dentate with reddish hydathode-tipped teeth, net-veined; terminal lft often 3-lobed. Stem lvs shortly petiolate. Petiole sheathing at base, with spreading or retrorse hairs, with 3 vb's and spiral fibres. Stipules to 1.2 x 1cm, toothed. Stems to 70cm, hairy, round or angled. Shady habs. PL 19..*Wood Avens* **Geum urbanum**

Terminal lft cordate-rounded at base, much wider than long. Mar-Oct

Basal lvs with (1)3-6 prs lfts; lfts 2-20mm, dark green, ± dentate, shortly hairy both sides, ciliate, dentate with reddish hydathode-tipped teeth, net-veined. Petiole sheathing at base, retrorsely hairy, with 3(5) vb's. Stipules to 2 x 2cm, toothed. Stems to 50cm, hairy, round. Damp habs...*Water Avens* **Geum rivale**

Basal lvs with (5)7-11 prs lfts; lfts 5-25mm, yellow-green and often ± rugose, weakly ± adpressed-hairy both sides, dentate with reddish hydathode-tipped teeth, net-veined; terminal lobe shallowly 5-7-lobed. Stem lvs 3-lobed, deeply dentate. Petiole sheathing at base, rough with swollen-based hairs, with 3(5) vb's. Stipules to 2 x 1.2cm, toothed. Stems to 50cm, hairy, round. VR alien, Wales, Scot.................................*Large-leaved Avens* **Geum macrophyllum**

■Basal lvs with terminal lft ± equal to at least uppermost pr of lateral lfts

Lfts silky-hairy at least below

Basal lvs 5-25cm; lfts 4-8(12) main prs, to 2.5cm, larger along rachis, ovate or oblong, ± folded when young, deeply serrate; intercalary lfts 2-5mm, (0)2-5-lobed. Petiole with long sheathing base, slightly channelled, with 3 vb's. Stipules papery-brown, never lf-like. Stolons developing in summer, reddish. ❷. Apr-Oct...*Silverweed* **Potentilla anserina***

Lfts spreading-hairy to hairless

Stipules absent

Rhizomatous per. Lvs to 25cm, deeply pinnately lobed, with tiny intercalary lobes; lfts up to 18 main prs, linear-oblong, gland-dotted, hairless or sparsely hairy (occ cottony), entire to deeply toothed. Upper lvs sessile, ± clasping. Petiole sheathing at base, often with tiny lf lobes at base, channelled, with 5 vb's. Stems 30-120cm, usu purplish, sparsely hairy to hairless, angled. ❸. Mar-Oct. PL 18..*Tansy* **Tanacetum vulgare**

Taprooted ann. Lvs 10-40cm; lfts 2-4 main prs, 5-8(10)cm, ovate or oblong, sessile or stalked, strongly odorous, hairy (occ strigose) esp below, glandular-hairy on rachis, ciliate, entire to lobed, often 2-pinnate, not or weakly net-veined, with stomata both sides. Petiole 2-5cm, glandular-hairy. Stems 60-200cm, sprawling, shortly hairy, with ± sessile yellow-tipped glandular hairs and sparse long simple septate swollen-based hairs, ± round, not or hardly ridged. Apr-Oct..*Tomato* **Lycopersicon esculentum**

Stipules (or pseudostipules) entire, papery or herbaceous, not lft-like

Tufted per, with taproot. Lfts 2(3) main prs (and 2 prs smaller lfts), to 5 x 4cm, folded when young, hairy both sides (hairs with minutely swollen bases), deeply crenate-serrate with 7-8 teeth per side, with stomata below only. Petiole spreading to retrorsely hairy, channelled, with 1(3) vb's. Stipules papery or herbaceous, with free part to 1cm, lanc-ovate, reddish, hairy ....*Bastard Agrimony* **Aremonia agrimonioides**

Rhizomatous per, with tubers. Lfts 2-4(6) main prs, 4-7 x 4cm, ± ovate, shiny dark green, with antrorsely curved hairs both sides, occ with ± sessile glandular hairs on rachis when young, ± entire, with stomata both sides, net-veined. Petiole 2.5-5cm, antrorsely crisped-hairy, with 3 vb's. Pseudostipules herbaceous, to 25 x 15 mm, ± crescent-shaped. clasping the stem. Stem 30-80cm, erect or sprawling, hairless or sparsely hairy (occ glandular), often sinuate-winged, with small hollow..*Potato* **Solanum tuberosum**

Stipules toothed or lobed, herbaceous, lft-like

Lfts 8-20 main prs, 5-15mm, v reduced proximally and distally on lf (like an ostrich feather)

Lfts ± oblong, folded-plicate when young, hairless, with antiseptic odour, pinnately lobed, with reddish hydathodes, with stomata below only; terminal lft resembling 3 fused lfts; intercalary lfts 1-4 prs between each pr of main lfts, 1-3mm, often entire. Petiole with 1(3) vb's. Stipules adnate to petiole for >½ length, toothed, green nr apex. ❶. All yr. Usu calc turf...*Dropwort* **Filipendula vulgaris**

Lfts 3-6 main prs, 20-60mm, becoming larger along rachis

Rhizomes often long. Lfts ± elliptic, with numerous small sessile glands (esp below), strongly aromatic, green above, often greyish below, mostly with short hairs above, less hairy below, acutely serrate or serrate-dentate. Petiole hairy, with 1(3) vb's. Stems 50-100cm, with abundant long hairs ≥2mm and tiny minute glandular hairs. Petals usu notched ....*Fragrant Agrimony* **Agrimonia procera**

Rhizomes short or absent. Lfts ± elliptic, with few or no glands, weaky aromatic, green above, often greyish below, with long hairs above, usu densely hairy below, deeply serrate or serrate-dentate, with stomata below only; intercalary lfts 2-3 prs between each pr of main lfts. Petiole hairy, with 1(3) vb's. Stems 30-60cm, hairy, often v glandular, hairs often swollen-based (rough to touch). Petals entire. Apr-Oct. Usu calc habs ....*Agrimony* **Agrimonia eupatoria**

Group SL – *Tansy-scented tufted per or bi, often with cottony or silky hairs. Lfts not or weakly gland-dotted. (*Tanacetum vulgare *would key out here but has strongly gland-dotted lfts and intercalary lfts)*

■Stems 60-120cm tall

Lvs 10-18cm, with 5(7) prs free lfts along winged rachis and tiny lobes at base, pale grey-green esp below, softly ± adpressed-hairy both sides, with stomata below only. Stems hairy, round but weakly angled..*Rayed Tansy* **Tanacetum macrophyllum**

■Stems 20-60cm tall

Lvs 3-8cm, 1-pinnate, with 1-2(5) prs of lobed lfts, yellow-green, with spreading or adpressed hairs both sides or ± hairless, minutely ciliate, with main veins raised both sides; lfts 2.5-3cm, the ultimate lobes c6 x 6mm, obtuse but apiculate with hydathodes. Petiole dilated at slightly sheathing base, hairy, ± triangular, not channelled, with flat arc vb and 2 rib bundles. Stems sparsely hairy, round but ridged. Per. All yr. PL 19....................*Feverfew* **Tanacetum parthenium**

Lvs 3-7cm, 1-3-pinnate, with many prs of pinnate lfts and smaller lobes nr base, grey-green, adpressed-hairy with silky long hairs esp below, occ ± hairless or cottony, ciliate, with obscure veins; lfts 1.5-2.5, with bristle-tipped teeth. Petiole with 3 vb's. Stems usu branched, tough, purplish, cottony and often with swollen-based septate hairs (hair base elongate and turning purple-black), ridged. Bi or per. All yr.................................*Yellow Chamomile* **Anthemis tinctoria**

Group SM – *Herb or stoloniferous soft-wooded dwarf shrub. Stipules present. Hairs unicellular (septate on stems of* Erodium moschatum*)*

■Petiole sheathing at base, with adnate stipules. Lfts net-veined

Lfts >2.5cm. Per

Bogs

Lfts 1-2(3) prs, usu opp, often appearing palmately arranged, 3-6cm, oblong, folded when young, dark blue-green above, pale or purple-glaucous below, hairless or with sparse glandular hairs above, silky-hairy to ± hairless below, ciliate, serrate with 7-9 red hydathode-tipped teeth per side, with stomata below only; terminal lft often 3-lobed. Petiole > lf, with long hairless sheathing base, often glandular-hairy, weakly channelled, with 3 vb's. Stipules on stem lvs only, papery, turning brown, sparsely minutely glandular-hairy, entire. Stems to 50cm. Rhizomes long, round, solid. Mar-Oct................*Marsh Cinquefoil* **Potentilla palustris***

Gsld, often damp

Lfts 3-7 prs, ± opp, 2-5 x 3cm, oblong-ovate, obtuse, cordate at base, stalked (6-35mm), folded when young, dark green above, ± glaucous below, weakly cucumber-scented, hairless, dentate with 9-20 reddish hydathode-tipped teeth per side, with 2° veins hardly raised below, with stomata below only. Petiole to 30cm, 1-2mm diam, narrowly sheathing at base, often reddish esp nr base, channelled, with 3-5(7) vb's and spiral fibres. Stipules largest on stem lvs, lft-like green. Stems to 120cm, hairless, angled. Tufted. Apr-Oct ....*Great Burnet* **Sanguisorba officinalis**

Mtn rocks

Lfts 2-4(8) prs, 2-6cm, rarely intercalary, ± ovate, obtuse, short-stalked, dark green above, paler below, sparsely hairy above (and with minute glandular hairs), hairy on veins below, ciliate, ± 2-dentate with 12-30 red hydathode-tipped teeth per side, with 2° raised below, with stomata both sides. Petiole channelled, with 3 vb's. Stipules green, scarious at base, entire. Stems to 50cm. Tufted. Mar-Oct. Sch8............................*Rock Cinquefoil* **Potentilla rupestris**

Lfts <2.5cm

Per herb, to 90cm tall. Stipules of basal lvs tiny, toothed, those of stem lvs large, lft-like ....*Fodder Burnet* **Sanguisorba minor** ssp **muricata**

Stoloniferous dwarf shrub, to 10cm tall. Lfts opp (alt if intercalary lfts are present), increasing in size along rachis, cuneate or rounded at base, sessile, folded when young, deeply crenate-serrate, with 2° veins ending in a white or reddish hydathode at tooth apex, with stomata below only. Petiole with 3 vb's and spiral fibres. Stipules small, green. All yr

Lfts green both sides (usu paler below), hairless above, 3-4(5) prs

Lfts 10-30 x 10mm, oblong, dull green above, sparsely silky-hairy below, with 6-10 teeth per side, the terminal tooth equal to the laterals. Stipules (3)5-lobed. Stems green to reddish, ± adpressed-hairy. VR alien...*Two-spined Acaena* **Acaena ovalifolia**

Lfts 5-10(18) x 4-6(8)mm, oblong, shiny green above, paler below, with long adpressed hairs on veins below, with 7-8(12) teeth per side (often recurved), the terminal tooth narrower but same length as the laterals. Stipules entire to 5-lobed. Stems reddish, ± adpressed-hairy...*Pirri-pirri-bur* **Acaena novae-zelandiae**

Lfts glaucous and purplish-brown on one or both sides. VR alien

Lfts hairless above, hairy on veins below, 5-7 prs, 3-7 x 3-7mm, obovate to orb (distal lfts as wide as long), glaucous or purplish-brown above, with 7-12 obtuse teeth per side. Petiole sparsely hairy. Stipules linear, entire. Stems brown.........*Spineless Acaena* **Acaena inermis**

Lfts silky- or patent-hairy above (occ sparsely so), silky-hairy below, cilia often tufted at tooth apices, 4-6 prs, 4-17 x 2-9mm, oblong light dull green to glaucous above, brownish at least on margins and veins, with 7-15 teeth per side (margins often recurved),. Stipules 3-8-lobed. Stems brown...*Bronze Pirri-pirri-bur* **Acaena anserinifolia**

Lfts pale glaucous both sides (without purple or brown coloration), hairless above

Lfts silky-hairy below, cilia often tufted at tooth apices, 5-7 prs (lower often v reduced), to 8 x 6mm, with 5 teeth per side. Stipules entire. Stems usu green. Hortal .... *Glaucous Pirri-pirri-bur* **Acaena magellanica**

■Petiole not sheathing at base, with free stipules. Lfts not net-veined

Lvs all on stems (basal rosette absent). Lfts shallowly toothed. Ann with fibrous roots

Plant glandular-hairy. Lfts 5 prs, to 10mm, ovate-oblong, acute, ± sessile, folded when young, yellowish-green, with 3-4 teeth per side in distal ½. Stipules lft-like, toothed. Stems ± square. Apr-Aug. VR casual..*Chick Pea* **Cicer arietinum**

Lvs in basal rosette. Lfts deeply incised. Stems often absent. Ann (bi) with long taproot

Lvs 10-45cm. Lfts 4-7 prs, 1.5-4cm, ovate-oblong, weakly musk-scented to odourless, with minutely gland-tipped hairs above, sparsely hairy on veins below, incised-dentate or pinnately lobed usu <½ way to midrib, with hydathode-tipped teeth often turning reddish, with translucent 2° veins, with stomata both sides. Petiole hairless or densely (glandular-) hairy esp along weak groove above, slightly flattened, with 4 vb's and spiral fibres. Stipules scarious, whitish, with green veins, entire. Stems with septate glandular hairs. All yr ....*Musk Stork's-bill* **Erodium moschatum**

Lvs 4-12(18)cm. Lfts 4-7 prs, to 1.5cm, oblong, odourless, with white stout hairs, incised-dentate usu >½ way to midrib, with few obscure 2° veins, with stomata both sides. Petiole usu hairy all round, flattened, not channelled, with 4 vb's and spiral fibres. Stipules scarious, often reddish, entire. All yr

Plant not or sparsely glandular-hairy. Acidic habs....*Common Stork's-bill* **Erodium cicutarium**

Plant densely glandular-hairy, often greyish-green. Dunes, S & W Br, Ire ....*Sticky Stork's-bill* **Erodium lebelii**

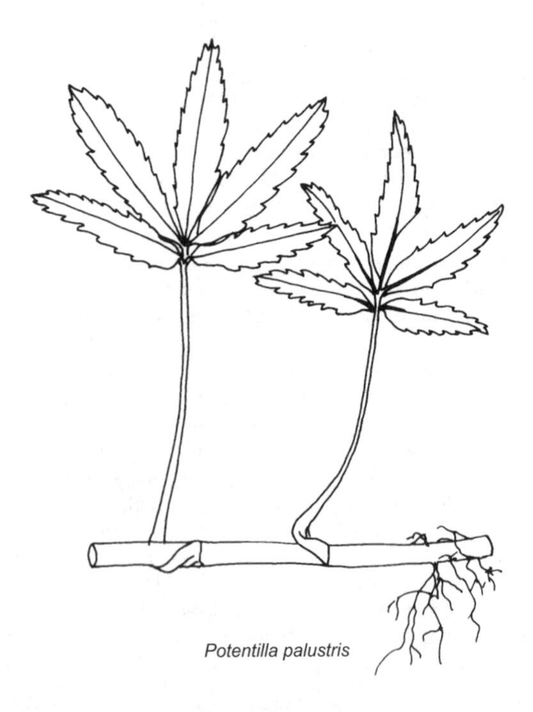

Potentilla palustris

Group SN – *Herb. Petiole with sheathing base, channelled. Lfts hairy above (often densely so)*

■Lvs with lowest pr of lfts stalked. Petiole with white latex

Lfts 1-2 prs, to 50cm, occ hairless above, usu minutely hairy below, scabrid-ciliate, net-veined, with stomata both sides. Petiole often purple-hispid. Stems to 4(6)m, ridged, purple-blotched, odorous, hollow. Usu monocarpic per. Feb-Oct......*Giant Hogweed* **Heracleum mantegazzianum**

Lfts 2(3) prs, 6-10cm, hairy both sides (often stiffly so but usu flattened on drying), lobed, serrate with semi-sunken white hydathodes, net-veined, with stomata both sides; terminal lft 3-lobed. Petiole hairy, with ciliate sheathing base overlapping itself, with irreg small hollows and scattered vb's. Stems to 1.8m, retrorsely hairy, ridged, hollow. Bi or polycarpic (occ monocarpic) per. Feb-Oct (all yr)..*Hogweed* **Heracleum sphondylium***

Lfts 3 prs, 2-5(10) x 3cm, hairy or hairless above.........*Greater Burnet-saxifrage* **Pimpinella major**

■Lvs with all lfts sessile

Petiole with white latex around margin (often sparse), parsnip-scented. Per

Lvs often basal. Lfts 3-5 prs, ovate, often yellowish, hairy (often hairless in var *hortensis*), minutely scabrid-ciliate, lobed, irreg serrate, net-veined, with stomata below and along veins above. Petiole tough, hardly inflated at sheathing base, occ purplish at base, laterally flattened, developing irreg hollows, with scattered vb's. Stems to 180cm, furrowed, hollow or solid. Mar-Oct (all yr)...*Wild Parsnip* **Pastinaca sativa***

Petiole without latex, weakly aromatic (tansy-scented). Shortly rhizomatous per

Basal lvs in a rosette, to 15(30)cm, with proximal lfts much reduced; lfts >10 prs, thread-like, linear, apiculate, long-hairy. Petiole tough, with wide channel above, solid, with 7 vb's. Stems to 50cm, ± woolly, furrowed, solid. All yr..*Yarrow* **Achillea millefolium**

Petiole without latex, odourless. Ann (bi)

Basal or lower lvs not in rosette, to 30cm, ± simple; lfts 2(3) prs, broadly ovate to ± orb, yellow-green, densely hairy both sides, occ lobed, crenate, with stomata both sides. Upper lvs lanc, obtusely serrate. Petiole hardly sheathing at base, with 7 vb's around small hollow. Stems to 100cm, rough with spreading hairs, striate to weakly ridged, with small hollow. VR, Essex ....*Hartwort* **Tordylium maximum**

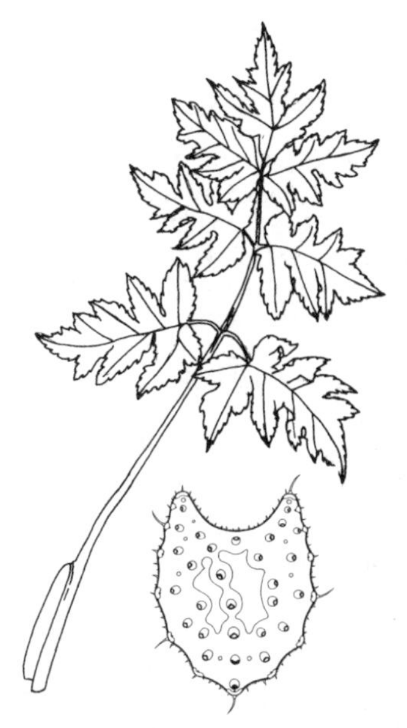

Heracleum spondylium with petiole TS

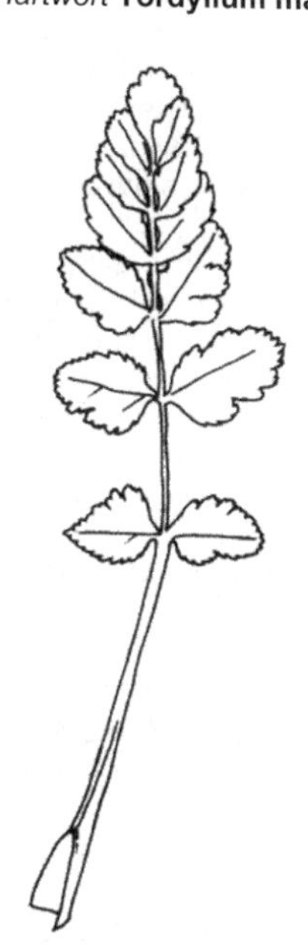

Pastinaca sativa

Group SO – *Herb. Petiole with sheathing base. Lfts hairless (or with sparse scattered unicellular hairs) above.* Apiaceae *have opp lfts and the teeth with a white hydathode.* ❶ *Lfts glaucous below, with cucumber odour.* ❷ *Petiole with strong celery odour*

■Plant usu in dry habs

Petiole with sparse white latex

Petiole usu hollow. Tufted per

Basal lvs 1-2-pinnate; lfts (2)4-5 prs, to 2.5cm, ovate to linear-lanc, (±) sessile, sparsely to densely hairy, scabrid-ciliate, serrate (to pinnately lobed), with stomata both sides. Stem lvs 2(3)-pinnate. Petiole long, often purplish, hairless or with sparse hairs. Stems to 70cm, slender, often purplish, tough, hairless to retrorsely hairy, ± round, slightly ridged, solid (but often hollow in fr). All yr..*Burnet-saxifrage* **Pimpinella saxifraga**

Basal lvs 1(2)-pinnate; lfts 3 prs, 2-5(10) x 3cm, ovate, truncate to ± cordate at base, lowest pr short-stalked (5-15mm), hairy to hairless (often hairy on veins below), scabrid-ciliate, serrate, with stomata below only (rarely above); terminal lft occ ± 3-lobed; rachis (±) hairless, channelled. Stem lvs with narrower lfts, sessile. Petiole long, not purplish, hairless, angled or triangular, channelled (occ slightly so), with 5(7) vb's and latex around margin. Stems to 100cm, stout, often reddish nr base, brittle, hairless, ridged or angled, hollow. Mar-Oct ....*Greater Burnet-saxifrage* **Pimpinella major**

Petiole solid

Rhizomatous per. Lfts 2 prs......................................*Ground-elder* **Aegopodium podagraria***

Bi. Lfts 2-5 prs, 3-7cm, oblong-ovate, (±) sessile, usu with nauseous tar-like odour, serrate with antrorse curving cartilaginous apices, often incised, net-veined, with stomata below only. Petiole with hairless sheathing base, ± laterally flattened, channelled (occ deep), with 9 vb's and latex mostly around margin. Stems to 100cm, round, finely striate, solid. All yr ....*Stone Parsley* **Sison amomum***

Bi. Lfts 4-12 prs, 0.5-3.5cm, held in a ± horizontal plane, ovate, sessile, with weak celery odour, dark green above, serrate with antrorse curving cartilaginous apices, occ lobed, with stomata below but confined to vein margins above. Petiole with long hairless sheathing base, channelled, with 5-9 vb's along lower margin and latex turning brown. Stems to 80cm, round, striate, solid. All yr...*Corn Parsley* **Petroselinum segetum**

Petiole without latex, solid

Ann. Lvs yellow-green both sides, (±) odourless

Lvs often in a basal rosette, with auriculate sheathing base, resembling *Cardamine impatiens*; lfts 5-6 prs, alt to opp, to 3cm, lanc, short-stalked, hairless or with sparse hairs above, incised, with indistinct veins, with stomata both sides. Upper lvs not sheathing at base. Petiole weakly channelled, solid or with several hollows, with 1 vb. Stems 15-50cm, slightly flattened, solid, with abundant sap. Mar-Oct (all yr)...............................*Meadow-foam* **Limnanthes douglasii**

Per. Lfts usu dark green above, ± glaucous below, cucumber-scented. ❶

Lfts 3-7 prs, ± opp, 2-5 x 3cm, oblong-ovate, cordate at base, stalked (6-35mm), folded when young, hairless, dentate with 9-20 reddish hydathode-tipped teeth per side, net-veined, with 2° veins hardly raised below, with stomata below only. Petiole to 30cm, 1-2mm diam, often reddish esp nr base, channelled, with 3-5(7) vb's and spiral fibres. Stipules largest on stem lvs, lft-like. Stems to 120cm, hairless, angled. Apr-Oct ....*Great Burnet* **Sanguisorba officinalis**

Lfts (4)7-12 prs, ± opp, 0.5-2cm, ovate-orb, rounded at base, short-stalked (0.5-2mm), folded when young, hairless above, sparsely hairy below, with reddish hydathode-tipped teeth, net-veined, with 2° veins not raised below, with stomata below only. Petiole to 25cm, 1mm diam, occ with long hairs, channelled, with 1(5) vb's and spiral fibres. Stipules occ present on basal lvs, lobed, lf-like. Stems hairy to hairless, weakly ridged. All yr

Lfts short-stalked to 0.5mm, dark green above, toothed. Stems to 40cm. Calc gsld ....*Salad Burnet* **Sanguisorba minor** ssp **minor**

Lfts stalked to 2mm, often pale green above, deeply toothed. Stems to 90cm. Mostly sown gsld...*Fodder Burnet* **Sanguisorba minor** ssp **muricata**

▪Plant in damp or wet (occ saline) habs (always hairless). Petiole with or without latex

Petiole with pale ring-mark (or remote pr of reduced lfts nr base)

Lfts 3-7 prs, rarely held in horizontal plane, to 12 x 4cm, lanc to ovate, sessile or v short-stalked to 2.5mm, mid-green, with distinct cartilaginous margins and teeth, with stomata below and along veins above. Submerged basal lvs 2-3-pinnate. Petiole with V-shaped sheathing base, 5-angled but not channelled above, v hollow, strongly paraffin- or petrol-scented, without latex. Stems to 200cm, strongly 7-ridged, hollow. Stolons absent. Calc aquatic, R ....*Greater Water-parsnip* **Sium latifolium**

Lfts 5-10 prs, often held in horizontal plane, to 4(5) x 2cm, ovate-lanc, ± clasping at sessile base, dull yellow-green above, often shiny below, with narrow cartilaginous margins and teeth, with stomata below and along main veins above. Submerged basal lvs 1-pinnate. Petiole with purplish sheathing base, round, soon hollow (with partial septum), sweetly celery-scented, without latex. Stems to 80(100)cm, with slight groove or channel, hollow. Stolons developing in spring. Apr-Oct (all yr). Aquatic, esp calc water.................*Lesser Water-parsnip* **Berula erecta***

Petiole without ring mark

Stems rooting at least at lower nodes, usu prostrate

Stems solid. Petiole without latex...................................*Marsh Cinquefoil* **Potentilla palustris***

Stems hollow. Petiole without latex

Lfts (2)3-6 prs, 0.5-6(10)cm, lanc to ovate, sessile (but lowest pr often short-stalked to 12mm), shiny, crenate or shallowly lobed, with hyaline margins v minutely scabrid to entire, with stomata below and along veins above. Petiole with obtuse auricles at sheathing base, ± round (with obscure channel above), developing large hollow (often retaining partial septum), with c10 vb's around margin. Stems to 60cm tall, rooting only at lower nodes, sweetly celery-scented. All yr. Aquatic........................*Fool's-water-cress* **Apium nodiflorum**

Lfts 5-11 prs, 0.5-1.2cm, ± orb, sessile, shiny, ± 2-lobed, with hyaline margins v minutely scabrid to entire, with stomata all over both sides. Stems to 10cm tall, rooting at most nodes. All yr. Muddy habs, VR. Sch8..*Creeping Marshwort* **Apium repens**

Stems not rooting at nodes, erect

Lfts thread-like, appearing whorled

Lfts 15-30 prs, to 1cm, increasing in size along rachis, sessile, lobed; lobes to 5 x 0.5mm, without cartilaginous margins, with purple apiculus to 0.5mm, with stomata both sides. Petiole striate, round, hollow, weakly sweetly celery-scented. Stems to 60cm, purplish nr base, often pruinose, striate, hollow. Mar-Oct...........*Whorled Caraway* **Carum verticillatum**

Lfts linear, deeply lobed or entire (not toothed)

Upper lvs 1-pinnate with lfts 0.5-2cm, distant, obtuse (but with a greenish-white hydathode), often cylindrical and hollow, with smooth to weakly scabrid non-cartilaginous margins, sweetly celery-scented, with midrib sunken above and raised below, with 2° veins usu obscure. Lower (and basal) lvs 1-2-pinnate. Petiole of basal lvs longer than lf, with green sheathing base, round, hollow. Stems to 80cm, to 8mm diam, often constricted at nodes, striate, thin-walled, v hollow. Often aquatic.......*Tubular Water-dropwort* **Oenanthe fistulosa**

Lfts broader, toothed, never deeply lobed

Petiole without latex

Lfts 1-3 prs, 1-5cm, broadly ovate to rhombic, stalked (uppermost sessile), shiny green esp below, without cartilaginous margins, (3-) lobed, serrate, with stomata below and usu confined to veins above. Upper lvs ternate. Petiole tough, sharply 5-7-angled, shallowly channelled, solid when young, becoming hollow, with 5-7 vb's around margin, strongly celery-scented. Stems to 80cm, strongly ridged, often purplish, solid. ❷. All yr. Often coastal...*Wild Celery* **Apium graveolens***

Petiole with white latex (often sparse)

Bracts few or several. Frs 2.5mm, ovoid, with styles 1mm. All yr. Usu damp brackish gsld ....*Parsley Water-dropwort* **Oenanthe lachenalii**

Bracts several. Frs 3mm, cylindrical (± straight-sided), with styles 2-3mm. Damp to dry gsld ....*Corky-fruited Water-dropwort* **Oenanthe pimpinelloides**

Aegopodium podagraria petiole TS

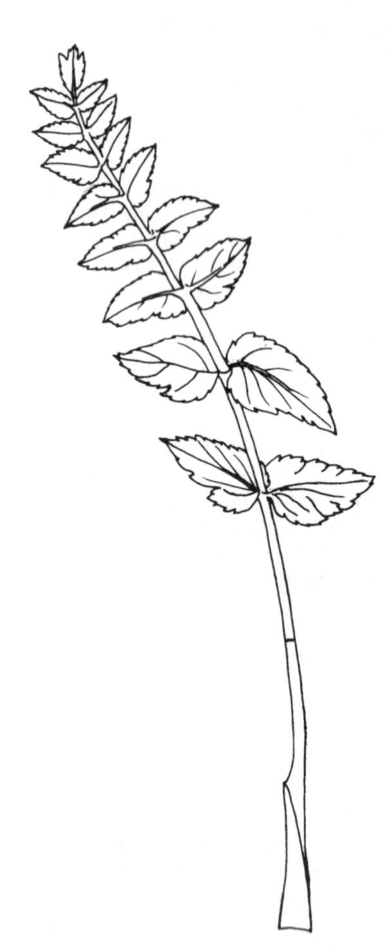

Berula erecta

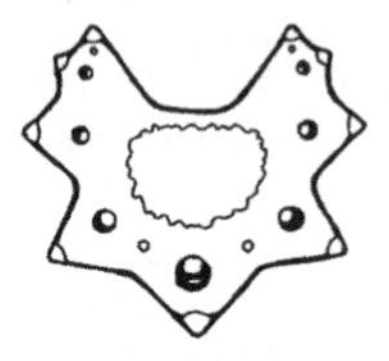

Apium graveolens petiole TS

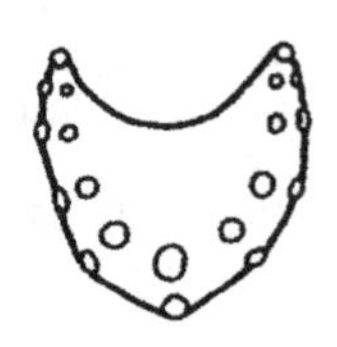

Sison amomum petiole TS

Group SP – *Herb. Petiole without sheathing base. (*Senecio jacobaea *and* Acanthus mollis *may rarely key out here)*

▪Plant with abundant yellow latex

Lvs 15-35cm. Lfts 4-7 prs, ovate to oblong, the terminal often 3-lobed, the laterals usu with a stipule-like lobe at base, all dull grey-green and ± hairless above, ± glaucous and sparsely hairy below, crenate. Stems to 90cm, brittle, ± glaucous, hairy. Per. All yr ....*Greater Celandine* **Chelidonium majus**

▪Plant with sparse white latex (at least when young)

Basal lvs to 20cm, limp, dull pale green to ± glaucous, (±) hairless above, often hairy on veins below, sparsely ciliate, with stomata below only. Petiole often long, hairy at base, channelled, with 3-5(7) vb's. Mar-Oct. PL 18..*Welsh Poppy* **Meconopsis cambrica**

▪Plant without latex

Stems rooting at least at lowest nodes. Petiole with (1)3-7 vb's. Terminal lft often larger than laterals. Aquatic or damp habs

Stems angled, soon hollow, hairless, with hot water-cress taste

Lfts (0)2-4(5) prs, the terminal 1-2.5cm, broadly elliptic to ± orb, all rolled when young, entire to sinuate-toothed with hydathodes, with ± indistinct veins, with stomata both sides. Petiole clasping at base, auriculate on upper lvs, channelled, solid, strongly cress-scented, with (1)3(5) vb's. Stems to 30cm tall, ascending. All yr (but lvs often simple in winter). Aquatic

Plant green in autumn. Seed pod 10-18mm; seeds in 2 rows, with 25-50 coarse reticulations per face..*Water-cress* **Rorippa nasturtium-aquaticum**

Plant purplish in autumn. Seed pod 17-30mm; seeds mostly in 1 row, with c100 moderate reticulations per face.................................*Narrow-fruited Water-cress* **Rorippa microphylla**

Stems round, solid, often shortly hairy below, with hot water-cress taste

Lfts 2-3 prs, 1-2cm, the lowest pr often remote, ovate or orb to lanc, ± cordate at base, short-stalked, light green, hairless, ± crenate, with obscure hydathodes, with opaquely netted veins, with stomata both sides. Petiole cress-scented. Stems to 50cm, 3-4mm diam. All yr. Wet shady habs...*Large Bitter-cress* **Cardamine amara**

Stems round, solid, hairless, with v mild taste

Lvs to 40cm, with 1-3(6) prs of lfts, occ with terminal and single lateral lft only; terminal lft 8 x 8cm, orb, sinuate-crenate, with ± opaque veins, with stomata both sides. Petiole channelled, solid, strongly water-cress-scented, with 5(7)vb's. Stems to 70cm, erect. Rhizomatous per. All yr. Damp shady habs, VR alien.......................*Greater Cuckooflower* **Cardamine raphanifolia**

Stems not rooting at nodes (or absent). Petiole with 1(3) vb's. Terminal lft (of lower lvs) occ larger than laterals. Wet to dry habs

Lvs emerging singly

Basal lvs with 1-2(3) prs of lfts; lfts often opp, to 5cm, hairless or occ with sparse adpressed hairs above, finely ciliate, with 5-7 acute hydathode-tipped teeth (often orange) per side (more deeply toothed in f. *ptarmicifolia*), with stomata below only; terminal lft 5-7cm, orb. Stem lvs similar but uppermost simple. Petiole channelled, with 1(3) vb's. Stems purplish at base, round below, slightly ridged above, with purplish bulbils in upper lf axils. Rhizomes with white swollen petiole bases. Apr-Sept. Woods, also hortal......*Coralroot* **Cardamine bulbifera**

Lvs not emerging singly, usu forming a rosette

Hairs septate (often v sparse)

Dry or coastal habs. Stem lvs always absent.......*Buck's-horn Plantain* **Plantago coronopus**

Damp or wet habs. Stems brittle, usu with sparse hairs at least in lf axils, ± grooved with 4-5 angles, solid, with abundant sap. Lvs limp, revolute when young, with strongly recurved margins at maturity, weakly radish- or dill-scented, often reddish, with all veins obscure, with stomata below only. Petiole 5-sided, weakly channelled, with 1-3 vb's

Per, with v stout rootstock. Stems 1-many from base, decumbent, densely branched. Lvs to 3cm; lfts to 4mm, deeply dentate. Apr-Oct (all yr). Damp hths ....*Lousewort* **Pedicularis sylvatica**

Ann (bi), with slender taproot. Stem 1, erect, loosely branched. Lvs (2)4-8cm, weak cucumber and dill scent; lfts to 15mm, dentate to pinnately lobed. May-Sept. Bogs, fens ....*Marsh Lousewort* **Pedicularis palustris**

Hairs unicellular or absent

Plant odourless when crushed. Ann

Lvs in basal rosette and/or on stems, 4-15cm, 1(2)-pinnate or pinnately lobed, limp, resembling *Pedicularis*; lfts 3-5 prs, 1.3-4cm, sparsely hairy, ciliate, lobed, with margins slightly recurved, with main veins visible, with stomata both sides. Petiole dilated at base but hardly sheathing, ciliate at base, channelled, with arc vb and 2 tiny rib bundles. Stems erect, minutely hairy, with sparse longer (often hispid) hairs, round, brittle, usu with small hollow. Mar-Sept. Arable..*Phacelia* **Phacelia tanacetifolia**

Plant cress-scented when crushed

Per to ann, with taproot, or perennating with a short underground stolon or rooting at lf tips

Lfts >2cm

Lf margin with any cilia confined to teeth and lobe apices (cilia soon falling)

Basal lvs persisting, to 15cm, dark glaucous-green above, ± fleshy, usu sparsely hairy on rachis only, with 4-5 prs deeply toothed or lobed lobes (occ reflexed). Stem lvs 0-2. Petiole sparsely hairy, channelled, with 3 vb's. Stems to 50cm, purplish at base, pruinose, (±) hairless, round. Bi. All yr. Dunes, W Br ....*Isle of Man Cabbage* **Coincya monensis** ssp **monensis**

Basal (and lower) lvs persisting, to 10cm, green to ± glaucous above, not fleshy, sparsely hispid both sides, with 3-5 prs of deeply toothed or lobed lobes (occ reflexed). Petiole retrorsely hairy, with 3 vb's. Stem lvs 3-6. Petiole sparsely hairy, channelled, with 3 vb's. Stems to 70cm, purplish at base, ± glaucous above, ± hispid with scattered spreading or retrorse hairs below, ± hairless above, round. Bi (ann). All yr. Ruderal....................*Wallflower Cabbage* **Coincya monensis** ssp **cheiranthos**

Lf margin ciliate or not (but cilia never confined to teeth)

Basal lvs persisting, to 60cm, lyrate-pinnatifid with a large rounded terminal lobe and usu 1-8(11) prs of lobes (occ alt in size, or basal reflexed and v small), teeth tipped with green hydathodes, ± sharply hispid with stout unequal 0.2-1mm hairs. Stem lvs not clasping, decurrent. Petiole with 5(7) vb's. Stems to 100cm, with spreading or reflexed hispid hairs (esp below), pruinose or glaucous above. All yr

Bi or per. Coastal....................*Sea Radish* **Raphanus raphanistrum** ssp **maritimus**

Ann..................................*Wild Radish* **Raphanus raphanistrum** ssp **raphanistrum**

Lfts <1cm

Ann but often perennating by rooting at lf tips. Lfts 1-3 prs, to 4 x 4.5mm. Garden or nursery weed, R (increasing)............*New Zealand Bitter-cress* **Cardamine corymbosa**

Per, with a short underground stolon

Lfts 1-7 prs; terminal lft 0.5-1.5cm, ovate to orb or reniform, often cordate at base, stalked, all occ minutely hairy to hairless above, often 3-dentate with white or purple hydathodes, with mild cress taste, with stomata both sides (often obscure above); lateral lfts usu smaller, ± entire or distantly toothed. Upper lvs sessile, with oblong ± entire lfts. All yr. Damp gsld................................*Cuckooflower* **Cardamine pratensis**

Ann to per, easily uprooted, with fibrous roots

Lf base with ciliate acute auricles clasping stem

Basal lvs in a rosette, dead by fl, strongly cress-scented, long-petiolate; lfts 2-4 prs, opp to alt, ± ovate, hairless, deeply toothed with bristle-tipped hydathodes, with stomata both sides; terminal lft largest, 1-4cm, ± lobed; rachis narrowly winged. Stem lvs ± sessile, with 6-9 prs of narrow toothed or entire lfts. Stems to 70cm, purplish at base, hairless, ridged, hollow. Bi. All yr ....*Narrow-leaved Bitter-cress* **Cardamine impatiens**

Lf base without auricles, not clasping stem

Basal lvs in false rosette. Stems pruinose

Basal lvs soon dying, petiolate, mid-green, occ sparsely hairy, ciliate, lyrate with toothed ovate to linear-lanc lfts. Stem lvs sessile, 1-2-pinnately lobed, often glaucous. Stem 1, 20-40cm, pruinose, hairless, round. Ann. Apr-Oct .... *Garden Cress* **Lepidium sativum**

Basal lvs in true rosette. Stems not pruinose

Lfts orb to ovate

Damp (often shady) habs. Per (ann). Basal lvs up to 12 in rosette, to 7(10)cm, with 3-6 prs of ovate to reniform lateral lfts and a larger terminal lft; terminal lft to 2.5cm, sparsely hairy above, ciliate, ± lobed or angled, sparsely toothed with green hydathodes, often opaquely net-veined, with stomata both sides. Stem lvs 4-10, with smaller narrower lfts (often longer than lfts of basal lvs), short-stalked or sessile. Stems c2mm wide, with short hairs to 0.5mm esp below, deeply grooved. Fls with (4)6 stamens. All yr......................... *Wavy Bitter-cress* **Cardamine flexuosa**

Dry sunny habs. Ann. Basal lvs often >20 in rosette, to 7(10)cm, with 1-3(5) prs of obovate or orb lfts and a larger terminal lft; terminal lft to 2cm, sparsely hairy above, ciliate, ± lobed or angled, sparsely toothed with green hydathodes, often opaquely net-veined, with stomata both sides. Stem lvs 2-4, with smaller narrower lfts, ± sessile. Stems c1mm wide, hairless or with v short hairs to 0.1mm, slightly ridged. Fls with 4(6) stamens. Oct-Jun (all yr)............. *Hairy Bitter-cress* **Cardamine hirsuta**

Lfts lanc

Basal lvs dying at fl, to 10cm, obovate or oblanc, not fleshy, bright green, strongly fetid, hairless, sparsely ciliate at base, pinnately lobed to 2-pinnate, with 3-5(7) prs of lfts and long narrow terminal lobe, all lobes with fragile bristle at apex and indistinct veins. Stem lvs pinnately lobed with narrow ± entire lobes. Petiole ± flattened, not channelled, with 1(3) vb's. Stems ± prostrate. Ann or bi. All yr .... *Lesser Swine-cress* **Coronopus didymus**

Basal lvs dying at fl, to 10(20)cm, obovate or oblanc, ± fleshy, dull grey- or bluish-green, cress-scented, hairless, sparsely ciliate at base, pinnately lobed, each lobe with short lobes esp on distal side. Upper stem lvs narrower, ± entire. Petiole ± flattened, not channelled, with 3 vb's. Stems ± prostrate to ascending. Ann. Mar-Oct .... *Swine-cress* **Coronopus squamatus**

Key to Groups in Division T

(Lvs 2-4-pinnate. Lfts all opp)

Mostly Apiaceae. *Hairs unicellular (but septate in some* Geranium *spp)*

■Tree, shrub or woody climber...**TA**
■Herb
  Stipules present...**TB**
  Stipules absent
    Petiole with fimbriate sheathing base, tough. Stipels occ present (*Thalictrum*).........................**TC**
    Petiole with sheathing base entire or absent, often fragile. Stipels absent
      Lfts hairy at least below
        Petiole (and stems) with white latex (often v sparse and confined to young growth)...........**TD**
        Petiole (and stems) without latex...**TE**
      Lfts hairless
        Petiole channelled above, or triangular in TS
          Petiole solid...**TF**
          Petiole hollow...**TG**
        Petiole not channelled above (may have narrow slit), round or laterally flattened
          Petiole solid, never with latex..**TH**
          Petiole hollow (with or without latex), OR solid with latex (occ v sparse)
            Aquatic to damp habs ....**TI**
            Dry habs, occ drier parts of saltmarshes or rocky coasts...**TJ**

Group TA – *Tree, woody shrub or climber. R alien*

▪Tree or tall shrub. Lvs alt, 2-pinnate. Twigs hard-wooded

Trunk and twigs spiny. Deciduous

Lvs to 75cm, with bristles in lf axils; lfts 6 prs, with an extra 1 basioscopic lft at pinna base, to 12 x 5cm, ovate, acuminate, often short-stalked to 5mm, dark green above, ± glaucous below, bristly on veins both sides, 1-2-serrate with 40-50 teeth per side, with 8-10 prs of 2° veins, with stomata below only. Petiole spiny, swollen and sheathing at base, v weakly channelled nr base. Lf scars U-shaped, extending ½ way around twig, with >30 traces. Trunk often >2m ....*Japanese Angelica-tree* **Aralia elata**

Trunk and twigs not spiny. Evergreen

Lvs to 16cm, grey-green; lfts 15-30 prs, 0.5-6cm, each lft divided into 15-70 prs of lobes; lobes 0.7-6 x 0.4-1mm, linear, minutely hairy, with ± circular pits (extra-floral nectaries) at base (occ absent from basal pr). Twigs pruinose, hairy, ridged, angled nr apices. Bark smooth, deeply fissured with age, grey-green or brown to ± black. Suckering shrub or tree to 30m ....*Mimosa* **Acacia dealbata**

▪Woody climber, with spiralling opp petiolules. Lvs opp, 1-2-pinnate. Stems pith-filled. Deciduous

Stems with 14 shallow ridges, hairless. Lfts 2-6 x 0.5-3cm, rhombic-ovate to narrowly ovate, rounded to broadly cuneate at base, green, (±) hairless, entire or deeply 3-lobed. Petiole 2-6cm, channelled...*Virgin's-bower* **Clematis flammula**

Stems with (10)12 ridges, hairy to hairless. Lfts 1-6 x 0.5-2(2.8)cm, rhombic-ovate to narrowly ovate, cuneate to ± cordate at base, glaucous grey-green, hairy at least when young, toothed, often lobed nr base. Petiole 2-6cm, channelled............*Orange-peel Clematis* **Clematis tangutica**

Stems with 6-10 ridges, hairless to sparsely hairy. Lfts 1.2-5 x 0.5-2.5cm, narrowly ovate to or linear-lanc, rounded to broadly cuneate at base, glaucous grey-green, hairless or hairy below, usu entire (occ 1-2-denticulate) but often 3-lobed nr base. Petiole 2.5-6cm, channelled ....*Oriental Clematis* **Clematis tibetana** ssp **vernayi**

Group TB – *Herb. Lvs alt. Stipules present*

▪Stipules ± herbaceous, red or green. Lfts 3-5, palmately arranged............(*Geranium*) Go to **GER A**

▪Stipules scarious, brown

Lvs 2-ternate

Lfts 5-10 x 4-6cm, ovate, acute to acuminate, cordate at base, shiny green above, dull glaucous below, hairless or with sparse hairs below, with short glandular bristles in axils, ± spiny-serrrate, net-veined, with anastomosing ± parallel or palmate veins, with stomata below only. Petiole reddish, dilated at base. Stems 30-45cm. Rhizomatous. *E. grandiflorum* x *pinnatum* ....*Persian Barrenwort* **Epimedium** x **versicolor**

Lvs 3-pinnate

Lfts 2-8 x 1-4cm, ovate to elliptic, acuminate, rounded at base, shiny green both sides, sparsely glandular-hairy on veins both sides, 2-serrate with bristle-tipped teeth, net-veined, with 2° veins weakly raised both sides (occ reddish). Petiole red, dilated at base, round, hairless (exc for long brown bristles in axils). Stems 50-100cm, red, hairless (exc for long brown bristles in lf axils), round, solid. Rhizomatous. Hortal. *A. ?chinensis* x *japonica* ....*Red False-buck's-beard* **Astilbe** x **arendsii**

Group TC – Thalictrum. *Herb. Lvs alt, basal and on stems (mostly or all basal in* T. alpinum*). Petiole sheathing at base with fimbriate margins at least on upper lvs, wiry. Lfts net-veined, usu with 3 veins converging at each hydathode at lobe apex. Stems tough. Apr-Oct.* ❶ *Stipels (stipule-like outgrowths below each lft) often present on upper lvs*

■Lfts 2-4(5)cm. Stems to 120cm

Lfts glaucous both sides. VR alien, hortal...........*Dusky Meadow-rue* **Thalictrum speciosissimum**

Lfts dark green above, paler (to ± glaucous) below. ❶

Lvs 2-3-pinnate, the upper sessile; lfts longer than wide, obovate, truncate, cuneate at base, hairless above, (±) hairless below, 3-4-lobed at apex, with 2-3° veins raised below, with stomata below only. Petiole ± round, hollow but with many scattered vb's. Stems hairless, ridged, hollow. Open damp habs...............................*Common Meadow-rue* **Thalictrum flavum**

Lvs 2-5-pinnate, the upper sessile; lfts usu wider than long. Shady, often damp, habs. N & W Br ....*Lesser Meadow-rue* (ssp) **Thalictrum minus** ssp **major**

■Lfts <2cm

Stems 20-80(120)cm

Lvs to 5-pinnate; lfts 7-12 x 6-15mm, usu wider than long, dull green to glaucous grey-green above, hairless or with stalked glands below, 3-5(7)-lobed or deeply dentate, with 2° veins raised below, with stomata both sides or below only. Petiole minutely glandular-hairy or hairless, occ weakly channelled, solid or hollow. Stems often wavy, green or pruinose, often purplish, minutely glandular-hairy, ± round to furrowed, usu hollow. ❶ ....*Lesser Meadow-rue* **Thalictrum minus** ssp **minus***

Stems usu <15cm

Lvs often at end of short stolons, 2-ternate, petiolate; lfts to 6 x 6mm, usu as wide as long, ovate to ± orb, shiny dark green above, shallowly lobed or crenate, with stomata below only. Petiole to 5cm, thread-like (0.4mm diam), round. Mtns or rocky habs. N Br, Ire ....*Alpine Meadow-rue* **Thalictrum alpinum**

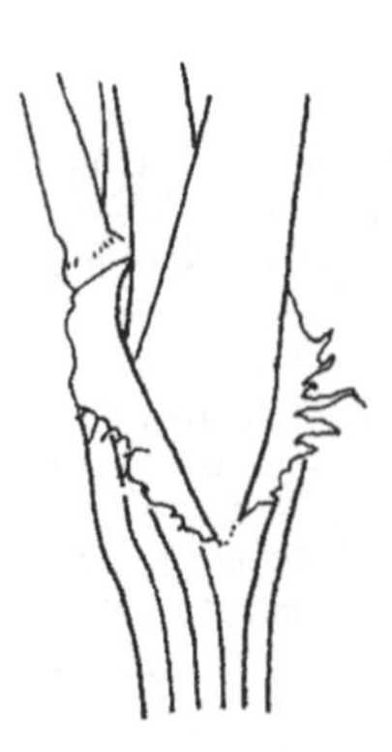

Thalictrum minus ssp *minus*
sheathing petiole base

Group TD – *Herb. Lfts hairy at least below, with stomata below only (but scattered or confined to veins above in* Pimpinella saxifraga*). Petiole (and stems) with white latex (often v sparse and confined to young growth).* ❶ *Aniseed-scented*

■Petiole (and stems) soon hollow (occ pith-filled)

Lfts (and petiole) strongly aniseed-scented ❶

Per. Lfts pale green and often white-flecked above, v shortly hairy above, more densely so with longer hairs below, lobed and serrate. Petiole stout, with long patent hairs but hairless at sheathing base, narrowly channelled. Stems 60-180cm, hairy, ± grooved. Apr-Oct ....*Sweet Cicely* **Myrrhis odorata**

Ann. Lfts pale green, sparsely hairy esp below. Petiole hairless but often ciliate nr apex of sheathing base, channelled, with 5-7 vb's. Stems 30-70cm, hairless or sparsely hairy above nodes, striate....................*Garden Chervil* **Anthriscus cerefolium**

Lfts with sour parsley odour or odourless

Per

Lfts often purplish, 3-9cm. Stems to 200cm....................*Wild Angelica* **Angelica sylvestris**

Lfts green, usu 1-4cm, ovate-lanc, occ white-flecked above, with sour parsley odour or odourless, hairless or with short patent hairs above, usu hairy below, sparsely scabrid-ciliate, lobed. Petiole often ciliate nr apex of sheathing base, triangular or channelled, with hollow longer than broad, with 7(11) vb's. Stems to 100cm, hairy below, hairless above, furrowed ....*Cow Parsley* **Anthriscus sylvestris***

Ann. Lfts green (but lf rachis purplish nr base), usu <1cm, odourless, ovate, hairless above, with scattered patent stiff hairs below (and on rachis), ciliate, lobed. Petiole with long-ciliate sheathing base, channelled, with small hollow broader than long. Stems to 40cm, often purplish nr base, hairless to sparsely hairy, striate, ± thickened below nodes. Dry sandy habs ....*Bur Chervil* **Anthriscus caucalis**

■Petiole (and stems) solid

Lfts roughly antrorsely adpressed-hairy. Stems not purple-blotched (may have purple ridges), not swollen below nodes, usu roughly retrorsely adpressed-hairy

Tall ann (bi), >50cm when mature, with 1 erect main stem

Lvs 1-3-pinnate, the basal >10cm; lfts 1-3cm, ovate to lanc, hairy above, lobed, the lobes serrate. Petiole retrorsely adpressed-hairy, ± hairless at sheathing base (not ciliate), channelled, with 5(7) vb's. Nov-Sept....................*Upright Hedge-parsley* **Torilis japonica**

Low ann, <30cm, with several stems

Petiole retrorsely adpressed-hairy, with hairless sheathing base, channelled, with 3(5) vb's. Lvs 2-3-pinnate, to 8.5cm; lfts 0.5-3cm, ovate to lanc, hairy esp below, pinnatifid into linear-lanc lobes (not bristle-tipped); terminal lobe not acuminate. Stems ± prostrate, sparsely retrorsely adpressed-hairy, striate. Sept-Jul....................*Knotted Hedge-parsley* **Torilis nodosa**

Petiole usu ± hairless, with hairless sheathing base, channelled, with 7 vb's. Lvs 1-2-pinnate, to 13cm; lfts 0.5-3(5)cm, lanc, hairy both sides, ciliate, pinnatifid into serrate lanc lobes (bristle-tipped); terminal lobe acuminate. Stems erect, hairless or with sparse adpressed retrorse hairs, round below, ridged above. Oct-Sept (all yr). Arable, R, S Eng ....*Spreading Hedge-parsley* **Torilis arvensis**

Lfts not roughly hairy or, if so, the hairs not adpressed

Stems purple-blotched, swollen below nodes, ± retrorsely hairy. Odourless bi

Lfts dull dark green, with unequal ± adpressed hairs, broadly ovate, acuminate, sweetly celery-scented, lobed, the lobes serrate. Petiole with ciliate sheathing base, hairy, with hairs to 0.5mm, triangular or rounded, weakly channelled, often sharply angled below, with 5(7) vb's. Stems to 100cm, with short stiff retrorse hairs and longer patent hairs, ± grooved. Sept-Jul....................*Rough Chervil* **Chaerophyllum temulum**

Lfts usu yellow-green, with equal ± adpressed to ± patent hairs (often sparse above), lanc, acuminate, sweetly celery-scented, lobed, the lobes serrate or ± entire. Petiole with auriculate hairy sheathing base often purplish nr apex, with dense ± patent to retrorse short hairs to 0.3mm, slightly channelled, with 12 scattered vb's. Stems to 100cm, ± rough, with minute retrorse hairs, ± grooved, occ with small hollow. Sept-Jul. VR alien....................*Golden Chervil* **Chaerophyllum aureum**

Stems not purple-blotched (may have purple ridges), not swollen below nodes

Carrot-scented bi. All lvs 2-3-pinnate. Lfts bristle-tipped, 0.5-3cm, sparsely bristly-hairy below (occ densely hairy both sides), usu antrorsely scabrid at least nr apex. Petiole not reddish, with ciliate sheathing base, hairy to hairless, channelled, with 7(9) vb's around margin and 1(3) in centre, with sparse latex. Stems to 100cm, striate or ridged. All yr .... *Wild Carrot* **Daucus carota***

Odourless per. Lower stem lvs (and those of mature rosettes) 2-pinnate, with 4-5 prs lfts. Lfts apiculate (but not bristle-tipped), densely short-hairy or sparsely long-hairy to hairless, scabrid-ciliate. Petiole reddish at base, with weakly inflated ciliate sheathing base, hairy to hairless, channelled, occ hollow, with 5(7) vb's (only 3(5) vb's in 1-pinnate lvs). Stems to 70cm, striate. All yr... *Burnet-saxifrage* **Pimpinella saxifraga**

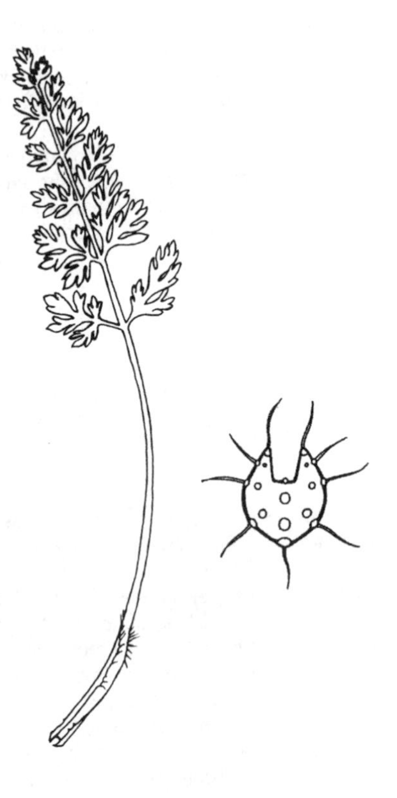

Daucus carota & petiole TS

Group TE – *Herb. Lfts hairy at least below. Petiole (and stems) without latex*

■Petiole solid

Lvs 30-100cm. Rhizomatous per

Lvs to 100cm, 2-3-pinnate, swollen at pinnate divisions, with minute glandular bristles in axils; lfts 7-9cm, elliptic, acuminate, petiolate, hairless above, sparsely hairy below, ciliate, folded when young, 2-serrate, net-veined, with 8-11 prs of 2° veins strongly sunken above and raised below (ending in teeth), with stomata below only. Petiole dilated at base, with abscission layer, hairless, channelled (occ weak), with many vb's around margin. Stems to 200cm, hairless, round. Apr-Oct..*Buck's-beard* **Aruncus dioicus**

Lvs to 50cm, 3-pinnate, with long brown bristles in axils; lfts 2-8 x 1-4cm, ovate to elliptic, acuminate, rounded at base, petiolate, shiny green both sides, sparsely glandular-hairy on veins both sides, 2-serrate with bristle-tipped teeth, net-veined, with 2° veins weakly raised both sides (occ reddish). Petiole dilated at base, red, hairless, round. Stems 50-100cm, red, hairless exc for long brown bristles in lf axils, round, solid. Hortal. *A. ?chinensis* x *japonica* ....*Red False-buck's-beard* **Astilbe** x **arendsii**

Lvs <20cm

Shortly rhizomatous per. Basal lvs to 15(30)cm, with proximal lfts much reduced; lfts >10 prs, opp below, alt above, linear, apiculate or bristle-tipped, weakly tansy-scented, long-hairy, with stomata above only. Petiole tough, with wide channel above, with 7 vb's. Stems to 50cm, ± woolly, furrowed, solid. All yr..*Yarrow* **Achillea millefolium**

Ann. Basal lvs to 10(15)cm, with proximal lfts longest; lfts 3-6 prs, all opp, 4-5 x 0.5-1mm, linear-lanc, bristle-tipped, odourless, hairless or with sparse stiff hairs, antrorsely scabrid, with stomata both sides. Petiole with reddish and ciliate sheathing base, with 3 vb's. Stems to 50cm, ± hairy with short scattered hairs, striate, hollow when old. Mar-Oct (all yr). Arable ....*Shepherd's-needle* **Scandix pecten-veneris**

■Petiole becoming hollow. Per

Plant fetid

Basal lvs to 30cm, long-petiolate, 2-ternate or with 2 prs of 2-pinnate divisions; terminal lft ovate, often 3-lobed, acute, with channelled stalk, dark green above, paler below, hairless above (or with minute curved hairs on veins), sparsely long-hairy on veins below, minutely ciliate (hairs curved), often 3-lobed, incised-serrate, with pale hydathodes visible above, with 2(3)° veins raised below, with stomata below only. Stem lvs 1-4, much smaller. Petiole weakly channelled to triangular, with large irreg hollow, with >20 vb's (turning brown) around margin. Stems to 60cm, minutely hairy, ± round, solid to hollow. Mar-Oct. Limestone habs, N Eng ....*Baneberry* **Actaea spicata**

Plant odourless

Basal lvs (1)2-ternate, with 2 prs lfts; lfts 3-6cm, ovate, often hairless above, glaucous and softly hairy below, irreg 3-lobed and crenate with white hydathodes in retuse apices (visible end-on or from above), ± net-veined, with veins not or hardly raised below, with stomata below only. Petiole often purplish, usu softly hairy, with sheathing base, round, with 10-12 vb's around margin. Stems to 100cm, hairy, round. Mar-Oct

Fls with strongly hooked spurs. Petiole with hairy sheathing base ....*Columbine* **Aquilegia vulgaris**

Fls with ± straight spurs (or hooked at end). Petiole with hairless sheathing base. Hortal, alien ....*Garden Columbine* **Aquilegia vulgaris** cv

Basal lvs 2-3-pinnate, with 3-5 prs lfts; lfts divided into many linear to linear-lanc lobes, silky-hairy or with long white patent hairs, with main veins visible only, with stomata both sides. Stem lvs (bracts) 3-whorled, sessile, with long-linear segments. Petiole hairy, persistently silky-hairy at sheathing base, channelled or weakly so, with several vb's around margin. Stems 5-20cm, with fibrous lf remains persisting for 4 yrs. Mar-Sept. Calc turf ....*Pasqueflower* **Pulsatilla vulgaris**

Group TF – *Herb. Lfts hairless. Petiole channelled or triangular, solid.* (Crithmum maritimum *and* Apium graveolens *may rarely key out here).* ❶ *Upper lvs perfoliate*

■Petiole with white latex (often v sparse)

Lfts >15mm wide, lanc to ovate

Rhizomatous per. Lvs 1-2-ternate; lfts 3-8cm, acuminate, often oblique or fused at base, stalked or sessile, dull green above, irreg serrate, with scabrid-ciliate narrow cartilaginous margins, net-veined, with stomata below only. Petiole to 15cm, sheathing at base with overlapping auricles, with 7 vb's and celery-like odour, the latex mainly confined to canals around margin. Stems to 100cm, ridged, hollow.................................... *Ground-elder* **Aegopodium podagraria***

Tufted bi or per, with swollen taproot. Lvs 2-ternate, with hairs in lf axils; lfts 2-6 x 5cm, ovate, light green, with stomata both sides but often confined to vein margins above. Upper lvs perfoliate, usu with stomata below only. Petiole with ciliate sheathing base, with dense tuft of minute bristles in lf axils, channelled to flat above, with 7 vb's, odourless. Stem usu 1, to 120cm, round below, with 2 ciliate cartilaginous ridges above, pith-filled or hollow. ❶. Feb-Aug. VR alien.. *Perfoliate Alexanders* **Smyrnium perfoliatum**

Lfts <3mm wide, linear-lanc. Tufted bi or per

Lvs glaucous. Stems to 20cm, with persistent fibrous dead remains of lvs at base

Lvs to 15cm; lft lobes 5-10(15) x 0.5mm, thread-like, acute, with stomata both sides. Petiole channelled, with 3 vb's and v sparse latex, odourless. Stems several, branched from base, hairless, deeply grooved, solid. Per. Mar-Oct. Limestone turf, SW Eng..*Honewort* **Trinia glauca**

Lvs not glaucous. Stems to 100cm, without fibrous remains at base

Petiole with ciliate sheathing base. Bi.. *Wild Carrot* **Daucus carota***

Petiole with hairless sheathing base. Per

Lft lobes 10-20 x 1-4mm, linear-lanc, apiculate with purple hydathode, mid-green, with serrulate narrow cartilaginous margins, with 2° veins partly translucent, with stomata below only. Petiole purple at base, with sheathing base not inflated, with 7(9) vb's around margin and several scattered in the centre, with weak celery-like odour and peppery taste. Stems striate, solid. Apr-Oct. Damp gsld.. *Pepper-saxifrage* **Silaum silaus**

Lft lobes 7-18 x 1.5-3mm, linear-lanc (to ovate), long-apiculate with white hydathode, dark bluish-green above, scabrid-serrulate, with 2° veins (±) obscure, with stomata below only. Petiole purplish at base, with 7(9) vb's, sweetly celery-scented, with mild taste. Stems with strong acute cartilaginous ridges, solid. Fens, VR, Cambs. Sch8....*Cambridge Milk-parsley* **Selinum carvifolia**

■Petiole without latex

Basal lvs arising singly from tuber, without sheathing base. Stem lvs with a sheathing base

Petiole of basal lvs slender, partly below ground, whitish, channelled, with 3 vb's. Basal (and lower) lvs few, 3-pinnate, broadly triangular, often dead by fl; lft lobes 5-10mm, spathulate to linear, with an obtuse cartilaginous apex, the margins antrorsely scabrid or weakly so. Petiole reddish nr base. Stem 1, to 50cm, striate, round, solid. Mar-Jun...................... *Great Pignut* **Bunium bulbocastanum**

Basal lvs not arising singly from tuber. All lvs with sheathing base

Petiole with ciliate sheathing base

Ann. Lfts lobes 4-5 x 0.5-1mm, bristle-tipped, odourless, hairless or with sparse stiff hairs, antrorsely scabrid, with stomata both sides. Petiole with reddish and ciliate sheathing base, with 3 vb's. Stems to 50cm, ± hairy with short scattered hairs, striate, hollow when old. Mar-Oct (all yr). Arable.. *Shepherd's-needle* **Scandix pecten-veneris**

Petiole with hairless sheathing base

Rhizomatous per. Lvs light green.. *Moschatel* **Adoxa moschatellina**

Tufted bi or per. Lvs light to dark green

Lfts parsley-scented, usu strongly crisped

Lfts 1-3cm, shiny bright green, with acute or apiculate teeth, with stomata both sides or below only. Upper stem lvs often ternate. Petiole channelled (occ weakly so), solid or with small hollow, with 7 vb's. Stems to 75cm, with strictly ascending branches, striate below, ridged above, solid. Bi. All yr.................................. *Garden Parsley* **Petroselinum crispum**

Lfts (±) odourless, not crisped

Lvs 2-pinnate; lfts 5-15mm, ovate, mucronate with long-acute purplish hydathodes, narrowly cuneate at base, ± shiny above, hairless, with antrorsely scabrid margins, lobed, with veins partly translucent in interrupted pattern, with stomata below and at least along veins above; rachis channelled. Petiole hairless to hairy, weakly channelled, with 9(11) vb's around margin. Stems to 100cm, (±) hairless, with sheathing base not auriculate, ± ridged, solid. Bi (occ short-lived monocarpic per), with persistent fibrous lf remains at base. Mar-Sept. Calc gsld, VR...*Moon Carrot* **Seseli libanotis**

Lvs 2-ternate or 3-pinnate; lfts 15-50mm, ovate, acute, cuneate at base, dull above, minutely hispid-scabrid on larger veins both sides, with antrorsely scabrid margins, laciniate, with main veins raised above, with stomata below only; rachis double-channelled. Petiole hairless, with sheathing base not auriculate, purplish and round nr base, channelled distally, with c17 scattered vb's. Stems to 120cm, hairless, striate to ribbed (esp above), solid. Per, with short scaly rhizomes. Mar-Sept. Woods, SW Eng and Bucks ....*Bladderseed* **Physospermum cornubiense**

Ann. Lvs glaucous or dark bluish-green

All lvs 2-3-pinnate, divided into small lfts. Lft lobes 0.5-1.5cm, ovate to lanc, fetid, dark bluish-green, lobed, with v weakly antrorsely scabrid (to entire) margins, with veins hardly raised above, with stomata below only; terminal lobe long. Petiole with rectangular channel, slightly ridged to ± angled below, solid, with 12 vb's around margin (often obscure). Stem(s) to 100cm, ± glaucous, finely striate, usu solid. Apr-Oct (usu) ....*Fool's Parsley* **Aethusa cynapium**

Basal lvs 1-2-pinnate, coarsely divided into large serrate lfts. Stem lvs usu 2-pinnate, finely divided. Lft lobes 1-3 x 0.6cm, lanc (to ovate), odourless, ± glaucous, the teeth ending in long cartilaginous points, with stomata both sides. Petiole channelled, often hollow. Stem(s) to 100cm, ± glaucous, striate to weakly ridged, hollow. May-Oct. R casual ....*Bullwort* **Ammi majus**

TF

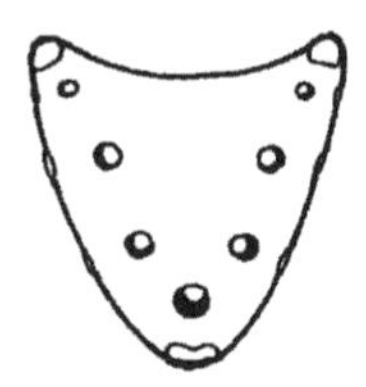

Aegopodium podagraria petiole TS

TG

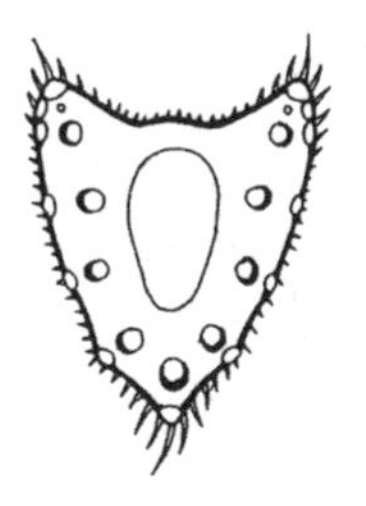

Anthriscus sylvestris petiole TS

Group TG – *Herb. Lfts hairless. Petiole channelled or triangular, hollow. Stems hollow*

■Plant with white (or cream) latex (often sparse). Per

Basal lvs arising singly from tuber, without sheathing base. Stem lvs sheathing

Petiole of basal lvs slender, partly below ground, whitish, with sparse latex (in 7 canals around margin) and 3 vb's. Basal lvs 2(3)-pinnate, 5-15cm, soon withering, the lowest pr long-stalked, odourless; lft lobes 3-10 x 1.5mm, linear-lanc, acute, antrorsely scabrid-ciliate, with stomata below only. Stem lvs ± sessile. Stems to 40cm, finely striate. Mar-Jun ....*Pignut* **Conopodium majus**

Basal lvs not arising singly from tuber. All lvs with sheathing base

Lft lobes >5mm wide, ± ovate

Lvs with purplish pinna junctions and hairs in lft axils; lfts 3 prs, 3-9 x 3.5cm, ovate (to lanc), asymmetric at base, rarely with short stiff hairs above and on veins below, scabrid-ciliate, with narrow cartilaginous margins (occ purplish), 1-2-serrate, net-veined, often with purplish veins, with stomata below only. Petiole purplish at sheathing base, often laterally flattened, celery-scented, with hollow usu broader than long and >20 vb's, with latex canals scattered throughout and around margin, the latex occ cream. Stems to 200cm, usu purplish, striate. All yr. Damp habs..*Wild Angelica* **Angelica sylvestris**

Lvs not purplish; lfts 3-6 prs, usu 1-4cm, ovate-lanc, with sour parsley odour or odourless, shiny green above, occ white-flecked, occ hairy, sparsely scabrid-ciliate, without cartilaginous margins, lobed, coarsely serrate, with stomata below only. Petiole often ciliate nr apex of sheathing base, triangular, usu channelled, with hollow longer than broad, occ pith-filled, with 7(11) vb's and latex canals around margin. Stems to 100cm, hairy below, hairless above, furrowed. All yr...*Cow Parsley* **Anthriscus sylvestris***

Lft lobes 2mm wide, linear-lanc

Lfts 4-5 prs, with lobes 5-20mm, acute, not or weakly odorous, with scabrid-ciliate cartilaginous margins, with translucent veins opaque when dried, with stomata below only (at least in sunny habs). Petiole with purple sheathing base, 7-angled. Stems to 150cm, strongly ridged, hairless. Mar-Nov (but single lf early in season). Fens ....*Milk-parsley* **Peucedanum palustre**

■Plant without latex (may have clear sap)

Hortal or casual alien

Lfts usu strongly crisped, parsley-scented. Bi.................*Garden Parsley* **Petroselinum crispum**

Lfts not crisped, odourless

Basal lvs 1-2-pinnate, coarsely divided into large serrate lfts. Stem lvs usu 2-pinnate, finely divided; lft lobes 1-3 x 0.6cm, lanc (to ovate), ± glaucous, the teeth ending in long cartilaginous points, with stomata both sides. Petiole channelled, often hollow. Stem(s) to 100cm, ± glaucous, striate to weakly ridged. Roots odourless. Ann. May-Oct. R alien ....*Bullwort* **Ammi majus**

All lvs 2-3-pinnate, finely divided, often held in horizontal plane; lft lobes 4-6 x 0.6-1mm, linear-lanc, acute, mid-green, with stomata below only, lobed to entire. Petiole channelled, soon with small hollow, with 5 vb's. Stem(s) to 60cm, striate. Roots carrot-scented. Bi or monocarpic per. Often all yr..*Caraway* **Carum carvi**

Hay-meadows, N Br

Lvs mostly basal, 4-5-pinnate, parsley-scented; lobes 3-5(8) x 0.1-0.3(0.5)mm, thread-like, apiculate (x20), with translucent main veins, with stomata both sides. Upper lvs with broad dilated sheathing base. Petiole laterally flattened (teardrop-shaped), narrowly channelled, with 10-20 vb's around margin. Stems to 60cm, to 3mm diam, 13-striate. Per. Apr-Oct ....*Spignel* **Meum athamanticum**

Aquatic or wet habs

Lvs 1-pinnate but lfts deeply incised (appearing 2-pinnate); lf lobes 2-8 x 1mm, (±) linear, with smooth margins. Submerged/lower lvs 2-3-pinnate, flaccid, often without stomata; lobes linear or thread-like. Upper and aerial lvs 1-pinnate, with oblanc to obovate (often 3-lobed) lfts, with stomata both sides. Petiole weakly channelled to ± triangular, hollow. Stems rooting at nodes. Per. All yr. Aquatic, usu partly submerged........................*Lesser Marshwort* **Apium inundatum**

Group TH – *Herb. Lfts hairless (but hispid-scabrid on veins of* Physospermum cornubiense*). Petiole not channelled, round or laterally flattened, solid, without latex*

▪Petiole not sheathing at base

Lvs 1-2-ternate; lfts to 10cm, broadly ovate, net-veined, with reddish 2° veins raised below, with stomata both sides; rachis to 10cm. Petiole to 16cm, often reddish above, round, slightly ridged, pith-filled, with strong antiseptic odour and 7-11 vb's around margin. Stems to 60cm. Roots tuberous. Mar-Oct..*Peony* **Paeonia mascula**

▪Petiole sheathing at base

Lvs glaucous

Lvs fleshy, strongly aromatic, 3-pinnate; lft lobes 2-5cm x 3-4mm, linear, acute, with 2° veins obscure, with stomata all round. Petiole with long auriculate overlapping sheathing base, round, striate, with deep narrow channel above. Stems to 45cm, pruinose. Rhizomes absent. All yr (v reduced in winter). Coastal...*Rock Samphire* **Crithmum maritimum**

Lvs not fleshy, odourless, 1-2-ternate; lfts 5-30cm, linear-lanc to linear, ± curved, with strongly serrate (± spinescent) cartilaginous margins, with midrib raised below, with anastomosing 2° veins and with ± parallel lateral veins each side of midrib, with stomata both sides. Petiole to 15cm, with long non-auriculate sheathing base, channelled at least nr base, with 5 vb's. Stems to 90cm. Rhizomes swollen. Mar-Oct (all yr). R alien........................*Longleaf* **Falcaria vulgaris**

Lvs green, not fleshy. Rhizomes absent

Stems prostrate, rooting at nodes. Lvs 1-8-ternate. Wet habs ....(*Ranunculus aquatilis* agg*) Go to **RAN-BAT**

Stems erect, not rooting at nodes. Dry habs

Lvs 2-ternate or 3-pinnate, dull green above, odourless; lfts 15-50 x 10-15mm, ovate, acute, cuneate at base, minutely hispid-scabrid on larger veins both sides, with antrorsely scabrid margins, laciniate, with main veins raised above, with stomata below only; rachis double-channelled. Petiole hairless, with sheathing base not auriculate, purplish and round nr base, channelled distally, with c17 scattered vb's. Stems to 120cm, hairless, striate to ribbed (esp above), solid. Per, with short scaly rhizomes. Mar-Sept. Woods, SW Eng and Bucks ....*Bladderseed* **Physospermum cornubiense**

Lvs 3-6-ternate, bronzey-green when young, parsley-scented; lft lobes to 40 x 2mm, linear, v acute and bristle-tipped, with v narrow cartilaginous margins often serrulate nr apex, with translucent main vein, other veins ± obscure. Petiole round, with c10-15 striations. Stems to 200cm, striate, solid. Per, with persistent fibrous remains of lvs at base. Coastal, SE Eng ....*Hog's Fennel* **Peucedanum officinale**

Group TI – *Plant hairless. Petiole not channelled, round or laterally flattened, hollow (with or without latex), or solid with latex. Per (occ ann in* Oenanthe aquatica*). Aquatic to damp habs*

■Plant with white latex (occ v sparse) AND/OR with tuberous roots

Lvs strongly parsley-scented. Petiole without latex

Petiole with auriculate sheathing base, slightly laterally flattened (0-shaped), with minute slit above, with c10 striations, solid or hollow. Lvs 3-4-pinnate; lfts deeply lobed; lobes 2-5(8) x 1mm, lanc to ovate, acute, v thin, with entire or weakly scabrid margins, with opaque veins when dried, with stomata obscure above. Submerged lvs stiffer with flattened lobes. Stems to 150cm, round (finely striate), with large hollow, with dead lvs persisting at base. Tubers slender. Rarely all yr..*Fine-leaved Water-dropwort* **Oenanthe aquatica**

Lvs weakly parsley- or celery-scented (occ fetid). Petiole with white latex (occ obscure)

Petiole solid, with latex drying orange-brown. Stems round (finely striate), to 150cm

Lfts lanc to ovate, shiny, with narrow erose cartilaginous margins, toothed, with stomata both sides. Petiole with auriculate sheathing base, often with obscure slit above, with fetid celery odour. Stems stout, pith-filled or hollow. Tubers 3-5, 6-10cm, spindle-shaped. Jan-Jul ....*Hemlock Water-dropwort* **Oenanthe crocata**

Petiole hollow or solid, with latex (often v sparse) not drying orange-brown

Stems round (finely striate), to 100cm

Basal (and lower) lvs (1)2-pinnate; lft lobes 0.4-2cm, linear-lanc to narrowly ovate-lanc, mucronate, often cylindrical and hollow, with smooth to weakly scabrid non-cartilaginous margins, with midrib sunken above and raised below, with 2° veins usu obscure. Upper lvs 1-pinnate with lfts 0.5-2cm, distant, obtuse (but with a greenish-white hydathode). Petiole of basal lvs longer than lf, with long green sheathing base (± not auriculate), round, hollow, sweetly celery-scented. Stems to 8mm diam, purplish nr base, thin-walled, often constricted at nodes, v hollow, the latex often obscure. Stolons developing from mid-summer. Tubers small. All yr. Often aquatic.................................*Tubular Water-dropwort* **Oenanthe fistulosa**

Stems ridged (often strongly so)

Stems to 12mm diam, ± pruinose, hollow. Lvs 2-4-pinnate, the lower soon withering; lobes 3-15 x 1(5)mm, linear-lanc, acute with hydathode often purplish, with erose to minutely scabrid margins, with opaque veins when dried. Petiole 3mm diam, round, striate. Tubers often globose, nr to rootstock, occ spindle-shaped or remote. Infl rays hollow. Bracts few or absent. Frs 4mm, ± cylindrical, with styles 1-2mm. Damp alluvial gsld ....*Narrow-leaved Water-dropwort* **Oenanthe silaifolia**

Stems 2.5-5mm diam, not pruinose, solid to hollow. Infl rays solid. Bracts few or several. Frs 2.5mm, ovoid, with styles 1mm........................*Parsley Water-dropwort* **Oenanthe lachenalii**

■Plant without latex or tuberous roots

Stems usu >20cm tall, erect

Lvs 1-pinnate but lfts palmately lobed into thread-like lobes (appearing 2-pinnate and whorled). Stems to 60cm. Mar-Oct..*Whorled Caraway* **Carum verticillatum**

Lvs 2-3-pinnate, to 30cm; lfts all stalked (the lowest pr longest); lobes 3-9cm, linear-lanc, unequal at base, with narrow entire or weakly scabrid cartilaginous margins, serrate, with stomata below only. Petiole stout, occ reddish nr base, weakly ridged, hollow, celery-scented. Stems to 150cm, round, striate, v hollow. Per. Apr-Oct. Usu aquatic.....*Cowbane* **Cicuta virosa**

Stems usu <15cm tall, (±) prostrate

Lvs 1-pinnate but lfts deeply incised (appearing 2-pinnate); lf lobes 2-8 x 1mm, (±) linear, with smooth margins. Submerged and lower lvs 2-3-pinnate, flaccid, often without stomata; lobes linear or thread-like, without apical bristles. Upper and aerial lvs 1-pinnate, with oblanc to obovate (often 3-lobed) lfts, with stomata both sides. Petiole weakly channelled to ± triangular, hollow. Stems creeping, hollow. All yr. Aquatic, usu partly submerged ....*Lesser Marshwort* **Apium inundatum**

Lvs 1-8-ternate; lfts with apical bristles.....................(*Ranunculus aquatilis* agg*) Go to **RAN-BAT**

Group TJ – *Plant hairless (exc occ for lf axils). Petiole round or laterally flattened, not channelled, soon hollow (remaining solid in* Ferula communis *and some* Oenanthe *spp). Tufted per or bi (ann). Dry habs, occ brackish coastal gsld or rocky coasts*

■Plant with latex (esp in petiole)

Petiole with dense tuft of minute bristles in lf axil and at base of each pr of lfts

Lvs 1-2-ternate; lfts 4-10cm diam, broadly ovate, dark yellow-green, with narrow cartilaginous margins (entire or remotely scabrid), irreg lobed, the lobes serrate with shortly aristate teeth, with stomata below and occ scattered along veins above. Petiole usu reddish at base, with minute papillae-like hairs nr lf axils, round, shallowly ridged, with vb's scattered around 2 vertical hollows, with abundant greenish-cream latex (turning brown). Stems 30-100cm, ridged, hollow, pine- or celery-scented. Tufted per. Mar-Oct. N Br ....*Masterwort* **Peucedanum ostruthium**

Lvs 1-3-ternate; lfts to 4 x 3cm, ovate to rhombic, often asymmetric at base, shiny dark green both sides, crenate-serrate, with white protruding hydathodes (unlike other *Apiaceae*), net-veined, the veins sunken above but with raised cartilaginous ridges, with stomata below only. Petiole often auriculate and ciliate at base, with minute hairs in lf axils, laterally flattened, with (6)11 vb's around hollow, with white latex (occ sparse). Stems 50-150cm, furrowed, solid, later hollow, celery-scented. Tufted per (bi). Nov-Jul. Often nr coast...........*Alexanders* **Smyrnium olusatrum***

Petiole without bristles

Lfts >50mm wide

Lfts to 12 x 7cm, ovate, asymmetric at base, occ 3-lobed (terminal always so), serrate with whitish teeth (each with obscure sunken hydathode), net-veined, usu with stomata below only. Petiole with purple sheathing base, slightly dorsally flattened, sweetly celery-scented. Stems to 200cm, usu green. Monocarpic per. All yr............*Garden Angelica* **Angelica archangelica***

Lfts <3mm wide

Giant per, >150cm. Roots not tuberous. Petiole (and stems) solid

Lvs 5-6-pinnate, not all in 1-plane; lobes v like *Foeniculum vulgare*, to 5 x 0.5-1.3mm, linear, mucronate, dark grey-green, with stomata both sides. Petiole with auriculate sheathing base, with many scattered vb's and abundant latex, odourless. All yr. VR alien ....*Giant Fennel* **Ferula communis**

Medium per, to 100cm. Roots tuberous. Petiole (and stems) solid to hollow (petiole hollow esp nr base)

Basal (and lower) lvs 1-2-pinnate, soon withering; lft lobes 2-4(25) x 2.5mm, occ spathulate, acute with green apiculus, with weakly scabrid to entire margin. Upper lvs with lobes 30-55 x 1-2.5mm, spathulate or linear. Petiole with sheathing base ± not auriculate, ridged, with 10-12 vb's around margin, sweetly celery-scented. Stems reddish nr base, ridged. Tubers cylindrical, elongate (like pipe-cleaners), occ spindle-shaped. Infl rays solid. Bracts few or several. Frs 2.5mm, ovoid, with styles 1mm. All yr. Usu damp brackish gsld ....*Parsley Water-dropwort* **Oenanthe lachenalii**

Basal (and lower) lvs 2-3-pinnate, soon withering; lft lobes (3)5-6(9) x 1-2.5mm, ovate to linear-lanc, acute, with v minutely scabrid margins. Upper lvs with lobes 20-55 x 1-2.5mm, spathulate or linear. Petiole with sheathing base auriculate only on basal lvs, ridged, with 10-12 vb's around margin, weakly parsley-scented. Stems reddish nr base, ridged. Tubers globose, remote from plant. Infl rays solid. Bracts several. Frs 3mm, cylindrical (± straight-sided), with styles 2-3mm. Nov-Aug. Damp to dry gsld ....*Corky-fruited Water-dropwort* **Oenanthe pimpinelloides***

■Plant without latex, totally hairless

Petiole and stems purple-spotted. Plant fetid

Lfts 1-2cm, gland-dotted when dry, with (±) entire non-cartilaginous margins, with stomata below only. Petiole slightly flattened, striate, v hollow. Stems to 200cm, weakly ribbed, 5-angled to round, hollow. Bi. All yr..........*Hemlock* **Conium maculatum***

Petiole and stems not purple-spotted. Plant not fetid

Lfts >10mm wide, ovate or ± so

Coastal, Scot. Lvs 1-2-ternate; lfts 1 pr, 2-5 x 3cm, ovate to rhombic, cuneate at base, sessile, shiny bright green, the terminal with channelled stalk, the laterals sessile, serrate distally with cartilaginous teeth apices occ purple, occ lobed, with stomata below only. Petiole purplish at auriculate base, laterally flattened, striate, with sweet celery odour. Stems to 50cm, often purplish nr base, ± round, striate, hollow. Per. Apr-Oct....*Scots Lovage* **Ligusticum scoticum**

Hortal. Lvs 2-pinnate; lfts 1-4 prs, 3-11cm, trullate to rhombic, cuneate at base, the lowest stalked, shiny green esp below, lobed or sparsely deeply serrate, with veins raised above, with stomata both sides. Petiole often with purplish auriculate sheathing base, round, strongly celery-scented. Stems 80-180cm, hollow, striate. Apr-Oct.........*Lovage* **Levisticum officinale**

Lfts to 1mm wide, thread-like

Aniseed-scented per, to 250cm. Lvs 4-5(6)-pinnate, not all in 1-plane; lobes 5-50 x 0.5mm, acute (often mucronate), with stomata both sides (exc midrib). Petiole with auriculate sheathing base, round or laterally flattened, striate, soon hollow. Stems shiny green, occ pruinose when young, striate, solid, soon hollow. Feb-Sept

Young lvs green..*Fennel* **Foeniculum vulgare***

Young lvs purplish................................*Bronze Fennel* **Foeniculum vulgare 'Purpurascens'**

Parsley-scented per, to 60cm...*Spignel* **Meum athamanticum**

Dill-scented (± fetid) ann, to 60(90)cm. Lvs 4-5(6)-pinnate, finely divided; lobes 10-20 x 1mm, acute (with mucro), dark green. Petiole with long non-auriculate sheathing base, round, hollow. Stems occ pruinose, striate, hollow. Mar-Oct (all yr).............*Dill* **Anethum graveolens**

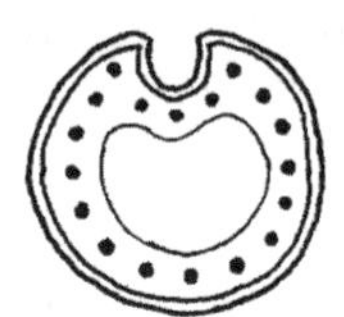

Angelica archangelica lft margin & petiole TS

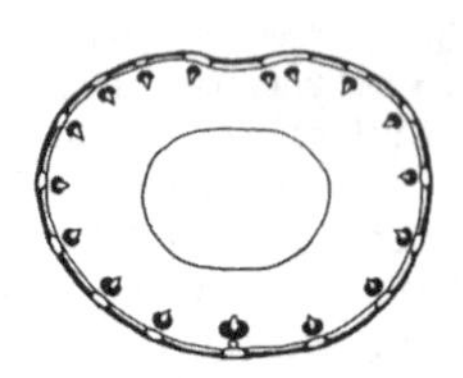

Conium maculatum petiole TS

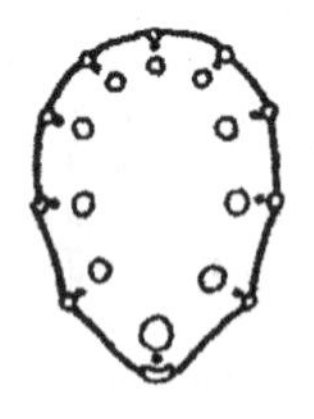

Foeniculum vulgare petiole TS

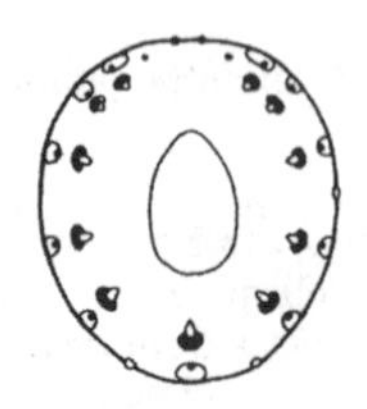

Smyrnium olusatrum petiole TS

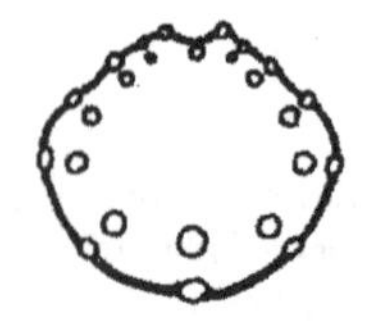

Oenanthe pimpinelloides petiole TS

Key to Groups in Division U

(Lvs 2-4-pinnate. Lfts mostly alt)

Herbs or low shrubs; if intercalary lfts present Go to **SK**

■Lvs strongly aromatic or fetid. Plant densely to sparsely hairy (rarely hairless)............................**UA**
■Lvs odourless or weakly aromatic. Plant with at least some hairs when young............................**UB**
■Lvs odourless. Plant always hairless..**UC**

Group UA – *Lvs strongly aromatic or fetid. Plant densely to sparsely hairy (always hairless in* Ruta graveolens*).* ❶ *Stems prostrate, rooting at nodes.* ❷ *Stipules present*

■Lvs densely hairy (often grey-cottony, woolly, or white-silky)

Absinth-scented. Soft-wooded shrub or shrub-like per

Lft lobes 1.5-4mm wide, densely white-silky (hairs adpressed)

Stems to 100cm, with crown of overwintering lvs below 25cm. Buds to 3mm. All yr ....*Wormwood* **Artemisia absinthium**

Lft lobes <1.5mm wide, grey-cottony to -woolly (hairs slightly spreading)

Lvs grey-cottony (to woolly) or sparsely so, 3-6cm, usu without lft lobes at base; lobes linear, obtuse, strongly aromatic, with stomata both sides. Upper lvs 2-3-pinnate. Petiole channelled. Stems to 120cm, woody, hairless, round, with crown of overwintering lvs. Buds to 3mm, dark. All yr. Hortal....................*Southernwood* **Artemisia abrotanum**

Lvs densely grey-woolly, 2-5cm, usu without lft lobes at base; lobes linear, obtuse. Upper lvs 1-pinnate, sessile. Stems 20-50cm, woody-based, usu woolly, weakly ridged, without overwintering lvs. Apr-Oct. Upper saltmarshes............*Sea Wormwood* **Artemisia maritimum**

Tansy-scented. Herbaceous per

Stems 60-120cm

Lvs 1-2-pinnate, 10-18cm, with 5-8 prs free lfts along winged rachis and tiny lobes at base, pale grey-green esp below, softly ± adpressed cottony-hairy both sides. Stems hairy, round but weakly angled. R alien....................*Rayed Tansy* **Tanacetum macrophyllum**

Stems to 60cm

Lvs 2-4-pinnate, to 15(30)cm, with proximal lfts much reduced, not gland-dotted. Stems not woody, ± woolly, furrowed. All yr....................*Yarrow* **Achillea millefolium**

Lvs 1-2-pinnate, to 12cm, weakly gland-dotted below, grey-cottony; lobes 2mm wide, ± acute. Stems woody below, white-cottony, often with short glandular hairs, angled. Hortal, R alien ....*Sicilian Chamomile* **Anthemis punctata** ssp **cupaniana**

■Lvs (and stems) always hairless. Low semi-evergreen shrub to 80cm

Lvs 6-12cm, translucent gland-dotted, strongly aromatic, glaucous ....*Common Rue* **Ruta graveolens**

■Lvs hairy to sparsely so. Herb. Petiole often with linear lobes at extreme base

Per (bi)

Stems prostrate, rooting at nodes. Lvs pleasantly chamomile-scented (occ scentless in winter); lft lobes 2.5-4 x 0.5mm, linear-subulate, fleshy, greyish-green, sparsely hairy to hairless, with stomata more abundant above. ❶. All yr. Short acidic turf....*Chamomile* **Chamaemelum nobile**

Stems (if present) not prostrate or rooting at nodes

Stipules present. Lvs ± fetid. ❷....................(*Geranium robertianum*) Go to **GER A**

Stipules absent. Lvs tansy-scented

Lf lobes c6mm wide, broad.

Lvs 3-8cm, 1-pinnate, with 1-2(5) prs of lfts. All yr. PL 18 ....*Feverfew* **Tanacetum parthenium**

Lf lobes <2mm wide, linear

Lvs to 15(30)cm, 2-4-pinnate, with many prs of lfts, the proximal lfts much reduced. All yr ....*Yarrow* **Achillea millefolium**

Lvs to 7cm, 1-3-pinnate with many ± pinnate lobes, with smaller lobes nr base. All yr ....*Yellow Chamomile* **Anthemis tinctoria**

Ann, with taproot. Stems ± erect, to 60cm

Plant pleasantly aromatic

Lvs 15-35mm, not flat in 1 plane, with lfts irreg pectinate (comb-like), with long weakly adpressed hairs, soon hairless; lobes overlapping, to 5 x 1.2mm, oblong, bristle-tipped, yellowish-green. Stems several, branched at base, long-hairy (to woolly), striate, solid. Arable, R....................*Corn Chamomile* **Anthemis arvensis**

Plant tansy-scented

Lvs 1-3-pinnate, flat in 1 plane, with lfts regularly pectinate (comb-like), usu hairy or woolly; lobes well-separated, to 5 x 0.8mm, linear or narrowly so, apiculate but not bristle-tipped, greyish-green. Stems much-branched, usu ± adpressed-hairy, striate, solid. Frequent sown in wildflower mixes..*Austrian Chamomile* **Anthemis austriaca**

Plant fetid (vomit-scented)

Lvs 1-3-pinnate, not flat in 1 plane, ± hairless or sparsely hairy, occ with sparse cottony hairs; lobes 3-7 x 0.5-1.2mm, narrowly linear, apiculate but not bristle-tipped, ± grey-green. Stems branched from base, sparsely long-hairy to hairless, weakly ridged, solid. Plant cottony when young. Arable chalk and clay..*Stinking Chamomile* **Anthemis cotula**

Group UB – *Lvs odourless or weakly aromatic. Plant with at least some hairs when young.* ❶ *Hairs stellate.* ❷ *Tendrils present*

■Per, with yellow latex throughout..*Greater Celandine* **Chelidonium majus**

■Per, without latex

Woody-based per (often shrub-like)

Basal and lower lvs 2-3-pinnate, long-petiolate, clasping at base. Upper lvs shortly petiolate, less divided, the uppermost lvs sessile, linear, entire. Lf lobes 0.5-1mm wide, linear, acute or mucronate, sparsely silky unicellular-hairy both sides, soon hairless, with obscure veins, with stomata both sides. Petiole with 1 or 3(5) flat vb's. Stems to 60cm, tough, branched, reddish-brown, ± hairless, ± round. All yr. E Anglian brecks. Sch8 ....*Field Wormwood* **Artemisia campestris** ssp **campestris**

Strongly rhizomatous per

Lvs densely white-felted both sides, revolute when young, 2-pinnate or 2-pinnately lobed; lobes 5-10mm wide, obtuse. Petiole with 1(3) vb's. Stems 30-60cm, densely white-felted, angled. Mar-Oct. VR alien..*Hoary Mugwort* **Artemisia stelleriana**

Herbaceous per (shortly rhizomatous in *Achillea*)

Petiole often hollow. Lvs without recurved margins, odourless. *Pasqueflower* **Pulsatilla vulgaris**

Petiole solid. Lvs without recurved margins, weakly tansy-scented. Stems 1-several, erect, solid. All yr..*Yarrow* **Achillea millefolium**

Petiole solid. Lvs with recurved margin, with weak cucumber/dill scent. Stems >1, decumbent, densely branched, solid. Often all yr. Damp hths.....................*Lousewort* **Pedicularis sylvatica**

■Ann, without latex

Hairs stellate (actually dendritic) ❶

Basal lvs soon dying. Stem lvs to 8cm, 2-3-pinnately lobed; lobes usu 1-2mm wide, linear-lanc, acute, greyish-hairy, with main vein visible only, with stomata both sides. Stems to 100cm, 1.5-2mm diam, purplish at base, greyish-hairy, solid. Taproot weak. May-Sept. Sandy arable habs ....*Flixweed* **Descurainia sophia**

Hairs simple, those on stem with swollen gland at base

Basal lvs long-petiolate or absent. Stem lvs 1-5cm, ± orb, deeply divided into numerous narrow-linear acute lobes (often appearing palmate), hairy. Stem usu 1, to 80cm, erect, ± glandular-hairy above, the hairs crisped with swollen yellow glandular bases...*Larkspur* **Consolida ajacis**

Hairs all simple, without swollen gland at base

Stems scrambling. Lvs without sheathing base. Acidic shady (or recently cleared) habs

Lvs 1-2-pinnate, developing 3-7-branched tendril at tip; lfts alt or subopp, 5-12mm, elliptic, mucronate, greyish-green, entire, ± parallel-veined, with stomata below only. Petiole often sparsely unicellular glandular-hairy. Stems to 70cm, reddish at base, occ sparsely hairy, ± trigonous. ❷. Mar-Sept....................................*Climbing Corydalis* **Ceratocapnos claviculata**

Stems not scrambling. Lvs with sheathing base. Arable..........(*Ranunculus arvensis*) Go to **RAN**

Stems not scrambling. Lvs without sheathing base

Bogs, fens. Lfts with recurved margins. Stem 1, erect, loosely branched, solid. Apr-Oct ....*Marsh Lousewort* **Pedicularis palustris**

Arable. Lfts without recurved margins. Stem 1-several, ascending to erect

Stems hollow

Lvs basal or on stems, 3-pinnate, sparsely hairy, with v divided pinnae at petiole base; lobes linear, acute. Stems to 50cm, v sparsely hairy (at least near base), ridged. Apr-Oct. Arable, VR...*Pheasant's-eye* **Adonis annua**

Stems solid

Lvs 3-15cm, usu (±) hairless; lobes 4-20 x 0.3mm, not or weakly fleshy, acute or bristle-tipped, yellowish-green, with stomata both sides. Stems with sparse hairs when young, with stomata all round. Apr-Sept..........*Scentless Mayweed* **Tripleurospermum inodorum**

Lvs to 4cm, (±) hairless, occ with scattered hairs on rachis; lobes 3-10 x 0.3-1.5mm, linear, usu crowded, ± fleshy, bristle-tipped, yellowish-green, with stomata both sides. Stems hairless. Fls pineapple-scented. Often all yr...............*Pineappleweed* **Matricaria discoidea**

Group UC – *Lvs odourless. Plant always hairless.* (Asparagus *spp may key out here in error).* ❶ *Lft lobes fleshy and ± cylindrical.* ❷ *Tendrils present*

■Rhizomatous per. Stems and petioles brittle, often with abundant sap

Petiole with 1(3) vb's

Lvs all basal, fern-like, to 30cm. Stems to 20cm ....*Fern-leaved Corydalis* **Corydalis cheilanthifolia**

Lvs all basal, 2-ternate, to 12cm. Stems to 12cm......................*Moschatel* **Adoxa moschatellina**

Petiole with 5(7) vb's

Lvs all basal, to 30cm, 2-4-ternate, with 1st pr of lfts often opp, with rachis winged distally (becoming deeply pinnately lobed nr apex); lobes 2.5-3mm, Fumaria-like, often glaucous, mucronate or cuspidate, thin, with weakly translucent veins, with stomata below only. Petiole channelled, solid. Mar-Oct. Shady habs.................................*Bleeding-heart* **Dicentra formosa**

■Tufted per. Stems and petioles ± brittle, with sap sparse or absent

Lvs basal and on stems, 3-10(15)cm, fleshy, bright green, hairless, odourless; lobes ± cylindrical, obtuse, sparsely pitted, with stomata all round; rachis flat above but ribbed below. Petiole with lobes at dilated base, with 1(3)vb. Stems to 50cm, usu decumbent, often purplish, rarely with a few scattered hairs. ❶. All yr. Coastal....................*Sea Mayweed* **Tripleurospermum maritimum**

■Tufted per. Stems and petioles brittle, with abundant sap

Plant with orange latex confined to taproot. Lvs basal, ± glaucous

Lvs pinnately or ternately lobed; lfts to 10 x 2mm, linear, flattened, ± acute with red hydathodes, with erose hyaline margins, with obscure veins, with stomata both sides. Petiole to 12cm, slightly sheathing at base, weakly triangular or ± flat, solid, with (1)3-5 vb's and spiral fibres. Stems to 30cm, erect, branched, ridged or striate to weakly angled, with stomata all around exc on cartilaginous ridges, solid or with small hollow. Apr-Oct ....*Californian Poppy* **Eschscholzia californica**

Plant without latex

Plant with a tuber. Lvs all basal, thin, dull pale grey-green, 2-ternate, with cuneate lfts and lobes to 6mm wide, with stomata below only. Petiole channelled, with 1(3) vb's. Stems to 30cm, round, hollow. Mar-May

Stems with 1(3) ovate scale-lvs at base. Bracts deeply lobed. Tuber solid, globose ....*Bird-in-a-bush* **Corydalis solida**

Stems without scale-lvs. Bracts entire. Tuber hollow, ellipsoid.......*Hollow-root* **Corydalis cava**

Plant fibrous-rooted. Lvs basal and on stems, thin, usu glaucous at least when young

Lfts occ all opp, distant, long-stalked, ternately or pinnately divided into 3-5 lobes, mucronate, with stomata below only; lobes obovate to elliptic, obtuse but apiculate (hydathode often reddish), pinnate-veined, with entire hyaline margins, usu 2-3-lobed nr apex. Petiole reddish, sharply triangular, narrowly winged, with 1(3) vb's. Stems 15-30cm, much-branched, 4-angled, solid. Dec-Oct (all yr). Often on walls

Fls yellow. Lft lobes 8-20 x 8mm, usu green at maturity ....*Yellow Corydalis* **Pseudofumaria lutea**

Fls white. Lft lobes to 12mm, glaucous below at maturity..*Pale Corydalis* **Pseudofumaria alba**

■Ann

Lvs with minutely scabrid margins. Stems and petioles with sparse sap

Basal and lower lvs long-petiolate; lobes all 0.5-1.5mm wide, linear, acute, light green, with midrib visible or obscure, with stomata absent above (or confined to margins). Petiole 0-10cm, weakly sheathing at base, channelled, hollow, with 5(7) vb's. Stems 10-75cm, ridged, pith-filled or hollow. Taproot yellowish. All yr.......................................*Love-in-a-mist* **Nigella damascena**

Lvs with smooth margins. Stems and petioles with sparse sap

Lvs to 5cm; lft lobes hairless, to 5 x 1mm, linear, lax, ± fleshy (0.4mm), bristle-tipped, yellowish-green, with stomata both sides. Apr-Oct. Usu arable.......*Scented Mayweed* **Matricaria recutita**

Lvs to 4cm, (±) hairless, occ with scattered hairs on rachis; lft lobes 3-10 x 0.3-1.5mm, linear, usu crowded, ± fleshy, bristle-tipped, yellowish-green, with stomata both sides. Stems hairless. Fls pineapple-scented. Often all yr...................................*Pineappleweed* **Matricaria discoidea**

Lvs with smooth margins. Stems and petioles with abundant sap

Stems brittle. Petiole triangular, brittle, with (1)3-7 vb's around small hollow. Stems 5-angled, solid or hollow. Often all yr. (*Fumaria* and *Ceratocapnos*)

Lfts with ultimate lobes 2.5-5mm wide

Plant v robust, often climbing to 100cm. Fls 12-15mm

Fl- and fr-stalks erect to spreading (occ recurved). Frs 3mm, wrinkled (at least when dry), without a 'neck'. Fls usu white to pink; upper petal with pink wing (with white margin), becoming pink with purple blotch at base; sepals 3-5.5 x 1.5-4mm, weakly toothed. Lvs ± glaucous; lobes 10-15mm, oblong-lanc. R, Cornwall, Scilly Is ....*Western Ramping-fumitory* **Fumaria occidentalis**

Plant robust, often climbing to 80cm. Fls 9-14mm

Fl- and fr-stalks mostly strongly recurved. Frs often with a distinct 'neck' at join with stalk

Fls whitish, becoming pinkish; sepals ovate, toothed; upper petal folded upwards but only ½ way to keel. Frs (±) smooth. Lvs yellowish-green; lobes oblong or wedge-shaped, mucronate, limp, the ultimate lobes linear, to 8mm. Stems 2(2.5)mm diam ....*White Ramping-fumitory* **Fumaria capreolata**

Fls purple, in loose infl; sepals oblong, toothed; upper petal folded upwards to cover most of keel. Bracts foliaceous. Lvs bluish-green; lobes narrow-oblong or wedge-shaped, the ultimate lobes broadly obtuse. R, mostly W Br ....*Purple Ramping-fumitory* **Fumaria purpurea**

Fl- or fr-stalks erect to spreading (occ recurved). Frs without a 'neck'. Sepals 3-5.5 x 1.5-4mm, weakly toothed

Fls 9-12mm, usu pink to purplish. Frs 2-2.5mm, usu ± smooth. Lvs grey-green (occ ± pruinose); lf lobes oblong, obtuse, often fleshier than other spp, the ultimate lobes 2-6mm, occ with stomata sparse along midrib above ....*Common Ramping-fumitory* **Fumaria muralis**

Lfts with ultimate lobes <2.5mm wide

Plant slender to robust, often climbing to 40(70)cm

Lvs developing 3-7-branched tendril at apex ❷

Lvs 1-2-pinnate; lfts alt or subopp, 5-12mm, elliptic, mucronate, greyish-green, entire, ± parallel-veined, with stomata below only. Petiole often sparsely unicellular glandular-hairy. Stems scrambling, reddish at base, occ sparsely hairy, ± trigonous. Fls yellow. Mar-Sept. Acidic shady (or recently cleared) habs ....*Climbing Corydalis* **Ceratocapnos claviculata**

Lvs never developing tendrils

Fls 9-14mm

Fl- and fr-stalks mostly strongly recurved. Frs usu with a distinct 'neck' at junction with stalk

Sepals 3-5 x 3mm, ovate, ± entire. Fls pink, tipped blackish-red; upper petal folded upwards to cover most of keel. Frs acute. Infl >> than its stalk. Lf lobes oblong or wedge-shaped, the ultimate lobes 3 x 2mm. VR, Cornwall, Isle of Wight ....*Martin's Ramping-fumitory* **Fumaria reuteri** ssp **martinii**

Fl- and fr-stalks erect to spreading (occ recurved). Frs without a 'neck'

Sepals 2-3 x 1.5mm, strongly toothed. Fls pale pink, the lateral petal purple-tipped. Frs 2-2.5mm, wrinkled. Sepals persisting after fl has dropped. Lf lobes oblong, 6mm, the ultimate lobes 6 x 1.5mm.......................*Tall Ramping-fumitory* **Fumaria bastardii**

Fls 5-8(9)mm. Fl- and fr-stalks erect to spreading

Sepals 2-3.5 x 1-3mm. Frs truncate or notched at apex, 2-3mm diam, usu wider than long, slightly wrinkled or shallowly pitted. Fls whitish to deep pink, purple-tipped, 3x as long as sepals, with 2.5mm spur; bracts ≤ than fr-stalks. Lf lobes lanc to linear-oblong, oblanc, the ultimate lobes 3-3.5 x 1.2-2.2mm.... *Common Fumitory* **Fumaria officinalis**

Plant slender, not climbing, usu <30cm tall. Fls 5-8(9)mm

Sepals 2-3.5 x 1-3mm

Frs rounded at apex, 2-3mm, ± as wide as long. Fls pink or purple, tipped blackish-purple, 2x as long as sepals; bracts > fr-stalks. Lf lobes linear or linear-oblong, ± fleshy, glaucous, obtuse, often parsley-like, the ultimate lobes clustered, 2-4 x 0.5-1mm, oblanc, often shallowly channelled.........................*Dense-flowered Fumitory* **Fumaria densiflora**

Sepals 0.5-1.2 x 0.5mm

Bracts mostly ½-¾ as long as fr-stalks. Fls pink (-purple) with dark tips, with 1-1.5mm spur. Frs usu minutely apiculate (or obtuse), slightly wrinkled or shallowly pitted. Lf lobes linear-oblong or lanc, flat..................................*Few-flowered Fumitory* **Fumaria vaillantii**

Bracts mostly ≥ fr-stalks. Fls white to pale pink. Frs usu v obtuse. Lf lobes linear or subulate, usu shallowly channelled..................*Fine-leaved Fumitory* **Fumaria parviflora**

Keys to Selected Groups

(Typically representing genera or subgenera)

Group ALC – Alchemilla. *Per herb. Lvs 1-15cm diam, plicate when young, dull, palmately 5-11-lobed, net-veined; lobes serrate, the teeth with a tuft of hairs (even on otherwise hairless lvs) and a pale sunken hydathode (turning reddish). Petiole usu long, round to weakly channelled, with 3 vb's. Stipules ± ovate-oblong, papery, turning brown (reddish in* A. filicaulis*), pinnate-veined. Hairs unicellular. Apr-Oct. Mostly N & W Br (exc* A. mollis*)*

▪Lvs deeply lobed >½ way to base, often divided into lfts, silvery silky-hairy below, with stomata below only ....**ALC A**

▪Lvs lobed to ½ way to base, never divided into lfts, not silvery below, with stomata both sides

Lvs hairy both sides (rarely sparsely so above)..**ALC B**

Lvs (±) hairless above (may have a few adpressed hairs)...**ALC C**

ALC A – *Lvs deeply palmately lobed >¾ way to base, often divided into lfts (or ± so), dark green above, silky-hairy below, with midrib raised below, with stomata below only. Petiole sparsely silky adpressed-hairy, trigonous, slightly channelled*

▪Lvs cut to ¾ way to base into 7(9) lobes; lobes 3-5(8)cm x 6-15mm, obovate or oblong lobes, serrate in distal ½. Stems to 40cm tall. R alien................*Silver Lady's-mantle* **Alchemilla conjuncta**

▪Lvs cut to base into 5-7 lfts (or lobes); lfts 1-3(4)cm x 3-9mm, oblong-oblanc, serrate only at or nr apex. Stems to 15cm tall. Mtns...*Alpine Lady's-mantle* **Alchemilla alpina**

ALC B – *Lvs hairy both sides (rarely sparsely so above), with terminal tooth usu equal to laterals. Petiole with ± dense spreading or retrorse hairs*

■Lvs shallowly lobed to ¼ way to base, mid-green. Hortal

Lvs (7)10-14cm diam, with dense long (0.6-1mm) hairs both sides, 9-11-lobed, each lobe with 7-9 wide equal teeth per side, the basal lobes often overlapping. Petiole with dense spreading hairs. Hortal...*Soft Lady's-mantle* **Alchemilla mollis**

Lvs 3-10cm diam, with dense short (<0.5mm) hairs both sides, usu shallowly 9-lobed, each lobe with 6-7 narrow ± equal teeth per side, the basal lobes not overlapping. Petiole with dense retrorse hairs. R alien, S Scot.................................*Crimean Lady's-mantle* **Alchemilla tytthantha**

■Lvs lobed ≥¼ way to base

Lvs glaucous, densely softly silky-hairy. Petiole and stipules not reddish at base

Lvs to 6cm diam, orb, with basal lobes close or ± overlapping, with long (0.7-1mm) adpressed hairs both sides; lobes usu 7-9, 1-2x as wide as long; teeth on middle lobe 4-5 per side, ± straight and obtuse. Limestone gsld......................*Silky Lady's-mantle* **Alchemilla glaucescens**

Lvs green to bluish-green (occ glaucous in *A. filicaulis*), often roughly spreading- or patent-hairy

Petiole and stipules usu reddish at base

Lvs 3-6cm diam, usu reniform, with wide basal sinus, usu with long (0.7-1mm) spreading hairs, lobed to ⅓ way to base; lobes 7(9), with convex sides, each usu with 11 acute ± equal teeth

Stems with lowest internode(s) hairless, internodes becoming progressively more hairy up stem...*Slender Lady's-mantle* **Alchemilla filicaulis** ssp **filicaulis**

Stems with lowest internode(s) hairy, internodes becoming progressively less hairy up stem ....*Hairy Lady's-mantle* **Alchemilla filicaulis** ssp **vestita**

Petiole and stipules never reddish at base.

Lvs with ± adpressed hairs

Petiole densely ± adpressed-hairy. Lvs 5-15cm diam, densely hairy above, less hairy below, with wide basal sinus; lobes usu 9, usu rounded, often 2x as wide as long; teeth of the middle lobe 6-7 per side, wide, those each side of apical tooth the largest and those nr the lobe base the smallest. Plant robust or medium-sized. Uplands or mtns, R ....*Clustered Lady's-mantle* **Alchemilla glomerulans**

Lvs with (±) patent hairs

W Yorks. Lvs usu to 3cm diam, lobed ⅓ to base, hairy only on folds above and on veins below...*Least Lady's-mantle* **Alchemilla minima**

Co Durham

Lvs 8-15cm diam, lobed ⅓-½ way to base, with dense long (1mm) hairs both sides

Lvs orb, with 7-10 teeth per side of middle lobe, the basal sinus open. Petiole with patent hairs..*Starry Lady's-mantle* **Alchemilla acutiloba**

Lvs <10cm diam, lobed to ⅓ way to base, with dense short (<0.7mm) hairs both sides

Lvs orb, the basal sinus closed; lobes 9-11, obtuse, with 15-19 teeth. Co Durham ....*Velvet Lady's-mantle* **Alchemilla monticola**

Northumberland

Lvs <8(10)cm diam, ± reniform, the basal sinus open; lobes 7-11, obtuse, with 15 acute teeth. Petiole with erecto-patent or weakly adpressed hairs. Northumberland ....*Shining Lady's-mantle* **Alchemilla micans**

ALC C – *Lvs (±) hairless above (may have a few adpressed hairs)*

■Lvs hairless below, 7-9-lobed, the middle lobes with 13-15 teeth, the terminal tooth often < than laterals. Petiole with sparse adpressed hairs

Lvs usu 5-10(15)cm diam, with rounded or straight-sided lobes, rarely with basal lobes overlapping (usu 60-180°), veins hairy below distally

Lobes up to 2x as wide as long, the sinus between them often wide; teeth of the middle lobe usu 9 per side, ± wide, those nr middle larger and wider. Stems with lowest internode(s) sparsely adpressed-hairy, becoming progressively hairless up stem ....*Smooth Lady's-mantle* **Alchemilla glabra**

Lvs to 8cm diam, with rounded lobes, often with basal lobes overlapping and veins hairy below

Lobes up to 2x as long as wide, the sinus between them v narrow or ± closed; teeth of the middle lobe 7-9 per side, narrow, equal. Petiole usu with sparse adpressed hairs. Stems often with lowest internode(s) densely adpressed-hairy. Moist calc gsld, damp rock ledges, VR ....*Rock Lady's-mantle* **Alchemilla wichurae**

■Lvs spreading-hairy below, 7-9-lobed, the middle lobes with 13-15 teeth, often with terminal tooth > laterals. Petiole with dense hairs

Petiole with spreading hairs. Lvs reniform, 1-10cm diam, yellow-green, usu strongly plicate, hairless above, sparsely hairy esp on veins below, lobed ≤⅓ way to base; lobes 7-9, ± long, usu with 9 teeth per side..........*Pale Lady's-mantle* **Alchemilla xanthochlora**

Petiole often with some retrorse hairs. Lvs ± orb, 1-6cm diam, green, usu strongly undulate and plicate, with upturned basal lobes, sparsely hairy above, more densely hairy below, lobed ≤⅓ way to base; lobes 7-9, ± wide, usu with 8 wide teeth per side. VR, Co Durham (Teesdale), Northumberland..........*Large-toothed Lady's-mantle* **Alchemilla subcrenata**

■Lvs densely ± adpressed-hairy below, 9-11-lobed, the middle lobes with 17-19 teeth, the terminal tooth often < than laterals. Petiole with adpressed hairs (often dense)

Lvs to 12cm diam, with ± wide basal sinus. VR hortal.....*Boreal Lady's-mantle* **Alchemilla venosa**

Group AMAR – Amaranthus. *Lvs with a fragile 0.5-1mm mucro (actually excurrent midrib), cuneate at base, folded when young, entire but usu with erose cartilaginous margins, opaquely net-veined with Kranz venation (the thin spaces between the Kranz veins often appearing as translucent dots), often with 2° veins strongly sunken above and raised below, with stomata both sides. Petiole channelled, with several vb's usu forming arc and with spiral fibres. Stem(s) round to weakly ridged, with scattered vb's, often odorous. Hairs septate. May-Oct. Ruderal or arable weeds.* ❶ *Infl nodding*

■Stem usu 1, erect. Lvs usu with strongly sunken 2° veins above. Ann

Stem 3-10mm diam. Infl in terminal spike

Stem ± hairless

Lvs often reddish-yellow esp when young, 5-10(15) x 8cm, ovate, truncate-retuse, hairless above, sparsely hairy below. Petiole ≤ lf length, sparsely hairy. Stem 15-80cm, green, sparsely hairy. Infl purplish. ❶................................*Love-lies-bleeding* **Amaranthus caudatus**

Lvs mid-green, 3-10 x 5cm, ovate, retuse, (±) hairless. Petiole often > lf length, with long wavy hairs. Stem 15-90cm, reddish esp above, sparsely hairy at least below. Infl greenish ....*Green Amaranth* **Amaranthus hybridus***

Stem hairy (often densely so), at least near infl

Lvs 3-15 x 5cm, ovate to ovate-oblong, obtuse to retuse, hairy esp below, with translucent dots at maturity. Petiole ≥½ lf length. Stem 15-60cm. Infl brownish, erect ....*Common Amaranth* **Amaranthus retroflexus**

Stem 2-4mm diam, ± hairless. Infl in lf axils

Lvs 3-6cm, obovate to spathulate, obtuse, occ retuse, hairless. Petiole c½ lf length (longer in young lower lvs). Stem usu 10-40cm, widely branched, whitish, rarely reddish. Infl greenish ....*White Pigweed* **Amaranthus albus**

■Stems several, ± prostrate to erect. Lvs usu with weakly sunken 2° veins above

Per. Stems hairy at least nr infl

Lvs 1-2 x 0.5-1cm, rhombic-ovate to lanc, obtuse, occ retuse, occ weakly undulate, often with yellowish hairs on veins below. Petiole ≤ lf length. Stems 20-50cm ....*Perennial Pigweed* **Amaranthus deflexus**

Ann. Stems hairless

Lvs deeply retuse, 1-6 x 0.5-4cm (larger nr infl), ovate or obovate, undulate, mid-green, never with chevron, with ± translucent dots. Petiole (2)3-6cm (longer nr infl). Stems 10-60cm ....*Guernsey Pigweed* **Amaranthus blitum**

Lvs retuse to deeply so, (1.5)2-4(5) x 1-3cm, rhombic-ovate to ± linear, with entire margins, rarely undulate, occ purplish with green chevron, without translucent dots. Petiole 0.3-5cm. Stems 10-90cm...*Short-tepalled Pigweed* **Amaranthus graecizans**

Lvs shallowly retuse to ± truncate, 1-2(4) x 0.4-1.5cm, obovate to elliptic or spathulate, with narrow silvery hyaline margins, green with whitish chevron, without translucent dots. Petiole 1-1.5cm. Stems 10-60cm...*Prostrate Pigweed* **Amaranthus blitoides**

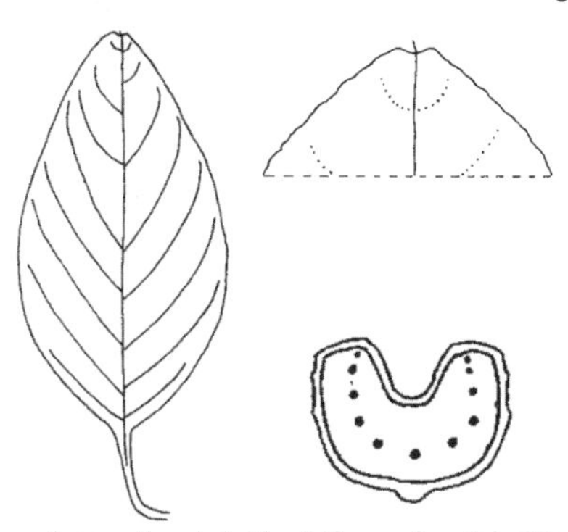

Amaranthus hybridus lf, lf apex & petiole TS

Group AST – Aster *(N American taxa only). Lvs ± linear to lanc, acute, often sessile and clasping at base, rolled when young, dull above, with antrorsely scabrid-ciliate cartilaginous margins (cilia often longer nr base), entire to weakly serrate esp in distal ½ (teeth tipped with dark hydathodes), net-veined, midrib raised below and often purplish above, with 2° veins hardly raised either side. Stems to 150cm, erect, purplish, round to weakly angled, brittle, solid. Rhizomes often purplish with white apices. Hairs septate. Apr-Oct (wintergreen as basal rosettes). Apart from* A. novae-angliae, *a ± continuous variation of morphology exists because of fertile hybrids.* ❶ *Lvs strongly auriculate*

▪Stems roughly hairy with white hairs, often with stalked glands. Lvs with stomata below only

Lvs to 6 x 1cm, linear, clasping at base with rounded auricles, hairy both sides, ciliate ....*Hairy Michaelmas-daisy* **Aster novae-angliae**

▪Stem hairless or with sparse hairs (esp above). Lvs with stomata both sides or below only

Lvs dull glaucous both sides, often pruinose when young. Fls blue

Lvs 5-14 x 3cm, the lower narrowed to petiole-like base, the mid and upper clasping with auricles, hairless. Stems pruinose when young. ❶.....*Glaucous Michaelmas-daisy* **Aster laevis**

Lvs dark green, occ ± glaucous. Fls blue

Lvs 7-14 x 3cm, clasping at base. *A. laevis* x *novi-belgii* ....*Late Michaelmas-daisy* **Aster** x **versicolor**

Lvs mid-green, never glaucous

Fls blue. Infl not lfy. Lvs 1.2-1.6(2.2)cm wide

Lvs 8-15cm, lanc or narrowly so, weakly auriculate at base. The commonest taxon encountered. *A. lanceolatus* x *novi-belgii*.........*Common Michaelmas-daisy* **Aster** x **salignus***

Lvs 8-12cm, lanc or narrowly so, usu strongly auriculate and ± clasping at base (some forms lack the auricles). ❶..*Confused Michaelmas-daisy* **Aster novi-belgii***

Fls white. Infl lfy

Lvs 10-15 x 1-1.6cm, narrowly lanc, usu weakly clasping at base (occ weakly auriculate), more strongly narrowed to petiole-like base than in *A.* x *salignus* ....*Narrow-leaved Michaelmas-daisy* **Aster lanceolatus**

Lvs (6)10-15 x (1)1.6-2.2cm, lanc, not clasping at base (without auricles). VR alien ....*Narrow-leaved Smooth-aster* **Aster concinnus**

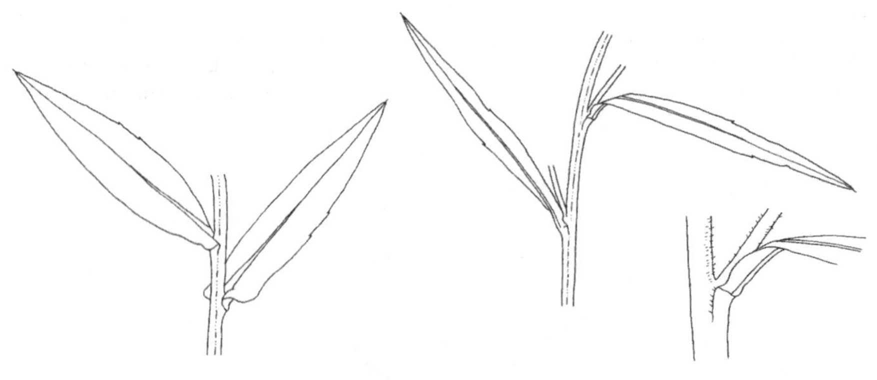

Aster novi-belgii lvs (showing auricles)

Aster x *salignus* lvs & (weak) auricles

Group AVENA – Avena *(Oats). Lvs to 40cm x (5)8-30mm, twisted, green to bluish-green (occ pruinose), often rough at least below, hairless or hairy above, often ciliate esp nr base, hardly ribbed in TS, with discernible midrib, other veins regular and unequal (2-4 minor veins between main veins), with stomata both sides. Ligule (1.5)2-8mm, truncate to acute, ciliolate-fimbriate, toothed. Sheaths open, usu smooth, hairless to hairy, the lowest often hairless and purple-striped. Ann crop relics or weeds (but occ resembling a per when overwintering)*

▪No florets disarticulating at maturity (each floret without a bearded basal callus)

Lemma without fine bristles..........*Oat* **Avena sativa**

Lemma ending in 2 fine bristles 3-9(12)mm. VR, mostly N & W Br ....*Bristle Oat* **Avena strigosa**

▪Only lowest florets disarticulating at maturity (each floret with a bearded basal callus). Lemma without a fine bristles..........*Winter Wild-oat* **Avena sterilis** ssp **ludoviciana**

▪All florets disarticulating at maturity (each floret with a bearded basal callus)

Lemma ending in 2 fine bristles 3-9(12)mm. R..........*Slender Oat* **Avena barbata**

Lemma without fine bristles..........*Wild-oat* **Avena fatua**

Group BAM – *Bamboos. Lvs evergreen for 2 yrs, attached to the lf sheath by a 1-2mm pseudopetiole, rolled when young, unicellular-hairy or hairless, with translucent tessellate veins (unlike other grasses), the midrib raised below, with stomata below only. Ligule 0.5-3mm, truncate or oblique. Ligule whiskers (also known as oral setae), when present, are long bristles adjacent to the ligule (often also found at culm sheath apices). Culms woody, attaining their maximum height in 1st yr and never increasing in diam after emerging from ground, hollow. The culm sheaths are often diagnostic but they often quickly drop off. Strongly rhizomatous spp have culms well-spaced; tufted (or shortly rhizomatous) spp have culms clumped together*

▪Lvs mostly >3cm wide. Main culms round ....**BAM A**

▪Lvs all <3cm wide

Main culms round..........**BAM B**

Main culms round but flattened or grooved on one side at least above..........**BAM C**

Main culms (±) square..........**BAM D**

BAM A – *Lvs >3cm wide. Main culms round*

▪Lvs ¾ glaucous below (only the outer ¼ not glaucous). Culms with 0-2 branches per node

Pseudopetioles usu greenish. Lvs rarely with extensively withered margins

Lvs 15-30 x 2-5cm, acuminate (apex often withered), deep green above, sparsely hairy to (±) hairless, with 5-9 main veins each side of midrib. Ligule 1.5-3mm, minutely hairy or sparsely so. Ligule whiskers usu present. Auricles absent or small. Culms 2.5-5m, 10-20mm diam. Culm sheaths densely roughly hairy when young, usu without whiskers. Strongly rhizomatous ....*Arrow Bamboo* **Pseudosasa japonica**

Lvs 12-30(40) x 3.5-9cm, acuminate, shiny bright green above, rarely hairy below, with 7-14 main veins each side of midrib. Ligule 2mm, minutely hairy. Ligule whiskers absent. Auricles absent. Culms (0.5)2-3m, 7-10mm diam. Culm sheaths often chalky-pruinose and white-hairy when young, without whiskers. Strongly rhizomatous..........*Broad-leaved Bamboo* **Sasa palmata**

Pseudopetioles often purplish. Lvs soon with broad white withered margins

Lvs 10-25 x 2.5-6cm, abruptly tapered, dull above, variegated, often shortly hairy below, with (5)6-9 main veins each side of midrib. Ligule to 1.6mm, dark, minutely hairy. Ligule whiskers short, purplish, minutely scabrid, soon falling. Culms 0.5-1.5m, 5-7mm diam. Culm sheaths often chalky-pruinose and white-hairy when young..........*Veitch's Bamboo* **Sasa veitchii**

▪Lvs green below (paler in one ½). Culms with ≥3 branches per node

Lvs 15-30 x 1-2.5(3.5)cm, sharply acute, occ variegated, rarely ½ glaucous below, with 4-7 main veins each side of midrib (7 minor veins between main veins). Pseudopetiole hairy above. Ligule 1mm, minutely hairy. Ligule whiskers long, white, smooth. Culms 3-5(8)m, 20-30mm diam. Culm sheaths hairless or v shortly hairy nr base, often with a few whiskers when young ....*Simon's Bamboo* **Pleioblastus simonii**

BAM B – *Lvs <3cm wide. Main culms round*

■Lvs hairy at least below. Strongly rhizomatous

Lvs 1-3(3.5)cm wide

Lvs light to yellow-green above, with darker green stripes of various widths

Lvs to 15 x 1.8-2.2cm, hairy both sides, with 5-6 main veins each side of midrib. Ligule whiskers often absent. Culms <2m, with 1(2) branches per node ....*Kamuro-zasa* **Pleioblastus auricomus**

Lvs dark green above, not striped but may have withered margins

Lvs 8-25 x 1-3.5cm, v shortly acute, hairy at least below. Ligule whiskers white, (±) smooth. Culms 0.5-1.5m, 3-8mm diam, with 1-2(4) branches per node. Culm sheaths soon hairless, with few whiskers...*Hairy Bamboo* **Sasaella ramosa**

Lvs <1.2cm

Lvs 8-16 x 0.6-1(1.2)cm, linear-lanc, occ minutely hairy below, with 3 main veins each side of midrib. Ligule whiskers smooth or obscurely rough. Culms 2-3m, usu with 3 branches per node, occ with aerial roots. Culm sheaths ciliolate....*Marbled Bamboo* **Chimonobambusa marmorea**

■Lvs (±) hairless

Culms usu with 1-2(3) branches per node. Strongly rhizomatous

Lvs ½ glaucous below

Lvs 8-20 x (0.3)0.9-2.5cm, lanc or linear-lanc, narrowly pointed at apex, bright green above, paler below (on one or both sides of midrib), occ hairy below nr base, with 2-7 main veins each side of midrib. Ligule hairy. Ligule whiskers scabrid. Culms 0.6-2m, 3-10mm diam. Culm sheaths soon hairless, ciliate, with small auricles and a few whiskers ....*Dwarf Bamboo* **Pleioblastus humilis**

Lvs ¾ glaucous below...*Arrow Bamboo* **Pseudosasa japonica**

Lvs weakly glaucous all over below

Culms 2-3m, usu with 3 branches per node. Ligule whiskers smooth or obscurely rough

Lvs 8-16 x 0.6-1(1.2)cm, linear-lanc, occ minutely hairy below, with 3 main veins each side of midrib. Culms 10-15mm diam, occ with aerial roots. Culm sheaths ciliolate ....*Marbled Bamboo* **Chimonobambusa marmorea**

Culms to 0.75(1.2)m. Ligule whiskers v obvious, smooth

Lvs 2.5-6(8) x 0.3-0.9(1.5)cm, abruptly acuminate, pale green above, hairless or hairy below, with 2-3(7) main veins each side of midrib. Culm sheaths occ hairy on veins, ciliate, with few whiskers. Culms 1-3mm diam.........................*Pygmy Bamboo* **Pleioblastus pygmaeus**

Lvs mid-green below. Ligule whiskers obscurely rough, white

Lvs 5-25 x (0.3)0.8-2.5cm, tapered to an acute apex, green above, occ cream-striped, hairless or hairy on one side below, with (2)5-7 main veins each side. Auricles absent. Culms 1.2-2(3)m, (6)10-15(20)mm diam. Culm sheaths usu ± softly hairy when young, soon hairless and ciliate...*Maximowicz's Bamboo* **Pleioblastus chino**

Culms with >3 branches per node

Lvs ¾ glaucous below. Ligule whiskers rough or obscurely so. Tufted

Lvs 5-9(13) x 0.5-1cm, with 2-4 main veins each side. Ligule 1mm, dark, ciliolate. Auricles present or absent. Culms >3m. Culm sheaths hairy to hairless, ciliate ....*Pea-stick Bamboo* **Thamnocalamus spathiflorus**

Lvs uniformly glaucous below

Ligule whiskers smooth. Tufted

Lvs (3.5)6-10 x 0.5-1.3cm, long-acuminate, usu with 2-4 main veins each side of midrib. Ligule 1mm, minutely hairy. Auricles usu absent. Ligule whiskers translucent. Culms 4-6m, 10-15mm diam, often purplish (or pruinose in shade). Culm sheaths densely bristly ....*Chinese Fountain-bamboo* **Fargesia spathacea**

Ligule whiskers scabrid. Strongly rhizomatous

Lvs 6-12(16) x 0.5-1.2(2)cm, linear-lanc to narrowly oblong, gradually narrowed to apex, thin, brilliant green above, often pruinose below, mostly with 2-4 main veins each side of midrib. Ligule 0.5-1mm. Culms 3-6m, 7-20mm diam, purplish-green. Culm sheaths hairless, ciliate, with obvious opaque whiskers.............................*Indian Fountain-bamboo* **Yushania anceps**

Ligule whiskers strongly scabrid. Tufted

Lvs 6-10(12) x 0.8-1.3cm, with long fine apex, with one margin minutely serrate, mostly with 2-4 main veins each side of midrib. Ligule 1mm, pale, hairless. Auricles absent. Ligule whiskers yellow-brown, rarely absent. Culms (1)3-5m, 5-14mm diam. Culm sheaths hairless, ciliate, with v short ligule, with whiskers......................*Umbrella Bamboo* **Fargesia murieliae**

BAM C – *Lvs <3cm wide, ¾ glaucous below (often completely glaucous below in* Phyllostachys bambusoides*), with (4)5(6) main veins each side of midrib. Main culms flattened or grooved on one side at least above, hollow, occ with aerial roots below (*Phyllostachys *spp also have young branches flattened on 1-side). Strongly rhizomatous*

▪Culms often flattened or grooved only at the upper internodes, mostly with 3-5 branches per node

Lvs 8-20 x 1.5-2.5(4)cm, narrowly lanc, often purplish, hairless, occ hairy below nr base. Ligule truncate. Ligule whiskers translucent, smooth (to opaque and minutely scabrid). Culms 3-8m, 15-40(80)mm diam, often purplish. Culm sheaths v large, usu hairless, shiny cream ....*Narihira Bamboo* **Semiarundinaria fastuosa**

▪Culms often flattened or grooved on one side throughout, mostly with 2(3) branches per node

Culms with lowest nodes clearly shorter than those above

Lvs 10-13 x 1.8cm, oblong, acuminate. Ligule hairy. Ligule whiskers long, scabrid ....*Golden-bamboo* **Phyllostachys aurea**

Culms with lowest nodes not clearly shorter than those above

Culms ± zig-zagging. Ligule whiskers (and auricles) usu absent ....*Zigzag Bamboo* **Phyllostachys flexuosa**

Culms straight. Ligule whiskers (and auricles) usu present (at least when young)

Culms blotched with red-black, becoming black when old

Lvs 6-11 x 1.3-1.7cm. Culm wall 3mm thick. Culm sheaths usu without spots ....*Black Bamboo* **Phyllostachys nigra**

Culms not blotched (occ streaked)

Culms pruinose at least below nodes, the wall 4-8mm thick. Lvs 9-13 x 1.2-1.8(3.8)cm. Ligule whiskers to 20mm (often much shorter or absent), scabrid, fragile. Auricles usu present, often purplish...........*Greenwax Golden-bamboo* **Phyllostachys viridiglaucescens**

Culms not pruinose (rarely a narrow pruinose band below nodes), the wall 4-8mm thick. Lvs 12-16 x 1.5(4.5)cm. Ligule whiskers to 12mm, occ absent. Auricles usu absent ....*Japanese Timber-bamboo* **Phyllostachys bambusoides**

BAM D – *Lvs <3cm wide. Main culms (±) square. Strongly rhizomatous*

▪Lvs 8-30 x 1-3cm, elliptic to lanc, acute, green (not glaucous below), occ sparsely hairy below, with 5-14 main veins each side of midrib. Ligule minutely hairy, ciliolate. Ligule whiskers long, white-reddish, smooth. Culms 5-8m, 20-40mm diam, occ rough, much-branched with >3 branches per node, with root buds at lower nodes, with hollow ⅓ diam. Culm sheaths sparsely scabrid, with erect whiskers.................................*Square-stemmed Bamboo* **Chimonobambusa quadrangularis**

Group BER – Berberis. *Shrub. Lvs alt (often in fascicles on short shoots), often spine-tipped, entire or with spiny teeth, with stomata below only. Branches with spines confined to lf axils (the spines are modified stipules), often with short shoots, the wood beneath the bark yellow. Buds small, ovoid, the c7 acute or acuminate scales tipped with a mucro (actually a vestigial petiole) and often reflexed. Roots yellow. Many more spp are grown in gardens and may rarely be bird-sown*

■Lvs with >6 teeth per side, spine-tipped. Twigs hairless. Spines 3-partite

Lvs net-veined, deciduous. Twigs ridged

Lvs 2-6cm, elliptic to obovate, cuneate at base, dull above, paler below, serrate with up to 40 teeth per side. Petiole <1cm...*Barberry* **Berberis vulgaris**

Lvs usu not net-veined, evergreen

Twigs round (occ ridged when young)

Lvs (3)6-10 x 1-1.5cm, linear to linear-lanc, often with recurved margins, shiny green both sides (dull above in var *lanceifolia*) but paler below, with 15-21 teeth per side. Spines 14-25mm, patent, yellow, not channelled. Twigs yellow to reddish ....*Gagnepain's Barberry* **Berberis gagnepainii**

Twigs ridged

Lvs 6-11 x 2.5cm, narrowly elliptic to lanc, often with slightly recurved margins, shiny green above, paler below, with 6-15 teeth per side, occ ± net-veined. Spines to 6cm, vicious, ± round (slightly channelled below)..*Chinese Barberry* **Berberis julianae**

■Lvs with 0-6 teeth per side. Twigs hairless or hairy

Spines 1 (simple). Lvs deciduous

Lvs toothed, net-veined, spine-tipped

Lvs (1)1.5-4 x 0.7-1.3cm, oblong to obovate, reddish, hardly shiny above, with 0-3(5) teeth per side. Twigs sparsely hairy to hairless, strongly angled. Spines usu 14-16mm. Hybrids with *B. wilsoniae* frequently occur and also key out here....*Clustered Barberry* **Berberis aggregata**

Lvs entire, not net-veined

Lvs spine-tipped, 1-4.5cm, obovate, green to purple. Petiole to 2cm. Spines to 10mm. Twigs reddish-brown, hairless, angled or ridged....................*Thunberg's Barberry* **Berberis thunbergii**

Lvs not spine-tipped, 2-4 x 2.5cm, ovate to ± orb, green or purplish. Petiole to 2cm. Spines 7-13mm, channelled. Twigs often reddish, hairless, angled. *B. thunbergii* x *vulgaris*. PL 11 ....*Ottawa Barberry* **Berberis** x **ottawensis**

Spines 3-7-partite

Lvs >3cm, ± net-veined, semi-evergreen

Lvs 2.5-6(8) x 2cm, oblanc to obovate, spine-tipped, ± shiny dark green, with 0-5 teeth per side. Spines 3-partite, the terminal longest, to 10mm, round, not channelled. Twigs hairless, round..*Great Barberry* **Berberis glaucocarpa**

Lvs <3cm

Lvs net-veined, deciduous (occ semi-evergreen)

Lvs 1-2.5 x 0.7cm, oblanc to obovate, apiculate, light dull green to red. Spines 3-partite, to 15mm, channelled. Twigs yellowish to reddish, sparsely hairy to hairless, slightly ridged ....*Mrs Wilson's Barberry* **Berberis wilsoniae**

Lvs not net-veined, evergreen

Lf margins strongly revolute margins (occ appearing cylindrical)

Lvs 1.3-2.5 x 0.5cm, often in fascicles of 5-6, narrowly elliptic, spine-tipped, leathery, dark green above, paler below, with 0-1 teeth per side, usu with obscure 2° veins. Spines 3-partite, 3-6mm, occ unequal or curved, reddish-brown, hairy, channelled below. Twigs reddish, brownish-hairy. *B. darwinii* x *empetrifolia*....................*Hedge Barberry* **Berberis** x **stenophylla**

Lf margins flat or weakly revolute

Twigs densely brown-hairy, with ± shaggy hairs >0.5mm

Lvs 1-3 x 1.2cm, usu obovate (occ oblong or oblanc), spine-tipped, shiny dark green above, with 2-3 teeth per side nr apex. Spines 5-6-partite, to 3mm ....*Darwin's Barberry* **Berberis darwinii**

Twigs not brown-hairy

Lvs pruinose below

Lvs with 1-2 teeth per side, to 1.5 x 0.8cm, ovate, spine-tipped, leathery, occ with recurved margins. Spines 3-partite (occ with 2 more v tiny peripheral spines), to 17mm long yellow, patent, hairless, channelled below. Twigs densely minutely papillate, yellow, ± round. Hortal, VR..................................*Warty Barberry* **Berberis verruculosa**

Lvs with 2-4 teeth per side. VR hortal....................*Whitish Barberry* **Berberis candidula**

Lvs green below

Lvs 1-3 x 1cm, ovate to obovate, spine-tipped, dull dark green above, with flat margins, usu entire (occ 3-dentate at apex). Spines 3-5-partite, occ lf-like on vigorous shoots, 8-12mm, channelled below. Twigs hairless, reddish-brown, ridged ....*Box-leaved Barberry* **Berberis buxifolia**

Group BET – Betula. *Deciduous tree or shrub. Lvs usu triangular-ovate, folded-plicate when young, usu shiny dark green above, paler below, with veins weakly raised both sides, with stomata below only. Petiole channelled. Stipules 2mm, ovate, soon falling. Twigs ± straight, slender, round. Buds with several scales. Bark red-brown on young trees, often silver-white when mature*

▪Twigs (±) hairless (hairy on saplings and sucker growth), often with glandular warts, pendent

Lvs ± truncate (to broadly cuneate) at base, hairless, sparsely ciliate

Lvs to 6 x 5cm, acuminate, viscid when young, with resin glands both sides, 2-serrate with c26 teeth per side, with main teeth prominent and ± curved towards apex, with 6-8 prs of 2° veins. Petiole 10-20mm, 1-2mm wide, usu hairless. Twigs ≤2.5mm diam, slender, ± shiny dark brown. Buds 6mm, acute, not viscid, green and brown, sparsely ciliate, with 4-7 scales. Bark silver-white, peeling, with black diamond-shaped lenticels. PL 14, 20.......*Silver Birch* **Betula pendula**

Lvs v cuneate at base, usu with adpressed hairs on veins below (rarely above)

Lvs to 6(10) x 4(8)cm, v acuminate, ± not viscid when young, with resin glands both sides, deeply sharply 1-2-serrate with c13 teeth per side (none in proximal ⅓), with 4-8 prs of 2° veins. Petiole 2cm, ± hairless, with orange glands. Twigs with orange lenticels. Buds 6mm, acute, viscid, green, hairless. Bark silver-white, not or hardly peeling, ± smooth. Hortal ....*Grey Birch* **Betula populifolia**

▪Twigs usu hairy, not or hardly pendent

Twigs with strongly raised glandular warts. *B. pendula* x *pubescens*......................**Betula** x **aurata**

Twigs without warts (but may have slightly raised resin glands)

Tree. Lvs 2-10cm

Lvs with 4(6) prs of 2° veins

Lvs often cuneate at base, ± not viscid when young, hairy (esp on veins below) or sparsely so, ciliate, with glands along veins both sides, 1-2-serrate (-dentate), with main teeth less prominent and not curved towards apex. Petiole 10-18mm, 1-2mm wide. Twigs dull dark brown or blackish, with long and/or short hairs. Buds 6mm, obtuse, viscid or not, ciliate. Bark turning grey, hardly peeling, ± smooth

Lvs 3-5(6)cm. Twigs usu without obvious resin glands ....*Downy Birch* **Betula pubescens** ssp **pubescens**

Lvs 2-3cm. Twigs with obvious resin glands. Often shrubby. N Br ....*Downy Birch (ssp)* **Betula pubescens** ssp **tortuosa**

Lvs with 7-9 prs of 2° veins

Lvs 5-10 x 3-7cm, not v acuminate, rounded to ± cordate at base, with resin glands esp below, sparsely adpressed-hairy above and below esp on veins, ± serrate with c17 teeth per side. Petiole 2-3cm, hairy. Twigs viscid with resin glands, soon hairless. Buds 10-12mm, viscid, green-brown, hairless, ciliate. Bark bright silver-white, often peeling, ± smooth. Intergrades with *B. utilis*. Hortal............................*Jacquemont's Birch* **Betula jacquemontii**

Lvs with 10-12(14) prs of 2° veins

Lvs 4-9 x 3-6cm, acuminate, rounded to ± cordate at base, with resin glands esp below, sparsely hairy above, hairy on veins below, with white or buff hairs in vein axils below, 2-serrate. Petiole 0.8-2cm, hairy. Twigs viscid with resin glands, with minute and longer hairs. Buds 10-12mm, viscid, green, hairless, ciliate. Bark silver-white or orange, often peeling, ± smooth. Hortal...*Himalayan Birch* **Betula utilis**

Shrub to 1m. Lvs 0.5-1.2(2)cm

Lvs orb or obovate-orb, with sparse orange resin glands both sides, hairless, crenate-serrate with 4-8 teeth per side, with 2-3(4) prs of 2° veins. Petiole to 3mm. Twigs dull grey-brown, minutely hairy, often with numerous resin glands. Buds 2mm, globose, ciliate. Uplands, mostly Scot...*Dwarf Birch* **Betula nana**

Group CAL – Callitriche. *Lvs opp, entire, often of two types (submerged linear lvs and broader floating lvs). Floating lvs often in a rosette, broadly spathulate to rhombic or ± orb, usu obtuse (rarely notched), with a swollen white hydathode at apex, (1)3-5-veined. Submerged lvs linear, usu notched, 1-veined. Stems round, branched, often with 2-4 roots at lower nodes. Often all yr (all spp, exc* C. stagnalis, *capable of overwintering as linear-lvd per). Wet, often aquatic habs*

■All lvs translucent, not connate, submerged, 1-veined, without stomata. Stems without scales

Lvs 5-11 x 0.8-1.3mm, linear (± parallel-sided), truncate or shallowly notched at apex, dark green, with minute blue flecks (x10), with axillary glandular scales. Stems often reddish. Apr-Oct (fragments may overwinter). Ann (per). Usu in base- or nutrient-rich water, rarely terrestrial on mud. R, S Br...*Short-leaved Water-starwort* **Callitriche truncata**

Lvs 8-18 x 1(2)mm, linear-lanc (widest at the base), with shallow wide notch at apex, pale green. Stems not reddish. Ann (per). Never terrestrial. Mostly N Br ....*Autumnal Water-starwort* **Callitriche hermaphroditica***

■All lvs opaque, connate at base. Stems often with minute peltate scales nr apices

Submerged linear lvs absent (3-veined floating rosette or submerged broad lvs only)

Lvs 4-16mm, ± orb to elliptic (usu <4x as long as wide), often bright apple-green, pitted both sides, often weakly 1-3-ridged above. Frs 1-2mm wide, greyish, broadly winged. Stamens 2x as long as fr. Ann (but aquatic plants are per). Often all yr. Often on wet mud ....*Common Water-starwort* **Callitriche stagnalis**

Submerged linear lvs present (3-veined floating rosette lvs present or absent)

At least some submerged linear lvs with deeply notched apex and 2 claw-like apical teeth

Submerged lvs to 20mm, narrowing evenly towards apex, then abruptly widened generally to a spanner-like apex. Floating lvs to 20mm, often rhombic, weakly ridged. Frs 1.4mm wide, brownish, narrowly winged, often stalked to 12mm, with the styles strongly recurved and adpressed to the fr. Ann to per. Usu acidic water ....*Intermediate Water-starwort* **Callitriche brutia** var **hamulata***

At least some submerged linear lvs with shallowly (to deeply) notched apex (no claw-like teeth)

Floating rosette lvs strongly veined, with midrib forming a weak ridge above (esp nr apex)

Submerged lvs to 20mm, parallel-sided, obtuse or notched, often with 2 unequal short apical teeth. Floating lvs to 20mm, often rhombic, mid-green. Frs 1.4mm wide, brownish, narrowly winged, ± sessile, with the styles strongly recurved and adpressed to the fr. Ann to per. Usu ephemeral ponds..........................*Pedunculate Water-starwort* **Callitriche brutia** var **brutia**

Submerged lvs to 20mm, parallel-sided, shallowly notched. Floating lvs to 16 x 6mm, rhombic, greyish-green, slightly waxy. Frs 1.5mm wide, brownish, with blunt unwinged lobes, sessile, with the styles erect at least when young. Per. Still or slow-flowing water ....*Blunt-fruited Water-starwort* **Callitriche obtusangula**

Floating rosette lvs weakly veined, with flat or sunken midrib above

Submerged lvs 10-30 x 1-2mm, ± linear, shallowly notched. Floating lvs 10-15 x 2-5mm, spathulate-elliptic, usu dull green. Frs 1.4-1.8mm wide, brownish, narrowly winged, (±) sessile, the styles erect at least when young. Per. Often in eutrophic water ....*Various-leaved Water-starwort* **Callitriche platycarpa**

Submerged lvs (4)5-10 x 0.5-1.2mm, linear to narrowly expanded, shallowly notched. Floating lvs 3-10 x 1.2-4.5mm, elliptic to ± orb. Frs 1-1.4(1.8)mm wide, blackish, narrowly winged (at apex only), sessile, the styles ± erect. Ann, germinating late summer. Often on wet mud, VR, Scot, Ire..................................*Narrow-ed Water-starwort* **Callitriche palustris**

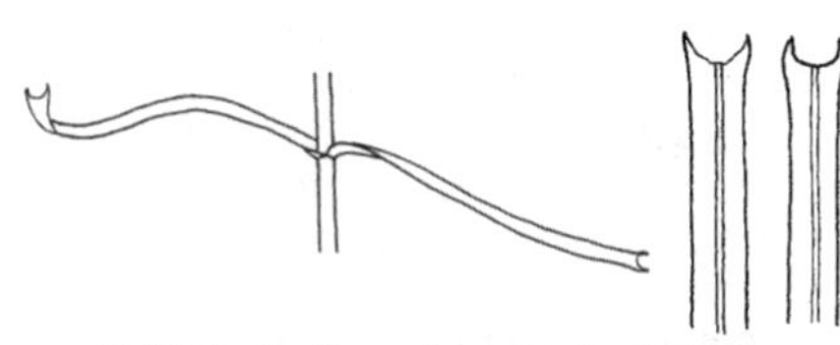

Callitriche brutia ssp *hamulata* lvs & lf apices

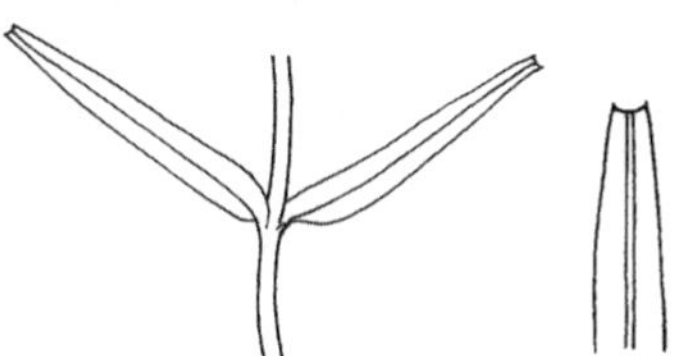

Callitriche hermaphroditica lvs & lf apex

Group CHAT – Chenopodium, Atriplex. *Ann herbs (per in* C. bonus-henricus*). Plant with bladder-like hairs which give the plant a mealy appearance (at least when young but a little meal is usu retained in the lf axils on older plants). Lvs all on stems (occ in basal rosette in winter plants of* C. bonus-henricus*), alt but the lowest opp, never connate at base, applanate when young, with stomata both sides. Petiole usu channelled. Stems occ ± woody at base, frequently longitudinally striped with white, red or green, round to ridged or 4-angled, solid but pith-filled, with scattered vb's. Jun-Oct (Mar-Oct in* C. bonus-henricus*)*

■Bladder hairs spherical. Stems without large lf in axils, small axillary lvs occ present ....(*Chenopodium*) **CHAT A**

■Bladder hairs slightly flattened to sausage-shaped. Stems with 1 shoot and 1 lf in axil, small axillary lvs occ present............(*Atriplex*) **CHAT B**

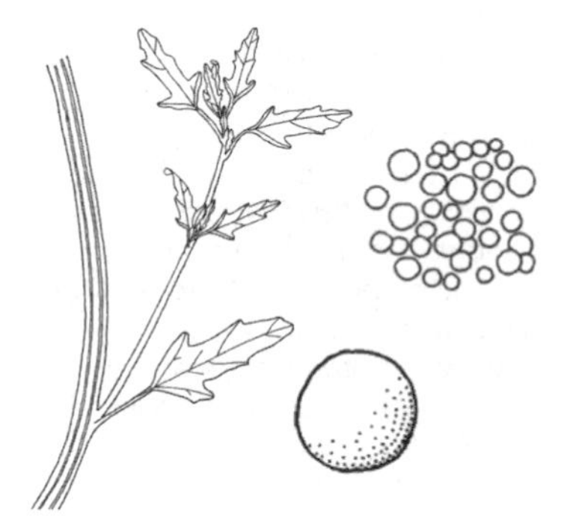

Chenopodium ficifolium (showing lf axils) & hairs

Atriplex prostrata (showing lf axils) & hairs

CHAT A – Chenopodium. *Bladder hairs spherical. Stems without a large lf in most axils (but small axillary lvs occ present). Lvs often minutely pitted above.* ❶ *Lvs net-veined (not net-veined in other spp).* ❷ *Plant stinking of rotten fish (which transfers to fingers on touching)*

■Lvs usu entire or with 1 tooth per side

Per, odourless. Lvs 5-10 x 3-9cm, broadly ovate, acute, truncate to cordate or hastate at base, mealy esp below when young, entire but sinuate, 3-pli-veined, with anastomosing 2° veins. Petiole to 12cm, often hollow, with 5 vb's. Stems 25-75cm, erect to ascending, unbranched, mealy, occ viscid. ❶............*Good-King-Henry* **Chenopodium bonus-henricus**

Ann

Plant stinking of rotten fish. Lvs and stems mealy (often strongly so). ❷

Lvs alt, 1-2.5cm, rhombic or ovate, acute, grey-green, entire or with a single tooth-like angle on one or both sides nr base, with few indistinct veins. Stems 5-60cm, 1.5mm diam, prostrate to ascending. Often nutrient-rich coastal shingle. Sch8............*Stinking Goosefoot* **Chenopodium vulvaria***

Plant not stinking of rotten fish. Lvs (±) hairless

Lvs 1.5-6(10)cm, ovate to elliptic, obtuse to ± acute, rounded to cuneate at base, thin, with erose (or v minutely crenulated) purple cartilaginous margins, usu dark green above, often purplish below, occ with sparse meal when young, (±) entire, with 3-5 prs of anastomosing 2° veins. Petiole to 2cm. Stems 15-50(100)cm, erect or decumbent, branched, often purplish, hairless............*Many-seeded Goosefoot* **Chenopodium polyspermum**

Lvs 1-2.5(4)cm, broadly ovate, slightly wider than long, usu obtuse, truncate or broadly cuneate at base and decurrent to base of petiole, ± fleshy, with minutely crenulate (to ± entire) cartilaginous margins, dark green above, violet below when young, entire or slightly toothed on vigorous growth, with indistinct to obscure 2° veins. Petiole to 0.8cm. Stems 10-50cm, prostrate to decumbent, usu hairless. Saltmarshes, VR, Kent, Essex ....*Saltmarsh Goosefoot* **Chenopodium chenopodioides**

■Lvs toothed (often regularly). Ann

Lvs ± fleshy and/or glaucous

Lvs glaucous

Lvs much longer than wide, never 3-lobed

Lvs 1-4 x 0.6-2cm, oblong-ovate to lanc, obtuse to ± acute, cuneate at base, fleshy, without cartilaginous margins, ± green and hairless above, glaucous (often purplish) and mealy below, dentate to ± entire (occ weakly lobed), with indistinct or obscure 2° veins. Petiole 0.5-1cm. Stems 20-40cm, decumbent, often reddish or with red and green stripes. Infl lfy .... *Oak-leaved Goosefoot* **Chenopodium glaucum***

Lvs ± as wide as long, often 3-lobed

Lvs 1-3.5(4.5) x 1-3(4)cm, broadly ovate or rhombic-ovate, similar to *C. murale*, obtuse, broadly cuneate to truncate at base, grey-green, often glaucous below, usu densely mealy both sides, lateral lobes (when present) ± equal to terminal lobe, dentate to ± entire, the teeth shortly mucronate. Petiole 0.5-2(3)cm. Stems 30-100(150)cm, erect to ascending, occ decumbent, usu much-branched, densely mealy. VR alien............. *Grey Goosefoot* **Chenopodium opulifolium***

Lvs green or reddish, not glaucous

Lvs often becoming reddish, hairless (but occ meal in lf axils)

Lvs (2)4-8 x 2-6cm, broadly ovate to rhombic-ovate, acuminate, cuneate at base, ± fleshy, shiny mid-green both sides, (±) hairless, strongly to weakly toothed, rarely entire, the teeth in 3-5 prs (usu ± incurved), with 2° veins not or weakly anastomosing (± not raised below). Petiole 0.5-3(5)cm. Stems 30-80cm, usu erect or ascending, light green to reddish, hairless. Nutrient-rich habs, often ruderal............................... *Red Goosefoot* **Chenopodium rubrum***

Lvs not reddish, mealy (at least when young)

Lvs 0.8-4(9) x 0.4-6(8)cm (a small-lvd form is widespread), ovate or rhombic-ovate, acute to acuminate, cuneate to rounded at base, slightly fleshy, with v narrow entire cartilaginous margins, dark (occ pale) green, irreg dentate with c12 deep teeth per side, the teeth incurved, with few anastomosing 2° veins (veins shiny below). Petiole 1-2.5cm, occ equal to lf. Stems 10-60(100)cm, erect or with spreading branches from the base, occ red-striped, hairless (to sparsely mealy when young)... *Nettle-leaved Goosefoot* **Chenopodium murale***

Lvs not fleshy or glaucous

Lvs with long terminal lobe (lvs widest 1/3 from base)

Plant stinking of rotten fish. Lvs distinctly 3-lobed, the central lobe the widest and usu longer than laterals. Stems to 100cm, erect to ascending. ❷. VR alien .... *Foetid Goosefoot* **Chenopodium hircinum**

Plant odourless. Lvs 2.5-5(8) x 1-3.5cm, ovate-oblong, obtuse to ± acute, mucronate, mealy or sparsely so, usu 3-lobed, the terminal lobe 2-3x longer than lateral lobes and ± parallel-sided, ± entire to sinuate-dentate, the lateral lobes usu with 1 tooth nr base, with 4 prs of 2° veins ending at lobe apices. Upper lvs weakly lobed to ± entire. Petiole 0.5-5cm. Stems 20-50(90)cm, erect, occ decumbent, green-striped................ *Fig-leaved Goosefoot* **Chenopodium ficifolium***

Lvs without long terminal lobe

Lvs v mealy at least below

Lvs violet below at least when young

Stem 1, to 300cm, stout (to 50mm diam), erect, much-branched above, tough, reddish-green or purple-striped. Lvs to 20 x 16cm, rhombic to ovate, usu obtuse, broadly cuneate at base, dark green above, paler below, irreg serrate-undulate. Upper lvs ovate to ovate-lanceolate, serrate to entire. Petiole to 10cm...... *Tree Spinach* **Chenopodium giganteum***

Lvs not violet below

Stems mostly >10mm diam

Lvs v similar to *C. album* (below), yellow- to dark green, with 1-2 shallow teeth in proximal ½. Stems to 150cm, sparsely mealy. Infl orange or red. Crop plant, often sown in wildflower mixes.. *Quinoa* **Chenopodium quinoa**

Stems mostly <5mm diam

Lvs 3-6 x 2.5-5cm, rhombic-ovate to broadly lanc, 1-2x as long as petiole, ± obtuse to acute, cuneate to broadly so at base, ± dull deep green, irreg serrate, with 3 prs of 2° veins fading before margin. Petiole to 5cm. Stems 15-150cm, erect, much-branched, usu with ± short strict branches (occ spreading), often reddish or red-striped ....*Fat-hen* **Chenopodium album***

Lvs hairless or sparsely mealy

Lvs 6-15 x 5-13cm, broadly ovate to ovate-triangular, acute or acuminate rounded to ± cordate at base, bright green both sides, with reddish erose cartilaginous margins, palmately lobed to deeply erose-dentate with a few v large teeth (maple-like). Petiole 2-7cm. Stems 20-100cm, erect, sparsely branched above, yellowish or purple-striped ....*Maple-leaved Goosefoot* **Chenopodium hybridum***

Lvs 3-8(14) x 3-8cm, usu triangular or rhombic (the upper often lanc), acute or ± obtuse, truncate to broadly cuneate at base, occ slightly fleshy, similar to *C. rubrum* but mid-green both sides, usu toothed, the teeth often long and hooked, sinuate to irreg dentate (occ ± entire), often with obtuse to acute outward-pointing lobes at base. Petiole 1.5-4cm. Stems (15)30-100cm, erect, rarely branched. VR alien ....*Upright Goosefoot* **Chenopodium urbicum**

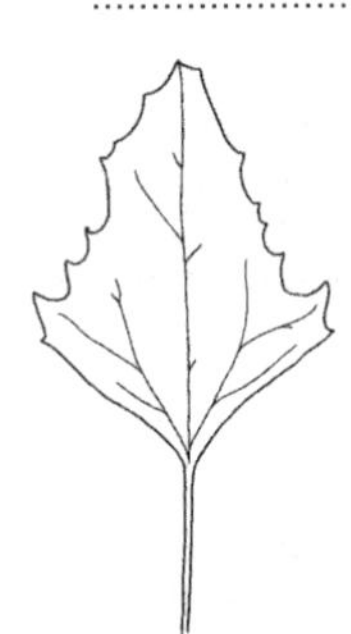

Chenopodium album

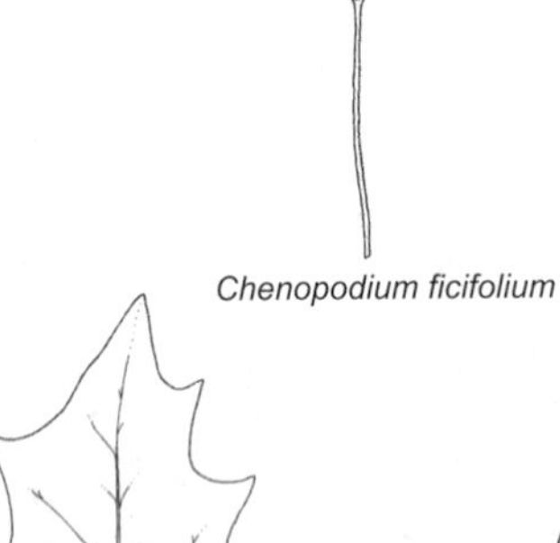

Chenopodium ficifolium

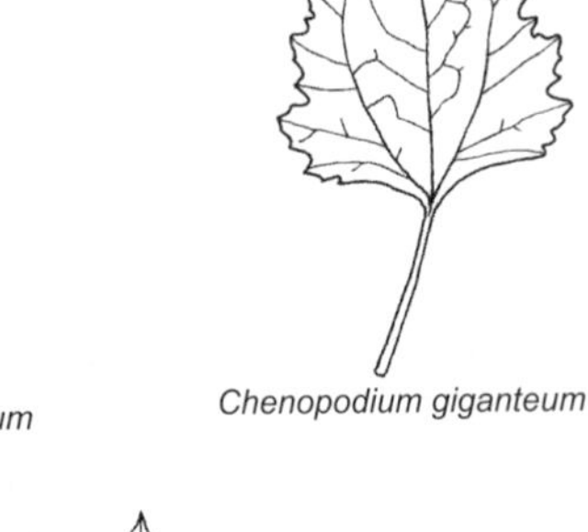

Chenopodium giganteum

Chenopodium glaucum

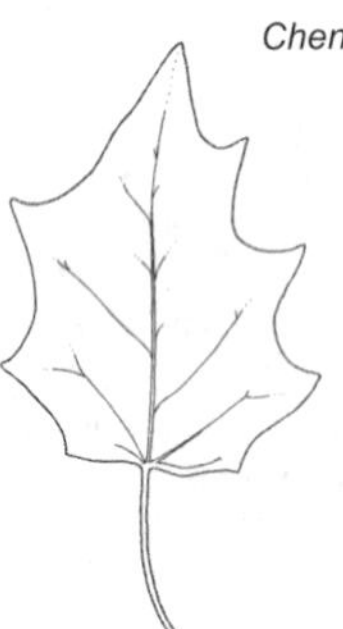

Chenopodium hybridum

Chenopodium murale

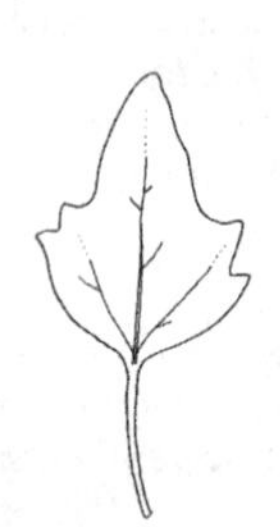

Chenopodium opulifolium

Chenopodium polyspermum

Chenopodium rubrum

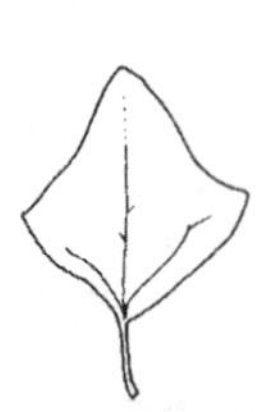

Chenopodium vulvaria

CHAT B – Atriplex. *Bladder hairs slightly flattened to sausage-shaped, collapsing to give a scurfy or frosted (salt-encrusted) appearance to foliage. Stems with 1 shoot and 1 large lf in most axils (below branch), small axillary lvs occ present. Ann*

▪Lvs with long terminal lobe, not hastate at base

Lvs 4-8cm, linear to linear-oblong, the lower shortly petiolate, the upper sessile, fleshy, dark bluish-green, hairless, all usu dentate distally. Petiole 1-1.5cm, often purplish. Stems to 100cm, ± erect, much-branched, often purplish, ± mealy..................*Grass-leaved Orache* **Atriplex littoralis**

▪Lvs without long terminal lobe, usu hastate at base

Lvs with Kranz venation

Lvs usu 1.5-4.5cm, rhombic to ovate or lanc, acute to obtuse, cuneate to a short petiole, sinuate-dentate, ± fleshy and v scurfy. Stems to 30cm (rarely taller), ± prostrate, much-branched, usu reddish. Sand and shingle beaches................*Frosted Orache* **Atriplex laciniata**

Lvs without Kranz venation

Lvs v fleshy (snapping along midrib)

Lvs to 10 x 6cm, triangular to hastate, (±) truncate at base, slightly scurfy, toothed (to entire), usu with 2° veins fading out before margins. Stems to 20cm (rarely taller), forming large prostrate mats, often reddish. Beaches...................*Babington's Orache* **Atriplex glabriuscula**

Lvs to 2cm, elliptic to oblong, sessile or shortly petiolate, obtuse but apiculate, cuneate at base, fleshy, silvery-mealy, entire, with midrib visible only. Stems to 15(30)cm, erect. Dry saltmarshes, VR, Essex. Sch8.....................*Pedunculate Sea-purslane* **Atriplex pedunculata**

Lvs not or weakly fleshy

Lvs often >10cm

Lvs 5-25 x 3-18cm, ovate-oblong to ovate-triangular, ± obtuse, hastate to broadly truncate at base, slightly scurfy above, mid-green both sides, often purplish-red, entire or irreg serrate. Petiole 0.4-5cm. Stems to 200cm, erect, with oblique or spreading branches ....*Garden Orache* **Atriplex hortensis**

Lvs <10cm

Lvs strongly toothed

Lvs 2.5-5(10) x 1-5(10)cm, triangular-hastate, acute or acuminate, (±) truncate at base, like *A. patula* but soon hairless above, ± mealy below, irreg serrate or with 1-3 prs of serrate lobes below middle, with translucent 2° veins (often weakly anastomosing) and raised below. Petiole 1-3cm. Stems to 100cm, erect, with obliquely spreading branches, often purplish, ± hairless. PL 14...................................*Spear-leaved Orache* **Atriplex prostrata***

Lvs (±) entire or with weak teeth

Bracteoles sessile or stalked to 0.4cm. Lvs 4-6 x 0.7-3cm, ovate-triangular to linear-lanc, shortly acuminate, broadly cuneate at base, sparsely mealy, usu scurfy below, entire or toothed with a pr of large teeth (or lobes) nr base. Petiole 0.5-1.2cm. Stems to 100(150)cm, prostrate to erect, much-branched, with obliquely spreading branches, usu green, mealy or scurfy, distinctly ribbed with coloured striae above, the nodes usu purplish ....*Common Orache* **Atriplex patula**

Bracteoles stalked to 2.5cm. Lvs 2.5-5 x 1-5cm, lanc to rhombic, hastate or not at base, fleshy to ± thin, sparsely mealy when young, slightly scurfy, (±) entire. Stems (15)30-60cm. Upper saltmarshes, R (but widespread)..................*Long-stalked Orache* **Atriplex longipes**

Group COT – Cotoneaster. *Shrub (tree in* C. frigidus *and occ so in* C. x watereri*). Lvs alt, simple, shortly petiolate, usu apiculate, folded when young, entire, with anastomosing veins. Stipules 3-5mm, subulate, soon falling. Twigs usu round, rarely angled, with adpressed stout hairs (hairs S-shaped at base), often with short shoots. Buds small, with several scales or with 2 scales always parted. Examine young lvs for wool/hair colour. All taxa are garden escapes (exc* C. cambricus*); additional spp rarely occur*

■Lvs mostly >5cm

Lvs woolly or densely hairy below..**COT A**

Lvs (sparsely) hairy to hairless below..**COT B**

■Lvs 3.5-5.5cm..**COT C**

■Lvs all <3.5cm

Lvs woolly below..**COT D**

Lvs densely hairy to hairless below

Tall shrub (usu >1m)..**COT E**

Low shrub (usu <1m)..**COT F**

COT A – *Lvs >5cm, woolly or densely hairy below, net-veined, often with 2° veins sunken above. Tall shrub or tree*

■Lvs mostly ± lanc. Petiole 4-6mm

Arching shrub

Lvs glaucous below, 3-10 x 1-2(3.5)cm, narrowly lanc, occ bullate, (±) shiny above, ± dull glaucous and sparsely adpressed-hairy above, densely hairy below (± woolly when young), with 2° veins sunken above (more strongly so than *C. lacteus*). Frs 4-5mm, bright red. Evergreen shrub to 5m.............................*Willow-leaved Cotoneaster* **Cotoneaster salicifolius**

Lvs not glaucous below, 6-14 x 2.5-5cm, lanc to elliptic-lanc, dull (to ± shiny) and hairless above, white-woolly (to sparsely hairy) below, with 2° veins often weakly (or not) sunken above. Frs 5-8mm, yellow to bright red. Semi-evergreen shrub to 8m. *C. frigidus* x *salicifolius* ....*Waterer's Cotoneaster* **Cotoneaster** x **watereri**

Erect shrub (or tree)

Lvs (6)10-17 x 1.5-4.5cm, elliptic to narrow-lanc, dull and hairless above, densely adpressed-hairy below (woolly when young), with 2° veins weakly sunken or flat above. Frs 4-6mm, usu bright red. Deciduous or semi-evergreen shrub or tree to 18m ....*Tree Cotoneaster* **Cotoneaster frigidus**

Lvs 3-8 x 3cm, ovate to elliptic-lanc, ± shiny and hairless above, densely whitish-grey hairy or woolly below, with 2° veins sunken above. Frs 5-6mm, bright red. Evergreen shrub to 3m. VR ....*Hylmö's Cotoneaster* **Cotoneaster hylmoei**

■Lvs ovate to elliptic. Petiole 6-12mm

Evergreen spreading shrub to 5m. Lvs 5-8 x 3.5cm, not bullate, ± shiny or dull dark green and (±) hairless above, ± white- or yellow- woolly below, with 2° veins sunken above. Frs 5(6)mm, bright to dark red..*Late Cotoneaster* **Cotoneaster lacteus**

■Lvs ± ovate. Petiole 2-4(6)mm

Deciduous arching shrub to 4m. Lvs 3.5-7 x 4cm, bullate, shiny and ± hairless above, densely hairy below, with 2° veins sunken above. Frs 6-8mm, shiny bright red ....*Hollyberry Cotoneaster* **Cotoneaster bullatus**

COT B – *Lvs >5cm, hairy to sparsely so below (occ hairless), net-veined. Tall shrub or tree.* ❶ *Lvs glaucous below*

▪Lvs strongly bullate, with 2° veins sunken above

Lvs 3.5-7 x 4cm, ± ovate, shiny above. Frs 6-8mm, shiny bright red. Petiole 3-6mm. Calyx hairy. Deciduous shrub to 4m..*Hollyberry Cotoneaster* **Cotoneaster bullatus**

Lvs 5-15 x 7cm, thick, shiny to dull above, hairy above. Petiole 2mm. Frs 8-11mm, shiny dark red. Calyx hairless (exc margins). Deciduous shrub to 5m....*Bullate Cotoneaster* **Cotoneaster rehderi**

▪Lvs weakly bullate, without 2° veins sunken above (or weakly so)

Lvs 4-12 x 3.5cm, lanc, ± shiny above. Petiole 4-5mm. Frs 4-5mm, bright red. Semi-evergreen shrub to 5m..*Henry's Cotoneaster* **Cotoneaster henryanus**

▪Lvs not bullate but with 2° veins sunken above

Lvs with reddish midrib below. Frs red or yellow

Lvs 6-14 x 2.5-5cm, lanc to elliptic-lanc, dull (to ± shiny) above. Petiole 5-6mm. Frs 5-8mm, yellow to bright red. *C. frigidus* x *salicifolius*.......*Waterer's Cotoneaster* **Cotoneaster** x **watereri**

Lvs 3-10 x 1-2(3.5)cm, narrowly lanc, (±) shiny above. Petiole 4-6mm. Frs 4-5mm, bright red. ❶ ....*Willow-leaved Cotoneaster* **Cotoneaster salicifolius**

Lvs with midrib not reddish below. Frs shiny black

Lvs 5-11 x 5cm (often variable on same shoot), ovate, thin, dull to ± shiny above, hairless above, v sparsely hairy to hairless below, with 2° veins sunken above. Petiole 5-7mm. Frs 9-13mm. Deciduous erect shrub to 5m. VR...........*Hummel's Cotoneaster* **Cotoneaster hummelii**

▪Lvs never bullate (margins occ undulate), with 2° veins not or weakly sunken above

Tree. Frs red, not changing colour......................................*Tree Cotoneaster* **Cotoneaster frigidus**

Shrub

Frs red, not changing colour................................*Waterer's Cotoneaster* **Cotoneaster** x **watereri**

Frs red becoming bluish-black or purple. Deciduous (often evergreen in *C. cooperi*)

Lvs 4-10 x 5cm, elliptic to elliptic-ovate or ovate, ± dull dark green above, with spreading yellow hairs below (occ sparse), with 2° veins not or weakly sunken above (but weakly raised below). Petiole 5-6mm. Frs 7-9mm, red becoming bluish-black. Shrub or tree to 8m. R ....*Purpleberry Cotoneaster* **Cotoneaster affinis**

Lvs 4-8 x 4cm, v similar to *C. affinis* but smaller, lanc, flat, dull and (±) hairless above, sparsely hairy below, with 2° veins not sunken above. Petiole 5-6mm. Frs 6-10mm, dull red becoming dark purple. Shrub to 8m. VR..............*Cooper's Cotoneaster* **Cotoneaster cooperi**

Lvs 5-10 x 4cm, lanc to ovate, similar to above spp but relatively narrower, flat, dull above, v sparsely hairy to hairless below, with 2° veins not sunken above. Petiole 7mm. Frs 6-10mm, red becoming brownish-black, with slight white bloom. Shrub to 5m. VR ....*Godalming Cotoneaster* **Cotoneaster transens**

Frs bluish-black. Deciduous shrub to 5m

Petiole 3-6mm

Lvs 4-7 x 3.2cm, ovate, dull and sparsely hairy above, hairy below, with weakly sunken 2° veins above. Frs 8-10mm, shiny black. VR ....*Kangting Cotoneaster* **Cotoneaster pseudoambiguus**

Petiole usu >6mm

Lvs 4-6(8) x 3cm (often variable on same shoot), elliptic, flat, dull above, hairy below. Petiole to 12mm. Frs 7-8mm, bluish-black. VR.........*Black-grape Cotoneaster* **Cotoneaster ignotus**

Lvs 2.5-6 x 3cm, obovate, ± undulate, dull above, hairy to sparsely so below. Petiole to 8mm. Frs 7-9mm, bluish-black. VR..........................*Lindley's Cotoneaster* **Cotoneaster ellipticus**

Lvs 5-8 x 4.5cm, elliptic to broadly ovate, flat, light green and hairless above, ± hairless below. Petiole to 8mm. Frs 6-8mm, black. VR...*Dartford Cotoneaster* **Cotoneaster obtusus**

COT C – *Lvs 3.5-5cm, woolly or hairy below. Shrub usu <3m.* ❶ *Lvs rugose*

■Lvs woolly below. Evergreen. Widespread

Lvs with 2° veins clearly sunken above

Lvs 2.5-5 x 3cm, ovate, shiny dark green and sparsely long-hairy above, densely white-yellowish woolly below, net-veined. Petiole 3-6mm. Frs 8-10mm, orange-red. Erect shrub ....*Stern's Cotoneaster* **Cotoneaster sternianus**

Lvs 3-5 x 2cm, ovate, shiny dark green and sparsely hairy above, silvery- to yellowish-woolly below, net-veined, with 2° veins sunken above. Petiole 4-5mm. Frs 6-9mm, orange-red. Erect shrub..*Ward's Cotoneaster* **Cotoneaster wardii**

Lvs with 2° veins weakly (or not) sunken above

Lvs 2-4 x 1-2cm, elliptic to ovate, ± dark shiny green and sparsely adpressed-hairy above, white-grey woolly below, net-veined. Petiole 3mm. Frs 6-9mm, orange-red. Arching shrub ....*Franchet's Cotoneaster* **Cotoneaster franchetii**

Lvs 1-3(4) x 1-1.5cm, ovate, (±) dull and hairless above, whitish- to silvery- woolly below, ± net-veined. Petiole 4mm. Frs 5-8mm, dull red. Erect shrub ....*Silverleaf Cotoneaster* **Cotoneaster pannosus**

■Lvs densely to sparsely hairy below. Deciduous (exc *C. tengyuehensis*). VR

Lvs with 2° veins sunken above

Lvs ± rugose, otherwise similar to *C. franchetii*, 3-5cm, ovate-elliptic, dull above, densely hairy below, net-veined. Petiole 3-4mm. Frs 6-8mm, maroon. Erect shrub. ❶ ....*Obscure Cotoneaster* **Cotoneaster obscurus**

Lvs not rugose, 3-5.5(6)cm, ovate or elliptic to lanc, ± shiny yellow-green and hairy above, hairy below (to sparsely so), (±) net-veined. Petiole 3-4mm. Frs 7-9mm, orange-red. Erect shrub ....*Bois's Cotoneaster* **Cotoneaster boisianus**

Lvs with 2° veins weakly (or not) sunken above

Lvs 3-5 x 2.5cm, elliptic, ± undulate, dull above, hairy to densely so below. Frs 8-11mm, maroon, with 1 stone. Arching shrub.......*One-stoned Cotoneaster* **Cotoneaster monopyrenus**

Lvs 3.5-5.5 x 3cm, ovate, flat, shiny above, densely spreading yellowish-hairy below, weakly net-veined. Petiole 4mm. Frs 7-10mm, bright red, with 3-4(5) stones. Evergreen or semi-evergreen erect shrub..............................*Tengyueh Cotoneaster* **Cotoneaster tengyuehensis**

COT D – *Lvs <3cm, woolly below*

■Lvs white-grey woolly below (more silvery in *C. pannosus* and *C. amoenus*)

Lvs (±) net-veined. Evergreen

Lvs 2-4 x 1-2cm, elliptic to ovate, ± dark shiny green and sparsely adpressed-hairy above, net-veined, with 2° veins weakly sunken above and raised below. Petiole 3mm. Frs 6-9mm, orange-red. Arching shrub to 3m....................... *Franchet's Cotoneaster* **Cotoneaster franchetii**

Lvs 1-3(4) x 1-1.5cm, ovate, (±) dull and hairless above, ± net-veined, with 2° veins not or weakly sunken above. Petiole 4mm. Frs 5-8mm, dull red. Erect shrub to 3(5)m .... *Silverleaf Cotoneaster* **Cotoneaster pannosus**

Lvs not net-veined (occ weakly so in *C. cambricus*)

Shrub usu >1m

Lvs 1.5-3.5 x 1.5cm, ovate, v like *C. franchetii*, ± shiny dark green above, greyish- or greenish-woolly to densely hairy below, with 2° veins sunken (or weakly so) above. Petiole 3mm. Frs 6-8mm, bright red. Deciduous arching shrub to 2(3)m .... *Diels' Cotoneaster* **Cotoneaster dielsianus**

Lvs 1-2.3 x 1.3cm, elliptic, dull above, white- to silvery-grey woolly below, with 2° veins not or v weakly sunken. Petiole 4mm. Frs 5-6mm, bright red. Evergreen erect shrub to 1.5m. VR .... *Beautiful Cotoneaster* **Cotoneaster amoenus**

Shrub usu <1m

Lvs crowded, 0.6-1 x 0.5cm, elliptic, apiculate, shiny and hairy above. Twigs persistently hairy. Frs 8-10mm, ± shiny red. Evergreen arching shrub to 0.3m. VR .... *Starry Cotoneaster* **Cotoneaster astrophoros**

Lvs often sparse, 1-4 x 2.5cm, ovate to orb, not apiculate, dull and (±) hairless above, woolly to v densely hairy below, with 2° veins not or hardly sunken above. Petiole 6mm. Twigs soon hairless. Frs 7-11mm, bright red. Deciduous spreading shrub to 0.5(1.5)m. VR, N Wales (Great Orme)... *Wild Cotoneaster* **Cotoneaster cambricus**

■Lvs yellow-woolly below

Lvs net-veined

Lvs 2-4 x 2cm, ovate, not apiculate, dull (occ weakly shiny) and sparsely hairy above, ± yellow-woolly to densely hairy below, with 2° veins sunken (or weakly so) above. Petiole 3mm. Frs 7-9mm, bright red. Deciduous erect shrub to 3m. VR .... *Cherryred Cotoneaster* **Cotoneaster zabelii**

Lvs not net-veined

Lvs 1.4-2.5 x 1.5(2)cm, orb-ovate, apiculate, dull and sparsely hairy above, with 2° veins sunken above and weakly raised below. Petiole 3mm. Frs 8-11mm, orange to orange-red. Deciduous erect shrub to 1.5m. VR....................... *Showy Cotoneaster* **Cotoneaster splendens**

COT E – *Lvs <3cm, densely hairy to hairless below.* ❶ *Commonest sp encountered (other spp R or VR)*

■Lvs densely hairy below (hairs often less dense in *C. sherriffii*)

Lvs with 2° veins sunken above, not net-veined (*C. dielsianus* may rarely key out here)

Lvs (1)1.5-2.5 x 1.5cm, ovate, strongly apiculate, thin, like *C. splendens*, not v shiny above, sparsely hairy above, densely (white-) yellow-hairy (to ± woolly) below. Petiole 2.5mm. Frs 6-10mm, bright red. Evergreen erect shrub to 3m. VR ....*Engraved Cotoneaster* **Cotoneaster insculptus**

Lvs with 2° veins v weakly sunken above, usu net-veined

Lvs 1-4 x 3cm, elliptic-ovate to ovate, thin, dull hairless or v sparsely hairy above, densely hairy below (the hairs often spreading, occ white- to grey- woolly). Petiole 4mm. Frs 7-9mm, bright red. Deciduous erect shrub to 3m. R......................*Fang's Cotoneaster* **Cotoneaster fangianus**

Lvs with 2° veins not sunken above, not net-veined

Lvs orb. Deciduous

Lvs 1.5-2.5(4)cm. Shrub to 2m. VR........*Circular-leaved Cotoneaster* **Cotoneaster hissaricus**

Lvs elliptic. Evergreen

Lvs 1-2.5 x 1cm, elliptic, thick, ± dull and sparsely hairy above, densely adpressed-hairy to woolly below, often spreading-ciliate. Petiole 3.5mm. Frs 5-8(10)mm, dark red. Erect or ± prostrate shrub to 2(3)m. R.............................*Fringed Cotoneaster* **Cotoneaster marginatus**

Lvs 0.5-1.5 x 1cm, elliptic, ± thin, ± dull dark green above, adpressed-hairy below (not densely so), spreading-ciliate. Petiole 1.5mm. Frs 7-10mm, orange. Erect to spreading shrub to 2(3)m. VR..*Sherriff's Cotoneaster* **Cotoneaster sherriffii**

Lvs 0.5-2 x 1.1cm, elliptic, thin, dull mid-green above, like *C. sherriffii* but more densely hairy below. Petiole 1.5mm. Frs 6-9mm, shiny bright red. Erect to spreading shrub to 2(3)m. R ....*Tibetan Cotoneaster* **Cotoneaster conspicuus**

■Lvs sparsely hairy to hairless below, with 2° veins not or weakly sunken above (not raised below), usu not net-veined (small-lvd forms of *C. salicifolius* key out here but are strongly net-veined)

Lvs 1.5-3cm

Lvs to 1.5cm wide, ovate, thin, ± shiny dark green above with sparse long adpressed hairs, shiny paler green and ± hairy or sparsely so below, not net-veined. Petiole 2.5-4mm. Frs 6-11mm, orange-red, with 3-4 stones. Deciduous or semi-evergreen erect shrub to 3.5m. ❶ ....*Himalayan Cotoneaster* **Cotoneaster simonsii**

Lvs 0.5-2.5cm

Lvs 0.8-2.5cm, lanc, shiny above, ± hairless, ± not net-veined. Petiole 1-2(3)mm. Frs 7-12mm, often cylindrical, bright red, with 2 stones. Deciduous shrub to 2m. R ....*Spreading Cotoneaster* **Cotoneaster divaricatus**

Lvs 0.8-2.2 x 2cm, elliptic to orb, dull to shiny above, hairless above, sparsely adpressed-hairy below, occ ± net-veined. Petiole 5-6mm. Frs 7-9mm, black, with 2(3) stones. Deciduous shrub to 3.5m. VR...*Few-flowered Cotoneaster* **Cotoneaster nitens**

Lvs 0.5-1.3 x 1.2cm, orb, shiny above, ± hairless, not net-veined. Petiole 1-3mm. Frs 7-11mm, bright- to orange-red, with 3 stones. Evergreen or semi-evergreen shrub to 3m. VR ....*Distichous Cotoneaster* **Cotoneaster nitidus**

COT F – *Lvs <3cm, sparsely hairy to hairless below. Low ± horizontal shrub (occ erect in* C. hjelmqvistii*), <1m tall, sprawling*

■Lvs (±) net-veined. Evergreen

Lvs mostly >1.5cm. Shrub to 0.5m tall (branches to 3m). Widely planted

Lvs strongly net-veined, 1.5-3.5 x 1.6cm, elliptic to obovate, with recurved apiculus, minutely rugose, shiny dark green above, hairless below, with 2° veins weakly or not sunken above (weakly raised below). Petiole 8mm. Frs 6-8mm, bright red ....*Bearberry Cotoneaster* **Cotoneaster dammeri**

Lvs ± net-veined or weakly so, 1-2.5 x 1.5cm, elliptic, v like *C. apiculatus*, with recurved apiculus, not rugose, shiny dark green above, sparsely hairy below, with 2° veins never sunken above (rarely raised below). Petiole 2.5-5mm. Frs 6-8mm, ± bright red. *C. conspicuus* x *dammeri*..*Swedish Cotoneaster* **Cotoneaster** x **suecicus**

Lvs mostly <1.3cm, weakly net-veined. Shrub to 0.2m tall. VR

Lvs 0.8-1.5 x 1.1cm, apiculate, not rugose, shiny green above, like *C. apiculatus* but hairy below (esp when young). Petiole 6mm. Frs 9-10mm, bright red ....*Procumbent Cotoneaster* **Cotoneaster prostratus**

Lvs 0.5-1.4 x 0.9cm, elliptic, apiculate, not rugose, shiny dark green above, hairless to sparsely hairy below. Petiole 3mm. Frs 6-10mm, bright to dark red ....*Yunnan Cotoneaster* **Cotoneaster cochleatus**

■Lvs not net-veined

Lvs obovate to oblong. Evergreen

Lvs 0.5-1.4 x 0.9cm, obovate, occ weakly rugose, with flat margins, dull green above, v sparsely hairy below, with 2° veins often sunken above. Frs 5-8(10)mm, dark red. Shrub to 0.7m. R..*Congested Cotoneaster* **Cotoneaster congestus**

Lvs 0.5-0.8(1.2) x 0.8cm, obovate to oblong, never rugose, with recurved margins, v shiny dark green above, adpressed-hairy below, with 2° veins never sunken above. Frs 8-10mm, dark red. Shrub to 1m. Widely planted and well-naturalised ....*Entire-leaved Cotoneaster* **Cotoneaster integrifolius**

Lvs ± orb. Deciduous

Lvs mostly 1.5-2cm. Branches not herringbone-like

Lvs 1.5-2 x 1.8cm, thin, ± shiny dark green and ± hairless above, shiny paler green and v sparsely adpressed-hairy below, with 2° veins weakly sunken above (weakly raised below). Petiole 3mm. Frs 6-8mm, orange-red, with 2 stones. Occ erect shrub. R ....*Hjelmqvist's Cotoneaster* **Cotoneaster hjelmqvistii**

Lvs 1-2 x 1.5cm, v like *C. nitens*, ± thin, ± flat, shiny dark green above, dull paler green below, hairless above, v sparsely adpressed-hairy below, with 2° veins not or weakly sunken above (weakly raised below). Petiole 2.5mm. Frs 10-12mm, bright red, usu with 3 stones. R ....*Apiculate Cotoneaster* **Cotoneaster apiculatus**

Lvs mostly <1.5cm. Branches herringbone-like (± in a flat plane)

Side branches regular and spreading at >70° to main branch. Lvs 0.6-1.3 x 0.9cm, densely arranged, ± thin, flat, shiny dark green above, shiny paler green and v sparsely adpressed-hairy below, with 2° veins never sunken above or raised below. Petiole 1-2mm. Frs 4-6mm, orange-red, with 3 stones....................................*Wall Cotoneaster* **Cotoneaster horizontalis**

Side branches irreg and spreading at <65° to main branch. Lvs 0.9-1.3 x 1.4cm, densely arranged, ± thin, undulate, shiny dark green above, shiny paler green and ± hairless below, with 2° veins never sunken above or raised below. Petiole 1-2mm. Frs 6-9mm, orange-red, with 2-3 stones..............................*Purple-flowered Cotoneaster* **Cotoneaster atropurpureus**

DOCK / DOCK A

Group DOCK – Polygonaceae. *Lvs revolute when young, usu with mildly acidic taste. Ochreae (fused stipules) always present at least when young, whitish, silvery, or turning brown and papery*

■Lvs mostly basal (smaller lvs on stem if present) but often dead by fr. Ochreae parallel-veined

Lvs palmately veined (or 3-5-veined at base), usu orb or wider than long.......................... **DOCK A**

Lvs pinnately veined, not orb

Lvs hastate AND/OR petiole deeply channelled.. **DOCK B**

Lvs not hastate (occ fiddle-shaped). Petiole weakly channelled

Lvs glaucous or hairy below (often minutely or sparsely so), net-veined (*Persicaria*)... **DOCK C**

Lvs green to grey-green, hairless (occ papillate), rarely net-veined (*Rumex*)............... **DOCK D**

■Lvs usu on stem only, basal leaves (if present) much smaller. Ochreae pinnate-veined

Stems climbing, or ± prostrate/trailing along ground.. **DOCK E**

Stems erect (not climbing or trailing)... **DOCK F**

DOCK A – *Lvs basal (smaller lvs on stem, if present), usu orb or wider than long, palmately veined (or 3-5-veined at base). Petiole solid. Ochreae brown or whitish, scarious, hairless, never ciliate, parallel-veined*

■Lvs >20cm diam, (±) net-veined. Shortly rhizomatous per

Lvs 40-100cm diam, cordate at base, undulate, v shallowly lobed and obscurely toothed, dull green above, papillate only on minor veins below (but not midrib or most 2° veins), often papillate on margins, with stomata both sides, with 5 main veins (anastomosing). Petiole to 100cm, stout, reddish, not papillate, semi-cylindrical, with many scattered vb's, rhubarb-scented. Stems to 150cm. Apr-Nov.. *Rhubarb* **Rheum** x **hybridum**

Lvs 20-60cm diam, deeply 7-lobed (the lobes deeply dentate or lobed), ± cordate at base, shiny green above, shortly papillate at least below on 2° veins, with 5 main veins (each to lobe apex), with stomata below only (or both sides). Petiole to 15cm, occ red-spotted, papillate, not channelled, with many scattered vb's, odourless. Stems to 200cm, papillate, not ridged .... *Ornamental Rhubarb* **Rheum palmatum**

■Lvs <7cm diam, not net-veined. Tufted per

Lvs basal and on stem, to 5(7)cm diam, obtuse, hastate with spreading lobes, ± glaucous, with crenulate cartilaginous margins, ± palmately veined, with 2° veins visible nr base becoming obscure nr margin, with stomata both sides. Petiole shallowly channelled, with 1 central vb and 3(5) vb's around margin. Ochreae whitish, occ with red veins. Stems many, 10-50cm, erect to ± prostrate, much-branched. Mar-Oct (all yr). R alien........................ *French Sorrel* **Rumex scutatus**

Lvs usu all basal, to 3cm diam, reniform-orb, obscurely angled, with crenulate cartilaginous margins, ± fleshy, palmately veined, with ± obscure weakly translucent veins forking 3-4x to margin, with stomata more abundant below. Petiole channelled, with several tiny vb's. Ochreae whitish-scarious (occ with green or red veins), thick, adnate to petiole for ½ length. Stems 1-4, 10-20cm, erect, often unbranched. Mtns.. *Mountain Sorrel* **Oxyria digyna**

DOCK B – *Lvs usu hastate (and/or petiole with deep channel), not orb, never as wide as long, pinnately veined, with stomata both sides*. Ochreae brown, torn

■Lvs to 4-15cm x 20-50mm. Petiole with U-shaped channel

Lvs oblong-lanc, usu with backward-pointed lobes, green, pitted both sides, with anastomosing 2° veins not or hardly raised below (veins weakly translucent or opaque) but slightly raised above. Petiole c2mm wide, angled, with c5 vb's. Stems 20-50cm, erect, hollow. Tufted per. Var *hirtulus* has long papillae on lvs (at least below) and petioles. All yr. PL 24 ....*Common Sorrel* **Rumex acetosa**

■Lvs to 2(4)cm x 2-6mm. Petiole with V-shaped channel

Lvs lanc to linear, usu with spreading or forward-pointing lobes, often purplish, not pitted, with obscure 2° veins. Stems 5-20cm, ± prostrate, solid. Rhizomatous per. PL 24 ....*Sheep's Sorrel* **Rumex acetosella**

DOCK C – *Lvs not orb, never as wide as long, hairy (often minutely or sparsely so) and glaucous below, net-veined. Petiole pith-filled. Stipules papery.* ❶ *Stems rooting at nodes*

■Lvs ovate, acuminate, cordate at base and slightly decurrent down petiole. Tufted

Lvs 10-15 x 9cm, occ hispid hairs on veins above, white-hairy on veins below, ciliate, with crenulate margins, with veins curving up and thickening at extreme margin, with stomata below only (occ both sides). Upper lvs sessile, clasping. Petiole to 25cm, winged nr apex, occ reddish nr base, hairy to ± hairless, channelled, with 9-13 vb's around margin. Stems to 130cm. Apr-Oct. Shady habs, R alien...*Red Bistort* **Persicaria amplexicaulis**

■Lvs linear-lanc to oblong, acute, long-cuneate to truncate at base

Shortly rhizomatous

Lvs to 7 x 1.5cm, linear-lanc, long-cuneate to truncate at base, with crease lines and slightly recurved margins, dark green above, hairless or hairy below, with midrib raised below, other veins thickened and slightly raised at margin, with stomata both sides. Upper lvs sessile, clasping. Petiole to 5cm, unwinged, with 3 main vb's. Ochreae obliquely truncate at apex. Apr-Oct. Mtn turf..*Alpine Bistort* **Persicaria vivipara**

Stoloniferous

Lvs to 8 x 2cm, oblong, rounded-cuneate at base, dull green above, orange-brown and persistent when dead, hairless, appearing serrate with slightly thickened recurved margins, with stomata below only. Petiole to 1.5cm, channelled, with 3(7) vb's. ❶. Apr-Oct. Hortal or VR alien..*Bronze Bistort* **Persicaria affinis**

■Lvs broadly ovate-oblong, obtuse, truncate (to ± cordate) at base and decurrent down petiole (to 6cm). Rhizomatous

Lvs to 18 x 9cm, with crease lines, occ with recurved margins, minutely white-hairy on veins below, minutely ciliate, the veins not thickening at margin, with stomata both sides or below only. Petiole to 25cm, triangular, reddish nr base, with 7-13 vb's around margin. Ochreae obliquely truncate, brown, sparsely hairy or hairless, not ciliate. Stems to 80cm, round, striate. Apr-Jul. Damp gsld..*Common Bistort* **Persicaria bistorta**

DOCK D – Rumex. *Lvs not orb, rarely as wide as long, never hastate (occ fiddle-shaped), hairless (but may be papillate), often minutely pitted both sides, opaquely net-veined (translucent in* R. pseudoalpinus *and often so in* R. obtusifolius*), with abundant stomata obvious both sides (sparse or obscure above in* R. sanguineus*). Petiole with scattered vb's (and occ spiral fibres). Hybrids often occur.* ❶ *Lvs fiddle-shaped*

■Lvs gradually tapered to petiole (long-cuneate), papillate or not on veins below

Rhizomatous per. Lvs 5-12 x 3-7cm. Sand dunes, VR alien, SW Eng, S Wales

Lvs often all on stems, narrowly obovate, ± obtuse, occ rounded at base, leathery, strongly crisped-undulate, often with sparse papillae on veins below. Stems to 60cm, unbranched, reddish nr base. Rhizomes yellow in TS. Jan-Oct................*Argentine Dock* **Rumex frutescens**

Tufted per. Lvs 30-100 x 10-25cm. Aquatic

Lvs mostly basal lanc to ovate, acute or acuminate, held erect, often ± undulate, dull grey-green, occ with papillae on veins below. Petiole purplish at base, with many scattered purple vb's and sparse spiral fibres. Stems 80-200cm. Apr-Oct...................*Water Dock* **Rumex hydrolapathum**

Ann. Lvs to 10 x 1cm. Marshy muddy habs

Lvs mostly on stems, with midrib papillate both sides (esp nr petiole), linear-lanc, yellowish, with weakly crisped margins, with 2° veins not or weakly raised below. Petiole papillate, aerenchyma-filled, with indistinct vb's. Stems 40-90cm, papillate. Plant golden-yellow in fr ....*Golden Dock* **Rumex maritimus**

Lvs mostly on stems, without papillae, linear-lanc, yellowish, with crisped margins, with 2° veins not or weakly raised below. Petiole not papillate, aerenchyma-filled, with indistinct vb's. Stems 40-90cm, not papillate. Plant yellowish-brown in fr........................*Marsh Dock* **Rumex palustris**

■Lvs usu deeply cordate at base, strongly papillate on veins below (exc *R. aquaticus*)

Rhizomatous. Usu upland gsld, N of Derbs

Basal lvs 20-50 x 20-40cm, broadly ovate, ± undulate, dark green, with dense hair-like papillae on veins below, with 2° veins raised both sides. Petiole to 50cm, purple-striped, channelled, not papillate. Stems to 80cm. Mar-Nov...............................*Monk's-rhubarb* **Rumex pseudoalpinus**

Rhizomatous. VR alien, Kent

Basal lvs to 20(30) x 15(25)cm, broadly ovate, occ as wide as long, grey-green (to yellowish), hardly crisped, papillate on veins both sides, with 2° veins raised above. Petiole shallowly channelled, strongly ridged below, papillate, with 30 scattered vb's. All yr ....*Russian Dock* **Rumex confertus**

Tufted (or with short rhizomes)

Lvs 25-35 x 9-15cm, ovate, often crisped-undulate, mid-green. Petiole 9-25cm. Stems 80-200cm. Mar-Sept. Wet habs, VR, Scot (Loch Lomond)............*Scottish Dock* **Rumex aquaticus**

Lvs 10-25(40) x 5-15cm, ovate-oblong, undulate or weakly crisped, dark bluish-green, with 2° veins not or weakly raised above (rarely papillate above). Petiole 6-12cm, ridged, sparsely or not papillate. Stems 50-100cm, furrowed. Taproot yellow in TS. Often ruderal ....*Broad-leaved Dock* **Rumex obtusifolius**

Lvs 5-10(15) x 2-4(5)cm, oblong or ovate-lanc but usu fiddle-shaped, mid-green, with 2° veins not sunken above. Petiole to 6cm, with papillate channel. Stems 20-50cm, with wavy branches diverging at 70-90°. Taproot yellow in TS. ❶. All yr. Dry gsld, S of Severn-Wash line. PL 24 ....*Fiddle Dock* **Rumex pulcher**

■Lvs usu rounded to slightly cordate at base, not or weakly papillate on veins below

Lvs crisped-undulate (often strongly so), 8-30 x 1.5-5(6)cm (>4x as long as wide), greyish-green, thickish

Tepals ovate-triangular. Lvs lanc, cuneate to ± cordate at base. Stems 40-100cm. All yr ....*Curled Dock* **Rumex crispus**

Tepals oblong. Lvs oblong or fiddle-shaped, rounded to ± cordate at base. Stems 40-70cm. ❶. All yr. Coastal. VR, SW Eng, Wales. Sch8....................................*Shore Dock* **Rumex rupestris**

Lvs not or weakly crisped-undulate

Lvs 20-50cm. Taproot yellow in TS
SE Eng, dry ruderal habs
Lvs 10-15cm wide, ovate or oblong-lanc. Petiole to 30cm, flat above, often reddish at base. Stems 50-200cm..*Patience Dock* **Rumex patientia**
Lvs 7-10cm wide, broadly lanc to oblong-lanc. Petiole to 30cm, flat above, often reddish at base. Stems 50-200cm...*Greek Dock* **Rumex cristatus**
N Br, often disturbed habs
Lvs 7-15cm wide (3-4x as long as wide), ovate to lanc. Stems 50-120cm ....*Northern Dock* **Rumex longifolius**
Lvs (5)10-25 x 2-6cm. Taproot red in TS
Infl (and branches) lfy to ± apex. Tepals 3, each with a large tubercle. Lvs with abundant obvious stomata both sides, oblong-lanc to oblong or slightly fiddle-shaped, ± acute, thickish, with crease-lines, occ crisped, ± dull green above. Petiole to 10(15)cm. Stems 30-90cm, wavy, with branches diverging at 30-90°. Sunny wet habs ....*Clustered Dock* **Rumex conglomeratus**
Infl (and branches) not lfy to apex. Tepals 3, only one with a tubercle. Lvs with sparse or obscure stomata above, ovate-lanc, occ fiddle-shaped, ± acute, ± limp, with crease-lines, occ crisped, ± dull green above, occ with reddish veins. Petiole to 10(15)cm, often reddish at base. Stems 30-80cm, straight, with branches diverging at 15-25(45)°. Shady dry habs ....*Wood Dock* **Rumex sanguineus**

DOCK E – *Lvs usu on stem only, basal leaves (if present) much smaller. Lvs not long-petiolate. Ochreae pinnate-veined. Stems climbing or ± prostrate/trailing, occ ascending.* ❶ *Lvs with purple-black chevron*

■Woody-stemmed per vine or climber to 3(5)m tall. (*Fallopia japonica* climbs after cutting)

Evergreen. Lvs widely-spaced along stems, 0.4-1.2(2)cm, usu orb (slightly angular), leathery, dark green above, ± entire (occ lobed), not net-veined, with indistinct anastomosing 2° veins, with stomata sparse or absent above. Petiole 4-6mm, without an extra-floral nectary. Stems scrambling, interwoven, 1-1.5(3)mm diam, red-black, tough, round, minutely scabrid (x20). Ochreae soon disintegrating, brownish, scabrid.................... *Wireplant* **Muehlenbeckia complexa**

Deciduous. Lvs often in fascicles, 3.5-6 x 4cm, ovate, acuminate, rounded to cordate at base (occ sagittate), weakly undulate, with reddish erose cartilaginous margins, green, weakly net-veined, with anastomosing 2° veins raised below, with stomata below only. Petiole 10-30mm, with an extra-floral nectary at base (below abscission layer), channelled, with 5-7 vb's. Stems clockwise-climbing, woody, green or reddish, soon ashy grey, ± round, with striations and faint lenticels, solid to hollow. Ochreae persistent, greenish, soon brownish, smooth........... *Russian-vine* **Fallopia baldschuanica**

■Twining or ± prostrate bindweed-like ann

Lvs 3-10cm, ovate, acuminate, truncate at base, with minutely papillate scarious margins, dark bluish-green with sunken glands above, paler and often with minutely papillate veins below, with stomata below only. Petiole to 3cm, jointed at base, channelled, with several vb's. Stems 1-2m, clockwise-climbing, often papillate on obscure angles, solid or hollow. Jun-Oct. Shady habs .... *Copse-bindweed* **Fallopia dumetorum**

Lvs 2-6cm, ovate, acuminate, cordate-sagittate at base, without papillate margins, dull green above, often with reddish tinge, mealy and rarely pitted below, with 2° veins minutely papillate both sides, with stomata both sides. Petiole to 6cm, jointed at base, channelled, with several vb's. Stems usu <0.5m, clockwise-climbing or ± prostrate, with papillate angles, solid or hollow. Jun-Oct. Arable or ruderal.. *Black-bindweed* **Fallopia convolvulus**

■Prostrate (or ± so) ann or per (occ woody at base). (*Persicaria affinis* may key out here in error)

Petiole with 2 auricle-like lobes at base. Ochreae with hairs or cilia, entire. Lvs ± ovate, with translucent pinnate veins, with stomata both sides

Per. Lvs 1.5-5 x 1-3.5cm, acute, cuneate at base, dull dark green (to purplish) above, with 8-10 prs of sunken veins (raised below), with reddish erose cartilaginous margins, sparsely unicellular-hairy both sides (occ adpressed-hispid), ciliate. Petiole 4-8mm. Ochreae hairy, with cilia to 1mm, with parallel green veins. Stems rooting at nodes, often purplish, hairy (rarely glandular-hairy or hairless), round but grooved distally, solid. ❶. Apr-Nov .... *Pink-headed Knotweed* **Persicaria capitata**

Ann. Lvs 1.5-5 x 1-4cm, acute, rounded to truncate at base, mid-green, hairless above, long-hairy and gland-pitted below, often scabrid-ciliate. Petiole 0-30mm, winged to base. Ochreae usu with bristle-like hairs nr base, not ciliate. Stems occ rooting at lower nodes, hairless but with stout retrorse whitish hairs at nodes, striate, solid .... *Nepal Persicaria* **Persicaria nepalensis**

Petiole without lobes at base (but with jointed abscission layer). Ochreae hairless, silvery when young, laciniate. Lvs usu oblong to linear, 1-veined or with weakly translucent pinnate veins, with stomata both sides

Lvs ± glaucous to strongly so. Sand or shingle beaches

Per. Lvs with strongly revolute margins, 1.5-2.5 x 0.6cm, oblong, ± acute, glaucous, without papillae. Petiole 0-1mm. Ochreae v obvious and much overlapping nr stem apex, turning reddish, with 8-12 branched veins, usu persistent. Stems (±) woody, stout, much-branched, without papillae. All yr. S Eng, Ire. Sch8.................... *Sea Knotgrass* **Polygonum maritimum***

Ann. Lvs with recurved to flat margins, 1.5-3.5 x 0.6(10)cm, lanc, ± acute, ± glaucous, with low papillae above and on margins. Petiole 0-1mm. Ochreae often not overlapping, with 4-6 unbranched veins, not persistent. Stems becoming ± woody at base, slender (1.5-2mm diam), forming a prostrate circular mat, with internodes ≥1cm, with low papillae. Apr-Oct .... *Ray's Knotgrass* **Polygonum oxyspermum** ssp **raii**

DOCK E

Lvs not or rarely glaucous. Usu ruderal or arable (rarely beaches). Ann. Stems usu >10cm, tough, striate

Lvs ± all same size. Stems usu forming prostrate mats

Lvs 7-14 x 4-5mm, elliptic-lanc, ± glaucous to green, with minutely erose-crenate cartilaginous margin. Petiole <3mm. Ochreae 2-3mm ....*Equal-leaved Knotgrass* **Polygonum arenastrum**

Lvs distinctly longer near stem apices (smaller on branches). Stems usu ascending or erect

Lf length usu <4x width

Petiole 4-8mm. Lvs 20-50 x 1-25mm, oblong-obovate to spathulate, green to glaucous. Ochreae 3-7mm. N Br.......................................*Northern Knotgrass* **Polygonum boreale**

Petiole <3mm. Lvs 10-40 x 3-13mm, lanc to ovate-lanc. Ochreae c5mm ....*Knotgrass* **Polygonum aviculare**

Lf length often >6x width

Petiole <3mm. Lvs 15-35 x 1-4mm, linear to linear-lanc. Ochreae often c10mm, with reddish-brown veins. Stems slender, wavy. Calc arable or disturbed habs ....*Cornfield Knotgrass* **Polygonum rurivagum**

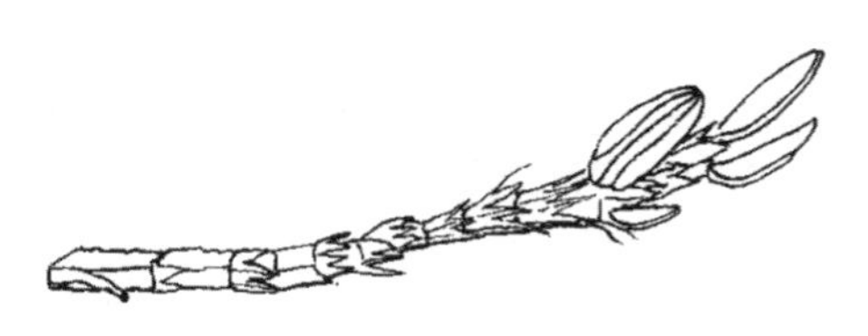

Polygonum maritimum

DOCK F

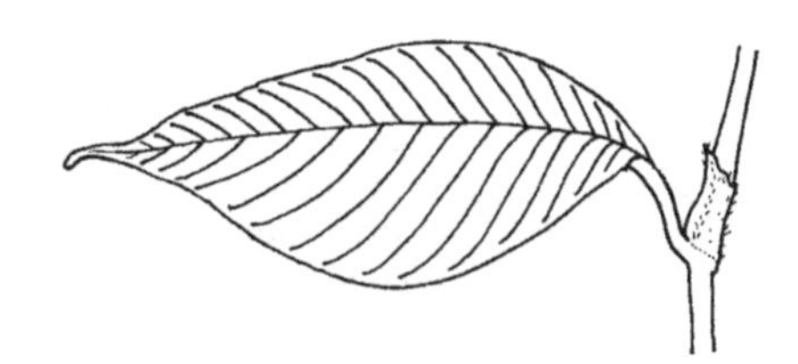

Persicaria campanulata

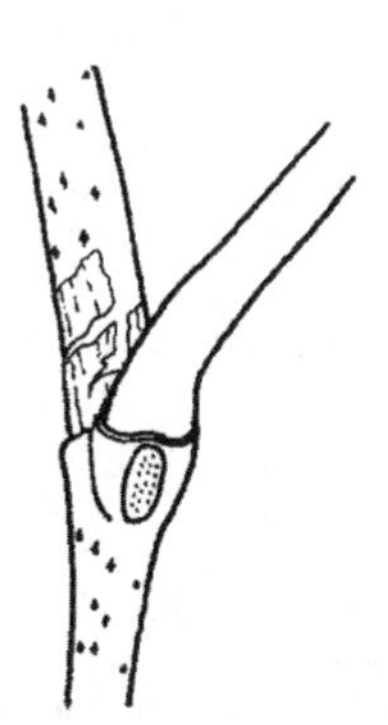

Fallopia japonica var *japonica* petiole base (showing extra-floral nectary)

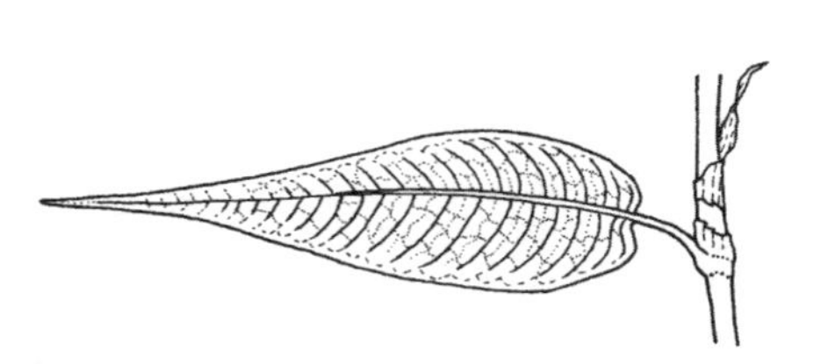

Persicaria wallichii

DOCK F

DOCK F – *Lvs usu on stem only (basal leaves, if present, much smaller), not long petiolate. Ochreae pinnate-veined. Stems erect, (±) round.* ❶ *Lvs with palmate veins.* ❷ *Lvs with hot peppery taste*

■Lvs densely white-hairy below, net-veined or not. Rhizomatous. Apr-Oct

Lvs 7-20 x 8cm, ovate-lanc, acuminate, not rugose, dark green and adpressed unicellular-hairy above with ± stiff white hairs, white-cottony below (hairs often turning brownish), ciliate, rarely net-veined but 9-22 prs of 2° veins strongly sunken above and raised below (anastomosing at margin), with stomata below only. Petiole to 1cm, flattened, with c13 vb's around margin. Ochreae hairy or hairless, often ciliate to 1.5mm. Stems to 1.5m, stout, brittle, hairless, v hollow ....*Lesser Knotweed* **Persicaria campanulata***

Lvs 8-17 x 2-5cm, broadly to narrowly ovate, acuminate, rugose, dull green above, white-woolly below, net-veined. VR alien..........*Chinese Knotweed* **Persicaria weyrichii**

■Lvs not densely white-hairy below, net-veined. Rhizomatous. Mar-Dec

Lvs cuneate at base. Petiole base without extra-floral nectaries

Lvs 12-17 x 7cm, mid-green and adpressed-hairy above, paler and densely long-hairy to ± hairless below, with non-swollen adpressed hairs on midrib below, finely sinuate, with 2° veins anastomosing once at or nr margin. Petiole to 1cm. Ochreae ± adpressed-hairy. Stems to 2m, much-branched, sparsely hairy, pith-filled to hollow......*Soft Knotweed* **Persicaria mollis**

Lvs to 14 x 3cm, *Chamerion*-like, light green, sparsely hairy above, short-hairy below, with swollen-based adpressed hairs on midrib below, densely ciliate, entire, with 2° veins hardly raised below and anastomosing several times, with stomata both sides or below only. Petiole to 1.5cm, slightly winged distally, with flat arc vb and several obscure round vb's. Ochreae ± adpressed-hairy. Stems to 1m, much-branched, occ reddish, hairless, pith-filled to hollow ....*Alpine Knotweed* **Persicaria alpina**

Lvs rounded to ± cordate at base

Petiole base with several extra-floral nectaries visible as c5 semi-circular pits around each node

Lvs to 30(38) x 10-20(28)cm, ovate-oblong, acute to obtuse, cordate at base, glaucous and pitted below, often hairy on veins above, with scattered 0.2-0.6mm wavy septate hairs below, with stomata below only. Petiole 1-4cm. Ochreae 6-12mm. Stems to 4m, green, hollow. Fls from Sept onwards..........*Giant Knotweed* **Fallopia sachalinensis**

Petiole base with 1 extra-floral nectary (visible as a circular pit on the underside just below abscission layer)

Stems 1-2(4)m

Lvs 16-23 x 15cm, lanc-ovate, cuspidate, truncate to cordate at base, with crinkled margins, glaucous and pitted below, with short stout pointed hairs on veins esp below, with stomata below only. *F. japonica* x *sachalinensis*..........*Hybrid Knotweed* **Fallopia** x **bohemica**

Lvs to 16(21) x 10(14)cm, broadly ovate, cuspidate, truncate to ± cordate at base, with scabrid-papillate margins, glaucous and pitted below, hairless (occ blunt papillae on veins below), with stomata below only. Petiole to 3cm, red, channelled, solid, with central vb and c7 small vb's around margin. Ochreae 6-10mm, brown, usu torn. Stems erect (climbing when cut!), stout, glaucous, reddish-spotted, hollow. PL 10 ....*Japanese Knotweed* **Fallopia japonica** var **japonica***

Stems <1m

Lvs 7-11 x 10cm, ± orb, ± leathery, with crinkled margins, hairless below, occ with stomata above..........*Japanese Knotweed* (var) **Fallopia japonica** var **compacta**

Petiole base without extra-floral nectaries

Lvs to 25 x 8cm, lanc to narrowly ovate, acuminate, rounded to ± cordate at base, hairless above (hairy in var *pubescens*), ± hairless to densely hairy below, ciliate, with reddish midrib, with stomata sparse above but abundant below. Petiole 3cm, reddish, not channelled, with several vb's. Ochreae hairless to hairy, rarely ciliate. Stems to 1.5m, unbranched below, solid, with roots (or root primordials) below lower nodes. Apr-Oct ....*Himalayan Knotweed* **Persicaria wallichii***

■Lvs not densely white-hairy below, not net-veined

Lf veins palmate. Ochreae not ciliate

Lvs broadly ovate to ± orb, acuminate, sagittate or hastate at base, petiolate, dull above, often papillate on veins below, with 5 translucent to opaque palmate veins, with stomata both sides. Upper lvs sessile, clasping. Petiole to 6cm, channelled, solid, with scattered obscure vb's. Ochreae hairless. Stems to 60cm, v sparsely papillate (esp at nodes), often with swollen internodes, hollow. Ann. ❶. Jun-Oct.................................*Buckwheat* **Fagopyrum esculentum**

Lf veins pinnate. Ochreae ciliate, usu hairy

Lvs roughly antrorsely adpressed-hispid above, cordate at base. Rhizomatous per

Lvs to 15 x 3cm, oblong-lanc, cordate at base, usu with black blotch above, with 3 crease lines each side, adpressed-hairy below (usu hairless in aquatic habs), ciliate, often with 2° veins slightly raised both sides, usu with stomata below only. Petiole 0.5-1cm (to 4cm in aquatic habs), with several vb's. Ochreae antrorsely adpressed-hispid, often long-ciliate. Stems to 60cm. Apr-Oct. Wet habs or ruderal..........*Amphibious Bistort* **Persicaria amphibia***

Lvs white-woolly or sparsely so below (occ hairless), cuneate at base. Ann

Lvs to 10 x 2cm, dull green above, often with black blotch, often with sessile glands below, ± adpressed hispid on midrib below, ciliate, pitted both sides. Petiole hispid. Ochreae often hairless, with v short cilia to 0.2mm. Stems to 80cm, usu greenish, ± swollen above nodes, glandular-papillate nr infl. Jun-Oct. Ruderal or damp habs ....*Pale Persicaria* **Persicaria lapathifolia**

Lvs sparsely hairy below, or with adpressed hairs on midrib below, cuneate at base. Ann

Lvs often black-blotched. Usu dry ruderal habs (or dried mud)

Lvs to 12 x 2.5cm, lanc, ± sessile, undulate, occ adpressed-hairy above, often sparsely hairy below, shortly ciliate, with stomata both sides. Ochreae ± adpressed-hairy, with cilia to 2mm. Stems to 80cm, reddish, hairless, swollen above nodes. Jun-Oct ....*Redshank* **Persicaria maculosa**

Lvs not black-blotched. Damp habs

Lvs 2-5x longer than wide, 2-8 x 0.3-2cm, mid-green above, hairless exc for adpressed hairs on midrib below and long cilia to 0.4mm, with stomata both sides. Ochreae with strongly adpressed (or fused) hairs, with cilia 1-4mm. Stems to 60cm. Plant often more reddish than *P. hydropiper*. Jun-Oct...*Tasteless Water-pepper* **Persicaria mitis**

Lvs 7-10x longer than wide, 2-8 x 0.2-0.8cm, dark green above, occ reddish, hairless below exc for sparse adpressed hairs on midrib, antrorsely ciliate, with stomata both sides. Ochreae sparsely adpressed-hairy, with cilia (1)2-3mm. Stems to 30cm. Jun-Oct ....*Small Water-pepper* **Persicaria minor**

Lvs hairless below (even midrib), cuneate at base, with hot acrid taste ❷. Ann

Lvs to 10cm, lanc or narrow lanc, ± undulate, with short cilia <0.1mm, with stomata both sides. Ochreae hairless, with cilia 1-2.5mm. Stems to 75cm, reddish at least below nodes (nodes often with a reddish ring). Jun-Oct. Damp habs............*Water-pepper* **Persicaria hydropiper***

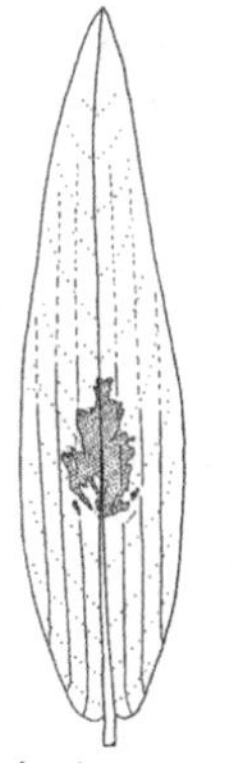

Persicaria amphibia

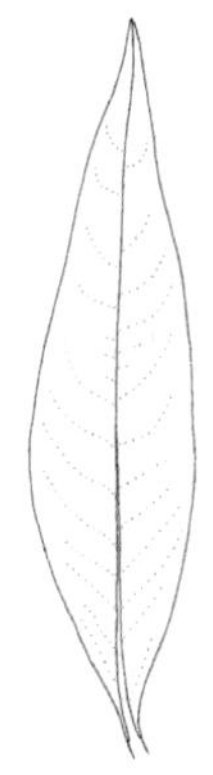

Persicaria hydropiper

Group EPIL – Epilobium. *Per (ann). Lvs opp below, usu alt above, occ with fragile brown scale at apex when young, occ with translucent dots (esp* E. ciliatum *and* E. parviflorum*), not net-veined, the translucent midrib often fading before apex and the 2° veins often fading nr margins (eucamptodromus). Petiole, when present, with flat arc vb. Stems usu branched, pith-filled. Hairs unicellular, whitish.* ❶ *Lvs often entire*

■Stem lvs densely patent-hairy (basal lvs hairless). Stems round

Stem lvs ± clasping, sessile, slightly decurrent down stem. Basal rosette lvs reflexed nr apex

Lvs alt above first few nodes, 6-18 x 1.5-2.5cm, oblong-lanc, acute, with 2 prominent hydathodes and a brown fragile scale at apex, rolled when young, with hairs to 1.3mm both sides, often with short glandular hairs, ciliate, minutely serrate, with stomata below only (stomata both sides on rosette lvs). Stems to 150cm, 8-15mm diam, densely hairy with 1-2mm hairs, glandular-hairy above (often viscid). Rhizomatous, with long thick lfy stolons in late winter. All yr...*Great Willowherb* **Epilobium hirsutum***

Stem lvs not clasping or decurrent, sessile or shortly petiolate. Basal rosette lvs not reflexed

Lvs mostly opp, 2-12 x 0.8-1.8cm, oblong-lanc, ± acute, with 2 ± prominent hydathodes but without a brown scale at apex, often held erect, ± applanate when young, softly greyish-hairy (occ reddish) with spreading 0.2-0.4mm hairs both sides, ciliate, minutely denticulate, with stomata below only (stomata both sides on rosette lvs). Stems to 75cm, 3-8mm diam, densely hairy with 0.7-1.5mm hairs, occ with short glandular hairs to 0.3mm above. Tufted, with short lfy stolons in late winter. All yr......................................*Hoary Willowherb* **Epilobium parviflorum***

■All lvs with sparse minute adpressed hairs (hairless TNE). Stems round

Lvs entire or obscurely denticulate. Wet habs

Lvs 2-7 x 0.4-1cm, lanc to linear-lanc, obtuse (without hydathode or brown scale at apex), cuneate and (usu) connate at base, ± sessile or uppermost v shortly petiolate, decurrent into v weak stem lines, dull green, occ purplish esp below, ± hairless both sides (occ with crisped hairs), with crisped cilia, with stomata below only. Stems to 60cm, 2-3mm diam, occ purplish, often with short antrorse crisped unicellular hairs, often glandular-hairy above. Rhizomatous. ❶. Apr-Oct (overwintering as a bud at end of thread-like stolons).......*Marsh Willowherb* **Epilobium palustre***

Lvs distinctly toothed. Dry habs

Lvs rounded at base, with ± adpressed crisped hairs both sides, ciliate, occ 3-whorled below, usu alt above, usu 4-7 x 1.5-3cm, ovate to ovate-lanc, acute, often with brown scale at apex, serrate, with stomata below only. Petiole 1-6mm. Stems to 75cm, often reddish, with short antrorsely crisped hairs, glandular-hairy above. All yr, occ overwintering with underground stolons. Usu shady habs, often humus-rich......*Broad-leaved Willowherb* **Epilobium montanum**

Lvs broadly cuneate at base, with strongly adpressed hairs above, ciliate, 1.5-5 x 0.6-1.6cm, elliptic-lanc to lanc, obtuse, often reddish, denticulate, with stomata both sides. Petiole 3-8mm. Stems to 60cm, often reddish, with short antrorsely crisped hairs, not glandular-hairy. All yr, overwintering as short stolons ending in lf rosettes. V dry sunny habs ....*Spear-leaved Willowherb* **Epilobium lanceolatum**

■All lvs hairless (may be ciliate) or with hairs confined to veins. Stems round to square

Plant in lowland habs, rarely climbing mountains. Stem usu 1, erect, often >20cm tall

Stem lvs with distinct petiole (basal may be sessile)

Lvs hairless exc for minute crisped cilia, 2-5(10) x 0.7-1.5(3)cm, all but uppermost opp, ovate-lanc, acute, without scale at apex, rounded to cuneate at base (rarely ± cordate), held erecto-patent, with translucent dots, weakly serrate, with stomata below only. Petiole 1.5-4mm. Stem 10-75cm, spreading- or crisped-hairy all round (but often hairless below) or with 2(4) rows of short crisped hairs, with spreading glandular hairs at least above, round or with 4 raised lines. Plant usu reddish, producing short leafy stolons and rosettes. All yr...............*American Willowherb* **Epilobium ciliatum**

Lvs with crisped hairs on veins both sides (occ hairless), 3-8 x 1.5-3cm, all but uppermost opp, ovate-elliptic to lanc-elliptic, acute, without scale at apex, cuneate at base, held patent or drooping, without translucent dots, sharply toothed, with stomata below only. Petiole 3-20mm. Stem 10-75cm, hairless below, with crisped hairs and spreading glandular hairs above, ± square with 2 distinct and 2 indistinct lines. Plant rarely forming thin stolons. All yr ....*Pale Willowherb* **Epilobium roseum**

Stem lvs sessile or ± so (lowest may have v short petiole)

Stems 4-angled below

Lvs 2-8 x 0.3-1cm, lanc, obtuse with 3 hydathodes at apex (no brown scale), not connate at base, clearly decurrent down stem and running into 2 stem ridges (check several nodes), often held erect, shiny green above, hairless exc for minute crisped cilia, denticulate with hydathodes, with stomata both sides. Stem to 75cm, 2mm wide, crisped-hairy esp on angles. Infl without glandular hairs. Plant without stolons but forming rosettes in autumn. All yr...*Square-stalked Willowherb* **Epilobium tetragonum**

Stems ± round below but with 2-4 raised lines (at least above)

Lvs usu 3-8 x (0.8)1.2-2.8cm, ovate-lanc, usu obtuse with a single hydathode at apex (no brown scale), rounded at base, the lower connate, the upper not connate, decurrent into raised lines on stem, dull green above, hairless or occ hairy on veins, with crisped cilia, remotely denticulate. Stem 20-80cm, 4-6mm diam, hairless below, often ± adpressed-hairy above. Infl with glandular hairs. Plant producing elongated lfy stolons from May (not ending in distinct rosettes). All yr (overwinters as lfy stolons) ....*Short-fruited Willowherb* **Epilobium obscurum**

Stems round along length, occ with 2 raised lines

Lvs 2-7 x 0.4-1cm, lanc to linear-lanc, obtuse (without apical hydathode or brown scale), cuneate and (usu) connate at base, ± sessile or uppermost v shortly petiolate, decurrent into v weak stem lines, dull green, occ purplish esp below, ± hairless both sides (occ with crisped hairs), with crisped cilia, with stomata below only. Stems to 60cm, 2-3mm diam, occ purplish, often with short antrorse crisped unicellular hairs, often glandular-hairy above. Rhizomatous. ➊. Apr-Oct (overwintering as a bud at end of thread-like stolons) ....*Marsh Willowherb* **Epilobium palustre***

Plant in damp mtn habs only. Stems usu several, not erect, <20cm tall

Lvs distantly toothed, 2-4 x 1.2-2.5cm, ovate to ovate-lanc, rounded or broadly cuneate at base, ± bluish-green, (±) hairless, often crisped-ciliate, with stomata both sides. Petiole 1(6)mm. Stems 1.5-3mm diam, strongly ascending from decumbent base, with 2 rows of hairs down the 2 faint ridges and a few scattered crisped hairs, occ glandular hairs above. Stolons often underground, slender, yellowish, with distant prs of yellowish scale-lvs. Apr-Oct ....*Chickweed Willowherb* **Epilobium alsinifolium**

Lvs weakly toothed to entire, 1-2.5 x 1cm, lanc or elliptic-lanc, rounded or broadly cuneate at base, often yellowish-green, ± hairless, occ crisped-ciliate, with stomata both sides. Petiole 1(6)mm. Stems 1.5-3mm diam, ± prostrate or gradually ascending, with 2 rows of hairs down the 2 faint ridges, occ glandular hairs above. Stolons above-ground, slender, often with distant prs of small green lvs. ➊. All yr...........................*Alpine Willowherb* **Epilobium anagallidifolium**

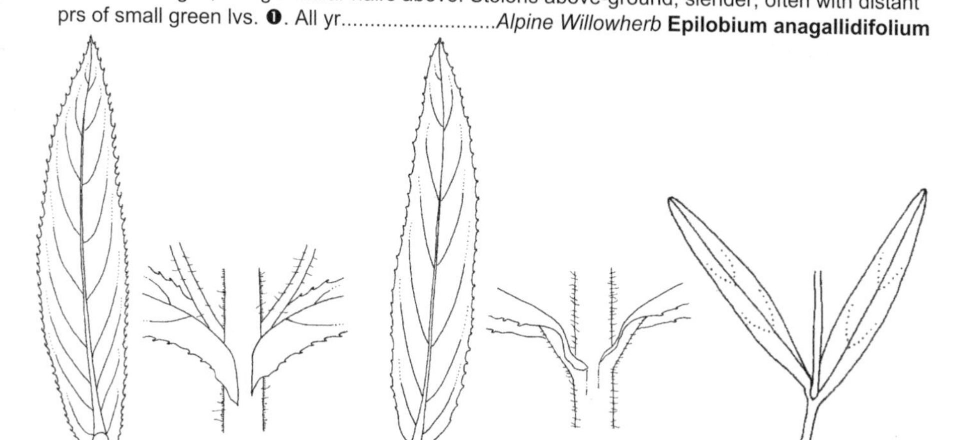

Epilobium hirsutum with lf bases

Epilobium parviflorum with lf bases

Epilobium palustre

Group EUPH – Euphrasia *(Eyebrights). Ann. Lvs opp below, alt above, 4-15mm, ± ovate to linear-lanc, shortly petiolate to ± sessile, with 1-4 teeth per side, with 2° veins ending in sinus (and raised below), with stomata both sides but often obscure. Stem(s) to 20cm, slender, erect, round. Roots ill-developed (semi-parasitic on grasses, etc). Hairs septate, minutely tuberculose (sculptured). Jun-Sept. Fertile hybrids frequently occur. This genus appears to be actively evolving and the specific boundaries are often obscure. The following key should be treated as tentative*

■Plant with at least some long patent glandular hairs >0.2mm

Stems with some long patent glandular hairs

Usu lowland damp acid gsld, rarely dry chalk gsld.. **Euphrasia anglica**

Usu lowland gsld, R, Wales, N Br, Ire.......................... **Euphrasia rostkoviana** ssp **rostkoviana**

Upland gsld, VR, Wales, N Eng...................................... **Euphrasia rostkoviana** ssp **montana**

Stems only with short retrorse nonglandular hairs

Damp coastal hths, Devon, Cornwall.. **Euphrasia vigursii**

Mtn damp gsld and flushes, Wales, N Eng (Lake District)............................ **Euphrasia rivularis**

■Plant with hairs v short, to 0.1(0.2)mm, occ glandular. Stems with short retrorse hairs

W Ire, limestone habs. Stems to 8cm, much-branched. Lvs often copper-coloured, with long-acuminate teeth... **Euphrasia salisburgensis**

Scottish coasts

Exposed sea-cliffs, N Scot. Lvs densely hairy, ± fleshy................................ **Euphrasia marshallii**

Sea-cliffs, N'most Scot. Lvs densely hairy esp below, ± fleshy.................. **Euphrasia rotundifolia**

Damp coastal hths, Outer Hebrides (Lewis). Lvs hairy esp distally, ± fleshy .... **Euphrasia campbelliae**

Wet gsld on exposed sea-cliffs, upper saltmarshes, N Scot. Lvs hairless to sparsely hairy, fleshy .... **Euphrasia foulaensis**

Saltmarshes, NW Scot. Lvs hairless to sparsely hairy, fleshy......... **Euphrasia heslop-harrisonii**

Mtns

Gsld, N Wales. Fls white or purplish, with yellow blotch; calyx-teeth broadly triangular. Lvs broad, sparsely hairy, bluntly toothed.. **Euphrasia cambrica**

Wet cliff ledges. Fls white or purplish, small, the lowest often at 2nd or 3rd stem node from base. Lvs broad, usu hairless, with v obtuse teeth.. **Euphrasia frigida**

Lowlands, may climb mtns

Mostly sea cliffs and sand dunes in W Br

Lvs fleshy, hairless to densely hairy (occ glandular-hairy), with obtuse teeth. Infl dense, 4-ranked.. **Euphrasia tetraquetra**

Dry calc habs (usu)

Widespread (the commonest sp in most of Br). Lvs shiny dark green, usu hairless, rarely densely hairy, with acute to aristate teeth. Stem branched above the middle, the branches straight and shorter than the stem.. **Euphrasia nemorosa**

N & W Br, often exposed or bare habs, VR. Lvs densely hairy. Stems simple to much-branched.. **Euphrasia ostenfeldii**

S Eng, W Wales, gsld, rarely fens, R. Lvs narrow, finely toothed, at least the upper hairless. Stems usu much-branched. Fls large, with elongated corolla tube, late (July onwards) .... **Euphrasia pseudokerneri**

Dry acidic or inland sandy habs

Widespread but mostly N & W Br, occ on limestone. Stem usu with slender wavy branches. Lvs rarely purplish. Fls often purplish.. **Euphrasia confusa**

Dry hths, mostly N & W Br. Stem usu with strict branches. Lvs usu purplish at least above. Fls purplish.. **Euphrasia micrantha**

Damp habs

Mostly N & W Br, usu lowland damp gsld (often acidic)

Stems robust, often with long upcurved branches (occ equalling main stem). Lvs usu with short glandular hairs (the commonest sp in much of N & W Br) .... **Euphrasia arctica** ssp **borealis**

Mostly N & W Br, usu lowland damp gsld (often acidic)

N & W Br, wet peaty flushes. Stem slender, erect, usu not or weakly branched. Fls usu white, with short lower lip. Lvs often purplish below. (The hybrid with *E. confusa* is common in more basic flushes and may be recognised by its larger fls (with longer lower lip), a more flexuous stem and lvs with ± aristate teeth)..**Euphrasia scottica**

Orkney, Shetland. Similar to *E. arctica* ssp *borealis* but stem initially ± prostrate or flexuous at base...**Euphrasia arctica** ssp **arctica**

Group GENT – Gentianaceae. *Hairless herbs. Lvs opp, decussate, usu sessile, applanate when young, often purplish in gentians, with narrowly hyaline margins (smooth to papillate), with (1)3-5(7) parallel veins, with sparse non-viscid sap and bitter taste (often odorous, esp gentians). Stem(s) round or ridged. Fls closed unless sunny.* ❶ *Lvs net-veined.* ❷ *Stem lvs perfoliate.* ❸ *Stems decumbent (erect in other spp)*

■Lvs mostly >5cm. Basal lvs usu absent

Lvs 3-8(12) x 1-2.5cm, lanc (to ovate), ± acuminate, with scabrid margins, net-veined, usu with 3-5(7) veins raised below, with stomata below only. Stems 30-60cm, erect to arching, unbranched, usu with erose cartilaginous ridges. Shortly rhizomatous per. ❶. Apr-Oct. R alien ....*Willow Gentian* **Gentiana asclepiadea**

■Lvs <5cm. Basal lvs several (in rosette), with 3-5(7) raised veins below (midrib strongly so)

Basal lvs not persistent at fl, 1-5 x 1-2cm, obovate, obtuse, often truncate and shortly petiolate at base, with entire hyaline margins, green to glaucous, with stomata both sides. Stem lvs ovate, acute, glaucous, with midrib raised below. Stem 1, to 45cm, often branched, round. Ann, with simple taproot. Dec-Oct (all yr). Calc gsld, dunes. PL 24.........*Yellow-wort* **Blackstonia perfoliata**

Basal lvs usu persistent at fl, 1-5 x 0.8-2cm, obovate or elliptic-oblong, usu obtuse, sessile, with entire or minutely scabrid hyaline margins, green, with stomata both sides. Stem lvs ovate to elliptic, ± acute, green, with 3(5) veins raised below. Stems 1-many, to 50cm, branched at base, 6-ridged. Bi, with branched taproot. ❷. All yr................*Common Centaury* **Centaurium erythraea**

■Lvs <5cm. Basal lvs with 0-1 raised veins below, or absent

Lvs with stomata both sides

Per to ann, with basal rosette at fl (>3 basal lvs). Lf margins often scabrid/crenulate (x20)

Mtns or uplands (or lowland in W Ire)

Rhizomatous per, with several rosettes. Basal lvs usu present at fl 8-15(20) x 5-7mm, ovate to elliptic, obtuse to acute, sessile, often purplish below, weakly channelled, with crenulate-papillate margins, 1-veined (lateral veins obscure). Stem lvs 2 prs, to 20mm. Stems 4-8cm. All yr. Limestone gsld, N Eng, W Ire. Sch8............................*Spring Gentian* **Gentiana verna**

Ann (bi), with 1 rosette. Basal lvs present or absent at fl, (2)5 x 5mm, ovate-orb to obovate, obtuse, ± sessile, green, ± entire, 1-3-veined. Stem lvs 1-2 prs, to 10mm, elliptic-lanc, ± acute, with weakly crenulate margins, 1-3-veined. Stems 3-6(15)cm, with purplish ridges. Mtns, VR. Sch8...*Alpine Gentian* **Gentiana nivalis**

Lowland, mostly coastal

Per. Stems ± prostrate to decumbent, several, branched from base, usu 4-angled. Basal lvs present or absent. Stem lvs 5-15 x 4-7mm, ± orb (on non-fl stems) to lanc-oblanc (on fl stems), obtuse to acute, dark green, shortly petiolate to sessile, often held erecto-patent, with scabrid margins (on fl stems), 1-3-veined (veins raised or flat above), with additional pinnate veins. ❸. All yr. Coastal cliffs, R, W Br. Occ hortal ....*Perennial Centaury* **Centaurium scilloides**

Bi. Stem(s) erect (or absent), 1-several, branched above, 4(6)-ridged (ridges minutely papillate). Basal and stem lvs 10-20 x 3-5mm, ± linear and parallel-sided (to narrowly oblong), obtuse, dark green, with scabrid to entire margins, (0)1(3)-veined (lateral veins indistinct, only the midrib raised below). Sand dunes, saltmarshes, Wales, N Br ....*Seaside Centaury* **Centaurium littorale**

Ann, without basal rosette at fl (or 2 basal lvs). Lf margins usu smooth to erose

Stem(s) 4-angled

Lower lvs usu shorter than upper lvs, 6-17 x 3-6mm, ovate to oblanc, obtuse, patent to erect, with ± smooth margins, weakly 1-3-veined, with veins (esp laterals) not raised below. Mid and upper lvs (2)6-20 x 2-10mm, often acute. Stem(s) 2-15cm, with 2-4 internodes, simple or branched above base at 30-45°. May-Oct........*Lesser Centaury* **Centaurium pulchellum**

Lower lvs usu longer than upper lvs, 8-17(20) x 5-10mm, ovate to oblanc, obtuse, erect, with ± smooth margins, (1)3-veined (laterals often opaque or obscure). Mid and upper lvs 4-15 x 5-10mm, often acute. Stem(s) 3-35cm, with 5-9 internodes, not branched below, branched above at 20-30°. Clay cliffs, Dorset. Sch8..........*Slender Centaury* **Centaurium tenuiflorum**

Stem round

Stem divaricately branched, 3-10cm, slender (0.5mm diam), ± prostrate to ascending. Lvs often numerous, 4-7mm, linear. Damp dune slacks, Guernsey ....*Guernsey Centaury* **Exaculum pusillum**

Stem unbranched (rarely branched above), 2-8cm, slender, erect. Lvs (0)1-2 prs, 2-6mm, linear. Damp hths, W Br, Ire..*Yellow Centaury* **Cicendia filiformis**

Lvs with stomata below only, with scabrid margins

Stem unbranched at least nr base. Wet hths

Basal lvs usu absent at fl. Stem lvs to 35 x 5mm, linear, obtuse, green to purple, with acutely scabrid-papillate recurved margins, with 1 translucent midrib raised below and 2 weak laterals. Stem erect, ± round, with 4 v weakly raised lines. Per. Apr-Oct ....*Marsh Gentian* **Gentiana pneumonanthe**

Stem usu branched nr base. Usu dry gsld. Basal lvs usu dead at fl, obtuse to acute, dark green, often purplish, with minutely crenulate to papillate-serrate margins often weakly recurved, 3(5)-veined (at least proximally), with midrib raised below (at least nr base). Stem(s) (4)6-ridged, the ridges erose to papillate-serrate

Usu acidic or neutral gsld. Calyx with 4 lobes, the outer 2 much wider than the inner 2

Basal lvs 1-2.5cm, ovate, lanc or spathulate, tending to fold in half, with lateral veins more visible in distal ⅔ of lf. Stem with v ridges. Bi (ann)....*Field Gentian* **Gentianella campestris**

Calc gsld. Calyx with 4-5 lobes, all ± equal

Basal lvs 2-4cm, broadly ovate to spathulate, with all veins sunken above and often fading out nr apex. Corolla (15)25-35mm. Ann (bi). VR, S Eng ....*Chiltern Gentian* **Gentianella germanica**

Basal lvs 1-3cm, obovate to spathulate or lanc, with lateral veins not sunken above and more visible in distal ⅔ of lf. Corolla 12-20mm. Ann (bi) ....*Autumn Gentian* **Gentianella amarella**

Basal lvs 1-2cm, spathulate. Stem lvs lanc to oblong- or linear-lanc. Corolla 12-22mm. Ann (bi). VR, S Eng, S Wales. Sch8..*Early Gentian* **Gentianella anglica**

Dune slacks, R, Wales, N Devon, W Scot. Calyx with 4-5 lobes, all ± equal, the outer often widely spreading

Basal lvs 1-2cm, lanc. Stem lvs ovate to ovate-lanc. Ann (bi). Sch8 ....*Dune Gentian* **Gentianella uliginosa**

Group GER – Geranium. *Lvs palmately lobed, with hydathode-tipped teeth. Petiole long, with 3-9 vb's (often indistinct) and spiral fibres. Stipules entire. Stem hairs are usu v similar in type, size and distribution to those of the petiole*

■Lvs 3-5-lobed (± composed of lfts), often fetid...**GER A**

■Lvs (3)5-9-lobed, odourless to strongly aromatic

Lvs usu 7-17cm diam. Per...**GER B**

Lvs usu <7cm diam. Ann (per in *G. sanguineum* and often in *G. pyrenaicum*)......................**GER C**

GER A – Geranium *subgenus* Robertium *section* Ruberta. *Lvs basal (alt) and on stem (opp), usu palmately divided into 3-5 lfts, often polygonal in outline and held horizontally from a ± erect petiole; lfts dark green to reddish, the teeth obtuse and usu tipped with purple hydathodes; rachis channelled. Petiole to 15cm, round, usu solid, with 3-5 vb's (many in* G. maderense*). Stipules ovate-triangular, acute. Plant often fetid. Hairs septate*

■Lvs >25cm diam. Petiole minutely nonglandular-hairy (hairless TNE)

Lvs minutely adpressed-hairy (hairless TNE), usu with reddish veins, with stomata below only. Petiole >30cm, reddish. Stem to 1(2)m, 5cm diam. All yr. R alien, mostly SW Br ....*Giant Herb-Robert* **Geranium maderense**

■Lvs ≤10cm diam. Petiole with patent hairs (often gland-tipped), or hairless

Lvs with stomata both sides. Bi. Anthers orange or purple. Petals 18-22mm

Lfts to 6cm, light green, with sparse patent stout hairs scattered both sides, often with dense minute curled hairs on veins above. Petiole to 15cm, reddish. Stems often hairless. Plant often odourless. All yr. VR alien..*Greater Herb-Robert* **Geranium rubescens**

Lvs with stomata below only

Bi (ann). Anthers orange or purple. Petals 8-14mm. Petiole sparsely to densely hairy, with red-tipped glandular hairs. Lfts with sparse stout hairs both sides. Stems with 0.4-1mm ± patent glandular hairs. Plant usu reddish. All yr. PL 16................*Herb-Robert* **Geranium robertianum**

Ann. Anthers yellow. Petals 5-9mm. Petiole hairless to sparsely hairy. Lfts hairless. Stems hairless or hairy. Plant often reddish. Oct-Aug (all yr). Rocky or shingle shores, railways, R ....*Little-Robin* **Geranium purpureum**

GER B – *Per. Lvs basal and on stems, usu 7-17cm diam, net-veined, with stomata below only. Petiole with 4-9 vb's. Hairs unicellular*

■Petiole hairless or with strongly adpressed retrorse 0.5mm hairs (± equal)

Lvs 3-5-lobed, antrorsely adpressed-hairy, net-veined. Petiole red. Stems adpressed-hairy to hairless, swollen at the nodes. Mar-Oct (all yr)...............*Knotted Crane's-bill* **Geranium nodosum**

■Petiole with strongly retrorse hairs (all ± equal)

Petiole with hairs 0.5-1mm. Lvs alt (the upper occ opp), cut to ⅔ way into 7 lobes, hairy both sides; lobes shallowly toothed with teeth 1.5-2x longer than wide, each tooth with a purple hydathode. Apr-Oct. Usu upland areas............................*Wood Crane's-bill* **Geranium sylvaticum**

Petiole with hairs 0.2-0.5mm. Lvs opp (but often unequal in size), cut to 4/5 way into 7 lobes, hairy both sides; lobes deeply toothed with teeth 3-4x longer than wide, each tooth with a purple hydathode. Apr-Oct. Gsld...*Meadow Crane's-bill* **Geranium pratense**

■Petiole with spreading hairs (often v unequal)

Plant strongly cat-scented, with viscid glandular hairs. Stems unbranched. Strongly rhizomatous

Lvs 5-10cm diam, dull or shiny above, softly hairy with long nonglandular hairs and with minute glandular hairs and sessile glands. Stem lvs 0-2. Petiole to 25cm, with patent hairs and minute sessile glands. All yr. PL 16.....................................*Rock Crane's-bill* **Geranium macrorrhizum**

Lvs to 5cm diam, dull above, with minute curled hairs above and abundant minute ± sessile hairs below. Stem lvs absent. Petiole to 25cm, with v minute curled hairs and minute glandular hairs (hairless TNE). Hortal, VR alien. *G. dalmaticum* x *macrorrhizum*
....**Geranium** x **cantabriense**

Plant odourless, with hairs not or hardly glandular. Stems usu branched. Tufted or shortly rhizomatous

Petiole with some hairs >1mm. Lvs not brown-blotched

Stems (and often petioles) red-spotted. Petiole sparsely long-hairy

Lvs alt, 8-15cm diam, 7(9)-lobed, hairy both sides, occ glandular-hairy below. Petiole with hairs patent to slightly retrorse, with sparse unequal long hairs 2(2.5)mm, more frequent hairs to 0.5mm and abundant v short hairs 0.1-0.2mm (occ glandular). Often all yr. PL 16
....*Dusky Crane's-bill* **Geranium phaeum**

Stems and petioles never red-spotted (but occ reddish). Petiole densely long-hairy

Lvs opp, to 12cm diam, 7-9 lobed, softly hairy both sides, weakly rugose. Petiole often reddish nr base, with long hairs to 2(3)mm (occ glandular and often v unequal). Hortal. *G. ibericum* x *platypetalum*. Apr-Oct. PL 16............*Purple Crane's-bill* **Geranium** x **magnificum**

Petiole hairs ≤1mm. Lvs occ brown-blotched

Lvs <10cm diam, 5(7)-lobed, hairy, the lobes toothed. All yr

Lvs shiny dark green above. Petiole hairs unequal, mostly 0.5-1(2)mm, with v short (0.1mm) hairs (glandular short hairs occ or absent)

Fls pink with faint purple veins.................................*French Crane's-bill* **Geranium endressii**

Fls pink with faint or bold purple veins. *G. endressii* x *versicolor*. PL 16
....*Druce's Crane's-bill* **Geranium** x **oxonianum**

Lvs ± shiny light green above. Petiole often less densely hairy than the above taxa

Fls white to pale pink with bold purple veins......*Pencilled Crane's-bill* **Geranium versicolor**

GER C – *Ann (but per in* G. sanguineum *and often in* G. pyrenaicum*). Lvs usu <7cm diam, with stomata below only. Petiole usu with 4 vb's. Often all yr.* ❶ *At least some hairs septate (all unicellular in other spp)*

▪Petiole ± hairless (occ with sparse patent hairs)

Lvs opp at least above, 1.5-4cm, cut ⅔ way to base into 5-7-lobes, with ± square terminal lobe, often with recurved margins, shiny green or red both sides (or with red spots in sinuses and vein junctions), hairless or with short stout hairs above. Petiole often red. ❶. PL 16 ....*Shining Crane's-bill* **Geranium lucidum**

▪Petiole with adpressed (strongly retrorsely curved) nonglandular hairs (hairs 0.3-0.6mm)

Lvs opp at least above, 2.5-5cm diam, often cut ± to base, with strongly adpressed hairs both sides. Stems with strongly adpressed hairs. Apr-Oct. Often disturbed calc habs ....*Long-stalked Crane's-bill* **Geranium columbinum**

▪Petiole with retrorse nonglandular hairs (hairs 0.3-0.8mm, slightly unequal)

Lvs usu opp above, 2-7cm diam, often cut ± to base *Cut-leaved Crane's-bill* **Geranium dissectum**

▪Petiole with ± patent hairs (hairs glandular or not)

Petiole usu with unequal red-tipped glandular hairs to 0.5(1)mm. Lvs opp, often with red sinuses

Lvs to 6cm diam, usu lobed ≤½ way to base, with central tooth of each lobe usu square. Petiole with additional ± sessile glandular hairs.....*Round-leaved Crane's-bill* **Geranium rotundifolium**

Petiole without red-tipped glandular hairs. Lvs opp or alt, without red sinuses

Rhizomatous per with horizontal rootstock. Stems geniculate. Petiole 1mm diam. Lvs opp

Lvs 2-6cm diam, cut ± to base into 5-7 lobes, with unequal stiff antrorsely curved white hairs both sides. Petiole with ± patent long (to 1.5(2.5)mm) hairs (often unequal), and medium (c0.5mm) hairs, and short hairs and minute glandular hairs (occ viscid) and ± sessile glands. All yr (reduced to small rosette in winter)...............*Bloody Crane's-bill* **Geranium sanguineum**

Tufted per (ann/bi) with taproot. Stems ± erect. Petiole 1.5-2mm diam. Lvs opp

Lvs 3.5-7(9)cm diam, cut to ½(⅔) way to base. Petiole with short ± equal glandular hairs, and short retrorse/patent hairs and occ long bristly hairs to 1.5mm. All yr ....*Hedgerow Crane's-bill* **Geranium pyrenaicum**

Ann. Stems erect. Petiole 1(1.5)mm diam

Lvs alt, 1-3(4)cm diam, cut to ⅔ way to base. Petiole with short (0.2-0.5mm) unequal glandular hairs, and abundant short (0.1-0.4mm) nonglandular hairs, and sparse long hairs to 1.5mm..*Dove's-foot Crane's-bill* **Geranium molle**

Lvs opp, 1-4cm diam, cut to ⅔ way to base. Petiole with short (0.2mm) ± equal patent hairs (occ retrorse or curved), occ with ± sessile glandular hairs. PL 17 ....*Small-flowered Crane's-bill* **Geranium pusillum**

Group IMP – Impatiens. *Hairless ann, with taproot. Basal lvs absent or not persisting, at least lowest stem lvs opp, all acuminate, usu weakly involute when young, (±) net-veined. Stipules absent or represented by glands. Stem usu 1 (but often many in* I. balfourii*), translucent, brittle, with wide hollow and abundant sap, without stomata. All vegetative parts with orange dye (often exuded on drying). May-Oct*

■Lvs with >15 teeth per side. Stipular glands present. Stem unbranched below

Lvs opp or 3(4)-whorled below

Lvs 6-15cm, lanc to elliptic, rounded or cuneate at base, serrate with (12)24-75 teeth per side, at least the lower teeth tipped with red glands, with stomata both sides. Petiole to 3.5cm, hollow. Stipular glands long, red. Stem to 200cm, stout, reddish, ridged, with ± swollen nodes, with reddish roots present at lower nodes (usu with antiseptic odour). Damp or wet habs. PL 10 ....*Indian Balsam* **Impatiens glandulifera**

Lvs alt (exc lowest pr)

Lvs serrate with (15)20-30 teeth per side (excluding 5-6 glandular cilia nr base), 5-15cm, elliptic-ovate, cuneate at base. Petiole to 6cm, winged nr apex. Stem 30-60cm, green, slightly ridged, with swollen nodes. Dry shady habs.........................*Small Balsam* **Impatiens parviflora**

Lvs serrate with 40-50 teeth per side (excluding occ glandular cilia nr base), 3-10cm, ovate to ovate-oblong, rounded at base. Petiole to 4cm, winged. Stem(s) to 100(200)cm, green, round, often reddish at nodes. VR alien..*Kashmir Balsam* **Impatiens balfourii**

■Lvs with <15 teeth per side. Stipular glands absent. Stems usu branched below

Lvs dull light green, crenate-serrate with 6-11(15) teeth per side (diminishing to glandular cilia nr lf base), 4-12cm, ovate-oblong, obtuse to acute, cuneate (occ ± cordate) with 2-5 strong cilia-like teeth at base, with stomata below only or sparse above. Petiole to 4cm. Stems 40-70cm, yellow-green, often reddish, often swollen above nodes, without white dots on inner surface. Fls yellow. Shady damp habs, mostly N Br.............................*Touch-me-not Balsam* **Impatiens noli-tangere**

Lvs dark (bluish-) green, serrate with (6)8-12 teeth per side (lower teeth glandular-ciliate), 3-9cm, ovate-oblong to elliptic-oblong, cuneate (to rounded) with 1-2(4) glands at base, with stomata below only. Petiole to 5cm. Stems 40-70cm, brownish, occ purplish, with strongly swollen nodes, slightly angled, with white dots on inner surface. Fls orange. Usu by rivers, S Br ....*Orange Balsam* **Impatiens capensis**

Group IVY – Hedera. *Evergreen shrub, climbing or trailing by means of adventitious rootlets at and between nodes. Stellate hairs or peltate scales present at least on young growth. Lvs mostly palmately 3-5 lobed, with white hydathode at each lobe apex, with veins raised above, with stomata below only. Petiole channelled or round, with 7-8 vb's in ring, the sap usu odorous. Buds to 4mm, reddish, with several scales. In* H. helix *(both ssp) the lvs of fl shoots are rhombic, entire and 3(5)-pli-veined from base (often with ± dichotomous veins)*

▪Stellate hairs spreading (occ adpressed on older lvs)

Lvs 4-8(10)cm, cordate at base, dark green above (often with paler veins), often hairless, usu 3-5-lobed to >½ way. Petiole usu bronzy-green. Stems usu green (occ red), densely hairy when young, with weakly acrid-scented sap. Hairs with (3)8-21 rays. PL 10, 15, 17 ....*Common Ivy* **Hedera helix** ssp **helix***

▪Stellate hairs adpressed, usu brownish (rays connate only at extreme base)

Lvs to 10 x 7-13cm, cordate at base, ± shiny green above, sparsely hairy when young, usu 3-5-lobed to <½ way. Petiole often red. Stems sparsely hairy when young, with sweetly pine-scented sap. Hairs with (6)8-25 rays

Plant climbing. W Br..................*Atlantic Ivy* **Hedera helix** ssp **hibernica***

Plant not climbing (or weakly so). Hortal..................*Irish Ivy* **Hedera 'Hibernica'**

▪Peltate scales adpressed, orange-brown (rays connate for ¼-⅓ length)

Lvs 6-20 x 8-20cm, ± cordate at base, often dull green above (or variegated), usu with scales below, usu not or weakly 3-5-lobed, entire or denticulate. Petiole green or reddish. Stems green. Scales with 15-25 rays..................*Persian Ivy* **Hedera colchica***

Lvs 8-15 x 6-15cm, rarely cordate at base, shiny light yellow-green above (or variegated), hairless, some usu shallowly 3-lobed, entire. Petiole reddish, hairless. Stems reddish, sap with acrid odour. Scales with (8)10-15(18) rays. VR hortal..................*Algerian Ivy* **Hedera algeriensis**

Hedera colchica scales

Hedera helix ssp *helix* hairs

Hedera helix ssp *hibernica* hair

Group LIM – Limonium. *Per. Lvs leathery usu with bristle-like mucro below at apex (at least when young), often viscid at weakly sheathing base of petiole, rolled when young, with cartilaginous margins, pitted with salt glands both sides, with midrib raised below.* Limonium binervosum *agg (rock sea-lavenders) have dense rosettes of lvs (with flat petioles) close to the ground, often forming low cushions (many of these taxa are endemic or of uncertain taxonomic status). Coastal, mostly cliffs, saltmarshes, shingle and dune slacks.* ❶ *Lvs without stomata above*

■Lvs 1-veined. S Wales (St David's Head, Pembroke)

Lvs 3-5 x 1.8(2)cm, oblanc or oblong-spathulate, obtuse. Petals 2-2.5mm wide ....*Rock Sea-lavender* (sp) **Limonium paradoxum**

Lvs 2-5 x 1.2cm, narrowly oblanc or linear-oblong, obtuse. Petals 1.2-1.5mm wide ....*Rock Sea-lavender* (sp) **Limonium transwallianum**

■Lvs 3-veined (at least proximally)

Lvs with additional weakly translucent pinnate veins. Evergreen

Lvs to 6cm, broadly obovate-spathulate, often without mucro, often with recurved margins. ❶. All yr. VR alien, E Sussex, Dorset..................*Rottingdean Sea-lavender* **Limonium hyblaeum***

Lvs with parallel veins only

Deciduous (Apr-Sept)

Lvs dying at fl, 2-4cm, obovate or lanc-spathulate, obtuse, often with mucro, 1-3(5)-veined. Stems <10cm, decumbent, dichotomously branched from nr base, scabrid. Dryish saltmarshes, Norfolk, Suffolk............................*Matted Sea-lavender* **Limonium bellidifolium**

Evergreen (*L. binervosum* agg)

S. Wales. Stems smooth below..............................*Rock Sea-lavender* (sp) **Limonium parvum**

Cornwall. Stems scabrid below..........................*Rock Sea-lavender* (sp) **Limonium loganicum**

Widespread. Stems scabrid below

Lvs (2)3-8cm, narrowly obovate-spathulate, obtuse, mucro often absent, with v weak lateral veins. Petiole with 3 vb's.....................*Rock Sea-lavender* (sp) **Limonium recurvum** (4 ssp)

Widespread. Stems smooth below

Lvs 2-12 x 0.5-2cm, obovate-spathulate to narrowly oblanc, with 0.1-0.6mm mucro ....*Rock Sea-lavender* (sp) **Limonium binervosum** (6 ssp)

Lvs 2-6 x 0.5-2.5cm, oblanc, with mucro to 0.6mm ....*Rock Sea-lavender* (sp) **Limonium procerum*** (3 ssp)

Lvs 1-6 x 0.5-2cm, obovate-spathulate, with mucro to 0.3mm ....*Rock Sea-lavender* (sp) **Limonium britannicum** (4 ssp)

■Lvs 5-9-veined (at least proximally). Evergreen

Lvs 6-12(20)cm, 5-9-veined, glaucous. Channel Is

Outer bract 2-2.5mm. Stems to 30cm. Lvs broadly obovate-spathulate, often viscid. Petiole broadly winged. Jersey.................*Broad-leaved Sea-lavender* **Limonium auriculae-ursifolium**

Outer bract 3-4mm. Stems to 20cm. Lvs broadly obovate-spathulate, often viscid. Petiole broadly winged. Alderney, Jersey..................*Alderney Sea-lavender* **Limonium normannicum**

Lvs to 5cm, 5-veined, not glaucous. Dorset

Lvs obovate-spathulate, with short (0.3mm) mucro. Stems rough or smooth ....*Rock Sea-lavender* (sp) **Limonium dodartiforme**

■Lvs pinnate-veined (occ obscure but never with parallel veins). Deciduous (Apr-Sept)

Lvs >2cm

Lvs hairy at least on veins below. Stems often hairy above. Hortal, R alien

Lvs 15-30 x 8cm, elliptic to obovate, shiny dark green above, with opaque veins. Petiole to 14cm. Stems 50-90cm...................................*Florist's Sea-lavender* **Limonium platyphyllum**

Lvs hairless. Stems hairless. Saltmarshes

Lvs 4-15(25)cm, elliptic to oblong-lanc, with recurved mucro to 5(8)mm, with 2° veins obscure and opaque, with stomata not visible to the naked eye. Petiole to 10cm, with ≥10 scattered vb's. Stems branched usu well above the middle. Absent from Ire ....*Common Sea-lavender* **Limonium vulgare***

Lvs 4-12(20)cm, oblong-lanc, with recurved mucro to 5(8)mm, with 2° veins obscure and opaque, with stomata larger and visible to the naked eye (esp below). Petiole to 10cm, with ≥10 scattered vb's. Stems branched from below the middle .... *Lax-flowered Sea-lavender* **Limonium humile**

Lvs <2cm

Lvs to 1.5 x 0.4cm, obovate or spathulate, with short (0.2mm) mucro, with recurved margins, papillate above. Petiole with 3 vb's. VR alien, E Sussex .... *Dwarf Sea-lavender* **Limonium minutum**

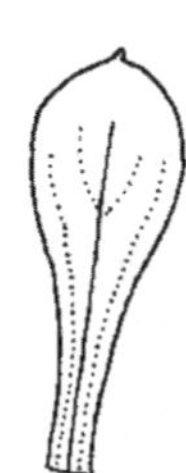

Limonium hyblaeum

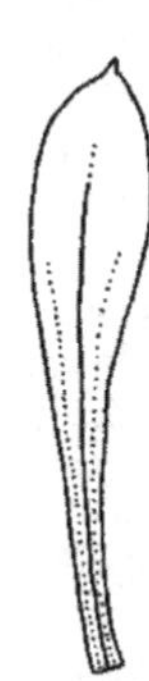

Limonium procerum

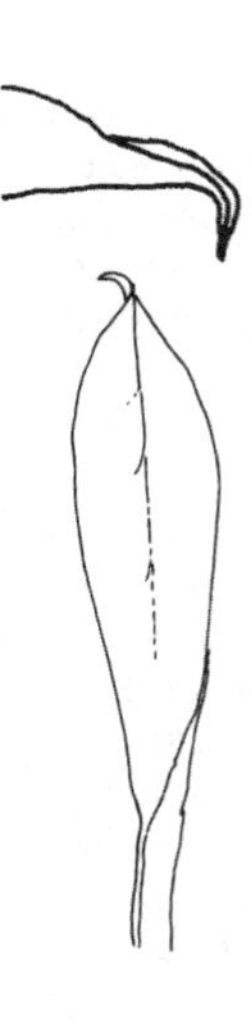

Limonium vulgare

Group MIM – Mimulus. *Per. Lvs opp, 1-11cm, ovate or orb-ovate, the lower petiolate, the upper sessile or perfoliate, all toothed, often net-veined, with 5-7 ± parallel veins (anastomosing) sunken above and raised below, with additional pinnate veins, with stomata both sides. Petioles connate at base (strongly so above), flat, with 5-7 vb's in a flat arc. Stems to 80cm, usu rooting at lower nodes and ascending, hairless at least below, often hairy above, round or with 2-4 ridges, often with swollen nodes, usu becoming hollow, with stele present as a broken ring. All yr (overwintering as rosettes).* M. moschatus *keys out elsewhere due to the densely glandular-hairy lvs and stems*

■Lvs with twisted teeth, with undulate margins. Fls yellow

Infl v sparsely glandular-hairy. Fls fertile. Usu lowland

Lvs narrow, acuminate, usu with few regular narrow teeth. Stems hairless. Fls usu with a red-brown blotch on the lower lobe (rarely on all five lobes). Bracts >1cm, ± lanc, toothed. Fl stalks hairless..*Blood-drop-emlets* **Mimulus luteus** var **rivularis**

Infl sparsely to densely glandular-hairy. Fls sterile. Often upland in N & W Br

Lvs usu longer than wide, 1-2-serrate with 7-9 long main teeth per side. Stem often reddish. Fls usu with unblotched lobes, the lowest lobe angled downwards. Bracts usu >1cm, ovate-lanc, toothed. Fl stalks usu ± hairless. Often upland in N & W Br. *M. guttatus* x *luteus* ....*Hybrid Monkeyflower* **Mimulus** x **robertsii**

■Lvs with teeth not twisted, margins undulate or not

Infl (esp fl stalks) densely glandular-hairy. Fls yellow, fertile. Usu lowland

Lvs to 10 x 8cm, undulate, 1(2)-serrate or dentate with teeth often longer and more irreg nr base (often ± lobed). Stems with yellow-tipped glandular hairs above (and nonglandular hairs). Fls usu with unblotched lobes, the lowest lobe held horizontally when young. Bracts <1cm, ± orb, mucronate, entire. Fl stalks densely hairy..........................*Monkeyflower* **Mimulus guttatus**

Infl ± hairless. Fls orange, sterile. Usu lowlands. *M. cupreus* x *guttatus* ....*Coppery Monkeyflower* **Mimulus** x **burnetii**

Infl sparsely glandular-hairy. Fls orange, partially fertile. Usu lowlands in N & W Br. *M. cupreus* x *luteus*...*Scottish Monkeyflower* **Mimulus** x **maculosus**

Group MYO – Myosotis. *Herb. Lvs with swollen white hydathode below at apex (dark when mature), often indistinctly petiolate, folded when young, with translucent midrib raised below, with stomata both sides. Stems strongly adpressed-hairy nr infl, round or angled. Hairs unicellular, white, usu swollen-based*

■Dry habs. Stems not rooting at nodes

Per. Lvs not opaquely net-veined, without submarginal vein

Basal lvs 1.5cm, oblong to oblong-lanc, acute to ± acute, antrorsely ± adpressed-hairy above, hairs often retrorse below (occ hairless), retrorsely ciliate. Upper lvs to 3cm, linear-lanc, sessile. Petiole 0.5-2cm. Stems 5-15cm, with spreading white hairs below. All yr. Calc mtn rocks ....*Alpine Forget-me-not* **Myosotis alpestris**

Bi (ann). Lvs obscurely opaquely net-veined, with opaque submarginal vein

Lvs in basal or aerial rosette, 4-8 x 2cm, ovate-spathulate, obtuse, bluish-green, spreading-hairy both sides. Stems 15-45cm. Usu shady habs.....*Wood Forget-me-not* **Myosotis sylvatica**

Ann. Lvs not opaquely net-veined, without submarginal vein

Lvs 1-6.5cm x 7-15mm, linear-oblong, obtuse (upper acute), with ± flat or recurved margins, hairy both sides. Stems 15-40cm, >1mm diam, with spreading or erecto-patent hairs. Oct-Aug (all yr)..*Field Forget-me-not* **Myosotis arvensis**

Lvs 1-4cm x 4-7mm, oblong-lanc, obtuse (upper acute), flat, hairy both sides. Stems to 10(30)cm, 1-2mm diam, with erecto-patent or adpressed hairs nr base. Oct-Jun ....*Changing Forget-me-not* **Myosotis discolor**

Lvs 0.5-1.5cm x 3-6mm, ovate-spathulate, obtuse, flat, hairy both sides. Stems to 8cm, 1mm diam, with spreading hairs nr base. Oct-Jun..........*Early Forget-me-not* **Myosotis ramosissima**

■Wet habs. Stems often rooting at nodes

Stems with spreading hairs (to 1.5mm) at least below (but above water-line). Stoloniferous per

Lvs with strongly adpressed short antrorse hairs esp above (occ ± hairless), ciliate, 4-10 x 1-2cm, oblong-lanc, usu obtuse, often indistinctly opaquely net-veined, with submarginal vein. Stems to 60cm, often sparsely hairy. Bracts all below lowest fl. All yr ....*Water Forget-me-not* **Myosotis scorpioides***

Lvs with weakly adpressed to spreading hairs both sides (occ sparse), ciliate, 2-5 x 0.5-1.5cm, ovate-spathulate to oblanc, obtuse, occ indistinctly opaquely net-veined, with submarginal vein. Stems to 50cm, often densely hairy. Bracts usu above lowest fl. All yr. Acidic habs. ....*Creeping Forget-me-not* **Myosotis secunda**

Stems with antrorsely adpressed hairs below. Lvs with strongly adpressed hairs (exc *M. sicula*)

Stoloniferous per. Lvs usu ≤2x longer than wide

Lvs 1-2 x 1cm, ovate-lanc, obtuse or emarginate, dark bluish-green. Stems to 20cm. All yr. Uplands, N Eng, S Scot...*Pale Forget-me-not* **Myosotis stolonifera**

Ann (occ bi in *M. laxa*). Lvs usu ≥3x longer than wide

Lvs 1-8 x 1cm, lanc, obtuse, hairy both sides. Stems to 40cm. Bracts usu above lowest fl. Often all yr...*Tufted Forget-me-not* **Myosotis laxa** ssp **caespitosa**

Lvs 1-2(6) x 0.8cm, oblong-spathulate, sparsely hairy above, (±) hairless below. Upper lvs hairy both sides. Stems to 15cm. Bracts all below lowest fl. Jersey, VR ....*Jersey Forget-me-not* **Myosotis sicula**

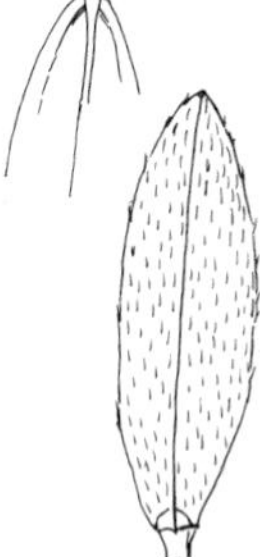

Myosotis scorpioides lf underside & upperside

Group OENO – Oenothera. *Taprooted bi (per in* O. rosea*). Basal lvs dead at fl, cuneate at base, often undulate or rugose, odourless, hairless or unicellular-hairy, remotely denticulate or hydathode-toothed, usu with whitish 2° veins fading nr margins (eucamptodromus) and raised below, with stomata both sides. Petiole often slightly winged, not or hardly channelled, with flat arc to involute vb. Stem usu 1 (often several in* O. rosea*), ± round. Fls yellow, often fading to orange-red (always pink in* O. rosea*). All yr. Dry habs. Fertile hybrids frequently occur*

■Lvs usu 10x as long as wide. Stems without red bulbous-based hairs

Lvs to 18 x 1.8cm, linear-lanc, often undulate, minutely crisped-hairy both sides to hairless, ciliate, with reddish midrib. Stems 25-100cm, red esp above, ± densely hairy with long patent and short crisped hairs. Mostly coastal, esp dunes.................*Fragrant Evening-primrose* **Oenothera stricta**

■Lvs <6x as long as wide

Stems with red bulbous-based hairs

Stems to 180cm, often red, with long patent and short crisped (often antrorse) hairs. Infl rachis reddish above, glandular-hairy. Lvs 7-25 x 2.5-5cm, mostly elliptic, often twisted or crinkled, with minute adpressed hairs above, ± patent-hairy below, with whitish veins (occ reddish). Petiole to 7cm. Sepals reddish..........*Large-flowered Evening-primrose* **Oenothera glazioviana**

Stems usu <100cm, green, with long patent and short crisped (often antrorse) hairs. Infl rachis green above, glandular-hairy. Lvs 7-15 x 2-5cm, obovate to linear-lanc, flat, with antrorsely adpressed or curved hairs both sides, often with reddish veins (esp on basal lvs). Petiole to 5cm. Sepals green................................*Small-flowered Evening-primrose* **Oenothera cambrica**

Stems without red bulbous-based hairs

Stems to 150cm, >5mm diam, erect, not or sparsely branched, antrorsely crisped-hairy or sparsely so. Infl rachis green above, glandular-hairy (esp above). Lvs 7-18 x 2-4cm, usu elliptic to elliptic-lanc, often twisted or undulate, shortly hairy both sides, with whitish veins (occ reddish). Petiole to 5cm....................................*Common Evening-primrose* **Oenothera biennis**

Stems to 50cm, 2mm diam, decumbent to ascending, much-branched, sparsely antrorsely crisped-hairy. Lvs 2-5 x 0.5-2.5cm, ovate-elliptic, resembling *Epilobium* spp, minutely adpressed-hairy both sides (occ sparsely so), with 10 teeth per side, with 5 prs of whitish veins. Petiole to 2cm. Per. ❶. All yr. VR alien........................*Pink Evening-primrose* **Oenothera rosea**

Group POP – Populus. *Deciduous tree. Lvs involute when young (weakly so in white poplars). Lf characters refer to those of short shoots unless otherwise stated (those of long shoots can differ dramatically). Petiole >2cm. Twigs monopodial, round or angled, with pith 5-angled in TS; lf scars with 3 prs of traces. Terminal bud always present and larger than laterals (unlike* Salix *spp where the terminal bud is often absent or poorly developed); the terminal buds can be dissected to determine the sex of fls long before they have opened*

▪Lvs white-felted below, without cartilaginous margins, with stomata below only (*White poplars*) .... **POP A**

▪Lvs sparsely hairy to hairless below

Lvs green below (but paler than above), with obvious cartilaginous margins, with stomata both sides (below only in *P. tremula*) (*Black poplars* & *aspen*).. **POP B**

Lvs whitish below (much paler than above with strongly contrasting green veins), with indistinct cartilaginous margins, with stomata below only (*Balsam poplars*)... **POP C**

POP A – *White poplars. Lvs white-felted below. Petiole laterally flattened (or ± so). Twigs round, white-felted (at least nr apex of long shoots when young). Trunk white or yellowish, with dark diamond-shaped lenticels. Suckering*

▪Lvs palmately 3-5-lobed, ± persistently white-felted below

Lvs 4-8 x 3-8cm, ovate to orb, white-felted both sides, soon shiny dark green and hairless above, with weakly dentate lobes. Petiole 3-6cm, white-felted. Twigs becoming hairless and dark brown after 1st yr. Buds 4-5mm, ovoid, ± acute, woolly esp nr base, soon hairless and reddish-brown

Crown spreading.. *White Poplar* **Populus alba**

Crown fastigiate... *White Poplar* (cv) **Populus alba 'Pyramidalis'**

▪Lvs usu deeply toothed, white-felted below when young, soon ± hairless. Crown spreading

Lvs 3-8(10)cm, ovate to orb-ovate, dark green and hairless above, obtusely dentate with 4-7 teeth per side, occ shallowly lobed. Petiole to 3cm. Twigs soon hairless dark grey-brown, smooth. Buds 6-10mm, ovoid, ± acute, woolly at base, slightly viscid. Branches often ± drooping. *P. alba* x *tremula*... *Grey Poplar* **Populus** x **canescens**

POP B – *Black poplars & aspen. Lvs shiny green above, slightly paler below, turning black when dried, with obvious cartilaginous margins, serrate, with stomata above at least along veins above (stomata below only in* P. tremula*). Petiole laterally flattened at least nr apex. Buds viscid (at least when opening), often scented (but rarely as strong as balsam poplars). Cultivars, exc 'Plantierensis', 'Vereecken', 'Italica' and 'Gigantea', belong to the* P. x canadensis *(*P. deltoides x nigra*) group*

■Twigs (or petiole/lvs) at least sparsely hairy when young. (*P.* 'Marilandica' may rarely key out here)

Lvs with >12 teeth per side, with stomata above (at least along vein margins)

Petiole often with glands nr apex. Trunk without burrs. Crown narrow and regular

Lvs appearing v early, 7-10 x 10cm, truncate to shallowly cordate at base, bronze-red when young, hairless, ciliate, serrate with slightly hooked teeth. Petiole with minute (to 0.4mm) patent hairs on vigorous shoots. Twigs with minute patent hairs in 1st-2nd yr, round to angled. Buds 10-20mm, viscid, scented. *P. deltoides* 'Cordata' x *nigra* 'Plantierensis'. ....**Populus 'Robusta'**

Petiole without glands

Trunk usu with burrs. Crown spreading, with downcurved boughs upswept at tips

Lvs 5-10 x 9cm, broadly ovate to rhombic, acuminate, cordate to broadly cuneate at base, green or bronze when young, soon shiny dark green above, hairy to hairless, often ciliate, obtusely serrate with weakly hooked teeth. Petiole 3-7cm, with unequal patent hairs to 0.4mm, soon hairless. Twigs hairless to densely hairy when young, round. Buds to 10mm, the lateral adpressed, all narrowly ovoid, acuminate, shiny dark brown, hairless, viscid or not, balsam-scented. Lowland floodplains. PL 14 ....*Black-poplar* **Populus nigra** ssp **betulifolia**

Trunk usu fluted (without burrs)

Crown strictly fastigiate. Lvs 5-8 x 9cm, rhombic, often wider than long, hairy below, ciliate or not. Twigs hairy..**Populus nigra 'Plantierensis'**

Crown cylindrical, not strictly fastigiate. Lvs 5-9 x 7cm, rhombic, longer than wide, sparsely hairy below, sparsely ciliate. Twigs sparsely hairy to hairless.....**Populus nigra 'Vereecken'**

Lvs with 9-14 teeth per side, without stomata above. Petiole without glands. Trunk with diamond-shaped lenticels

Lvs 1.5-8(12)cm, ± orb to broadly ovate, usu obtuse, truncate to shallowly cordate at base, dull dark green above, paler below, occ silky-hairy when young, soon hairless, coarsely sinuate-dentate. Petiole 1.5-4(7)cm, hairless or sparsely hairy. Twigs persistently hairy, greyish, round, with flat longitudinal lenticels. Buds 5-10(14)mm, conical, sharply acute, shiny brown, hairless, viscid when opening. Suckering...*Aspen* **Populus tremula**

■Twigs (and petiole/lvs) hairless

Petiole without glands. Lvs not ciliate. Twigs round. Lateral buds mostly <8mm, adpressed

Crown narrowly fastigiate. Trunk usu fluted. Lvs usu 5-7.5cm, strongly rhombic, often wider than long, acuminate, usu cuneate at base, with c26 slightly hooked teeth per side. Petiole 5cm, flattened. Buds hairless. Fls male. VR planted.............*Lombardy-poplar* **Populus nigra 'Italica'**

Crown broadly fastigiate. Trunk usu fluted. Lvs >8cm, usu more triangular, strongly acuminate, usu cuneate at base. Fls female.................*Female Lombardy-poplar* **Populus nigra 'Gigantea'**

Petiole often with glands nr apex. Lvs ciliate when young. Twigs usu angular or ribbed. Lateral buds usu ≥10mm, spreading

Petiole with (1)2-3(4) glands. Lvs 8-14cm, usu longer than wide ....*Eastern Cottonwood* **Populus deltoides**

Petiole with 0-2 glands (often obscure). Lvs 4-10cm, as wide as long

Lvs 7-10cm, truncate to ± cordate at base. (*P.* 'Robusta' may key out here if hairs absent or overlooked)..**Populus 'Heidemij'**

Lvs mostly <7cm

Lvs mostly cuneate to broadly so at base

Fls female. Lvs 4-8cm, often as wide as long, broadly ovate, with twisted apex, cuneate at base (occ shallowly cordate nr twig apex), ± green when young. Crown wide and rounded. Branches spreading, wavy, pendent but upturned at apices. Twigs yellowish-grey, often strongly ribbed, rarely shortly hairy when young. Buds slightly viscid. *P. nigra* ssp *nigra* x 'Serotina'...**Populus 'Marilandica'**

Fls male. Lvs 4-8cm, longer than wide, ovate, with apex not twisted, broadly cuneate. Crown narrow, symmetrical. Branches initially ascending then curving gently upwards, not pendent. Twigs yellowish-grey, strongly ribbed. Buds often v viscid. *P.* 'Regenerata' x *nigra* 'Italica'..*Carolina Poplar* **Populus 'Eugenei'**

Lvs mostly truncate to ± cordate at base

Fls male

Crown spreading, usu symmetrical. Lvs appearing late but 1-2 weeks earlier than *P.* 'Serotina'

Lvs bronze when young, soon light to mid-green, with sharply twisted apex. Twigs light brown. Branches slightly wavy, ascending to begin with then curving downwards, or continuing to grow upwards at a wide angle to the vertical.................**Populus 'Gelrica'**

Crown spreading, often symmetrical, with diverging branches forming a fan- or goblet-shape. Lvs appearing v late

Lvs 6-10cm, broadly ovate, with apex not twisted, truncate to shallowly cordate at base, bronze when young, serrate with strongly hooked teeth. Petiole (4)6-10cm. Twigs shiny olive-grey. Buds 1-2cm, narrowly ovoid, acute, greenish-brown, viscid. Branches initially ascending (exc in lower crown), then curving upwards so that young branches are ± vertical. *P. nigra* ssp *nigra* x *deltoides*

Lvs pale yellow-green........................*Golden Italian-poplar* **Populus 'Serotina Aurea'**

Lvs shiny dark green above............................*Black Italian-poplar* **Populus 'Serotina'**

Crown strictly fastigiate. Trunk straight, usu vertical. Lvs similar to *P.* 'Serotina' ....**Populus 'Serotina de Selys'**

Fls female

Lvs shallowly cordate to truncate at base, to 7 x 7cm, glandular-serrate. Petiole c4cm. Twigs strongly ribbed. Buds v viscid. Crown spreading sideways and becoming asymmetric with age. Branches wavy and at wide angle to the trunk. *P.* 'Marlandica' x 'Serotina'...*Railway Poplar* **Populus 'Regenerata'**

Lvs cordate to cuneate at base, to 6 x 8cm, glandular-serrate. Petiole c3cm. Twigs ± round. Buds slightly viscid. Crown broad but remaining symmetrical. Branches ascending throughout their length...**Populus 'I-78'**

POP C

POP C – *Balsam poplars. Lvs whitish or greyish below (much paler than above with strongly contrasting green veins, occ rusty-brown in* P. balsamifera *and* P. trichocarpa*), turning black when dried, with indistinct cartilaginous margins, with stomata below only. Petiole ± round, never strongly flattened. Buds 1-2cm, narrowly ovoid, acuminate, shiny, viscid, strongly scented.* ❶ *Lvs rugose.* ❷ *Buds light green (brown in other spp)*

■Petiole hairless

Lvs 8-16cm, broadly ovate to ovate-lanc, rounded to broadly cuneate at base (occ ± cordate), ± leathery, dark green above, whitish-green or silvery below (occ rusty-brown), ciliate, midrib not reddish. Petiole 3-4cm, with 1-2 glands at apex. Twigs usu round when young, hairless. Buds 2cm. Suckering..*Eastern Balsam-poplar* **Populus balsamifera**

Lvs 7-12(15)cm, triangular-ovate, cordate to truncate at base, thin, light green above, pale green below, ciliate, often with reddish midrib. Petiole 3-7cm, often reddish, slightly flattened, with 1-2 glands at apex. Twigs hairless or with sparse minute hairs, usu ribbed or angled when young. Not suckering. *P. deltoides* x *trichocarpa*..................................*Generous Poplar* **Populus** x **generosa**

■Petiole with v minute hairs <<0.1mm (x20)

Petiole densely hairy, often with 2 minute glands nr apex. Crown not fastigiate. Branches usu long, with pendent twigs

Lvs 5-15(23)cm, ovate (-elliptic), acute, usu truncate at base, dark green above, occ rusty-brown below, sparsely hairy to hairless below, ciliate, shallowly obtusely glandular-serrate, the teeth not hooked. Petiole to 6cm, ± round to channelled above. Twigs sparsely hairy to hairless, usu ribbed or angled when young. Buds 10-15mm. Rarely suckering ....*Western Balsam-poplar* **Populus trichocarpa**

Petiole sparsely hairy, without glands. Crown narrowly fastigiate. Branches short, strongly ascending, with few pendent twigs

Lvs to 10 x 12cm, broadly ovate, acuminate, shiny dark green above, sparsely minutely patent-hairy below. Twigs angled or ridged when young. *P. balsamifera* x *trichocarpa* ....*Hybrid Balsam-poplar* **Populus 'Balsam Spire'**

■Petiole with longer hairs to 0.5mm

Petiole usu with 2 glands nr apex (or at base of lf), with dense short hairs and sparse longer hairs

Lvs 5-15cm, broadly ovate, abruptly cuspidate with flat apex, rounded to cordate at base, ± leathery, dark green above, sparsely minutely hairy below, ciliate, obtusely serrate. Petiole 3-7cm, slightly flattened. Twigs minutely hairy (occ hairless), angled when young. Buds 10-15mm, ciliate. Suckering. *P. balsamifera* x *deltoides*............................*Balm-of-Gilead* **Populus** x **jackii**

Petiole without glands, with sparse long hairs only

Lvs rugose, 5-10(17)cm, elliptic-oblong, with an abruptly twisted apex, ± cordate at base, ± leathery, dark green above, grey-green below, hairy along veins both sides. Petiole 2-4cm, round. Twigs hairy, round when young. Usu suckering. ❶......*Doronoki* **Populus maximowiczii**

Lvs not rugose, 4-12cm, ovate to ovate-rhombic, with apex not twisted, rounded to cuneate at base, thin, light green above, pale greyish-green below. Petiole >3cm, slightly flattened. Twigs often hairy, usu ribbed or angled when young. Not suckering. ❷. *P. laurifolia* x *nigra* 'Italica' ....*Berlin Poplar* **Populus** x **berolinensis**

Group PRU – Prunus. *Deciduous tree or shrub. Lvs toothed, the teeth tipped with a minute fragile claw-like hydathode, with stomata below only. Petiole often with glands (extra-floral nectaries) nr apex (occ on lowest lf teeth), channelled, with arc vb and spiral fibres. Stipules soon falling. Twigs often with short shoots. Buds often clustered at twig apex, hairy or hairless, with erose (or retuse) scales.* ❶ *Buds ciliate (may be hairy or hairless).* ❷ *Lvs strongly almond-scented when crushed*

■Lvs rolled when young (or vernation not apparent). Thorns occ present
Twigs hairy
Thorns often present. Suckering. Lvs 2-4cm....................................*Blackthorn* **Prunus spinosa**
Thorns absent (rarely present). Not suckering
Lvs 4-10cm, ovate or elliptic, acuminate, cuneate or rounded at base, dull or shiny above, patent-hairy at least below (esp along midrib proximally) to ± hairless, crenate-serrate with 35-50 teeth per side. Petiole (0.5)1.5-2cm, hairy, often with 1 pr of yellow-green glands (occ on lowest lf teeth). Stipules 8mm, linear. Twigs stout, usu with short shoots, green but turning purplish or grey/brown, ridged, with peeling bark. Buds 2-4mm, conical, acute, often hairy. ❶ ....*Wild Plum* **Prunus domestica***
Twigs hairless. Thorns absent. Not suckering
Lvs 3-7cm, ovate-elliptic or obovate, acute, rounded or broadly cuneate at base, purplish in some cultivars, hairless above, sparsely hairy along lower part of midrib below, crenate-serrate with c70 teeth per side. Petiole 0.5-1cm, usu with 1-2 glands nr apex (or on lowest lf teeth). Twigs shiny green, occ reddish above. Buds usu in terminal or lateral clusters, hairless exc at extreme base..*Cherry Plum* **Prunus cerasifera**

■Lvs folded when young. Thorns always absent
Twigs with strongly odorous inner bark. Twigs hairless (occ minutely hairy when young in *P. padus*)
Lvs 5-10cm, elliptic or obovate, acuminate, rounded or cordate at base, dull above, usu with minute translucent dots (x20, HTL), hairless (exc below with white hair-tufts in vein axils or sparse white hairs on midrib), serrate with 80-110 teeth per side (lowest not glandular), with 10-13 prs of distinct 2° veins raised below. Petiole 1-2cm, with 1 pr of small yellow-red glands at apex. Stipules lanc, fimbriate. Twigs shiny dark reddish-brown, round, often with short shoots, with lenticels. Buds single, adpressed, 2-10mm, conic-ovoid, acute, reddish-brown, hairless. Trunk with brown peeling bark. Not suckering......................*Bird Cherry* **Prunus padus**
Lvs 5-12cm, oblong-ovate to -lanc, cuneate or rounded at base, shiny above, without translucent dots, hairless but often with narrow row of orange hairs along midrib below proximally, shallowly serrate with 40-60(75) teeth per side (glands often on lowest teeth), with 9 prs of ± indistinct 2° veins not or hardly raised below. Petiole 1-2.5cm, with 1-3 unequal prs of small red glands. Stipules lanc, fimbriate. Twigs v dark, without short shoots, round, with horizontal flat lenticels. Buds single (occ in prs), 4mm, ± acute, dark reddish-brown, hairless. Not suckering ....*Rum Cherry* **Prunus serotina**
Twigs with odourless inner bark (lvs almond-scented in *P. dulcis*)
Twigs densely minutely hairy when young
Lvs to 5 x 4cm, broadly ovate to ± orb, rounded at base, usu with some hairs along midrib proximally below, shallowly crenate-serrate with 45-65 orange gland-tipped teeth per side. Petiole to 2cm, hairless, with 0-2 green or brown glands nr apex. Stipules 0.5-1.5mm, ovate, glandular-fimbriate. Buds often clustered at twig apex, 3.5mm, brown, hairy and green nr apex. Not suckering. VR alien..*St Lucie Cherry* **Prunus mahaleb***
Twigs hairless
Lvs with sparse adpressed or spreading hairs below (esp on veins and in vein axils below)
Lvs 6-16 x 5-8cm, obovate-elliptic, acuminate, cuneate or rounded at base, slightly viscid when young, often drooping, mid- to dark green above, dull at least below, hairless above, 2-serrate with 38-55 teeth per side, with 11-15 prs of 2° veins. Petiole 2.5-5cm, with 1(2) prs large red glands at apex. Stipules 6-12mm, linear, glandular-fimbriate. Twigs stout, 6-8mm diam, with stout short shoots. Buds in dense clusters of up to 9, 5-9mm, ovoid, acute, brown, hairless. Tree, non-suckering. Pl 9, 22............................*Wild Cherry* **Prunus avium**

Lvs 5-9 x 3.5-6cm, obovate-elliptic, acuminate, cuneate or rounded at base, viscid when young, spreading or erect, dark green above, shiny at least below, hairless or sparsely hairy below, 1-2-serrate, usu with <8 prs of 2° veins. Petiole 1-3cm, with 0-2 yellow or red sessile disc-like glands. Stipules to 10mm, linear, glandular-fimbriate. Twigs slender, 3-5mm diam, grey-brown, often drooping, usu without short shoots. Buds single or in clusters of 2-3 at twig apex, to 5(8)mm, ovoid, obtuse, chestnut-brown, hairless. Shrub, suckering .... *Dwarf Cherry* **Prunus cerasus**

Lvs with sparse adpressed hairs confined to midrib below

Lvs to 14 x 4cm, lanc, shiny above, serrate-crenate with c50 teeth per side, often gall-infested. Petiole to 1.5cm, with 1-3 sunken disc-like or curved glands on petiole/lower lf. Stipules 1cm, lanc, fimbriate. Twigs green, soon purple. Not suckering .... *Peach* **Prunus persica**

Lvs totally hairless

Lvs strongly almond-scented when crushed ❷

Lvs 7-12 x 3.5cm, dark shiny green above, dull below, shallowly bluntly crenate-serrate with 45-95 teeth per side, with midrib white below. Petiole 1.5-2cm, with 2-7 small reddish glands (occ orange or green when young). Stipules lanc, fimbriate. Twigs reddish on upper side, occ with short shoots. Buds 4mm, hairless, with free scale tips .... *Almond* **Prunus dulcis**

Lvs odourless when crushed

Lvs 9-18 x 5-9cm, obovate, acuminate, shiny above, v sharply serrate with c80 acuminate teeth per side, with >10 prs of 2° veins, midrib often reddish below. Petiole 2-4cm, with 1(2) prs large red glands (rarely disc-like). Stipules to 20mm, lanc, fimbriate, often lobed. Twigs stout, >5mm diam, with short shoots. Buds clustered at twig apex, to 8mm. Not suckering... *Japanese Cherry* **Prunus serrulata***

Lvs 5-9 x 3.5-6cm, obovate-elliptic, acuminate, shiny at least below, 1-2-serrate, usu with <8 prs of 2° veins. Petiole 1-3cm, with 0-2 yellow/red sessile disc-like glands. Stipules to 10mm, linear, glandular-fimbriate. Twigs slender, 3-5mm diam, usu without short shoots. Buds single or in clusters of 2-3 at twig apex, to 5(8)mm. Suckering .... *Dwarf Cherry* **Prunus cerasus**

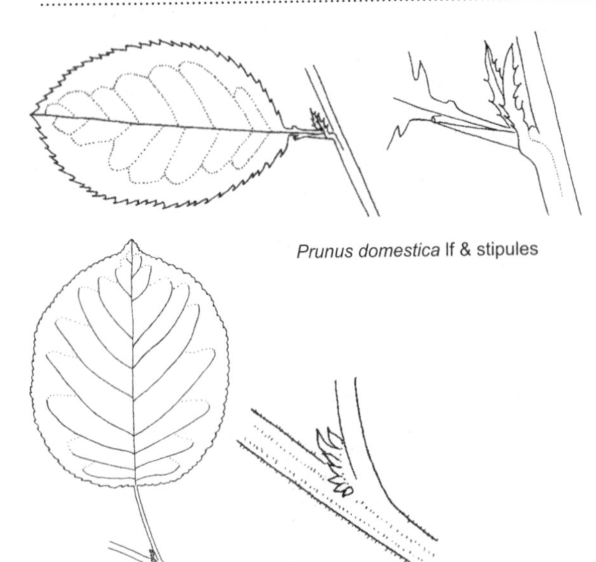

Prunus domestica lf & stipules

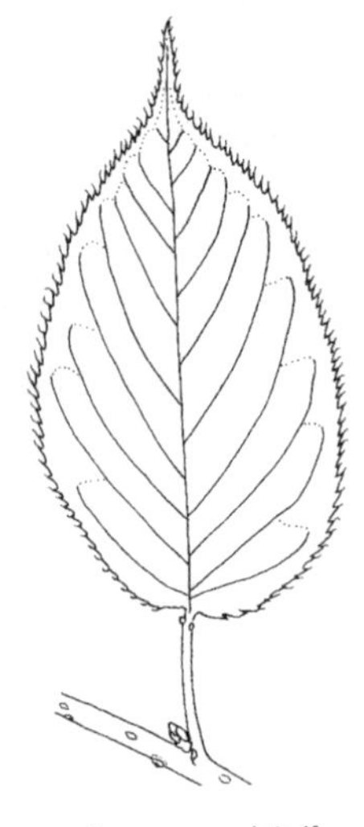

Prunus serrulata lf

Prunus mahaleb lf & stipules

Group PRUNE – Prunella. *Per. Lvs 1-8.5cm, ovate, truncate to cuneate (occ slightly winged) at base, applanate when young, dull green above, not or hardly pitted, with swollen bases of hairs visible as dark dots when HTL, with 0-8 teeth per side, with stomata both sides. Petiole to 2.5cm, with 1 vb. Stems to 30cm tall, rooting at lower nodes, deeply channelled on 2 sides. All yr*

■Lvs 4-8.5cm. VR alien..*Large Selfheal* **Prunella grandiflora**

■Lvs 1-5cm

Upper lvs entire to shallowly dentate, 2-5 x 3cm, ovate, sparsely hairy. Lower lvs ovate, entire or toothed. Stems sparsely hairy. PL 11..*Selfheal* **Prunella vulgaris**

Upper lvs with 1(2) prs of lobes nr base, 1-2.5 x 1cm, lanc, densely hairy esp below. Lower lvs oblong-lanc, often entire. Stems densely hairy....................*Cut-leaved Selfheal* **Prunella laciniata**

Group RAN – Ranunculus, Trollius, Caltha. *Herb. Lvs entire, toothed or with 3-7 main lobes (often ± divided to base). Petiole sheathing at base, usu channelled, with vb's usu turning brown. Sap acrid and often caustic to touch*

■Lvs lanc to ovate, not orb, unlobed but weakly toothed with hydathodes................................**RAN A**
■Lvs orb AND/OR lobed
Plant (±) hairless..**RAN B**
Plant hairy to densely so...**RAN C**

RAN A – *Lvs lanc to ovate, not orb, unlobed, weakly toothed with white (turning dark) hydathodes sunken along margins above (best viewed end-on). Stems hollow. Roots fibrous. Wet habs.* ❶ *Lvs grey-green*

■Per
Lvs 8-25 x 3-5(8)cm. Stems soon rooting at lower nodes, ascending to 120cm tall
Basal lvs (visible Oct-Apr) often submerged, ovate to ovate-oblong, cordate at base, long-petiolate, rolled when young, shiny green, remotely denticulate with hydathodes, with opaque or translucent pinnate or parallel veins (occ obscure), with midrib raised below. Stem lvs 2-ranked, oblong-lanc, acute to acuminate, ± sessile, ± clasping, grey-green, often reddish below, often adpressed-hairy on veins below, adpressed-ciliate. Petiole laterally flattened, channelled, becoming hollow, with 5 tiny vb's. Stems reddish-green mottled, occ flattened. ❶. Oct-Sept ....*Greater Spearwort* **Ranunculus lingua**
Lvs 2-6 x 0.5-2cm. Stems occ rooting at lowest nodes, ascending to 50cm tall
Basal lvs subulate to ovate, cuneate to cordate at base, hairless or adpressed-hairy below, often sparsely ciliate, occ opaquely net-veined, with 3-7 opaque to translucent ± parallel main veins (not anastomosing), also pinnate veins, with midrib raised below, with stomata both sides. Petiole 2-8cm, slightly laterally flattened or grooved above, with 3(5) indistinct vb's within aerenchyma. Stems purplish, round. All yr................*Lesser Spearwort* **Ranunculus flammula***
Lvs 0.7-3 x 0.15cm. Stems soon rooting at nodes, prostrate, to 10cm tall. VR, Scot
Lvs linear to filiform. Stems 0.2-1mm diam, thread-like. All yr
....*Creeping Spearwort* **Ranunculus reptans**
■Ann. Lvs 1-2 x 1.2cm. Stems often rooting at lower nodes, to 30cm tall. VR, Gloucs
Basal (or submerged) lvs ovate to ± orb, cordate at base, long-petiolate, pale yellow-green, often obscurely toothed with 3-5 hydathodes, occ 3-5-lobed, with obscure veins. Stem lvs lanc to linear-lanc. Petiole channelled, with 2 small hollows. Jun-Sept. Sch8
....*Adder's-tongue Spearwort* **Ranunculus ophioglossifolius**

Ranunculus flammula

RAN B – *Plant (±) hairless*

■Per

Petiole with 1 large hollow (occ pith-filled in *Caltha*)

Wet sunny or shady habs. Basal lvs 3-30cm diam, ± reniform, petiolate, shallowly lobed, ± shiny green above, paler below, hairless, crenate, rarely serrate-dentate, with white hydathodes visible above or end-on, often net-veined (veins translucent or opaque), with translucent 2° veins occ raised below, with stomata both sides. Petiole purplish at base, sheathing base inflated like a ligule or stipules but forming an ochrea on stem lvs, not or weakly channelled above, with c12 vb's. Stems to 40cm, hollow. Shortly rhizomatous. Usu all yr ....*Marsh-marigold* **Caltha palustris**

Wet mtn pastures. Basal lvs 3-11cm diam, pentagonal, petiolate, palmately 3-5 lobed, dark green above, paler below, hairless, the lobes ± deeply cut and serrate, often net-veined, with 2° veins raised below, with stomata below only. Stem lvs ± sessile, usu narrowly 3-lobed. Petiole not purplish at base, round, obscurely channelled, with 6-10 vb's. Stems to 60cm, hollow. Tufted. Mar-Oct..........*Globeflower* **Trollius europaeus**

Petiole with 2 hollows. Shady habs, often damp

Lvs basal and/or on stems, 1-4cm diam, ± entire to angled or weakly lobed, crenate, hairless, cordate at base, occ ± fleshy, shiny dark green above often with darker or pale markings, paler below, with white hydathodes above or end on, occ net-veined, with stomata both sides. Petiole with translucent channel. Stems to 25cm, hollow, often rooting at decumbent base. Roots with spindle-shaped tubers. Jan-Jun. PL 13..........*Lesser Celandine* **Ranunculus ficaria**

Petiole pith-filled or with 1-several hollows. Shady dry calc habs

Lvs mostly basal, 1.5-5cm diam, reniform to ± orb, petiolate, undivided or palmately divided into 3(5) lfts or deep lobes; lfts ± dull dark green above, paler below, hairless or with a few hairs, minutely ciliate, often deeply lobed, crenate, not net-veined (but occ opaquely so), with indistinct 2° veins not raised below, with stomata both sides or below only. Stem lvs sessile, ± deeply 3-5(7)-lobed. Petiole sparsely hairy to hairless, channelled. Stems to 40cm, hollow. Rootstock stout, with many fibrous roots. (Nov) Mar-Jun ....*Goldilocks Buttercup* **Ranunculus auricomus**

■Ann

Dry arable habs

Basal lvs 1.5-4cm, orb-ovate, (±) hairless, shortly ciliate, 3-lobed to ≤½ way, with lobes toothed. Stem lvs similar to basal. Petiole 3-10cm. Frs with 0.5-1mm spines. VR, W Cornwall, Scilly Is ....*Rough-fruited Buttercup* **Ranunculus muricatus**

Basal lvs 1.5-3(4)cm, obovate, sparsely hairy to hairless, usu shortly ciliate, with 2-3 distal teeth per side, occ appearing parallel-veined. Stem lvs with 3 lfts, the lfts lobed or toothed, occ 2-3-ternate with linear lfts. Petiole to 5cm, channelled, usu hollow. Frs with 1-2mm spines. R ....*Corn Buttercup* **Ranunculus arvensis**

Damp muddy habs (the first ± unlobed lvs of *R. sardous* may key out here)

Basal lvs 4-12 x 7cm, reniform or pentagonal, cordate to truncate at base, shiny mid-green both sides, hairless, 3(5)-lobed (lowest lvs often ± entire), with ± obscure opaque veins often slightly raised both sides, with stomata both sides; lobes often 2-3-lobed, crenate with sunken white (purple-edged) hydathodes often in retuse indentations. Stem lvs often 3-foliate. Petiole round but channelled, solid, with 3(7) small vb's in aerenchyma. Stems to 60cm, rarely rooting at lower nodes, occ furrowed, hollow. (Nov) Mar-Nov ....*Celery-leaved Buttercup* **Ranunculus sceleratus**

Basal lvs to 1.5(2) x 1.2cm, ovate to ± orb, cordate at base, pale yellow-green, (±) hairless, entire or with 3-5 indistinct dark hydathodes along margins, occ 3-5-lobed, with obscure opaque veins, with stomata both sides. Stem lvs elliptic, toothed. Petiole channelled, with 2 small hollows. Stems to 60cm (usu much less), often rooting at lower nodes, round, hollow. Jun-Sept. VR. Sch8..........*Adder's-tongue Spearwort* **Ranunculus ophioglossifolius**

RAN C – *Plant hairy to densely so*

■Tufted per with globose bulb-like tuber (exc in 1st yr)

Basal and lower stem lvs 2.5-10cm, with ± adpressed or spreading antrorse hairs (hairs never bulbous-based), usu with 3 main lfts, the middle lft long-stalked, the laterals short-stalked or sessile, all further divided or lobed. Petiole with long adpressed hairs, channelled, soon ± with 1-2 hollows, with 3 vb's. Stems to 40cm, branched, often adpressed-hairy. Oct-Sept ....*Bulbous Buttercup* **Ranunculus bulbosus**

■Tufted per with spindle-shaped tubers

Lvs mostly basal, ± adpressed-hairy, shallowly 3-lobed. Stem lvs 3-foliate (occ pinnately lobed), the middle lft long-stalked, all divided into narrow dentate lobes. Petiole channelled, solid. Stems to 40cm, simple or weakly branched, silky-hairy. Damp gsld, Jersey. Oct-Jun ....*Jersey Buttercup* **Ranunculus paludosus**

■Per or ann without tubers (stolons developing in *R. repens*)

Per

Stolons always absent. Lvs 2.5-10cm, palmately 3-7-lobed (often resembling *Geranium* spp), the lobes further cut or lobed, without pale sinus marks, spreading-hairy. Petiole 3-20cm, channelled, soon hollow, with 5 vb's. Stems to 75cm, with ± dense spreading or reflexed hairs below; all hairs turning brown. Oct-Sept. PL 17................*Meadow Buttercup* **Ranunculus acris**

Stolons developing (Jun-Sept). Lvs 1.5-10cm, divided into 3 lfts or lobes, with pale sinus marks, spreading-hairy, the middle lft often long-stalked, laterals short-stalked or sessile, all lobed, occ net-veined, with stomata both sides. Petiole with spreading or ± antrorse hairs, hairless at sheathing base, soon with 1-2 irreg hollows, with 3(5) vb's. Stems to 50cm. All yr. PL 17 ....*Creeping Buttercup* **Ranunculus repens***

Ann, difficult to uproot

Damp habs. Basal lvs 2-6cm, 3-foliate or 3(5)-lobed (the first lvs simple and toothed) and lobed or toothed again, densely ± adpressed-hairy to hairless, with stomata both sides, the middle lft long-stalked, often ± deeply 3-lobed and toothed. Petiole with antrorse or spreading hairs, usu hairy at sheathing base, with 0-2(4) small hollows, with 3 vb's. Stems 3-5(7), to 40cm, ascending or decumbent, with spreading or adpressed hairs. Oct-Aug ....*Hairy Buttercup* **Ranunculus sardous**

Ann, easily uprooted (recently introduced forms of *R. arvensis* may key out here)

Dry disturbed habs. Lvs 0.5-2.5cm, (3)5-lobed or shallowly so and deeply toothed (first lvs lobed only), rounded-cordate at base, yellowish-green, softly hairy, not net-veined. Petiole spreading-hairy, hairy or ciliate at sheathing base, shallowly channelled, solid or with 2 hollows. Stems 4-6, to 20(30)cm, often prostrate. Oct-Aug ....*Small-flowered Buttercup* **Ranunculus parviflorus***

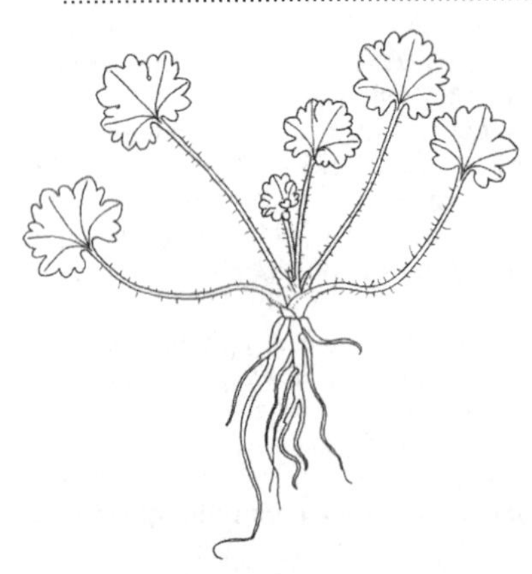

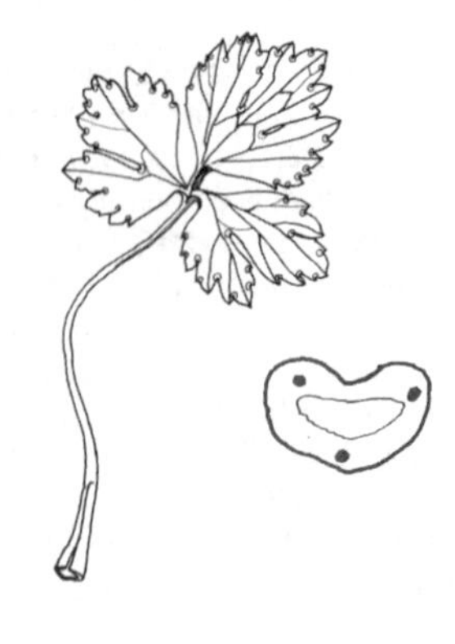

Ranunculus parviflorus

Ranunculus repens lf & petiole TS

Group RAN-BAT – Ranunculus *subgenus* Batrachium *(*Ranunculus aquatilis *agg). Lvs often of two kinds; floating and submerged. Floating lvs palmately lobed, reniform to ± orb, with obscure veins not sunken above or raised below, with stomata above only. Submerged lvs thread-like, ternately divided into 3's, the segments 1mm diam, obtuse, usu tipped with 1-several bristles. Intermediate lvs (when present) are usu submerged and ± palmately lobed. Petiole with stipule-like auriculate sheathing base (tubular when young), round to triangular, ± solid to hollow. Stipules absent. Stems (±) round, soon developing 2-4(8) roots at nodes*

▪Both submerged and floating lvs present (only submerged lvs in winter). Stems hollow
 Floating lvs 3-lobed (cut >⅔ way to base), usu hairy below. (*R. trichophyllus* may key out here)
  Floating lvs 2-5cm, reniform to ± orb; middle lobe often shorter (and narrower) than lateral lobes, cuneate at base, crenate to dentate. Intermediate lvs occ present. Petals 5-10mm. Ann to per. Often all yr. Often brackish water......................*Brackish Water-crowfoot* **Ranunculus baudotii**
  Floating lvs 0.5-2cm, usu reniform; middle lobe usu narrower than lateral lobes, cuneate at base, entire or crenate. Intermediate lvs often present. Petals <6mm. Fr stalks erect, ± straight. Ann to per. New Forest. *R. omiophyllus* x *tripartitus* ....*New Forest Water-crowfoot* **Ranunculus** x **novae-forestae**
  Floating lvs 0.5-1.5cm diam, usu reniform; middle lobe narrower than lateral lobes, cuneate at base, entire or crenate. Intermediate lvs usu absent. Petals <5mm. Fr stalks strongly recurved. Ann. Muddy habs...*Three-lobed Crowfoot* **Ranunculus tripartitus**
 Floating lvs usu 5-lobed, usu hairless below
  Streams or ponds. (*R.* x *novae-forestae* may occ key out here)
   Floating lvs 3-5cm, reniform to orb; lobes broadly cuneate at base, usu cut <⅔ to base, crenate or dentate. Submerged lvs (3)4-6x divided, the segments usu divergent, rigid or flaccid, with 2 minute bristles at apices. Intermediate lvs occ present. Ann or per
    Petals usu >10mm, with pear-shaped nectary pit. PL 10 ....*Pond Water-crowfoot* **Ranunculus peltatus**
    Petals <10mm, with circular nectary pit........*Common Water-crowfoot* **Ranunculus aquatilis**
  Acidic rapid-flowing rivers. R, mainly W Eng, Wales, Ire
   Floating lvs 1-2.2cm, reniform to orb; lobes broadly cuneate at base, usu cut <⅔ way to base, crenate. Submerged lvs 7-22cm, longer than adjacent stem internode, with 100-150 segments, the segments ± parallel, flaccid. Intermediate lvs absent. Per ....*Stream Water-crowfoot* **Ranunculus penicillatus** ssp **penicillatus**

▪Only floating lvs present. Stems hollow
 Lvs usu lobed <½ way, hairless below. Ann (per)
  Lvs 1.5-4cm diam, reniform to ± orb, usu 3-5-lobed (lobes almost touching), with lobes narrowest at base (broadest above, with narrow acute sinus). Petiole often 3-6x lf length, solid. Ann (per). Mar-Oct (all yr). Muddy habs or shallow water ....*Round-leaved Crowfoot* **Ranunculus omiophyllus**
  Lvs 0.4-1.8cm diam, ivy-shaped to reniform, usu 3-5-lobed, with lobes widest at base (with broad ± obtuse sinus). Petiole to 3x lf length, with 1-2 hollows. Ann (per). Mar-Oct (all yr). Muddy habs...*Ivy-leaved Crowfoot* **Ranunculus hederaceus**
 Lvs usu lobed >½ way, usu hairy below. Ann
  New Forest. Floating lvs 0.5-2cm, usu reniform, 3-5-lobed (occ <½ way); middle lobe usu narrower than lateral lobes, cuneate at base, entire or crenate. Intermediate lvs often present. Petals <6mm. Fr stalks erect, ± straight. *R omiophyllus* x *tripartitus* ....*New Forest Water-crowfoot* **Ranunculus** x **novae-forestae**
  Muddy habs, VR. Floating lvs 0.5-1.5cm, usu reniform and 3-lobed; middle lobe narrower than lateral lobes, cuneate at base, entire or crenate. Intermediate lvs usu absent. Petals <5mm. Fr stalks strongly recurved......................................*Three-lobed Crowfoot* **Ranunculus tripartitus**

▪Only submerged lvs present
 Plant often in brackish water. Stems hollow
  Lvs 3-5x divided, the segments rigid, divergent, with 2-3 bristles at apex ....*Brackish Water-crowfoot* **Ranunculus baudotii**

Plant in still or slow-moving freshwater. Stems hollow
  Petiole usu >0.5cm. Lvs ± orb in outline, with segments not in 1-plane
    Lvs (3)4-6x divided, the segments usu divergent, rigid or flaccid, with 2 minute bristles at apices
      Petals usu >10mm, with pear-shaped nectary pit. PL 10 ....*Pond Water-crowfoot* **Ranunculus peltatus**
      Petals <10mm, with circular nectary pit........*Common Water-crowfoot* **Ranunculus aquatilis**
    Lvs 4-6x divided, with short (1-2cm) rigid divergent segments, occ sparsely bristly, with 2-4 bristles at apices. Petals 3-6mm. Ann or per ....*Thread-leaved Water-crowfoot* **Ranunculus trichophyllus**
  Petiole <0.5cm. Lvs ± orb in outline, with rigid segments in 1-plane (like wheel-spokes)
    Lvs 3-4x divided, each segment with 2-3 bristles at apex. Per. All yr (lvs flaccid in winter). Permanent water, often eutrophic............*Fan-leaved Water-crowfoot* **Ranunculus circinatus**
Plant in moderate to fast flowing rivers. Stems solid or with small hollow
  Lvs mostly ≥5x divided
    Lvs <8(18)cm, usu shorter or ± equal to adjacent stem internode; segments 30-350, rigid or flaccid, divergent or ± parallel, sparsely bristly when young; ultimate segments to 2.5cm, 0.3-0.4mm diam, with (1)2-4 bristles at apices. Petiole 1.5-7(15)cm, sparsely bristly when young, often laterally flattened. Per. All yr
      Calc streams....*Stream Water-crowfoot* (ssp) **Ranunculus penicillatus** ssp **pseudofluitans**
      Acidic rapid-flowing rivers. R, mainly W Eng, Wales, Ire ....*Stream Water-crowfoot* **Ranunculus penicillatus** ssp **penicillatus**
  Lvs ≤4x divided
    Non-chalky rivers. Lvs to 30cm, often longer than stem internodes; segments few, mostly >8cm, firm to flaccid, ± parallel. Petiole to 22cm. Per ....*River Water-crowfoot* **Ranunculus fluitans**

Group SAL – Salix. *Deciduous shrub. Lvs often revolute or rolled when young (rolled lvs may appear revolute on drying!). Petiole short, with flat arc vb. Stipules persistent or soon falling (rarely absent), often larger and more persistent on vigorous shoots. Twigs sympodial, without short shoots (unlike* Prunus*), round or angled, usu flexible, with circular pith in TS (unlike* Populus*); lf scars narrow, with 3 traces. Buds usu adpressed to twig, ovoid, usu obtuse, with 1 scale that falls off as a conical cap, the terminal bud often absent or poorly developed, often 2 collateral buds on 2nd yr twigs.*

Many hybrids occur – only some of the more distinctive ones appear in the key. The following hybrids may readily be encountered; they are usu intermediate between the parents in all respects, or they may resemble one parent more closely: S. aurita x cinerea *(***S**. x **multinervis***)*, S. caprea x cinerea *(***S**. x **reichardii***)*, S. caprea x viminalis *(***S**. x **sericans***) and* S. cinerea x viminalis *(***S**. x **smithiana***)*

■Tree or tall shrub ≥1m, not rooting at nodes. Lvs usu >3cm
  Lvs ≤3x as long as wide, suborb to oblong.. **SAL A**
  Lvs >3x as long as wide, elliptic to linear.. **SAL B**
■Low shrub <1m, rooting at nodes. Lvs usu <3cm.. **SAL C**

SAL A – *Tree or tall shrub >1m, not rooting at nodes. Lvs ≤3x as long as wide, usu >3cm, ± orb to oblong, with stomata below only. To check for striae on wood, peel the bark off a 2nd yr twig*

■Lvs densely hairy (or with sparse rusty hairs) at least below

Lvs broadly ovate to suborb (≤1.5x as long as wide). Buds spreading from twig

Lvs often obtuse, rounded to ± cordate at base, ± undulate, thick, dull green. Petiole 0.8-1.5(2.5)cm, hairy, not or hardly channelled. Twigs greenish, sparsely minutely hairy when young, ± round, without striae on wood. Buds 5mm, shiny yellow or reddish, soon hairless

Lvs 5-12 x 2.5-8cm, shortly hairy to ± hairless above, shortly crisped grey-hairy to woolly below, weakly glandular crenate-serrate. Stipules 8-12mm, auricle-like, ± cordate at base, soon falling. Tree or shrub..*Goat Willow* **Salix caprea** var **caprea***

Lvs 3-7cm, silky-hairy both sides, entire, with brownish blotches. Stipules absent or soon falling. Small gnarled shrub or tree. Mtns, N Eng, Scot ....*Gangrene Willow* **Salix caprea** var **sphacelata**

Lvs usu obovate to oblanc (usu 2-3x as long as wide). Buds ± adpressed to twig

Lvs usu >3cm, occ rugose when young, usu with flat or recurved margins, usu with hairs below turning rusty-brown at maturity

Lvs 2-8(16) x 1-3(5)cm, rarely ± orb, often (±) acute, occ obliquely twisted at apex, cuneate at base, revolute when young, shiny green above, slightly glaucous below, soon hairless above, shortly hairy below, minutely serrate. Petiole to 1cm. Stipules usu small, auricle-like, often persistent. Twigs densely minutely hairy when young, with long weak scattered striae. Buds yellow or red, rusty-hairy...*Grey Willow* **Salix cinerea***

Lvs usu <3cm, always rugose, often with undulate margins, with hairs below not turning brown

Lvs 1.5-3(6) x 1-2.5cm, occ obtrullate, obtuse, often with twisted apiculate apex, cuneate at base, revolute when young, often with recurved margins at maturity, dull dark green above, green to glaucous below, sparsely hairy above, usu hairy below, serrate. Petiole 0.3-0.8cm. Stipules large, ± cordate to reniform, persistent. Twigs often branching at 90°, dark reddish-brown, minutely hairy, soon (±) hairless, with prominent striae. Buds occ slightly spreading esp nr twig apex, often reddish, minutely hairy to hairless. Often bogs ....*Eared Willow* **Salix aurita**

■Lvs (±) hairless below

Lvs glaucous below, rolled when young. Petiole without glands. Shrub

Stipules large, ovate, persistent (occ small or absent). Lvs 2-6.5 x 1.5-3.5cm, obovate to elliptic or oblong, usu acute, cuneate or rounded at base, ± thin, ± shiny dark green above, turning blackish, sparsely hairy when young, ± serrate (rarely ± entire). Petiole to 1cm. Twigs dull brown or greenish, densely hairy when young, soon (±) hairless, occ with a few distinct striae. Buds usu hairy...*Dark-leaved Willow* **Salix myrsinifolia**

Stipules usu v small or absent. Lvs 2-6 x 1-5cm, oblong to elliptic (rarely ± orb), acute to obtuse, cuneate or rounded at base, ± thick, leathery, shiny bright green above, not turning blackish, often sparsely hairy when young, serrate. Petiole to 1cm. Twigs shiny brown, usu hairless, usu without striae. Buds hairy...*Tea-leaved Willow* **Salix phylicifolia**

Lvs pale green below, involute when young. Petiole with 3-15 small sessile glands nr apex. Tree

Lvs 5-12 x 2-5cm, ovate to obovate-lanc, acute to shortly acuminate, rounded to cuneate at base, ± leathery, viscid and weakly aromatic when young, dark shiny green above, turning black, glandular-serrulate with 50-60 teeth per side. Petiole 0.5-1cm. Stipules minute, ± ovate, glandular-serrate, soon falling. Twigs shiny green to brown/reddish, hairless, usu without striae. Buds dark brown, viscid..*Bay Willow* **Salix pentandra**

Salix caprea var *caprea* bud

Salix cinerea bud

SAL B – *Tree or tall shrub >1m, not rooting at nodes. Lvs >3x as long as wide, usu >3cm, oblong to linear.* ❶ *Twigs pruinose.* ❷ *At least some lvs opp or subopp*

■Lvs hairless both sides at maturity

Petiole with 3-4 small glands at apex (at least on some lvs)

Lvs 4-10(15) x 1-3(4)cm, oblong-lanc to narrowly elliptic, acute to acuminate, rounded to cuneate at base, rolled when young, shiny (or dull) green above, green or glaucous below, glandular-serrate with 25-45 teeth per side, with stomata below only. Petiole 0.8-2cm, sparsely hairy. Stipules 5-10mm, ± ovate, often persistent. Twigs shiny green to olive-brown, hairless, often strongly angled or ridged when young, with inner bark tasting of rosewater. Buds to 6mm, flattened, yellow-green (occ reddish), soon hairless. Trunk with bark often flaking..................*Almond Willow* **Salix triandra**

Petiole without glands

Twigs pruinose ❶

Lvs (4)7-12(14) x 1-3(4)cm, oblong to linear-lanc, acuminate, cuneate, folded when young, shiny green above, glaucous below, with stomata both sides (often confined to vein margins above), shallowly glandular-serrate with 20-40 teeth per side. Petiole 0.7-2cm, hairy or hairless, channelled. Stipules narrowly ovate, glandular-serrate. Twigs shiny green or reddish, hairless, round, with bitter-tasting inner bark. Buds to 15mm, hairless or hairy

Lvs silky-hairy below when young. Stipules to 2cm, persistent. Buds dark red ....*European Violet-willow* **Salix daphnoides** var **daphnoides**

Lvs hairless (bronze when young). Stipules to 0.4cm, soon falling. Buds black ....*Siberian Violet-willow* **Salix daphnoides** var **acutifolia***

Twigs not pruinose

At least some lvs opp or subopp (at least proximally on twigs) ❷

Lvs 2-8(10) x 0.5-2(3)cm, linear or linear-oblanc, acute, cuneate at base, rolled when young, dark green above, usu glaucous below, often turning black on drying, occ sparsely woolly below when young, nonglandular serrate with c13 teeth per side nr apex, with stomata both sides. Petiole to 1cm. Stipules absent. Twigs yellowish or greyish, occ purplish, flexible, hairless, with bitter-tasting lime-yellow inner bark. Buds yellowish or reddish, hairless. Trunk with smooth greyish bark. PL 10...*Purple Willow* **Salix purpurea**

All lvs alt

Lvs to 12 x 3cm, oblong-lanc, ± acuminate, cordate at base, ± folded when young, occ purplish, (±) hairless both sides, shallow glandular-serrate with c70 teeth per side, with stomata below only. Petiole 2cm, hardly channelled. Stipules auricle-like, often persistent. Twigs hairy when young, round, with bitter-tasting inner bark. Buds hairless. VR alien ....*Heart-leaved Willow* **Salix eriocephala**

■Lvs with at least some hairs below at maturity

Petiole with 2-5 small glands nr apex (at least on some lvs). Lvs rolled when young, with stomata both sides. Tree

Lvs hairless above at maturity

Lvs 8-15 x 1.5-3cm, lanc, acuminate, cuneate at base, shiny dark green above, glaucous with sparse adpressed silky hairs below, glandular-serrate with 50-60 teeth per side. Petiole 0.5-1.5cm, hairless or sparsely hairy, channelled. Stipules 3-8mm, ± lanc, soon falling. Twigs olive-brown to yellowish, fragile at branch junctions, minutely hairy when young, soon hairless, angled, bitter-tasting. Buds to 6mm, flattened, reddish or yellowish, hairy to hairless. Trunk often pollarded, with furrowed bark..*Crack-willow* **Salix fragilis**

Lvs with sparse silky hairs above at maturity (dense when young)

Branches not drooping (but twigs may droop), not fragile

Twigs densely silky-hairy, soon hairless and shiny olive-brown, round. Lvs 5-12 x 0.5-2.2cm, lanc, acuminate, cuneate at base, usu glaucous below, densely silky-hairy below, shallowly glandular-serrate with up to 50 teeth per side (to ± entire). Petiole to 1.5cm, hairy, channelled at least nr base. Stipules to 5mm, linear-subulate, glandular-serrate, soon falling. Buds 8mm, dark brown, reddish or yellow, adpressed-hairy. Trunk often pollarded, with deeply fissured greyish-brown bark. PL 9..*White Willow* **Salix alba**

Branches strongly drooping ('weeping')

Twigs adpressed-hairy, yellowish, round. Lvs 6-15 x 1-2.5cm, lanc, acuminate, cuneate at base, glaucous below, sparsely silky-hairy below, shallowly glandular-serrate with 45-70 teeth per side along thickened margin. Petiole 0.8-1.3cm, often hairy, channelled. Stipules absent, or small and soon falling. Buds 7mm, pale green or yellow, adpressed-hairy. Trunk never pollarded, with rough fissured bark

Twigs fragile at branch junctions. R planted. *S. babylonica* x *fragilis* ....*Weeping Crack-willow* **Salix** x **pendulina**

Twigs not fragile. *S. alba* x *babylonica*..........................*Weeping Willow* **Salix** x **sepulcralis***

Petiole without glands. Lvs revolute when young, with stomata below only. Shrub

Lvs silky-hairy (with ± parallel hairs) below, 4-17(25) x 0.6-1.5(2.5)cm, linear to linear-lanc, acuminate, cuneate at base, with recurved and often undulate margins, dull or shiny green and sparsely minutely hairy above, entire. Petiole 3-10mm, hairy, channelled. Stipules to 10mm, linear to linear-lanc, often curved, soon falling. Twigs silky-hairy, soon hairless and ± shiny yellow- or olive-brown, round or ± so. Buds often obscured by swollen petiole, yellowish or reddish-brown, shortly hairy, soon (±) hairless. Bark dark greyish-brown, fissured ....*Osier* **Salix viminalis**

Lvs white-woolly (with tangled hairs) below, 5-15 x 0.4-0.8cm, linear, acuminate, cuneate at base, with revolute margins, shiny dark green above, turning black on drying, sparsely hairy to hairless above, slightly glandular-toothed (to ± entire). Petiole 3-5mm. Stipules usu absent. Twigs whitish-woolly, soon hairless and yellowish-brown or reddish, angled. Buds often obscured by adpressed petiole, yellow-green, sparsely hairy. Bark greyish-brown, fissured. Hortal, alien..*Olive Willow* **Salix elaeagnos**

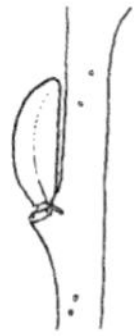

Salix daphnoides var *acutifolia* bud

Salix x *sepulcralis* bud

SAL C – *Low shrub <1m, rooting at nodes. Lvs usu <3cm, (±) net-veined (exc* S. arbuscula*).* ❶ *Buds balsam-scented*

■Lvs with stomata above (upper lf surface ± hairless)

Lvs dull glaucous below

Lvs 1.5-3(5) x 1-1.5(3)cm, ovate to elliptic, usu acute, cuneate at base, ± folded when young, shiny green above, turning black when dried, occ hairy above, densely adpressed-hairy below when young, soon hairless, obtusely glandular-serrate with c25 teeth per side. Petiole to 5(8)mm, hairy. Stipules usu absent. Twigs ± shiny dark reddish-brown, sparsely patent-hairy, soon hairless. Mtns, Scot..*Mountain Willow* **Salix arbuscula**

Lvs shiny green below

Lvs 2-7 x 0.5-3cm, oblong to obovate, acute to obtuse, rounded or cuneate at base, channelled when young, shiny dark green both sides but paler below, turning black (and persisting over the winter), adpressed-hairy at least on midrib below, ciliate, glandular-serrate with 20-35 teeth per side. Petiole to 10mm, channelled, soon hairless. Stipules small, oblong, usu persistent. Twigs green, soon shiny reddish-brown and hairless. Shrub <40cm tall. ❶. Uplands, Scot ....*Whortle-leaved Willow* **Salix myrsinites**

Lvs 0.3-2(3) x 0.3-2(3)cm, obovate to suborb, obtuse to retuse (rarely ± acute), usu rounded at base, ± channelled when young, shiny dark green both sides, sparsely white-hairy below, soon hairless, serrate with 12-16 hooked teeth per side. Petiole to 4mm, channelled, white-hairy, soon (±) hairless. Stipules minute or absent. Twigs dark shiny brown or reddish, sparsely ± adpressed-hairy, soon hairless. Buds sparsely hairy to hairless. Shrub <10cm tall. Mostly mtns ....*Dwarf Willow* **Salix herbacea**

■Lvs without stomata above (upper lf surface may be obscured by hairs)

Lvs white-woolly (with tangled hairs), at least below, when young

Lvs rugose. Extinct in wild. Hortal

Lvs to 2 x 2cm, orb, retuse, cordate at base, rugose, with recurved margins, shiny dark green above, ± glaucous below, sparsely hairy above, entire. Buds reddish. Slow-growing gnarled shrub. *S. lapponum* x *reticulata*..*Boyd's Willow* **Salix 'Boydii'**

Lvs not rugose. Mtns

Lvs 3.5-7 x 3-6.5cm, suborb to broadly ovate, usu obtuse (occ shortly mucronate), often cordate at base, ± flat, often soon ± hairless, greyish-green above, ± glaucous below, (±) entire. Petiole to 1.5cm. Stipules to 2cm, broadly ovate, persistent. Twigs sparsely woolly when young. Buds dark reddish-brown, soon hairless.....................*Woolly Willow* **Salix lanata**

Lvs 1.5-4(7) x 1-2.5cm, usu lanc to narrowly obovate, acute, cuneate or rounded at base, occ ± undulate, with recurved margins when young, adpressed-woolly above, occ ± hairless both sides, (±) entire. Petiole to 5(10)mm. Stipules small or absent. Twigs ± shiny dark reddish-brown, sparsely woolly to hairless. Buds shiny dark brown, hairless ....*Downy Willow* **Salix lapponum**

Lvs silky-hairy (with adpressed hairs) at least below, not rugose

Lvs 1-3.5 x 0.4-2.5cm, lanc to ovate-oblong, obtuse to acute or shortly mucronate (occ twisted at apex), cuneate to rounded, with recurved margins, turning black on drying, entire or with glandular teeth. Petiole usu <4mm. Stipules usu absent. Twigs yellowish- or reddish-brown, densely silky-hairy when young. Buds 2-3mm, yellow to dark red, silky-hairy..........*Creeping Willow* **Salix repens**

Lvs hairless or ± so below, strongly rugose

Lvs 1.2-4(5) x 1-2.5(4)cm, ovate to ± orb, obtuse, usu rounded at base, with ± recurved margins, dark green above, whitish-grey below, soon hairless above, entire to obscurely glandular crenate-serrate. Petiole 0.7-4cm, reddish, (±) hairless, channelled. Stipules usu absent. Twigs dark reddish-brown, with sparse long silky hairs when young. Buds reddish-brown, densely hairy, soon hairless. Mtns...*Net-leaved Willow* **Salix reticulata**

Group SED – Sedum *section* Sedum. *Evergreen low per herbs (exc* S. villosum, *which is a non-wintergreen per or bi). Lvs ± cylindrical and fleshy, imbricate at least at stem apices, entire.* ➊ *Shoots with hot peppery taste (tasteless in other spp)*

■Lvs glandular-hairy

Lvs alt, 6-12 x 2mm, linear-oblong, obtuse, flattened above, not glaucous, usu reddish, persisting. Stems to 7cm, unbranched, glandular-hairy. Jun-Sept. Mtns.....*Hairy Stonecrop* **Sedum villosum**

Lvs opp, 3-5 x 4mm, ovoid or obovoid, obtuse, slightly flattened above, glaucous, often pinkish, soon falling. Stems to 5cm, branched, papillate and glandular-hairy. Old walls ....*Thick-leaved Stonecrop* **Sedum dasyphyllum**

■Lvs hairless, alt

Stems papillate (often looking like glandular hairs). Lvs without spur

Lvs 6-12mm, ovoid to cylindrical (occ slightly flattened above), obtuse, spreading, bright green, occ reddish; lower lvs soon falling. Stems to 20cm....................*White Stonecrop* **Sedum album**

Stems not papillate. Lvs with short spur at base (<0.3mm, push lf against stem to see it)

Lvs 7-20mm, apiculate, linear to oblong-lanc

Lvs 2-2.5mm wide, reflexed (esp lower lvs), fleshy, slightly flattened above, slightly glaucous, not papillate; lower lvs soon falling. Stems to 30cm, nodding at apex ....*Reflexed Stonecrop* **Sedum rupestre**

Lvs 1mm wide, upswept or erect (esp upper lvs), ± fleshy, flat above, slightly glaucous, occ reddish, papillate esp on margins; lower lvs usu persisting as brown shrunken remains. Stems to 20cm, erect at apex.......................................*Rock Stonecrop* **Sedum forsterianum**

Lvs 3-6mm, obtuse

Lvs linear-cylindrical, green, ± spreading; lower lvs soon falling. Stems green. R alien ....*Tasteless Stonecrop* **Sedum sexangulare**

Lvs ovoid to ± globose

Lvs light green (occ reddish), imbricate; lower lvs persisting as whitish shrunken remains. Stems to 5cm, green. ➊. Basic habs..*Biting Stonecrop* **Sedum acre**

Lvs ± pale glaucous to reddish, spreading; lower lvs soon falling. Stems to 10cm, reddish. Acidic habs, often coastal or uplands.............................*English Stonecrop* **Sedum anglicum**

Group SOL – Solanaceae. *Plant with fetid roast beef odour (occ weak). Lvs all on stems, alt (oblique to branch axil) or in unequal-sized prs above (both lvs originating from same side of node, unlike truly opp-leaved plants). Petiole with 1 flat arc vb (unless otherwise stated). Stems often much-branched with widely spreading outline. Hairs septate or unicellular. (*Solanum tuberosum *and* Lycopersicon esculentum *key out under Group SK)*

■Scrambling or sprawling woody-based per

At least some lvs developing 2(4) lobes or lfts at base (otherwise entire), 4-8(11) x 2-8cm, ovate, acute-acuminate, rounded to cordate at base (occ hastate), slightly decurrent down petiole, occ undulate, occ purple-edged, hairless or shortly unicellular-hairy, ciliate, ± net-veined. Petiole 1-2cm. Stems 0.5-3(7)m, scrambling, hairy to hairless, ridged or angled. Buds 1-2mm, spherical to ovoid, brown. Var *marinum* has ± fleshy lvs and prostrate stems and is confined to shingle beaches in S Eng. Apr-Nov (usu overwintering as woody stems). PL 19 ....*Bittersweet* **Solanum dulcamara**

All lvs entire, to 5 x 4cm, ovate, ± obtuse, asymmetrically rounded at base, occ undulate, shortly unicellular-hairy both sides when young, occ with crisped cilia, not net-veined, with 2° veins weakly raised below, with stomata both sides. Petiole 1-2cm, crisped-hairy. Stems to 1.5m, sprawling, often with swollen-based crisped hairs, often square, with cartilaginous ridges. Buds absent. Rhizomatous. Mar-Oct. VR alien...............................*Cock's-eggs* **Salpichroa origanifolia**

■Erect per herb (occ woody at base)

Stems and lvs glandular-hairy. Mar-Oct

Lvs 8-20 x 5-8cm, ovate, acuminate, ± folded when young, sparsely minutely glandular-hairy (occ obscurely so) at least on veins below, rarely hairless, occ ciliate, ± net-veined, with white ± opaque midrib and 6-8 prs of translucent anastomosing 2° veins raised below, with stomata both sides. Petiole to 6cm, minutely glandular-hairy, slightly winged. Stems to 150cm, purplish at least nr base, glandular-hairy, round or slightly angled, solid, with stomata (absent in similar-looking *Phytolacca* spp). Usu calc scrub.........................*Deadly Nightshade* **Atropa belladonna**

Stems and lvs nonglandular-hairy to hairless

Rhizomatous. May-Nov

Lvs broadly cuneate to ± cordate at base, to 8 x 6cm, broadly ovate, acuminate, not rugose, mid-green, hairy (occ sparsely so), ciliate, entire to dentate, (±) net-veined. Petiole v slightly winged, channelled. Stems to 60cm, occ branched nr base, yellow-green, sparsely hairy, angled, with swollen nodes above................................*Japanese-lantern* **Physalis alkekengi***

Lvs cordate at base (often strongly so), to 6 x 4cm, ovate, shortly acuminate, rugose, dull dark green, densely minutely hairy, entire to dentate, net-veined, with purplish midrib and 2-3° veins raised below. Petiole v slightly winged, channelled. Stems to 100cm, purple-striped, densely minutely hairy, angled, with nodes not swollen. R alien ....*Cape-gooseberry* **Physalis peruviana***

Tufted. Lvs broadly cuneate at base. May-Nov (all yr)

Lvs to 12 x 7cm, light green, densely hairy with stout hairs, sinuate-toothed. Petiole to 5cm, winged, not channelled, with flat arc of several vb's and occ scattered vb's. Stems to 160cm, 6-8mm diam, ± woody at base. R alien..................*Tall Nightshade* **Solanum chenopodioides**

■Erect or ± prostrate ann. Jun-Oct

Stems and lvs nonglandular-hairy (at least when young)

Lvs deeply pinnately lobed

Lvs 2-5cm, ± thick, often with recurved margins, hairy below, sparsely ciliate, with translucent main veins, with stomata both sides. Petiole 0.5-1cm, with many scattered vb's. Stems 20-80cm, with stout antrorse unicellular or septate hairs (hairs usu v swollen and multicelled at base), ± round. VR alien..*Small Nightshade* **Solanum triflorum**

Lvs unlobed

Lvs usu with scattered dark purple-based pustulate bristles above

Lvs 5-20 x 2-14cm, broadly ovate, acuminate, folded when young, soon limp, dull above, usu hairless below, with purple-based scabrid cilia, irreg dentate, often net-veined, with 2° veins raised below, with stomata both sides. Petiole 2-6cm, slightly winged, minutely hispid above. Stem usu 1, 5-furrowed, usu with small hollow..............*Apple-of-Peru* **Nicandra physalodes**

Lvs without scattered purple-based bristles

Stems hollow, with swollen nodes. Lvs with stomata below only

Lvs to 8 x 5cm, ovate, cordate at base, hairless (exc for main veins esp above), limp, v thin, irreg toothed, not net-veined, with translucent 2° veins raised below. Petiole to 4cm, shortly hairy above, occ long-hairy, v narrowly winged, channelled, with 3 vb's (forming flat arc). Stems 15-60cm, with swollen nodes, shortly hairy (± hispid), angled, with cartilaginous ridges..*Tomatillo* **Physalis philadelphica**

Stems solid or pith-filled, without swollen nodes. Lvs with stomata both sides

Lvs 8-20 x 10cm, broadly ovate to elliptic, acuminate, often asymmetric at base, dull dark green and minutely hairy above, irreg sinuate-dentate. Petiole to 7cm, weakly channelled. Stems 30-100cm, >5mm diam, tough, round, solid...............*Angel's-trumpets* **Datura ferox**

Lvs 2.5-5(10) x 4(7)cm, ovate or rhombic to lanc, acute, cuneate at base (decurrent down petiole), occ hairless, entire or sinuate-dentate. Petiole 2-5cm, winged. Stems to 60cm, to 4mm diam, hairy to hairless (hairs occ crisped), (2)4-ridged esp below nodes, pith-filled ....*Black Nightshade* **Solanum nigrum** ssp **nigrum**

Stems and lvs glandular-hairy

Lvs yellow-green, densely glandular-hairy, 3-7 x 3.5cm, rhombic-ovate, cuneate at base, with stomata both sides. Petiole to 2cm. Stems 25-50(80)cm, prostrate to ascending, minutely glandular-hairy, ± winged (wings often sinuate-toothed). *Green Nightshade* **Solanum physalifolium**

Lvs mid-green, glandular-hairy (occ sparsely so). Stems to 60cm, ± erect, glandular-hairy at least above, with stomata both sides. . *Black Nightshade* (ssp) **Solanum nigrum** ssp **schultesii**

Stems and lvs (±) hairless

Lvs deeply pinnately 3(5)-partite with long terminal lobe

Lvs 10-25cm, cuneate at base, folded when young, dull dark bluish green above, paler below, with stomata both sides. Petiole to 1.5cm. Stems to 150(300)cm, ridged. A shrub in warmer climes!..*Kangaroo-apple* **Solanum laciniatum**

Lvs unlobed

Stems slender, <5mm diam............................*Black Nightshade* **Solanum nigrum** ssp **nigrum**

Stems stout, >5mm diam

Lvs 8-20 x 10cm, broadly ovate to elliptic, acuminate, often asymmetric at base, ± shiny dark green above, irreg sinuate-dentate, with stomata both sides. Petiole to 7cm, weakly channelled. Stems 30-100cm, tough, round.......................*Thorn-apple* **Datura stramonium**

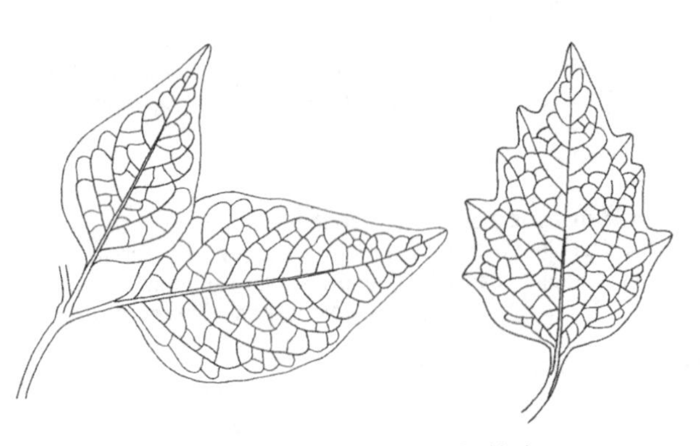

Physalis alkekengi (entire & toothed lvs)

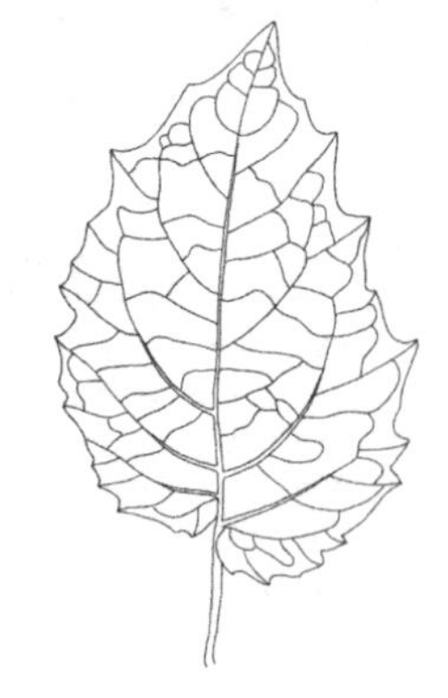

Physalis peruviana

Group SOR – Sorbus. *Deciduous tree or shrub. Lvs alt, ± folded when young. Petiole 1-3(4)cm. Stipules present on long shoots (absent from short shoots), soon falling. Twigs stout (3-5mm diam), with lenticels; lf scars narrow, with 3 traces. Buds held erect, to 2cm, acute, greenish, hairless to woolly, the scales ciliate and with broad purplish scarious margins. (*S. torminalis *keys out under Group PA; the truly pinnate-leaved spp (*S. aucuparia, S. cashmiriana, S. domestica, S. pseudohupehensis*) key out under Group SD). More spp are currently being described*

▪Lvs with deepest lobes >⅓ way to midrib (partially pinnate)..**SOR A**
▪Lvs with all lobes <⅓ way to midrib (toothed to lobed)
Lvs widest above middle (fold in half to check), usu broadly obovate ....**SOR B**
Lvs ± orb...**SOR C**
Lvs widest at or below middle, usu ovate to elliptic ....**SOR D**

SOR A – *Lvs widest at or below middle, ovate to elliptic, with deepest lobes >⅓ way to midrib*
▪Widespread, often planted. Tree >2m. (Some lvs of *S. latifolia* may key out here)
Lvs 9-15cm, ovate-oblong, rounded at base, dark green above, grey-woolly below, partially pinnate, usu with (1)2-3(5) prs of free or adnate well-separated lfts, with 10-14 prs of 2° veins. *S. aucuparia* x *aria*...*Bastard Service-tree* **Sorbus** x **thuringiaca**
Lvs 7-12cm, ± elliptic, rounded or broadly cuneate at base, dark yellowish-green above, ± sparsely yellowish-grey woolly below when young, with lobes ⅓-½ way to midrib (those of vigorous shoots occ lobed ± to midrib or rarely with a free lft), with 7-8 prs of 2° veins ....*Swedish Whitebeam* **Sorbus intermedia**
Lvs 7-10(12)cm, ± elliptic, rounded at base, dark green above, grey-woolly below, partially pinnate with (1)2(3) prs of free or adnate well-separated lfts, with 8-10 prs of 2° veins ....*Swedish Service-tree* **Sorbus hybrida**
▪W Scot (Arran)
Tree or small shrub. Lvs 6-9cm, elliptic-ovate, rounded at base, ± shiny yellowish-green above, whitish-grey woolly below, partially pinnate with (1)2-3 prs of free or adnate lfts (usu overlapping), serrate, with 7-9(10) prs of 2° veins.............................*Arran Service-tree* **Sorbus pseudofennica**
Tree 2-4m. Lvs 6-9cm, elliptic or elliptic-ovate, long-cuneate at base, ± shiny yellowish-green above, whitish-grey woolly below, lobed ½-¾ way to midrib (occ ± to base, rarely with free lft), sharply serrate..*Arran Whitebeam* **Sorbus arranensis**
Tree 2-4m. Lvs 7-13cm, elliptic, ± rounded at base, dark green above, paler and sparsely hairy below, partially pinnate, with 4-5 prs of free lfts (the lowest pr sessile or clearly stalked), serrate with acuminate teeth..*Catacol Whitebeam* **Sorbus pseudomeinichii**
▪S Wales (Brecon)
Shrub ≤2m. Lvs 7-9cm, ± ovate, obtuse or ± acute, cuneate at base, dull dark yellowish-green above, ± unevenly grey-woolly below, lobed ⅓-¾ way to midrib (occ to base), serrate, with 6-10 prs of 2° veins..*Ley's Whitebeam* **Sorbus leyana**

SOR B – *Lvs widest above middle (fold in half to check), usu broadly obovate*

▪Old Red Sandstone, N Devon

Lvs 8-11cm, obovate, rounded distally to an obtuse apex, tapered from the middle to a cuneate base, yellow-green above, white-woolly below, irreg 2-serrate, (±) entire in the basal ¼-⅓, with 6-9 prs of 2° veins. Small tree..*Bloody Whitebeam* **Sorbus vexans**

▪Limestone habs

Widespread (but esp W Br), R

Lvs 8-14.5cm, obovate or oblanc, rounded distally to (usu) an obtuse apex, usu tapered from middle (occ above) to a cuneate base, dark green above, ± densely white-woolly below (but often patchy with age), coarsely and unequally serrate, the teeth acute or ± acuminate and much smaller in basal ⅓ (occ ± absent), with 7-9 prs of 2° veins. Petiole usu turning reddish. Shrub, occ small tree..*Rock Whitebeam* **Sorbus rupicola**

SW Eng, Wales

Lvs to 9cm, usu obovate, cuneate from nr the middle (occ more rounded on fl shoots), shiny bright green above (and occ slightly woolly), greenish- or white-woolly below (occ sparse), unlobed, strongly 2-serrate in distal ½, ± entire for at least basal ⅓, with 8-11 prs of 2° veins. Shrub or occ small tree................................*Erect-leaved Whitebeam* **Sorbus porrigentiformis**

Avon Gorge

Lvs to 12cm, obovate, broadly cuneate (to ± rounded) at base, ± shiny bright yellowish-green, grey-woolly below, shallowly lobed ≤1/6 way to midrib (esp above middle), serrate, with (7)8-9(10) prs of 2° veins. Tree...*Bristol Whitebeam* **Sorbus bristoliensis**

Lvs to 14cm, obovate to obtrullate, cuneate at base, mid-green above, greenish-woolly below, usu unlobed but 1-2-serrate, with 8-11 prs of 2° veins. Shrub or tree ....*White's Whitebeam* **Sorbus whiteana**

NW Eng

Lvs to 10cm, obovate, cuneate in basal ⅓, dark green above, greyish- or white-woolly below, ± unlobed, 2-serrate (the largest to 1/9 way to midrib), (±) entire in the basal ¼, with 8-10 prs of 2° veins. Shrub or small tree................................*Lancaster Whitebeam* **Sorbus lancastriensis**

SOR C – *Tree. Lvs ± orb (those of non-fl shoots often smaller and narrower), regularly 2-serrate*

▪Planted street tree

Lvs to 16 x 16cm, broadly cuneate to rounded at base, dark green above, white-woolly below, regularly 2-serrate, with 11-16 prs of 2° veins (some forking 3x). Petiole 2.5-4cm, white-woolly. Buds 5-9mm, acute, viscid, green (to reddish)......................*Tibetan Whitebeam* **Sorbus thibetica**

▪Limestone habs

Lvs to 10 x 10cm, broadly cuneate to rounded at base, shiny bright green above, greenish-white woolly below, with 10-11 prs of 2° veins....................*Round-leaved Whitebeam* **Sorbus eminens**

SOR D – *Lvs widest at or below middle, ovate to elliptic, lobed <⅓ way to midrib.* ❶ *Lvs often with intercalary veins (distinct lateral veins between the 2° veins)*

▪Lvs with ≥11 prs of 2° veins

Tree to 15m

Lvs 15-20cm, elliptic to obovate, dark green above, adpressed white-woolly below, sparsely serrate, the teeth often cuspidate. VR alien.....................*Himalayan Whitebeam* **Sorbus vestita**

Lvs 7-13cm, elliptic to ovate, rounded to cuneate at base, dark green above, grey-woolly below, lobed to 1/5 way to midrib, lobes (or deep teeth) decreasing towards apex, serrate, with 10-13 prs of 2° veins. R alien..*Sharp-toothed Whitebeam* **Sorbus decipiens**

Lvs 5-12cm, ovate to elliptic (rarely obovate), rounded or cuneate at base, dull yellow-green above, densely white-woolly below, lobes not more than 1/6 way to midrib (apical tooth hardly projecting beyond its neighbours), 1-2 serrate, entire for the basal ≤1/5, with 10-14 prs of 2° veins..*Common Whitebeam* **Sorbus aria**

Shrub

Lvs usu 9-12cm, broadly elliptic to obovate, cuneate at base, yellowish or dark green above, sparsely greenish-white woolly below, lobed to ¼ way to midrib, entire for the basal ≤1/5, 2-serrate (teeth sharper and coarser than in *S. aria*), with 11-12 prs of 2° veins. VR, Wales ....*Thin-leaved Whitebeam* **Sorbus leptophylla**

▪Lvs with (7)9-11 prs of 2° veins. Shrub or small tree to 3m

Lvs 7-11cm, broadly elliptic to obovate, usu cuneate at base, yellowish or dark green above, ± sparsely whitish-grey woolly below, lobed 1/6-¼ way to midrib, serrate, with of 2° veins. ❶. R, W Br, Ire..*English Whitebeam* **Sorbus anglica**

▪Lvs with ≤10 prs of 2° veins

Lvs with acuminate teeth. W Eng, SE Ire. Large shrub or tree

Lvs 7-12cm, ovate to oblong-ovate, occ obovate, acute to ± acuminate, rounded at base, dark green above, sparsely greenish-grey woolly below, shallowly lobed to ⅛(¼) way to midrib, 2-serrate, with 7-9 prs of 2° veins....................................*Devon Whitebeam* **Sorbus devoniensis**

Lvs with non-acuminate teeth (*S. porrigentiformis* may key out here)

Widespread, R. Tree

Lvs to 13 x 12cm, broadly ovate, rarely ± orb, rounded to broadly cuneate at base, shiny bright green above, sparsely grey-woolly below, shallowly lobed to ¼ way to midrib (esp above middle), 1-2-serrate, with 7-9(10) prs of 2° veins. Alien ....*Broad-leaved Whitebeam* **Sorbus latifolia**

Lvs to 10 x 8cm, ovate to broadly so, occ elliptic or rarely obovate, rounded to broadly cuneate at base, shiny bright green above, sparsely grey-woolly below, occ shallowly lobed to 1/10 way to midrib, 1-2-serrate, with 9(11) prs of 2° veins. Alien ....*Orange Whitebeam* **Sorbus croceocarpa**

Lvs to 12 x 8cm, ovate to elliptic, broadly cuneate at base (occ rounded), lobed from 1/7 to over ¼ way to midrib, serrate, with 7-10 prs of 2° veins. *S. aria* x *torminalis* ....*Wye Whitebeam* **Sorbus** x **vagensis**

Avon. Small tree or shrub

Lvs 7-12cm, elliptic to obovate-elliptic, acute to obtuse, cuneate at base, ± shiny bright green above, greenish-white woolly below, shallowly lobed (1/10 way to midrib), sharply 2-serrate, entire for basal ≤1/5, with 8-9(10) prs of 2° veins....*Wilmott's Whitebeam* **Sorbus wilmottiana**

N Devon, S Somerset. Tree

Lvs 7-10.5cm, narrowly ovate or elliptic, acute, cuneate to ± rounded at base, bright green above, whitish-grey woolly below (± creamy when young), lobed to (1/7)1/6-¼(⅓) way to midrib, sharply serrate, with 8-9 prs of 2° veins. ❶. Old Red Sandstone ....*Somerset Whitebeam* **Sorbus subcuneata**

S Wales (Brecon). Shrub to 3m

Lvs 6-8cm, elliptic or oblong-elliptic, acute or ± acute, cuneate to ± rounded at base, dull green above, sparsely grey-woolly below, shallowly lobed 1/5-⅓(½) way with obtuse or acute lobes, serrate (teeth mostly curved and directed towards lobe apex), with (7)8-9(10) prs of 2° veins..*Least Whitebeam* **Sorbus minima**

Ire. Tree

Lvs 8-10cm, ovate to obovate, obtuse, broadly cuneate to rounded at base, dull green above, densely whitish-green woolly below, 2-serrate, entire in distal 1/5, the teeth longer and narrower towards the apex (teeth ending 2º veins are more prominent, at least nr lf apex), with 9-11 prs of 2º veins...*Irish Whitebeam* **Sorbus hibernica**

Group SPI – Spiraea. *Deciduous shrub. Lvs with stomata below only. Petiole to 5mm. Twigs brittle. Buds 3mm, pale brown, hairless or hairy, ciliate.* ❶ *Lvs 3(5)-lobed distally (unlobed in all other taxa)*

■Mature lvs woolly below

Lvs white- to grey-woolly below, 4-11 x 3-4cm, oblong, obtuse, with margins often recurved, entire in at least proximal ½, serrate distally with 3-8 unequal teeth. Twigs reddish, round, striate (without raised lines). Erect shrub to 2m. Strongly suckering ....*Steeple-bush* **Spiraea douglasii** ssp **douglasii**

Lvs yellowish-grey woolly below, 3-7 x 1-3cm, ovate, obtuse, irreg serrate ± to base. Twigs brown-hairy when young, angled. Shrub to 2m, ± erect. Strongly suckering..*Hardhack* **Spiraea tomentosa**

Lvs greyish-woolly to hairy below, 1.5-4 x 1.5cm, elliptic-oblong, obtuse, toothed only nr apex. Twigs angled. Shrub to 2m, ± erect or arching. Weakly suckering. VR hortal. *S. canescens* x *douglasii*..*Lange's Spiraea* **Spiraea** x **brachybotrys**

■Mature lvs hairy (to sparsely so) below

Twigs ± round. Erect shrub to 1.5m. Lvs often >6cm

Lvs (2)5-8(12)cm, elliptic to narrowly so, acute, occ ± glaucous below, hairless or sparsely hairy on veins below, coarsely serrate. Twigs ± round or striate (occ flattened or angled) ....*Japanese Spiraea* **Spiraea japonica**

Twigs angled or slightly so (at least distally). Erect shrub to 2m. Lvs 2-6(8)cm

Pollen fertile

Lvs 4-8cm, oblong, ± hairless to hairy below, entire in at least proximal ½, serrate distally with 3-8 unequal teeth. Twigs reddish. Strongly suckering ....*Steeple-bush* (ssp) **Spiraea douglasii** ssp **menziesii**

Pollen mostly sterile

Lvs to 8 x 1.5-3.5cm, acute, usu slightly dark glaucous below, ± irreg serrate with up to 28 teeth per side, entire nr base. Twigs soon hairless. *S. alba* x *douglasii* ....*Billard's Bridewort* **Spiraea** x **billardii**

Lvs to 6 x 1.5-3cm, ± obtuse, mid-green below, sparsely hairy below, regularly serrate, entire or toothed nr base. Twigs soon hairless. *S. douglasii* x *salicifolia* ....*Confused Bridewort* **Spiraea** x **pseudosalicifolia**

Twigs strongly angled. Arching shrub to 2m. Lvs 1-2(3)cm

Lvs 1-2.7 x 0.5-1.5cm, ovate to obovate, obtuse, dull dark grey-green and shortly grey-hairy above, often hairless below, entire to shallowly serrate-dentate with 3-4 teeth in distal ½. Twigs reddish-brown, shortly hairy when young.......................*Himalayan Spiraea* **Spiraea canescens**

■Lvs (±) hairless below (occ ciliate, or hairy at base of midrib below) (examine mature lvs)

At least some lvs 3(5)-lobed distally. Twigs round, hairless. Arching shrub to 2m

Lvs 1.5-4 x 3cm, ovate to obovate, ± glaucous below, with veins hardly raised below. Twigs reddish-brown. ❶. *S. cantoniensis* x *trilobata*..........*Van Houtte's Spiraea* **Spiraea** x **vanhouttei**

All lvs serrate, never lobed

Twigs (±) round (occ flattened or angled), usu shortly hairy when young, reddish-brown. Lvs (3)5-12cm, elliptic to narrowly so, acute, occ ± glaucous below, occ sparsely hairy on veins below. Infl umbel-like. Erect shrub to 1.5m..........................*Japanese Spiraea* **Spiraea japonica**

Twigs (±) angled, usu hairless. Lvs 3-8cm

Erect shrub to 2m

Pollen fertile

Lvs oblanc-oblong to elliptic-oblong, acute to ± obtuse, minutely hairy when young, 1(2)-serrate. Twigs yellowish-brown. Infl panicle dense, cylindrical, or with few ascending branches from base. VR (much over-recorded)........................*Bridewort* **Spiraea salicifolia**

Lvs narrowly lanc, acute to ± obtuse, hairless, 1(2)-serrate. Twigs yellowish-brown. Infl panicle open, conical..*Pale Bridewort* **Spiraea alba**

Pollen sterile

Lvs 4-8 x 3cm, elliptic-oblong to lanc, acute, sharply serrate exc at extreme base. Twigs yellowish-brown. *S. alba* x *salicifolia*....................*Intermediate Bridewort* **Spiraea** x **rosalba**

Arching shrub to 2m. VR hortal

Lvs 2-7cm, ovate, (±) hairless, coarsely serrate exc nr base. Twigs pale grey. Pollen fertile ....*Elm-leaved Spiraea* **Spiraea chamaedryfolia** ssp **ulmifolia**

Lvs 2-5cm, elliptic-oblong, (±) hairless, serrate in distal ½. Twigs reddish- or warm-brown. Pollen fertile..*Russian Spiraea* **Spiraea media**

Lvs 1.5-4cm, oblanc to oblong-obovate, hairless at maturity, serrate. Pollen sterile. *S. multiflora* x *thunbergii*...*Bridal-spray* **Spiraea arguta**

Group THIS – *Thistles and thistle-like herbs. Lvs alt (the upper stem lvs often opp in* Acanthus*), spiny. Stems often with spiny wings*

▪Basal lvs persistently white-cottony at least below, occ with a long petiole............................**THIS A**

▪Basal lvs with at least some hairs either side (but cottony hairs sparse or absent) at least when young...**THIS B**

▪Basal lvs totally hairless, glaucous..**THIS C**

THIS A

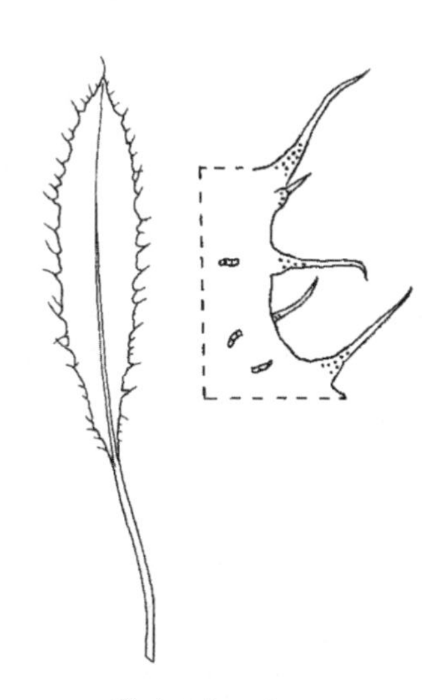

Cirsium dissectum

THIS A – *Basal lvs persistently white-cottony at least below, often with additional septate hairs*

▪Basal lvs persistently white-cottony both sides. Tufted

Lvs 75-200cm, oblong-lanc, pinnately lobed; lobes lobed again, with occ dark short spines esp below nr rachis; rachis winged and ridged below. Petiole distinct on lower lvs, ridged below, with many scattered vb's. Stem(s) to 250cm, unwinged. Per. All yr. R alien .... *Globe Artichoke* **Cynara cardunculus**

Lvs 10-50cm, broadly elliptic, lobed <½ way, with short (0.5-1mm) spiny margins. Petiole indistinct, with spines to base. Stem(s) 50-15(300)cm, with continuous broadly spiny wings, white woolly. Bi or per. All yr.. *Cotton Thistle* **Onopordum acanthium**

▪Basal lvs persistently white-cottony below only (occ sparsely so in *Cirsium dissectum*)

Basal lvs with long distinct petiole. Shortly rhizomatous

Basal and lower stem lvs 15-40 x 4-8cm, elliptic-lanc, cuneate at base, revolute when young, shiny and hairless above, finely 1-2-serrate with spine-tipped teeth, with 2° veins ± herringbone-like and anastomosing nr margins, with stomata both sides or below only. Upper lvs clasping, with auricles. Petiole often >10cm, reddish nr base, triangular, channelled, with small hollow and 7 vb's. Stems to 100cm, ridged. Apr-Oct. Uplands, N Br .... *Melancholy Thistle* **Cirsium heterophyllum**

Basal lvs 6-15(25) x 1-3cm, elliptic-lanc, cuneate at base, rolled when young, with soft marginal prickles, dull and septate-hairy above, with weak spines along margins with elongated swollen purple bases, entire (exc for spines) or toothed, with midrib raised below, with anastomosing 2° veins, with stomata both sides. Stem lvs sessile, ± clasping with auricles. Petiole to 10cm, often purplish, ± round, channelled. Stems to 15cm, round. Mar-Oct. Fens, wet hths, mostly S Br, Ire .... *Meadow Thistle* **Cirsium dissectum***

Basal lvs sessile or with indistinct petiole

Strongly rhizomatous per, forming false rosettes (stem always present but often v short). Mar-Oct

Lvs 8-20cm, oblong, grey-green above, cobwebby below, with long septate hairs both sides, weakly lobed or toothed, weakly spiny with 1.5-3mm spines along margins. Stem lvs sessile, ± clasping, hairless or cottony below. Petiole green, with spines to extreme base. Stems 20-120cm, ridged.. *Creeping Thistle* **Cirsium arvense**

Tufted per to ann, forming true rosettes at least in 1st yr. All yr

Lvs white-felted below, 10-40 x 2-10cm, 1(2)-pinnately lobed, the ± triangular lobes ending in a strong spine and with weakly spiny-toothed margins, with stomata below only; rachis winged. Stem lvs clasping with spiny auricles. Stems 50-200cm, ridged

Lvs densely glandular-hairy and sparsely bristly above. Fls white or greyish. Per .... *Glandular Globe-thistle* **Echinops sphaerocephalus**

Lvs sparsely septate-hairy above, the hairs obscurely glandular with withering tips. Fls blue. Per (bi).. *Blue Globe-thistle* **Echinops bannaticus**

Lvs hairless or with sparse strigose hairs above. Per. Fls white or greyish .... *Globe-thistle* **Echinops exaltatus**

Lvs white-cottony below (at least when young)

Bi. Lvs to 1.5cm wide, linear

Lvs to 10cm, dull grey, cottony but soon ± hairless above and sparsely hairy below, undulate, weakly toothed when young, becoming ± lobed, the lobes narrow and at 90° to midrib, the spines weak, 1.5mm, amber-coloured, not in 1-plane. Petiole ± absent, with spines to base, with 3 vb's. Stem lvs broadly ± clasping at base, ± hairless. Stems 5-50cm. All yr.. *Carline Thistle* **Carlina vulgaris**

Bi (ann). Lvs 2-8cm wide, elliptic to oblanc

Lvs to 15cm, ± shiny above, often appearing to have variegated margins, v sparsely cottony above, usu (±) cottony below, with septate hairs esp above, pinnately lobed (even when young, unlike *Cirsium vulgare* which is also bullate), spiny margins with 2mm sharp spines, with 2° veins sunken above. Stems 5-80cm, ± cottony, not ridged. Taproot slender. All yr

Stem with continous spiny wings 5-10mm wide........ *Slender Thistle* **Carduus tenuiflorus**

Stem with interrupted spiny wings (to 5mm wide) absent below some fl heads. Coastal cliff-tops, VR alien, Devon................*Plymouth Thistle* **Carduus pycnocephalus**

Per. Lvs 2-15cm wide, elliptic to broadly oblong

Basal lvs 5-40 x 4-15cm, viciously spiny, pinnately lobed, with lobes entire and not markedly in 1-plane, with recurved margins, dark green and ± hispidly septate-hairy above with some additional spines, white-cottony or -woolly and septate-hairy below, with midrib often purplish nr base, with stomata both sides; spines (4)7-9(11)mm, not or hardly swollen at base. Stem lvs ± clasping, prickly-hairy above, cottony below, the lobes 2-lobed with 1 lobe directed upwards, the other down. Petiole often purplish, with sparse moniliform hairs and spine-like cilia, with 5 main vb's. All yr (but overwintering as a tiny rosette). Calc habs...*Woolly Thistle* **Cirsium eriophorum**

Basal lvs 10-25 x 2-8cm, weakly spiny, septate-hairy above (hairs with withering tips); spines short (1.5mm), with swollen base (occ purplish). Stem lvs ± sessile, ± clasping with auricles small or absent, sparsely cottony both sides. Petiole 5cm; spines absent at extreme base, with 5(7) vb's. Apr-Oct. Calc habs, VR, S Br ....*Tuberous Thistle* **Cirsium tuberosum**

THIS B – *Basal lvs with at least some septate hairs either side (but cottony hairs sparse or absent).* ❶ *Spines with swollen yellowish base*

■Basal lvs with long petiole, with stomata below only

Lvs 20-60cm, shiny dark green above, occ weakly marbled, paler below, sparsely hairy on veins both sides, shortly ciliate, teeth tipped with 0.5mm spine-like bristles, not net-veined. Petiole to 25cm, ± round, hairless, with 1 large vb and 2 rib bundles. Per. Mar-Oct. R alien, hortal ....*Spiny Bear's-breech* **Acanthus spinosus**

■Basal lvs sessile (or petiole short and spiny), with stomata both sides

Lvs spiny to touch (can pierce skin)

Lvs with white marbling

Lvs 15-60cm, broadly oblanc, hairless above (exc nr base), often hairy below, sinuate-lobed, with main veins ending in a spine. Stem lvs clasping with spiny-ciliate auricles, shiny pale green, hairless. Stems to 250cm, not spiny. Bi (ann). All yr.....*Milk Thistle* **Silybum marianum**

Lvs without marbling

Lvs with weak spines above, bullate

Lvs dull grey-green above, hairy esp below (1st yr lvs ovate, not deeply lobed, pustulate above); spines along margins 3mm, with swollen yellowish base. Petiole not purplish at lf base, weakly spiny, with 3(5) vb's, occ with sparse latex (also in rootstock). Stems to 150cm, with interrupted spiny wings. Bi. ❶...*Spear Thistle* **Cirsium vulgare**

Lvs without spines above, not bullate

Stem to 100cm, with interrupted spiny wings (and absent for some distance below fl heads), ± cottony. Lvs 5-30cm, dark green above, paler below, sparsely hairy both sides or soon hairless above, cottony hairs sparse or absent, pinnately lobed; spines to 5mm, vicious, occ purplish nr base. Petiole purplish at base, often spiny to winged base, septate-hairy and usu with some cottony hairs. All yr. Usu calc gsld..................*Musk Thistle* **Carduus nutans**

Stem usu ± absent. Lvs 4-15(20)cm, shiny green and soon ± hairless above (hairs mostly on midrib), hairy on veins below (with withering tips), cottony hairs always absent, pinnately lobed; spines 3-4.5mm. Petiole purplish with few or no hairs or spines at base, with 3-5 main vb's. Apr-Nov. Calc gsld..*Dwarf Thistle* **Cirsium acaule**

Lvs weakly spiny to touch (unable to pierce skin)

Spines along lf margin 3-5mm. Lvs often ± shiny green above

Lvs 10-50cm, oblanc to broadly so, septate-hairy both sides, pinnately lobed, margins turning purplish. Stem lvs decurrent down stems, hairy above. Petiole indistinct, purplish, with purple (often bulbous-based) spines to extreme base. Stems to 170cm, with continuous spiny wings. Bi...*Marsh Thistle* **Cirsium palustre**

Spines along lf margin 3mm. Lvs dull green above
Lvs 10-50cm, elliptic, sparsely hairy both sides. Stems to 140cm, with narrow spiny wings, cottony. All yr..*Welted Thistle* **Carduus crispus** ssp **multiflorus**

Spines along lf margin 1mm (excluding herbaceous base). Lvs dull greyish-green above
Lvs 5-50cm, elliptic to ovate, with short septate hairs above and on veins below, inrolled and ± silvery above when young, soon hairless. Stem lvs sessile, clasping, not or hardly decurrent, limp, ± hairless, with large rounded auricles. Petiole often weakly spiny-winged to base, channelled. Stems to 180cm, ± hairless, not spiny, ridged, hollow. Per. All yr. VR alien ....*Cabbage Thistle* **Cirsium oleraceum**

THIS C – Eryngium *(*Apiaceae*). Basal lvs totally hairless, glaucous, rolled when young, net-veined. Tufted per*

▪Basal lvs 10-30cm, pinnate with 3-5 prs of spiny lobes, stiff, with cartilaginous margins, with citrus-pine odour; rachis winged-spiny. Stem lvs clasping, pinnate. Petiole flattened, dilated at base, stout, occ slightly channelled above, rounded below, not spiny, with 15 obscure vb's. Stems to 70cm, striate, repeatedly branched in 2's or 3's, spineless. Mar-Dec. Calc gsld, S Eng. Sch8 ....*Field Eryngo* **Eryngium campestre**

▪Basal lvs 5-15cm diam, ± orb, palmately 3-lobed, spiny-toothed (tips of spines yellow), leathery, with thick cartilaginous margins, odourless, with veins often purplish, with stomata both sides. Stem lvs sessile, palmate. Petiole channelled, dilated at base, not spiny, with 8-13 vb's in an arc. Stems to 50cm. Apr-Oct. Littoral sand or shingle..*Sea-holly* **Eryngium maritimum**

Group TIL – Tilia. *Deciduous tree. Lvs broadly ovate to ± orb, abruptly acuminate, often cordate at base, rolled when young, hairy or hairless but always with longer tufts of hairs in vein axils below, toothed, often with 3-5 main veins from nr base (appearing ± palmately veined). Petiole weakly dilated at base. Stipules soon falling (leaving scars). Twigs sympodial, with sparse small raised lenticels; lf scars ± round, with 3 traces. Buds lopsided, with 3 scales (1 smaller, making bud asymmetric), not ciliate.* ❶ *Trunk often with abundant epicormic shoots (dense twiggy outgrowths)*

▪Lvs with abundant white stellate hairs below, hairless above

Petiole (4)6-12cm, >½ lf length, stellate-hairy to hairless. Lvs 6-12 x 12cm, dark green above. Twigs stellate-hairy or hairless. Buds to 4mm, green, ± hairless. Branches pendent ....*Pendent Silver-lime* **Tilia 'Petiolaris'**

Petiole to 5cm, <½ lf length. Lvs 5-12 x 10cm, dark green above. Twigs usu densely short-hairy. Buds to 4(8)mm, brown, hairy. Branches hardly pendent....................*Silver-lime* **Tilia tomentosa**

▪Lvs without stellate hairs below, or confined to vein axils

Lvs with at least some hairs below (with longer whitish hair-tufts in vein axils)

Lvs 6-16cm, broadly ovate, obliquely cordate or truncate at base, usu dull green and often hairy above, ± shiny green below, ciliate, dentate, with 3° veins raised below. Petiole 1.5-4.5cm, usu <⅓ lf length, 1.5-2mm diam. Stipules 10mm, green, hairy. Twigs hairy at least when young. Buds (4)7-10mm, reddish, hairless

Lvs densely hairy below. Petiole hairy. Native, also alien ....*Large-leaved Lime* (ssp) **Tilia platyphyllos** ssp **cordifolia**

Lvs hairy only on veins below. Petiole hairy. Alien (from central Europe) ....*Large-leaved Lime* **Tilia platyphyllos** ssp **platyphyllos**

Lvs ± hairless below. Petiole hairless. VR alien (from E Europe) ....*Large-leaved Lime* (ssp) **Tilia platyphyllos** ssp **pseudorubra**

Lvs hairless both sides (exc for tufts of hairs in vein axils below)

Lvs mostly 7-12(15)cm, broadly ovate, with 3° veins slightly raised below. Stipules green

Lvs dull green above, ± shiny green to dull greyish-green below, cordate to obliquely truncate at base, with white or buff hair-tufts below, serrate. Petiole 2.5-5cm, sparsely hairy or hairless. Stipules to 20mm, hairless. Twigs usu hairless. Buds 5-8mm, reddish, hairless. ❶. *T. cordata* x *platyphyllos*. Pl 10, 20......................................*Common Lime* **Tilia** x **europaea***

Lvs shiny dark green above, greyish-green below, turning yellow in autumn, obliquely cordate at base, with obvious orange-buff hair-tufts below, sharply serrate with mucronate teeth. Petiole 3.5-7cm, hairless. Stipules to 20mm, ovate. Twigs hairless. Buds 4-5mm, reddish or yellow-red (occ green), hairless. *T. dasystyla* x *platyphyllos* ....*Caucasian Lime* **Tilia** x **euchlora**

Lvs mostly 2.3-6(8)cm, ± orb, without 3° veins raised below. Stipules pinkish

Lvs abruptly long-acuminate, usu cordate at base, dull green above, (±) glaucous with reddish hair-tufts below (hair-tufts occ pale brown to colourless), dentate with 20-60 obtuse teeth per side. Petiole 1.5-4(6)cm, hairless or with sparse hairs. Stipules 8 x 1.5mm, hairless. Twigs soon hairless. Buds to 5(8)mm, ovoid, shiny green or reddish-green, hairless ....*Small-leaved Lime* **Tilia cordata**

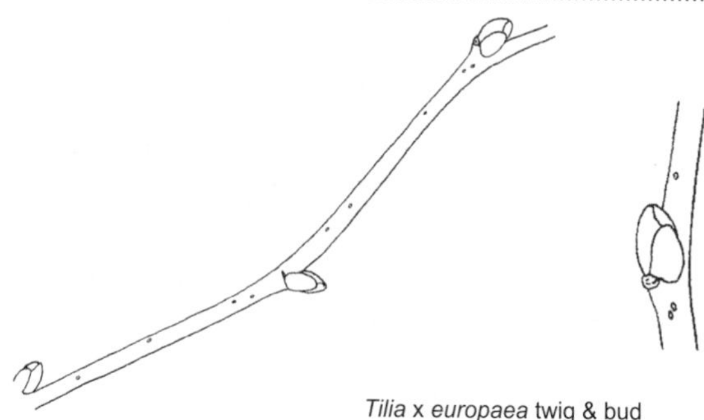

Tilia x *europaea* twig & bud

Group ULM – Ulmus. *Tree or shrub. Lvs deciduous, often asymmetric at base, 2-serrate. Stipules to 6mm, linear-lanc, soon falling. Twigs often with short shoots (with 3-5 lvs); lf scars with 3 traces. Buds to 6mm, usu ovoid (fl buds to 2.5mm, globose). Suckering (rarely so in* U. glabra*)*

■Lvs mostly >7cm (length x width >28)

Lvs v rough above. Petiole usu ≤3mm, mostly overlapped by lf base. Buds often with reddish cilia

Lvs 7-16cm, acuminate to cuspidate, strongly asymmetric with rounded auricle at base, usu v rough above, hairy below, with >65 teeth per side, with 12-18 prs of 2° veins. Twigs not corky, hairy when young. Buds dark brown

Lvs broadly obovate. Mostly S Br......................................*Wych Elm* **Ulmus glabra** ssp **glabra**

Lvs narrowly obovate. Mostly N & W Br...............*Wych Elm* (ssp) **Ulmus glabra** ssp **montana**

Lvs ± rough to smooth above. Petiole usu >5mm, not or partly overlapped by base of lf

Buds ciliate, dark brown. Lvs with 60-80 teeth per side. Trunk without burrs

Petiole 8-11mm. Lvs 7-12 x 6cm, ± 2x as long as wide, acuminate to cuspidate, dull dark green above, with 12-18 prs of 2° veins. Twigs sparsely hairy, not corky-winged. Tree broadly obovate to orb or fan-shaped in outline, with long straight branches from low down. *U. glabra* x *minor*...*Huntingdon Elm* **Ulmus** x **vegeta**

Petiole 5-8(10)mm. Lvs (5)7-10 x 7cm, <2x as long as wide, acute to shortly acuminate, ± shiny dark green, with 9-14 prs of 2° veins. Twigs sparsely hairy, some becoming corky winged. Tree obovate in outline, with spreading crooked branches above. *U. ?glabra* x *minor*, or *U. glabra* x *minor* x *plotii*..*Dutch Elm* **Ulmus** x **hollandica**

Buds not ciliate, chestnut-brown. Lvs with <60 teeth per side. Trunk usu with burrs

Petiole (2)5-10mm. Lvs 4-12 x 5cm, obovate to broadly so, usu strongly asymmetric at base, smooth above, softly hairy (esp below) to hairless, with 10-18 prs of 2° veins. Twigs minutely hairy. Buds 2-5mm, narrowly ovoid. R hortal......................*European White-elm* **Ulmus laevis**

■Lvs mostly <7cm (length x width <28)

Lvs rough to smooth above. Buds 3-6mm, ovoid

Lvs 4-7(9) x 5cm, ovate (-elliptic) to ± orb (length/width >0.75), slightly asymmetric at base, dark green, usu rough above with short curved hairs, hairy below, usu with <55 teeth per side, with 8-12 prs of 2° veins. Petiole 3-8mm. Twigs hairy when young, some becoming corky-winged. Buds purple-black, often with reddish cilia. Tree obovate to oblong in outline, with strong branches at all levels. Commonest sp in hedges. PL 20..................*English Elm* **Ulmus procera**

Lvs 3-5(7)cm, narrowly elliptic (length/width <0.75), strongly asymmetric to ± equal at base, usu light green, slightly rough or smooth above, usu with <70 teeth per side, with 7-10 prs of 2° veins. Petiole 5-10mm. Twigs hairy when young. Buds blackish-brown, with whitish cilia or hairless; leading shoot drooping, long shoots conspicuous in mature canopy. Tree narrow but irreg in outline, with leading shoot arching or drooping, without strong branches (all partly pendent), the short shoots mostly continuing growth as long shoots.......*Plot's Elm* **Ulmus plotii**

Lvs smooth above. Buds 2-3mm, ovoid, blackish-brown. Trunk with or without burrs. Tree with erect leading shoot, usu with some strong branches, only the lower pendent (if at all), the short shoots rarely continuing growth

Lvs usu strongly asymmetric at base. Tree ± spreading in outline, with main branches in lower ½ of tree, at least some of which become horizontal or drooping, the canopy composed of short shoots..*Small-leaved Elm* **Ulmus minor** ssp **minor**

Lvs usu weakly asymmetric at base. Tree v narrow (or broad at top) in outline, always with all main branches ascending, with no or v few main branches in lower ½ of tree

Tree pyramidal, with pointed crown. Trunk persisting to tree apex, with burrs below. Branches numerous, slightly ascending. Lvs shiny dark green above, with c50 teeth per side, with 8-12 prs of 2° veins. Petiole to 10mm...........................*Guernsey Elm* **Ulmus minor** ssp **sarniensis**

Tree narrow in outline, with flat crown. Trunk ending short of tree apex, without burrs. Branches few, the lowest steeply ascending. Lvs shiny bright green above, with <35 teeth per side, with 8-12 prs of 2° veins. Petiole 4-7mm......*Cornish Elm* **Ulmus minor** ssp **angustifolia**

Group VAL – Valerianella. *Ann. Lvs opp, the basal often in a rosette, 2-9 x 0.4-1.2cm, spathulate to linear-lanc, obtuse with a tiny white hydathode at apex (best viewed end-on), narrowly cuneate at base, connate (unusual for ridged stem), weakly rolled when young, soon limp, weakly aromatic, bright green, hairless or with scattered hairs above, scabrid-ciliate esp nr base, with teeth often obscure to absent, with up to 4 white hydathodes per side along upper margin (occ protruding), not not-veined, pinnate-veined (occ 3-pli-veined or with 2° veins obscure), with midrib and 2° veins sunken above, midrib only raised below, with stomata both sides. Petiole often indistinct, with 1 vb. Stems 3-15(40)cm, dichotomously branched, the axils without a lf, brittle, 6-7-ridged, with retrorse hairs on ridges, hollow. Bracts with pinnate lobes at base. Rootstock aromatic. Hairs stiff, white, unicellular. Oct-Jun. Reliable vegetative characters appear not to exist*

▪Ruderal or man-made sites (rarely arable, often dunes in *V. locusta*). Widespread. Calyx above fr minute to absent

Frs 1.8-2.5 x 1-2mm, 2x as thick as wide, hardly longer than thick, flattened on 2 sides, ± round in side view, shallowly grooved below, the cell with seed inside has a thick corky wall. ....*Common Cornsalad* **Valerianella locusta**

Frs 2.0-2.7 x 0.8-1.4mm wide, as ± wide as thick, much longer than wide or thick, ± square, oblong in side view, with a deep groove below, the cell with seed inside has a thin wall ....*Keeled-fruited Cornsalad* **Valerianella carinata**

▪Arable, R. Calyx above fr ¼-½(⅔) as long as rest of fr, <½ as wide as fr, usu with 1 tooth

Frs 2 x 1.5mm, not flattened, inflated. Calyx tooth with short (±) entire teeth. S Eng ....*Broad-fruited Cornsalad* **Valerianella rimosa**

Frs 2 x 0.8-1mm, flattened on 1 side, not inflated. Calyx tooth usu denticulate with ≥2 teeth. Usu chalky soils, mostly S Eng..........*Narrow-fruited Cornsalad* **Valerianella dentata**

▪Eroding stony ground, usu on maritime cliff tops, R, S Eng. Calyx above fr >⅔ as long as rest of fr, ± as wide as fr, with 5-6 teeth. Frs 1.2 x 0.9mm..........*Hairy-fruited Cornsalad* **Valerianella eriocarpa**

Group VIO – Viola. *Herb. Lvs involute when young, toothed, each tooth usu with a pale hydathode on the distal side, with stomata both sides (but often confined to veins above in* V. canina*). Petiole slightly channelled, often with a central raised rib nr apex, with 1 vb. Stipules persistent, usu toothed or lobed. Hairs unicellular*

■Stipules weakly toothed to entire, never lobed or lf-like. Stems solid (*Violets*)

Petiole densely hairy. All yr

Petiole with 0.3-1mm patent or weakly retrorse hairs, 2.5-7cm. Lvs in basal rosette, 1.5-4cm, ovate-lanc or oblong-ovate, obtuse to ± acute, cordate at base, yellowish-green, densely hairy, crenate (-serrate) with 16-30 teeth per side. Stipules lanc, weakly toothed, the teeth usu gland-tipped. Calc turf........*Hairy Violet* **Viola hirta**

Petiole with 0.1-0.3mm weakly retrorse hairs. Lvs usu in rosettes at end of each stolon, 3-7cm, ovate-orb, cordate at base, shiny dark green, hairy to ± hairless, crenate-serrate with 20-25 teeth per side. Petiole 4-15(25)cm. Stipules ovate-lanc, glandular-fimbriate. PL 13, 23 ....*Sweet Violet* **Viola odorata**

Petiole with <0.1(0.3)mm usu retrorse hairs (if hairless, then lvs also hairless unlike *V. riviniana*), 0.4-1(2.5)cm. Lvs in basal rosette, 0.8-1.5cm, ovate, obtuse with v wide terminal tooth, cordate to ± truncate at base, hairless or shortly hairy, with 7-11 teeth per side. Stipules ovate-lanc, fimbriate. Limestone gsld, N Eng........*Teesdale Violet* **Viola rupestris**

Petiole sparsely hairy or hairless

Lvs strongly cordate at base

Dry habs. Lvs 3-many, in basal rosette or on stems. Roots often budding giving rise to 2-3 rosettes. All yr

Lvs 2.5-8cm, orb-ovate to ovate, shiny or dull green, often sparsely hairy above, crenate with 13-17 teeth per side, ± net-veined. Stipules << petiole, with patent cilia-like gland-tipped teeth

Fls with spur usu paler than petals. Lvs ± constant size throughout season, acute or obtuse ....*Common Dog-violet* **Viola riviniana**

Fls with spur usu darker than petals. Lvs increasing in size throughout season, obtuse to acuminate........*Early Dog-violet* **Viola reichenbachiana**

Damp habs. Lvs 1-3(5) per node along stolons/rhizomes. Roots not budding. Apr-Oct

Lvs 3-7cm diam, orb-reniform, dull above, shiny below, hairy to hairless, crenate, ± net-veined. Petiole to 16cm, hairless or with 0.5mm hairs. Stipules lanc to ± ovate, obscurely toothed to entire........*Marsh Violet* **Viola palustris**

Lvs slightly cordate to cuneate at base, all on weakly prostrate stems

Lvs 2-3 x 1.5-2.5cm, dark green, opaque, ± thick, hairless. Usu dry habs

Lvs usu slightly cordate at base, not purplish below, ovate to ovate-lanc, obtuse to ± acute, crenate with 10-15 teeth per side. Petiole 1.5-3cm. Stipules to 10 x 2mm, <½ petiole length, lanc, sparsely fimbriate or toothed. Apr-Oct........*Heath Dog-violet* **Viola canina**

Lvs usu truncate at base, occ purplish below, ovate-lanc, ± acute, shallowly crenate-serrate with 10-14 teeth per side. Petiole 1.5-3cm. Stipules 5-7 x 0.5-0.8mm, upper often > petiole, ovate-lanc to thread-like, fimbriate. Apr-Oct. Hths, SW Br........*Pale Dog-violet* **Viola lactea**

Lvs 2.5 x 1.2cm, light green, translucent, v thin, hairless. Fens (or turloughs)

Lvs ovate-lanc, cordate at base but slightly decurrent down petiole, shallowly crenate-serrate with 13 teeth per side. Stipules to 8 x 1mm, usu <½ as long as petiole, laciniate (occ entire). Apr-Oct. VR, Eng, Ire. Sch8........*Fen Violet* **Viola persicifolia**

■Stipules deeply pinnately lobed nr base, often lf-like. Stems hollow (*Pansies*)

Stipules with large toothed terminal lobe (lf-like). Ann

Lvs sparsely minutely hispid-hairy, 0.5-5cm, ovate (-elliptic), crenate-serrate. Upper lvs oblong-lanc. Stems 5-30cm, much-branched, retrorsely hairy, with strong antiseptic odour. Mar-Oct ....*Field Pansy* **Viola arvensis***

Lvs densely grey-hairy with short crisped or retrorse hairs, 0.3-1(2)cm, ± orb, crenate-serrate, occ purplish. Upper lvs oblong-lanc. Stipules 0.5-1cm. Stems to 3cm, simple or little-branched, retrorsely hairy, ± odourless. Oct-Jun. Scilly Is........................... *Dwarf Pansy* **Viola kitaibeliana**

Stipules with long entire terminal lobe (not lf-like)

Per, with abundant above-ground stolons or rhizomes

Lvs to 1.5(3)cm, ovate to lanc (narrower upwards), crenate-serrate. Petiole <10mm, scabrid-ciliate. Stipules with long narrow lanc terminal lobe to 1cm and 2 prs of lateral lobes. Stems to 30cm, branched, minutely (0.1mm) retrorsely hairy, with strong antiseptic odour. Rhizomatous. Mar-Oct... *Wild Pansy* (ssp) **Viola tricolor** ssp **curtisii**

Lvs 1.5-3cm, ± orb to ovate, crenate-serrate. Petiole 7-13mm, sparsely scabrid-ciliate. Stipules serrate or lobed ½ way to midrib, often appearing 3-lobed with short entire end-lobe (often broad). Stems to 30cm, branched, hairless or with a line of hairs. Rhizomatous, or with slender stolons, forming dense tufts. All yr. R alien...................... *Horned Pansy* **Viola cornuta**

Per, with underground rhizomes. Upland gsld

Lvs 1-2cm, ovate or oblong, with sparse short stiff white hairs above, with 2-5 teeth per side. Stipules ciliate. Apr-Oct (all yr)... *Mountain Pansy* **Viola lutea**

Ann to per, tufted

Lvs 2-5cm, ovate-oblong, v shallowly crenate with 6 broad teeth per side. Petiole 10-30(50)mm, with retrorse scabrid cilia. Stipules with ovate-lanc terminal lobe. Stems with retrorse scabrid cilia, with strong antiseptic odour. All yr. *V. altaica* x *lutea* x *tricolor* .... *Garden Pansy* **Viola** x **wittrockiana***

Lvs to 1.5(3)cm, ovate to lanc, crenate-serrate. Petiole scabrid-ciliate. Stipules <10mm, with long (to 1cm) narrowly lanc terminal lobe and 2 prs of lateral lobes. Stems to 30cm, branched, minutely (0.1mm) retrorsely hairy, with strong antiseptic odour. Mar-Oct .... *Wild Pansy* **Viola tricolor** ssp **tricolor**

Viola arvensis
(showing stipules)

Viola x *wittrockiana*
(showing stipules)

SYMBOLS & ABBREVIATIONS

< less than; **<<** much less than
> greater than; **>>** much greater than
± more-or-less (qualitative); approximately (quantitative)
° order of branching, e.g. 2° = secondary, 3° = tertiary; degrees of an angle, e.g. 90° = 90 degrees
***** see illustration(s)
x10, **x20**, etc hand lens / microscope magnification recommended
agg aggregate
alt alternate
ann annual
bi biennial
Br Britain
c circa (approximately)
calc calcareous
cv cultivar
Eng England
exc except
fl flower
fr fruit
gsld grassland
habs habitats
hths heaths
HTL hold to the light
infl inflorescence
Ire Ireland
irreg irregular
Is Island(s)
lanc lanceolate
lf leaf
lft leaflet
lvs leaves
mtn mountain
N, **E**, **S**, **W** points of compass
nr near
oblanc oblanceolate
occ occasionally
opp opposite
per perennial
PL plate(s)
pr pair
R rare
Sch8 Schedule 8 (of the Wildlife & Countryside Act 1981 (as amended)
Scot Scotland
sp (plural **spp**) species
ssp subspecies
TNE to the naked eye
TS transverse section
usu usually
v very
var variety
vb (plural **vb's**) vascular bundles
VR very rare
yr year

GLOSSARY

SUFFIXES

-fid divided to, or less than, half-way
-foliate a leaf made up of leaflets, e.g. a 3-foliate leaf consists of three leaflets (*see inside front cover*)
-lvd leaved e.g. small-leaved
-partite deeply divided to near the base into ± equal parts
-pli-veined with the main veins originating from near the leaf base and appearing ± parallel

ORDER OF VISIBILITY

distinct clearly visible
indistinct visible with careful observation
obscure difficult to see

TERMS

acicle a very slender prickle
acroscopic on the side towards the apex
acuminate narrowing gradually to a point
acute sharply pointed
adnate fusing together of two different organs e.g. stipules adnate to petiole
adpressed lying flat; closely pressed against a surface
adventitious (of roots) arising in an unusual position, e.g. along a stem
aerenchyma tissue with tiny air-hollows, a common character of aquatic plants
aggregate a group of very closely related species
anastomosing with the 2° veins forming a closed loop
annual a plant that completes its life cycle from germination to death within twelve months
anther the part of the stamen that produces pollen
anti-clockwise twining to the right with stem in front of the support, growing from lower left to upper right (*see inside back cover*)
antrorse pointing forward or upwards towards the apex
apiculate with an apiculus
apiculus a short sharp point
applanate flattened out
arcuate curved, usu fairly strongly
arista an awn or bristle
aristate with an arista (*see PL 22*)
ascending growing obliquely at first but becoming erect
auricle a small lobe or ear-shaped appendage
auriculate having auricles (*see inside back cover*)
awl-shaped narrow and sharply pointed; gradually tapering from base to a slender or stiff point
awn a bristle-like projection
awned with an awn
axil the junction of the upper side of a leaf and the stem
axillary in or pertaining to an axil
basioscopic on the side towards the base
biennial a plant that completes its life cycle in two years, usually forming a basal rosette in the first year
bifid divided to, or less than, half-way into two parts
bipartite deeply divided to the near the base into two parts
bract a much-reduced leaf, usu on the upper stem
bracteole small bract subtending a flower, usually one in monocots and two in dicots
bulb a swollen rootstock consisting of fleshy scale leaves, e.g. in *Allium* species
bulbil a small bulb

bulbous having a bulb
bullate with blister-like swellings on the surface; blistered or puckered
calyx the outer perianth, composed of free or united sepals
cartilaginous tough and without chlorophyll; resembling cartilage in consistency (*see* PL *23*)
cataphyll a reduced or vestigial leaf at base of a stem
ciliate fringed with hairs (**cilia**), usually referring to a leaf margin (*see* PL *23*)
ciliolate minutely ciliate
circinnate coiled in a flat spiral, with the apex innermost, as in many unrolling fern fronds
cladode a green leaf-like lateral organ or shoot; a branch resembling a leaf in form and function
clathrate resembling a lattice; used to describe scales with a single layer of translucent cells with dark cell walls (only occurring in ferns)
cleistogamous self-pollinating, without the flower ever opening
clockwise twining to the left with stem in front of the support, growing from lower right to upper left (*see inside back cover*)
collar (of grasses) connective tissue between leaf margin and sheath margin
collateral lying side-by-side
colleters minute papillae-like glands found in the leaf axils of *Galium* species
compound leaf leaf that is divided to the rachis into discrete segments that individually resemble leaves
conical cone-shaped
connate (of similar organs) joined or fused together
cordate heart-shaped; with base rounded and prominently notched (*see inside back cover*)
corm a bulb-like swollen underground stem, but solid and not composed of fleshy scale leaves
cormous having a corm
cotyledon a seed leaf; one of the first leaves of the embryo of a seed plant (typically one in monocots, two in dicots, and two or more in conifers)
crenate with rounded teeth or lobes; scalloped (*see* PL *23*)
crenulate minutely crenate
culm the flowering stem of a grass
cultivar a distinct garden selection maintained by vegetative propagation, or a true-breeding race developed in cultivation
cuneate wedge-shaped, with straight sides converging at the base (*see inside back cover*)
cuspidate ending abruptly in a sharply pointed tip
decumbent lying along the ground with the apex ascending
decurrent extending downwards beyond the point of attachment
decussate four-ranked, in opposite pairs with each pair at right angles to the next
dendritic branched like a small tree
dentate toothed, with acute teeth facing outward (*see* PL *22, 23*)
denticulate minutely dentate (*see* PL *23*)
dichotomous with two branches
dicot dicotyledon; a flowering plant usually with two cotyledons
digitate with lobes like the fingers of the hand
disarticulating separating at a joint or articulation
distal situated away from the point of attachment; not proximal
distichous arranged in two opposite vertical rows
divaricate widely spreading
divergent spreading away from each other, usually at a wide angle
domatia (singular **domatium**) tiny pockets in the axils of veins on the lower surfaces of leaves
eccentric off-centre; having the axis not centrally placed
ellipsoid (of a 3-dimensional body) elliptical in outline
elliptic widest in the middle, tapering equally at both ends
emarginate deeply and broadly notched at the apex; deeply retuse
entire with margin unbroken by teeth, lobes or other indentations; uninterrupted; whole (*see inside front cover,* PL *22*)
epicormic sprouting from dormant buds on the trunk of a tree or shrub

equitant leaves overlapping in two ranks and folded (and mostly fused) along the midrib, e.g. in the leaves of *Iris* species
erecto-patent between spreading and erect, ± at 45°
erose having an irregularly minutely eroded margin; appearing gnawed
excurrent extending beyond the apex, as in a midrib developing into a mucro or awn
explicative with the leaf margins in the bud folded back
extravaginal (of a grass shoot) arising from an axillary bud which breaks through the sheath of the subtending leaf and thus giving rise to loose tufts; not intravaginal
fascicle a cluster or bundle
fascicled arranged in clusters or bundles (fascicles)
fastigiate with branches erect and more or less adpressed to the vertical
fimbriate fringed; having the margin cut into long, slender lobes
floccose covered with tufts of soft woolly hairs which rub off readily with age
geniculate abruptly bent; knee-like
gland an organ that produces a secretion
glandular possessing glands
glaucous bluish-white
globose spherical or globe-shaped
glume a bract at the base of a spikelet in grasses (usually occurring in pairs)
hastate widely spearhead-shaped, with basal lobes spreading outwards at a wide angle (*see inside back cover*)
herb a non-woody plant, or one that is woody only at the base
herbaceous composed of green non-woody tissue
hispid having stiff bristly hairs, often piercing to the touch
hortal a plant of garden origin
hyaline very thin, colourless and transparent
hydathode a gland that exudes water (occasionally with lime, salt, etc), usually confined to the apex and teeth of a leaf
imbricate overlapping, like roof tiles
incurved curved inwards
inflorescence the arrangement of flowers on the floral axis; a flower cluster
intercalary leaflets/lobes leaflets/lobes alternately large and small
intercalary veins distinct lateral veins between the secondary veins
internode the part of a stem between two nodes
interpetiolar between two opposite leaf bases
intravaginal (of a grass shoot) arising from an axillary bud which does not break through the sheath of the subtending leaf giving rise to dense tufts; not extravaginal
involucre a ring of bracts surrounding or subtending an inflorescence
involute (of leaf in bud) with each margin rolled equally inwards (upwards)
Kranz venation opaquely and thickly net-veined; wreath-like
laciniate jagged, torn
lacunae air-hollows
ladder-fibrillose the ladder-like pattern of fibrillae (small fibres) best seen on older sheaths in some *Carex* species
lanceolate lance-shaped; at least 3 times longer than wide, widest below the middle and tapering to apex
lateral at the side
latex coloured juice (sap), usually white, but may be cream, green, yellow, orange or red
leaflet a leaf-like segment of a compound leaf
leaf-opposed at a node on the opposite side from a leaf
ligule membranous or scarious projection from the top of a leaf sheath in monocots; an appendage near the base of leaves in *Selaginella* and *Isoetes*
linear many times longer than wide, with the margins nearly parallel
lobe a division of an organ; cut deeper than ½ way to the midrib
lobed having one or more lobes (*see inside front cover, PL 22*)
lobulate having small lobes

lyrate pinnatifid, with the terminal lobe much larger than the others and usually rounded
mealy covered with a flour-like powder
medifixed (of a hair) attached at the middle
megaspore a very large spore
membranous membrane-like; thin, dry and translucent; slightly thicker than hyaline
midrib the main or central vein of a leaf
moniliform (of a hair) resembling a string of beads
monocarpic flowering and fruiting once only before dying
monocot monocotyledon; a flowering plant having one cotyledon
monopodial with a simple main stem or axis, growing by apical extension
mucro a short stiff point, often an extension of the midrib
mucronate with a mucro
multiseriate (of a hair) consisting of 2 or more rows of parallel cells
net-veined having veins that join up to form a network
node the point on a stem from which one or more leaves are borne
obcordate inversely heart-shaped, broadest towards the notched apex and tapering to the base
obdeltoid inversely triangular, with the apex truncate and tapering to the stalk
oblanceolate inversely lanceolate, widest above the middle and tapering to the base
oblique asymmetric; having unequal sides
oblong longer than broad with nearly parallel sides and rounded at both ends
obovate inversely ovate, widest above the middle and tapering to the base
obovoid (of a 3-dimensional body) inversely ovoid (egg-shaped), attached at the narrower end
obtrullate inversely trullate; shaped like a bricklayer's trowel with the two short sides at the apex
obtruncate inversely truncate
obtuse blunt; with a more or less rounded apex (at an angle >90°)
obvolute (of leaves in bud) with half of one leaf wrapped round half of another leaf
ochrea (plural **ochreae**) a sheath formed from two stipules fused into a tube, e.g. in Polygonaceae (*see inside back cover*)
orbicular circular in outline
ovate egg-shaped, less than 3x as long as wide, widest below the middle and tapering to apex
ovoid (of a 3-dimensional body) egg-shaped
palmate radiating from a central point; with more than three segments or leaflets arising from a single point, as in the fingers of a hand (*see inside front cover*); having veins radiating from the end of a leaf stalk to the tips of the lobes
panicle a much-branched inflorescence
papillae small rounded or pimple-like protuberances
papillate bearing papillae
patent spreading widely and straight; at ± 90º to a surface
pedicel the stalk of a single flower
peduncle the stalk of an inflorescence (which may be reduced to a single flower)
peltate shaped like a disc with stalk attached centrally (i.e. not at the margin). Peltate leaves are found in *Hydrocotyle vulgaris*, *Tropaeolum* and *Umbilicus*
perennial with a life cycle lasting three or more years
perfoliate (of leaf) completely encircling the stem
petiolar relating to a petiole
petiole a leaf stalk
petiolule the stalk of a leaflet in a compound leaf
phyllaries (singular **phyllary**) involucral bracts that surround the head of flowers in Asteraceae
phyllode a flattened and expanded petiole having the function and appearance of a leaf
pinna (plural **pinnae**) primary division or leaflet of a compound leaf (which may be further divided)
pinnate compound, with leaflets or pinnae arranged on opposite sides of a common stalk, with or without a single terminal leaflet (*see inside front cover*); having veins along each side of the mid-rib of a leaf
pinnatifid pinnately lobed, the lobes extending from a quarter to half-way to the midrib (or rachis)
pinnatisect deeply pinnately lobed almost to the midrib (or rachis)

pinnule the second or third order divisions of a pinnate leaf (bi- or tripinnate); the smallest (lowest-rank) foliar unit of a divided pinna
plagiotropic growing at an angle to the sun (usually horizontally)
plicate folded more than once into pleats; fan-like
polychotomous with several branches
prickle a hard sharp-pointed outgrowth (lacking vascular tissue) from the epidermis of a stem or leaf
pricklet a small prickle
prostrate lying on the ground
proximal situated near to the point of attachment of an organ; not distal
pruinose covered with a whitish bloom that is easily rubbed off
pseudowhorled (of leaves) alternate but closely spiralling around a stem so as to appear whorled
raceme an elongated inflorescence of stalked flowers on a common rachis and opening progressively upwards
rachis (plural **rachises**) the axis (excluding petiole) of either a compound leaf or an inflorescence
ray floret one of the outer, irregular tiny flowers in the flower heads of some plants in the Compositae
recurved bent or curved downwards or backwards
reflexed bent abruptly backwards
reniform kidney-shaped
reticulate marked with a network pattern of veins
retrorse (of a hair) bent or curved backwards or downwards
retuse shallowly notched at a rounded apex
revolute (of a leaf in bud) with both margins rolled equally downwards
rhizome a root-like stem, usu lying horizontally under the ground
rhombic having the shape of a rhombus (an equilateral parallelogram)
rootstock junction of stem(s) and roots
ruderal growing in waste places; a plant that grows in waste places
rugose markedly wrinkled
runcinate (of leaf) with lobes directed backwards towards the base
sagittate shaped like an arrowhead, with two backward-directed lobes (*see inside back cover*)
saprophytic living on decaying organic matter
scabrid rough, often to the touch
scape a leafless stalk, arising from the rootstock which bears an inflorescence
scarious thickly membranous
scurfy covered with small flake-like scales
sepal a single segment of the calyx
septa (singular **septum**) partitions
septate divided into compartments by septa
serrate saw-toothed, with acute teeth pointing towards the apex (*see Pl 22, 23*)
serrulate minutely serrate
sessile without a stalk; attached directly to an axis or organ
sheath a tubular structure surrounding an organ or part of an organ
sheathing (of leaf or petiole) with at least the basal portion dilated and/or forming a sheath
shrub a woody plant smaller than a tree, with several stems arising from ground level
silky having a covering of soft fine straight adpressed hairs
sinuate wavy at margins only (2-dimensional, unlike **undulate**) (*see Pl 22*)
sinuous shallowly curved
sinus (plural **sinuses**) the gap between two teeth, lobes or segments of an organ
sori (singular **sorus**) an aggregation of spores (sporania)
spathulate spoon-shaped, usually with a rounded apex
spikelet a unit of the inflorescence in grasses, consisting of one or more florets and usually subtended by a pair of glumes
spindle-shaped swollen in the middle and tapering at both ends
spine a stiff sharp-pointed structure originating as a modified leaf, stem or stipule
spinescent ending in a spine or sharp point
spinose spiny

spinulose bearing small spines
spiral fibres tightly coiled fine strands of xylem contained within veins and vascular bundles visible as fine threads when tissue is pulled apart
stamen one of the male sex organs, usually consisting of an anther and filament
stele the central core of the stem of some plants; a cylinder of vascular strands
stellate (of a hair) star-shaped
stigma the apex of the style, usually enlarged, on which pollen grains germinate
stipel a stipule-like outgrowth at the base of a leaflet
stipoid glands glands in the position where stipules would normally originate
stipular relating to a stipule
stipule a small herbaceous (or rarely spiny) appendage, normally in pairs at the base of the petiole
stolon an above-ground stem usually rooting at the nodes and capable of giving rise to new plants
stoloniferous bearing stolons
stomata (singular **stoma**) pores found in the leaf (and often stem) epidermis used for gas exchange
striae fine longitudinal lines, grooves or ridges
striate having striae
strigose with sharp, adpressed, stiff hairs or bristles (often basally swollen)
stylar orifice (of *Rosa* hips) the opening at the top of the fruit from which the styles protrude
style the stalk of a stigma, usually borne at apex of ovary
subopposite almost opposite
subshrub a low shrub, often with herbaceous stems
substrate the surface to which a plant is anchored
subtending adjacent to (usually beneath); originating from the same point
subulate awl-shaped; long, narrow and gradually tapering to a fine point
sucker a vigorous vegetative shoot of underground origin
superposed (of buds) positioned one above the other at one node
surcurrent running up, as when the base of the leaf is prolonged up the stem as a wing
sympodial with the main stem or axis ceasing to elongate, growth being continued laterally
taproot a swollen or well-developed main root with much smaller lateral roots
tendril part or all of a stem, leaf or petiole modified to form a slender twining appendage
tepals sepals or petals (or both) - used when the distinction is not obvious
ternate arranged or divided into threes
thorn a short, modified branch with a sharp, hard point
translucent allowing light to pass through, but not transparent; semi-transparent
transverse crosswise
tree a tall woody plant with a single main trunk
trichotomous with three branches
trifid divided to, or less than, half-way into three parts
trigonous three-angled, the angles blunt
tripartite deeply divided to near the base into three ± equal parts
triquetrous three-angled, the angles sharp
trullate shaped like a brick-layer's trowel; rhombic (with 4 straight sides)
truncate appearing as if transversely cut off (*see inside back cover*)
tuber an underground stem or root that is swollen and contains food reserves
tubercle a small wart-like or knobbly projection
tubercled with tubercles
tuberous with tubers
tufted having rhizomes short or absent
turion a winter-bud formed of tightly-packed leaves by which some aquatic plants propagate; a small rootless daughter plant in *Lemna*
twiner a climbing plant that supports itself by winding around an object
umbel an inflorescence in which all the pedicels arise from the same point; umbrella-like
undulate with a wavy or corrugated surface or margin (*see Pl 22*)
uniseriate (of a hair) consisting of a single row of cells (i.e. simply septate)

utricle a bladder-like structure, especially the membranous sac of the fruit of *Carex* species that contains a nutley within
valvate (of leaves in bud) meeting without overlapping
vascular bundle the thread-like primary conducting tissue in higher plants consisting of xylem (water-conducting tissue) on the inside and phloem (food-conducting tissue) on the outside, separated by a layer of cambium
vein an assemblage of strands of vascular tissue in a leaf or similar structure
veinlet a small vein
venation the arrangement of veins, especially in leaves or leaf-like structures
vernation the arrangement of unexpanded leaves in a vegetative bud (still discernible as the young leaves emerge and sometimes much later)

EXPLANATORY NOTES ON SOME IMPORTANT CHARACTERS & GLOSSARY TERMS

The following notes briefly explain some of the important characters that are useful in vegetative identification. They expand in more detail some of the terms to be found in the glossary; they are a précis of a series of articles in *BSBI News* (Poland, 2005-2008). Additionally, details may be searched for in a range of botanical textbooks that cover the anatomy and morphology of plants.

Some of these characters may be used as a shortcut to the identification of an unknown plant. Remember that the characters are not mutually exclusive, and may only be present when very young. Families or genera in brackets (...) indicates that not every species shares that particular character.

HAIRS

The nature and type of hairs present on the vegetative parts of plants can give reliable clues to the identification of many species. Aquatic plants are never hairy (like Olympic swimmers!) as the hairs would soon clog with mud and other particles. In a few species, emergent stems and leaves of aquatics can become hairy, for example *Ludwigia grandiflora*.

Septate hairs (major families given only) - Asteraceae, Lamiaceae.

Septate (or multi-cellular) hairs are recognisable as the septum of each individual cell is normally discernible. This is clearly seen with a x20 lens, although x10 will usually suffice

Unicellular hairs (major families given only) - Apiaceae, Boraginaceae, Brassicaceae, Fabaceae, Ranunculaceae, Rosaceae.

Each hair simply consists of a single cell; no septa are present. The hairs of most Brassicaceae are unicellular, even when branched.

Hispid hairs - (Boraginaceae)

Some (usually unicellular) hairs can be sharply bristly (hispid) and almost piercing to the touch. Hispid hairs are particularly frequent in Boraginaceae, *Echium vulgare* being one such example.

Unicellular hair (*Arabidopsis thaliana*)

Septate hair (*Thymus polytrichus* ssp *britannicus*)

Hispid hair (*Echium vulgare*)

Glandular hairs (many families and genera, variable even within a species)

Glandular hairs can be unicellular or septate but they always have a swollen gland at the apex (sometimes minute and often disappearing as the volatile oil contained within escapes).

Glandular hair (*Spergula arvensis*)

Forked hairs - (Asteraceae), (Brassicaceae)

Within the Brassicaceae, forked hairs are mostly confined to the genera *Erophila*, *Draba* and *Arabis*. The basal leaves of *Hesperis matronalis* can look remarkably similar to some Boraginaceae, but at least some hairs are forked on the former.

Forked hair (*Leontodon saxatilis*)

Stellate hairs - *Artemisia dracunculus*, (Brassicaceae), *Buddleja*, (*Hedera*), *Helianthemum*, (*Hieracium*), *Lavandula*, Malvaceae (usu present), *Parrotia*, *Pilosella*, *Pterocarya*, (*Quercus*), (*Sorbaria*), (*Tilia*), *Tuberaria*, (*Viburnum*)

These are branched hairs where the rays or arms radiate out like a star.

Dendritic hairs - *Erica arborea*, *Descurainia*, *Dryas*, (*Mentha* - much-branched), *Phlomis*, *Platanus*, (*Ribes*), *Verbascum*

These are multi-branched hairs. They are a feature of *Verbascum* and a few other species. The petiole and lower side of the leaves of *Platanus* x *hispanica* may also have some dendritic pubescence.

Dendritic hair (*Verbascum* sp)

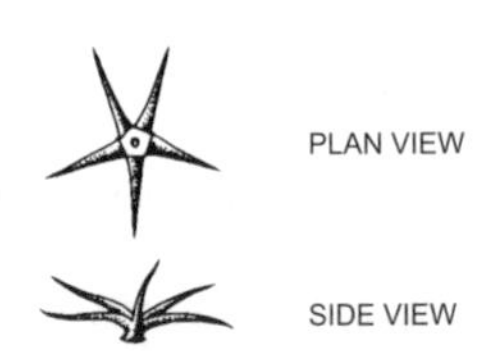

Stellate hair (*Capsella bursa-pastoris*)

Hooked hairs - *Parietaria*, *Picris*, *Symphytum*

Hairs strongly recurved at the apex, forming a hook or barb. The simple hooked hairs of *Symphytum* are unlike those of *Picris* which are minutely bifid or trifid (like miniature grapnels). Hooked hairs are often not easily seen but can easily be detected as the leaves readily stick to woolly clothing!

Hooked hair (*Symphytum* sp)

Hooked hair (*Picris echioides*)

Scabrid hairs - *Hieracium, Papaver, Pilosella*

In these hairs the surface is distinctly rough.

Scabrid hair (*Papaver* sp)

Medifixed hairs - *Cornus*, *Erysimum*, *Lobularia*

These hairs are positioned flat against the leaf surface and are 'bolted' in the middle, giving a distinctive appearance.

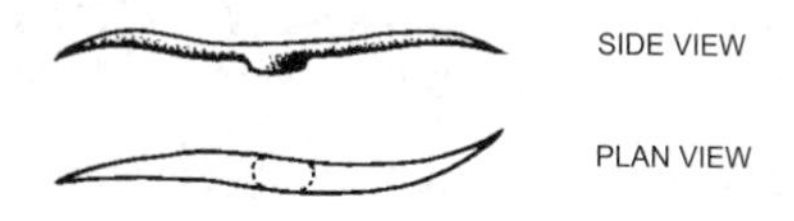

Medifixed hairs (*Lobularia maritima*)

Cobwebby & cottony hairs - (Asteraceae)

Lumped here for simplicity, these hairs are a feature of many Asteraceae including *Tragopogon pratensis*, *Pulicaria dysenterica* and thistles in the genera *Carduus*, *Carlina*, *Cirsium* and *Onopordum*. However, strands of mildew mycelia can look remarkably similar - observers beware!

Anvil-shaped hairs - *Humulus lupulus*

These hairs undoubtedly act in lieu of tendrils in order to climb up to the light.

Anvil-shaped hair (*Humulus lupulus*)

Mealy (& bladder) hairs - *Atriplex*, *Chenopodium*, (*Primula*)

Mealy hairs may be found covering the leaves of some *Primula* species. Bladder hairs uniquely characterise the genera *Atriplex* and *Chenopodium* (although they are absent from some species). These hairs, appearing like rather large sessile glands, often soon wear off so it is best to look at the youngest leaves. The shape is subtly different between the two genera, typically spherical in *Chenopodium* and often, more or less, flattened in *Atriplex*. Bladder hairs have probably evolved to rid the plant of excessive amounts of salt in the tissues.

Bladder hair (*Chenopodium album*)

Peltate scales (not strictly a hair) - (*Callitriche* - minute and confined to stems), *Deutzia*, *Elaeagnus*, (*Hedera*), *Hippophae*

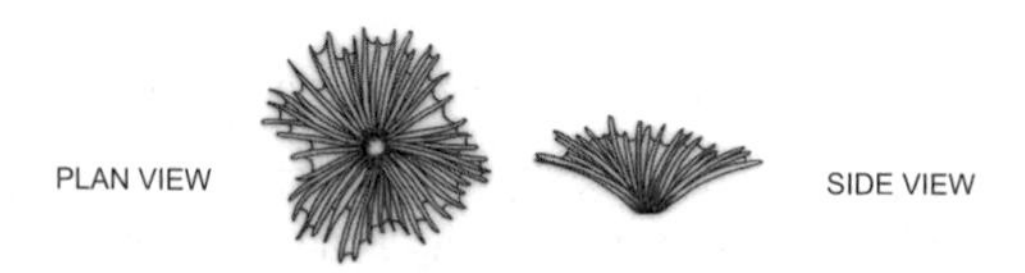

Peltate scales (*Hippophae rhamnoides*)

LATEX

White latex - (Aceraceae), (Alismataceae), (Apiaceae), Asteraceae, tribe Lactuceae, Campanulaceae (see also green/cream latex below), Convolvulaceae, *Ficus*, *Morus*, Papaveraceae (often turning yellow or drying red), *Rhus*, *Vinca*

Green or cream latex - *Angelica sylvestris*, *Campanula rhomboidalis*, *Campanula trachelium*, *Peucedanum ostruthium*, *Phyteuma spicatum*

Yellow, orange or red latex - *Eschscholzia* (confined to rootstock), (Papaveraceae), *Scorzonera*

Latex is simply a coloured juice which exudes from broken canals or lactifers within the plant tissues. Latex is usually white (at least initially) and is often present only in the young tissues of a plant. A cross-section of young petioles (and rootlets) is a good means of locating latex although it can still be sparse and easily overlooked, thus confirmation often demands a good hand lens.

STOMATA

Stomata are pores on the leaf (or stem) surface used for respiration during photosynthesis. Contrary to popular belief, stomata are indeed a field character - not just a microscopic one - and they can normally easily be seen as little white dots under a hand lens (ideally x20). It is primarily the larger sized pair of guard cells that surround the pore that we actually see. The presence of stomata on the underside of leaves is generally often little use for identification (except for their absence in some grasses) and the structure of the cuticle can make them appear obscure or absent. Hence, it is thus the presence, or absence, of stomata on the upperside of the leaves that is the useful character.

The majority of our vascular plants are **amphistomous** (stomata present on both sides of the leaf, although often in unequal proportions) or **hypostomous** (stomata on the lower (abaxial) side of the leaves only). **Hyperstomous** species (those with stomata on the upper (adaxial) side of the leaves only) are generally rare although some grasses, and those aquatic plants with floating leaves, qualify in this category. Totally submerged aquatic plants do not develop stomata since they would effectively drown. As a general rule, stomata are arranged in parallel lines in monocots (with many exceptions!) and are randomly scattered in dicots.

VASCULAR BUNDLES (vb's)

Many similar species can be separated solely and reliably by observing the vascular bundles visible in a cross-section of a petiole. Vascular bundles are tightly packed strands of xylem and phloem visible (as a whole) to the naked eye. The number, shape and arrangement of the vascular bundles within the petioles often provide clues to the identity of many species. Although cross-sections are best done using a razor blade, sharp scissors or even a clean break will suffice for rapid identification in the field. It is important to observe the cross-section in the mid-section of the petiole, as the structure may vary at the proximal and distal ends.

HYDATHODES

Hydathodes (*see Pl* 22) are modified stomata forming pores that are found at the apex and along the margins of some leaves. They act as a mechanism for the removal of excess water in a process called guttation. Although hydathodes are probably present in virtually all plants they are often much-reduced, obscure or invisible even to a strong hand lens. The shape, position and pigmentation of hydathodes are all useful characters in vegetative identification.

BUDS

Buds are useful aids to identification at all times of year. Even during late spring when buds are bursting into 'leaf', a few dormant or moribund buds can usually be found - one of the few instances where dead twigs may be more helpful than living ones!

Buds normally occur in the leaf axils (a lateral bud) or at the tip of the stems (a terminal bud). Consequently, the distribution of buds on the twig is identical to that of the leaves - alternate, opposite, or whorled. Similarly, predominantly opposite twigs indicates opposite leaves. These features are helpful when identifying twigs during the winter.

As the bud develops, the bud scales may enlarge before dropping off, leaving scars indicating the age of young twigs. However in some families, especially Caprifoliaceae (e.g. *Symphoricarpos* and *Lonicera*), the bud scales persist on the twig and provide a useful indication of the family.

RHIZOMES

Plants with long underground rhizomes form large spreading patches; these may constitute a mono-culture with many shoots closely set, e.g. *Phragmites australis*. Those with short rhizomes will generally appear as tufts of several shoots; and those without rhizomes will appear as obviously single plants (although a plant may have more than one stem). Experience should enable the presence or absence of rhizomes to be accurately assessed simply by looking carefully at the ground encompassing the plant.

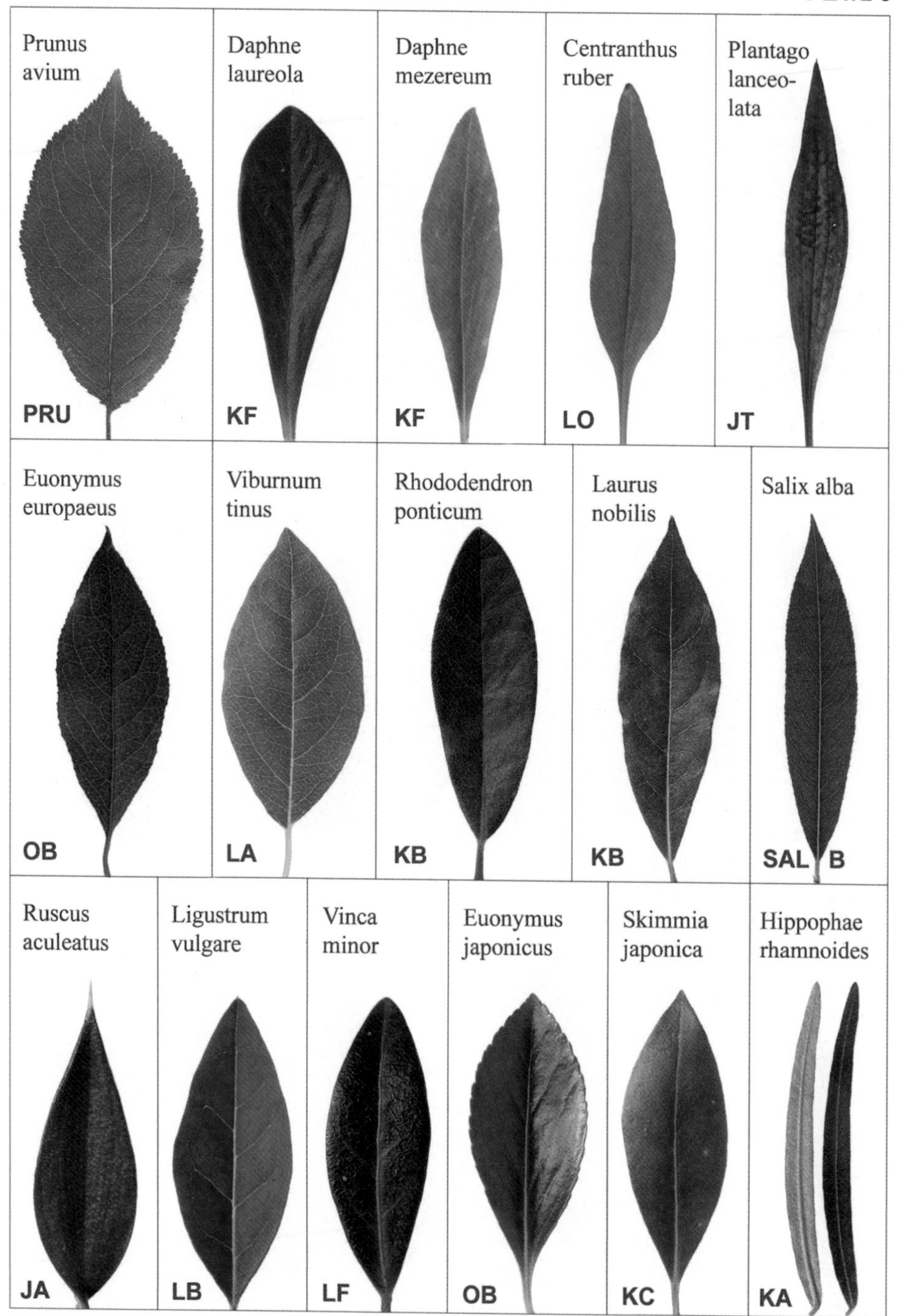
Prunus avium
PRU
Daphne laureola
KF
Daphne mezereum
KF
Centranthus ruber
LO
Plantago lanceo-lata
JT
Euonymus europaeus
OB
Viburnum tinus
LA
Rhododendron ponticum
KB
Laurus nobilis
KB
Salix alba
SAL B
Ruscus aculeatus
JA
Ligustrum vulgare
LB
Vinca minor
LF
Euonymus japonicus
OB
Skimmia japonica
KC
Hippophae rhamnoides
KA

PLATE 10

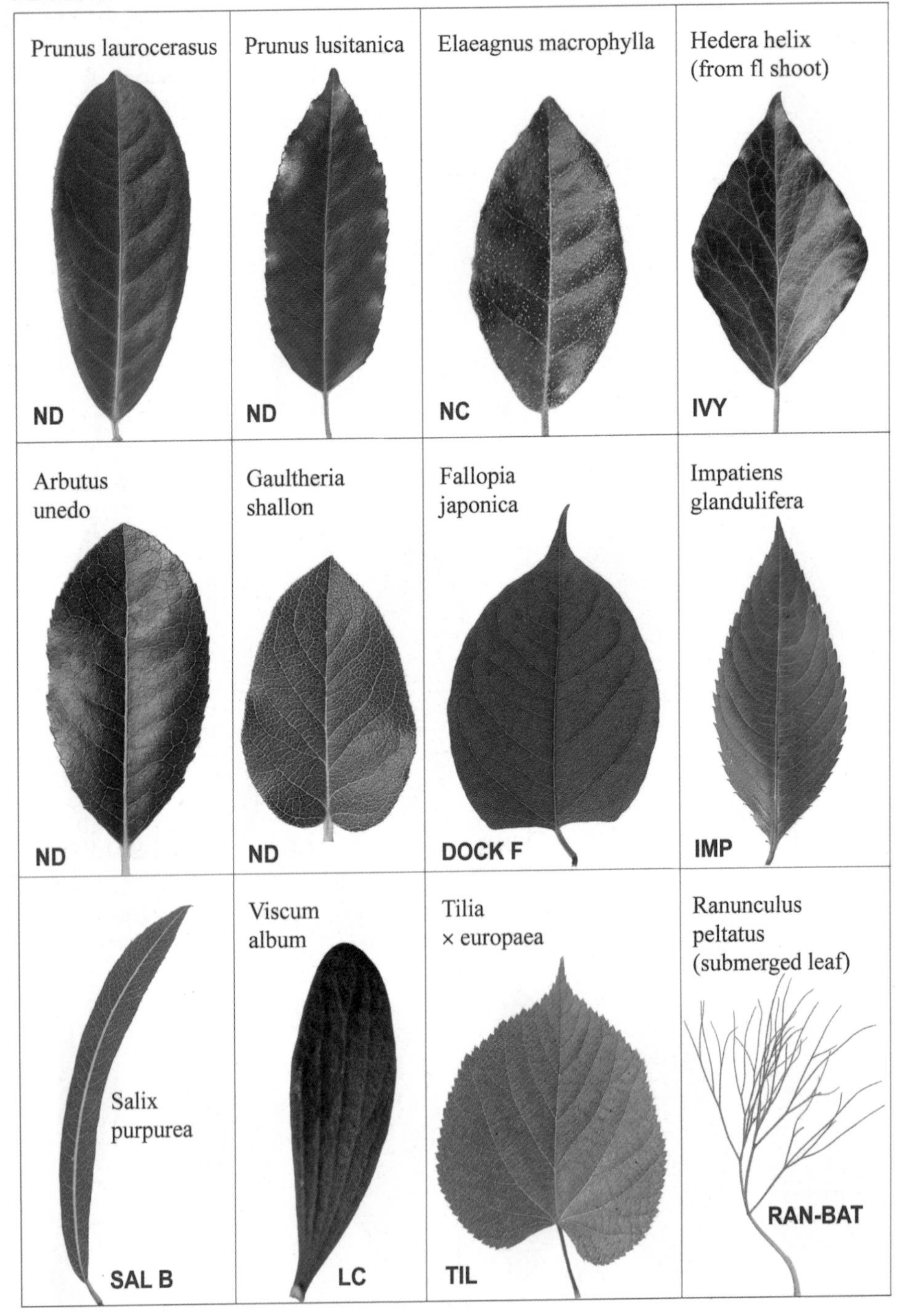
Prunus laurocerasus
ND
Prunus lusitanica
ND
Elaeagnus macrophylla
NC
Hedera helix
(from fl shoot)
IVY
Arbutus
unedo
ND
Gaultheria
shallon
ND
Fallopia
japonica
DOCK F
Impatiens
glandulifera
IMP
Salix
purpurea
SAL B
Viscum
album
LC
Tilia
× europaea
TIL
Ranunculus
peltatus
(submerged leaf)
RAN-BAT

NT
Bellis perennis
NR
Saxifraga spathularis
BER
Berberis x ottawensis
NQ
Primula vulgaris
NQ
Primula veris
PRUNE
Prunella vulgaris
NB
Dryas octopetala
LS
Silene dioica
LK
Claytonia perfoliata

PLATE 12

Alnus glutinosa
NE
Griselinia littoralis
KC
Petasites fragrans
NO
Cotinus coggygria
KB
Rhamnus cathartica
OB
Viburnum lantana
OA
Corylus avellana
NE
Carpinus betulus
NE
Fagus sylvatica
KB

Asarum europaeum
LT
Aristolochia clematitis
KD
Tamus communis
KD
Glechoma hederacea
OK
Alliaria petiolata
NQ
Lunaria annua
OK
Ranunculus ficaria
RAN B
Viola odorata
VIO
Lamium album
OJ

PLATE 14

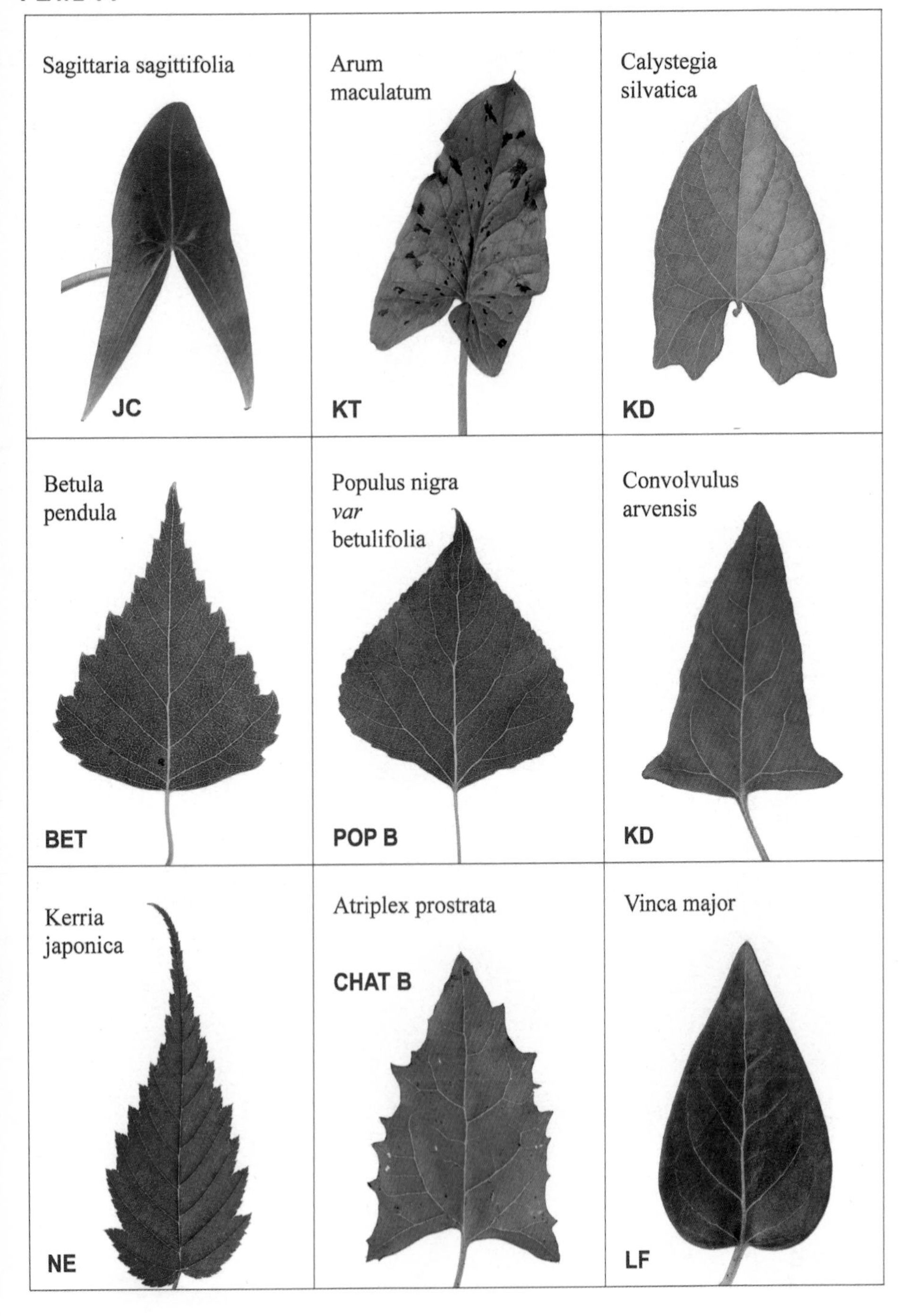

Sagittaria sagittifolia
JC
Arum maculatum
KT
Calystegia silvatica
KD
Betula pendula
BET
Populus nigra *var* betulifolia
POP B
Convolvulus arvensis
KD
Kerria japonica
NE
Atriplex prostrata
CHAT B
Vinca major
LF

Hedera helix
(juvenile)
IVY
Bryonia
dioica
PC
Malva
sylvestris
PI
Ribes
uva-crispa
PB
Ribes
nigrum
PB
Acer
campestre
QA
Ficus carica
PB
Acer platanoides
QA
Acer pseudoplatanus
QA

PLATE 16

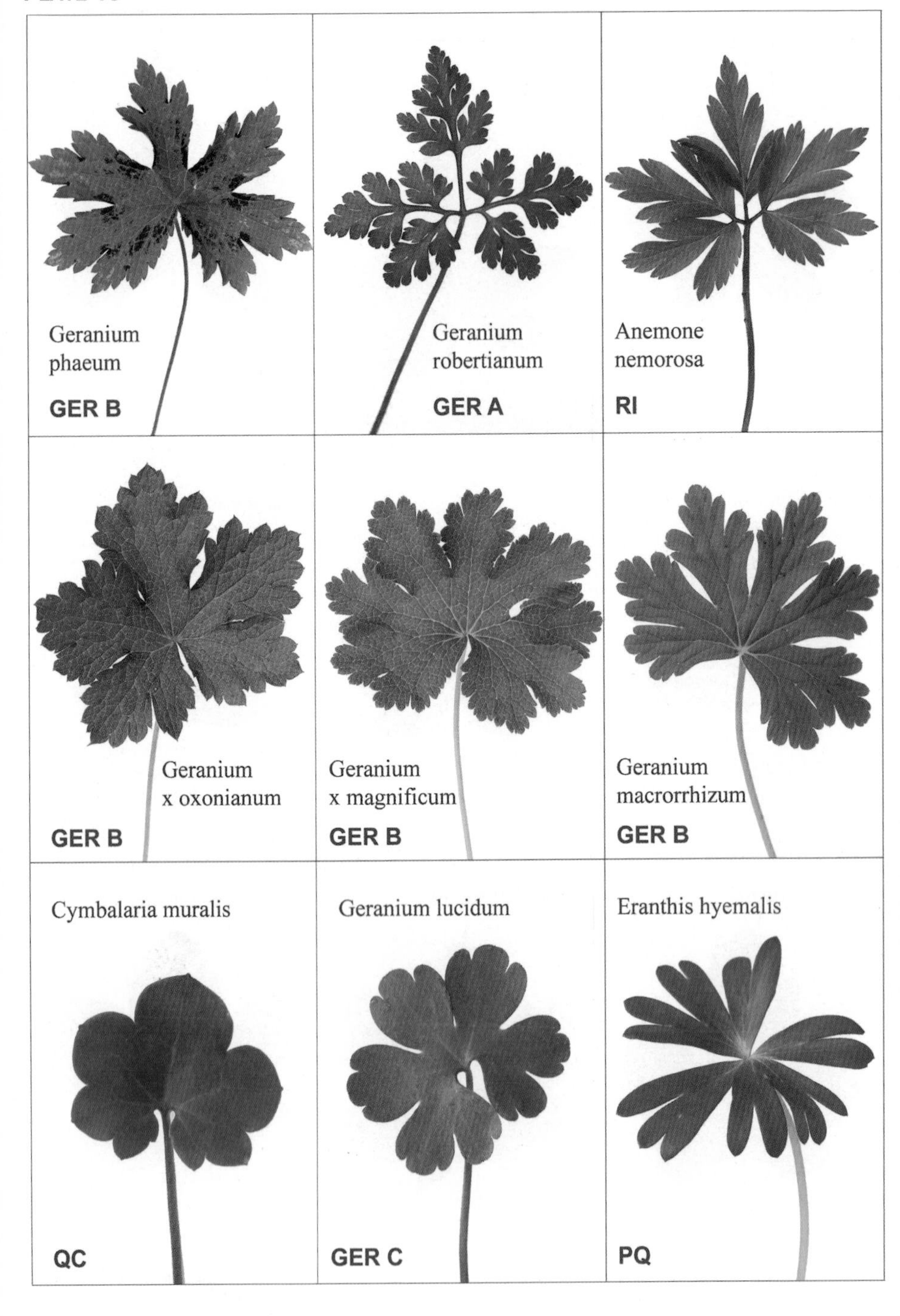
Geranium phaeum
GER B
Geranium robertianum
GER A
Anemone nemorosa
RI
Geranium x oxonianum
GER B
Geranium x magnificum
GER B
Geranium macrorrhizum
GER B
Cymbalaria muralis
QC
Geranium lucidum
GER C
Eranthis hyemalis
PQ

Jasminum nudiflorum
RH
Fragaria vesca
RO
Trifolium repens
RN
Ranunculus acris
RAN C
Ranunculus repens
RAN C
Oxalis acetosella
RC
Geranium pusillum
GER C
Hedera helix (juvenile)
IVY
Humulus lupulus
QB

PLATE 18

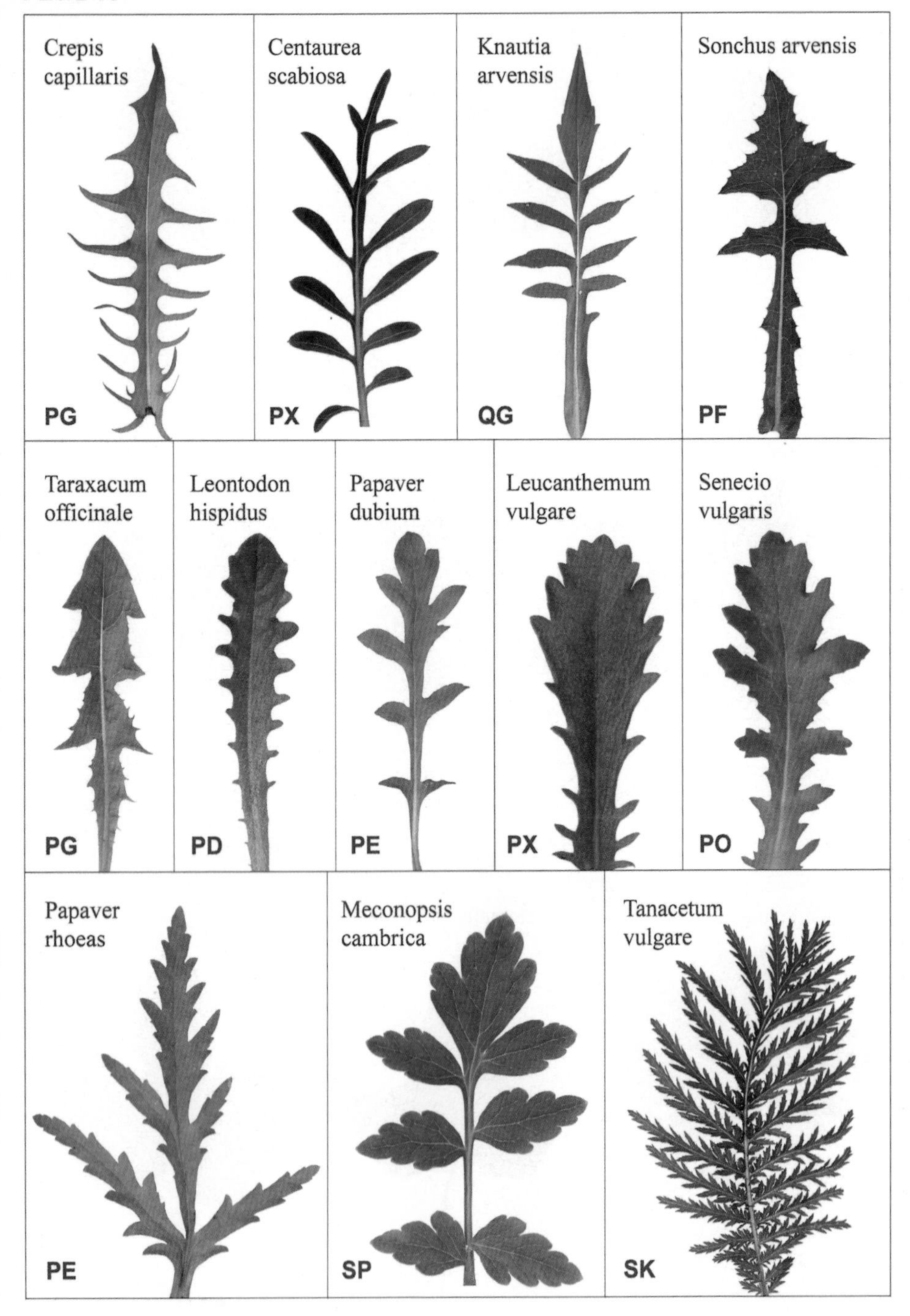

Crepis capillaris
PG
Centaurea scabiosa
PX
Knautia arvensis
QG
Sonchus arvensis
PF
Taraxacum officinale
PG
Leontodon hispidus
PD
Papaver dubium
PE
Leucanthemum vulgare
PX
Senecio vulgaris
PO
Papaver rhoeas
PE
Meconopsis cambrica
SP
Tanacetum vulgare
SK

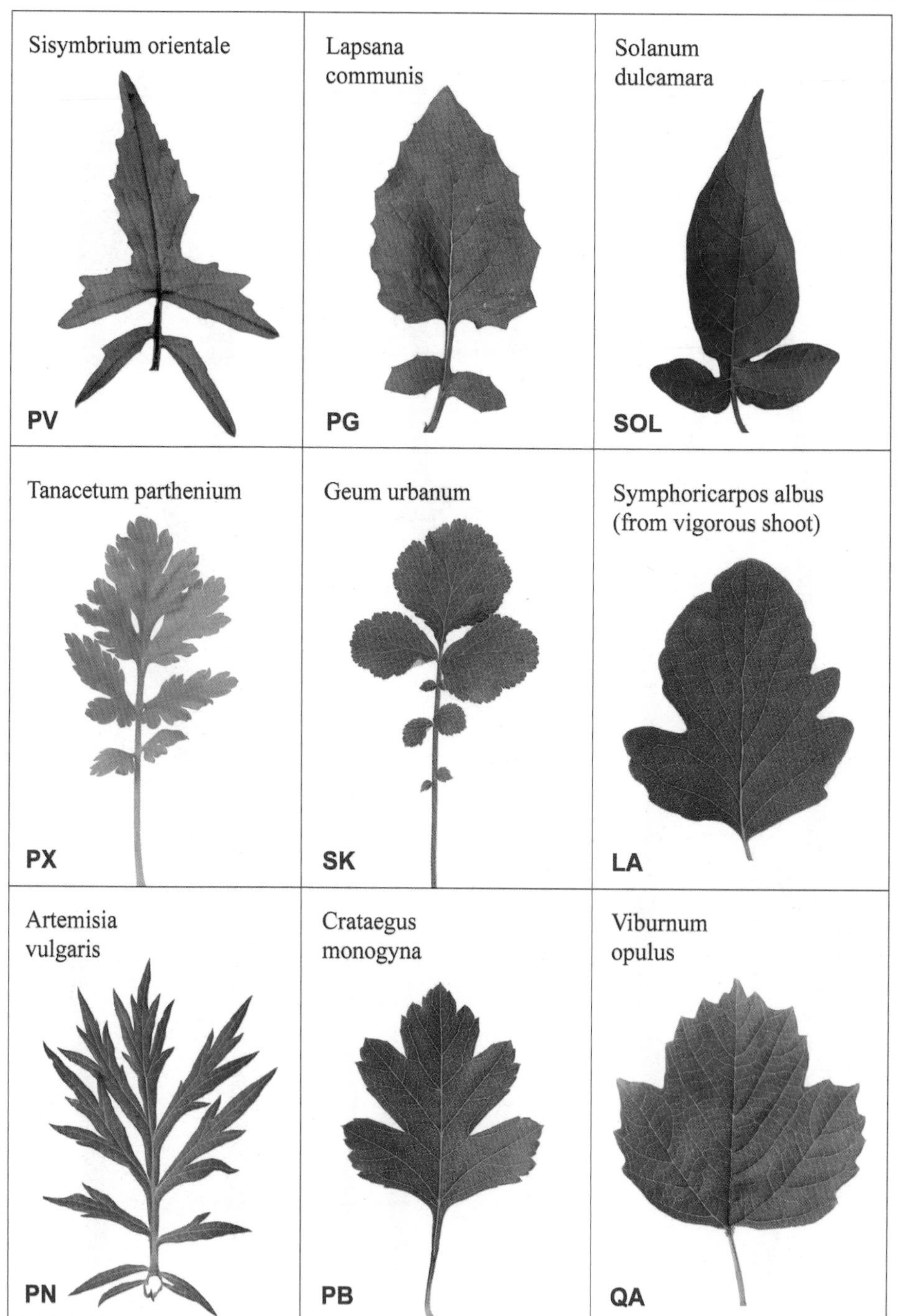
Sisymbrium orientale
PV
Lapsana communis
PG
Solanum dulcamara
SOL
Tanacetum parthenium
PX
Geum urbanum
SK
Symphoricarpos albus (from vigorous shoot)
LA
Artemisia vulgaris
PN
Crataegus monogyna
PB
Viburnum opulus
QA

PLATE 20

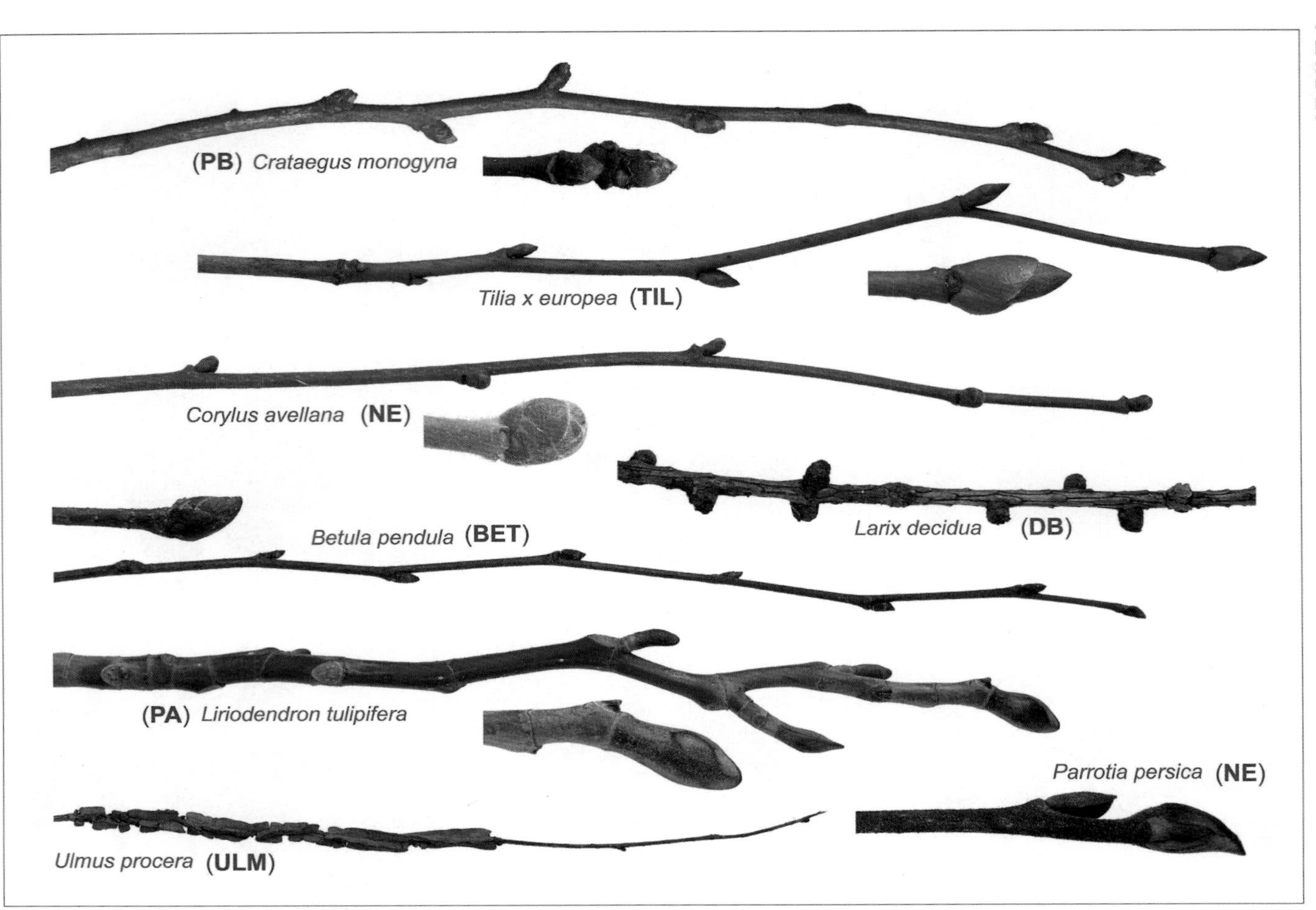

(PB) Crataegus monogyna
Tilia x europea (TIL)
Corylus avellana (NE)
Larix decidua (DB)
Betula pendula (BET)
(PA) Liriodendron tulipifera
Parrotia persica (NE)
Ulmus procera (ULM)

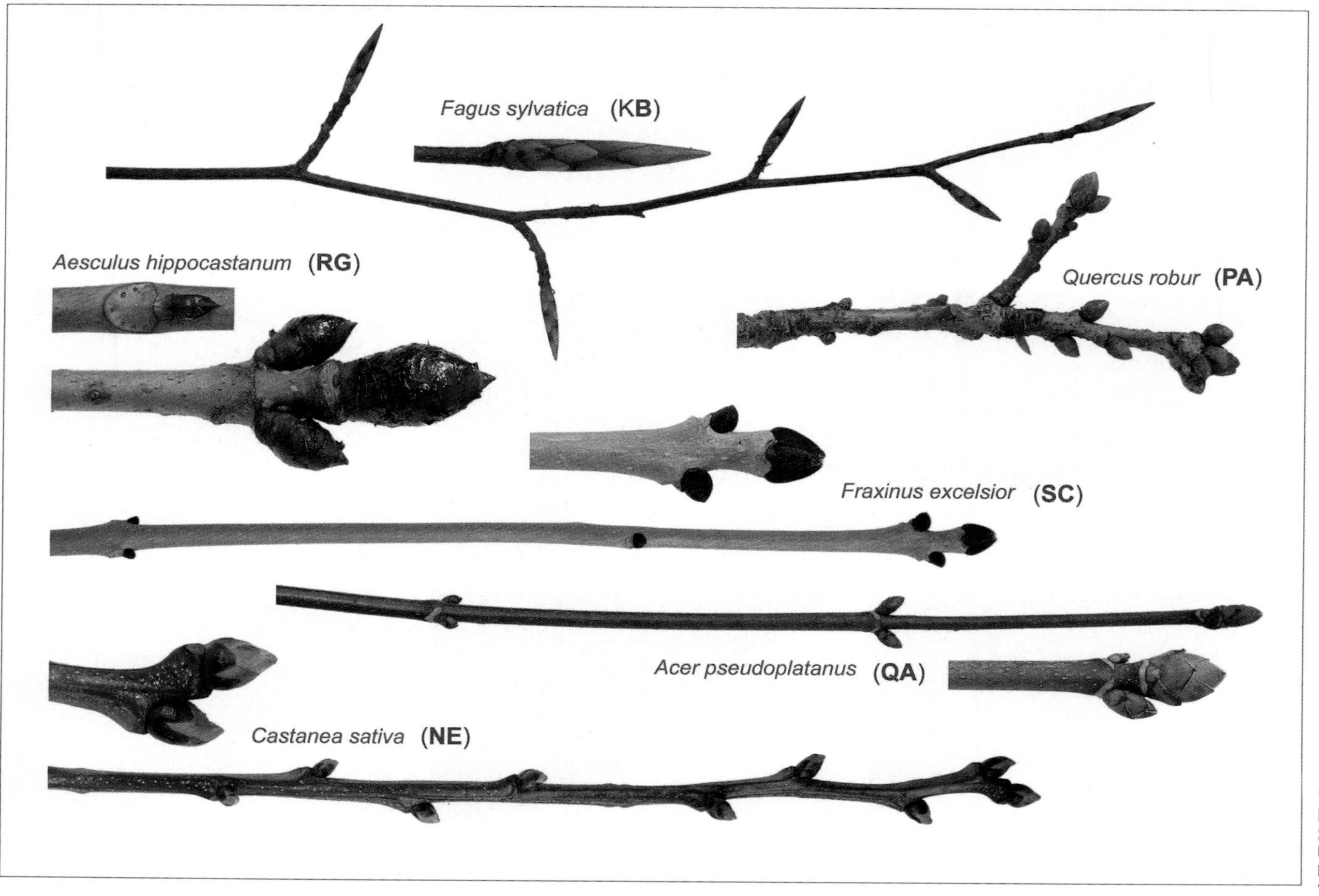
Fagus sylvatica (KB)
Aesculus hippocastanum (RG)
Quercus robur (PA)
Fraxinus excelsior (SC)
Acer pseudoplatanus (QA)
Castanea sativa (NE)

PLATE 22

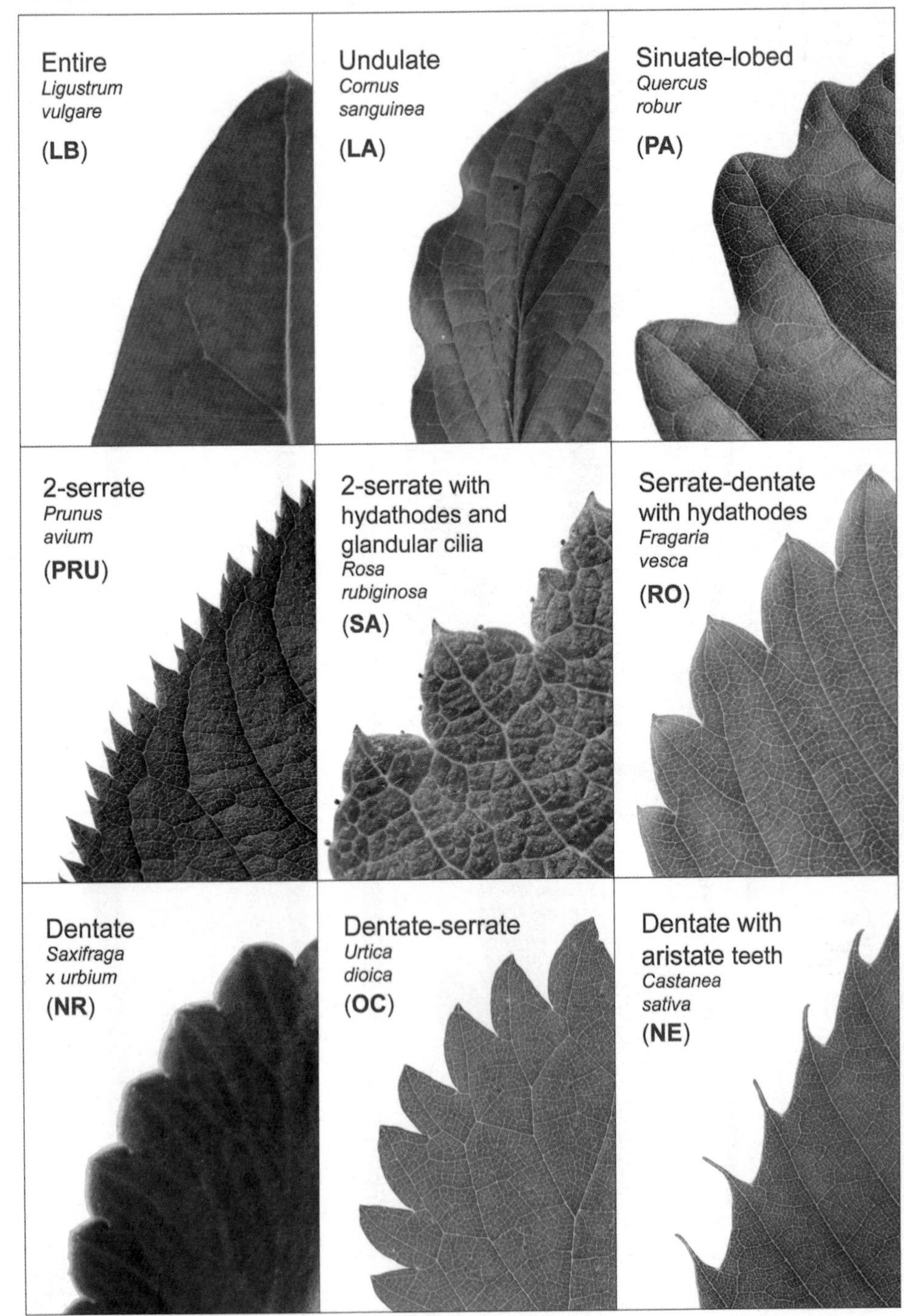

Entire
Ligustrum vulgare
(**LB**)
Undulate
Cornus sanguinea
(**LA**)
Sinuate-lobed
Quercus robur
(**PA**)
2-serrate
Prunus avium
(**PRU**)
2-serrate with hydathodes and glandular cilia
Rosa rubiginosa
(**SA**)
Serrate-dentate with hydathodes
Fragaria vesca
(**RO**)
Dentate
Saxifraga x urbium
(**NR**)
Dentate-serrate
Urtica dioica
(**OC**)
Dentate with aristate teeth
Castanea sativa
(**NE**)

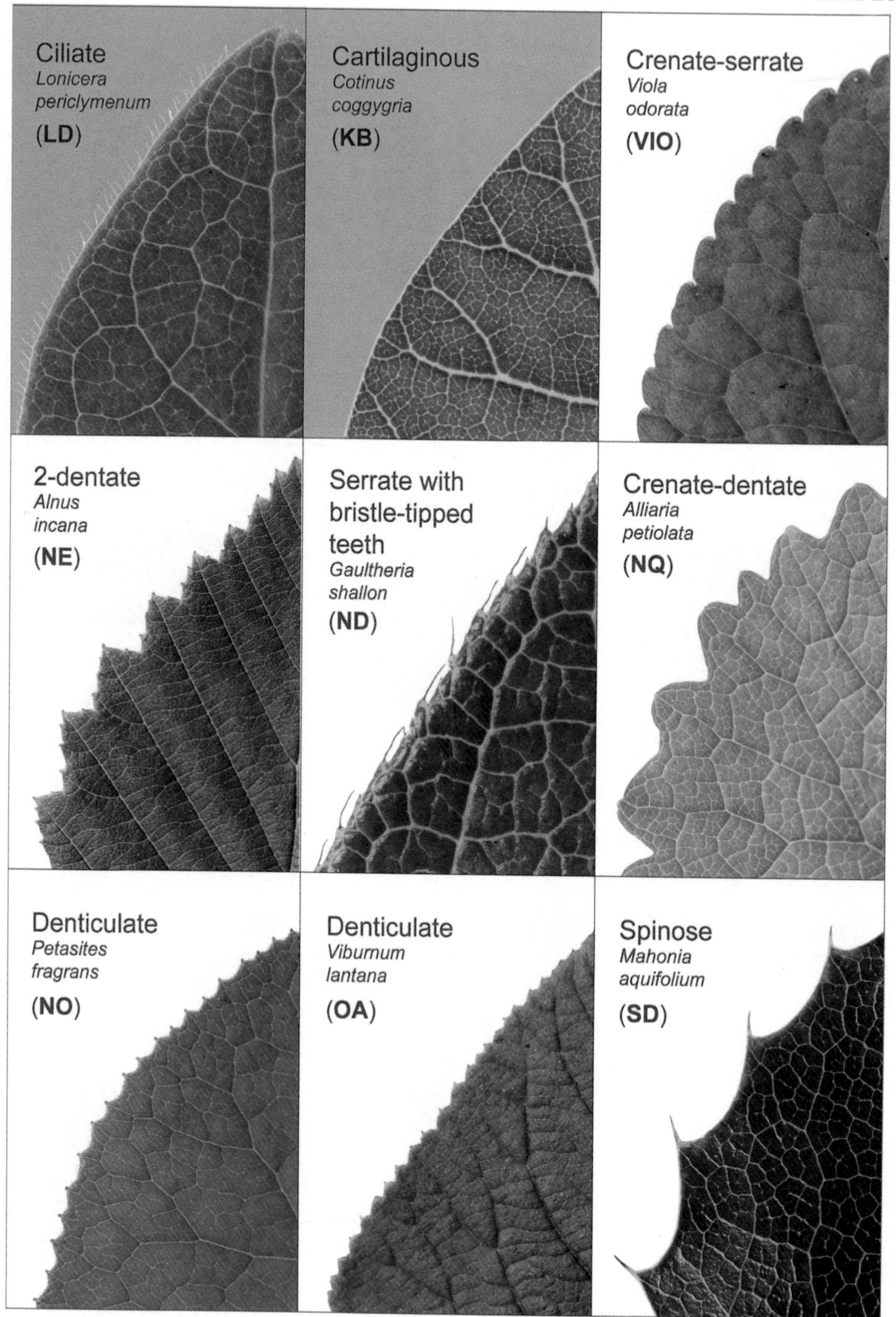
Ciliate
Lonicera periclymenum
(LD)
Cartilaginous
Cotinus coggygria
(KB)
Crenate-serrate
Viola odorata
(VIO)
2-dentate
Alnus incana
(NE)
Serrate with bristle-tipped teeth
Gaultheria shallon
(ND)
Crenate-dentate
Alliaria petiolata
(NQ)
Denticulate
Petasites fragrans
(NO)
Denticulate
Viburnum lantana
(OA)
Spinose
Mahonia aquifolium
(SD)

PLATE 24

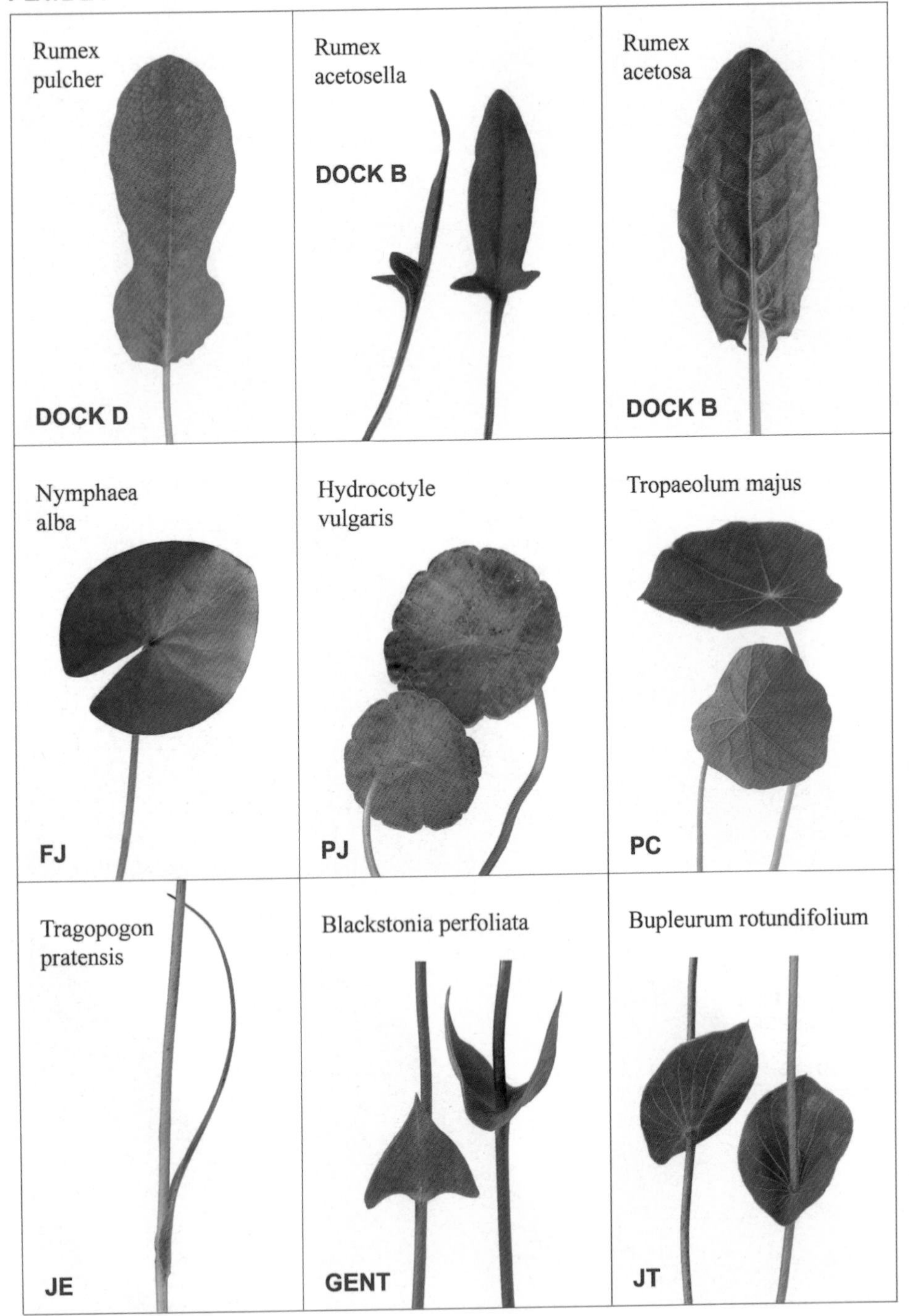
Rumex
pulcher
DOCK D
Rumex
acetosella
DOCK B
Rumex
acetosa
DOCK B
Nymphaea
alba
FJ
Hydrocotyle
vulgaris
PJ
Tropaeolum majus
PC
Tragopogon
pratensis
JE
Blackstonia perfoliata
GENT
Bupleurum rotundifolium
JT

INDEX

Groups are referenced instead of conventional page numbers and the main entry or entries (where each species is most likely to key out) are highlighted in bold. To get a complete description (if required) all entries should be read. An asterisk (*), attached to the scientific name, indicates that a line drawing occurs in a position at or near to the indicated *Group(s)*. Plate numbers are also included.